architecture
Design • Engineering • Drawing

Sixth Edition

architecture
Design • Engineering • Drawing

Sixth Edition

William P. Spence

GLENCOE/McGRAW-HILL
A Macmillan/McGraw-Hill Company
Mission Hills, California

Send all inquiries to:
Glencoe/McGraw-Hill
15319 Chatsworth Street
P.O. Box 9509
Mission Hills, CA 91395-9509

ISBN 0-02-677123-3 (Text)
ISBN 0-02-677124-1 (Problems and Quizzes)
ISBN 0-02-677125-X (Teacher's Guide)

1 2 3 4 5 6 7 8 9 10   94 93 92 91 90

Chapter 24, "CADD in Architecture," was written by Terry T. Wohlers of Wohlers
Associates, a CAD Consulting firm.

Introductory photos: Brent Phelps, Ann Garvin

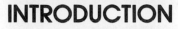

INTRODUCTION

Architecture: Design-Engineering-Drawing provides an introduction to the diverse and complex field of architecture in a comprehensive, organized manner. Information is divided into units: History of Architectural Styles; Residential Design and Construction; Electrical and Mechanical Systems; Architectural Design for Energy Conservation; Finance and Presentation; Commercial Design and Construction; Computer-Aided Design and Drafting; Careers; and Developments in Architecture and Construction. In addition, there are four appendices that provide useful supplementary data.

The planning and design of residences, both single and multifamily, are covered in great detail, including the efficient use of space in room planning and in the design of the total house. Interior design solutions are then related to exterior design, showing how the two work together to form an integrated whole. An introduction to basic drafting procedures is presented, along with a detailed explanation of how to understand and draw architectural plans. Computer-aided design is discussed, and information is supplied on such topics as building codes, finance, and insurance. The basic methods of residential foundation, floor, wall, and roof construction are covered in detail, including instructions for designing footings and determining beam, column, joist, and rafter sizes. Together, these many topics provide a well-rounded explanation of home design and construction.

Information about mechanical and electrical systems enables students to understand the systems currently in use and helps them apply this knowledge in the design of new systems.

Today, strong emphasis is being placed on the efficient use of energy. One unit is devoted to this important topic. The discussion includes basic designs for active and passive solar systems, as well as a detailed section on earth sheltered housing.

Architects must be able to present their designs well. To this end, an extensively illustrated section on architectural rendering and model-building techniques is provided.

Since metric standards in the field of building design and construction are presently being developed, the U.S. Customary system is used as the primary measure and its standards are treated as basic. Throughout the text, however, metric units are used where applicable. This, together with the text's conversion charts and explanation of the SI metric system, will help students gain a working understanding of metrics.

The commercial design and construction section of the text explains commercial working drawings — what they should include and how they should be drawn. Typical methods of commercial construction are discussed, including the use of pilings and various concrete, steel, and wood structural systems. Among the illustrations are detail drawings and design data.

One unit presents planning principles for a wide range of commercial enterprises. These establishments must accommodate a wide variety of people, including those with physical impairments. A section within this unit deals with the adaptation of commercial design for people with special needs.

The career opportunities in architecture and related construction occupations are many and diverse. Students will become aware of career possibilities, and they can use the sources listed to obtain further information. New in this edition is a section about architectural firms. The discussion includes typical methods of organization and the various types of occupations involved.

The fields of architecture and construction are dynamic. A new unit on recent developments has been added to this edition. For example, new technology is being used in the reproduction of drawings. Basic processes as well as new processes are explained.

Major health and safety factors to consider during the design process are presented. This section includes a detailed listing of design and construction techniques for radon prevention.

New approaches to assembling wood-framed buildings are becoming accepted. Some of the most successful designs are discussed and illustrated.

The extensive content of this text will enable advanced students as well as beginners in the study of architecture to utilize their abilities to the maximum. For example, there is sufficient material for a two-year course in architectural drawing, covering both residential and commercial design.

Considerable effort was put forth to make this text comprehensive, practical, and useful. The drawings have been selected with care. Color has been used to emphasize important points. The end result is a text that is a creative, effective learning tool for students in the field of architecture.

CONTENTS

architecture *Design Engineering Drawing*

UNIT ONE

History of Architectural Styles

Chapter 1
*Residential Architecture —
Past and Present*

Chapter 1

Residential Architecture — Past and Present

A study of the development of the various architectural styles reveals that the story of buildings cannot be separated from the story of the people who built them and lived in them.

The study of ancient architecture is a study of large buildings — cathedrals, palaces, tombs, and public buildings. Ancient people lived in huts of no particular style or description. Their homes were not a part of the architectural heritage. The following discussion is centered upon later events which influenced dwellings.

Old World Residential Architecture

Much of the influence on American residential architecture came from Europe. Several periods in European history produced architectural styles that eventually influenced the houses built in America. These styles can be classified as:

English

Old English or Cotswold
Tudor
Elizabethan or Half-Timbered
Jacobean
Georgian
Regency

Mediterranean

Italian Renaissance Villas and Farm Houses
Spanish Renaissance Farm Houses

From these European influences, there developed in America several architectural styles reflective of the European heritage of the people.

American Styles

The Early 17th Century New England House
The Southern Colonial House
The Georgian Style
The Cape Cod Style
The Dutch Colonial
The French Colonial
The Federal House
The Greek Revival Style
The Victorian Style
Spanish Architecture in Florida
Spanish Architecture in the Southwest

English House Styles

Following the Norman conquest of England in 1066, English country houses and farm houses developed distinguishing characteristics. These cottages influenced early American architecture. They are commonly called Old English or Cotswold, since many were built in the Cotswold area of England. Sometimes this style is simply called Norman, because these houses closely resembled the houses of the Normans from Northern France. This type of house was popular until about 1500.

Old English Style or Cotswold

These cottages were built by the small merchants, farmers, and country squires. See Fig. 1-1. They were low in appearance, with the roof line at the head of the window and sometimes lower. They were built from stone, brick, stucco (a cement and lime mixture), and occasionally a bit of half timber was used. The roofs, generally, were tile or slate, while poorer families thatched their roofs. The roof has a very steep pitch to help shed snow and rain. The thatch was rolled under at the eaves giving a rounded effect. The casement windows were few in number and quite small. They were made from many small pieces of glass, sometimes diamond-shaped, because it was not yet possible to make glass in large pieces. They were placed at random in the exterior walls. Their location depended upon the placement of interior partitions or where the owner happened to want a window.

The Old English house was neither formal nor symmetrical. Its rambling, uneven plan was the result of being occupied for centuries. Wings were added by various occupants wherever room was needed. Frequently, the additions were from a material different from the original house. Thus, Old English houses may be part stone, part brick, and part half timber.

Tudor Style

Late in the fifteenth century, England was in the midst of civil wars called the Wars of the Roses. The House of Tudor came to power and, with the country at peace, home construction began again. Many of the wealthy had fled to Europe during the fighting. There they saw the beginning of the Renaissance, a period of culture. The architectural forms of France and Italy impressed these people. They brought these ideas back with them when they returned to England. Workers were even brought from Italy to assist in building construction.

Out of these influences came a style which was called Tudor after the reigning power of the time. A large Tudor house is shown in Fig. 1-2.

Tudor houses were large houses. Characteristic of this style were large, prominent gables: large, high chimneys with high, decorative chimney pots; and walls of masonry (brick, stone) or stucco. The casement-type windows were tall and narrow and occasionally included stone mullions and transoms (vertical and horizontal dividers). Windows were made of many small panes of glass, frequently with lead strips between them (leaded glass).

The Tudor house was usually a full two-story house and sometimes had a half-story in the attic. The roof was very high-pitched and steep, with the roof line generally on a level with the ceiling of the second floor. This style did not have the low, rambling appearance of Old English houses. It was not symmetrical, though it exhibited some formal planning and a semblance of symmetry.

The Bettmann Archive, Inc.

Fig. 1-2. A Tudor house — Hengrave Hall, Suffolk, 1538. A typical, A-shaped gable is at the far left.

H. Armstrong Roberts

Fig. 1-1. Old English, Cotswold or Norman farmhouse — The Ann Hathaway Cottage, Stratford on Avon, England. Notice the half-timber construction and the way the casement windows open out.

B. T. Batsford, Ltd.

Fig. 1-3.   An Elizabethan House — Speke Hall, Lancashire. The half-timber construction is the outstanding characteristic of this style.

Elizabethan or Half-Timbered

During the reign of Queen Elizabeth, the Tudor house underwent gradual styling changes until a new variety emerged to become a style of its own. The use of an exposed timber frame with stucco or brick filling between the timbers became popular. (Many Elizabethan houses had considerable stone on the exterior.) The exposed timber frame became the outstanding characteristic of the Elizabethan style. See Fig. 1-3.

Most other characteristics were similar to the Tudor. The Elizabethan house had many prominent gables, was usually a full two stories and had tall, narrow, casement windows. The windows were fitted between the structural timbers, and the sashes contained many small panes of glass.

Several large, prominent chimneys usually were a part of this style. Occasionally, the second floor would cantilever (overhang) the first floor, and the heavy corner timbers of the second floor would project down and be carved.

The exterior of the Elizabethan house was not symmetrical. It was an informal, rambling house. The rooms merely were placed where the owner wanted them.

Jacobean Period

After the reign of Queen Elizabeth, the House of Stuart ruled England. Few changes occurred in house styles for over one hundred years. No one style developed sufficiently to become recognized. This was the Jacobean Period.

When the Stuart line ended, a prince from Holland was invited to become the King of England. This was the beginning of the House of Hanover. During the reign of George III, a Hanover king, the American Colonies gained their freedom and formed the United States of America.

Georgian Style

After several hundred years of development, the influence of the Renaissance brought about a new style of house on the Continent and in England. This style came to be known as the Georgian, since it evolved during the reign of the Georges of the House of Hanover. By 1700, it was a well established style.

The Georgian house was very formal and symmetrical. It was usually rectangular. Reflected in the details were direct copies of the classical architecture of early Italy. The cornice (horizontal trim at the top of a wall) was usually of a classical design. A triangular decoration or pediment was frequently used over windows and doors.

Basically, the front entrance was exactly centered on the front elevation (view of the house), and the windows were balanced on either side. The house was usually two full stories with a high-pitched gable roof providing attic rooms if dormer windows were added. The second-floor windows were placed directly above those of the first floor.

The exterior was usually brick with decorative stone blocks called quoins set in at the corners. Large chimneys, generally at the ends of the house, were prominent.

Since the Georgian style was so formal and fixed, little freedom was left for interior planning or room arrangement. Typical of a Georgian house was a large hall through the center of the building, with the stairs to the second floor directly in front of the main entrance. The living room was usually on one side of this hall, and a dining room was on the other. Upstairs, the two main bedrooms were placed with a bath or sewing room between, to enable the builder to preserve the exterior balance of windows on the front elevation. Other less important rooms, such as the kitchen, pantry, or maid's room, were placed in the rear. Little attempt was made to keep this part of the structure symmetrical.

The Georgian is a beautiful house and is still very popular. See illustrations of the Georgian houses built in the American Colonies in the next portion of this chapter.

Regency Style

At the time of the American Revolution, King George III became unfit to rule. His son was appointed as a Regent to act in his place. During this period of history, the Georgian style was modified so that it was less formal and more refined. Much of the classical detail of the Italian Renaissance, which influenced the Georgian, was removed. Designers studied ancient Greek and Roman architecture for design ideas. Characteristic of a Regency house were a curved porch roof over the front entrance, octagonal windows, long shutters on the first-floor windows, and curved top edges of walls projecting from the corners of the house. The roof was usually rather steep and had four surfaces all sloping toward the center of the house in the hip style. The house was a full two stories. The windows were made of small panes of glass. They were double-hung and could be opened from the top or bottom.

England had developed trade with the Far East, and some people believe the curved roof over the front porch or over the front, bay windows was the influence of the Chinese.

Wrought iron balconies and porch railings were frequently used.

The Mediterranean Styles

The architectural styles of Italy and Spain found acceptance in the southern parts of the United States. Both countries have warm climates similar to that of our Southern states. These houses were designed to ward off heat rather than intense cold as were the English and French styles. It may frequently be difficult to separate Italian and Spanish houses, although there are some distinguishing features.

The Italian Styles

The architecture of Italy was influenced by the Italian Renaissance. This awakening initiated a movement that sought out and copied the details of the classic buildings of ancient Rome. It was this style that was used in the designing of Italian villas. The Italian style had a low-pitched, tile roof and classical design features such as balustrades (railings). It, generally, was a full two-story building, with a loggia having columns and arches. (A **loggia** is a roofed gallery within the body of a building. It is attached at the second floor and has three sides enclosed and one side open to the air.) The casement-type windows, generally, had a wide trim surrounding them. The exterior was usually stucco. See Fig. 1-4.

The Italian styles were brought to the United States by American tourists who visited the Italian villas, liked them, and copied them upon returning home.

The Spanish Styles

The developments of the Spanish Renaissance influenced the architecture in Spain. However, Spain had been conquered and occupied by the Moors from Africa, and this factor also exerted an influence. This style differed somewhat from the Italian. A low-pitched, gable roof was common, but it was not tile, as were Italian houses. Stucco was a common exterior material, and casement windows were used. The Spanish employed balconies with

H. Armstrong Roberts

Fig. 1-4. Grand Canal, Venice, Italy. The balustrades and the trim around the windows are typical of Italian styles.

decorative railings, Fig. 1-5. They also built around a patio, frequently enclosing it on all four sides by the house. This provided a cool, private place to relax. In an attempt to keep the houses cool, the Spanish built the walls very thick. Windows had to be located in a deep opening causing a dark, shadowy effect.

The Spanish styles were brought to this country by Spanish settlers who lived in Florida and southern California.

Early American Residential Architecture

The first settlers in the New World lived in crude shelters. These, usually, were dugouts, cabins, or wigwams. They were built from whatever material was close at hand and with no concern for style. Log cabin construction was introduced in the colonies in 1638 by the Swedes in Delaware.

Early 17th Century New England Houses

As the settlers prospered and towns grew substantially, well finished houses were con-

structed. These were usually made from the common materials available locally. See Fig. 1-6. Sawyers cut lumber where it was plentiful. If the soil was right, bricks were made and burned in kilns.

These houses were much the same as those lived in by the settlers in their mother country. There is a striking resemblance between the medieval houses of Old England and the Early Colonial houses of New England.

H. Armstrong Roberts

Fig. 1-5. Spanish style house. Notice the use of arches and delicate railings.

Essex Institute

Fig. 1-6. An early 17th century New England house in Salem, Massachusetts.

HABS, Library of Congress

Fig. 1-7. A 17th century cottage — The Peake House, Medfield, Massachusetts, 1680.

Old Sturbridge Village, Massachusetts

Fig. 1-8. Saltbox or Lean-To style, built 1750.

The houses were predominantly timber framed. Stone and brick were scarce and used only for the most expensive houses. The typical house used hewn oak posts and beams, joined with pegged joints. It was usually covered with clapboard siding. (Clapboard siding is made up of overlapping narrow boards each of which is thicker on one edge than on the other.)

Most houses were very simple. The people had to work to survive and did not have time to spend on ornate decorations. The houses, usually, were rectangular, with a large chimney in the center serving many fireplaces. See the Peake House, Fig. 1-7.

The rooms tended to be low, and the roofs were steeply pitched and often thatched. The doors and windows were placed where needed, with little consideration for exterior appearance. Windows were small, and if they contained glass, it was in small pieces, for glass was precious and expensive.

The second floor occasionally overhung the first floor. This overhang was an influence from the medieval towns of Europe where houses were crowded and ground was precious. A larger space was gained on the second floor without consuming more ground space.

Through the years, rooms were added on and roofed over. This led to a type of house referred to as the Lean-To or Saltbox. See the Richardson House, Fig. 1-8.

An example of a large farmhouse of the 1800's is shown in Fig. 1-9. The roof style is known as gambrel. Notice also the double-hung windows.

Fig. 1-9. A large New England farmhouse built in 1801.

Alpha Photo Associates, Inc.

Southern Colonial Architecture

A great region running from Delaware Bay to the Savannah River and inland to the Piedmont region became what was called Virginia. This area was settled mainly by the English. Here developed a style of architecture commonly known as Southern Colonial.

The early homes here were much the same as those in other parts of the colonies — crude huts or dugouts. As more colonists arrived, frame houses were built. These were of hand-hewn timber frames covered by clapboards. See Fig. 1-10. Brick houses were built early here, possibly in the early 1600's, but the frame house still predominated at this time. Construction methods were much the same as in the other colonies.

The soil in the South was suitable for brickmaking. With skilled brickmakers among the early settlers, brick fast became a popular building material. This was much more true in the South than in New England.

The typical Southern house was built on a central hall plan with chimneys located at each end of the house. It was usually rectangular, with a small projecting porch, and was commonly one and one-half story. Dormers were used a great deal to increase the livability of the second floor. The houses built without dormers relied only on small windows in the end walls for light and ventilation. See Fig. 1-11.

The Georgian Style in the Colonies

The Colonies looked to England for trade and fashions. As the Renaissance brought about the development of the Georgian house in England, eventually the English Georgian influenced house styling in the Colonies. The Georgian house became popular in the Colonies in the early 1700's.

This was a very formal style. The houses were usually rectangular, and the windows were placed in perfect symmetry. Interior rooms were arranged to permit a balanced exterior. The front entrance was placed in the exact center of the structure. The pediment of the Romans was used a great deal. Doorways were usually flanked with pilasters (wall columns) rising to a cornice or pediment.

Windows were almost universally of the double-hung type and were made up of many small panes of glass. Blinds of the louvered type (movable, horizontal slats) were very common. Some paneled blinds were used in the Middle Colonies.

Fig. 1-10.   Early Southern Colonial — the Blair House. Frame houses were the most common at this time.

Colonial Williamsburg

Fig. 1-11. A brick Southern Colonial — the Red Lion Inn.

Fig. 1-12.   Mount Vernon, the home of George Washington in Virginia.

Fig. 1-13.   A late Georgian House — the Governor's Palace, Williamsburg, Virginia.

Most Georgian houses were either frame or brick. Attempts were made to stucco over the brick or frame exterior and incise lines to make it resemble stone masonry, but this was esthetically unsuccessful. Mount Vernon is a classical example of a frame Georgian treated to resemble stone masonry, Fig. 1-12.

The roof was usually pitched about 30 degrees, which was lower than Colonial roofs. While many gable roofs were used, the hipped roof was the most popular. Some roofs were cut off below the ridge to form a flat deck which was enclosed with balustrades. See Fig. 1-13.

Many later Georgian houses had large pilasters extending the full height of the facade (front of the house). Some used a small entrance portico with pilasters and a pediment above.

Georgian houses were built with one and one-half to two and one-half stories. They had gable, gambrel, and hipped roofs. Dormer windows of many shapes were used. They were usually narrow, and, frequently, the gable end was treated as a pediment.

Georgian chimneys were usually simple, rectangular shapes with a small, molded cornice at the top.

The early Georgian house was much simpler than those built after 1750. It was a large, impressive house with a bold appearance. The roof tended to be pitched steeper than the later styles. Usually, it had no entrance pilasters or portico. The windows were built in smaller panes than the later Georgian, sometimes having 18 to 24 lights per window. The dormers had rectangular windows, while the later Georgian frequently had arched dormer windows. Later Georgian houses, also, commonly had a balustraded roof deck, while these were seldom used on earlier styles.

The Georgian houses built in the Middle Colonies were much like those of New England and the Southern Colonies, except they were usually stone. Some were built of brick, but stone was a readily available material. Throughout Pennsylvania, Georgian houses can be found.

Fig. 1-15. Row houses with wrought iron and small front yards, typical of Virginia.

As cities grew, row houses were built. They had common walls between them and were built right up to the front sidewalk. See Fig. 1-14. Sometimes, a small front yard was allowed. The styling was a direct influence of the homes of the time, and the Georgian influence was strong. The used of wrought iron railings and columns became common. See Fig. 1-15.

By the beginning of the 18th century, the deep South was beginning to build its system of large plantations. Rivers were the main method of transportation because much of the land was swampy and unsuitable for roads. Since wealth was beginning to come to plantation owners, they began to build large houses. These were of the Georgian style. Many of the largest were facing rivers, since this was the "highway" of the day. They were built miles from the nearest neighbor.

Fig. 1-14. **Captain's Row on King Street, Alexandria, Virginia. Built just before the Revolution.**

Fig. 1-16. Georgian house with portico on three sides, built in Louisiana.

Alpha Photo Associates, Inc.

Some plantation owners began adding a portico with two-story, slender columns. George Washington was among the first to do this at Mount Vernon. This portico permitted the exposure of all the windows on the second floor, yet it protected them from the hot sun and driving rain. Frequently, this portico was built facing the river. Some built it on three sides. Figure 1-16 illustrates a beautiful Southern house with columns on the front and sides.

While many call this "Southern Colonial," it is a Georgian house built during Colonial times. Porticos built in northern Colonies were definitely a Southern development.

Dutch Colonial Architecture

The Dutch and Flemish settled in Long Island and New Jersey. They brought with them the architectural style copied from the streets of Amsterdam, Leyden, and Utrecht, and from the farmlands of the Dutch and Flemish lowlands. Although, in the New World, this style developed into one that was distinctly American, it retained a great deal of the flavor of its origin.

As the colony of New Amsterdam thrived, new houses were built. Usually, these were wood, brick, or stone. Occasionally, all three materials were used in one house. A large number of the very early houses were frame, since the settlers were familiar with wood, and it was easily worked.

In the Hudson Valley, settlers found ample supplies of stone and built many fine stone houses. The walls, usually, were from one and one-half to three feet thick.

Another style also developed during the 18th century in this area. It is most commonly today as a Dutch Colonial, but it is more properly called a Flemish Colonial house. The most outstanding feature is the gambrel roof. The actual heritage of this roof cannot be traced to the Dutch or Flemish, since the English and Swedish also used the gambrel roof.

The roof was large, permitting almost a full second story, with the overhangs boxed in. Frequently, the overhang was large enough to cover a front porch. The roof was usually covered with wide wood shingles. Dormer windows were used to open up the living area on the second floor.

The chimney was always on the interior. See Fig. 1-17. This style of house was of frame and stone construction. Shutters on all windows were common. Frequently, wings were built as additional space was needed.

Cape Cod Colonial Cottage

This house style can now be seen throughout the country, but it seems to have originated in the Cape Cod area of Massachusetts.

They are one story or one and one-half story buildings. The roof is rather steep and is a simple,

H. Armstrong Roberts

Fig. 1-17. Flemish (Dutch) Colonial, Lexington, Massachusetts.

gable type. The eaves are plain and have little overhang. The eave line is near the top of the window, giving the house a low appearance.

The exterior siding is white, frame clapboard or brown (or gray), wood shingles. Shutters are used on all windows. The windows are the double-hung type and have many small panes. Dormer windows are not characteristic of this style, but many modern versions do add two single dormers on the front and, occasionally, a shed dormer at the rear. These should be small and in proportion to the house.

A large central chimney containing several flues is a distinct characteristic of the Cape Cod. Since it required several fireplaces, a large chimney was necessary.

French Colonial Architecture

The French Colonial empire in North America was immense. It extended from the Alleghenies to the Rocky Mountains and from the Gulf of Mexico

HABS, Library of Congress

Fig. 1-18. An early French Colonial house, 1737 — a re-erected structure in Cahokia, Illinois.

Alpha Photo Associates, Inc.

Fig. 1-19. A two-story French Colonial as found on plantations — Poente Coupee Parish, Louisiana, 1750.

to Labrador and Hudson's Bay. This remained under French control until the early 18th century. There were few towns in this vast area and only a few forts and Indian villages.

French explorers, trappers, missionaries, and some settlers worked their way up the St. Lawrence River and across the Great Lakes. They established forts at strategic points such as Sault Sainte Marie, Frontenac, and Duluth. They began establishing settlements along the Mississippi Valley. By 1682, French explorers had traveled south all the way to the Gulf of Mexico. Settlements were started at Mobile, Biloxi, Fort Toulouse, Natchez, and Natchitoches. In 1718, New Orleans was founded and, in 1764, St. Louis.

France lost this territory to England and Spain after the Seven Years War (1756-63). The division of this territory in 1763 gave all lands east of the Mississippi to England and all to the west to Spain. However, the territory remained culturally French, and strong elements of this can be found in places today.

The architecture of many of these cities became a blend of several cultures. The houses of New Orleans are an excellent example of this blending.

The French Colonial Style

The early French Colonial house was usually of half-timber construction. However, it differed considerably from that of England. Heavy cedar and cypress logs were placed vertically on the ground and set much as fence posts. They were spaced a few inches apart, and the spaces were filled with a clay and grass mixture. Some builders used Spanish moss or deer hair instead of grass with the clay.

Later, the same type of construction was improved by the use of a stone foundation with a wood sill. The upright logs were placed on this sill. Such an arrangement prevented the rotting of the logs, as occurred with those buried in the ground. See Fig. 1-18.

This house was one story, with several rooms all in a line, and a stone chimney either in the center or at one end. Generally, this structure was surrounded by a "galerie" or porch, which gave access to the rooms. Many slender posts were used to support the roof of the galerie. The roof shape in Fig. 1-18 is characteristic of this style. Usually, the main house was covered with a steep, hipped roof, and a lower-pitched roof was placed over the galerie.

As plantations grew, larger houses were built. They were most often two stories with stucco-covered brick walls and columns on the ground floor, and wood construction on the second floor. The galerie was extended across the front on both levels. Sometimes, it was also across the rear of the house, and a few houses had a galerie on all four sides. See Fig. 1-19.

The supporting posts for the first-floor galerie were also stucco-covered brick and were styled on classical lines. The second-floor posts, usually, were slender and made of wood.

Spanish Architecture in Florida

Spanish explorers penetrated the New World early. The first landing in Florida was in 1513 by Ponce de Leon. By 1528, Panfilo de Narvaez had landed in Tampa Bay and explored inland in the state of Florida. DeSoto left Tampa Bay in 1539 and traveled through Florida, Georgia, the Carolinas, Tennessee, Alabama, and almost to Ohio. He then went south into Arkansas and Oklahoma, returning to the Mississippi River where he died. Remnants of his party reached Mexico in 1543.

In 1565, the Spanish sent about 2,000 persons to establish a defense outpost at St. Augustine. They murdered a group of French settlers who had already made homes there.

A crude fort was built of earth and logs. In 1593, a group of Franciscan friars arrived and, over the years, established many Spanish missions in the area. The mission buildings were built of stone and wood.

Indian and pirate raids destroyed many of the forts. In 1672, the first stone fort, Castillo de San Marcos, was built at St. Augustine. It is one of the most impressive examples of Spanish architecture in Florida, Fig. 1-20.

Britain acquired control of Florida in 1763 and ceded it back to Spain in 1783. It remained Spanish until American occupation in 1821. Few buildings of this early Spanish period now exist. The present Spanish influence is due to a revival of this influence in the twentieth century.

Spanish Architecture in the Southwest

In 1609, the Spanish were building La Villa Real de Santa Fe de San Francisco, the capitol of the province of New Mexico. This was only two years after the English landing at Jamestown, Virginia. Coronado explored the mountains and plains of Arizona, New Mexico, Colorado, and Kansas. In 1542, Cabrillo sailed north beyond San Francisco. Mexico was by now settled by the Spanish, north of

today's Mexico City. The Franciscan and Dominican friars were establishing missions which were the main examples of Spanish architecture. The friars were responsible for their construction and used what they knew about architecture from Spain. The mission was more than just a place of worship. It was an entire community, containing sleeping rooms, kitchen, storerooms, and shops.

The missions, built by unskilled Indian labor at a time when tools were scarce, were rather crude. The walls were built of adobe bricks — a mixture of clay and straw, formed into blocks weighing about 50 pounds (22.68 kg). These were dried in the sun and stacked to form the walls.

Roof beams were hewn logs laboriously lifted into place. The roof was of rough-hewn planks. Over the planks, six inches (152 mm) of adobe clay was spread to form a solid slab roof.

When the Spanish settled the area now called Texas, missions were built as in New Mexico. The most famous of the missions is the Alamo, built in 1718 at San Antonio.

Another characteristic type of building had adobe walls with flat logs as roof supports. This type house is still built today in New Mexico.

The first California Spanish mission was established at San Diego in 1769. From this small, difficult beginning, the Spanish moved north, establishing many missions along their route. In 1776, as the thirteen English Colonies gained their independence, the Spanish established a mission at the Golden Gate.

Fig. 1-20. The Castillo de San Marcos, St. Augustine, Florida, built 1672.

The California missions were built mainly from adobe bricks, as in New Mexico. The walls were very thick and tapered to the top. They were coated with lime and sand to prevent erosion by rain. These missions differed from those in other areas in the Southwest because they used quite a lot of burned brick. This enabled the builders to reinforce arches and to build piers.

Red tiles formed into half cylinders were used for roofing, and wooden beams and planks made up the structure of the roof.

Since the builders had bricks, they could build arches. The arch is repeated in the long arcades of the missions, forming a major architectural feature.

Figure 1-21 is an example of a splendid mission, Mission San Carlos Borromeo, built in 1793. This was one of the first structures to use stone as the basic material. The facade was covered with stucco. Arches were used a great deal. The dome on the tower was directly influenced by the Moorish tradition in Spain.

The houses built by the Spanish at this time in the Southwest were of the same materials as the missions — adobe walls and wood roof planking with a tile outer cover. The roofs were usually the shed type. Architecturally, the houses were very simple. They were commonly built in a **U** shape, with a patio open on the fourth side. The veranda facing the patio was usually supported by wooden posts and not by heavy pillars and arches, as were common on the missions. Since the building was only one-room wide and had no inside halls, the **veranda** served as a covered corridor, giving access to each room. It also served as a lounging place.

In the northern part of the Spanish holdings in California, one of the most beautiful of all Spanish California houses developed. It was a two-story house with a long veranda on the ground level and a balcony above. Wood supporting posts were run from the ground to the roof. Since many of these houses were built in Monterey, the style is named after this town. See Fig. 1-22.

The ground level usually contained the living area, dining area, cooking and service areas, and a stairway leading to the second floor. This stairway was on the outside of the building and ran up from the veranda. The balcony and veranda frequently were covered on the ends by a framework of slats called a lattice.

The balcony railings were light and delicate. The roof was usually a gable style and was low-pitched. Frequently, it was hipped at the ends. Wood shingles were used extensively as a roofing material.

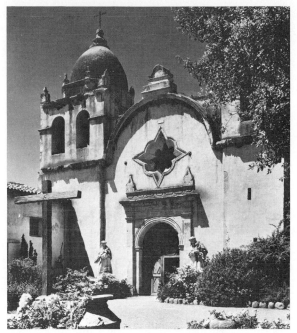

Joseph Muench Pictorial Photography

Fig. 1-21.   The Mission San Carlos Borromeo, built in 1793, located on the Monterey Peninsula, California.

HABS, Library of Congress

Fig. 1-22.   A Monterey style house — the Larkin House in Monterey, California.

The house was rectangular in shape and, therefore, did not encircle a patio. Adobe walls were built around an area to the rear of the house to form a private patio.

The exterior walls were mainly adobe, averaging three feet in thickness on the first story and two feet in thickness on the second story. The adobe exterior was covered with a mud plaster and whitewashed. Occasionally, the adobe walls were covered with wood planks placed either horizontally or vertically.

The windows were small and had small panes of glass. Since the walls were very thick, the windows were set deeply and the resulting window ledge formed window seats.

The Federal Style

After the Revolutionary War, houses began to change. The architecture of England had a great deal to do with these changes. By 1800, a style called the Federal Style was popular. See Fig. 1-23. The house was still a basic rectangle and usually two stories high. The symmetrical appearance of the English Georgian house was used. The detail of the house became delicate as smaller members replaced the heavy ornamentation of the earlier Georgian period. The hip roof and the gable roof were used, but they were not pitched as steeply as earlier styles. A roof style called the monitor was common, with small windows near the ridge on a hipped roof.

Another common characteristic of the Federal house was a delicate balustrade on the roof. See Fig. 1-24. Often, this was near the eave and practically concealed the roof.

The main entrance usually had an overhead, fan-shaped light (window) and glass side lights. If columns were used on the front, they were regularly spaced. Each column tapered gradually to a narrow extremity.

Greek Revival Style

In the early 1830's, interest in classical Roman and Greek architecture became great. This caused some abrupt styling changes in the houses of the time. Designers found in the early Greek buildings a masculine vigor that was a sharp contrast with the delicate, feminine Federal style. The revived interest took the country by storm.

Greek temples became examples of excellence. Massive columns supporting a portico were used to imitate these temples. The Doric order of Greek architecture was the most popular and was extensively copied. The building was comprised of large, plain surfaces and large mouldings dividing these surfaces into panels.

Full-length French windows were commonly used, and large panels of glass were installed.

The Greek Revival home had to be a large structure. The design was massive and entirely out of keeping for small houses. It was not popular long and by 1850 was being replaced by the Victorian style. See the Levi Lincoln house, Fig. 1-25.

Victorian Style

The Victorian period was one of confusion. Houses were large and heavily ornamented. Carvings and turnings were fastened on porch roofs,

The Bettmann Archive, Inc.

Fig. 1-23. The Corbit-Sharp House built in the Federal style in 1772. Notice the use of delicate features in the roof and window trim.

H. Armstrong Roberts

Fig. 1-24. A Federal Style — Longfellow's House, Cambridge, Massachusetts.

Old Sturbridge Village, Massachusetts

Fig. 1-25. Greek Revival Style — Levi Lincoln House, 1836.

eaves, and gables. Porch balustrades were very ornately carved. It was a period of poor taste, when many large, irregular, two-story houses were built. The heavy, fancy trimming of this style was given the name **gingerbread.** See Fig. 1-26. Designers paid little attention to pleasing proportions, balance or a feeling of unity or harmony.

By 1900, the people reacted against this style, and a rapid change was made to a style that did away with all forms of ornament — the bungalow. The form and decoration of the house was supplied by the structure. If something was needed for structural purposes, it was used, but nothing extra was used simply for decoration.

These later styles — the Greek Revival, the Victorian, and the bungalow — have had little influence on the American house of today. The earlier Colonial styles, however, have exerted considerable influence in the styling of today's house, and some are quite popular in a form closely resembling the original.

The Modern House

The style of house commonly referred to as "modern" or "modernistic" developed primarily in Europe during the early 1920's. Architects such as Le Corbusier, Gropius, Hoffmann, and Neutra rejected the traditional way to plan and build a house. Their approach was to plan the interior to suit the needs and way of life of the occupants. Then they covered this plan with a simple shell, devoid of unnecessary decoration. Its function was to shelter the plan from the elements of nature. The exterior, then, became simply the result of the developed plan.

The outstanding feature of the modern house was not its exterior appearance, but the interior plan of rooms. The house was designed to be functional and to reflect the modern way of living. Considerable emphasis was placed on working out the best room arrangement and use of space.

Building costs were getting higher, so houses were becoming smaller. The modern house attempted to get a feeling of spaciousness in a small floor area, by combining several rooms. It was called an **open plan.**

The use of large, glass areas in the exterior walls also contributed a great deal to the open feeling. The windows were placed where needed, with little regard to what this did to the exterior. The emphasis was on a livable, open, floor plan.

The large glass areas were usually placed away from the street and toward the best view. Since automobiles were becoming common, garages were placed on the front of the house. Roofs were commonly used as sun decks and for outdoor living.

An influential factor was the development of new materials and new construction techniques. The exterior wall was no longer needed to serve as the main support of the structure, because steel framing and columns could be used. Walls, then, became only thin skins to retard the elements. New materials began replacing brick and stone. Large glass panels, aluminum skins, and plastic materials were new exterior materials coming into use at that time.

The exterior style of the modern house can be easily recognized because of its box-like shape. Simplicity was demanded. Little decoration was used. The roofs were flat, and the exterior walls presented large, plain, unbroken areas. Long expanses of glass broke the wall on occasion, and corner windows came into use. A pleasing architectural effect was created by skillful use of proportioned, rectangular solids. The entire house portrayed simplicity and functionalism. Many people disliked it, and the style did not gain the acceptance in the United States that it did in Europe. However, the modern house did contribute much to residential

H. Armstrong Roberts

Fig. 1-26.   A Victorian House.

architecture. Progress made in the planning and construction of houses had lagged far behind that made in such fields as transportation, communication, and medicine. The modern house was a rich and genuine effort to enable people to live comfortably in a modern society. While the style is now out of favor, the lessons it taught are reflected clearly in houses of today.

Architectural Styles of Today

The Southwestern Ranch Style

On the open prairie in the Southwest, land was cheap and plentiful. Timber was scarce, the prairies flat and windy. It was hot in the summer and cold in the winter. The people settling the area could not build the two-story stone or frame Colonial that they were accustomed to seeing in the east coast colonies. Instead, they developed an entirely new style of house — the low, rambling, ranch house. See Fig. 1-27.

This house was built of adobe bricks and had a flat or low-pitched roof. Since it seldom rained in this area, the houses were built on the ground, and hard-packed, earth floors were common. If wood floors were used, they were built over a crawl space.

These houses were one-story high and had low ceilings. Since land was cheap and plentiful, it was

Fig. 1-27.   A modern version of the western ranch house. Notice the rambling plan, veranda, and private patio.

Fig. 1-28.   A typical midwest ranch house.

more economical to add rooms on the ground level than to build a second story. All that was necessary to add a room was to mix adobe bricks and let the sun bake them. The walls could be built right to the ground.

These houses were designed for outdoor living and, frequently, were built around a patio, as were the Spanish houses. Since all rooms were on the ground floor, most of them had doors opening onto the patio.

The Midwest Ranch House

The ranch house did not become popular until the 1940's. As it spread to the midwest, it was modified by builders to fit a more restricted building site. The modified ranch house still retains many of the features of the early California and prairie ranch houses. They are now brick as well as frame, and tend to fit an **L** shape rather than a winding, rambling **U** shape.

The rooms in the modernized version of the rambling ranch house are smaller and fewer in

Fig. 1-29.　A variety of the midwest ranch house.

number. Fireplaces are retained for decorative purposes, and garages have been added. The house, often, is built with a basement. Typical examples are shown in Figs. 1-28 and 1-29.

The Japanese Influence

The style of house sometimes called Japanese Modern has developed from the study and appreciation of Japanese architecture. The best way to understand what this is today is to go back a few hundred years and follow the story of the development of building in Japan. The Japanese had a strong culture and an architecture of their own. However, the architecture of China spread its influence into Korea, Indonesia, and Japan. Therefore, what we think of as Japanese architecture is really an assimilation of the Chinese influence by the existing Japanese architecture.

Climate is a big factor influencing the Japanese architecture, as is the rough landscape. Japan is a country of steep hills and mountains, rapidly falling valleys, and lakes, oceans, rivers, and waterfalls. The Japanese people have always been sensitive to the esthetic, and they take advantage of these natural beauties in their architecture. The use of the picturesque and the close relationship of a building to a slope, to water, to a striking tree, or to a hillside became a part of the architecture of Japan. The houses were not strictly symmetrical, as were the Chinese houses, and the Japanese design had more freedom and was not as formal and strict as the Chinese design. An American version is shown in Fig. 1-30.

Fig. 1-30.　Japanese modern home, built in Hawaii.

R. Wenkam, photographer and House and Home Magazine

The Japanese also try to make their houses part of the total landscape, Fig. 1-31. They use trees to balance their typically asymmetrical houses and relate the house and nature with stone lanterns, walls, and shrines. This composes a complete architectural picture that is related not only to the immediate surroundings but to the distant landscape. To view a Japanese home is the same as viewing a completed painting.

The Japanese built their houses primarily from wood. They used wood so that the natural beauty of the grain and color could be displayed. Usually, it was undecorated, but rubbed smooth. The houses were built on a rectangular framework of delicate appearing members, so that the entire effect was one of lightness and delicacy. Even the thatched roofs enhanced this effect. The Chinese house tended to portray solidity and formal dignity. See Fig. 1-32.

The interior and exterior of the Japanese Modern house are highly characteristic of the people. The interior partitions are sliding paper screens, and the exterior is composed of sliding wood shutters. The floors are covered with woven mattings. The wood panels are plain and unpretentious.

Freedom of arrangement is great. Often, the interior screen partitions can be moved, and the entire house can be opened into one large room.

Japanese houses are planned with freedom for expansion. As more space is needed, it is added on. Each addition is placed to take advantage of a view or is built with a private garden or court.

In the United States, successful attempts have been made to adapt the Japanese concepts of house planning and design to the American way of life. The United States version attempts to capture the open floor plan and spacious rooms, the simple, undecorated, exterior styling, the honest use of materials, and the use of large openings to take advantage of a view or a private courtyard.

The Hawaiian Influence

The architectural styles of the Japanese and Chinese, many of whom settled in Hawaii, appear predominately in the type of houses being built there. They all stress view and climate and try to make the most of these local advantages. It is difficult to separate some Hawaiian houses from Japanese. A modern example of the lingering Japanese influence in Hawaii is shown in Fig. 1-33.

Figure 1-34 shows a departure from the traditional Hawaiian architecture in a modern A-frame home.

C. W. Sorensen

Fig. 1-31.   A Japanese house built around existing trees.

R. Wenkam, photographer

Fig. 1-32.   Chinese influence on Hawaiian home.

R. Wenkam, photographer

Fig. 1-33.   A Hawaiian house. The exterior is more conventional, but interior shows Japanese influence.

R. Wenkam, photographer

Fig. 1-34.   A break with tradition in Hawaii.

Hutchinson Photographers

Fig. 1-35.   A Pacific style house.

The Pacific Influence

The Pacific style tends to be a blend of the traditional and the contemporary styles. The predominant influencing style is Japanese architecture. The Pacific style tends to use Japanese ideas for house arrangement, but does not use their design details. These are more traditional American. It is built on a 4'-0" × 8'-0" modular component.

Open room planning and outdoor living are part of this style. It establishes a pleasant, indoor-outdoor relationship through the use of glass and view; yet it retains some of the traditional style in overall appearance.

The roof has considerable overhang and usually has a small gable at the ridge. The slope changes at the eave and the ends of the rafters (lookouts) are exposed, Fig. 1-35. It has been influenced by Japanese styling and has an Oriental appearance. The style makes use of sliding screens and natural materials. Stained wood exteriors and wood shingle roofs are characteristic.

Private patios with trellises, pools, and gardens are used. Glass walls face this area, while the street side has few or no windows, thus insuring privacy.

Quarry tile is used on floors in some rooms, as well as on countertops. This is an Oriental influence.

The house is usually low, irregular in shape, without a basement. The delicate framing and rectangular shapes of the Japanese predominate.

The Florida House

Geographic locations often influence residential architecture. Such a case is the architecture in the extreme Southeast and especially in Florida. These structures are sometimes referred to as the Florida house, Fig. 1-36.

Fig. 1-36.   The Florida house includes large screened areas for outdoor living.

Photo by Philip H. Hiss

Fig. 1-37.   A large, private screened roof area off a bedroom.

Fig. 1-38.   A large, screened yard area. Photo by Philip H. Hiss

This type of house is open to the breeze, with many houses having every room with large screened openings. Figure 1-37 shows an area with screened roofing built off of a bedroom. Some houses have screened areas of grass outside. See Fig. 1-38. Even swimming pools are screened as shown in Fig. 1-39. Though many houses are air-conditioned, they still have large areas that can be opened to the air. Such a house has no definite boundary between indoors and outdoors. The Japanese call this characteristic "engawa," meaning an uncluttered space belonging inside and outside.

This house was developed in an attempt to meet the special climate conditions and to enable the occupants to get the most out of living in Florida. The design of the Florida house met the big problems of hot sun, steamy rain, and insects. It also met the demand for outdoor and informal living. The houses have large overhangs to shield windows from the sun, and they use sun screens and shrubs for additional protection. The large overhang, louvered sun screens, and awning-type windows enable the house to receive cooling breezes during the season of heavy rain. See Fig. 1-40.

Since the stress in Florida is on leisure and informal living, the Florida house is designed to be easily maintained. Much brick is used for exterior walls, with prefinished plywood popular for interior walls. Floors are commonly terrazzo and need practically no care. (Terrazzo flooring is made of chips of marble, granite, or other colored stone embedded in cement and polished smooth.) Aluminum and plastic panels are used on the exterior and interior, and much glass is evident.

The California House

Out of the architectural revolution on the Pacific Coast developed a type of house commonly referred to as the California house. It grew out of an awareness of environment and a way of life. The main concept is that of outdoor-indoor living.

Photo by Robert H. Ford

Fig. 1-39.   A screened indoor pool in a Florida house. Notice the informal atmosphere for relaxed indoor-outdoor living.

House and Home Magazine, Rudi Rada, photographer

Fig. 1-40.   Large overhang and louvered openings help weather condition a Florida house.

The California house is designed for an informal, relaxed way of life. It takes full advantage of California's bright sunshine and luxurious growth, but also provides shelter against the abundant rainfall. While outdoor living can take place year-round in much of California, provision is made for chilly evenings and nights.

The California house is related to its natural environment, as well as to the individual site, and is an attempt to provide a restful, yet exciting experience. See Fig. 1-41. It uses natural materials in

a natural setting. Stone, brick, and wood are characteristic. Glass is widely used in an attempt to relate the outdoors to the interior setting.

Many California cities are located upon mountainous terrain. These sites provide full and extended views, as well as opportunities to utilize the sun and wind. The site, as well as fences and walls, is used to insure privacy.

The California house is a simple house. The structure is prominently exposed and provides the character. Unnecessary embellishments are avoided.

The heat and glare of the sun are controlled by a large overhang. Natural light in each part of the house is controlled by clerestory windows, translucent panels, and louvered screens.

Considerable planting (interior as well as exterior) is characteristic of this house. It is considered as an integral part of the overall design.

The roof is flat or the shed type. Interior walls are commonly wood, using the intrinsic color and grain to give a feeling of naturalness.

Large, outdoor terraces that open to the sky are characteristic of the California house. These are usually paved with bricks, wood blocks, or other suitable materials. Again, plantings tie the house and terrace together as a unified whole.

Fig. 1-41. A sketch of a California house showing its relationship to simple mountainous terrain, the uti of the sun, and the simple functional use of each part.

Architecture Today

The house of today is designed for livability. It is designed and built to meet the needs of today's way of living. It is an attempt to look to the present and future needs of the family. As the workday shortens and the three-day weekend becomes a reality, the home will play an increasingly important role in leisure-time needs. It must include family recreation space for all kinds of activities, but must still provide a high degree of privacy for every member of the family. Such a house must be flexible so that rooms can be changed as needed.

The contemporary house is manifestly suburban and informal in character. It is less pretentious than the traditional house and is an outward expression of our industrial age and the American way of living.

The living areas face private gardens or a distant view rather than the street. The design is a supreme attempt to secure a close relationship between indoor and outdoor living and bring the outdoors into the house. Considerable indoor planting is a part of the contemporary house. This is not only attractive, but helps to maintain a healthful humidity level. In some designs, much of the exterior wall surface is glass, usually in the living areas. This is especially useful when designed as part of the solar heating system. The feeling desired is one of openness and

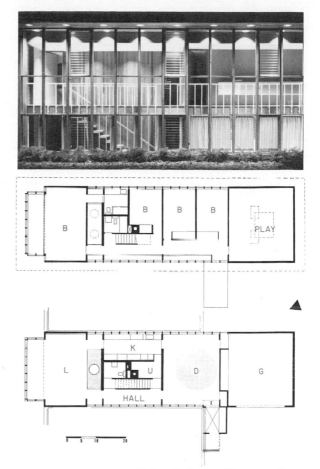

The Secondary Treasury of Contemporary Houses

Fig. 1-42.   A contemporary house with floor plan — the Kirkpatrick house, Kalamazoo, Michigan, Norman Carver, Jr., designer.

Western Wood Products Association

Fig. 1-43.   The natural color of exposed wood beams and roof decking provides a warm feeling.

California Redwood Association

Fig. 1-44.   Today's home is designed so the house and its surroundings blend together to form a unified whole.

Western Wood Products Association

Fig. 1-45. A wide cantilever gives this house a soaring appearance.

freedom. People should not feel enclosed by walls. The interior floor plan is an open plan with few partitions. Frequently, the kitchen, dining, and living areas are one large room with perhaps a screen or fireplace serving as a partial divider. See Fig. 1-42.

Exposed ceiling beams are common, and much emphasis is placed on the use of natural materials. The material is not painted, but is permitted to reflect its own characteristics of color and texture. See Fig. 1-43.

The house and its lot are treated as a whole. Together they form a single unit and present a feeling of belonging together. The house in Fig. 1-44 was designed in an irregular U-shape to avoid removing trees and natural vegetation. One side faces the pond. The other overlooks the ocean. The structure appears to have grown here along with the trees and plants.

A house may be anchored snugly to the ground or suspended above the site, depending upon the conditions found. See Fig. 1-45.

The exterior design is simple. Window sizes vary according to the needs of the occupants. Doors are placed where needed, rather than where they will produce a symmetrical appearance. The structural skeleton is not covered and becomes part of the character of the house. Materials requiring a minimum of maintenance are selected. See Fig. 1-46.

American Plywood Association

Fig. 1-46. Today's homes reflect simple, clean-cut exterior design. The structural frame of this house sets the design tone.

Fig. 1-47. The flat roof contributes to the lines, mass, and character of this house. The natural materials help blend it in with the surroundings.

The roof framing is commonly wood or steel beams, or trusses which are the rafters and the ceiling joists combined into one unit. The exterior wall panels may be hung on this skeleton. They can be glass, aluminum, brick, wood or any desired material. Complete freedom of interior arrangement is thus available, with flexible room dividers, such as large panels, cabinets, glass, etc. These partitions can be moved to new locations to permit rapid changes in the area of partially enclosed space. They sometimes extend to a height of six or seven feet, which is just sufficient to break the lines of sight. Even the closets are designed for flexible rearrangement of hanging rods and shelf space.

Interior materials are also selected for easy maintenance. Tile floors commonly replace hardwood flooring. Walls of glass, wood paneling, and plastic materials are common.

The house of today utilizes laborsaving and automatic devices extensively. Temperature control, stereophonic phonographs and speakers, television, and many kitchen devices are among the services built in.

A variety of roof styles are used. The flat roof is popular and contributes to the rectangular mass of a building. See Fig. 1-47. The redwood siding blends this house into its natural setting. The large glass panels take advantage of the beauty of the surroundings.

The two-story residence in Fig. 1-48 utilizes a large roof overhang to shield the large windows. These windows provide a substantial source of solar heat. This residence has the redwood siding bleached to give a weathered appearance. It blends in with the trees in its Virginia setting. The plan includes a living room, den, study, dining room, and three bedrooms. Dressing and laundry areas are provided.

An unusual design is in Fig 1-49. The elevation of the living area to provide auto storage below is

SECOND FLOOR

FIRST FLOOR

Fig. 1-48. A contemporary house with floor plan. The design of the roof creates a feeling of rectangular mass. The roof shelters the windows, yet allows the sun to heat the rooms.

Fig. 1-49. A contemporary residence using flush tongue and groove siding.

Fig. 1-50. This dramatic design uses wood shingles on the roof and walls.

unusual. The dramatic roof and dual chimneys provide the center of interest and give character to the building.

A dramatic design is in Fig. 1-50. The flaring roof and heavy wood shingles provide the character for the house. The woodshake siding ties together the roof and deck areas.

Special engineering problems had to be solved to build the house in Fig. 1-51. It extends over the waters of San Francisco Bay and is supported on concrete pilings and redwood timbers. Only the garage and front entry are over dry ground. Every room is oriented to the water. The redwood siding was left natural. A redwood deck extends on three sides providing the only exterior walking area. Inside, the ceiling, beams, and posts are redwood. Interior walls are painted white to provide contrast.

A wide variety of houses are being designed today. See Fig. 1-52. The ones shown represent current trends in architecture.

FIRST FLOOR   SECOND FLOOR

Fig. 1-51. This house extends over San Francisco Bay. The designer had to overcome special engineering problems.

Fig. 1-52. Modern architecture applies new uses of natural resources. This solar home uses the energy of the sun for heating and air conditioning.

Residential Design and Construction

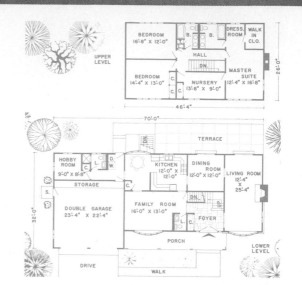

Chapter 2

Planning the Individual Rooms

A house is designed to meet the desires of the owner. Generally the individual rooms are planned first. These rooms are then related to one another to form the floor plan. (Floor plans are discussed in Chapter 4.)

When planning a room, the designer must make decisions concerning the furniture and fixtures to be used in each room. Furniture size and style will influence the room design. The room dimensions will be determined by the size and amount of furniture planned.

List the furniture to be in each room. For example, a dining room planning list could include the following: 1 dining table 48″ × 72″ (1219 mm × 1829 mm), 6 dining chairs 20″ × 21″ (508 mm × 533 mm), 1 buffet 16″ × 48″ (406 mm × 1219 mm). Similar lists should be made for the other rooms. The actual size of the furniture varies with the style. If possible, the style should be known so size requirements can be reasonably accurate. Typical furniture sizes are given as the individual room planning principles are presented in this chapter.

Furniture Templates

The use of furniture **templates** makes planning easier. Furniture templates are made by drawing the top view of each piece of furniture on cardboard. These show the amount of space required by the furniture.

Templates are made to the same scale used to draw the floor plan. For residential work this is usually 1/4″ = 1′-0″ or in metric 1:50 measured in millimeters. (Information on the metric system of measure is presented in Chapter 5.) Furniture templates for commercial work are usually 1/8″ = 1′-0″ or in metric 1:100. Each template should, when possible, have lettered on it the name of the

item it represents. Templates should be used to locate custom-built items such as wall-hung units or room dividers made of cabinets and bookshelves.

Using the Templates

Furniture templates are used to design a livable room. The principles of room planning must be observed and the desires of the owner of the house must be considered. Arrange the templates on a sheet of graph paper having squares representing the scale used to make the template. For residential work, this would normally be 1/4-inch squares. See Fig. 2-1.

Metric graph paper is divided into 1 mm squares with heavy lines every 5 mm. At a 1:50 scale, 1 mm would represent 50 mm. The 5 mm line would represent 250 mm. The items in Fig. 2-1 drawn to the metric scale 1:50 would appear as in Fig. 2-2.

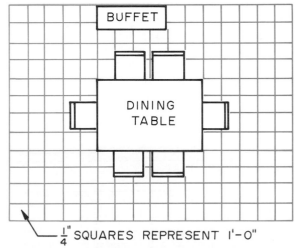

Fig. 2-1. Furniture templates placed in relationship with each other using inch measurements.

Notice on this trial plan that the room has no walls. This allows freedom in arrangement. Once a workable plan is developed, the walls can be drawn to establish room dimensions as shown in Fig. 2-3.

The room templates are then used to develop the floor plan. Doors and windows to each room are also planned. It may become necessary to change the furniture arrangement or the size of a room.

Typical Room Sizes

There are no perfect or ideal room sizes. These vary with the planned activities for the room. Figure 2-4 gives some suggested minimum sizes as an aid to planning. Remember, the larger the rooms, the more expensive the building becomes. Larger rooms also increase heating and air-conditioning costs. However, if a room is too small, it will not fully serve its planned purpose.

The size of some rooms can vary with the size of the house. For example, a house with two bedrooms will probably have fewer occupants than one with four bedrooms. Therefore, it is logical that the kitchen, dining area, and living room area should be larger for the four-bedroom house. There is no universal agreement on minimum room sizes.

A good way to check room sizes selected for a plan is to measure rooms in your present house. See how these sizes compare with those in your plan. Based on how well your present house suits your needs, decide whether the sizes you selected for the rooms in your new home are suitable.

Planning the Living Room

The living room is the most versatile room in the house. It serves as a sitting room, dining room, television room, extra bedroom (if you have a sofa bed), music room, den (if your desk is there), and library. Usually, it is the largest room in the house and the best furnished. When people come to visit,

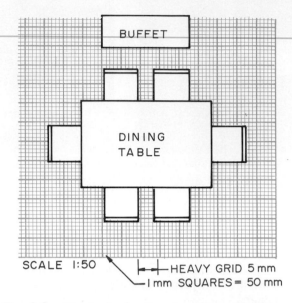

Fig. 2-2. Metric size furniture templates placed on a metric grid. Thick lines are every 5 mm and the thin lines are every 1 mm.

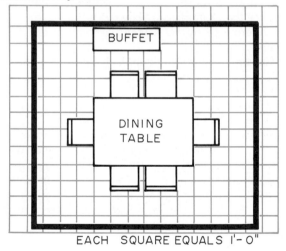

Fig. 2-3. Dining furniture located in a room with minimum dimensions.

Rooms	One- or two-Bedroom House		Three-Bedroom House		Four-Bedroom House		Least Room Dimension	
	Sq. Ft.	Sq. Meters	Sq. Ft.	Sq. Meters	Sq. Ft.	Sq. Meters	Lineal feet	Milli-meters
Living Room	175	16.3	200	18.6	240	22.3	11'-0"	3350
Dining Room	120	11.1	180	16.7	250	23.2	9'-0"	2740
Master Bedroom	140	13.0	120	11.1	120	11.1	9'-4"	2840
Other Bedrooms	110	10.2	110	10.2	110	10.2	8'-0"	2440
Kitchen	100	9.3	120	11.1	140	13.0	8'-0"	2440
Bath	40	3.7						

Fig. 2-4.   Suggest minimum room sizes.

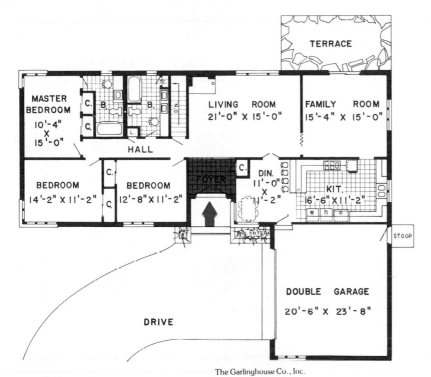

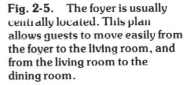

The Garlinghouse Co., Inc.

Fig. 2-5. The foyer is usually centrally located. This plan allows guests to move easily from the foyer to the living room, and from the living room to the dining room.

Fig. 2-7. (Below) The dining area of a combined living-dining room should be located near the kitchen.

American of Martinsville

Fig. 2-6. An open living-dining area. Notice the attention given to the use of color in the furniture.

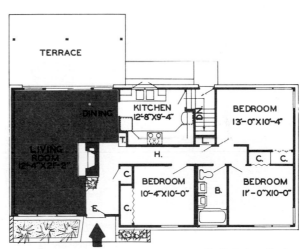

The Garlinghouse Co., Inc.

one of their first impressions of your house is made when they enter the living room.

Planning Considerations

1. Plan the living area to be in a central location near the front entrance.
2. Have the front entrance open into a **foyer,** not into the living room. Less expensive houses often omit the foyer as an economy. However, when the entrance opens into the living room, part of the room has to serve as a hall. This may interfere with the way people move about the house. A good plan is in Fig. 2-5.
3. Locate the living room and dining area near to each other. Guests normally visit in the living room and then move to the dining area for the meal. See Fig. 2-6. If the living and dining areas are in one room, the dining area should be near the kitchen. See Fig. 2-7.

4. Decide which of the two general types of living rooms is preferable — open or closed. An **open living room** has a minimum of walls. The living area, dining area, and sometimes the foyer and parts of the kitchen flow together. Often, some kind of low room divider or screen is used as in Fig. 2-8. A fireplace can be a divider. Even a high-pitched cathedral ceiling may be used. Folding doors make excellent room dividers.

A **closed living room** has walls to the ceiling which completely enclose it. It opens into other rooms through doors or small arches. A plan with a closed living room is shown in Fig. 2-9.

5. Locate the living room facing a pleasant view. See Fig. 2-10. If a good view is not available, an alternative is to build a landscaped, fenced area opening off the living room. A living room may also open onto a balcony, patio, or screened porch. This design is especially popular in warmer climates. See Fig. 2-11. If a pleasant view is not possible, however, avoid putting in a large window or glass wall.

PPG Industries

Fig. 2-8. A divider can be used to separate living and dining areas in an open plan.

PPG Industries

Fig. 2-10. Living rooms should face a pleasant view.

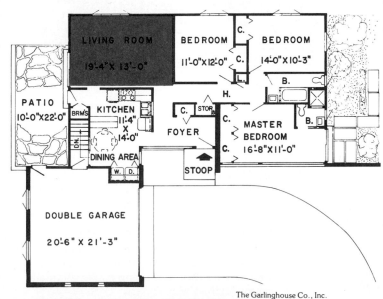

The Garlinghouse Co., Inc.

Fig. 2-9. A closed living room plan.

PPG Industries

Fig. 2-11. A living room often opens onto a patio or balcony.

Fig. 2-12. Furniture can be placed out in a room.

Fig. 2-13. A fireplace can serve as the center of attention in a living area.

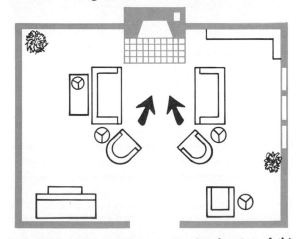

Fig. 2-14. The fireplace is the focal point of this room.

6. In a living room with large windows or a glass wall space, place furniture in open areas of the room away from the walls. See Fig. 2-12.

7. Consider using unusual design ideas. A sunken or elevated living room has identity and uniqueness.

8. Center the room about some focal point, such as a fireplace, large windows, or a special piece of furniture. See Fig. 2-13. Chairs and sofas normally are grouped about the focal point. See Fig. 2-14. A living room may have a second less important furniture grouping. A large room may have three to five groupings. See Fig. 2-15.

9. Consider the need for built-in units, such as bookcases, television and record cabinets, log boxes by the fireplace, and planters. These influence furniture arrangements and movement about the house.

10. Keep the decor of the living room consistent with the style of the house. If the style is contemporary, the interior design should reflect the clean-cut lines of this style. A Colonial style should have traditional wall covering patterns, moldings, and subdued use of color. The living room makes the greatest impression upon visitors. It is the showplace. It should be comfortable, inviting, and appear uncrowded and neatly arranged. The variety of interior wall finishes, lighting, and carpeting and other floor

Fig. 2-15. A formal living room with the furniture grouped around a table to form a conversation center. Notice the built-in bookshelves and how the fireplace with a balanced window design forms a focal point.

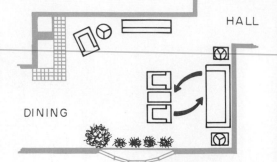

Fig. 2-17. A living room arranged for conversation.

Fig. 2-16. An exciting, comfortable living area. The glass gable end allows a wide view of the outside. The large overhang affords protection from the sun. The fireplace has a wide, raised hearth which serves as a bench. The artificial lighting provides a dramatic touch.

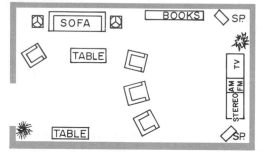

Fig. 2-18. This furniture is arranged for using the television, radio, and stereo in the entertainment center.

Fig. 2-19. This room receives many hours of exposure to the sun. It is bright and cheery all day.

covering material provides the designer with extensive choices. See Fig. 2-16.

11. Constantly be aware of patterns of movement in the living room. These can influence room size and shape as well as furniture arrangement. (A section on traffic patterns is included in Chapter 4.)

12. Arrange furniture to allow easy movement about the room. A comfortable major aisle should be about 3'-6" (1067 mm). A minor aisle should be about 2'-0" (610 mm). The aisle from the entrance of the living room to the major seating center should be from 3'-6" to 4'-6" (1067 mm to 1372 mm).

13. Space furniture so conversation is held easily. Refer back to Fig. 2-13 for an example. Furniture facing each other in a conversation grouping should be from 4'-6" to 6'-0" (1372 mm to 1829 mm) apart. Anything over 6'-0" makes conversation difficult, Fig. 2-17. If an activity such as TV viewing is important, special arrangements can be made. See Fig. 2-18.

14. Consider including a fireplace in the living room. Generally, it is used for the pleasure of watching the fire. More and more often it is used as a supplementary source of heat.

Azrock Floor Products

Fig. 2-20. The floor covering adds to the overall appearance of this divided living area.

American Olean Tile Company

Fig. 2-21. Ceilings can be of many types. This exposed beam ceiling is popular.

Special fireplace units are available which deliver considerable heat to the living area. (See Chapter 12, Heating and Air Conditioning.) The design of the fireplace, the mantel, and hearth should be in keeping with the style of the house.

15. If possible, orient the living room toward the south. The south wall receives many hours of exposure to the sun. South rooms are bright and well-illuminated all day. In the winter, they receive considerable solar heat. In the summer, the south wall needs protection from the long hours of hot sun. Window walls are effective on the south. See Fig. 2-19.

16. Select window styles for the living room that are consistent with the style of the house. Window styles and placement must allow for comfortable furniture arrangements. Think about privacy as you locate the windows.

17. Consider the design of the living room when selecting floor covering. A deep, rich carpet is good for one type of living room. See Fig. 2-20. A ceramic tile, flagstone, or vinyl floor covering is appropriate for other designs. The selection of a floor covering is very important to the finished results.

18. Decide which kind of ceiling to use. A variety of ceilings are available. Most common is the flat ceiling which is painted or papered. Contemporary houses often use an open-beam ceiling with exposed wood or composition decking. See Fig. 2-21. The ceiling is often a cathedral or shed ceiling. These open, higher ceilings make the rooms appear larger, but also make more air space to heat and cool.

19. Plan the lighting in the living room carefully. Some subtle form of general illumination is needed. Special features, such as a fireplace, might require spotlights. Provisions for reading can often be made by using furniture brought into the room.

20. Since the furniture in the living room is likely to be rearranged occasionally use special care in locating electrical outlets.

21. Remember, there is no best shape for a living room. However, it is easier to plan arrangements for a rectangular room. Arrangements for a square room are quite difficult to plan.

22. Make certain the room provides enough wall space to hold the furniture needed. A room with many doors and windows can be very difficult to arrange.

23. See Fig. 2-22 for selected living room furniture sizes and templates.

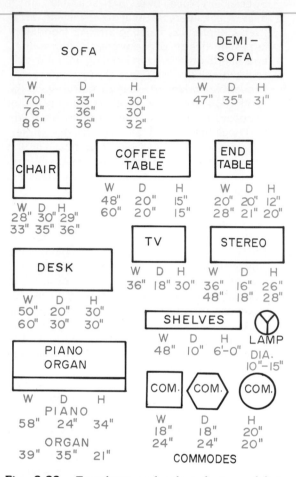

Fig. 2-22. Templates and selected sizes of living room furniture.

Fig. 2-23. A den should provide storage and a place to work.

area are needed. Writing requires a desk and good lighting. A person who does drafting will need drafting equipment.

If necessary, a corner of a living room can be used to serve, in a small way, as a den. It can be set up to provide the facilities for reading and study.

Planning the Family Room

The **family room** tends to be a second living room. It is one in which the family can take part in hobbies, sewing, music, television viewing, and other activities. It serves as the indoor play center for children. The family probably lives in this room more than any other. The formal living room is spared the wear and tear of day-to-day family activities.

The room must be planned for the activities enjoyed by the family. In addition, the room and furnishings should be composed of durable materials to withstand the hard use. Materials should also be easily cleaned. Usually, this room is cheerfully decorated and well lighted. Since noise is a problem, acoustical ceiling tile and other noise-control devices should be considered. (See the section on sound control in Chapter 13.)

This room may be located in any convenient place. Since it is a noisy area, it must be away from the quiet areas of the house. It does become messy due to the activity and needs to be out of the usual traffic pattern of visitors. A popular location is off the kitchen. Refer back to Fig. 2-5. Sometimes, the snack bar or informal eating area is between the kitchen and family room. In this location, the children can be supervised as meals are prepared.

Planning the Den

The **den** tends to be a quiet room where a person goes to study, read, or write. See Fig. 2-23. It can be designed to serve as a guest room, and should be located in a quiet area of the house. If the activities in the den are to be entirely private, it can be located in a basement or second floor room. If it serves as an office in the home to which clients come, it will need to be close to the entrance. Possibly it will need a separate outside entrance. A den should be comfortable, pleasant to work in, and utilize materials that reduce noise.

The furnishings found in dens depend upon the planned activity. If a den is to be used for reading and study, book storage and a comfortable reading

A family room and living room may be located back to back with some type of divider in between. This is useful when a large group is to be entertained. Both rooms can be used.

Because of the variety of activities, special attention must be given to providing adequate storage space. Closets, built-in cabinets, and furniture with storage capacity are needed.

Planning the Recreation Room

A **recreation room** provides space for more vigorous activities than those which generally take place in a family room. Typical activities include ping-pong, pool, dancing, and informal parties. Some activities, such as hobbies, that can occur in a family room can also be done here.

The recreation room should be located away from the day-to-day activities. It may be located in the basement. The basement area is usually of a nature to withstand the activity. This location also helps control noise.

Recreation rooms located on the first floor often open onto a patio or balcony. This allows expansion of the activities or party. It is especially helpful if a swimming pool is available.

Recreation rooms tend to be decorated in gay, cheerful colors. The materials used must be durable and easily cleaned. See Fig. 2-24.

Planning the Dining Area

The typical family has many different kinds of meals which occur at irregular hours. When family members maintain different schedules, a small eating area, preferably in the kitchen, is useful. If guests are entertained at rather formal dinners, a formal dining room is desirable. Informal entertaining on a porch, balcony, or patio requires food service to that area. The designer must consider the eating habits of the family as the dining areas are planned.

Planning Considerations

1. Consider the types of meals to be served. Formal dinners require a separate dining room and a large table with several chairs. See Fig. 2-25. Often, a formal dining room must serve eight to twelve persons. A buffet, serving cart, and china cabinets are required. Less formal meals are usually eaten in the kitchen. See the section on kitchen planning next in this chapter.

2. Locate the dining area near the kitchen. This location makes it easy to serve and to clean-up after meals. See Fig. 2-7. Provide a way for people to move to the dining area without going through the kitchen. There should also be

Fig. 2-24. A recreation room utilizes durable materials. The floor receives unusually hard wear.

American of Martinsville

Fig. 2-25. A formal dining room. The table can be expanded to serve eight people. The light above the table provides a subdued illumination.

a route to the kitchen without going through the dining room.

3. Locate the dining area near the living room for easy access by guests. Refer again to Figs. 2-5 and 2-6.

4. See Fig. 2-26 for the most common locations for dining areas. Remember, too, that snacks are usually a part of the activities in a recreation room, family room, or patio. The closer these areas are to the kitchen, the easier it is to serve food there.

5. When locating a dining area, consider the view. A pleasant view contributes to the pleasure of dining.

6. Design the dining area large enough for people to be seated and the food served. Spacing recommendations are in Figs. 2-27 and 2-28. Notice that at least 2'-6" (762 mm) is required between the back of a chair when a person is seated and the wall or a piece of furniture. If this is to be an aisle for serving, it should be 3'-6" (1067 mm).

7. Provide a way to separate the dining area from the kitchen. This can be some form of partial room divider, folding doors, or even a fireplace.

8. Design the dining area to be quiet and comfortable. A snack or light meal area should be cheery and bright. A formal dining room is more subdued and luxurious. In either case, the style of the dining room should be in keeping with the style of the house. The wall, ceiling, and flooring material must reflect the design. The furniture selected is critical to a proper setting and atmosphere. See Fig. 2-29.

9. Plan for the lighting in a formal dining room to be on a dimmer switch. This type of switch allows greater regulation of the amount of light. For example, adjustments can be made for a candlelight meal which requires only a minimum of electrical illumination. Lighting can be by the typical chandelier or other types of hanging fixtures. Usually it is placed directly over the dining table. See Fig. 2-29. Floor lamps and cove lighting are also effective.

10. Provide storage for linens, dishes, and silver very near the dining area.

11. Remember, there is no best size or shape for a dining area. In general, a rectangular area is

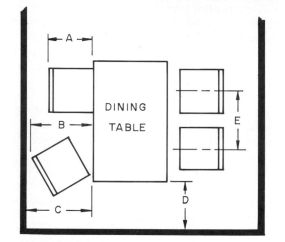

Fig. 2-27. Dining room space requirements.
  A. Seated person occupies 1'-6" to 1'-10".
  B. Average person requires 2'-6" to 3'-1" to rise from table.
  C. If an aisle for passage behind a chair is required, 3'-4" to 3'-6" space is needed.
     If an aisle for tray service is required, 4'-10" to 5'-4" space is needed.
  D. If only an aisle is needed, 2'-0" is sufficient. If a chair is to be at this end of the table, apply parts A, B, and C above.
  E. A minimum of 2'-3" is needed between the centers of chairs.

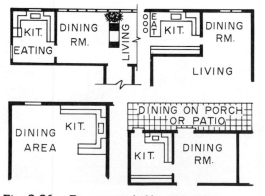

Fig. 2-26. Recommended locations for dining areas.

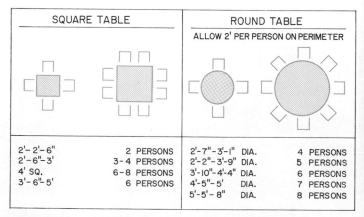

SQUARE TABLE		ROUND TABLE	
		ALLOW 2' PER PERSON ON PERIMETER	
2'- 2'-6"	2 PERSONS	2'-7"-3'-1" DIA.	4 PERSONS
2'-6"-3'	3-4 PERSONS	2'-2"-3'-9" DIA.	5 PERSONS
4' SQ.	6-8 PERSONS	3'-10"-4'-4" DIA.	6 PERSONS
3'-6"-5'	6 PERSONS	4'-5"-5' DIA.	7 PERSONS
		5'-5"-8" DIA.	8 PERSONS

Fig. 2-28.   Standard table sizes and seating requirements.

Fig. 2-29. Select furniture appropriate for the area in which it will be placed. Notice the difference between the furniture used in this dining room and that selected for the patio.

Western Wood Products Association

Armstrong Cork Company

Fig. 2-30. The vinyl floor requires little maintenance in this cheerful, pleasant dining room.

easier to arrange. If an expandable dining table is planned, the room should be large enough to hold it in its largest position.

12. Select the type of dining room which best meets the needs. A dining room can be closed or open. If closed, it is confined within the four walls of the room. See Fig. 2-5. If open, it flows into other rooms, generally the living room. See Fig. 2-6. The open plan provides greater flexibility for handling larger groups.

13. Select floor covering carefully. It must be easily cleaned, stand the abrasive action of chairs, and yet enhance the style of furniture to be used. See Fig. 2-30.

14. Remember as you plan, that a dining area at the end of a living room or kitchen will require less space than a formal dining room. People will not have to contend with four walls when leaving the table or serving food. Typical dining room plans are in Figs. 2-31 and 2-32.

Fig. 2-32. (Far right) A medium-sized dining room. Eight people can be comfortably seated.

Fig. 2-31. (Right) A small dining room. It is very confining for six people.

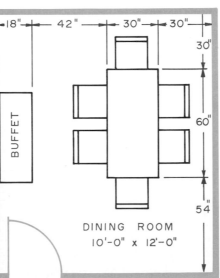

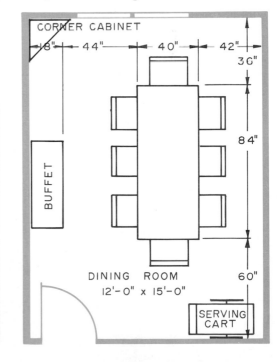

Notice the minimum for a small formal dining room is about 10'-0" × 12'-0" or 120 square feet (11.1 m²). In an open plan, the same activity can be held in about 80 square feet (7.4 m²). A room 12'-0" × 15'-0" or 180 square feet (16.7 m²) will accommodate six to eight people. Recommended dining area sizes are in Fig. 2-4.

15. See Fig. 2-33 for selected dining room furniture sizes and templates.

Planning the Kitchen

The kitchen has received considerable attention from home planners. It has become quite mechanized and is planned with close attention to efficiency and a minimum of walking. Kitchens are planned for food preparation, laundry activities, and sewing and mending. Frequently, it is combined with the family room, making a modern version of the former large, family kitchen with sofa, television, hobby space, and other provisions for family activity. The kitchen serves as a snack bar, food storage space, utensil storage space, freezer room, and informal location for buffet parties.

Efficient Kitchen Design

The basis of kitchen planning is a pattern of work flow for storing, processing, cooking and serving foods, and cleaning up after the meal. See Fig. 2-34. The kitchen should be arranged around three appliances—the refrigerator, cooking unit, and sink. These are sometimes referred to as the food storage center, cooking center, and preparation and cleanup center. See Fig. 2-35. Cabinets are built between these appliances to provide storage and a work surface.

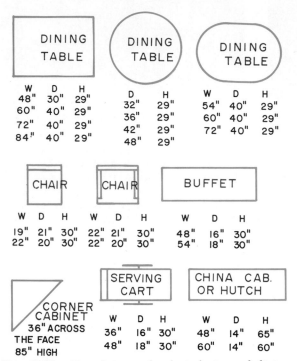

Fig. 2-33. Templates and selected sizes of dining room furniture.

Fig. 2-34. Pattern of work flow. Notice how the major appliances have been located for ease of work, with storage facilities for utensils and equipment near the point of use.

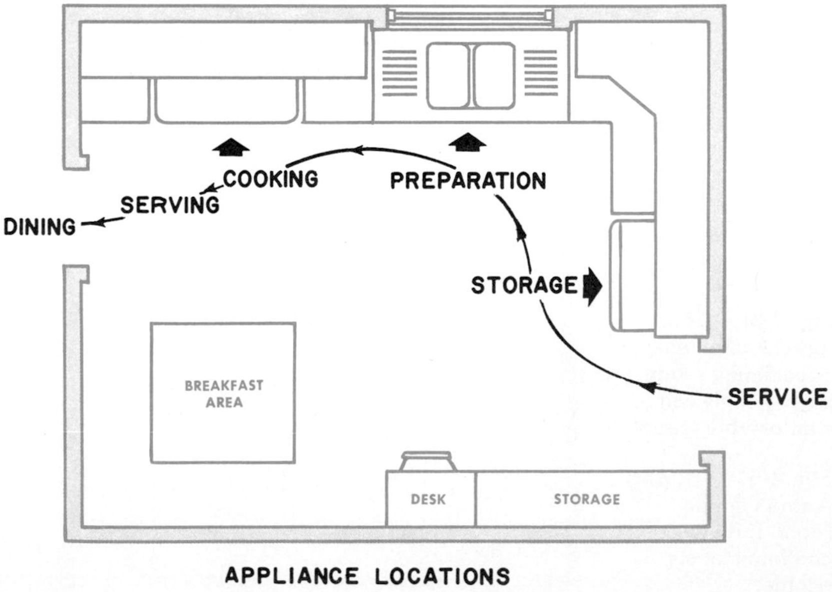

APPLIANCE LOCATIONS

PREPARATION AND   COOKING CENTER   FOOD STORAGE
CLEAN-UP CENTER   CENTER

Fig. 2-35. The three basic centers around which a kitchen is planned.

One way used to check the efficiency of a kitchen is to measure its **work triangle.** This is the total straight-line distance from the range to the refrigerator, the refrigerator to the sink, and the sink to the range. This is how far a person would have to walk to move from one center to another. A distance of 22 feet (6706 mm) is considered maximum. As a general rule, the distances between appliances should be as follows: sink to refrigerator 4 to 7 feet (1219 mm to 2134 mm), sink to range 4 to 6 feet (1219 mm to 1829 mm), and range to refrigerator 4 to 9 feet (1219 mm to 2743 mm). See Fig. 2-36.

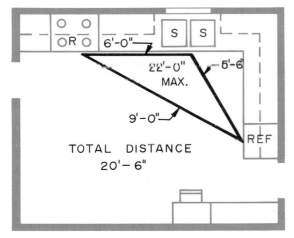

Fig. 2-36. The work triangle is a measure of the efficiency of a kitchen plan.

The Food Storage Center

The **food storage center** is designed to hold fresh, frozen, dried, and canned foods. It is located near the outside door for convenience when bringing in and storing groceries. The refrigerator-freezer is the heart of this center. This may be a single appliance or two separate units. The freezer can be located in a less desirable location because it is not used constantly. The refrigerator is used often and

requires convenient access. There should be at least 1'-0" (305 mm) of counter space beside the refrigerator to hold articles being taken from it.

The center includes cabinets to store fresh vegetables, dried foods, canned goods, and bakery products. Various special accessories are available for this storage, such as built-in bread boxes, sugar and flour bins, and sliding and revolving shelves for canned goods.

The Cooking Center

In the **cooking center,** the food is heated, broiled, baked, or fried. It is located near the dining area to make serving easier. The cooking units, oven, and possibly a microwave oven form the heart of this area. A variety of units are available. Some have the surface-cooking and oven as a single unit. Others have separate surface-cooking units and ovens. Since the surface-cooking unit is used often, it is given a central location not far from the food-preparation and cleanup center. Separate ovens can be placed in a less desirable location since they are not used as much. Allow 2'-0" (610 mm) or more of counter space beside surface-cooking units for placement of pots and pans. At least 2'-0" of counter space beside the oven is helpful when removing baked items.

Many microwave ovens are portable and are placed on a countertop. Space must be planned for this and a special 115 volt, 15 ampere electrical circuit should be provided.

A hood should be planned over the surface-cooking unit to remove cooking fumes and water vapor. Space for storage of cooking utensils is needed. Often, small appliances, such as toasters, coffeemakers, and broilers, are used in the cooking area. Countertop area and electrical outlets must be provided.

The Preparation and Cleanup Center

The **preparation and cleanup center** is designed for the preparation of foods for cooking and for cleaning up the dishes and utensils after the meal. The sink is the key unit; however, the disposal, dishwasher, and compactor are involved.

The sink is usually placed next to the food storage center. This arrangement makes it easy to get the food from storage, prepare it for cooking, and return it to storage until time to cook it. Some people prefer the sink to be by a window. While this is pleasant, it is not necessary. The sink can be placed on an inside wall, a peninsula, or an island cabinet. The important thing is to give it a central

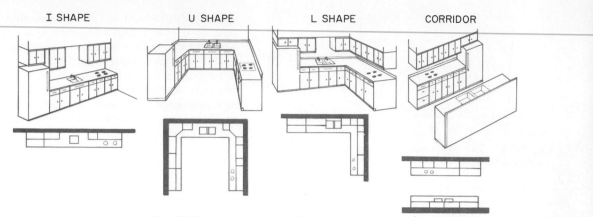

Fig. 2-37. Basic kitchen shapes.

location. In large kitchens, a second sink might be placed near the cooking center.

Allow 3'-0" to 4'-0" (914 to 1219 mm) counter space on each side of the sink. This is needed for stacking dishes or preparing foods.

The dishwasher is located next to the sink. For convenience, locate it left of the sink for right-handed persons or right of the sink for left-handed persons. The trash compactor is placed on the other side of the sink.

The disposal is placed beneath the sink used for food preparation and cleanup. The switch to operate it should be far enough away from it so that a person could not have fingers in the unit and turn it on at the same time.

Basic Kitchen Shapes

Kitchens are generally arranged in **I**, **L**, **U** and **corridor** shapes. See Fig. 2-37. In addition, **peninsulas** and **islands** are useful. See Fig. 2-38.

The **I-shaped kitchen** has cabinets and appliances along one wall. It is good for use in small

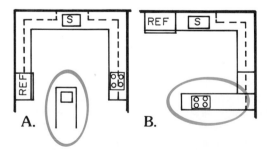

Fig. 2-38. Island and peninsula counters improve kitchen efficiency.
- A. A U-shaped kitchen with an island counter.
- B. An L-shaped kitchen with a peninsula counter. It serves as a room divider.

Connor Forest Products

Fig. 2-39. The I-shaped kitchen places all fixtures on one wall.

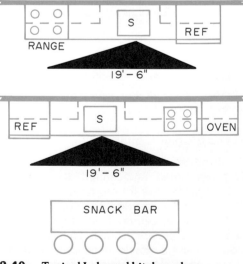

Fig. 2-40. Typical I-shaped kitchen plans.

Fig. 2-41. A minimum L-shaped kitchen. Notice that counter space is provided on each side of the cooking unit and sink.

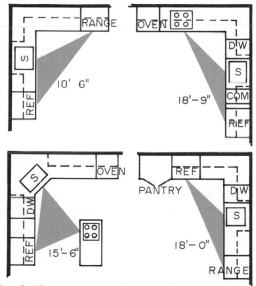

Fig. 2-42. A variety of L-shaped kitchen plans.

houses or apartments. See Fig. 2-39. Needed services can be planned in a minimum of space. Avoid making it too long, however. This design becomes difficult to use if too much walking is required. Typical arrangements are in Fig. 2-40.

The **L-shaped kitchen** has cabinets and appliances along two walls. See Fig. 2-41. This reduces the walking distance often found with I-shaped kitchens. If the cabinets on each wall are too long, however, the efficiency suffers. The L-shaped arrangement leaves a large, open floor

area which can be used for dining. Several L-shaped arrangements are in Fig. 2-42.

The **U-shaped kitchen** has cabinets and appliances arranged on three sides of the kitchen area. See Fig. 2-43. This design effectively reduces the walking distance between appliances. It is an efficient plan. The space between cabinets on opposite sides should be at least 4'-0" (1219 mm). If this is increased too much, efficiency is lowered. Several arrangements are shown in Fig. 2-44.

Fig. 2-43. This tight U counter puts storage, preparation, and serving areas close together.

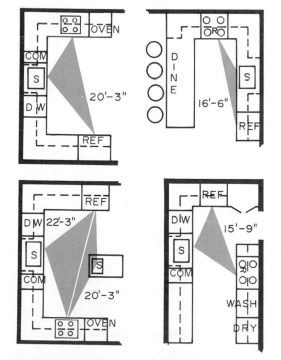

Fig. 2-44. A variety of U-shaped kitchen plans.

The **corridor kitchen** has cabinets and appliances on two opposite walls. It is used when a long, narrow area is available for a kitchen. It can be efficiently arranged. The space between the opposite rows of cabinets should be at least 4'-0" (1219 mm). Several examples are in Fig. 2-45.

A **peninsula** is a section of cabinet across the end of the kitchen that has no wall. It forms a divider to the next room yet is open on the top. It can be used for a sink, surface unit, food preparation area, eating counter or any other useful purpose.

An **island** is a section of base cabinet that is not connected to any wall or other cabinets. It serves the same purposes as a peninsula cabinet. See Fig. 2-46.

There is appeal in the large, old-fashioned, eat-in **country-style kitchen.** This design requires a large area because it includes all kitchen activities plus comfortable dining. See Fig. 2-47. Sometimes television, sewing, laundry, and/or game areas are included. Examples are in Fig. 2-48.

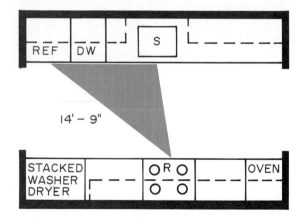
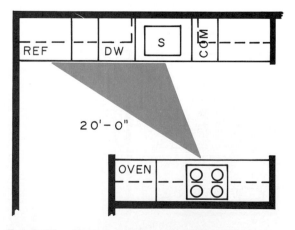

Fig. 2-45.   Typical corridor-type kitchen plans.

Location of the Kitchen

The kitchen must have an outside door for bringing in supplies and removing trash. Usually, the kitchen is planned to be next to a garage or carport and the area where trash is stored.

Locate the kitchen next to the dining room or dining area so that food may be served conveniently. See Fig. 2-49. Provide access to the kitchen without requiring persons to go through the dining room. Since food is often served on porches and patios, the kitchen should be closely related to them, if possible. Refer back to Figs. 2-5 and 2-9.

Other Planning Considerations

1. Most families desire a dining space in the kitchen. Figure 2-47 illustrates a spacious, well-planned kitchen with a nice dining area.
2. Helpful in a kitchen is a section of counter with leg room beneath so a person can sit to eat or work. See Fig. 2-50.
3. When a family has young children, the kitchen should be located to allow a view of the outside play area.
4. If much outdoor living is planned, facilities for simplifying the movement from kitchen to patio should be furnished.
5. If you desire an all-gas or all-electric kitchen, manufacturers of these items can provide considerable assistance for kitchen planning.

General Electric Company

Fig. 2-46.   **An island counter gives extra working and storage space.**

Fig. 2-47. The country-style kitchen includes a large eating area.

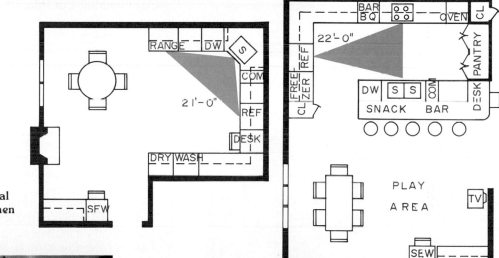

Fig. 2-48. Typical country-style kitchen plans.

Kitchen Kompact, Inc.

Fig. 2-49. This kitchen has the fixtures in an L-shaped plan with storage on a third wall. Notice the direct access to the dining area.

Coppes Napanee

Fig. 2-50. A counter provides a place where a person can sit to eat or work.

6. The kitchen should be bright and well lighted. See Fig. 2-51.
7. Electrical outlets should be plentiful, especially along the counter for use with minor appliances.
8. If it is difficult to work a satisfactory kitchen arrangement in the space allotted on the plan, the floor plan should be rearranged.
9. No thru traffic should be permitted through kitchen work areas.
10. The number of doors opening off the kitchen should be limited. Exterior doors opening into the kitchen can block cabinet doors and make them difficult to use.
11. Easy access to the front entrance from the kitchen should be provided, preferably not routing traffic through the dining room or the living room.
12. Enough cabinet space must be provided to store articles in use in each of the work areas of the kitchen. The area in which food is cooked and served should have space to store cooking utensils, silver, china, and spices.

13. Windows are important in a kitchen, but they reduce the amount of wall-cabinet storage space.
14. The average kitchen should have not less than 15'-0" (4572 mm) of free, clear, wall space available for use of cabinets and appliances. This does not include corners.
15. Standards for planning cabinets are shown in Fig. 2-52. Sizes vary depending upon the manufacturer. See Fig. 2-53.
16. A planning center in the kitchen is useful. Refer back to Fig. 2-34.
17. Laundry facilities are often located in the kitchen. See Fig. 2-54. Also see the section on planning laundry facilities, later in this chapter.
18. Templates and sizes of standard appliances are shown in Fig. 2-55.

Special Planning Considerations

Special designs may be required for a variety of reasons. For example, to design a kitchen that will comfortably accommodate a person in a wheelchair, consider the following:

1. Counter height 30 to 33 inches.
2. Keep 5 feet of clear space in front of appliance for wheelchair's turning radius.
3. Avoid smooth, slippery floor covering.
4. Provide knee space under sink.
5. Keep controls of cooking unit on front so person does not have to reach across burners.
6. Use front-loading dishwasher.
7. Set wall-mounted ovens low.
8. Use round dining table.
9. Use pegboard to hang cooking utensils, pots.
10. Provide low storage for foods.
11. Use a side-by-side refrigerator-freezer unit.

Fig. 2-51.   A cheerful, well-lighted kitchen.

Planning the Bedrooms

A large part of a house is devoted to the bedroom area. The number of bedrooms to be planned is important. Often, houses are described by the number of bedrooms. Sales material will read, "Three-bedroom Colonial style house." A three-bedroom house has become the norm. Two-bedroom houses are satisfactory for a couple with one child but are more difficult to sell. A three-bedroom house will accommodate a family with two children. This is the ideal situation because

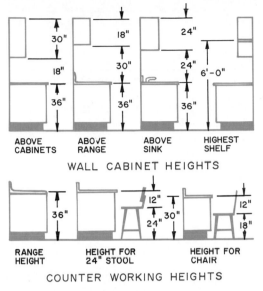

WALL CABINET HEIGHTS

COUNTER WORKING HEIGHTS

Fig. 2-52. Kitchen cabinet planning standards.

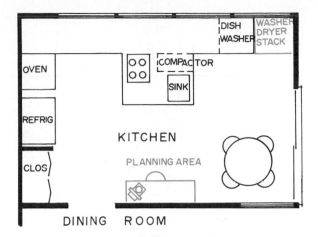

KITCHEN

PLANNING AREA

DINING ROOM

Fig. 2-54. Laundry facilities can be located in the kitchen. The planning area is a valuable feature to consider when designing kitchens.

Fig. 2-53. Typical kitchen cabinet sizes.

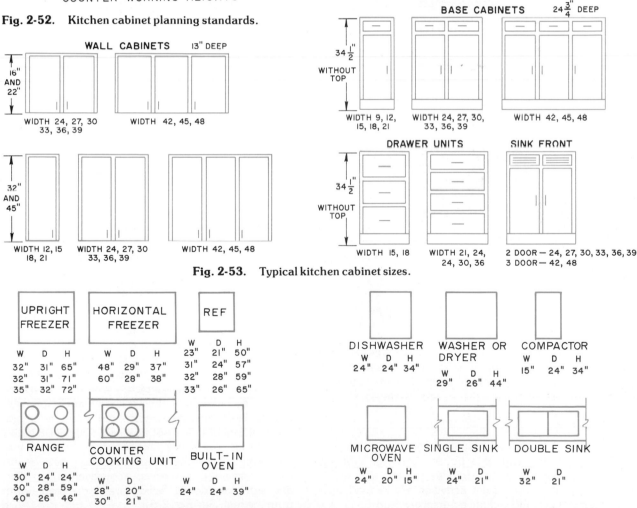

Fig. 2-55. Templates and selected sizes of kitchen appliances.

Home Planners, Inc.

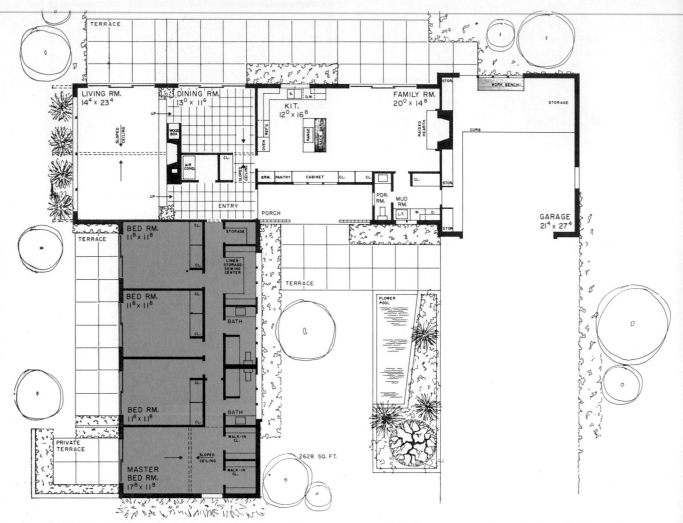

Fig. 2-56. Bedrooms are located together in the quiet section of the house. Notice how the entry hall provides a noise barrier.

each child has a private bedroom. For a larger family, the rooms can be made larger and can easily sleep two children per room.

The planning process for bedrooms is like that for other areas. Develop a list showing its uses. It will provide a sleeping area. In addition, it may serve as a place to sew, read, write, or relax to music.

Location

The bedrooms are usually located together in a wing on one end of a house. See Fig. 2-56. A second-floor location provides privacy and quiet. Rooms for other quiet activities are located with them. This could include a nursery or study.

Noise Control

As bedrooms are planned, the control of noise from within the house and outside should be considered. Placing closets on inside walls is effective. Walls can be insulated. Even better control is achieved if studs are staggered when the house is built. (This building method is shown in Chapter 7.) Locate bedrooms away from the street to avoid street noises. If this is not possible, consider a high, solid fence or heavy shrubbery in front of the bedroom area. Other noise reduction devices include carpeting, draperies, acoustical ceiling tile, and storm windows. Solid core doors stop noise from the hall. See Fig. 2-57.

Size

The size selection for the bedroom is most influenced by two major factors: the size of the furniture and the minimum spacing between furniture. Typical templates and furniture sizes are shown in Fig.

2-58. The sizes of bedroom suites vary considerably. Certain styles, such as Mediterranean, are large and massive and require a large room. Contemporary furniture is lighter, more delicate, and requires less space. See Fig. 2-59. Accurate planning requires some knowledge of the type of furniture to be used.

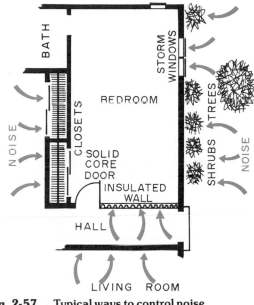

Fig. 2-57. Typical ways to control noise.

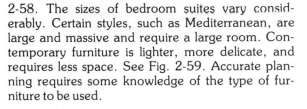

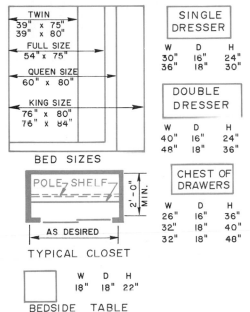

Fig. 2-58. Templates and sizes of bedroom furniture.

American of Martinsville

Fig. 2-59. A contemporary style of bedroom.

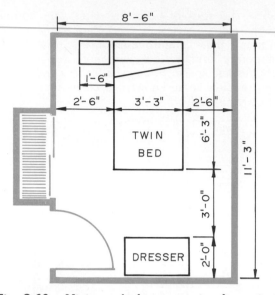

Fig. 2-60. Minimum bedroom spacing for a single occupancy room with one twin bed.

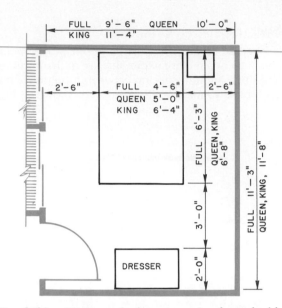

Fig. 2-61. Minimum bedroom spacing for a double occupancy room with a double bed.

The minimum spacing between furniture is shown in Figs. 2-60, 2-61, and 2-62. Remember that these are minimum sizes. Notice that a single twin bed and night table require 4'-9" (1448 mm) wall space. Twin beds with a night table need 8'-2" (2489 mm) wall space. A regular double bed with a night table requires 6'-0" (1829 mm) wall space, queen size — 6'-6" (1981 mm) and king size — 7'-10" (2388 mm).

A single occupancy bedroom requires a minimum of about 100 square feet (9.3 m²). A double occupancy with a full size bed requires 110 square feet (10.2 m²). A queen size bed requires 120 square feet (11.1 m²) and a king size bed needs 135 square feet (12.6 m²). A room with twin beds requires at least 140 square feet (13 m²).

Traffic aisles between furniture should be at least 2'-6" (762 mm). If furniture is to be near a door, enough space should be left clear for the door to open. Arrangements are easier to plan for a rectangular-shaped room than for a room that is square.

Storage

The furniture and closets in a bedroom are most often used to store clothing accessories. In a room designed for one person, a minimum of 3 lineal feet (914 mm) of closet space is needed. A double occupancy room should have 6 lineal feet (1829 mm). These are minimums and more should be provided if possible. Closet design sizes are given in the storage section in this chapter.

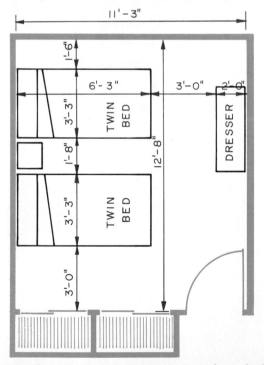

Fig. 2-62. Minimum bedroom spacing for a double occupancy with twin beds.

As the room is planned, the closets should be considered as the furniture is arranged. Closets consume long sections of wall space. Some designers make closet templates and move them around on the plan as if they were a piece of furniture.

Other Planning Considerations

1. Allow for two windows in every bedroom. Try to place them so the room has cross ventilation. Do not have drafts blowing across the bed. See Fig. 2-63.
2. Locate bedrooms so they are entered from a hall. Refer again to Fig. 2-56.
3. Consider using high windows with a sill about 5'-0" (1524 mm) from the floor. This increases privacy and allows easier arrangement of furniture. It may, however, prevent easy escape in case of fire.
4. Consider using sliding glass doors to a private patio. This opens up the room to the outside and makes a small room seem larger.

5. Coordinate the placement of bathrooms with bedrooms. Consider a private bath as part of the master bedroom. See Figs. 2-64 and 2-65.

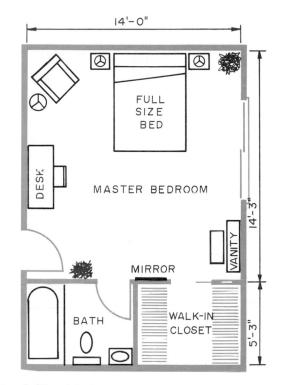

Fig. 2-65. A bedroom with private bath and walk-in closet.

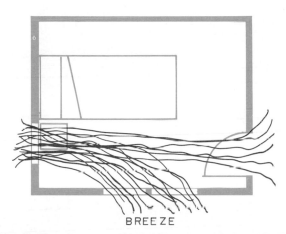

Fig. 2-63. Cross ventilation increases room comfort.

Fig. 2-64. A master bedroom with a private bath.

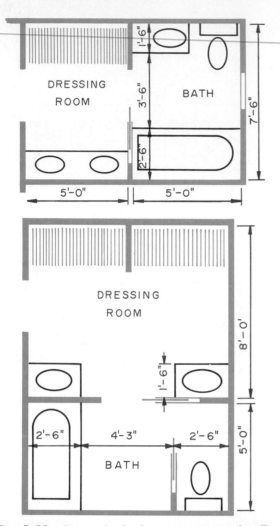

Fig. 2-66. Private bath, dressing room, and walk-in closet designs.

6. Locate mirrors and the dressing table to take advantage of the natural light.
7. In a more expensive house, design a private dressing room and walk-in closet off the master bedroom. See Fig. 2-66.
8. Locate beds so they have space on three sides for easy movement about the room. See Fig. 2-67.
9. Choose floor and wall coverings carefully. For example, those used in a child's room should be durable. See Fig. 2-68.

Azrock Floor Products

Fig. 2-68. A child's bedroom requires a durable floor covering. Bright colors add to the cheerful atmosphere.

Fig. 2-67. Properly located beds should have space on three sides.

PPG Industries

Fig. 2-70. A small sitting or reading area makes a bedroom more versatile.

Fig. 2-69. The use of high windows in a bedroom allows more wall space for placement of furniture and adds to the privacy of the room.

Fig. 2-71. A bedroom designed using large windows gives the room a light, cheerful atmosphere.

Fig. 2-72. A child's bedroom becomes a living area with imaginative design.

Planning a Bath

Most new houses have a bath and a half or two full baths. Many houses have a bath off the master bedroom and a second bath for the remaining bedrooms. A four-bedroom house should have two full baths in the sleeping area. A house built on several levels should have a half bath on the living level and at least one full bath on the sleeping level. It is desirable to have a half bath near the entertainment area; this could be the living room, family room, or a basement recreation room.

As a bath is planned, consider what functions it will serve. In addition to the normal uses, it may serve as a dressing area, a sauna, a laundry, exercise room, or sunbathing area.

Fixture Types and Sizes

The manufacturers of bath fixtures produce a wide range of units of various sizes. Some bathtubs are designed to be open on only one side. Others may be open on two and three sides. See Fig. 2-73. There are tubs that are almost square and fit in a corner. Sunken tubs of various sizes are used. Bathtubs may be made of steel, cast iron, or fiberglass. The fiberglass unit may have the shower walls as part of it.

Prefabricated shower stalls are generally made of metal or fiberglass. They are made as a single unit and simply set in place. Showers built on the job are often finished with ceramic tile, terrazzo, or marble. The sizes can be made to suit the designer.

Water closets are either wall-hung or floorstanding. The bidet is placed beside the water closet. It is used for personal hygiene. See Fig. 2-73.

Lavatories are either wall-hung or are built into a base cabinet. The wall-hung lavatory requires a minimum of space. However, the base cabinet provides storage.

Hydromassage baths use adjustable water jets to provide a swirling, relaxing, hot-water massage. These units are often located in the bathroom. See Fig. 2-74.

Templates and typical sizes of fixtures are in Fig. 2-75.

Location of the Bath

A house with one bath should be planned with the bath in a central location. When a house has more than one bath, the additional baths should be located in the areas they are to serve. One factor to consider is grouping the baths and possibly the kitchen to centralize the plumbing. See Fig. 2-76.

Kohler

Fig. 2-73. This bath has a spacious plan. Notice the bathtub is open on three sides. The two lavatories are built into a base cabinet.

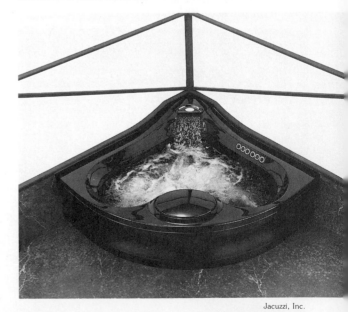

Jacuzzi, Inc.

Fig. 2-74. This hydromassage, or whirlpool, bath is designed to fit into a corner.

Plumbing costs are reduced if second-floor baths are located above first floor baths or the kitchen. However, if a more efficient location requires additional plumbing, the location should have priority.

A bath for use by those in several bedrooms should be entered from a hall. Privacy is difficult to maintain in a bath which can be entered directly from two bedrooms.

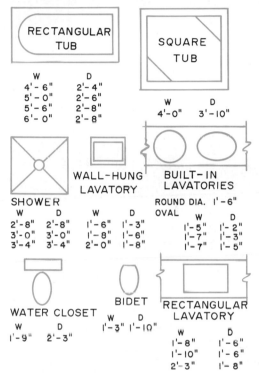

Fig. 2-75. Templates and selected sizes of bathroom fixtures.

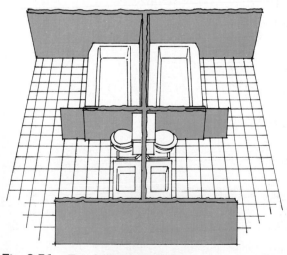

Fig. 2-76. Two full baths utilizing a common wall.

Fixture Size and Placement

Bath fixtures can be arranged effectively in many different ways. Basic design sizes are in Fig. 2-77.

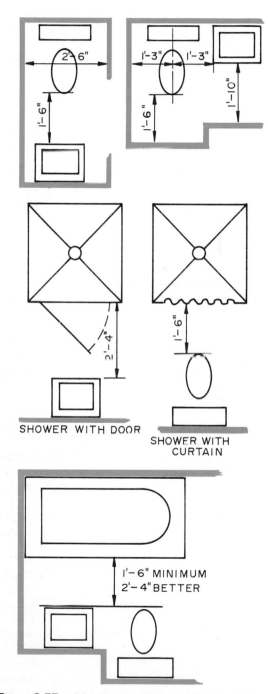

Fig. 2-77. Minimum space requirements for bathroom fixtures.

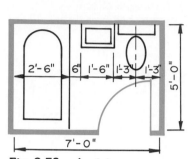

Fig. 2-78. A minimum bathroom plan.

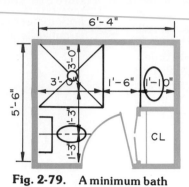

Fig. 2-79. A minimum bath with a shower.

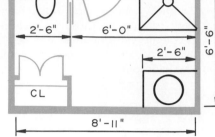

Fig. 2-80. A water closet can be placed in a private compartment.

A minimum bath includes the three basic fixtures — bathtub, lavatory, and water closet — placed in a minimum area. See Fig. 2-78. Some people prefer to use a shower instead of a tub. This minimum plan is shown in Fig. 2-79. From these basic plans, other arrangements are developed. The water closet can be located in a private compartment. This arrangement makes it possible for two people to use the bath at the same time. See Fig. 2-80. Two lavatories cost very little extra and allow two persons to groom simultaneously. See Fig. 2-81. Another efficient arrangement is to place the tub or shower and the water closet in one compartment. The lavatories can then be used while someone else is bathing. See Fig. 2-82.

A half bath usually contains a water closet and lavatory. A minimum example is illustrated in Fig. 2-83. When located in an area where formal entertaining occurs, it may need to be expanded to include an area for grooming. See Fig. 2-84.

Private baths are often located off master bedrooms. They may be connected with a dressing area. See those shown in the section on bedroom planning.

Kohler

Fig. 2-84. A large half bath used in a formal entertainment area. A comfortable seat and counter is provided for personal grooming. Notice the water closet in a separate compartment.

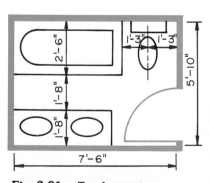

Fig. 2-81. Two lavatories increase the efficiency of a bath.

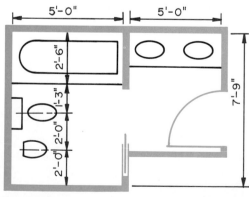

Fig. 2-82. The tub, water closet, and bidet can be located in a private compartment.

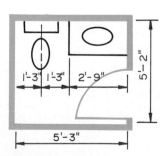

Fig. 2-83. A minimum half bath.

Kohler

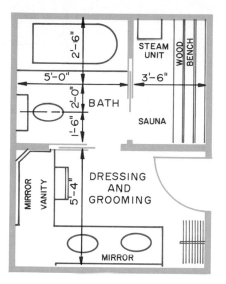

Fig. 2-85. A comfortable bath with a sauna, private bath, and a dressing and grooming area.

Fig. 2-87. A window is very desirable in a bath. It should not be directly above any of the fixtures. Notice the beauty of the natural light on the materials used. The grouping of these fixtures occupies a minimum of space.

A mirror should be above the lavatory. Electrical outlets are needed for hair dryers and electric razors. Good lighting is needed for grooming as well as for general illumination. It can also enhance the attractiveness of the room. Plants contribute to the appearance. See Fig. 2-86.

Other Planning Considerations

1. A sink and water closet may be placed in a "mud room" by the rear exit. This is especially useful in homes where there are children.
2. A window in a bath is highly desirable but not essential. See Fig. 2-87. A good exhaust fan can be used to change the air. Locate the fan near the tub or shower to help remove water vapor. The switch should be out of the reach of the person in the tub or shower.
3. The tub, water closet, and lavatory should not be directly in front of the window.
4. The water closet should be located so it cannot be seen from the next room when the bathroom door is open.
5. A means of rapidly increasing air temperature should be provided in the bathroom. A small electric heating unit is often used.
6. Since a bath is usually small and has many fixtures, doors must be planned carefully. In some cases, a door sliding into a pocket in the wall is suitable.
7. Storage space for towels and other accessories is needed. Medicines and items used daily should be stored in or near the lavatory. See again Figs. 2-79 and 2-80.

Kohler

Fig. 2-86. A mirror is located above the lavatory. The lighting is good and the decor is cheerful.

Expensive homes usually have more luxurious baths. An example is shown in Fig. 2-85.

Materials and Decoration

A bath should be a colorful, cheerful area. Materials used must withstand moisture and mold. They should be easily cleaned. Ceramic tile, marble, slate, special carpeting, vinyl materials, plastic and silk wall coverings, and moisture-proof wood paneling are typical examples.

Fixtures come in many colors. They should set the tone of the room. The colors of other materials must be coordinated with the fixture color.

Special Planning Considerations*

Special designs may be required for a variety of reasons. For example, to design a bath that will comfortably accommodate a person in a wheelchair, consider the following:

1. The water closet should be set 17 to 19 inches (430 mm to 485 mm) above the floor, measured to the top of the seat.
2. It is important to place grab bars next to the water closet. See Fig. 2-88.
3. Allow adequate clear floor area.

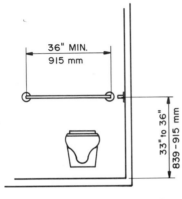

REAR WALL

Fig. 2-88. Location and minimum length of grab bars near a water closet.

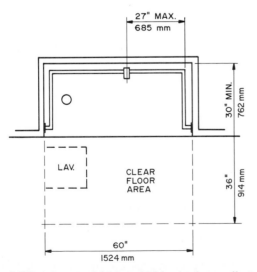

Fig. 2-89. A typical 30″ × 60″ bathtub installation.

*See Chapter 16 for more detailed information about architectural design for the orthopedically impaired.

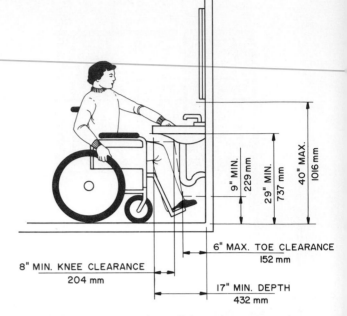

Fig. 2-90. Clearances for wall-hung lavatories.

4. Place grab bars on all walls around a bathtub. See Fig. 2-89. Showers with seats and grab bars are also widely used.
5. Allow adequate knee clearance below the lavatory. See Fig. 2-90.

Planning Laundry Facilities

The automatic washer and dryer are common appliances in today's homes. Space should be provided for these, as well as for sorting and ironing clothing. See Fig. 2-91.

A primary consideration is convenience. The most common location for laundry facilities is in a utility room or basement. Some homes have the washer and dryer in the kitchen or bathroom. These appliances could even be located in a recessed space in a hall. Folding doors can close them from view when not in use. In milder climates, the garage or carport serves as a laundry area.

Other Planning Considerations

1. The washer and dryer should be located side by side. This requires a 5′-6″ (1676 mm) wall space and a 3′-6″ (1067 mm) aisle. A washer and dryer located opposite each other require a 4′-0″ (1219 mm) passage. Standard appliance sizes and spacing requirements are shown in Fig. 2-92. Some washers and dryers can be stacked one above the other.
2. A sink is necessary in a laundry area. Again see Fig. 2-91.

General Electric Company

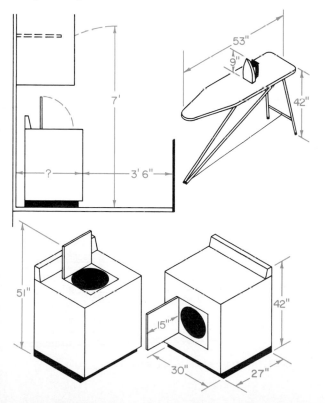

Fig. 2-91. The laundry area requires a sink and counter space.

3. A countertop is very helpful for sorting and arranging clothes. Wall cabinets are necessary for storage purposes.
4. The laundry facilities should be arranged in a sequence so the clothes move toward the place they are to be dried. The usual sequence is to have a sorting counter, then a sink for removal of difficult stains, then the washer, the dryer,

Fig. 2-92. Standard laundry appliance sizes and space requirements.

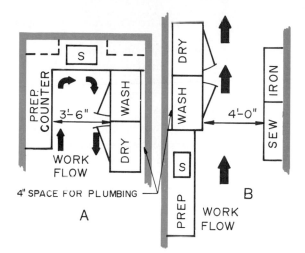

Fig. 2-93. Laundry facilities should be arranged with the work flow in a logical sequence.
   A. A U-shaped laundry
   B. A corridor-shaped laundry

and finally the ironing area. See Fig. 2-93. If the washer door opens to the left, it is convenient to have the dryer door open to the right.
5. Dryers are often placed on outside walls to aid in their venting.
6. If the laundry area is in the basement, a laundry chute is a real convenience.
7. Easy access to outdoor clothesline is desirable.
8. About 6 inches (152 mm) of space should be allowed between the wall and the laundry appliances for pipes and wiring.

Special Planning Considerations

Special designs may be required for a variety of reasons. For example, to design a laundry that will comfortably accommodate a person in a wheelchair, consider the following:

1. Laundry should be in or near the kitchen.
2. Use an ironing board that is adjustable from 28 to 32 inches in height.
3. Use front-loading washers and dryers.
4. Use appliances that have controls on the front.

Planning Utility Rooms

In homes without basements, a ground-level space for the furnace, laundry, and water heater is necessary. Such a room is called a utility room. It frequently has a lavatory and a water closet. The furnace can be closed off from the rest of the room. See Fig. 2-94.

Homes with basements often have a ground-level utility room for laundry and toilet facilities, with the furnace and water heater in the basement.

Planning Considerations

1. The utility room should be located next to the kitchen for convenience and economy.
2. It is desirable to have an outside door from the utility room.
3. If a furnace is included, the proximity of a chimney should be considered.
4. Some means of ventilation should be provided, either by a window or an exhaust fan.
5. A dryer must be placed where it can be vented.
6. Equipment should be located so it may be serviced, and room should be allowed around the furnace for cleaning and adjusting. Since laundry equipment is heavily used, it should be most accessible.
7. If laundry facilities are to be in the utility room, refer to previous section, "Planning Laundry Facilities."
8. A home workshop can be located in a utility room.

Planning Storage Facilities

An important and often overlooked part of house planning is storage. An adequate amount should be available in locations where it is most needed. Plan storage areas as you plan each room.

There are several types of storage. These include closets, built-in furniture units, cabinets, outside storage, and the furniture itself.

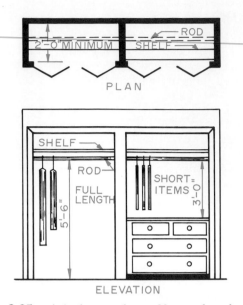

Fig. 2-95. A bedroom closet. Notice that drawers can be built in and hidden by the folding doors.

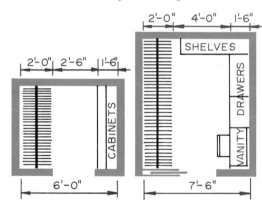

Fig. 2-96. Typical walk-in closet plans.

Lennox Industries, Inc.

Fig. 2-94. A furnace can be placed in a small closet off the utility room.

Fig. 2-97. A typical built-in wardrobe.

Closets

A closet is formed as the rooms are built. It generally is made of the same materials as the room. A typical closet, such as the one in Fig. 2-95, should be at least 2'-0" (610 mm) deep. Bedroom closets are usually 4'-10" (1473 mm) wide or wider. One person needs about 3 to 4 lineal feet (914 mm to 1219 mm) of closet space. Usually, closets have a rod upon which clothes hangers are placed. The rod should be 5'-6" (1676 mm) above the floor. One or more shelves may be above it. Each closet should have a light. The light can be operated by a pull-chain, but a wall switch is preferable. A closet may have built-in drawer units. Sliding or folding doors allow easier access to items in the closet.

Linen closets are often built near bedrooms or baths. They are usually 1'-0" to 1'-6" (305 mm to 457 mm) deep and 2'-0" to 4'-0" (610 mm to 1219 mm) wide. They usually have shelves their entire length since they hold towels and linens. This type of closet is often built as part of a bathroom.

A walk-in closet is large enough to hold many garments. See Fig. 2-96. Sometimes it is combined with a dressing area. (See the section on bedroom planning.) A typical size would be 6'-0" × 5'-0" (1829 mm × 1524 mm). A 2'-0" (610 mm) aisle is minimum. A dresser, chest, or built-in cabinets can be part of the storage capability of a walk-in closet.

Another type of storage is a built-in wardrobe. This unit is usually 12" to 18" (305 mm to 457 mm) deep. It has sections of shelving and drawers. Basically, it serves in place of a chest of drawers. See Fig. 2-97. If the doors are small, they may be hinged or folding. If they are large, they should be sliding doors. If this type of unit is to be used to

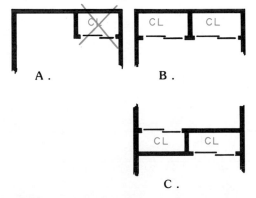

Fig. 2-98. Proper placement of closets.
A. Wrong — closets should not protrude into a room.
B. Closets can extend across the end of a room.
C. Closets can open into adjoining rooms.

Fig. 2-99. The shelves, cabinets, and drawers of this wall unit provide extra storage space.

hang clothing such as suits or coats, it will have to be 24" (610 mm) deep.

Closets should be planned so they do not protrude into a room. Either extend the closet the full length of the room or build another closet in the unused area. The second closet can open into the next room. See Fig. 2-98.

Other Storage Units

Wall units provide a considerable amount of storage space. Various types and sizes of items can easily be stored in units such as that shown in Fig. 2-99.

Storage is the key to kitchen planning. The size and location of base and wall cabinets are fundamental considerations. Many sizes of kitchen cabinets and many types of special-purpose cabinets are available. Review the section on kitchen planning.

A workshop requires tool storage. Paints, materials, and items being built must also be stored. Usually, wood or metal shelves can be used. However, flammables must be stored in fireproof metal cabinets.

A laundry area needs to store soiled clothing, washing materials, and clean clothes. See the section on laundry.

In a den, books, records, and papers must be stored.

The garage is usually used to store items needed for auto maintenance and the yard equipment. Sometimes, a small storage building is built for storing yard equipment and lawn furniture. Children's toys, including bicycles, must be stored. In a garage, a storage wall made up of shelving units is commonly used. Sometimes, doors are placed over the shelves for security and cleanliness.

Planning Porches and Patios

In most climates, there are several months when outdoor living is possible. In southern climates, it may be possible for most of the year. Provisions for outdoor living are desirable in a house plan.

Porches and Decks

Porches and decks are attached to the house and the ground. They often have a roof but this is not always the case. Sometimes, the roof of the house continues over the porch area. The porch may be open on the sides, screened, or have provisions to enclose it and make it weatherproof in cool or rainy weather. See Fig. 2-100.

Porches have been popular for years. Study the early architectural styles in Chapter 1, Residential Architecture-Past and Present. Notice the porches on the Georgian style, the early row houses in Virginia, and the French Colonial. They are also significant factors on southwestern ranch style houses, Monterey style, and the heavily ornamented Victorian style. These porches were designed mainly for sitting, but some were used for dining. Porches on houses today serve many purposes.

Porches are located wherever needed. If outdoor dining is desired, they are off the kitchen or dining area. If used for parties and entertaining, they may be off the living room. Porches off bedrooms provide places to relax and enjoy the evening in privacy. The size of the porch and the furniture used depends upon the activities planned. Also, the designer must always keep in mind what the porch does to the exterior design of the house.

Fig. 2-100. This porch opens off the house and is open to the weather.

Lee Woodard Sons, Inc.

Since porches are exposed to the elements, they are made of durable materials. Masonry, redwood, and cypress are often used.

A **balcony** is a form of a porch. It is attached to the house and may have some coverage by a roof. It extends into space and often is unsupported from the ground. It may have posts below to provide support. The improvement in structural members has made the construction of larger balconies possible. They can now extend farther out from the house. Balconies serve the same purposes as discussed for porches.

Balconies were used on many early homes. They are most often associated with Italian and Spanish houses.

Like porches, balconies can be located where needed. Typically, they are off bedrooms, living, and dining areas. They can be extended from any level of a house. Second-floor balconies give a feeling of great height and freedom.

GAF Corporation

Fig. 2-101. A front entrance can be sheltered with a small porch.

Western Wood Products Association

Fig. 2-102. A wood deck can be used to tie the outside to the interior of the home.

Fig. 2-103. This deck is used for informal entertaining.

The designer must consider how a balcony fits in with the style of a house. The influence on the mass and exterior detail must be studied. Often, the balcony could become the design factor or focal point of an exterior design.

Some houses have a small front porch with a small roof. These may be referred to as **stoops.** See Fig. 2-101. The front entrance to every house should have some type of porch. This gives callers protection from the weather as they wait for someone to answer the door.

A **breezeway** is a form of a porch. It is a covered area that connects a house to a garage or another building. In bad weather, it provides protection as a person moves from the house to the garage. A breezeway may be screened or weathertight. Often, it is furnished and serves as an outdoor living area.

A **deck** is an open, wood platform. A durable wood such as redwood, is generally used. The deck is usually connected to the house. See Fig. 2-102. It can be on several levels. It is used for entertaining and relaxation. See Fig. 2-103. Its appearance is enhanced by building it around existing trees and having planting boxes for flowers. See Fig. 2-104.

Patios or Terraces

A **patio** is sometimes called a **terrace.** It is a paved ground area near the house which performs the same functions described for a porch. A large patio can serve for outdoor entertaining. See Fig. 2-105. Small patios may be placed off a kitchen or

Fig. 2-104. Trees and flowers enhance the appearance of a deck.

Fig. 2-105. This patio is paved with bricks set in a bed of sand.

dining area for private outdoor meals. See Fig. 2-106. Each bedroom could have a patio. It can be screened with a fence and shrubs for privacy. Slatted walls permit air to circulate yet provide privacy and some sun control. Brick, stone, and wood can be used to form privacy walls. See Fig. 2-107.

Patios are paved with any durable material. Popular are stone, brick, slate, wood, and concrete. Patios should be sloped so they drain away from the house. Often, unpaved areas are left for planting trees, shrubs, and flowers.

A patio may have some form of roof, such as a slatted wood roof or fiberglass panels which transmit light. Vines can grow over the roof or upon screening fences. See Fig. 2-108.

Swimming Pools

If a swimming pool is built, it is often surrounded by a paved area which blends into a patio. See Fig. 2-109. There are many sizes and types of pools. All require equipment to filter the water. Usually, a small building must be planned to hold this equipment. It could also serve as storage for outdoor furniture. Swimming pools are most often installed by companies specializing in this type of construction. They install the pool and the filtering system. They also provide regular service to keep the system operating.

A pool should be fenced so it can be locked up when not in use. Small children can get into difficulty if a pool is left unguarded. See Fig. 2-110.

Courtyard

A **courtyard** provides an outdoor area of privacy. It is usually enclosed on all sides by the walls of the house and/or by fences. See Fig. 2-111. It includes areas paved with brick, stone, or other materials.

Lee Woodard Sons, Inc.

Fig. 2-107.   Walls are used to provide privacy for a patio.

Mid-State Tile Company

Fig. 2-108.   Patios can be screened to keep out insects.

Home Planners, Inc.

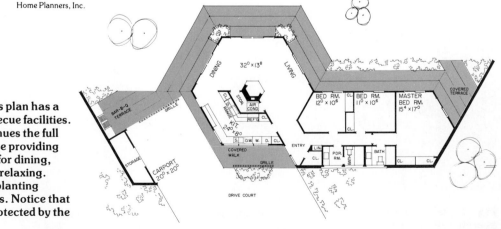

Fig. 2-106.   This plan has a terrace with barbecue facilities. The terrace continues the full length of the house providing outdoor facilities for dining, entertaining, and relaxing. Areas are left for planting shrubs and flowers. Notice that the entrance is protected by the main roof.

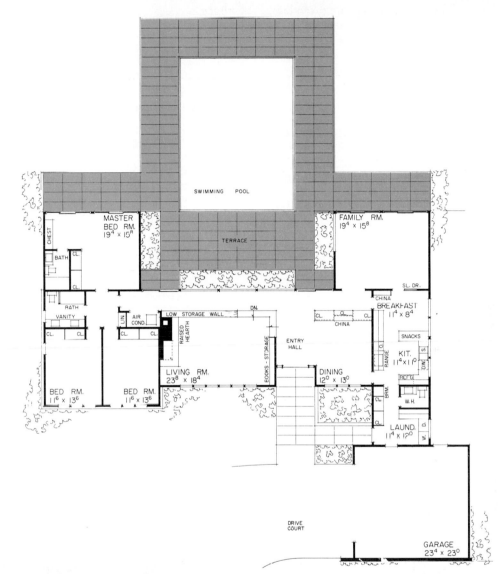

SWIMMING POOL

MASTER BED RM. 19⁴ x 15⁸

FAMILY RM. 19⁴ x 15⁸

TERRACE

CHEST

BATH

CL.

CL.

BATH

VANITY

CL.

CL.

LIN.

AIR COND.

CL.

CL.

LOW STORAGE WALL

RAISED HEARTH

DN.

CHINA

CL.

CL.

CHINA

SL. DR.

BREAKFAST 11⁴ x 8⁴

SNACKS

RANGE

KIT. 11⁴ x 11⁰

DW.

S.

REF'G.

W.R.

ENTRY HALL

BOOKS · STORAGE

LIVING RM. 23⁸ x 18⁴

DINING 12⁰ x 13⁰

BRM.

BED RM. 11⁶ x 13⁶

BED RM. 11⁶ x 13⁶

LAUND. 11⁴ x 12⁰

DRIVE COURT

GARAGE 23⁴ x 23⁰

Fig. 2-109. A swimming pool surrounded by a large, paved patio.

Home Planners, Inc.

California Redwood Association

Fig. 2-110. Pools are usually surrounded by a paved area. A fence to secure the pool is necessary.

COURTYARD   LOGGIA

Home Planners, Inc.

Fig. 2-111. A courtyard and loggia. These are shown on the plan in Fig. 2-112.

Areas for planting shrubs and flowers are provided. Special touches might include a fountain or small pool. See Figs. 2-111 and 2-112. Special consideration must be given to lighting. The courtyard should be usable at night with the lights blending into the surroundings.

Loggia

A **loggia** is a roofed, open gallery along the front or side of a building. It serves as a passage connecting areas of a house. See Figs. 2-111 and 2-112. Loggias may be on both the ground and second-story levels. Modern versions may have doors which can be used to enclose the area in bad weather.

Gallery

A **gallery** is an interior area which serves as a passageway. It often opens onto a courtyard as in Fig. 2-112. Usually, it connects with the foyer. One wall may be glass. Durable materials are commonly used. The gallery might contain potted plants and a few pieces of furniture.

Atrium

An **atrium** is an area, usually under a roof, in which plantings dominate. See Fig. 2-113. The roof admits light needed by the plants. The atrium might have paved paths and private places to sit and relax. Often, rooms open onto it or have glass

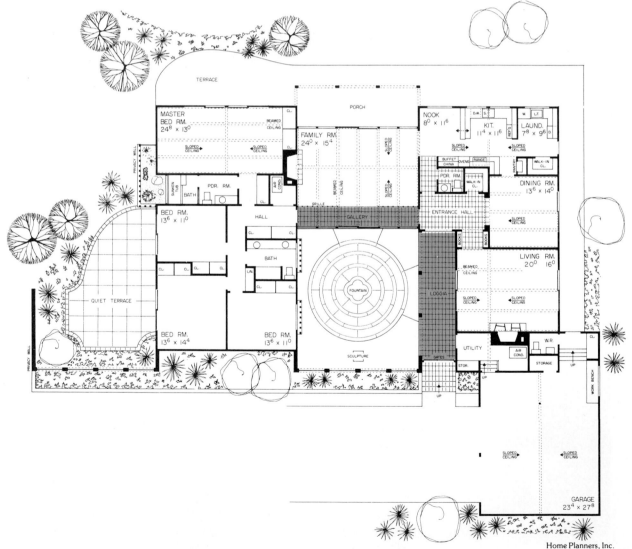

Home Planners, Inc.

Fig. 2-112.   The fountain is in the courtyard. The gallery and loggia are part of the traffic system in a house.

walls facing it. If totally enclosed, it provides year around greenery to the interior rooms. See Fig. 2-114.

Planning Considerations

1. A patio, porch, or balcony placed on the north side of a house will be in almost constant shade. One on the south will receive considerable sun. If sunbathing is important, the orientation of the area is important.

2. Since these outdoor areas are often used at night, lighting must be planned. This requires general illumination and local lighting.

3. Weatherproof electrical outlets should be provided. These are needed to operate cooking appliances, record players, televisions, and other such devices.

4. Plan for ease of maintenance. These areas are exposed to the weather and, in general, get very hard use. People are not as careful there as they are inside the house.

Home Planners, Inc.

Fig. 2-113. **An atrium is a roofed area containing plants.**

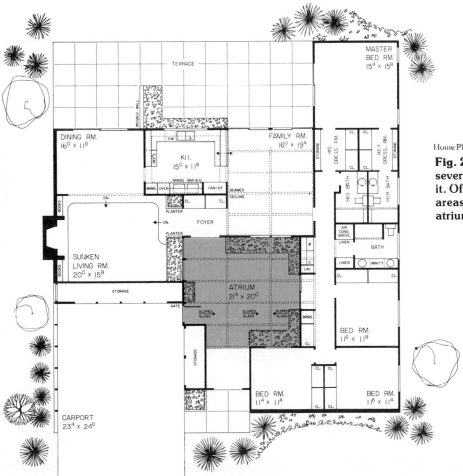

Home Planners, Inc.

Fig. 2-114. **An atrium has several rooms opening onto it. Often, rooms have large areas of glass facing the atrium.**

Western Wood Products Association

Fig. 2-115. The main entry creates the first impression of the home.

Stanley Door Systems

Fig. 2-116. Double doors help focus attention upon the entrance.

Stanley Door Systems

Fig. 2-117. Side lights add to the attractiveness of the entrance.

5. Consider any danger areas. Balconies need guard rails to keep people from falling. Steps need to be wide and carefully planned. Handrails are necessary. Light should be adequate to enable a person to see easily.
6. Allow areas for plantings. They add a great deal to the appearance and generally make it a more pleasant place to relax.

Planning Entrances

The Main Entrance

The main entrance to a house deserves special consideration in the design process. It often is the focal point of the building and sets the tone and reactions for those who enter. An entrance with a finely designed door and warm colors gives a visitor a different greeting from one which is all glass, metal framed, and uses cool colors. The entrance should reflect the style of the house. See Fig. 2-115.

The first reaction is to the exterior portion of the main entrance. This might be a ground-level, patio-type area or a small porch several steps above ground level. The main door becomes the focal point of attention. Many types of doors are available. Larger houses often use a double door. See Fig. 2-116. Side lights (panes of glass beside the door) add to the importance of the entrance. See Fig. 2-117. They are especially attractive at night.

It should be possible to see those at the door before opening it. This can be done by using side lights, lights in the door itself, or a peephole device.

The main entrance is usually centrally located on the plan. See Fig. 2-118. This location provides the shortest distance to the various parts of the house. When planning the main entrance, consider where guests will park their automobiles. A plan for guests to approach the main entrance is necessary. Sidewalks or a patio-type area is often used. In either case, the route to the main entrance should be clearly identifiable.

An entrance court adds to the attractiveness of an entrance. It can be paved and/or have plantings of all kinds, a fountain, and items of furniture. The court provides a measure of privacy for the front entrance.

Study the entrances in the floor plans in this chapter. Notice that they may go through the atrium, a covered walk, loggia, or courtyard.

The main entrance should have some means of protecting visitors from the weather. A small roof or an extension of the main roof may cover it. Sometimes the entrance is recessed into the building. See Fig. 2-117.

The main entrance should flow into the overall exterior design. The styling and materials must relate to those in the overall structure. Sometimes other architectural features, such as a chimney, a porch, or planters, become part of the overall entrance design. See Fig. 2-119.

Home Planners, Inc.

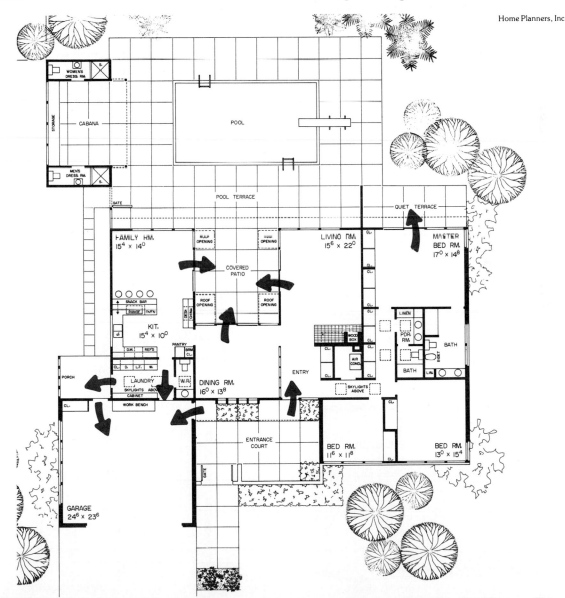

Fig. 2-118. The main entrance is centrally located and easily accessible for guests from the driveway. This entry is protected by the roof of the house extending over it.

Sun-Dor-Company

Fig. 2-119. The stone wall, concrete steps, and rock garden become part of the overall entrance design.

Azrock Floor Products

Fig. 2-120. The durable floor of this foyer flows smoothly into the interior design.

The Foyer

When the main door is open and the visitor steps into the foyer, the impression received is immediate. The foyer should complement the exterior entrance design and flow smoothly into the interior design. The floor must be a durable, easily-cleaned material. Ceramic tile, brick, stone, and slate are popular. See Fig. 2-120.

The lighting should be planned so that the visitor can see to enter, yet not be blinded by excess illumination. The parking area, exterior walk, and steps must be lighted. The interior lighting can be soft and direct the visitor to the living area.

The foyer itself should give a feeling of openness. Windows, mirrors, or a cathedral ceiling are ways this might be done. The foyer does not contribute

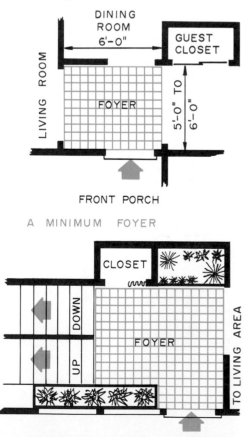

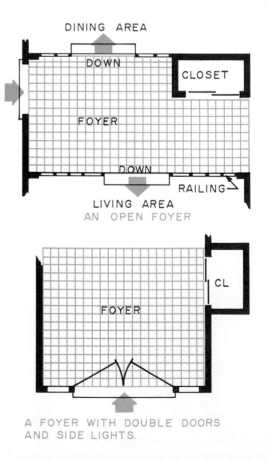

Fig. 2-121. Typical foyer designs.

to the living space and is often not included in lower cost homes.

The size and shape of the foyer can vary considerably. In expensive homes, it is large and might contain outstanding pieces of furniture, such as a rare table, chest, or chairs. In more conservative homes, it will be only large enough to fulfill its purpose. Typically, the foyer is square or rectangular, but any shape needed for it to perform its function is satisfactory. The key is to direct and control traffic flow. See Fig. 2-121.

It is convenient to locate a guest closet in or near the foyer. The guests can easily leave their coats and proceed to the living area.

Other Entrances

A house will have other entrances. See Fig. 2-118. Possibly the most used entrance is from the garage to the kitchen area. While this need not be as elaborate or expensive as the main entrance, it needs to be pleasing. It is important to design it so that the deliveries can be made easily.

A house may have a rear door for direct access to the yard. This might be from a kitchen or family room. Often, a small "mud room" with a lavatory is located near this entrance. Children can clean up a bit before entering the house.

Other rooms may have doors to porches, patios, or balconies. Sliding glass doors are popular.

Planning Stairs and Halls

Stairs and halls are the means utilized for assisting the flow of traffic through a house and between levels of a house. They are not classified as living space. Other than facilitating movement about a house, they are wasted space and,

therefore, should be kept to a minimum. See Fig. 2-122.

Halls

The average hall should be at least 3'-0" (915 mm) wide from finished wall to finished wall. A wider hall is nice, but becomes expensive because it serves as little more than a passageway. Sometimes, building in a storage wall or closets along one side of a hall makes it more useful.

Usually, a long hall becomes a rather dark, tunnel-like area. In flat-top contemporary homes, skylights are used to combat this, but most homes must rely on electric lighting. The best solution is to try to eliminate the long hall during the planning stage.

The foyer, entrance hall, and stairs are part of the traffic system. They are connected to any halls used to provide passage to the various rooms. The gallery also is a device for aiding in traffic flow. Study the floor plans in this chapter to see how the foyer, entrance hall, gallery, stairs, and halls form the traffic network.

Stairs

The stair is the part of the traffic system which allows people to move from one level to another. Its location must be decided as the halls and other traffic moving areas are planned.

Almost every house will have some type of stair. Most common are those to a basement or second floor. A split-level house will have several short stairs since levels are usually one-half story apart. See Fig. 2-123.

American Olean Tile Company

Fig. 2-122. A circular stair occupies a minimum of space.

PPG Industries, Inc.

Fig. 2-123. A short stair is used in a split-level house.

Fig. 2-124. A stairway elevator may be installed to accommodate special needs.

The design of stairs varies. Some stairs are very simple in design and serve no decorative purpose. Others form part of the interior style and are an important design feature. See Fig. 2-124.

Types of Stairs

The most commonly used types of stairs in homes are the **straight,** the **L,** and the **U.**

The **straight flight** is the least expensive to install and the easiest for moving large articles up and down. If room is available on the floor plan, a long flight of straight stairs can be improved by placing a landing halfway in the flight. This gives the person ascending a spot to pause and rest.

If floor space is not available for a straight stair, an **L** or **U stair** can be used. The location of the landing can be varied to suit the conditions of the house and to permit the necessary headroom.

Winders are triangular-shaped treads used to turn a corner in a stair without using a landing. They utilize less floor space than that required by a stair with a landing. Such stairs are not advised, however, because they are dangerous. The tread at the turn is narrow. A tread width of at least 7 ½" (190 mm) must be provided 12" (305 mm) from where the tread is narrowest. The rise must not exceed 9 ½" (241 mm). The stair must have at least 6'-6" (1981 mm) headroom. Common types of stairs are illustrated in Fig. 2-125.

Stair Design Features

The **tread** is the portion of the stair that is stepped upon. The **riser** is the vertical member running between the treads. See Fig. 2-126. The sizes of the tread and riser must be carefully planned. The maximum riser is 8" (203 mm) and the minimum tread is 9" (229 mm). The tread has a **nosing.** Again, see Fig. 2-126. The minimum nosing is 1" (25 mm). This is in addition to the tread size. For example, the stair with a 9 1/2" (241 mm) tread will have a walking surface of 10 1/2" (267 mm). Standard tread to riser sizes are shown in Fig. 2-127. The tread plus riser should equal approximately 18" (457 mm). For example, a 7" (178 mm) riser will require an 11" (279 mm) tread. All tread and riser sizes must be uniform in a stair.

Interior stairways between floors should have a minimum width of 2'-8" (813 mm) clear of the handrail. This will make the actual stair a minimum of 3'-0" (914 mm). See Fig. 2-128. If a stair is located where two people might pass frequently, 3'-6" (1067 mm) is needed. If possible, 4'-0" (1219 mm) should be used.

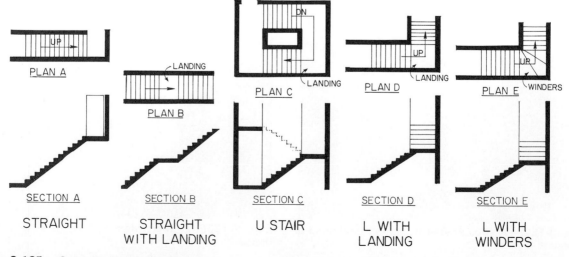

Fig. 2-125. Common types of stairs.

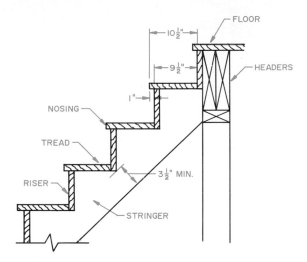

Fig. 2-126. The major parts of a stair.

Fig. 2-127.

Standards for Residential and Commercial Stairs

Type of Stair	Maximum Riser	Minimum Tread
Residential	8″ (203 mm)	9″ (229 mm)
Commercial	4″ to 7″ (100 to 178 mm)	11″ (279 mm)

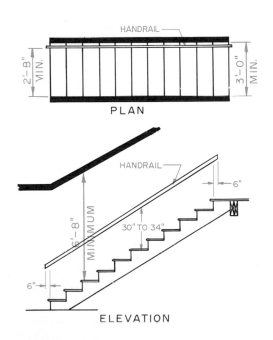

Fig. 2-128. Minimum standards for interior stairs in residences.

A main stair should have a minimum of 6′-8″ (2032 mm) clear headroom. See Fig. 2-128. A basement stair can have a minimum of 6′-4″ (1930 mm). The distance is measured vertically from the edge of the tread.

All stairs must have a handrail on at least one side. Codes require a stair to have at least three risers. Stairs with over eighteen risers or 12′ (3658 mm) total rise must have a landing in the middle of the stair. The landing must be the same width and length as the width of the stair. Also, doors must not open over a stair. A landing must be provided.

How to Figure a Stair

When developing a floor plan, space must be allowed for the stairs. This includes the width and length. The first thing that must be known is the **total rise.** See Fig. 2-129. This is the total vertical distance the stair will cover. Then select the riser size. Divide the riser size into the total rise to find the number of risers. The riser size must be adjusted until it divides evenly into the total rise.

There is always one more riser than tread in a stair. To find the **total run** (length) of the stair, select the tread width (Fig. 2-127) and multiply it by the number of treads. See Fig. 2-129.

Following is a typical problem. The procedure would be the same using metric measures.

Total rise	8′ 1½″
Riser size	0′-7½″
	(8′-1½″ ÷ 0′-7½″ = 13 risers)
Tread size	0′-9½″
	(12 treads × 0′-9½″ = 9′-6″)
Total run	9′-6″

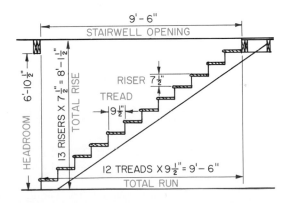

Fig. 2-129. When total rise and total run are known, tread and riser sizes can be figured.

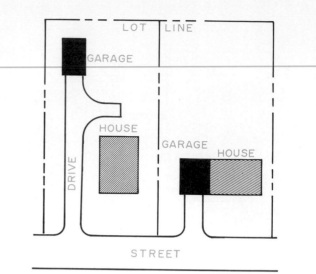

Home Planners, Inc.

Fig. 2-130.   A carport reduces the cost of construction.

Planning Garages and Carports

Climate influences the type of storage facility needed for automobiles. In warm, dry climates, a carport gives adequate shelter and provides space for the storage of yard tools and lawn furniture. A carport is a covered roof with one or more open sides or ends and no doors. See Fig. 2-130. In areas having cold weather, rain, or snow, an enclosed garage is desirable.

Consideration should be given to a two-car garage. Many families have two cars, or a boat or trailer. The extra money invested in the second garage space is quite wise. It also is an attraction to help sell a house.

In planning the garage, consider the following:
1. The garage is a good location for laundry equipment, a clothesline, the family workshop, or a play area for children.
2. The garage should be located on the side of the house receiving winter winds to provide a buffer against these winds and against snow.
3. Prevailing summer breezes should not be blocked by a garage.
4. A carport can be used if climate is mild and only shelter from sun is desired.
5. A carport is cheaper to build than a garage.
6. Building a one-car garage with a single carport beside it provides car shelter and also serves as a covered outdoor living area.
7. The garage should be located so one can reach the kitchen door and front entrance without going out into the weather.
8. If the garage is detached from the house, it should be connected to the house by a covered walk or breezeway.
9. The garage should be located so a long drive is not necessary. Turning room should be provided, if needed. See Figs. 2-131 and 2-132.

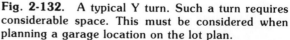

Fig. 2-131.   A long drive is expensive to build and maintain, and is inconvenient in bad weather. Notice the advantage of a short drive and attached garage.

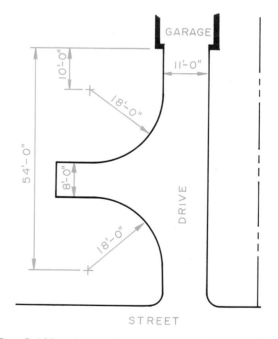

Fig. 2-132.   A typical Y turn. Such a turn requires considerable space. This must be considered when planning a garage location on the lot plan.

10. A parking area for guests' cars should be provided. See Fig. 2-133.
11. Minimum inside dimensions for a one-car garage should be 12'-0" × 22'-0" (3658 mm × 6706 mm), and for a two-car garage, 21'-0" × 22'-0" (6401 mm × 6706 mm).

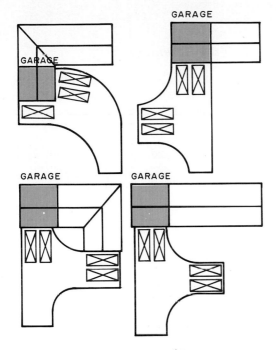

Fig. 2-133. Plans for guest parking.

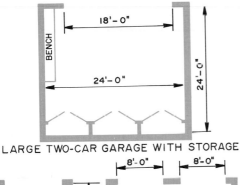

LARGE TWO-CAR GARAGE WITH STORAGE

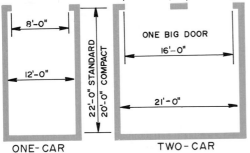

ONE-CAR

TWO-CAR

Automobile Size	Length		Width	
	in.	mm*	in.	mm*
Large	201-244	5100-6100	71-80	1800-2000
Intermediate	160-215	4000-5500	61-79	1500-2000
Small	150-160	3800-4000	60-65	1500-1600

*Rounded off to nearest hundred

Fig. 2-134. Recommended garage sizes.

These dimensions do not allow space for storage. See Fig. 2-134.

12. The garage door should be in keeping with the style of the house. Doors should be simple. They should not stand out or overbalance the design. Bright, contrasting colors are undesirable.

13. For best appearance, the garage should be constructed from the same material as the house; it should not be a sharp contrast.

14. Most contemporary house styles are enhanced by having a garage roof that is a continuation of the house. Lining up the ridge or the eaves gives the house a long, low, impressive appearance. However, a break in the roof line or eave line can make the house more interesting, if it is not too severe. Projecting the garage forward can be enhancing.

15. Standard sizes for garage doors are found in manufacturers' catalogs. Typical sizes include a height of 7'-0" and widths of 8'-0", 9'-0", 10'-0", and 16'-0".

16. The **type** of door should be considered. The use of one large door for a two-car garage is popular. Some prefer to use two single doors. This allows a car to be removed without exposing the various items stored in the garage.

17. An 8-foot (2439 mm) single garage door is minimum. Wider sizes are quite desirable. A 16-foot (4877 mm) width is minimum for two-car garages. Refer again to Fig. 2-134.

18. Overhead types of doors are very desirable. Swinging or folding doors are difficult to operate, take wall space, and frequently are blocked by snow.

19. A single 3'-0" × 7'-0" (914 mm × 2134 mm) door is desirable to allow persons to enter without having to open the large car door.

20. The wall between the attached garage and the house should be fireproofed. Metal lath and plaster or 5/8" gypsum board should be used.

21. The garage floor should be several inches above grade to prevent rain and snow from entering.

22. Concrete is the best material for garage floors.

23. The floor should slope to the door to allow drainage.

24. Lights and electrical outlets should be provided.

25. To cut costs, part of the basement can be used as garage space. This necessitates a sloping lot that will enable part of the basement to be exposed.

26. In very cold climates, garages can be heated. This is also important if the garage is used for a

laundry. Heat should be from the house furnace; open-flame heaters should never be used.

27. Houses without basements require storage space in garages. Consider making one wall a storage wall.

28. Electronically controlled garage doors may be conveniently opened or closed from the car or the house.

Planning the Basement

In warm climates or in areas with a high water table, basements are generally omitted. If a basement cannot be built because of water or rock problems, a ground-level addition can be built to compensate for this loss of space.

Many contemporary homes in all climates are built on a concrete slab. This enhances the long, low appearance of a house, but necessitates elimination of the basement.

In cold climates, basements are popular because so much of the year is spent indoors. The basement provides an excellent play space, recreation area, clothes drying area, and workshop or hobby area. See Fig. 2-135. The laundry and furnace can be located here. See Figs. 2-136, 2-137, and 2-138. A garage may also be included.

On a sloping lot, a basement can be included with little extra cost. On a flat lot, a basement will add significantly to the cost of a house. The big difference is that the house on the sloping lot requires less excavating and would require a high foundation anyway. This waste space underneath the house can be economically used for a basement. See Fig. 2-139.

Wood Conversion Company

Fig. 2-135. An ample basement workshop. Note pegboard tool panels and noise-deadening tile ceiling.

Bilco Company

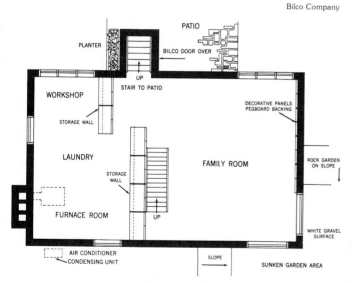

Fig. 2-136. Basement plans.
Interior stair running front to back. With a southeast sun exposure, the deep windows at the corners of the house catch the summer breezes. A storage partition dividing the recreation room from the workshop serves as an effective sound barrier. Doors provide easy access to basement entrance and the yard. Open planning ensures excellent ventilation.

Bilco Company

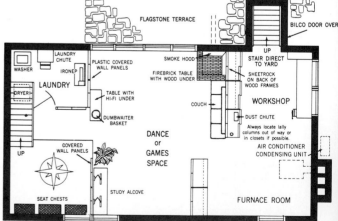

Fig. 2-137. Interior stair at end of plan. The family room adjoining a flagstone terrace and the unusual basement fireplace-barbecue arrangement offer economical indoor-outdoor enjoyment at a luxury level. Outside access to workshop, cleanup room, or dirt-associated areas makes good planning sense.

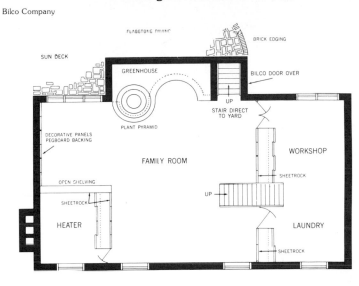

Bilco Company

Fig. 2-138. Interior stair running length of plan. Here, with the installation of a longer window and centralized basement entrance, the homeowner can create a blend of indoor-outdoor pleasures: barbecue, greenhouse, spacious functional basement. Further partitioning for hobbies, food storage, etc., should be done with concern for traffic and ventilation.

Andersen Corporation

Fig. 2-139. Good use is made of a sloping lot by exposing the basement to light and air.

Bilco Company

Fig. 2-140. A well-located laundry area and laundry chute. Notice the convenient outside door.

Other Planning Considerations

1. Interior stairs should be planned to give access to the basement. These are most satisfactory if they come down from the kitchen or family room on the first floor.
2. The stairs should be arranged so they do not spoil a large portion of the basement area by cutting into usable space. Refer again to Figs. 2-136, 2-137, and 2-138.
3. Having an outside entrance to the basement is important.
4. The furnace should be located near the chimney.
5. The laundry area should be separated by partitions and located near an outside door. The laundry chute should be in this area also. See Fig. 2-140.
6. The basement recreation room should be located so that one can reach it without going through the furnace or laundry room.
7. A fireplace in the basement should be below the fireplace on the first floor. The furnace flue and all fireplace flues should be in one chimney for most economical construction. A variation of this is shown in Fig. 2-141.
8. Bathroom facilities should be located near the laundry area. It is economical to locate these below a first-floor bath or kitchen.
9. Plenty of windows for light and ventilation are a necessity. Basement windows need not be small and widely spaced. Grouping windows

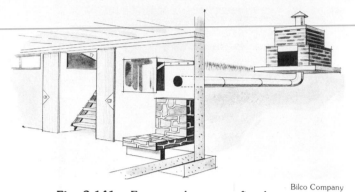

Bilco Company

Fig. 2-141. Economy basement fireplace.

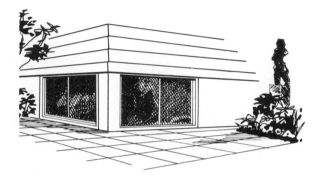

Fig. 2-142. Windows in series aid in lighting and ventilating a basement.

adds much to the brightness and ventilation. See Fig. 2-142.

10. Proper orientation can utilize breeze and sun.
11. A sloping lot will allow a large portion of the basement to be exposed. Refer again to Fig. 2-139.
12. Basement rooms should be kept in normal proportion.

Build Your Vocabulary

Following are terms that you should understand and use as part of your working vocabulary. Write a brief explanation of each term.

atrium	kitchen peninsula
closed living room	loggia
cooking center	open living room
courtyard	preparation and
den	cleanup center
family room	recreation room
food storage center	stair rise
foyer	stair run
gallery	template
kitchen island	

Class Activities

1. Make a drawing of the bathroom or kitchen in your home. List the ways it violates the principles of good planning.
2. Plan a new bath or kitchen for your home, using the existing space. Carefully apply the principles of good planning.
3. Make a notebook, using pictures of kitchens and bathrooms found in magazines and manufacturers' catalogs. Try to find examples that illustrate the principles of good planning. Devote one section to new ideas and new products for these areas.
4. Visit exhibit houses and write a report, analyzing the good and poor points in room planning. Explain what could be done to improve each item rated as poor.
5. Compute the cost of adding the following desirable features to a home: a foyer 4'-0" wide and 6'-0" long or 1200 mm wide and 1800 mm long; a second bathroom; a separate surface-cooking unit and oven, instead of the usual stove; an outside exit to a basement of a house on a flat lot; enlarging a minimum-sized two-car garage.
6. Compute the distance between the three major centers in your kitchen at home. Does the kitchen require unnecessary walking? What could be done to improve its efficiency?
7. Compute the amount of wall and base cabinet space in your kitchen at home. How does it compare with the minimum requirements for good kitchen planning?
8. Compute the square feet of closet space in your home. Considering the value of your home and the total square feet, how much did this space cost?
9. Write an evaluation of the laundry facilities in your home. Consider convenience of location, ease of working in the area, and accessibility for repairs and adjustments of equipment.
10. Develop a complete plan to convert your basement at home into useful living space. In addition to drawing a floor plan, figure the cost of the additions (walls, ceiling, floors, expanded electrical systems, etc.).

Chapter 3

Designing the Exterior

Before a final set of plans can be drawn, the elevations must be planned and sketched. Even a simple freehand sketch to an approximate scale is very helpful. While certain construction problems may be involved, the primary problem is that of developing a suitable exterior for the plan.

Evaluating Designs

Not everyone agrees on what is the most pleasing or best design. Certainly, people are entitled to their own value judgments and to like the things they find pleasing. However, just because someone likes a design does not mean it is a good design or in good taste. There are certain principles of design, reflective of good taste. These must be carefully considered by designers and purchasers of buildings.

The ability to examine and evaluate building designs is something intangible. It develops through study and practice, much as an artist's perceptions develop. It is not just a matter of liking a design. A well-trained designer may dislike a design and still appreciate that it is a good design, in good taste. Still, the human element is always present in judging the aesthetic. Designers' preferences lead them to certain choices. For example, some architectural designers say that there is only one way to design a

house: The house should appear to belong to the ground and the hills and trees surrounding it. Frank Lloyd Wright was one such designer.

Those who use this approach try in their exterior designs to achieve the feeling that the house was put there by nature — that it belongs, just as the rocks and hills belong. See Figs. 3-1 and 3-2. Figure 3-2 shows the most natural of all — an earth

Fig. 3-2. An earth sheltered design that retains the beauty of the natural surroundings.

Fig. 3-1. A low, stone house that appears to fit closely to the ground and surrounding vegetation.

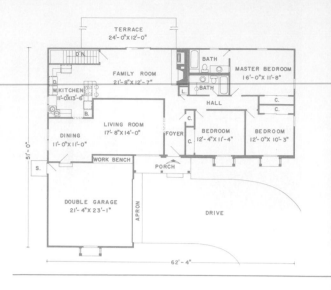

sheltered house. Earth is used on three sides and sometimes on the roof. Such a house fits comfortably into the side of a hill. The natural surroundings are retained and may be enhanced with plantings and rocks to help blend the house into the countryside. (See Chapter 13 for more about earth sheltered housing.)

Another group of designers might be classified as the traditionalists. They believe that the classical house styles are the best, and they reflect the design and details of the past in all their designs. The classical house must be modified so it is acceptable to modern society; yet it must retain the flavor of the past as authentically as possible. Certainly, a genuine 1800 New England colonial with its primitive kitchen, outside toilet, and poorly operating windows would not sell today. People may like traditional design, but they demand modern conveniences, such as central heat, plumbing, and a garage. It is possible to blend the best of the past with the conveniences of the present into an excellent, well-styled house. See Fig. 3-3.

The designer's style preferences should be clearly reflected in the design of a house. A house should never be "no style" or just a house. Neither should it be a mixture of styles, borrowing a part of one style and mixing it with parts of others. The end result is reflective of nothing; it is simply a nondescript hodgepodge. If a style is desired, it should be honestly reflected.

Today it is generally considered best practice to design a functioning floor plan and then to design a shell for this plan. Forcing a plan to fit a preselected exterior is extremely difficult. However, as the plan is being developed, the designer should constantly keep in mind what is happening to the exterior. The plan and the elevations must be developed together. Refer again to Fig. 3-3.

Basic Principles of Building Design

As the elevations are being designed, the composition, proportion, scale, contrast, rhythm, and unity of the building must be considered.

The **principle of composition** refers to the combining of the parts of a house to achieve a harmonious whole. All the principles of building design influence the composition of a building.

Symmetrical and Asymmetrical Designs

A building is of either symmetrical or asymmetrical composition. A **symmetrical** house has the same features on the right and left of its centerline. House A in Fig. 3-4 is an example of a symmetrical house. The garage projection on the left is balanced by the bedroom projection on the right. The roof design contributes to the symmetry. The brick wall and lights are in balance. The slight variance in symmetry does not spoil the overall feeling of precision, balance, and order. The plan of a house should not be forced to fit a predetermined symmetrical exterior, and the exterior should not control interior design.

House B in Fig. 3-4 is an example of a house with asymmetrical design. An **asymmetrical** design is one that is not balanced on a centerline. The right and left parts of the house are different. The designer must make the asymmetrical house appear to be balanced. This is done by grouping portions of the structure of varying sizes around a focal point, such as a front entrance. Although it is not mathematically symmetrical, the observer feels satisfied. The structure does not appear top-heavy or off balance.

In Fig. 3-4, the long left side of house B is balanced by the massive chimney on the right. This is necessary because the focal point, the front door, is to the right of the center of the building. To achieve a feeling of balance, the larger portions of a building should be located nearer to the focal point than the lower or smaller portions. Notice that the chimney is near the front door. Place a piece of paper over about two inches of the left side of house B. This design is now out of balance. The chimney overbalances the house. Place the paper over the right side covering the chimney. The long left side makes the building appear out of balance.

Fig. 3-3. A modern home using French Provincial styling.

Fig. 3-4. Building designs may be symmetrical or asymmetrical.

A. This house has a symmetrical design.

B. This design is in balance. It is an asymmetrical design.

Proportion in Design

Proportion must also be considered in designing exteriors. This refers to the relation of the size of one thing to the size of another or to the whole. Various designers and artists through the years have set up rules of proportions which they felt were most satisfactory or pleasing. However, the actual proportions that are pleasing vary with the individual and the culture. The proportions between the windows and the mass of the house which were pleasing in a colonial house in 1700 certainly are not pleasing in today's contemporary homes with their glass walls. In this sense, proper

81

proportion involves where a thing is used and with what it is associated. The entire effect can be altered by a simple change in the relationship of the size of associated elements. See Fig. 3-5.

A designer will consider the **scale** of the items involved in elevations. Such items as doors and stairs must be of such a size or scale as to allow easy use by the average person.

A frequent violation of the proper consideration of scale is a huge picture window placed in a small house. This window dominates the entire wall area and conflicts with the standard-size windows and door. It is **out of scale.** A more pleasing exterior would result if the picture window were smaller or, in other words, on the same scale as the other windows. This does not mean that all windows have to be the same size. Size variation is common and acceptable. If a window appears to be somewhat out of scale, the appearance can be helped by keeping the glass panes approximately the same size.

The fireplace, together with its chimney, is an example of another common item that frequently gets out of scale. Some are too large for the size of the house, while others are so slender that they look weak and unimportant.

Contrast in Design

The monotony of an otherwise satisfactory exterior can be relieved by **contrast** in the design. A house may have contrast in the mass of the structure. For example, a small wing or ell can break the monotony, Fig. 3-6.

Various materials may be used in developing contrast. See Fig. 3-10. A house with a part brick and part frame or stone exterior is more interesting than one constructed entirely of one material. Contrast may be obtained by the use of color or glass.

It is difficult to obtain contrast by using various geometric shapes, such as rectangular and semicircular masses and arches. In most cases, each should stand alone. It is usually best to vary the proportions of each geometric mass, rather than to introduce another form. For example, if a structure is basically built on a rectangular mass, it is unsatisfactory to introduce semicircular or triangular forms, such as an arch. See Fig. 3-7.

Even though a building is a stationary object, it can offer the viewer **a feeling of rhythm or movement.** A solid, unbroken wall does not offer this feeling. However, a wall that is broken by windows, shutters, trim, and other details which describe a definite, repetitive pattern introduces a feeling of rhythm. Rhythm can be subdivided into **pronounced patterns** and **secondary,** or more

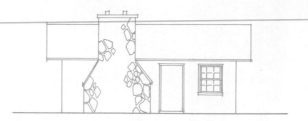

Fig. 3-5.   A massive chimney that is out of proportion to the house.

Fig. 3-6.   The house at the top leaves a feeling of monotony. This is relieved by adding an ell.

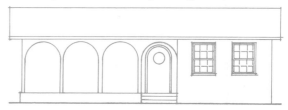

Fig. 3-7.   Introduction of a conflicting geometric shape—the round arch—has violated the basic rectangular mass of this house.

subtle **patterns.** To illustrate, a building may have an evenly spaced detail, such as columns, that is highly prominent. This detail sets the main theme or pattern of rhythm. A minor detail, such as a small amount of paneling, could also present a second pattern of rhythm, but be subservient to the main theme, the columns. This is much the same as writing music, with a main theme maintained throughout a piece, and a minor rhythm interjected in a subdued manner. See Fig. 3-8.

Unity of Design

A well-designed exterior must have **a sense of unity.** If a building is in balance, exhibits good pro-

Fig. 3-8. The exposed rafters give a sense of rhythm. The vertical V-groove siding provides a minor or secondary pattern of rhythm.

portion of its parts, is to scale and has a sense of rhythm, it is a unified, pleasing whole. In other words, the mass and details seem to belong together.

The violation of any of the principles of design will destroy the sense of unity. Dormers out of scale or the selection of conflicting geometric masses will damage the unity of the building. No two parts should be in conflict for domination. Certainly, an examination of the completed elevation with a consideration of the principles of design is a vital final step in the designing of elevations.

Identifying Characteristics

The purpose for which a building was designed can usually be guessed by examining the exterior. Buildings have characteristics that, because of their function and association with the past, make them different from other buildings. It is easy to identify a school or a church without ever going inside. You know what they are because they look like a church or school. This is their character.

Some of today's buildings are quite different from their counterparts of 200 years ago, and yet there is still something about them that is characteristic. Churches have changed as drastically as any building. Nevertheless, even though clothed in new materials which are assembled in radical manners, it is possible to tell that the building is a church.

Homes can have character, too. A low, charming Cape Cod leaves the visitor with entirely different reactions that a flat-top, glass-wall, contemporary house. Even the building materials influence the character of a house. The designer should consider this when planning exterior designs.

Mass, Texture, and Color

The designer cannot proceed to work on the elevations without considering the mass, surface texture, and color of the exterior.

The **mass** is the large geometric bulk of the structure, usually a rectangular form. The overall appearance probably is influenced more by the mass of a building than any other one thing. Compare your reaction to the house in Fig. 3-8 with your reaction to the A-frame in Fig. 3-9. Compare your reaction to the mass of a long, wide, low supermarket with that to a six-story office building. Your

Fig. 3-9. This building is a triangular mass.

first impression or reaction is to the mass. Then the factors of surface texture and color are noted.

The exterior surface **texture** stimulates sensory reactions. The various exterior materials evoke varied reactions. Glass and aluminum panels leave a cold feeling when compared with the warmth of a brick or stone exterior. A wood exterior is warm and conventional. Other materials, such as precast concrete, mosaic tile, and porcelain panels, have different textures and produce different feelings and reactions. As the character of a building is planned, consider its texture. See Fig. 3-10.

Anyone who has examined paintings knows the value of **color.** A feeling of warmth and friendliness can be engendered by using the warm colors. Other colors can provoke a feeling of coldness or loneliness. Interest can be excited, character spelled out, and good taste reflected by proper and careful selection and use of color.

Factors Influencing Styling

Before the plans can be drafted, the design of the roof and its relationship to the house proper must be determined. This includes the type of roof suitable for the style of house, as well as the height and pitch of the roof. The cornice details and amount of overhang should be decided.

Another consideration is the exterior materials to be used. While cost and availability of materials are important factors involved in this decision, certainly appearance should be paramount. The material must be suitable for the style of the house. The monotony of a large wall of one material can be relieved by a skillful blending of several materials. For example, frame blends well with brick or stone, stone and brick look well together, and large areas of glass can be used with any of these materials.

Refer again to Fig. 3-10.

As styling decisions are made, consider the appearance of the side and rear elevations. These are important to the overall effect created by the house. All sides of the house should be in character.

The shape and slope of a lot influence the type of house best suited for the site, and this, in turn, influences the styling.

Styling is influenced by the neighborhood. It would be unwise to have a house that is drastically different from those in the immediate neighborhood.

The likes and dislikes of the owners should be reflected in the styling. While this may produce a satisfied homeowner, if the design is not tempered by the judgment of a trained designer, a "no style" house may very well be created.

The amount of money available certainly will affect styling. This will be reflected in many ways, such as in the choice of exterior materials or in the choice of a gable roof instead of a more expensive hipped roof, or a square plan instead of an **L, T,** or **U** plan.

Building codes and deed restrictions frequently limit the style and type of house that can be built.

Climate has a lot to do with styling. Consider the effect that mild winters have on the way people live. Warmer climates with long growing seasons encourage designers to bring the outside right into the living area through the use of glass. Large patios and outdoor living are involved. Homes in areas of long, severe winters cannot use large glass areas because of the great loss of heat.

The section of the country in which the house will be built frequently dictates its style. Various sections look with favor upon particular styles. For example, in Virginia, colonial styling reigns supreme. Some extreme contemporary houses are built, but they are at a minimum. The people have lived in this

Fig. 3-10. The contrast of wood shingles and brick siding used on the exterior of this house relieves monotony and brings a warm sensory reaction to the viewer.   The Garlinghouse Co., Inc.

colonial atmosphere for generations, and this is what they desire when they build. Building a house which clashes with the local tastes could be a risky investment.

It is usually a mistake to build a house in one section of the country that was designed for another. For example, a New England Colonial house would be out of character in southern California. With its small windows, it would be difficult to ventilate and keep cool in the hot seasons. The many completely enclosed rooms and lack of contact with the outside give it a closed, restricted feeling. It would be difficult to live in such a house in an area with long, warm seasons.

Whether a house will have a basement influences its planning and styling. If a low appearance is desired, the basement could well be eliminated. This puts the finished floor almost on grade. If a basement is used on such a house, the windows are partially buried below the surface of the ground, permitting very little light and ventilation.

If a house is built on a slope, with a good portion of the basement wall exposed, the basement area is much more usable. However, this greatly affects the exterior styling since careful consideration must be given to making this foundation pleasing. See Fig. 3-8. A split-level house uses the foundation wall to good advantage by allowing the use of windows and doors.

Common Roof Types

No study of styling would be complete without consideration of the types of roofs in common use. Fig. 3-11 illustrates some common roof types.

Probably the most frequently seen roof is the **gable** or, as it is sometimes called, the **A roof.** This is an economical, easy roof to construct. It handles rain and snow easily and is adaptable to many styles of homes. It is easily vented by louvers at the peak of the gable end.

An adaptation of the gable roof is the **hip roof.** This roof is more expensive to build because of the labor involved in framing it. It tends to give a house a low appearance and does away with the gable end. This eliminates maintenance of the siding on the gable end. Houses that are almost square can use a hipped roof to advantage if all sides are brought to a common point in the center.

The **flat roof** is coming into more frequent use on contemporary homes. It requires the least cost for materials and labor, but demands special care in sealing against leaks. Some contemporary homes

use a **shed roof** to provide some slope to facilitate drainage. This is also inexpensive to construct. Both the flat and shed roofs need special consideration of structural strength in areas of heavy snows; considerable extra weight can accumulate on such a roof.

A variety of the shed roof found on some of today's homes is the **monitor.** It is a combination of two shed roofs, with skylights between the two roof levels.

The **gambrel** and the **mansard** are roof styles representative of some traditional house styles. The gambrel is characteristic of the Dutch Colonial house, while the mansard is found on French Provincial houses.

A roof style seen on contemporary homes is the **butterfly roof.** Since all water is shed to the center of the house, it must be built carefully to prevent leakage. It is not an expensive roof to build.

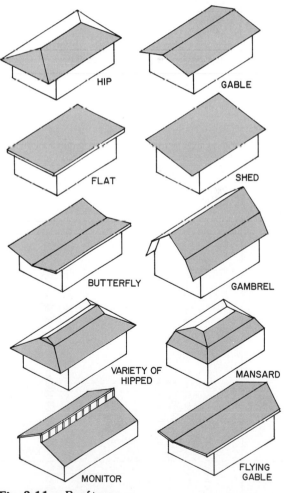

Fig. 3-11. Roof types.

Andersen Corporation

Fig. 3-12. The character of a house is strongly influenced by architectural detailing and window selection.

A styling feature used on gable roofs is to build out the overhang at the ridge and slant it to the eave. Such a feature is called a **flying gable.** It gives a house a rakish appearance and a feeling of increased length.

Window Planning

Windows are a very important part of both interior planning and exterior styling. Properly selected and located, they increase the usefulness of a room. See Fig. 3-12.

A pleasant view can either be enjoyed or it can be blocked, depending upon window selection. If a room has a pleasant view, the window sill should be low enough to enable a seated person to look out without straining. Windows with large, open glass areas are recommended for this purpose. Muntins across windows make viewing difficult. See Fig. 3-13.

Properly located windows can increase the possibilities for natural ventilation. Natural ventilation can be limited to a single room or it can come through windows in several rooms. See Fig. 3-14. The U. S. Department of Housing and Urban Development requires a minimum of 5% of the floor area to be available in open window area. Many homes have year around heating-cooling systems which keep the temperature comfortable at all

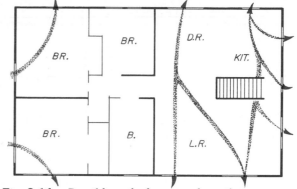

Fig. 3-14. Possible paths for natural ventilation.

times. Natural ventilation is less important in this case than natural light and view.

Possibly the most important function of windows today is to relieve the close, boxed-in feeling a person has in a room having few or no windows. The smallness of a room is relieved by tying it into the outdoors through windows. The use of natural light greatly helps in making a room pleasant. While artificial lighting is well-developed, natural lighting is still of great importance in designing a room that is pleasant to use. The minimum acceptable glass area in residential work is 10% of the floor area. Actually, it is not unusual to have window glass area equal to 20% or 25% of the floor area. Remember, however, that glass areas increase heating costs. Use double- or triple-paned windows to reduce heat loss and gain.

Windows can be planned to achieve a sense of privacy. Typical of this are the high narrow windows often used in bedrooms. See Fig. 3-15. They allow natural light and ventilation, yet increase the sense of privacy. They also increase the amount of wall space upon which furniture can be placed.

The location and size of windows has a great influence upon the furniture that can be used in a

Fig. 3-13. Choice of windows is important in room planning.

Fig. 3-15. High windows give privacy and allow wall space for furniture.

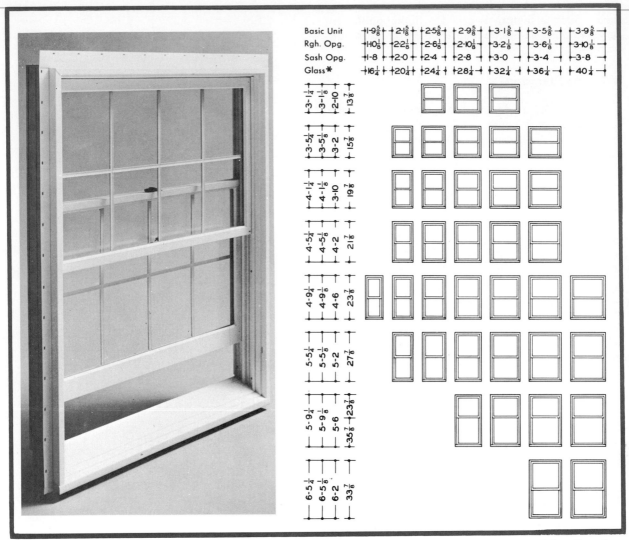

Andersen Corporation

Fig. 3-16.   Double-hung windows.
    Note: Several windows may be placed side by side to form a large window area. To find the overall unit dimension for a multiple window unit with a nonsupporting mullion, add the basic unit dimensions plus 1/2″ to this overall basic unit dimension. For openings requiring a supporting mullion, add 2″ for each supporting mullion. To find the rough opening, add 1/2″ to the overall basic unit size.

room. Windows with a sill near the floor remove that portion of the wall as a place to locate a piece of furniture. As window sizes are selected, keep in mind the standard furniture sizes.

The designer must always be aware of exterior design when selecting windows. Changes in window size influence the exterior appearance. The style of window must be in keeping with the style of house. For example, aluminum awning windows would be completely out of place on a Colonial style house.

Even the location of a window in a wall is important. After selecting the placement for the best interior use, the designer must see how the exterior appears with the window in that location.

Window Types

There are many types of windows available today. Those in most common use in residential construction are shown in Figs. 3-16 through 3-22. These are available in wood and metal.

UNIT DIM.	3'-0"	4'-0"	5'-0"	6'-0"
RGH. OPG.	3'-0½"	4'-0½"	5'-0½"	6'-0½"
GLASS*	14"	20"	26"	32"

Andersen Corporation

Fig. 3-17. Gliding windows.
Note: Multiple units of these sizes are available. To find the overall unit dimension which has a nonsupporting mullion, add the sum of the unit dimensions and subtract 2". The overall rough opening dimension is equal to 3/4" less than the overall unit dimension.

The **double-hung window** is most commonly used. They can provide ventilation at the top or bottom of the window opening. However, only half of the window can be open at any time. Most double-hung windows are held open with spring-loaded balances or a friction holding device. See Fig. 3-16.

A horizontal **sliding window** is much like the double-hung. The major difference is that the sash slides sideways rather than up and down. They run on nylon rollers. No spring balances are needed to hold them open. Sliding doors are also available. See Figs. 3-17 and 3-18.

The **awning window** hinges from the top of the sash. See Figs. 3-19, 3-20, and 3-21. It swings open to the outside of the house. Most types have a crank operating through gears and a lever to provide ventilation. It can be left open during a rain. They are sold in single units and multiple units. A multiple unit can be several windows high and wide.

Hopper windows are much like awning windows. The major differences are that they hinge from the bottom and open into the building. These windows deflect the incoming air toward the ceiling, thus reducing a direct draft. They are available in single and multiple units. See Fig. 3-21.

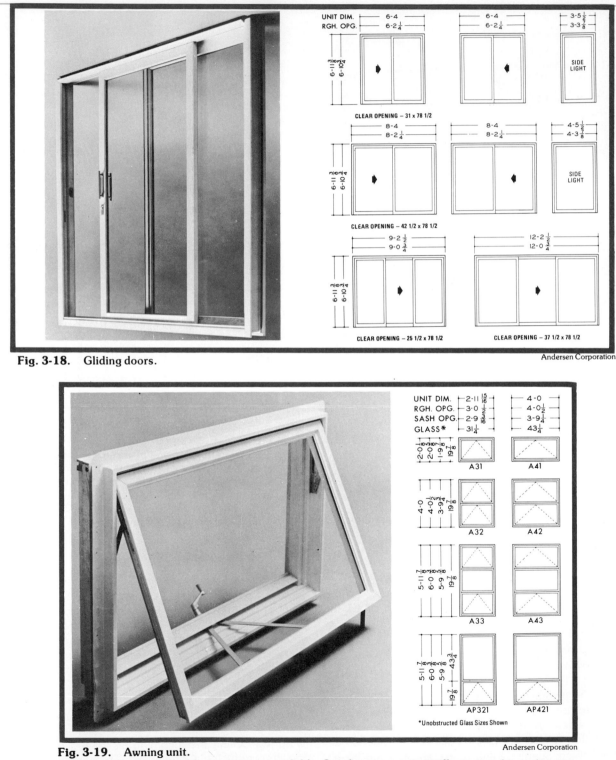

Fig. 3-18.   Gliding doors.

Andersen Corporation

Fig. 3-19.   Awning unit.
   Note: Multiple units of these sizes are available. See the casement unit illustration for explanation.

Andersen Corporation

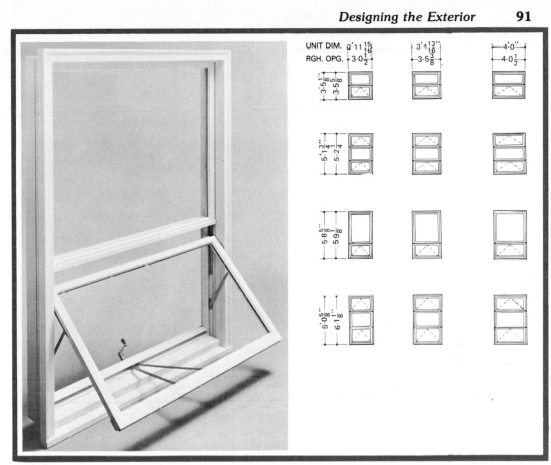

Fig. 3-20.   Fixed sash with awning sash.

Andersen Corporation

Note: Multiple units are available. To find the overall basic unit dimension, add the basic unit dimensions plus 2-7/8″. The rough opening size is the sum of the basic units plus 1/2″.

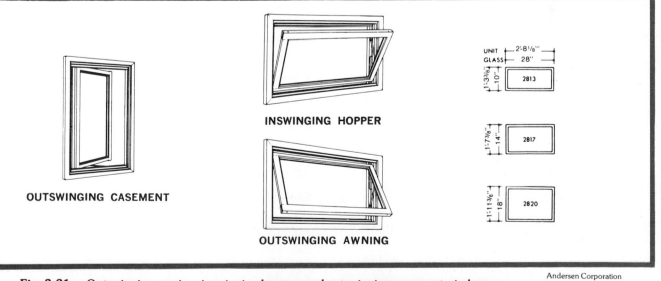

OUTSWINGING CASEMENT

INSWINGING HOPPER

OUTSWINGING AWNING

Andersen Corporation

Fig. 3-21.   Outswinging awning, inswinging hopper, and outswinging casement windows.

   Note: To find the overall basic unit width of multiple units, add the basic unit dimensions plus 2-7/8″ to the total. To find the rough opening width, add the basic unit width plus 1/2″.

   These units can be stacked vertically. The overall unit dimension for stacked units is the sum of the basic units plus 3/4″ for two units high, less 1/4″ for three units high, and less 1-1/4″ for four units high.

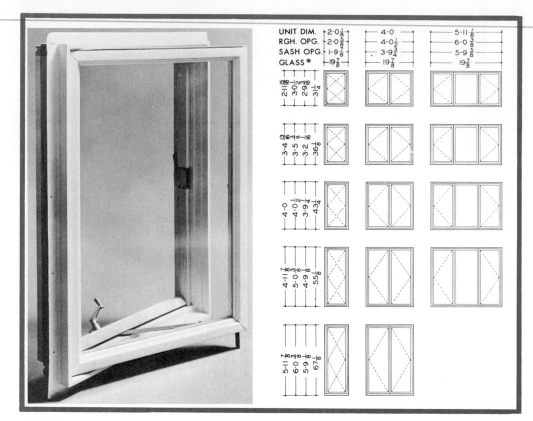

Andersen Corporation

Fig. 3-22.   Casement window.
   Note: Multiple units of these sizes are available. To find the overall unit dimension for a combination unit, add the unit dimensions together plus 1/8″ for each unit joining. To find the rough opening, add 1/2″ to the overall unit dimension. **Example:** If 2 unit C23 windows are joined to form a 4-window opening, the unit dimension is 4′-0″ + 4′-0″ + 1/8″ = 8′-0 1/8″ + 1/2″ or 8′-0 5/8″.

Casement windows hinge from the side of the sash. They open to the outside of the building. Usually a crank mechanism is used to open and close them. When open, the entire window area provides ventilation. They are sold as single or multiple units. Some are available with a fixed window unit. See Figs. 3-21 and 3-22.

Picture windows are large, fixed windows. They provide no ventilation. The glass is usually 1/4″ plate glass or an insulating glass. The insulating glass is two sheets of glass with a dead air space between. If ventilation is desired, other types of windows are placed next to the picture window.

Roof windows (often called skylights) are designed to open to permit ventilation. They provide natural light to rooms where conventional windows are not easily installed. See Chaper 27.

Build Your Vocabulary

Following are terms that you should understand and use as part of your working vocabulary. Write a brief explanation of each term.

asymmetrical	mass
contrast	principle of composition
double-hung window	proportion
gable roof	symmetrical
hip roof	unity

Class Activity

Have a class competition. Using the floor plan in Fig. 3-2, sketch three new exterior designs. Change the roof, windows, and exterior materials to suit each style. Select the sketch you like best and draw it to the scale 1/4″ = 1′-0″ or the metric scale 1:50. Have a local builder or architect select the best scale drawings for an exhibit.

Chapter 4

Getting Started on the Plan

The planning of a new home is usually the most important and exciting event in the life of the average family. Everyone has a "dream house" in mind and hopes to plan and build it some day. Families are willing to spend hours reading magazines and books and visiting homes to find ideas for their new house. However, the planning of a house it not this simple. Many factors must be considered before working drawings can be started.

General Considerations

A good way to start planning a house is to list all the features wanted. Specific items must be studied, such as the kinds of rooms, size of family, hobbies of family, personal likes and dislikes, and the style of the house. The lot upon which the house is to be built should be selected before planning progresses too far. The slope of the lot and the houses in the neighborhood influence the planning. The climate and the section of the country in which the house is to be built should be considered. Building codes, trends in style, advanced methods of construction, new building materials, and energy conservation all deserve consideration as a house is planned.

Another important consideration is the cost per square foot of the house. Some rooms are more expensive to build than others. Kitchens and bathrooms cost more than bedrooms or living rooms because of their special features, such as built-in cabinets and plumbing and wiring needs.

Perhaps more than any other single factor influencing house planning is the amount of money available. Information concerning financing can be found in Chapter 14.

Developing the Floor Plan

The following is a sample list of the things one family desired in a house.

> Three bedrooms
> Kitchen with dining area
> Living room at least 12 × 20 feet
> No dining room
> Full walk-out basement (lot slopes to rear)
> Recreation room in basement
> Fireplace in living room and recreation room
> One and one-half baths
> Two-car garage
> Prefer ranch-type home
> Price cannot be over $90,000 with the lot
> Brick exterior
> Gable roof, wide overhang
> Washer and dryer in basement
> Electric heat pump
> Room for a workshop

With the preliminary information concerning the house gathered, a floor plan can be developed.

Using an Existing Plan

Probably the most common method for people not trained in home planning and design is to use plan books, to examine magazines, or to look at plans already prepared. Certainly this is a sound approach. However, it is folly to copy a ready-made plan if it does not meet your needs. Try to "live" in a plan in your imagination. Always keep in mind the question, "Will it meet all of the needs of the family?" Usually it will not, and a process of revision must begin. This involves the application of the principles of room planning, traffic patterns, orientation of the house on the lot, as well as many personal considerations.

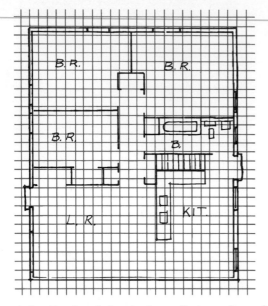

Fig. 4-1. Freehand sketch of a preliminary floor plan made by revising an existing plan. Use graph paper with ¼-inch squares. Allow each square to represent one square foot and represent walls with a single line.

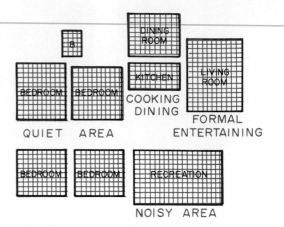

Fig. 4-2. The room templates are grouped according to activities.

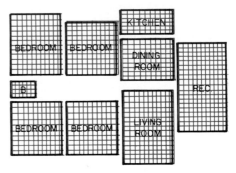

Fig. 4-3. A preliminary arrangement of the room templates into a possible floor plan.

The original plan can be revised by making a sketch of it. See Fig. 4-1. Place vellum (a strong, fine-grained paper) over the original and trace the parts to remain. Then sketch in the changes. It will probably take several sketches and much study before a plan is finalized. Be certain to utilize the principles of planning as changes are made. These principles are also used to develop original plans. The following section, Planning Using Templates, gives directions for developing your own plan.

Planning Using Templates

Another method used to develop the floor plan is to first develop each room using templates. This is described in Chapter 2, Planning the Individual Rooms.

The room templates form the basis for developing a floor plan. Remember that templates are drawn to scale. Usually a $1/4'' = 1'\text{-}0''$ or 1:50 metric scale is used. They are then arranged in various ways on graph paper until a desirable floor plan is developed.

Keep the principles of planning in mind while developing the floor plan. First, separate the rooms into basic areas. This is **zoning.** See Fig. 4-2. Templates may be moved about within these areas as consideration is given to other planning factors.

These may include orientation, view, conditions of the building site, and other outside factors. Inside factors may be needs for halls, stairs, and closets and other storage areas. Provisions must be made for laundry and heating and air-conditioning equipment.

If the house is on several levels, a plan is developed for each level. Each level must be carefully coordinated with the others.

The layout developed is a preliminary plan not a finished plan. See Fig. 4-3. It shows only the proposed relationship between rooms and areas. Adjustments must still be made. For example, this plan has an irregular exterior shape and the halls need to be developed. The plan can be adjusted by slightly changing the room sizes. Check the furniture arrangement in each room to see how the new size may change it.

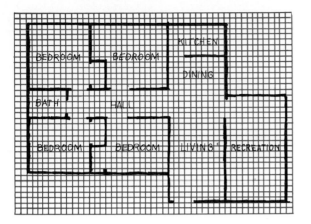

Fig. 4-4. The plan is sketched on graph paper. Closets and halls are added.

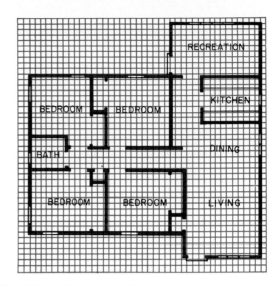

Fig. 4-5. The proposed final plan.

Next, sketch the plan on graph paper. Include closets, fireplaces, and interior walls in your plan. See Fig. 4-4. Usually, there will still be undesirable features. Make new sketches as you solve the problems. Tracing paper or vellum can be used over the original to copy the features to be retained. Apply the principles of modular design to the solutions. A section on modular planning is included later in this chapter.

Eventually, major problems will be solved and a final sketch can be made. See Fig. 4-5. Since the final sketch should be reasonably accurate, drafting tools might be used. Draw interior and exterior walls to indicate their proper thickness. Interior walls usually require 5 inches (127 mm) of space. Exterior frame walls are usually 6 inches (152 mm) thick, solid masonry walls - 8 inches (203 mm) and masonry veneer - 10 inches (254 mm). Adding wall thickness on the plan can reduce the size of a room 10 to 12 inches (254 mm to 304 mm). This may particularly affect closets. Next, accurately locate doors and windows. Sometimes, the final sketch will locate driveways, sidewalks, patios, and even trees and shrubs.

Exterior Styling

As the floor plan is developed, a new factor enters in — the exterior styling of the house. Many changes in locations of interior wall partitions, closets, and halls can be made without changing the exterior, but as soon as a window or fireplace is moved, the exterior appearance is affected. If the house is made larger or an ell added, the exterior

Fig. 4-6. Freehand sketch of a possible elevation.

styling is greatly changed. It is not a matter of whether the exterior or interior should be planned first — they must be planned together. Therefore, as each change in a floor plan is made, consider its effect on the exterior styling.

Make sketches of the exterior after a floor plan is developed but before it is finalized. See fig. 4-6. Any plan can have many different elevations. For more information, study Chapters 1 and 3.

It is important to work out roof details. Often, an unbalanced appearance is produced by an awkward roofing situation. Difficult and expensive roof framing situations can be spotted and corrected. This may require sketches of the elevations of all sides of the house. A cardboard model helps with roof planning and makes it easy to visualize how the roof will look.

Before proceeding with the final floor plan drawings, decide how the house will be constructed. A freehand sketch indicating the typical wall section is

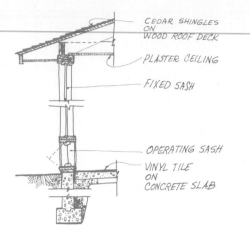

CEDAR SHINGLES
ON
WOOD ROOF DECK

PLASTER CEILING

FIXED SASH

OPERATING SASH

VINYL TILE
ON
CONCRETE SLAB

Fig. 4-7. Freehand trial sketch of a wall section.

essential. See Fig. 4-7. Such preliminary planning will frequently correct errors that will influence the elevations.

Modular Planning

As the width and length of the house and its rooms are being decided, use the principles of modular planning. A **module** is a standard unit of measure. Modular planning saves materials and reduces labor costs because most construction materials are made on a 4-inch standard module. Buildings are designed using 16 inches and 24 inches as **minor modules** and 48 inches as the **major module.** For example, sheets of sheathing material are 4'-0" × 8'-0". The 4'-0" width is three 16-inch or two 24-inch modules. When the construction industry sets metric standards, the module will probably be 100 millimeters.

If the house is to be constructed from factory-assembled modular wall components, design the length and width to use the modular measure. Details of a **modular system** are in Chapter 7.

Some houses are constructed from factory-assembled modular units. A **modular unit** is a room size or larger part of a building that is completely finished in a factory. It is moved to the building site much like a mobile home. When making a floor plan for such a house, the rooms have to be planned so the modular units will fit together.

After the floor plan is finalized, the elevations planned, and the wall section decided upon, it is safe to proceed with the drafting of the final floor plan.

Summary of Planning Steps

1. List all the things wanted in the house.
2. Examine existing plans for one near to what you want or develop a plan using templates.
3. Try to "live in" the plan.
4. Revise the plan to better meet your needs.
5. Study traffic problems.
6. Apply principles of planning to each room as explained in Chapter 2.
7. Consider the proper orientation of the house.
8. Check to see if stock parts can be used with little waste.
9. Sketch wall thickness on revised plan.
10. Sketch exterior elevations. (See Chapter 3.)
11. Examine roof for problems.
12. Ascertain wall construction. (See Chapter 7.)
13. If all checks, then begin drafting final floor plan.

Traffic Patterns

As the house is being planned, constant attention must be given to the flow of traffic through the house. Considerable inconvenience can be caused by poor room relationship. People must be able to move from room to room and from one part of the house to another with a minimum of congestion and unnecessary steps. Rooms can be rendered useless by having a main traffic aisle directed through them.

Following is a list of factors to observe as the flow of traffic between rooms is considered.

1. In most homes, the heaviest flow of traffic will occur between the kitchen and dining areas, the bedrooms and bath, and the living and dining areas.
2. The dining area should be next to the kitchen and easily accessible. This eases the task of serving a meal and cleaning up. No hall should separate these areas.
3. It should be possible to enter the kitchen without crossing the dining area. An aisle of traffic through a dining room could be completely blocked if the family were seated at the table.
4. The kitchen should have an outside door. This is necessary for removing trash. It also gives the children an entrance to use, so they do not soil the carpets.

5. The garage or carport should be near the kitchen and have an entrance permitting easy access into it. This facilitates bringing in supplies and entering the house in bad weather.

6. Each bedroom should have easy and quick access to a bath — no other room should come between these. The location of a bath between two bedrooms with entrances from each room is very unsatisfactory. See Fig. 4-8 for an example. This robs everyone of privacy. Fig. 4-9 illustrates a more satisfactory solution.

7. Each bedroom should have access to a hall. It should not be necessary to go through one bedroom to reach another.

8. It is desirable to be able to go to the front entrance without crossing through the dining area or living area. This principle of traffic flow is occasionally violated under the guise of economy. If the front entrance opens directly into the living area, the expense of an entrance hall or foyer is eliminated, but the effective space in the living area is also cut.

9. It is ideal to be able to enter a house and go to the bedroom area or bath without crossing through the living area. This is desirable for the sake of privacy.

10. An inside means of reaching the basement is vital. This is best if it can be built off the kitchen

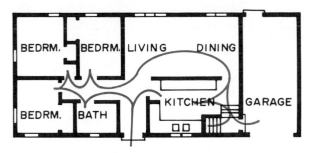

Fig. 4-10. A plan with a good traffic pattern.

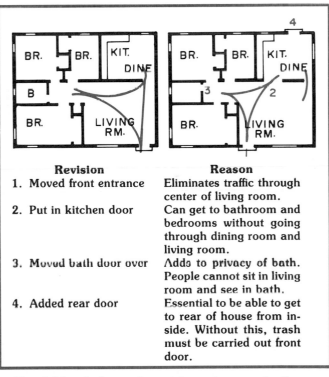

Revision	Reason
1. Moved front entrance	Eliminates traffic through center of living room.
2. Put in kitchen door	Can get to bathroom and bedrooms without going through dining room and living room.
3. Moved bath door over	Adds to privacy of bath. People cannot sit in living room and see in bath.
4. Added rear door	Essential to be able to get to rear of house from inside. Without this, trash must be carried out front door.

Fig. 4-11. A plan and a revision to improve the flow of traffic. Such revisions do not increase the cost of the house.

area. Also, a person should be able to go directly outside from the basement without passing through the living or dining areas or front entrance hall.

11. An outside entrance to a basement is desirable, but not essential.

Figure 4-10 illustrates a well planned house. Can you find any way to improve the traffic pattern?

Examine the floor plan and its revision in Fig. 4-11. Can you suggest any other improvements? Notice the reasons given for the changes made. Are they valid?

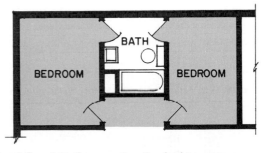

Fig. 4-8. A bathroom situation lacking privacy.

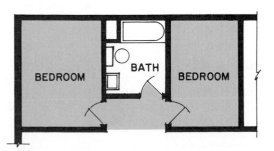

Fig. 4-9. A satisfactory solution allowing privacy in the bath.

Required Problem

Now try your hand at improving the following floor plans (Fig. 4-12). It probably will take several tries before you are satisfied with a plan. Remember it usually is necessary to revise a plan several times. Submit the revised plan to others for their opinions. Sometimes, another person can see things you have overlooked. Constantly apply the principles of planning for traffic movement as presented in this chapter.

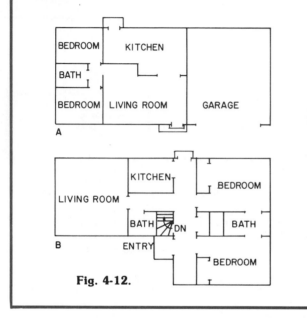

Fig. 4-12.

As you work on your plan revisions, consider designing several alternate kitchens, baths, or bedroom combinations, each following the principles of sound planning. Then, as you try to decide on a final plan, you will have a wider choice of ideas. Do not accept the first plan that satisfies the planning principles. Remember there are many variations of every plan. All variations could be good, but some will be more exciting, fresher, and more pleasing to you. This is your goal — a plan that is not just sound, but one that pleases you and is stimulating, fresh, and exciting.

Sketch your revisions to scale on squared paper. Be careful that the revised plan does not introduce poor factors.

Zoning the House

Zoning refers to the dividing of a house in areas according to the various activities of the family. It is an attempt to develop a plan that provides some privacy for each person. Obviously, some activities in a home conflict. Noise from children playing is not conducive to sleep. However, a house can usually be divided into the quiet area, including the sleeping and study areas, and the noisy area, such as the living room, recreation, and food areas. When the zoning of activities is considered, the bedrooms are usually placed in one end of the house or in a wing away from the living area. Placing a sound barrier between the sleeping area and the living area reduces transmission of noises. A stair, hall, or a closet provides an effective sound barrier to help isolate the sleeping area.

Usually, formal and quiet entertaining take place in the living room. This means that some provision should be made for the more active recreational activities. A basement recreation room is an excellent location for noisy activities, although a ground-level recreation room, located off or near the kitchen, is good.

If a family desires quiet and privacy for reading, a den or study should be provided. Again, this must

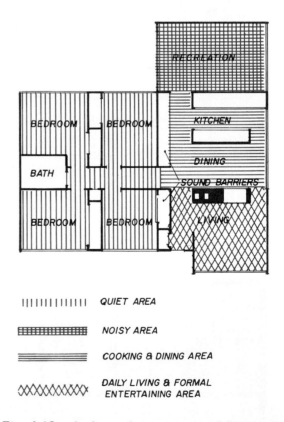

Fig. 4-13. A plan with activities carefully zoned to improve livability.

be away from the living area. In most houses, it is closely related to the sleeping area.

The kitchen and dining areas are best separated from the other areas. This reduces the annoyance of cooking odors and the disturbances from running water and the clatter of dishes.

The formal dining area usually is very near the kitchen, but separated from its view and traffic. This insures a relaxed and pleasant environment for dining. However, in many homes, these areas are combined.

Figure 4-13 illustrates a plan that takes advantage of zoning considerations.

Basic House Types

As choices are being made during the planning stages, an early, essential decision is the type of house to build. See Fig. 4-14. The thinking of the planner must be centered about a specific, basic type. Ideas for a good split-level type of house may not be suitable for a ranch-house type.

Fig. 4-15. A one-story house. This type of house may or may not have a basement.

One-Story

A one-story house has all the habitable rooms on one level. A ranch house is an example. Most long, low, contemporary homes are of this type. An advantage is that any room can be reached without ascending or descending stairs. Some one-story houses have a basement with a flight of stairs, while others are built on a concrete slab or over a crawl space. The attic space is small and not usable for expansion of living area, but is useful for limited storage. A house of this type usually has a low-pitched or flat roof. See Fig. 4-15.

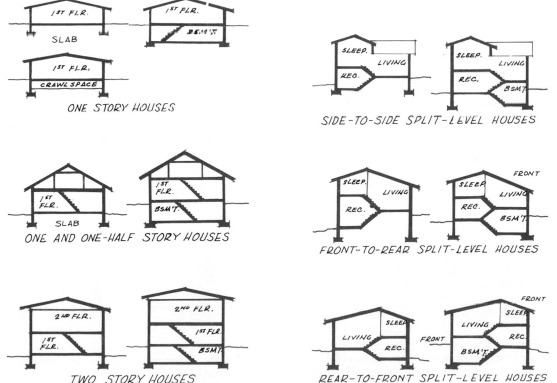

Fig. 4-14. Typical house types.

Fig. 4-16. A one-and-one-half story house is economical to build.

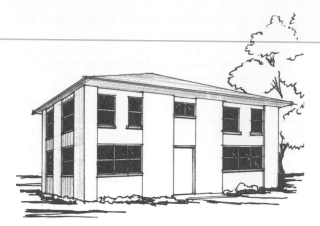

Fig. 4-17. A two-story house offers a large amount of living area.

One-and-One-Half-Story

The one-and-one-half-story house is similar to a one-story, except it has a high-pitched roof, making the attic space useful for living area. Generally, it has two bedrooms and a bath on the first floor and additional bedrooms and a bath on the second floor. About half of the total area of the attic floor is useful living area; space with less than 5-foot (1524 mm) headroom is not considered useful. The walls and ceiling slant, and knee walls are necessary. Dormers on the front and a shed dormer off the rear are common.

This is an economical house to build, because it offers a considerable amount of living area. The only extra cost is the higher roof, dormers, and stair. This type of home is desirable as an expansion house. The Cape Cod is a popular one-and-one-half-story house, Fig. 4-16.

Two-Story

The two-story house has been a favorite for years. It consists of a first floor and full second floor of the same area as the first floor. It affords certain economies, the most important being a large amount of living area with a minimum of foundation and roof being required. Any heat loss through the first-floor ceiling benefits the rooms on the second floor. Also, this type house can be built on a smaller lot than the long, rambling homes. See Fig. 4-17.

While most two-story houses are of a traditional style, architects have designed excellent contemporary two-story houses.

The long flight of stairs to the second floor is somewhat of a disadvantage, especially since all

bedrooms are on the second floor in most two-story houses. A half bath located on the first floor is an essential item.

The two-story house can be built with or without a basement.

Split-Level

A split-level looks like a one-story house which has had one portion of its basement pushed up about halfway out of the ground. The living area, generally, is on the ground level, with a deep basement below. The bedrooms are on the highest level, with a garage, utility area, recreation room, or additional bedrooms in the raised portion of the basement below. Such a plan converts what generally is a large, wasted, basement space into light, well-ventilated, living space. See Fig. 4-18.

The split-level house is admirably suited for sloping lots. Some people build this type on flat lots and fill in around them. This usually results in a high, off-balance exterior and should be avoided for the sake of aesthetics.

Fig. 4-18. Split-level house. This design makes good use of space.

There are a number of frequently used variations of the split-level. Most commonly used is the side-to-side split. This is well suited for lots sloping to the right or left. Such a house can have three or four levels.

The front-to-rear split is well suited for lots sloping away from the street. From the street, the house appears as a one-story, while from the rear, it appears as a two-story.

A similar split is from rear to front, giving the street elevation the appearance of a two-story house. This is well suited for lots sloping toward the street.

Common House Shapes

The shape of a house influences the layout of the floor plan, exterior styling, and cost.

The most common shapes are the square, rectangle, **L, U,** and **T.** Projections from a house, at right angles to the length of the main portion of the house, are called **ells.**

Ells or breaks in a long elevation do much to improve the apprearance, but they do increase the cost. Each break causes extra corners. The more the house deviates from the square shape, the more **lineal feet** of wall and footings are required. See Fig. 4-19. Notice that the square house has

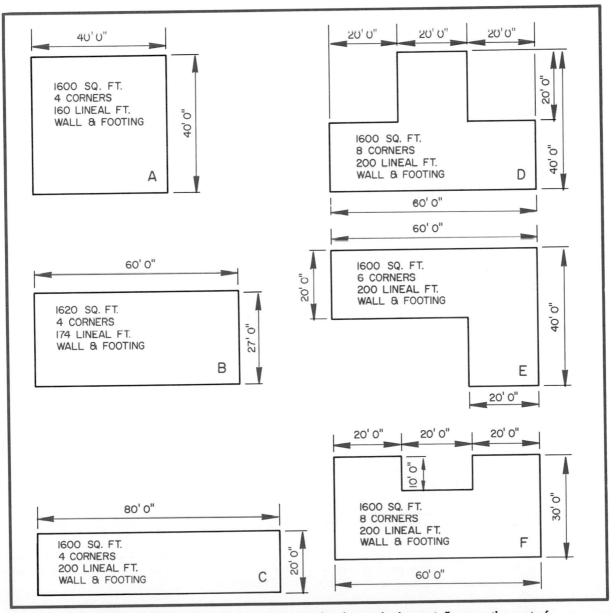

Fig. 4-19. Typical house shapes. Changing the shape of a house influences the cost of building.

1600 square feet (148.8 m²) of floor space, 4 corners, and 160 lineal feet (48 768 mm) of wall and footing. Both rectangular houses have the same square footage of floor space, but one has 174 lineal feet (53 035 mm) of wall and footing, and the other has 200 lineal feet (60 960 mm). The less square the house, the more expensive it will become. The **L** plan also has 1600 square feet (148.8m²) of floor space, but has 6 corners and 200 lineal feet (60 960 mm) of wall and footing. Thus can be seen the reason for considering cost as house shape changes.

The rectangular shape is used more frequently than the square space, because it makes it easier to develop a satisfactory floor plan. A good square floor plan is very difficult to devise.

The **L** shape costs a little more, mainly because of the extra corners and more expensive roof framing. However, the extra cost is well spent when the exterior styling is considered. This same reasoning can apply to houses built in other shapes; the cost is increased, but there is more freedom for interior planning and exterior styling.

Building Codes

Building codes specify the types of materials and construction methods acceptable. They specify allowable loads, stresses, mechanical and electrical requirements, and items pertaining to health and safety. Fire protection is an important example. Designers and builders must be certain to observe these laws. The city has an inspector who visits the building site to see that codes are being met.

Building codes vary from area to area. Some differences are due to geographic location. For example, earthquake problems are more likely to arise in certain parts of the country than in other parts.

In addition to local building codes, certain other organizations, such as the U.S. Department of Housing and Urban Development may have minimum standards to meet.

There are several standardized building codes which find wide acceptance in the United States. These are **The Standard Building Code** prepared by the Southern Building Code Congress, **The Basic Building Code** of the Building Officials and Code Administrators, International, Inc., **The National Building Code** published by the American Insurance Association, and the **Uniform Building Code** published by the International Conference of Building Officials.

Another possible place to look for building requirements is in the deed to the land upon which a house is to be built. A **deed** is a written contract which conveys the title of ownership of a piece of real estate to the purchaser. A deed frequently contains building restrictions. For example, it may require the house to be one story, or brick, or to have a basement. Or it may require all houses built on the land involved to be two stories high. The deed deserves careful examination.

The Building Site

When a home is planned, the selection of the lot is of primary importance. A house should be planned to fit a specific lot. Frequently, people plan a house first and then try to find a lot to fit it. It is very difficult to find a lot that meets all the requirements of a predetermined plan. In making a search for a lot, begin by locating the general area in which you would like to live. Investigate the building codes, for harsh or outdated codes do exist and you should be aware of them. A sound code is highly desirable and is for your protection. So beware of an area with no codes.

Examine the tax structure. Taxes vary widely between areas, even adjacent areas. Determine if you are going to get your money's worth in schools, streets, police protection, and utilities. Visit the local government office and try to find out what the future holds for taxes in that area. Low taxes do not necessarily indicate a desirable area in which to build.

Consider the **zoning regulations.** These are local laws specifying the types of buildings that can be built in an assigned area. Ascertain if an area is zoned for residential building only. Some zones permit erection of stores, factories, office buildings, apartments, and service stations. Examine the zoning restrictions of neighboring areas. You may border a factory area. A business built next to or near your house can cause a rapid loss of value.

Consider two points when studying zoning: (1) How are the zoning regulations amended — by vote of the residents in the affected zone or by a board? (2) Does the zoning require a minimum house which is too large for you to afford to build?

Examine neighboring areas. Drive around the streets bordering the area of interest to you. Look for objectionable factors such as noisy, busy thoroughfares, dumps, swamps, railroads, or substandard sections of houses. Is the neighborhood on the downgrade? All these factors will lower the future value of your house and the pride you will have living there.

Look for the conveniences you desire. These could include easy access to shopping centers, schools, bus service, and churches. The availability of utilities (gas, water, sewers, and electricity), as well as telephone service, should be considered. It is wise to check into such items as fire protection, garbage disposal, street paving, and streetlights. If these are not available, find out who is to pay for these improvements — you individually or the public.

If a thorough examination of the area proves it to be desirable, you are ready to find your lot. As you consider various sites, study the size and shape of the lots. Planning a house to fit a narrow, deep lot is quite difficult. For example, it is hard to locate the front entrance and a garage on such a lot. It also prevents the construction of certain styles of houses such as the ranch, contemporary, or colonial house. Such houses demand a lot with wide street frontage, but do not require great depth. A typical wide lot would range from 100 to 130 feet (approximately 30 m to 40 m) wide and 130 to 150 feet (40 m to 46 m) deep, while a small, narrow lot could be 40 feet wide by 125 feet deep (12 m wide by 38 m deep).

When considering lot size, examine the needs of your family and their interests. Some people like gardens and spacious lawns while others desire a minimum of yard work and would not be happy with a large yard. Children in the family may also influence the choice in lot size.

The type of houses in the neighborhood influences, somewhat, the type of house to be built. If colonial-style houses predominate, it would be out of keeping to place a flat-top contemporary house among them. The house being planned should blend with the others in the neighborhood. It need not be of identical style, but it should not clash.

The value of the surrounding houses warrants consideration. It would be a risky investment to build a house that is considerably more expensive than any other in the area. For example, if a contractor built a house valued at $75,000 in a neighborhood of $45,000 houses, it would be very difficult to sell, because the family able to afford the more expensive house would also prefer to live with other families in a similar income bracket. Many families in the lower income bracket would like to purchase the expensive house, but probably would have great difficulty in raising the down payment and making the monthly payments. Most loaning organizations would refuse to issue them a loan, because of their lower income.

Sloping land can present advantages as well as difficulties. A flat lot is easiest to build upon, but most difficult to landscape interestingly. A lot with some slope aids in drainage of the yard and could make the basement partially or fully exposed even to the point of having a walk-out basement. Steeply sloped lots frequently make retaining walls and expensive fills necessary. Low-lying lots usually should be avoided, because dampness problems frequently are encountered.

An examination of the portion of this chapter illustrating the basic house types shows why some types of homes are better adapted to lots with a particular slope. It would be a mistake to plan and draw a long, low, ranch-style house and then purchase a lot with a steep slope. This would not be in keeping with the character of the house.

The condition of the soil should be examined. A layer of topsoil at least a foot thick is excellent. Sand or clay below the topsoil is fine. Watch for indications of rock close beneath the surface. The occurrence of rock may eliminate the possibility of having a basement, and it may make footings expensive and septic tanks almost impossible. Recently filled lots should be avoided, because they will not support a house, and the walls will crack due to settling.

Examine the deed of the lot for building restrictions and possible easements. An **easement** is the right granted by the owner of land to utility companies to cross portions of this land with gas and water mains and electric lines, and it also grants them the right to come on the land to service or repair them. If there are easements, see if any will cause inconvenience, expense, or difficulty after the house is built.

Upon being satisfied that the lot is the one you want, make a minimum down payment and sign a sales agreement. Do not pay for the lot in full until you have a lawyer search the title and have the property surveyed. When you are certain the title is clear, with no prior liens, it is safe to pay the balance.

Accept only a warranty deed. A good safety device is to get a title guarantee from a guarantee insurance firm. Such a firm will guarantee the title to be clear and will pay any liens against it, should they occur.

Topography

Site Plans

Before buying or building on a piece of ground, the owner should have it surveyed. For a fee, a

Fig. 4-20. Surveyor's site plan.

This drawing was accompanied by signed statements that "the undersigned, a Registered Civil Engineer under the laws of the State of Michigan, certifies that he/she has made a survey of Lot 63, Plat of 'Broan Addition', Section 6, T.2 S., R. 10 W., Township of Comstock, County of Kalamazoo, State of Michigan, as recorded in the Office of the Register of Deeds for Kalamazoo County. Measurements were made and corners perpetuated in accordance with the true and established lines of the property as described, and the dimensions and lines of the property are indicated on the following Plat. The above Survey, Plat and Certificate are hereby certified correct as described. Surveyed [Date]; Survey No. 2929-97."

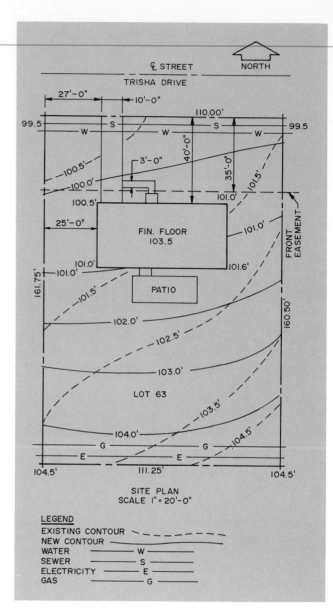

surveyor will come to the property and carefully ascertain the information desired. From this information, a site plan is drawn. A **site plan** is a top view of the lot, which usually shows the lot and block number, if the lot is in a development; the dimensions of the lot in decimal feet; the north point; the location of accessory buildings; the location of walks, driveways and approaches, steps, terraces, porches, fences and retaining walls; and the location and size of easements and required setbacks from the street, or lot boundries. A typical plan is shown in Fig. 4-20.

The site plan also should indicate the elevation above the local datum level of the first floor of the house, the garage or carport floor and other buildings, the elevation of the finished curb and crown of the street, and the finish grade level at each principal corner of the house. The **datum level** is an assumed basic level used as a reference for reckoning heights. In various areas, especially cities, this datum level is established as a part of the building code, and it controls all building standards for that area or city.

If special grading, drainage provisions, or foundations are necessary, their elevations must be specified on the site plan. Each corner of the lot should have the finish grade after the special grading is done, and the grade at each corner of the house should be indicated. The finish grade on both sides of retaining walls should be given.

If a private well is on the property or a private sewage disposal system is to be used, this should be indicated on the site plan. Often, the approximate locations of utilities are also indicated.

More complete information about the existing and finished slope of a lot can be obtained by having a surveyor establish contour lines. On a plan of a piece of land, **contour lines** connect points having the same elevation above or below an established level, such as the datum level or sea level. Refer again to Fig. 4-20.

Contour lines are very useful in orienting a house on a lot, for they indicate amount of slope. For example, widely spaced contour lines indicate a gradual slope while closely spaced lines define a steep slope.

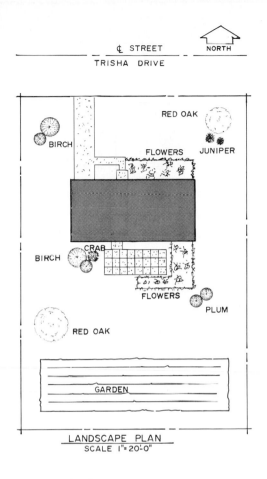

Fig. 4-21. A landscape plan shows the locations of buildings, plants, and other features on the lot.

Landscape Plans

A **landscape plan** indicates the location of the house, garage, and drives and walks. It also shows the location and, frequently, the species of the plantings that are to be made on the lot.

Such a plan is made by a landscape architect. These services are most often used by owners of expensive houses, because a fee is charged. Many nurseries offer free landscaping services and advice to their customers. This enables most persons to have some assistance in planning the plantings that will surround their house.

Since the plan as drawn by the landscape architect indicates the species of the plantings, it can be used to obtain competitive bids on the plantings and their installation. Figure 4-21 illustrates a simple landscape plan.

Prevailing Winds

A little time spent in a study to determine the direction from which the prevailing winds blow will pay big dividends.

In the summer, the air should blow through the living area and sleeping area. If something blocks these breezes, one must resort to fans or air conditioning. Examine your site to see if a hill or growth of trees will block the summer breeze. A house in a hollow may miss all breeze and be stifling.

Undesirable winter winds present an entirely different problem. Every attempt should be made to block them. See Fig. 4-22. Place the garage and utility room so they will receive the brunt of winter

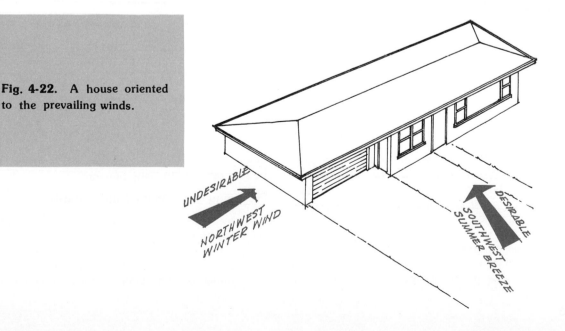

Fig. 4-22. A house oriented to the prevailing winds.

Climatic Atlas of the United States, Visher, Harvard University Press, Cambridge, Massachusetts

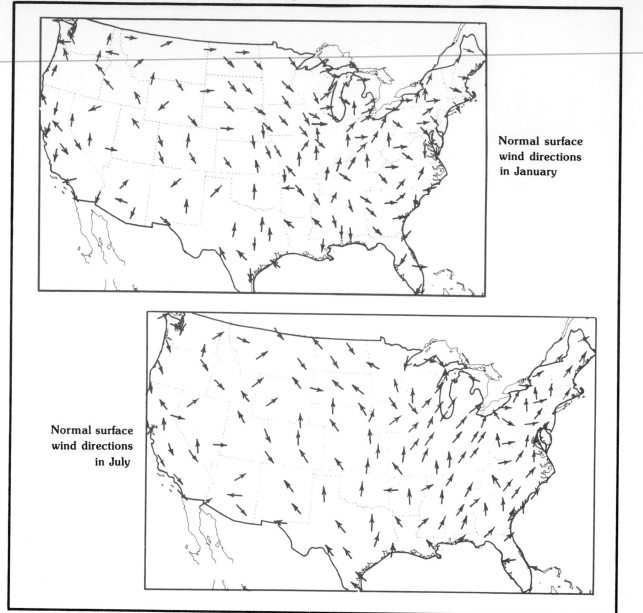

Normal surface
wind directions
in January

Normal surface
wind directions
in July

Fig. 4-23. Normal surface winds.

storms. This greatly increases the comfort and ease of heating a home. A planting of trees or tall shrubs helps break the force of the wind and settle the snow. A northwest wind driving against a garage door will pile it high with snow and increase the difficulty of moving the automobile.

Winds can bring smoke and odors from nearby factories and make living objectionable in what would otherwise be an acceptable area. In areas

where high winds are common, construction must be strengthened to resist damage.

Prevailing Winds in the United States

Figure 4-23 indicates the major normal surface wind directions in January and July. It presents an interesting study of the influence of mountains or bodies of water on the direction of prevailing winds. As a house is being planned for a particular loca-

106

tion, check with the local weather bureau for specific information concerning the directions and peculiarities of the local winds.

The View and Privacy

If a lot presents a pleasant view, the living area should be placed to take advantage of it. The master bedroom can be located to share the view or look upon another that is pleasing. Some families consider the view from the dining area of great importance. Occasionally, the best view will conflict with the orientation of the house in relation to the sun or the prevailing winds. A compromise must then be reached.

Many contemporary homes have few or no windows on the street side. A row of houses across the street usually is not the best view. A large glass area toward a street only exposes the activities in the house to the public. It is an intrusion upon the privacy of the occupants.

For the sake of privacy, fences can be built around glass window walls, or clusters of shrubbery can be planted.

Privacy in a bedroom can be attained by using long, narrow windows, placed approximately 5'-0" (1524 mm) above the floor. This allows some light and ventilation, but guarantees privacy. However, other plans include full-length glass walls for the bedrooms. Such a plan requires special consideration in lot selection and landscaping to insure privacy.

Orientation

The orientation of a house on a lot is of great importance. This is something that must be considered as the house is being planned. Orientation influences the location of the rooms on the plan as well as the positioning of the house on the lot.

Orientation involves a consideration not only of the prevailing winds, views, and privacy as already discussed, but also of the sun. Each of these affects the livability of the house. Generally, it is not possible to incorporate all the best factors involved, but each should be considered and some provision made to help ease a bad situation. For example, if the hot western summer sun will bear on the bedroom wing in the evening, and if it is not possible to reorient these rooms to face east, then some provision must be made for sheltering this area. A large overhang, trees, a fence, or some such device should be planned to break the sun's rays. Such

planning will help overcome disagreeable factors caused by unfortunate orientation.

The Sun

The comfort of a home depends a great deal upon a careful orientation of the house to the sun. In the summer, the sun can make a poorly oriented house unbearably hot. In the winter, a poorly oriented house could lose the benefit of free solar heat.

The **south wall** of a house receives a sustained and intense exposure to the summer sun. This means that this side of the house will be bright and well illuminated all day, receiving considerable heat from the direct rays of the sun.

Proper orientation requires that the living area face south. It is desirable for the master bedroom to face south also. It is necessary, however, to provide a deep overhang or some other means of breaking the sun's rays during the heat of the day.

This southern wall could use large areas of glass to advantage. In the winter, it will benefit from the low rays of the sun by using this light to make the living area cheery and by using the heat to warm the room. Such an orientation can reduce winter fuel bills.

The **north wall** of a house receives only a small amount of direct sunlight. It needs little overhang or shading devices. The northern light is uniform and free from the glare. For this reason, artists value northern light. Generally, the garage, kitchen, or utility room is placed to the north. Such rooms do not require a great deal of solar light or are affected considerably by solar heat. If a view is to the north, however, large areas of glass can be used to advantage to secure the even light, but there are few other advantages to this exposure.

The **east wall** of the house receives the early morning sunshine, which is usually moderate and pleasant. Bedrooms, kitchens, and dining areas can be placed to advantage using this exposure. A cheery morning sun starts the day right and brightens the breakfast table. It is wise to keep these windows high, so only the early morning rays will enter.

The **west wall** of the house receives intensive periods of heat from rays striking at low angles in the afternoon. As the sun sets, the angles get lower, causing the rays to penetrate any windows in this wall; therefore, this wall should have a minimum of windows, and they should be high, so only the lowest rays will enter. Any rooms facing west will be hot every afternoon all summer. Devices to control the rays of the sun are needed. A large porch is fre-

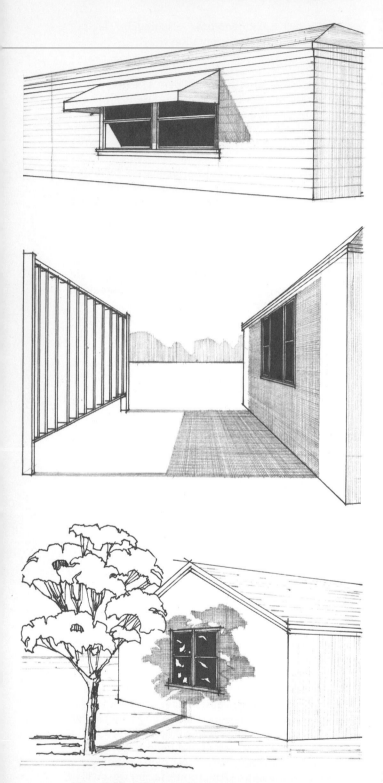

quently used as a protective device, as are trees, fences, awnings, and trellises. See Fig. 4-24. Deciduous trees are excellent, because they shade the house in the summer and shed their leaves in the winter, allowing the desirable winter sunlight to enter.

Figure 4-25 shows why the south wall receives sunlight all day, the east wall in the morning, and the west in the afternoon. It also shows why the north wall receives little or no sunlight at any time.

In the deep South and hot Southwest, the problem of orienting the house in relation to the sun is somewhat reversed. Usually, the living area faces north, and a wall with few or no windows faces south. A large roof overhang and trees are much desired on the hot, southern side.

There are two factors that directly influence the relationship between the sun and a house. One of these is the angle of the sun, called the **altitude,** and the other is the direction from which the sun rises and sets during the various seasons. This is called the **azimuth.**

The Altitude

The altitude is the angle formed by the rays of the sun and the earth's surface. Figure 4-26 illustrates the maximum and minimum angles of the sun for the 40° parallel of latitude (sighting on Milwaukee).

Fig. 4-24. Sun protective devices. A large porch also protects rooms from the sun's rays.

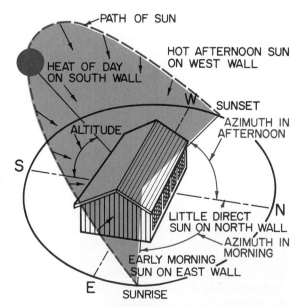

Fig. 4-25. Sun strikes the east side of the house in the morning, the south side through the middle of the day, and the west side in the late afternoon.

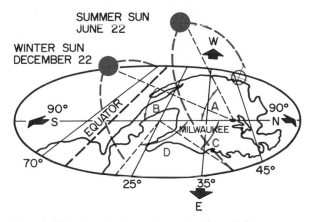

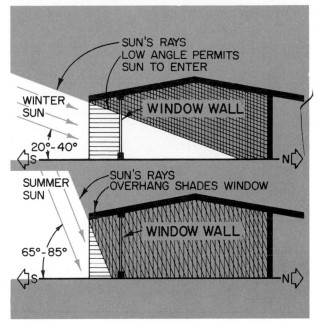

Fig. 4-26. Angle **A** is the altitude of the sun at Milwaukee at noon on June 22.

Angle **B** is the latitude on December 22.

Angle **C** is the azimuth at sunrise in Milwaukee on June 22.

Angle **D** is the azimuth on December 22.

Angle **A** is 73°-30′. This occurs at noon June 22. From this date, the angle gets lower each day. It reaches its minimum on December 22 when it is in position **B.** This angle would be 26°-30′ for the 40° parallel of latitude.

Maximum and minimum altitudes for selected parallels of latitude can be found in architectural standards books.

It can be seen that the altitude of the sun is considerably lower in the winter than in the summer. This fact enables a designer to utilize the sun to warm the house in winter, yet to use such devices as a large roof overhang to shield the interior of the house from the hot, summer sun. See Fig. 4-27.

Since the altitude varies with the parallel of latitude, the amount of overhang needed also varies. Figure 4-28 indicates the size overhang necessary to protect windows from the summer sun, but still allow the winter sun to enter. It can be seen that as the house is turned southeast or southwest, the amount of overhang must be increased to afford protection from the low-angle, early morning, east rays and the low-angle, late afternoon rays of the sun.

The Azimuth

The sun rises from an easterly direction and sets in a westerly direction. It does not rise due east and set due west every day. As the earth turns on its axis and revolves about the sun, the angle of inclination of the axis to the sun varies a little each

Fig. 4-27. Roof overhang shields house from hot summer sun, yet allows desirable winter sun rays to enter.

Fig. 4-28.
Overhang Width for Sun Protection

Latitude	South Windows	Windows 30° E or 30° W of South
45°	3′-6″	6′-0″
40°	3′-0″	5′-4″
35°	2′-5″	4′-8″
30°	1′-10″	4′-1″
25°	1′-3″	3′-8″

day. This causes the direction from which the sun rises and sets to vary a few degrees each day.

This angle is measured from the north-south axis and is called the azimuth. In the morning, the azimuth is measured from the north-south axis in an easterly direction, and in the afternoon, it is measured in a westerly direction from the same axis. In Fig. 4-29, selected azimuth readings are recorded for the 40° parallel of latitude at sunrise, 11:00 a.m., 1:00 p.m., and sunset on June 22.

A complete report of altitude and azimuth angles can be found in architectural standards books and the **American Nautical Almanac.**

Positioning the House

In the northern states, the long axis of a house should be along a northeast to southwest angle.

Usually, this angle is from 30° to 60°. This will cause the sunlight to strike the northern side of the house and help melt snow and ice. See Fig. 4-30.

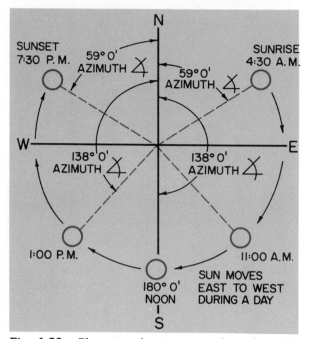

Fig. 4-29. Plan view showing azimuth readings for the 40° parallel of latitude on June 22.

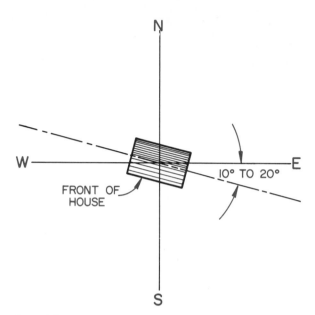

Fig. 4-31. Alternate position useful when orienting a house in northern climates.

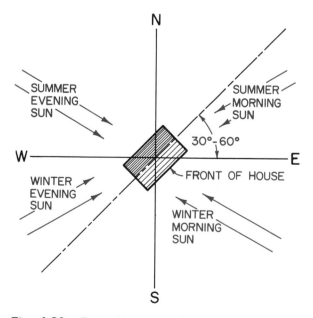

Fig. 4-30. Desirable orientation in relation to the sun in northern climates.

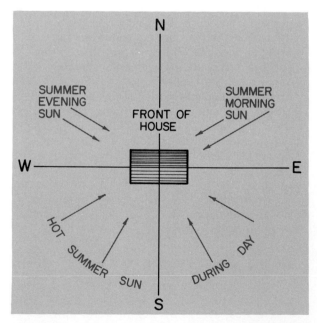

Fig. 4-32. Proper orientation of house in extreme South.

Figure 4-31 illustrates an alternate direction to orient a house in a northern climate. Such a location permits use of the southern exposure and takes advantage of some of the western heat in the winter. Any angle over 20° northwest to southeast would allow too much hot summer sun to hit the front of the house in the afternoon.

In the extreme South, the house can approach a true east-west direction with its major axis. This keeps the hot afternoon summer sun away from the living area. See Fig. 4-32.

How to Find the Shadow Cast by Overhang Using the Altitude and Azimuth

It is possible to locate the shadow cast by the overhang on a house graphically by using the altitude and azimuth of the house location.

For example, assume a cottage was built along the 45° parallel of latitude. To find where the

shadow will fall on a particular day, refer to a table of altitudes and azimuth angles for that day. Let us assume the day to be June 22, the longest day of the year, and the time 10 a.m. The tables indicate that the altitude is 57°-30', and the azimuth is 121°-30'. Now, follow these steps, shown in Fig. 4-33.

1. Lay out the plan of the building.
2. Locate the north-south axis.
3. Draw a line 121°-30' east of the north-south axis. It is east because it is in the morning. It would be west in the afternoon.
4. Draw altitude angle of 57°-30'.
5. On the plan, draw a ray line parallel to the 121° 30' azimuth line. This passes through the extreme corner of the roof overhang on the plan (labeled A).
6. Draw ray line parallel to altitude line, through same corner of the roof, in side view.

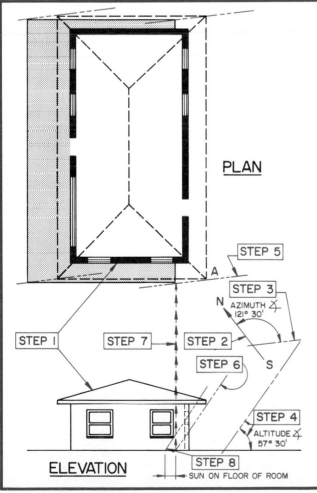

Fig. 4-33.   How to find the shadow cast by a house.

1. Draw plan view and side view.
2. Locate north-south axis.
3. Draw azimuth (in this problem 121° 30' E of N).
4. Draw altitude angle (in this problem 57° 30').
5. Draw ray line parallel to azimuth through the corner of roof in plan view.
6. Draw ray line parallel to altitude through the same corner of roof in side view.
7. Project the point of intersection between the altitude ray and the ground line to plan view intersecting the azimuth ray.

Note: The shadow cast by any point may be found similarly. All horizontal planes (or lines) cast a shadow parallel to themselves and all vertical lines or planes cast a shadow parallel to the azimuth; therefore, the shadow is constructed by the intersection of azimuth rays and parallel lines.

8. To locate where the sun will fall inside a room, draw only a simple wall section instead of a side view.

7. Project the altitude line from the point where it crosses the azimuth line until it cuts the azimuth line on the plan.

8. This intersection between the azimuth line and the altitude line on the plan locates the shadow line. The shadow runs through this point parallel to the roof edges.

If shadows for all sides of a building are desired, continue with the following steps:

9. Draw the front elevation. Project the shadow line into the front elevation to the ground line.

10. Connect the shadow line with the edge of the overhang. This gives the shadow angle.

11. Transfer this shadow angle to any other part of this elevation where finding another shadow location is desired.

12. Project this new shadow line from where it hits the ground line to the plan view. Where it cuts the azimuth line is another point of the shadow. This gives the location of the shadow on the plan.

Such a procedure even permits the location of the sun's rays inside a house. The length of the sun's rays inside the house is terminated by the head of the window casting a shadow on the floor.

If shadows are to be found for latitudes not given in the tables, refer to more elaborate tables, or use the **process of interpolation.** Following is an example:

Assume shadows were to be found for 43°-0′ latitude at 10 a.m., June 22. The tables indicate that for 45° latitude (the same time and date), the altitude is 57°-30′, and the azimuth is 121°-30′; for 40° latitude, the altitude is 60°-0′, and the azimuth is 114°-0′.

To find the altitude by interpolation, set up the problem as follows:

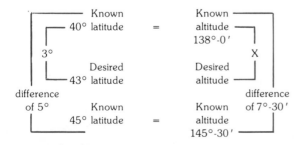

The altitude for 43° of latitude is desired. The 43° of latitude is 3/5 of the difference between 40° and 45° latitude. Therefore, the altitude angle is

3/5 of the difference between their known altitudes, 138°-0′ and 145°-30′. Let the unknown altitude be represented by the letter **X.**

Since the difference in latitudes is in the same proportion as the difference in altitudes, it can be said that 3 is to 5 as **X** is to 7°-30′. The solution follows:

$$\frac{3}{5} = \frac{X}{7°\text{-}30′}$$

$$5X = 22°\text{-}30′ \quad (60′ = 1°)$$

$$X = \frac{22°\text{-}30′}{5}$$

$$X = 4°\text{-}30′$$

Since 40° of latitude has an altitude of 138°-0′, 43° latitude would have an altitude of 138°-0′ plus the amount **X,** 4°-30′, or be equal to 142°-30′.

The same process can be used to find azimuth angles not given in tables.

Multifamily Housing

With the ever-increasing cost of land and housing, more families are living in multifamily housing units. Units that are sold are called **condominiums.** Units that are rented are called **apartments.** The principles of planning the individual living space are much the same as those used for planning individual homes. Often space is at a premium. Therefore, in planning, extra care is taken to see that every square foot is utilized.

A multifamily housing unit is shown in Fig. 4-34. This is a sixteen-unit building. The entrance to the public area, shown in Fig. 4-35, provides access to the main halls on the first and second floors. This unit was designed with a colonial-style exterior. Changes in exterior design contribute to the living environment and do much to reduce the monotony of long rows of identical exteriors. See Fig. 4-36.

This particular building has eight one-bedroom and eight two-bedroom living units. The two-bedroom units are available in interior and end-of-building designs; the one-bedroom units are all interior types. The overall plan for the building is found in Fig. 4-37. Notice that in addition to the main entrance there are stairs on each end of the hall. The laundry and storage room is centrally located. Details of each living unit are shown in

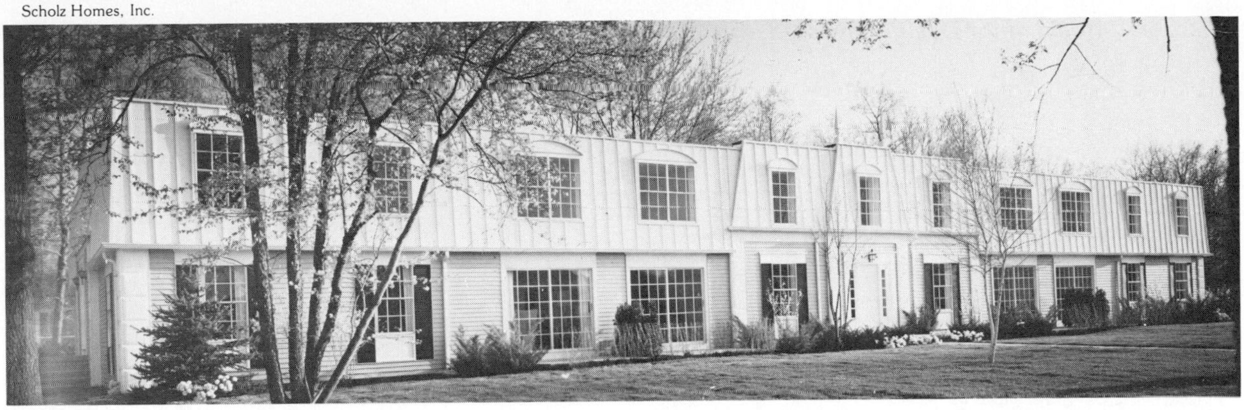

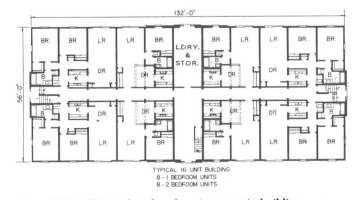

Fig. 4-34. This is a sixteen-unit multifamily housing unit. It contains eight one-bedroom and eight two-bedroom apartments.

Fig. 4-35. Multifamily housing unit public entrance and lobby.

Fig. 4-36. A colonial style exterior provides a pleasing change for the community.

TYPICAL 16 UNIT BUILDING
8 - 1 BEDROOM UNITS
8 - 2 BEDROOM UNITS

Fig. 4-37. Floor plan for the sixteen-unit building shown in Fig. 4-34.

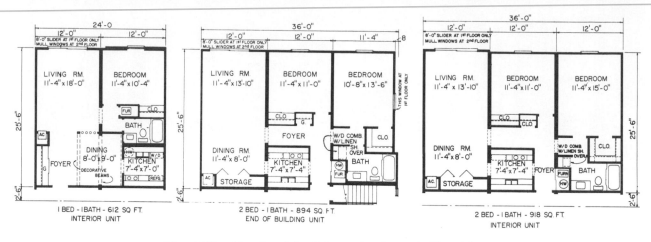

Fig. 4-38. Details of the living units shown in Fig. 4-37.

Fig. 4-38. The first-floor units have sliding windows to a patio. This feature enhances the quality of living in these units. See Fig. 4-39.

Detailed standards for designing these units can be found in various architectural standards books. A major source is the **Minimum Property Standards for Multifamily Housing**, published by the U.S. Department of Housing and Urban Development, Washington, D.C. (available from the Superintendent of Documents, U.S. Government Printing Office, Washington, D.C. 20402). These standards do not replace local building codes, however. Designers of multifamily units must observe local codes as well as the national standards. The following information includes only a few of the many design requirements in the standards.

Building Configuration

Many factors influence the shape and size of multifamily housing. Among these are the potential market, the site size, financing available, local building codes, and the projected cost of the project. It may be a simple one- or two-story building that houses ten to twenty families. It may have many buildings and cover many acres. Or it may be a high rise with eight or more stories of units.

There are many ways to lay out such buildings. Figure 4-40 shows only three basic approaches to relating the building to the site. There are other possibilities. The basic shape and plan can likewise vary. A few ideas are shown in Fig. 4-41. The ultimate shape and size, therefore, depend upon these and other local factors. The designer must study the entire situation carefully before deciding on the approach to use.

Fig. 4-39. The living room and dining room of a first-floor living unit. Notice the patio area made accessible by sliding glass doors.

Site Design

1. The site must be free from any hazards that could affect the health and safety of the occupants.

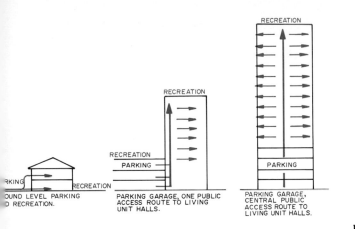

Fig. 4-40. Three basic approaches for relating a multifamily living unit to the site.

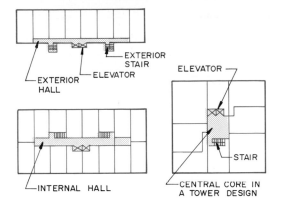

Fig. 4-41. A few examples of building layout and a public access system.

2. The site design for outdoor areas and facilities shall respond to the social and physical needs of the residents.
3. The design shall be arranged to utilize and preserve the favorable features of the site.
4. All elements of the site plan shall be designed to fit the natural contour of the land as closely as practicable.
5. Buildings should be located in areas having the least potential ground water hazard.
6. The design shall harmonize with and complement the appearance of surroundings.
7. Provision shall be made to store snow removed from public areas.
8. The buildings should be located to minimize sources of adverse noise.
9. Yard depth and width shall assure adequate distance between buildings. These distances are often specified by codes. See Fig. 4-42.
10. Access to property shall be designed to minimize all traffic hazards.
11. Adequate parking space shall be provided for residents, guests, and service vehicles. Five percent of this space should be arranged for wheelchair users.
12. Maximum walking distance from parking to building should be as follows: a. non-elderly, 250 feet (76.2 m) b. elderly, 150 feet (45.72 m) c. guests, 300 feet (91.44 m)
13. Walks shall be provided for safe, convenient access to all dwellings and for circulation throughout the development.

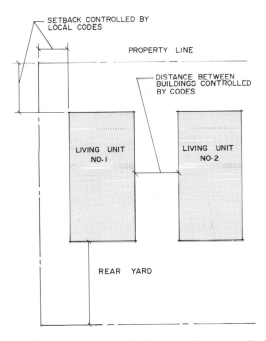

Fig. 4-42. Setbacks and distances between buildings are regulated by codes.

14. Exterior steps should be designed as follows: maximum riser 6 inches (152 mm), minimum riser 3 inches (76 mm), minimum tread 12 inches (305 mm), or stepped ramp with two easy paces between single risers.
15. Direct access shall be provided for all deliveries and services (ambulance, fire trucks, mail delivery, furniture delivery).

Building Design

Many projects are planned to accommodate all ages, from senior citizens to families with children. In these, it is desirable that housing units for senior citizens be located in separate floors or wings of the building, or in separate structures. Any indoor or outdoor community facilities shall also be separate.

1. Management and maintenance space should be adequate for the size of the development.
2. Community social rooms shall be designed to meet the needs of the residents. They shall have separate rest-room facilities for men and women.
3. Common laundry facilities shall be located near the elevators or another pedestrian traffic center.
4. Provide for the temporary sanitary storage of trash and for its disposal.
5. Special facilities for the elderly may include occupational or physical therapy rooms, an office for the dietitian, central dining, first aid and nursing facilities, and a medical office.

Living Area

Adequate space shall be provided in the living area to provide comfortable use and circulation (movement) while accommodating the following furniture:

1 couch, 3'-0" × 6'-0" (914 mm × 1828 mm)
Easy chairs — 2 for a two- or three-bedroom unit, 1 for an efficiency, and 3 for a four-bedroom unit
1 desk, 1'-8" × 3'-6" (508 mm × 1067 mm)
1 desk chair
1 television set
1 table, 1'-6" × 2'-6" (457 mm × 762 mm)

Dining Area

1. Each living unit shall contain space for dining. This area may be combined with the living room or kitchen, or it may be a separate dining room.
2. Space shall be provided to allow for proper circulation while accommodating the following sizes of dining tables, plus chairs:
   Efficiency or 1 bedroom, 2 persons: 2'-6" × 2'-6" (762 mm × 762 mm)
   2 bedrooms, 4 persons: 2'-6" × 3'-2" (762 mm × 965 mm)
   3 bedrooms, 6 persons: 3'-4" × 4'-0" (1016 mm × 1219 mm) or 4'-0" (1219 mm) round

4 bedrooms, 8 persons: 3'-4" × 6'-0" (1016 mm × 1829 mm) or 4'-0" × 4'-0" (1219 mm × 1219 mm)

Bedrooms

1. Each dwelling unit shall have space allocated to sleeping, dressing, and personal care.
2. Each bedroom shall allow comfortable use and circulation space while accommodating at least the following furniture:

Primary bedroom
   2 twin beds, 3'-3" × 6'-10" (991 mm × 2083 mm)
   1 dresser 1'-6" × 4'-4" (457 mm × 1321 mm)
   1 chair 1'-6" × 1'-6" (457 mm × 457 mm)
   1 crib 2'-6" × 4'-6" (762 mm × 1372 mm)

Secondary bedrooms
a. Double-occupancy bedroom
   1 double bed 4'-6" × 6'-10" (1372 mm × 2083 mm)
   1 dresser 1'-6" × 3'-6" (457 mm × 1067 mm)
   1 chair 1'-6" × 1'-6" (457 mm × 457 mm)
b. Single-occupancy bedroom
   1 single bed 3'-3" × 6'-10" (991 mm × 2083 mm)
   1 dresser 1'-6" × 3'-6" (457 mm × 1067 mm)
   1 chair 1'-6" × 1'-6" (457 mm × 457 mm)

Combined Spaces

1. Where required habitable rooms are combined into multi-use spaces, the furniture requirements and circulation space shall be applied to the multi-use space.
2. For efficiency apartments, the combined living-dining-sleeping space shall accommodate the living, dining, and sleeping requirements specified in those specific sections.

Minimum Room Sizes

Instead of designing according to use and furniture requirements, minimum room sizes can be used. When these sizes are followed, they must be used for all rooms of the project.

Minimum Room Sizes

A. Minimum Room Sizes for Separate Rooms

NAME OF SPACE[1]	MINIMUM AREA								LEAST DIMENSION	
	LU with 1-BR		LU with 2-BR		LU with 3-BR		LU with 4-BR			
	Ft.²	m²	Ft.²	m²	Ft.²	m²	Ft.²	m²		
LR	160	14.865	160	14.865	170	15.795	180	16.725	11'-0"	3353 mm
DR	100	9.290	100	9.290	110	10.220	120	11.150	8'-4"	2540 mm
BR (primary)[2]	120	11.150	120	11.150	120	11.150	120	11.150	9'-4"	2845 mm
BR (secondary)	NA	NA	80	7.435	80	7.435	80	7.435	8'-0"	2439 mm
Total area, BR's	120	11.150	200	18.580	280	26.015	380	35.305	—	—

B. Minimum Room Sizes for Combined Spaces

COMBINED SPACE[1]	MINIMUM AREA									
	LU with O-BR		LU with 1-BR		LU with 2-BR		LU with 3-BR		LU with 4-BR	
	Ft.²	m²	Ft.²	m²	Ft.²	m²	Ft.²	m²	Ft.²	m²
LR-DA	NA	NA	210	19.510	210	19.510	230	21.370	250	23.225
LR-DA-SL	250	23.225	NA	NA	NA	NA	NA	NA	NA	NA
LR-DA-K	NA	NA	270	25.085	270	25.085	300	27.870	330	30.660
LR-SL	210	19.510	NA	NA	NA	NA	NA	NA	NA	NA
K-DA	100	9.290	120	11.150	120	11.150	140	13.010	160	14.865

[1]Abbreviations:

LU = Living Unit	DR = Dining Room	O-BR = LU with No Sep-	K = Kitchen	BR = Bedroom
LR = Living Room	DA = Dining Area	arate Bedroom	NA = Not Applicable	SL = Sleeping Area

[2]Primary bedrooms shall have at least one wall space of at least 10 feet (3048 mm) uninterrupted by openings less than 44 inches (1118 mm) above the floor.

Kitchens

1. Each living unit shall include adequate space to provide for efficient food preparation, serving, storage, utensil storage, and cleanup activities.
2. Kitchen fixtures and countertops shall be provided as follows:

3. Required countertops may be combined when they are located between two fixtures, such as the stove, refrigerator, and/or sink.
4. One-third of the kitchen storage area shall be located in base or wall cabinets. At least 60 percent of the required area shall be enclosed by cabinet doors. Storage area requirements follow:

Countertops and Fixtures

WORK CENTER	MINIMUM FRONTAGES									
	0 Bedrooms		1 Bedroom		2 Bedrooms		3 Bedrooms		4 Bedrooms	
	Lineal Inches	mm	Lineal Inches	mm	Lineal Inches	mm	Lineal Inches	mm	Lineal Inches	mm
Sink	18	457	24	610	24	610	32[1]	813	32[1]	813
Countertop, each side	15	381	18	457	21	533	24	610	30	762
Range or Cooktop Space[2, 3, 6]	21	533	21	533	24	610	30	762	30	762
Countertop, one side[4]	15	381	18	457	21	533	24	610	30	762
Refrigerator Space[5]	30	762	30	762	36	914	36	914	36	914
Countertop, one side[4]	15	381	15	381	15	381	15	381	18	457
Mixing Countertop	21	533	30	762	36	914	36	914	42	1067

[1]When a dishwasher is provided, a 24-inch (610 mm) sink is acceptable.
[2]Where a built-in wall oven is installed, provide a counter 18 inches (457 mm) wide adjacent to it.
[3]A range burner shall not be located under a window nor within 12 inches (305 mm) of a window. Where a cabinet is provided above a range, 30 inches (762 mm) clearance shall be provided to the bottom of an unprotected cabinet, or 24 inches (610 mm) to the bottom of a protected cabinet.
[4]Provide at least 9 inches (229 mm) from the edge of a range to an adjacent corner cabinet and 15 inches (381 mm) from the side of a refrigerator to an adjacent corner cabinet.
[5]Refrigerator space may be 33 inches (838 mm) when the refrigerator door opens within its own width.
[6]When a range is not provided, a space 30 inches (762 mm) wide shall be provided.

117

Minimum Storage Area*

No. BR's	TYPE OF STORAGE			
	Shelf Area		Drawer Area	
	Ft.²	m²	Ft.²	m²
0-BR	24	2.230	4	0.375
1-BR	30	2.790	6	0.560
2-BR	38	3.530	8	0.745
3-BR	44	4.090	10	0.930
4-BR +	50	4.645	12	1.115

***Planning Considerations:** 1. A dishwasher may be counted as 4 sq. feet (0.372 m²) of base cabinet storage. 2. Wall cabinets over refrigerators shall not be counted as required shelf area. 3. Shelf area above 74 inches (1880 mm) shall not be counted as required area. 4. Inside corner cabinets shall be counted as 50 percent of the shelf area; where revolving shelves are used, the actual shelf area may be counted. 5. Drawer area in excess of the required area may be counted as shelf area if drawers are at least 6 inches (152 mm) in depth.

Baths

1. Each dwelling unit shall have one bathroom containing a bathtub with a minimum outside width of 30 inches (762 mm), a lavatory, and a water closet. In other bathrooms, a shower may be substituted for a tub. Bathrooms shall be convenient to bedrooms.
2. Bathrooms shall have the following: grab bar and soap dish at tub or shower, enclosure at shower, soap dish at lavatory, toilet paper holder, mirror and medicine cabinet or equivalent storage, one towel bar.
3. Shower stalls shall have a minimum area of 1024 square inches (0.661 m²) with the least dimension of 30 inches (762 mm).
4. Water impervious wainscot shall be provided at walls around showers or tub-showers to a height of 6'-0" (1829 mm).
5. Housing for the elderly requires grab bars along a bathtub wall and a seat and grab bars in a shower.

Laundry

If a common laundry is not furnished, space in each living unit for a clothes washer and dryer is required.

Closets and General Storage

1. Closets and storage space shall be provided for living and housekeeping items within each living unit.
2. Each bedroom should have the following minimum closets:

Primary and double occupancy:
   2'-0" × 5'-0" (610 mm × 1524 mm)
Single occupancy: 2'-0" × 3'-0"
   (610 mm × 914 mm)
3. Provide a 2'-0" × 2'-0" (610 mm × 610 mm) coat closet near the entrance.
4. Provide the following linen storage:
   10 square feet (0.930 m²) shelf area for 2 bedrooms or less
   15 square feet (1.395 m²) shelf area for 3 or more bedrooms
   Space shelving not less than 12 inches (305 mm) O.C.
   Shelving over 74 inches (1880 mm) above floor is not counted.
5. Provide general storage for each living unit as follows:

General Storage Requirements

No. BR's	VOLUME OF STORAGE			
	Entirely within Living Unit		At Least Half within Living Unit	
	Ft.³	m³	Ft.³	m³
0-BR	100	2.830	140	3.965
1-BR	150	4.245	200	5.660
2-BR	200	5.660	275	7.785
3-BR	275	7.785	350	9.905
4-BR +	350	9.905	425	12.030

6. Each living unit having one or more bedrooms shall have at least one closet for general storage within the unit. It must have at least 6 square feet (0.560 m²) of floor area.

Ceiling Heights

Clear ceiling heights under beams or other obstructions are as follows:

Minimum Clear Ceiling Heights

Habitable Rooms	7'-6" (2286 mm)
Halls within Living Unit, Baths	7'-0" (2134 mm)
Luminous Ceilings Within Living Unit Public Corridor	7'-0" (2134 mm) 7'-4" (2235 mm)
Sloping Ceilings	at least 7'-6" (2286 mm) for 1/2 the room with no portion less than 5'-0" (1524 mm)
Public Corridors	7'-8" (2337 mm)
Public Rooms	8'-0" (2438 mm)
Basements without Habitable Rooms	6'-8" (2032 mm)

Access and Circulation

1. Minimum doorway widths are as follows:

Minimum Doorway Widths

Public Doors	
Main Entrance to Bldg.	3'-0" (914 mm), 6'-0" (1829 mm) for double doors
Secondary Public Entrance to Bldg.	3'-0" (914 mm)
Service Entrance to Bldg.	2'-8" (813 mm)
Public Stairway	3'-0" (914 mm)
Private Doors	
Main Entrance to Living Unit	3'-0" (914 mm)
Secondary Entrance to Living Unit	2'-8" (813 mm) (5'-0" [1524 mm] sliding glass doors may be used)
Bathrooms, Toilets in Living Unit	2'-0" (610 mm), 2'-8" (813 mm) for elderly or wheelchair access
Habitable Rooms	2'-6" (762 mm), 2'-8" (813 mm) for elderly or wheelchair access

2. All exterior doors shall be equipped with a deadlock.
3. Exit doors, other than those from individual living units, shall swing outward.
4. Minimum clear widths of halls are as follows:

Minimum Clear Hall Widths

a. Public Halls:

Length	Width
Less than 10' (3048 mm)	3'-6" (1067 mm)
10' to 30' (3048 mm to 9144 mm)	4'-0" (1219 mm)
30' to 100' (9144 mm to 30 480 mm)	4'-6" (1372 mm)
More than 100' (more than 30 480 mm)	5'-0" (1524 mm)
Housing for elderly	5'-0" (1524 mm)

b. Exterior Access Corridors: 5'-0" (1524 mm)

c. Halls within Living Units: 3'-0" (914 mm)

d. Halls within Living Units for Wheelchair Access: 3'-4" (1016 mm)

5. Maximum hall lengths are as follows:
   a. Those providing access to stair or horizontal exit (on same level) in two directions, 100 feet from living unit to stair or exit. May be 150 feet if sprinklers are installed.
   b. Dead-end corridors 35 feet between door of living unit and exit.
   c. Within the living unit, 50 feet between exit and most remote room.
   d. From boiler room or other area of fire hazard, 50 feet to exit.
6. Designs for public stairways in the building are as follows:

Public Stairway Design
(Minimum Dimensions)

	Interior		Exterior	
Clear Headroom	6'-8"	2032 mm	6'-8"	2032 mm
Tread[1]	9"[2]	229 mm	11"	279 mm
Riser [1],[3]	7 3/4"	197 mm	7 1/2"	190.5 mm

[1]All treads shall be the same width and all risers the same height in a flight of stairs.
[2]Plus 1 1/8" (28.5 mm) nosing minimum on closed riser, plus 1/2" (13 mm) nosing minimum on open riser
[3]In buildings required to be accessible to the physically handicapped, the maximum riser is 7 1/2" (190.5 mm). Open risers shall not be used.

7. Minimum stairway widths are as follows:
   a. Building serving fifty or fewer people, 3'-0" (914.5 mm). Fifty or more people, 3'-8" (1118 mm). Handrails project not more than 3½ inches (89 mm) on each side.
   b. Landing size in direction of travel shall not be less than width of stairway.
8. Stairs must have a handrail on one side. Stairs 3'-8" (1118 mm) and wider need a handrail on both sides.
9. Install at least one elevator in each smoke compartment of the building.
10. Elevators are required as follows:
    a. In buildings five or more stories high
    b. In buildings three or more stories high if housing for elderly
    c. In buildings two or more stories high if occupants in wheelchairs live above first floor

NOTE: In all buildings requiring elevators, at least one elevator must accommodate a wheelchair.

Fire Protection

1. Planning and construction shall provide means of egress (exit) which will permit persons to leave the building safely and which will provide a high degree of property protection.
2. The building must meet all of the fire code regulations.

Exits

1. All means of egress shall provide a continuous and unobstructed path of travel from any point in the building to a public way.
2. Every living unit should have at least two separate exits which are remote from each other. There are exceptions to this in some standards.

Vertical Services

Provision must be made in the plan for all **vertical services**. These include plumbing, water, electrical, heating, cooling, mechanical ventilation, trash removal, and elevator services as well as stairs. These services require considerable floor area and must be carefully related to the living unit. They are all regulated by codes.

Build Your Vocabulary

Following are terms that you should understand and use as part of your working vocabulary. Write a brief explanation of each term.

altitude	deed
apartments	easement
azimuth	orientation
building codes	traffic pattern
condominiums	vertical services
contour lines	zoning

Class Activities

1. List all the factors that you will need to consider as you plan your house.
2. Examine local building codes and the Federal Housing Administration requirements to see how they will influence the plan.
3. Try to visit homes representative of the basic types mentioned in this chapter. After discussing the home with the occupant, write a brief report indicating the favorable and unfavorable aspects mentioned for each type.
4. Study the deed to your house. Look for building restrictions, and report these to the class.
5. Visit the local office of the Recorder of Deeds, and ask to read the restrictions placed on lots in subdivisions in your area.
6. Visit a subdivision under development. Have the realtor explain the merits and drawbacks of the various lots for sale. Based on the discussion, choose the lot upon which you would like to build your house. As you design the house, be certain it will fit the size and slope of this lot.
7. Design a house floor plan that will meet the needs set forth in Activity 1 above. Consider orientation and exterior styling as you proceed.
8. Sketch the floor plan of your house on graph paper. Criticize the arrangement of rooms as it influences the patterns of traffic flow. Improve the plan, if possible.
9. Examine the neighborhood in which you live. List the advantages and disadvantages of the location of the lot upon which your home was built.
10. List the principles of orientation that each house in Fig. 4-43 violates. Make drawings reorienting the houses.
11. Design a duplex having a maximum floor area of 140 square meters. One unit should contain the kitchen, dining, living, and sleeping activities in one room, plus a bathroom. The second unit should be a one-bedroom unit. Select materials to keep interior and exterior maintenance to a minimum.
12. Plan a four-family, two-story condominium. Two units should have two bedrooms and occupy 1100 square feet each. The other two units should have three bedrooms and occupy 1300 square feet each. These are expensive luxury units so include features common in luxury units, such as large bathrooms and a large kitchen. The back of the condominium faces a lake so plan for utilizing this view.

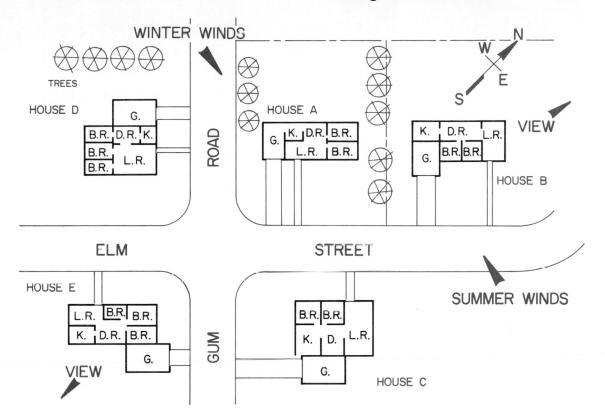

Fig. 4-43. Orientation problem.

Chapter 5

Drafting Tools and Techniques

After the process of planning the house is complete and scale sketches have been made and all ideas pertaining to the house have been recorded, the actual plans from which the house will be built can be drawn. Because clarity and accuracy are necessary in these drawings, **drafting tools and techniques** are used to prepare them. The basic drafting tools and techniques used in architectural drawing are presented in this chapter as well as a section on systems of measurement.

Drafting Tools

T-Square

The **T-square** is used to draw horizontal lines. It serves as the base for triangles and other tools. The head of the T-square is placed on the left edge of the drawing board for right-handed persons. Left-handed persons may prefer to place it on the right edge. See Fig. 5-1.

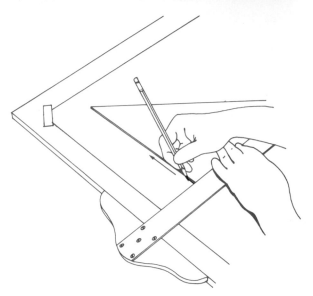

Fig. 5-2.   Vertical lines are drawn from bottom to top.

Horizontal lines are drawn from left to right along the top edge of the T-square. See Fig. 5-1. **Vertical lines** are drawn by placing a triangle on the top of the T-square and drawing from bottom to top. See Fig. 5-2. Triangles are also used to draw inclined lines. The direction in which to draw these lines is shown in Fig. 5-3.

The Parallel Straightedge

The **parallel straightedge** serves the same purposes as the T-square. It is used to draw horizontal lines, and triangles may be used with it in the same manner as described for the T-square. The straightedge is superior to the T-square because it

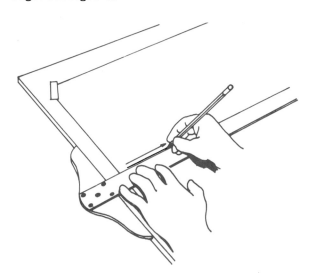

Fig. 5-1.   The head of the T-square is on the left edge of the drawing board. The T-square is held firmly to the board. Horizontal lines are drawn from left to right.

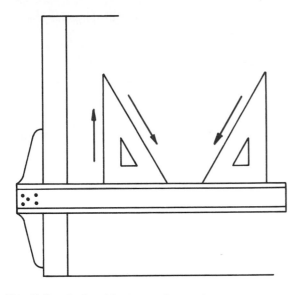

Fig. 5-3. Inclined lines are drawn downward toward the T-square.

Fig. 5-4. A parallel straightedge replaces the T-square and is more accurate.

always stays in a horizontal position. It is moved up and down the drawing board on wires running over pulleys. See Fig. 5-4.

Triangles

Triangles are made of transparent plastic. Those most often used are the 45 degree and the 30-60 degree types. See Fig. 5-5. They are made in many

sizes. The size of the 30-60 degree triangle is the length of the longer side forming the right angle. The size of the 45 degree triangle is the length of either side. Triangles are used in several ways in architectural drawing. A typical use of the 45 degree triangle is in section-lining wall sections to indicate where the wall has been cut.

Protractor

Angles can be laid out using a **protractor**. A protractor is a tool with degrees indicated on a semicircular edge. See Fig. 5-6.

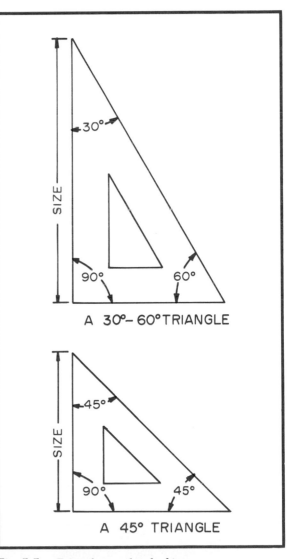

A 30°– 60° TRIANGLE

A 45° TRIANGLE

Fig. 5-5. Triangles used in drafting.

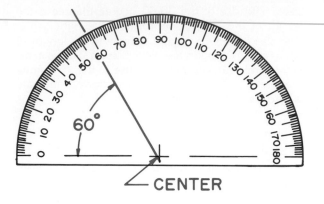

Fig. 5-6. A protractor may be used in drawing or measuring angles.

To draw an angle using the protractor, place the center point on a horizontal line where the lines forming the angle will converge. Align the horizontal line with the guidelines on the protractor. Read the degrees on the scale and mark the required number. Draw a line from the center to the desired degree mark.

Scales

A **scale** is a tool used to measure or lay out dimensions on a drawing. These dimensions may be either full size, or larger or smaller than full size.

The tool is available in flat and triangular shapes. See Fig. 5-7. The term **scale** also refers to the use of proportional measurements in which one distance represents another. (For example, to make furniture templates, the scale of 1/4" = 1' was recommended.) The two-bevel and opposite bevel scales (tools) have two different scale (measurement) markings on them. The four-bevel has four different scale markings, and the triangular scale has six.

These tools may be **open divided** or **fully divided.** Open divided scales have a subdivided unit on the end. Fully divided scales have the subdivisions running the entire length. See Fig. 5-8.

When marking distances with a scale, make certain it is flat on the paper. Always use a sharp pencil to make a short dash mark to locate the distance. After distances are located, draw the lines with a triangle, T-square, or other drafting tool designed for that purpose. Never use the edge of the scale for this purpose. The exception is the drafting machine. Its scales are designed for both measuring and drawing the line.

There are several types of scales available. The **architect's scale** is designed for preparing working drawings of buildings. The **engineer's scale** is used for making maps. The **metric scale** is gradually replacing these since the United States is converting to the metric stystem of measure. Eventually, everything will be in metrics.

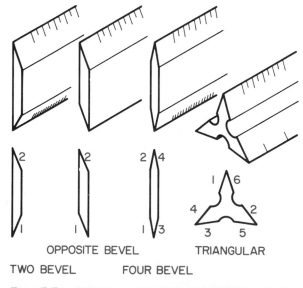

Fig. 5-7. Scales are made in triangular and flat shapes.

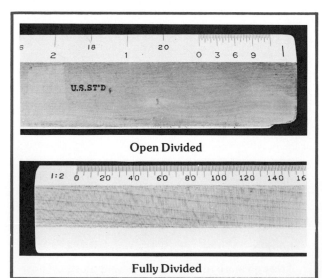

Fig. 5-8. Scales are available in open divided and fully divided types.

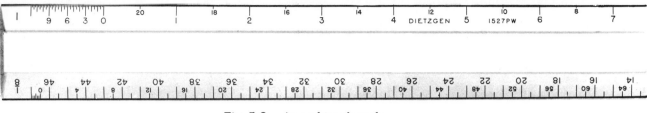

Fig. 5-9. An architect's scale.

Architect's Scale

The **architect's scale** is based on the **foot.** See Fig. 5-9. All of the measurement scales on the tool represent one foot. For example, the 1/4 scale means that 1/4 inch on the scale represents one foot on the drawing. The architect's scales commonly used include 3/32, 1/8, 3/16, 1/4, 3/8, 1/2, 3/4, 1, 1 1/2, and 3. At the end of each architect's scale, the unit used is divided into 12 parts. Each part represents one inch. See Fig. 5-10.

The architect's scale is divided into feet along its length. The number of feet are counted off the length beginning at the zero. If the distance contains inches, these are counted off the subdivided unit at the end of the scale. Look at Fig. 5-10. The distance to be drawn is 2 feet 6 inches at a scale of 1″ = 1′-0″. First find the correct scale. Then count two feet to the right of zero and 6 inches to the left. The total is 2 feet 6 inches. Notice the distance 3′-9″ marked on the scale 1/2″ = 1′-0″ in Fig. 5-10. Can you see how it was measured?

Engineer's Chain Scale

The **engineer's chain scale** is divided into **decimal parts.** The divisions are 10, 20, 30, 40, 50, and 60 parts to the inch. For example, the scale marked 20 means it has 20 divisions to the inch. See Fig. 5-11. This could be used to produce a drawing with the scale 1″ = 20′. Each division would represent one foot. Likewise, one inch could represent 200 feet and each division would represent 10 feet. If one inch represents 2000 feet, each division represents 100 feet. The other scales are used in the same manner.

Metric Scale

Metric scales are used for all drawings which use metric linear measurements. Metric scales are based on a **ratio** such as 1:10. This means that one millimeter on a drawing represents 10 millimeters of the actual object.

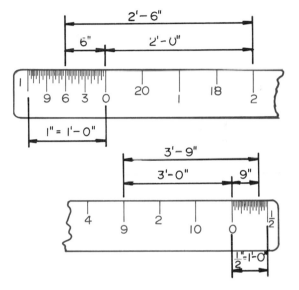

Fig. 5-10. The architect's scale is read in feet and inches. It may be read from either end depending on the measurement scale used.

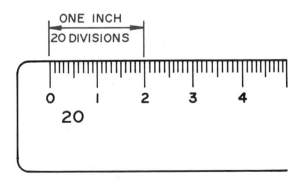

Fig. 5-11. The engineer's chain scale divides the inch into various subdivisions. This scale has the inch divided into 20 divisions.

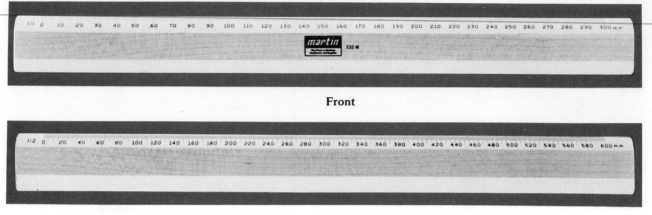

Front

Back

Fig. 5-12. A typical metric scale.

Scales which measure full size (actual) are marked 1:1. They are divided into millimeter divisions. Some may be marked in 0.5 mm divisions. The scale has every 10 mm marked. The typical metric scale is either 300 mm or 150 mm long. See Fig. 5-12.

Metric scales such as 1:1, 1:2, 1:5, and 1:10 are used to draw an object at a reduced size. Since the metric system is base 10, a scale can be used to measure several different reduction or enlargement ratios. For example, the scale 1:1 means one division represents one millimeter (mm). Each division on the scale is one millimeter so the drawing is actual size. However, it can be used to draw to scales of 1:10 or 1:100 if desired. The 1:10 ratio means the drafting technician would let one millimeter on

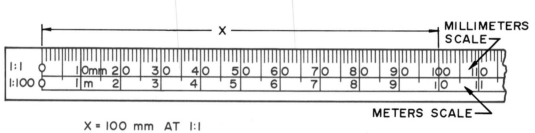

X = 100 mm AT 1:1
X = 1000 mm AT 1:10
X = 10 000 mm AT 1:100

Fig. 5-13. A metric scale showing the 1:1 and 1:100 ratios.

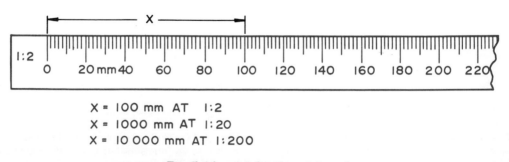

X = 100 mm AT 1:2
X = 1000 mm AT 1:20
X = 10 000 mm AT 1:200

Fig. 5-14. A 1:2 ratio metric scale.

Fig. 5-15. The drafting machine greatly speeds up the drafting process.

Keuffel and Esser Company

Fig. 5-16. This track-type drafting machine can operate in any position from horizontal to vertical.

the scale equal ten millimeters on the object. Likewise, one millimeter on the scale could represent one hundred millimeters on the object.

Since architectural drawings are often of very large objects, some dimensions will be in meters. One meter (m) equals one thousand millimeters. Suppose the 1:1 scale is used to make a drawing in meters. At a ratio of 1:100 (mm), ten millimeters on the scale represents one meter (1000 mm). At a ratio of 1:10, ten millimeters represents one-tenth of a meter (100 mm). See Fig. 5-13.

An example of the 1:2 ratio scale is shown in Fig. 5-14. The drawing produced by using this scale will be half the size of that using the 1:1 scale.

The Drafting Machine

The **drafting machine** replaces the T-square, triangles, protractor, and scale. See Fig. 5-15. There are many different kinds of machines. One example is shown in Fig. 5-16. The scales are set at a 90° angle to each other. They are used for measuring distances and drawing lines. The horizontal scale is used to draw horizontal lines. Vertical lines are drawn with the vertical scale. The scales are removable and may be replaced when a different scale is needed. The control knob is marked in degrees and can be rotated for drawing angles. See Fig. 5-17.

Most drafting technicians hold the control knob in the left hand. This hand moves the machine over the drawing. Machines are also available with the control knob in the opposite position.

Keuffel and Esser Company

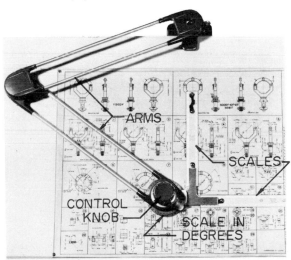

Fig. 5-17. The main parts of a drafting machine.

Irregular Curves and Adjustable Curve Rulers

Either of two kinds of tools may be used to draw irregularly curved lines. These are **irregular curves** and **adjustable curve rulers.** An irregularly curved line curves but is not a true circle. In architectural drawing, these lines may be used to indicate such things as driveways, swimming pools, or arches.

The **irregular curve** tool is used in the following manner. Find the points which form the irregularly curved line. Lay the tool up to the points. Move the tool about until it touches three or more points. Draw this section of the line. Then move the tool again until a part of it connects more points and draw that section. Continue the process until the curved line is complete. Irregular curves are

available in many sizes and shapes. One example is shown in Fig. 5-18.

Adjustable curve rulers are flexible and can be bent to fit irregularly curved lines. Several types are available. See Fig. 5-19.

Drawing Instruments

The compass and dividers are the two most commonly used instruments. See Fig. 5-20.

The **compass** is used to draw circles. It has a metal pin in one leg and a pencil lead in the other. To draw a circle adjust the legs of the compass to the radius of the desired circle. Hold the compass in

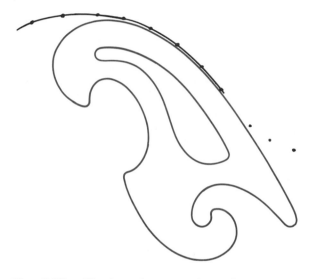

Fig. 5-18. The irregular curve is used to connect points forming irregularly curved lines.

Rola Tape Corporation

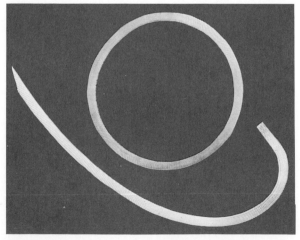

Fig. 5-19. Adjustable curve rulers can be formed to fit irregular curves.

Teledyne Frederick Post

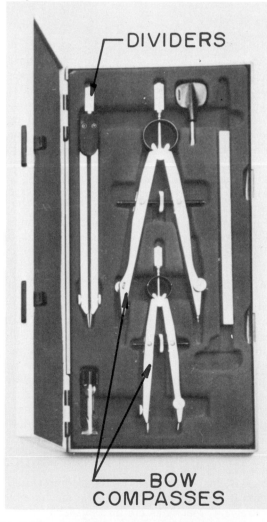

Fig. 5-20. A typical set of drafting instruments.

one hand. Rotate it in a clockwise direction. Lean it in the direction it is moving. See Fig. 5-21.

A **beam compass** is used to draw large circles. See Fig. 5-22. It is used in the same manner as the regular compass except it is held in both hands.

The **dividers** has metal pins in both legs. It is used to transfer distances from one place on a drawing to another. It is also used to "step off" equal distances. See Fig. 5-23.

Pencils

Drawing pencils are available as wood cased or mechanical. See Fig. 5-24. Both use the same type of lead. Lead is made in varying degrees of hardness. The **H** leads are hard. The **B** leads are soft. See Fig. 5-25. The hardness of the lead is printed on the leads used in the mechanical pencil and on the wood cased pencil. Do not sharpen the end with the hardness stamp. This would remove the stamp and make it difficult to tell the exact grade of hardness.

Fig. 5-21. A compass is rotated in a clockwise direction.

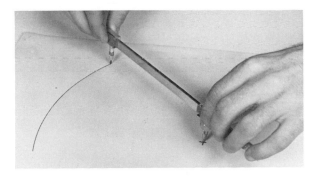

Fig. 5-22. A beam compass is used to draw large circles. It is rotated in a clockwise direction.

Fig. 5-23. A dividers can be used to lay out equal distances. It is held in one hand and the distances are stepped off.

Koh-I-Noor

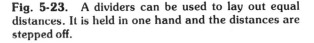

MECHANICAL

WOOD CASED

Fig. 5-24. Architects use both wood cased and mechanical drafting pencils.

Venus Drawing Pencils

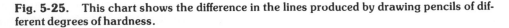

DEGREE CHART	Soft						Medium			Hard			Very Hard				
	6B	5B	4B	3B	2B	B	HB	F	H	2H	3H	4H	5H	6H	7H	8H	9H

Fig. 5-25. This chart shows the difference in the lines produced by drawing pencils of different degrees of hardness.

Fig. 5-26. A drafting pencil sharpener removes the wood but does not touch the lead.

Fig. 5-27. The lead is made pointed in a lead pointer.

The wood cased pencil is sharpened in a **drafting pencil sharpener.** This sharpener looks like a regular pencil sharpener. See Fig. 5-26. The difference is that the cutter removes the wood but does not sharpen the lead. The lead is formed to a conical point in a **lead pointer.** See Fig. 5-27. The sharpened point should be about 1 1/4 inches (32 mm) long. The exposed lead should be about 3/8 inch (10 mm) long. See Fig. 5-28.

The exact lead to use for various lines on a drawing will vary with the type of line and the preference of the technician. Most line work will be done with an H or 2H pencil. Thin lines will be made with a 3H or 4H pencil. Thick lines and sketching will be done with an F, HB or B pencil.

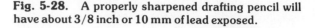

Fig. 5-28. A properly sharpened drafting pencil will have about 3/8 inch or 10 mm of lead exposed.

Vemco Corporation

Fig. 5-29. An electric eraser.

Pencils with a special plastic lead are used on polyester drafting film.

Erasers

Several types of **erasers** are available. A hard rubber eraser is used to remove dark, heavy lines. Light lines and smudges are removed by a soft rubber eraser. A vinyl eraser is used to remove plastic lead from polyester drafting film.

Erasers are sold in block and stick form. The stick form has a wood shell much like a pencil. It can be pointed in a pencil sharpener.

Electric erasing machines use long, round erasers. See Fig. 5-29. These erasers come in varying degrees of hardness.

Improper erasing can cause serious damage. Do not put too much pressure on the eraser. If you press too hard, the eraser will overheat and might leave a smudge on the drawing. Be especially careful when using the electric eraser. It can cut through the paper rapidly if improperly used. When erasing, stretch the paper between two fingers so it does not wrinkle.

Templates

A **template** used for drafting is a thin, transparent, plastic tool with openings of various shapes cut into it. See Fig. 5-30. The template is

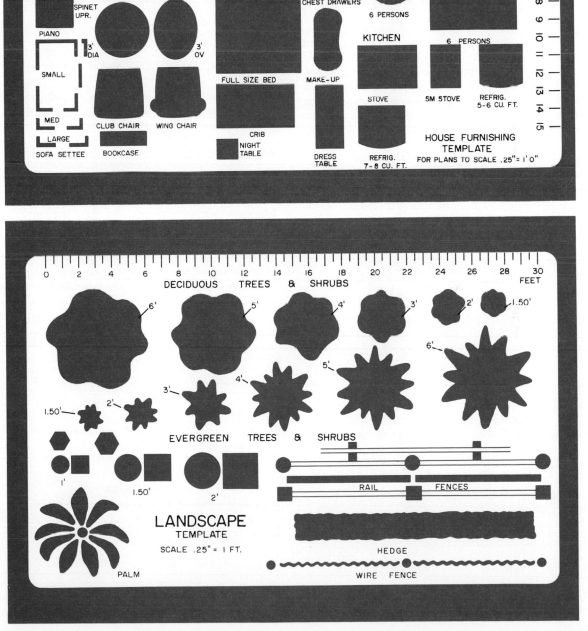

Fig. 5-30. Typical templates for use on architectural drawings.

placed on the drawing. A pencil is used to trace the shape of the opening onto the drawing. Templates are used a great deal because they are accurate and speed up the drafting process.

There are hundreds of different templates available. Many are of a general nature, such as a circle template. Others are specialized. Many are designed for use in architectural drawing. For example, one might contain scale openings representing the plumbing symbols used on drawings. Templates are also available for drawing the landscaping portions of a plan.

Pressure Sensitive Overlays

Pressure sensitive overlays are printed on a clear material and have an adhesive back. They include a wide variety of things, such as letters, symbols, and illustrations. The items wanted are removed from the large sheet and pressed into place on the drawing. Excellent copy can be produced quickly. Examples of the items available are shown in Fig. 5-31.

Drawing Papers and Film

There are a great many types of materials upon which drawings can be made. Most architectural drawings are made on tracing paper or vellum. These papers can be drawn upon with pencil or ink. **Tracing paper** is a thin, translucent, untreated paper. Treated tracing paper has a transparentizing agent applied and is called **tracing vellum.** It is more translucent than untreated tracing paper.

Vellum is made from 100 percent white rag stock. It is strong and lasts longer than untreated tracing paper. All papers deteriorate with age. If long-term storage of drawings is important, they should be drawn on polyester drafting film.

Polyester drafting film is translucent and very tough. It is stable and will not change size as humidity and temperature change. It can be drawn upon with pencil or ink. Pencils with a special plastic lead are used. Typical sheet thicknesses include 0.003″ and 0.005″ and 0.09 mm and 0.14 mm.

Papers and films are available in standard sizes and are also sold in rolls. Typical sizes are shown in Fig. 5-32.

Alphabet of Lines

Each line on a drawing has a special meaning. In order to help make and read drawings, **standard line symbols** were developed. The line symbols in Fig. 5-33 are the actual thickness they should be on a finished drawing. There are two thicknesses of lines: thick and thin. The thick lines are used for visible, cutting-plane, and short break lines. The thin lines are used for long break, hidden, center, section, extension, dimension, and phantom lines.

Fig. 5-31. Examples of the items available on pressure sensitive overlays.

Sheet Sizes

U.S. CUSTOMARY		INTERNATIONAL	
Designation	Inches	Designation	Millimeters
A	8.5 × 11	A4	297 × 210
B	11 × 17	A3	297 × 420
C	17 × 22	A2	420 × 594
D	22 × 34	A1	594 × 841
E	34 × 44	A0	841 × 1189

Roll Sizes

Width in Inches	Length in Yards
30	20 and 50
36	20 and 50
42	20 and 50

Fig. 5-32. Standard paper sizes.

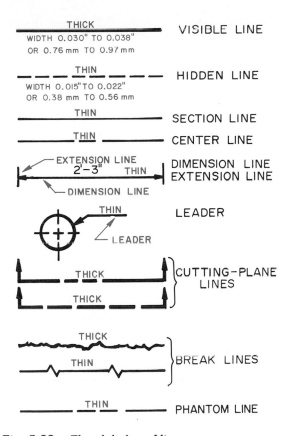

Fig. 5-33.   The alphabet of lines.

of long and short dashes. The long dashes can be from 3/4 inch (19 mm) to 1 1/2 inches (38 mm). The short dashes are about 1/16 inch (1.6 mm) long.

Section lines are used to show a surface that has been cut in a section view. They are thin lines drawn parallel and spaced 1/16 inch (1.6 mm) to 1/8 inch (3 mm) apart.

Cutting-plane lines show where a section has been taken on a drawing. They are thick lines with arrows on the end to show the direction in which the section was taken. Two symbols are in use. One is a series of dashes of equal length and the other is a series of long and short dashes.

Break lines are used to show the edge where part of the drawing has been removed. Short breaks are made with a thick, freehand, jagged line. Long breaks are made with a thin, solid line with a **Z** symbol located every several inches.

Phantom lines are part of the alphabet of lines though seldom, if ever, used in architectural drawing. They are used to show the alternate position of a part that moves. The part is drawn with visible lines. It is also drawn in its **moved location** using phantom lines. The symbol is a long dash followed by two short dashes. It is a thin line.

Applications for the various line symbols are shown in Fig. 5-34. **Visible lines** are used to show the main outline of a building and all interior walls. They are used for porches, patios, driveways, and construction details. All outlines of any major part which should stand out on the drawing should use a visible line.

Hidden lines show edges and surfaces which are not visible to the eye but are hidden below a visible surface. They are made of short dashes about 1/8 inch (3 mm) long. The spacing between dashes is about 1/32 inch (1 mm).

Dimension lines are thin, solid lines. They are used to show the distance represented by a size dimension. They usually have arrowheads on each end.

Extension lines are thin, solid lines used with dimension lines. They extend to a point on a drawing to which the dimension line refers.

Center lines are thin lines used to locate the center of holes or cylindrical solids. They are made

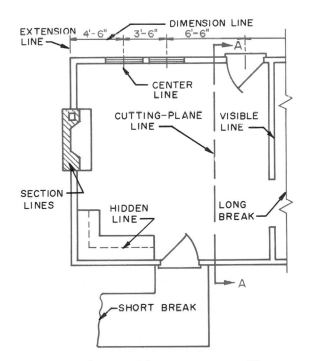

Fig. 5-34.   The use of the various types of line symbols.

Fig. 5-35. The recommended way to form vertical capital letters, and the order of strokes to follow.

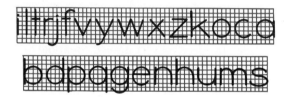

Fig. 5-36. Lowercase vertical letters.

Architectural Lettering

The purpose of lettering on a drawing is to communicate things which the drawing itself does not clearly show. Lettering must be easy to read. Experienced drafting technicians and architects generally have their own unique lettering styles. However, legibility is more important than decoration.

Beginners should follow the recommendations shown in Fig. 5-35 for using the capital, vertical Gothic style in lettering. Notice the size of the letters. Some are as wide as they are high. Others are a little narrower than they are high. What letter is wider than it is high? What letter is higher than it is wide? See how the letter sizes compare.

In Fig. 5-35, the recommended order of strokes is shown. Learn to form each letter in this order. Always form the letters the same way each time. Learning to letter properly requires a great deal of practice. Practice slowly at first. Build up speed after the order of strokes becomes habitual.

Lettering on architectural drawings is usually done in all capital letters. The only time lowercase

letters are used is when they have special significance, such as the symbol **mm** which means millimeters. Many metric symbols are lowercase letters. See Fig. 5-36.

Architectural lettering should "dress-up" the drawing. This means not only carefully formed letters, but also letters that are properly spaced. The space between letters must **appear** to be equal. The space cannot actually be equal because of the shape of the letters. Some letters must be closer together. Open letters such as the **A** or **V,** can be closer together because of the open space around them. Letters with vertical parts, such as the **I** or **H** should be farther from other letters because they appear to be closer. Spacing letters so they appear to the eye to be equally spaced even though they are not is called **optical spacing.** See Fig. 5-37. The space between words in a note should be equal to the size of the letter **O.** See Fig. 5-38.

Lettering should be horizontal and kept in straight lines. Horizontal guidelines can be drawn on the paper. They should be drawn so lightly that they do not reproduce when the drawing is used to run prints. Vertical guidelines may be drawn to help keep the letters straight vertically. See Fig. 5-39.

Most lettering will be 1/8 inch (3 mm) high. On certain parts of the drawing, it will be larger. Room identification and title block information may be in letters as large as 1/4 inch (6 mm).

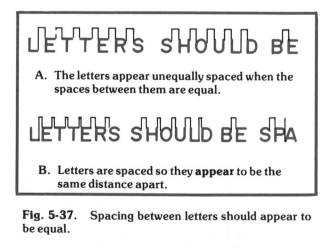

A. The letters appear unequally spaced when the spaces between them are equal.

B. Letters are spaced so they **appear** to be the same distance apart.

Fig. 5-37. Spacing between letters should appear to be equal.

THE␣SPACE␣BETWEEN WORDS␣IS␣EQUAL␣TO␣THE LETTER␣O.

Fig. 5-38. Spacing between words in a sentence.

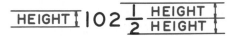

Fig. 5-39. Guidelines are used when lettering.

Fig. 5-40. Lettering fractions.

The lettering of numbers and fractions is shown in Fig. 5-40. Numerals are the same height as letters. Fractions have a total height of two letters. Since there is a bar between the numerals, they are actually smaller in size than letters and whole numbers. Notice the guidelines used to draw the fractions.

Once you can letter rapidly and accurately, you may want to vary your lettering somewhat from the Gothic style. Typical approaches are to condense or expand the letters a little. Some people change the height of the horizontal parts of the letters. Examples are in Fig. 5-41.

Computer-Aided Architectural Design

Engineering analysis, design, and drafting can be done with the assistance of the computer. Computer drafting facilities can produce drawings of all kinds. Some architectural firms already are doing a large part of their work in this way, and more and more will use computers in the future.

In order to fully utilize computer capabilities, the architect and architectural drafter must thoroughly

Fig. 5-41. Architectural drafters often develop their own lettering styles.

understand the principles of architectural drafting. A computer can do only what it is told. The person using it controls the information entered into the computer. That person must know the proper architectural graphic and engineering techniques.

The Computer

A **computer** is an electronic device that can store data and then follow instructions that enable it to operate upon the data to produce the desired results. A computing system includes hardware, data, and software.

Hardware

The equipment that makes up the computer system is called **hardware**. It includes input equipment, memory, a central processing unit, and output equipment. A simple computer is diagrammed in Fig. 5-42.

Fig. 5-42. The make-up of a simple computer.

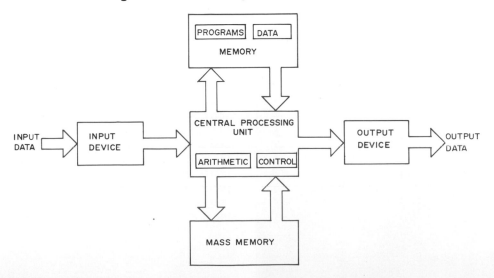

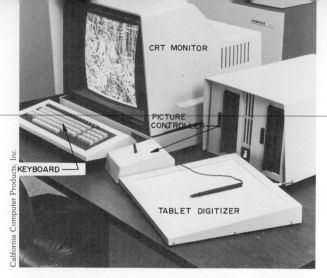

CRT MONITOR

PICTURE CONTROLLER

KEYBOARD

TABLET DIGITIZER

Fig. 5-43. A workstation where the operator produces the graphic image.

Fig. 5-44. The output is displayed on these monitors.

Fig. 5-45. Output can also be in the form of drawings produced on a plotter. These are beltbed plotters.

Data (facts) are entered through an **input device**. Typically this is a keyboard much like an electric typewriter. See Fig. 5-43. The input device encodes the data as electrical pulses. These are stored in the **memory**. Memory can consist of silicon chips, magnetic discs and tapes, or minute ceramic rings strung on a mesh of fine wires. The information to be taken from memory and used is transmitted to output devices in the form of electrical impulses. The **output device** decodes these impulses and produces the output in usable form. For example, output could be pictures on a cathode ray tube or an architectural drawing on a plotter. See Figs. 5-44 and 5-45.

The computer system is controlled by the **central processing unit** (or **CPU**). It has a **control section** and an **arithmetic and logic section**. The control section secures, interprets, and acts upon instructions from the program stored in the memory. The arithmetic and logic section performs the basic manipulations, such as adding or subtracting.

Data

Data are the pieces of information, such as sizes or weights, that make up the **input** entered into the computer. Input devices are used to enter these as electrical signals. Typical data input devices include a card reader and a keyboard. After data are processed by the central processing unit, the **output** is handled by an output unit. Typically this is an electric typewriter or a cathode ray tube display device.

Programs

The operations to be performed on the data by the CPU are detailed in a computer program. A **program** is a series of sequential instructions written in a computer language. Simple programs perform steps in the sequence in which they are written. Complex programs contain **loops** and **branches**. A **loop** is a sequence of instructions that is repeated a specified number of times within a program. In **branch instruction**, the specific instruction to be executed is determined by the result of the preceding specified operation.

Computer programs are referred to as **software**. Application software consists of programs written to solve specific problems. For example, a program may be designed to draw architectural perspectives and working drawings. Computer manufacturers and the companies that develop software have many programs available. Architectural firms needing additional programs hire programmers to develop these. Programs must be written in a computer language that the firm's computer will be able to use.

Computer-Aided Design and Drafting

In computer-aided design, the computer is used to produce solutions to design problems. Traditionally in an architectural office these solutions are found by architects and drafters working at a drawing board. They refer to standards manuals and record solutions on paper with a pencil. The use of a computer system to perform many of these processes speeds up the work and produces more accurate results at a lower cost.

The architectural designer can use the computer and computer graphics devices for many types of analyses. One type is **problem structuring**. The designer tries to identify all the elements of the design problem. These could include such things as goals, performance specifications, and restrictions. The designer gathers data about each element, then organizes the elements and the data into a complete problem description. The overall description is then broken down into sub-problems.

Another type of analysis is **feasibility analysis**. This is a study done to determine if a solution is practical. During this study, the design objectives and restrictions are set. Traditionally, the decision of feasibility has been based to a great extent upon the judgment of the designer. Now computers, using mathematical techniques, can assist in this process. It does not replace the final judgment of the designer, but provides an exhaustive analysis of the situation. The designer uses the information contained in this analysis in making a final judgment.

Space needs can be developed by gathering data about the people who will use the building, and about the equipment they will use and the activities they will perform. These data can be translated by the computer into a set of space requirements.

Many other functions can be analyzed. A building code check might be run to see if a design is in compliance with regulatory codes. An estimating and cost accounting study could be made. An analysis of the internal environment — including daylight factors, artificial lighting requirements, heat loss and gain, and heating and cooling curves — could be generated.

Much of the needed information on site development, such as mapping, drainage, and accessibility, can be generated. Drawings of floor plans, space with equipment located, and three-dimensional space analyses can be developed. Drawings of elevations, sections, and perspectives can be produced. See Fig. 5-46. Many detail drawings, such as electrical, plumbing, and heating layouts, can be drawn. Production schedules, cost control, worker time, and payroll records — as well as material in-

Fig. 5-46. A partial view of a perspective of a large commercial building drawn on a plotter.

ventory and purchasing operations — can be computerized. Chapter 24, "CADD in Architecture," provides a detailed description of the ways in which computers are used in architectural design and drafting.

Systems of Measurement

There are two systems of linear measurement in use in the United States. These are the **U.S. Customary** system and the **SI Metric** system.

The U.S. Customary system has been the nationally accepted system. The entire world uses the metric system; therefore, the United States is switching to the metric system. The change will be gradual, but eventually all architectural drawings and all construction materials will use metric measures and standards.

The U.S. Customary System

The U.S. Customary system is based on the **yard.** The yard is divided into three **feet.** Each foot is divided into 12 **inches.** See Fig. 5-47. The inch can be divided into fractional parts such as 1/4 inch or 1/8 inch.

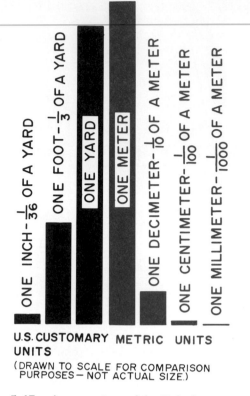

ONE INCH—$\frac{1}{36}$ OF A YARD

ONE FOOT—$\frac{1}{3}$ OF A YARD

ONE YARD

ONE METER

ONE DECIMETER—$\frac{1}{10}$ OF A METER

ONE CENTIMETER—$\frac{1}{100}$ OF A METER

ONE MILLIMETER—$\frac{1}{1000}$ OF A METER

U.S. CUSTOMARY METRIC UNITS
UNITS
(DRAWN TO SCALE FOR COMPARISON
PURPOSES—NOT ACTUAL SIZE.)

Fig. 5-47.   A comparison of the U.S. Customary and metric systems of linear measure.

The Metric System

The international system of metric units is known by the abbreviation **SI.** This stands for Système International d' Unités, the French name for International System of Units.

Of the many metric units, a selected few will be presented. These are the units most likely to be found in architectural and construction work.

Metric System of Linear Measurement

The basic SI unit of linear measurement is the **meter** (m). See Fig. 5-47. The meter is divided into 10 **decimeters** (dm), 100 **centimeters** (cm), or 1000 **millimeters** (mm). Since the metric system is based on ten, it is possible to convert from one metric measure to another by simply moving the decimal point. For example, 1.536 meters equals 15.36 decimeters, 153.6 centimeters, or 1536 millimeters.

The metric linear units, the symbols used to represent them on drawings, and conversion factors are given in Fig. 5-48. These units are used to measure both large and small dimensions.

Metric Measurement of Mass

The SI unit of mass is the **kilogram** (kg). This corresponds to our measure of weight. However, **mass** is the amount of substance in an object and is always the same regardless of the pull of gravity. The **weight** of an object varies according to the pull of gravity. For practical purposes in construction, the terms mass and weight can be used interchangeably. Many materials used in construction are now sold by weight.

Metric units of mass, the symbols used to represent them on drawings, and conversion factors are given in Fig. 5-49.

Metric Measurement of Volume

The SI unit that is used to measure volume is the **cubic meter** (m^3). The basic unit is a cube with each of its sides one meter in length. See Fig. 5-50. Possibly, concrete will be sold by the cubic meter rather than the cubic yard.

The metric units of volume, the symbols used to represent them on drawings, and conversion factors are given in Fig. 5-51. As in linear measurement, changing from one unit to another involves simply moving the decimal point. Move it to the left for larger units of measure and to the right for smaller units of measure. For example, 500 cubic centimeters equal 0.000 500 of a cubic meter, 0.500 of a cubic decimeter, or 500 000 cubic millimeters.

The **liter (L)** is a special term used to replace "cubic decimeter" (dm^3) when measuring liquids and gases. The only approved subdivision is the **milliliter** (mL), which is one-thousandth of a liter.

Fig. 5-48.

Metric Measures of Linear Distances

1 kilometer (km)	=	1000 meters (m)
1 hectometer (hm)	=	100 meters (m)
1 decameter (dkm)	=	10 meters (m)
1 meter (m)	=	1 meter (m)
1 decimeter (dm)	=	0.1 meter (m)
1 centimeter (cm)	=	0.01 meter (m)
1 millimeter (mm)	=	0.001 meter (m)

To Convert:	Multiply:	
inches to millimeters	inches by	25.4
millimeters to inches	millimeters by	0.0394
inches to centimeters	inches by	2.54
centimeters to inches	centimeters by	0.394
inches to meters	inches by	0.0254
feet to meters	feet by	0.3048

Metric Measurement of Area

The **square meter** (m²) is the SI unit for area measurement. It may be divided into smaller units. See Fig. 5-52. Areas must be figured to compute such things as cost per unit or to find the amount of floor space.

For large areas, as in land surveying, larger units than square meters are used. Refer to Fig. 5-53.

The **square kilometer** (km²) is used for large land areas. Smaller areas may be described by **square hectometers** (hm²). The term **hectare** is sometimes used in place of hectometer for land measurements.

Fig. 5-51.

Metric Measures of Volume

1 cubic meter (m³)	=	1000 cubic decimeters (dm³)
1 cubic meter (m³)	=	1 000 000 cubic centimeters (cm³)
1 cubic meter (m³)	=	1 000 000 000 cubic millimeters (mm³)

To Convert:	Multiply:	
cubic yards to cubic meters (m³)	cubic yards by	0.765
cubic feet to cubic meters (m³)	cubic feet by	0.028
gallons to cubic meters (m³)	gallons by	0.0038
quarts to cubic meters (m³)	quarts by	0.00095
ounces to cubic meters (m³)	ounces by	0.000029

Fig. 5-49.

Metric Measures of Mass

1 kilogram (kg)	=	1000 grams (g)
1 hectogram (hg)	=	100 grams (g)
1 decagram (dkg)	=	10 grams (g)
1 decigram (dg)	=	0.1 gram (g)
1 centigram (cg)	=	0.01 gram (g)
1 milligram (mg)	=	0.001 gram (g)

To Convert:	Multiply:	
pounds to kilograms	pounds by	0.454
kilograms to pounds	kilograms by	2.205

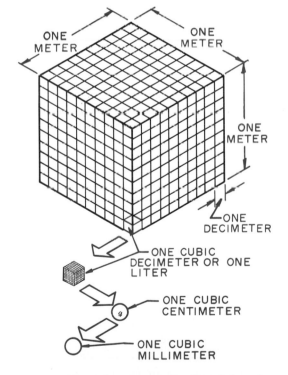

Fig. 5-50. The cubic meter is the SI unit for volume.
1000 cubic millimeters (mm³) =
1 cubic centimeter (cm³)
1000 cubic centimeters (cm³) =
1 cubic decimeter (dm³)
1000 cubic decimeters (dm³) =
1 cubic meter (m³)

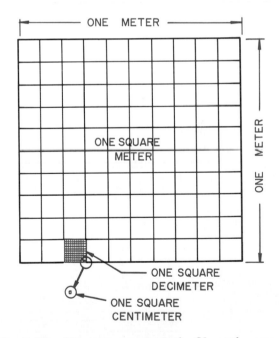

Fig. 5-52. The square meter is the SI unit for area.
100 square millimeters (mm²) =
1 square centimeter (cm²)
100 square centimeters (cm²) =
1 square decimeter (dm²)
100 square decimeters (dm²) =
1 square meter (m²)

Metric Measures of Area	Designations for Large Areas
1 square meter (m³) = 100 square decimeters (dm²) 1 square meter (m²) = 10 000 square centimeters (cm²) 1 square meter (m²) = 1 000 000 square millimeters (mm²)	1 square kilometer (km²) = 1 000 000 square meters (m²) 1 square hectometer (hm²) = 10 000 square meters (m²)

To Convert:	Multiply:	
square inches (in.²) to square millimeters (mm²)	in.² by	645.200
square inches (in.²) to square centimeters (cm²)	in.² by	6.452
square centimeters (cm²) to square inches (in.²)	cm² by	0.155
square feet (ft.²) to square meters (m²)	ft.² by	0.093
square meters (m²) to square feet (ft.²)	m² by	10.760
square yards (yd.²) to square meters (m²)	yd.² by	0.836
square meters (m²) to square yards (yd.²)	m² by	1.196
acres to square meters (m²)	acres by	4046.870
square miles (mi.²) to square meters (m²)	mi.² by	2 589 988
square miles (mi.²) to square kilometers (km²)	mi.² by	2.590

Metric Measurement of Electricity

The **ampere** is the SI unit of electrical measurement. It is the unit currently in use and measures the intensity or strength of an electric current. It is defined as the amount of current necessary in two parallel wires one meter apart that will result in a specific force between the wires.

Metric Measurement of Luminous Intensity

A **candela** is the unit of luminous intensity. It is used in measuring the amount of light given off by an object.

Metric Measurement of Temperature

Degrees **Celsius** (°C) is the SI unit used to measure temperature. There are one hundred degrees from the freezing point of water (0°C) to the boiling point (100°C). A normal room temperature (72° Fahrenheit) would be 22.2°C. The conversion factors are given in Fig. 5-54.

Fig. 5-54.

Temperature Conversion Factors

Fahrenheit to Celsius	=	(°F − 32) x .555
Celsius to Fahrenheit	=	(°C x 1.8) + 32

Build Your Vocabulary

Following are terms that you should understand and use as part of your working vocabulary. Write a brief explanation of each term.

architect's scale
°C
computer
CPU
data
dimension line
drafting machine
engineer's scale
extension line
hardware
input device
irregular curve
kilogram
m²
memory

meter
metric measurement
   system
metric scale
millimeter
output device
parallel straightedge
program
SI
software
template
U.S. Customary measurement system
vellum
visible line

Class Activities

1. Develop your drafting skills by copying some of the kitchen and bath plans shown in Chapter 2. Draw them four times as large as they are printed.
2. Draw light horizontal lines 1/8" apart on an 8 1/2" × 11" sheet of paper. Select a short poem you enjoy. Letter the poem on these lines. Skip every other drawn-in line.
3. Copy some of the wall sections shown in Chapter 7. Draw to the scale 3/4" = 1'-10". Carefully letter the identification of each part. If you have metric scales, draw to the scale 1:20.
4. Using a metric scale, measure the size of the top of your drafting table. Convert these measurements to inches, using conversion factors. Then measure the top with the architect's scale to see how close you came.

Chapter 6

The Working Drawings and Specifications

The actual plans from which the house will be built are called **working drawings.** A set of working drawings for the average house will include a plot plan, floor plans for each level or story, a foundation or basement plan, elevations of all sides of the house, wall sections to illustrate the method of construction, and details on which the builder will need special information.

No set of plans is complete without a set of specifications. **Specifications** detail all the points involved in the house construction that cannot be indicated on the working drawings.

As the working drawings are prepared, notes are made concerning the things that need to be clearly indicated. These are then written in a set form and become a part of the specifications for the house.

Preparing for Work

Before any drawing is started, several preliminary decisions must be made. First is the size of paper to use. If the house is to be small or average in size, usually **B** size (11″ × 17″) or metric A3 (297 × 420 mm) paper will be satisfactory. For larger houses, **C** size (17″ × 22″) or metric A2 (420 × 594 mm) will be needed. If the house is very large, it may be necessary to cut the sheet from a roll of vellum. All drawings in a set should be the same size. (Standard paper sizes are given in Chapter 5.)

Some prefer to place the floor plan, kitchen details, and door and window schedule on one sheet; the foundation plan, wall sections, and other details on a second sheet; and the elevations on third and fourth sheets, as required. Others prefer to use large sheets of vellum and place the floor plan, foundation plan, and details on one sheet and all the elevations on a second sheet. This is satisfactory; however, it is a little more difficult to keep the larger drawings clean.

Drawing to Scale

When the working drawings for a house are made, it is not possible to draw them full size. The drawings must be made using a small increment, such as 1/4 inch, to represent one foot of the actual house dimension. In metrics, 1 mm may represent 50 mm. This is called **drawing to scale.** A floor plan drawn to the scale 1/4″ = 1′-0″ reduces the house 48 times, so it will fit on a standard sheet.

It is necessary to decide to what scale each part of the working drawing will be drawn. Under normal conditions the following scales are used for residential work:

U.S. Customary Scales — inches	
Floor plan	1/4″ = 1′-0″
Foundation plan	1/4″ = 1′-0″
Elevations	1/4″ = 1′-0″
Construction details	3/4 to 1 1/2″ = 1′-0″
Wall sections	3/4 to 1 1/2″ = 1′-0″
Cabinet details	3/8 to 1/2″ = 1′-0″
Site plan	1″ = 20′ or 40′

ISO Metric Scales — millimeters	
Floor plan	1:50
Foundation plan	1:50
Elevations	1:50
Construction details	1:20 and 1:10
Wall sections	1:20 and 1:10
Cabinet details	1:50 and 1:25
Site plan	1:100

Larger scales are used if they will increase the clarity of the drawing.

Architectural Symbols

Since working drawings are made to a very small scale, it is not possible to indicate many parts of the house by actually drawing them as they appear to the eye. It is also not economical, considering the time it would take. Many parts of the house are, therefore, represented by standardized symbols.

Also, it is difficult to represent materials on a drawing, and symbols have been developed to indicate these. Figure 6-1 presents the more frequently used symbols. For a complete listing, consult architectural standards books.

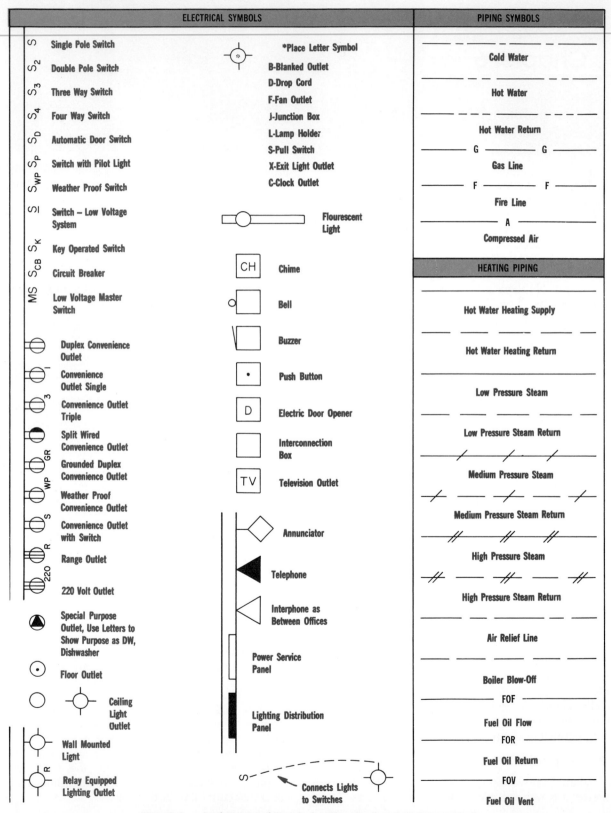

Fig. 6-1.   Architectural Symbols. Electrical and piping symbols.

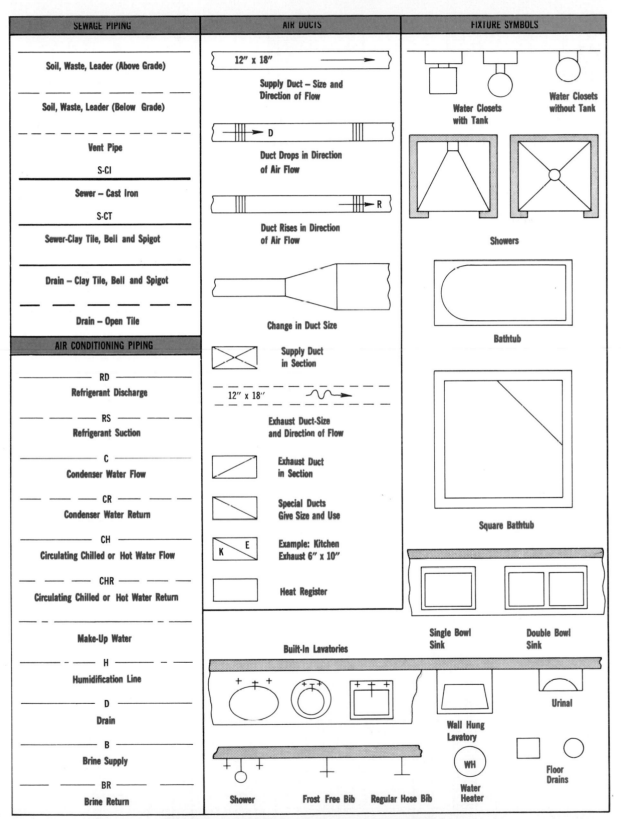

Fig. 6-1.   (cont.) Piping, air duct, and fixture symbols.

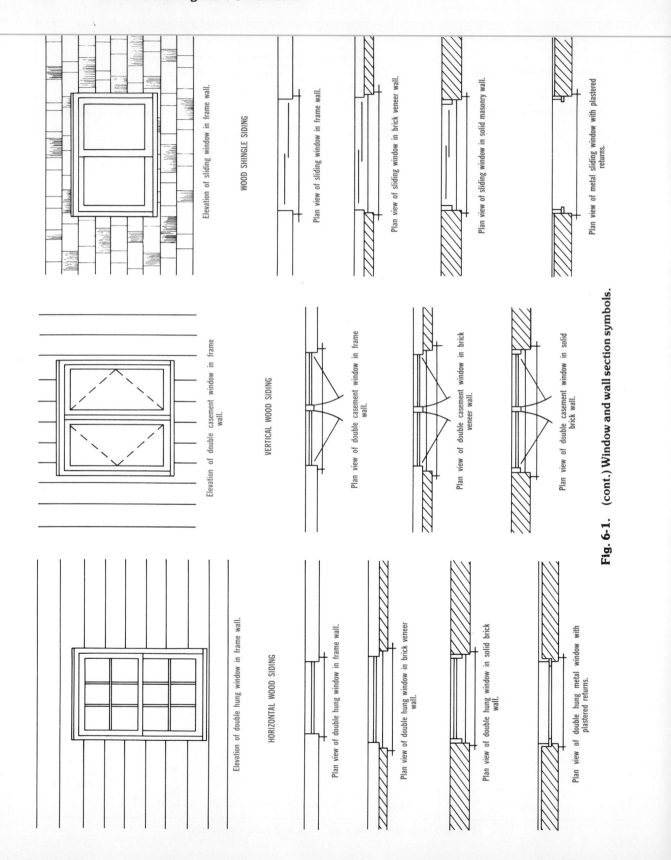

Elevation of sliding window in frame wall.

WOOD SHINGLE SIDING

Plan view of sliding window in frame wall.

Plan view of sliding window in brick veneer wall.

Plan view of sliding window in solid masonry wall.

Plan view of metal sliding window with plastered returns.

Elevation of double casement window in frame wall.

VERTICAL WOOD SIDING

Plan view of double casement window in frame wall.

Plan view of double casement window in brick veneer wall.

Plan view of double casement window in solid brick wall.

Elevation of double hung window in frame wall.

HORIZONTAL WOOD SIDING

Plan view of double hung window in frame wall.

Plan view of double hung window in brick veneer wall.

Plan view of double hung window in solid brick wall.

Plan view of double hung metal window with plastered returns.

Fig. 6-1. (cont.) Window and wall section symbols.

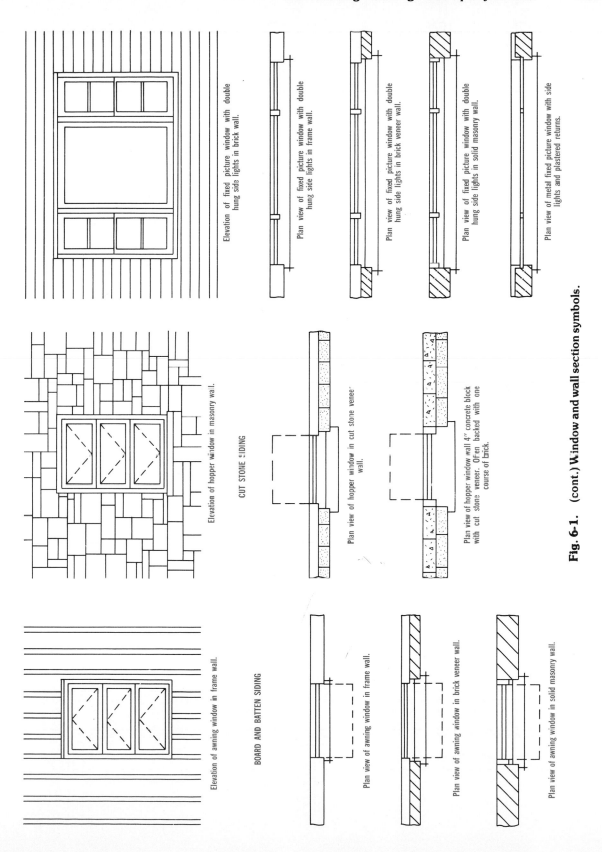

Elevation of fixed picture window with double hung side lights in brick wall.

Plan view of fixed picture window with double hung side lights in frame wall.

Plan view of fixed picture window with double hung side lights in brick veneer wall.

Plan view of fixed picture window with double hung side lights in solid masonry wall.

Plan view of metal fixed picture window with side lights and plastered returns.

CUT STONE SIDING

Elevation of hopper window in masonry wall.

Plan view of hopper window in cut stone veneer wall.

Plan view of hopper window wall 4″ concrete block with cut stone veneer. Often backed with one course of brick.

BOARD AND BATTEN SIDING

Elevation of awning window in frame wall.

Plan view of awning window in frame wall.

Plan view of awning window in brick veneer wall.

Plan view of awning window in solid masonry wall.

Fig. 6-1.   (cont.) Window and wall section symbols.

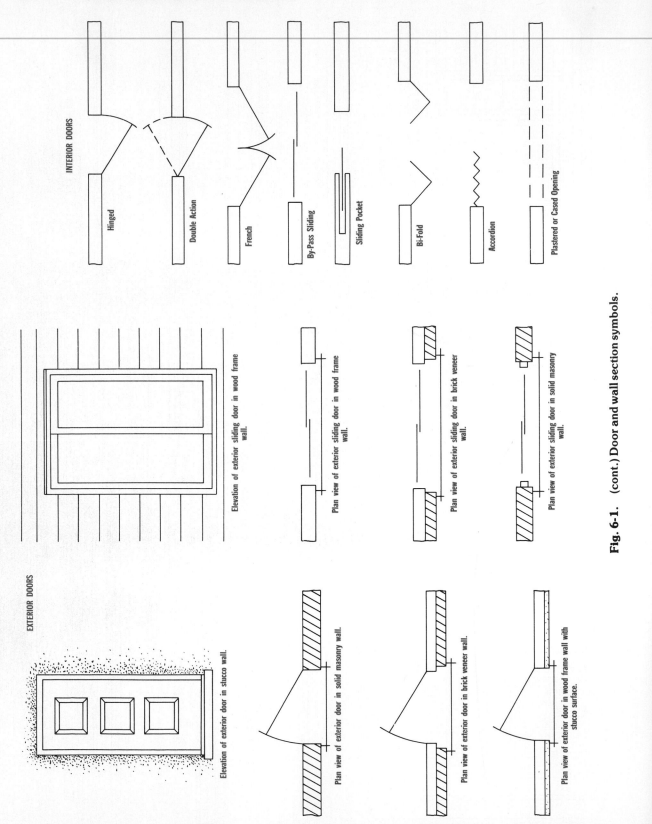

Fig. 6-1. (cont.) Door and wall section symbols.

ARCHITECTURAL SYMBOLS

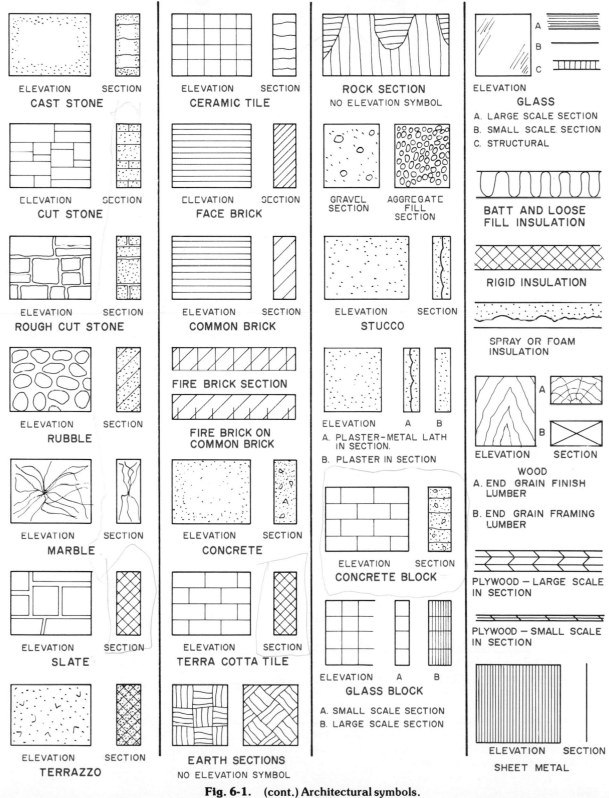

Fig. 6-1.   (cont.) Architectural symbols.

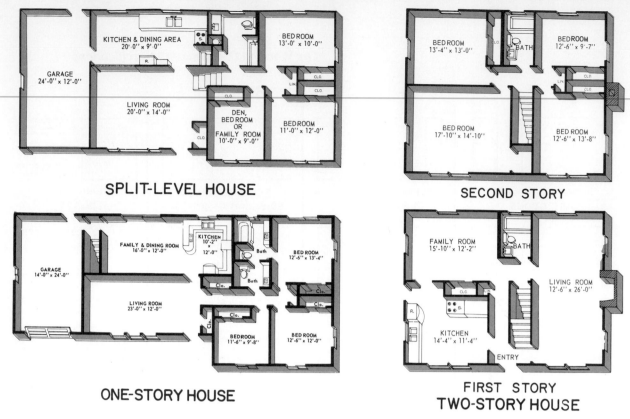

SPLIT-LEVEL HOUSE

SECOND STORY

ONE-STORY HOUSE

FIRST STORY
TWO-STORY HOUSE

Fig. 6-2. A floor plan is drawn as though the building has been cut through the windows and you are looking down on it.

Drawing the Site Plan

The two main purposes of a **site plan** are to locate the house on the lot and to show the slope of the land. A site plan is a protection for the owner because the builder must observe everything the plan specifies. Site plans are shown in Chapters 4 and 16.

Examine the site plan in Fig. 4-29. The house is located on the lot and dimensioned from two sides. The sidewalk and driveway are also located. Both are dimensioned in feet and inches or in millimeters. The size of the lot is lettered, parallel to each side, in decimal feet or in meters. The property line is drawn with long-dash - short-dash lines. The house outline and property lines are thick, and all other lines are thin.

The elevations of each corner of the lot and each corner of the house (ground level) are given, as is the floor elevation. **Contour lines** are located by the surveyor. These connect points of the same height above sea level. Contour lines showing the original slope of the land are drawn with dashed lines, while the new contours wanted after the house is built are shown with solid lines. Lines are broken to insert elevation figures.

Utilities are located using the proper symbols. Trees that are to remain are identified by space and trunk diameter.

Drawing the Floor Plan

Since the entire set of working drawings is based upon what develops as a floor plan, it is drawn first. The floor plan represents a top view of a house that is cut about halfway between the floor and ceiling. When the top half is removed, the view that remains represents the floor plan. It reveals the location of windows, stairs, cabinets, the fireplace, and many other items. See Fig. 6-2.

These items are located and their sizes are indicated by dimensioning this floor plan. Other details are shown also, such as the locations of electrical outlets, lights, bathroom fixtures, and the stove. Notes indicate the direction of floor joists and, sometimes, the type of material on the floor. A typical set of plans is in Fig. 6-29, pp. 169-178.

One-Story House

The preliminary layout should be drawn very lightly. Remember to allow room on the paper for such things as porches, patios, and dimensions. Following are the steps to draw the floor plan. See Fig. 6-3.

Fig. 6-3. Procedure for drawing floor plans.

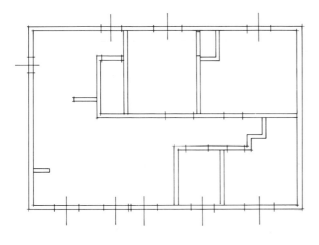

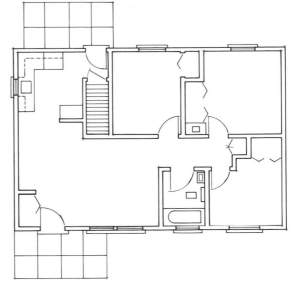

1. Using construction lines, lay out the width and length and the partitions.
2. Locate doors and windows. Be certain that locations give a pleasing appearance on the elevations.
3. Draw window and door symbols.
4. Draw interior details, such as fireplace, stairs, bath fixtures, and kitchen fixtures.
5. Draw porches and patios.
6. Darken all lines.

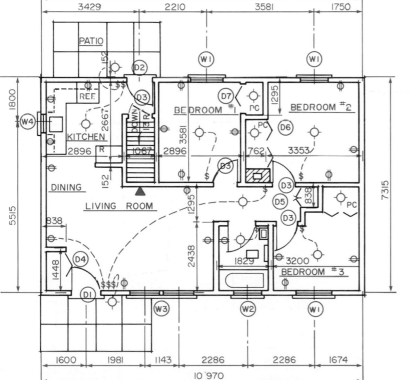

7. Indicate material symbols on the exterior walls.
8. Locate electrical fixtures and switches.
9. Dimension interior of house.
10. Dimension exterior of house. Locate doors, windows, and porches first. Indicate overall dimensions last.
11. Code doors and windows on plan, and complete door and window schedules in a clear corner of the paper. See Figs. 6-25 (p. 161) and 6-29 (pp. 169-178) for examples of schedules.
12. Label items needing identification, such as range and refrigerator, and add notes.

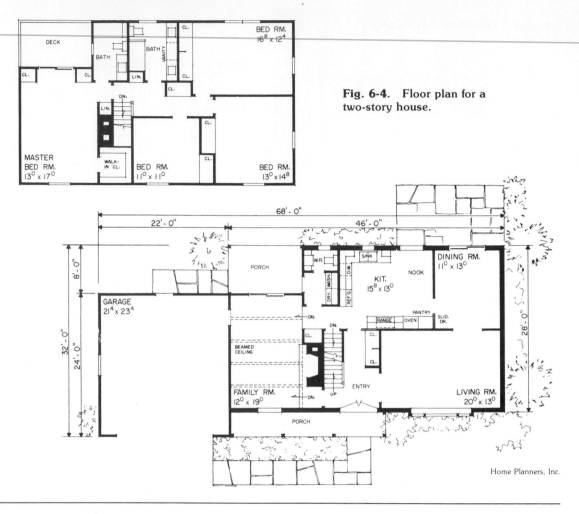

Fig. 6-4. Floor plan for a two-story house.

Home Planners, Inc.

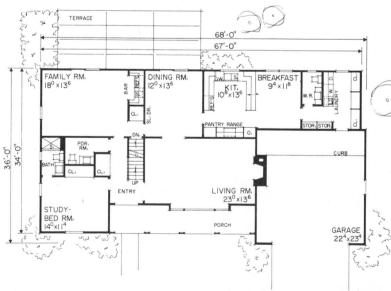

Fig. 6-5. Floor plan for a one-and-one-half-story house.

Home Planners, Inc.

Two-Story House and One-and-One-Half-Story House

The same procedure is used to draw floor plans for a two-story or one-and-one-half-story house as that used for a one-story house, except a floor plan for the second floor is necessary. This is the only additional drawing required. See Figs. 6-4 and 6-5.

The second floor can be developed by tracing the outline from the first floor plan. Stairs, bearing walls, and other features can also be traced.

Split-Level House

A split-level house is a little more difficult to draw than the house types just discussed. Usually, the rooms on levels four and three are placed together on one floor plan, and levels two and one form the basement plan. See Fig. 6-6. The actual layout and drawing are the same as for a one-story house. See Fig. 6-7.

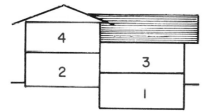

Fig. 6-6. Split-level house.

Drawing the Foundation or Basement Plan

A **basement plan** shows the basement layout and the footings. See Fig. 6-8. When no basement is used, the drawing shows the footings and foundation wall and is called a **foundation plan.** See Fig. 6-9.

The same procedure for drawing the floor plan can be followed for the foundation or basement plan. Much time can be saved by laying a piece of

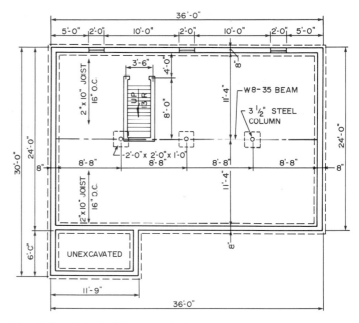

Fig. 6-8. A typical basement plan.

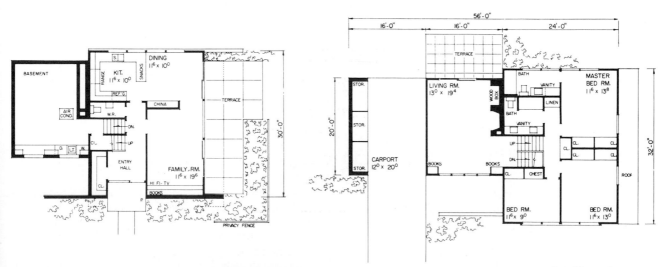

Home Planners, Inc.

Fig. 6-7. Floor plan for a split-level house.

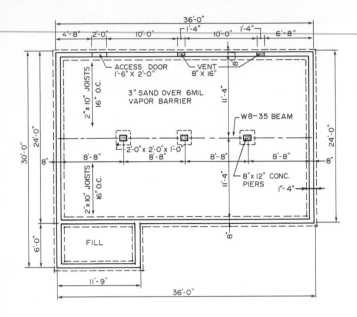

Fig. 6-9. A typical foundation plan for a house with a crawl space.

vellum over the floor plan and lightly tracing the exterior walls of the house and locating such items as fireplaces and stairs.

The basement plan includes windows and doors, columns and girders to support the house, a furnace, and footings. Perhaps laundry facilities, a recreation room, or a garage will be included. All these and other items are located and dimensioned in the same manner as the floor plan. The direction of floor joists is shown with a symbol and note.

Drawing the Elevations

An **elevation** is a view looking directly at a house. Four elevations are required to describe satisfactorily the exterior of the average house. These are views from the front, rear, right, and left sides.

The elevations, more than any other single factor, illustrate the taste of the owner and architect. They can be seen and understood by everyone. The designer must constantly keep in mind the style of house being developed, and apply the principles of design, balance, proportion, and color.

Many different elevations can be made from the same floor plan. Certainly, the designer should try many versions before the final decision is reached.

Suggested Procedure for Drawing Elevations

The preliminary layout must be drawn lightly because changes are usually necessary as the elevation is developed.

It is quite possible that some preliminary work on the elevations has been finished at this point. Generally, some of the elevations must be completed, at least partially, as the floor plan is developed, so the influence of such things as ells and window locations can be seen.

The elevation is laid out with dimensions from the floor plan and wall section. See Fig. 6-10. The floor plan is used to locate the doors, windows, porches, and other exterior details. The wall section gives the location of the roof ridge, ceiling line, window height, finished floor, finished grade, and top of the footing.

Suggested steps for drawing the elevation follow. See again Fig. 6-10.

1. Lay out the basic width and height of the house, using measurements from the floor plan and the wall section. Locate windows and doors with very light construction lines. The tops of the windows line up with the tops of the interior doors.
2. Draw roof details, overhang, and pitch.
3. Draw the grade line, foundation, and footings with a thin but dark line. Darken the outline using a thick dark line.
4. Draw the windows, doors, porches, and other exterior details using a thin, dark line.
5. Add material symbols using a thin line that is slightly lighter than that used for windows.
6. Letter any needed identification or notes. Common notes include identifying roof pitch; finished ceiling, floor, roof, and siding material; and the title and scale of each elevation.

When drawing material symbols, do not attempt to draw every brick or shingle. The symbol indicates the material. A common practice is to finish the front elevation entirely and record material symbols only partially on the other elevations. Do not try to get every line to meet at the corners, either. Architectural drafters tend to cross the lines in the corners.

Drawing the Sections

Wall construction is detailed by drawing a typical **wall section.** This is based on earlier sketches and on decisions that are made as the house is being planned. The scale used is usually $3/4'' = 1'-0''$ but can be as large as $1 \ 1/2'' = 1'-0''$. Metric scales are usually 1:20 and 1:10. All details of con-

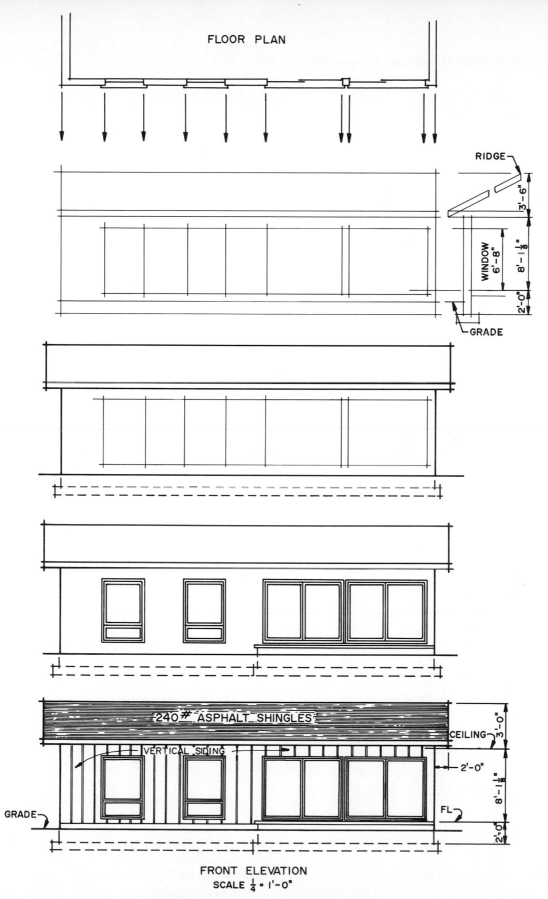

FLOOR PLAN

RIDGE

3'-6"

WINDOW 6'-8"

8'-1⅛"

2'-0"

GRADE

240 # ASPHALT SHINGLES

CEILING 3'-0"

VERTICAL SIDING

2'-0"

8'-1⅛"

GRADE

FL

2'-0"

FRONT ELEVATION
SCALE ¼ = 1'-0"

Fig. 6-10. Procedure for drawing elevations.

struction are shown, including the materials used and their sizes. Examples of typical wall sections are shown in Figs. 6-11 and 6-12.

If a house has some unusual feature, the designer may make longitudinal and transverse sections. A **longitudinal section** runs through the length of the building. See **A** in Fig. 6-13. A **transverse section** is through the entire building, running the narrow width. See **B** in Fig. 6-13. These sections are commonly used in commercial buildings. Longitudinal and transverse sections are usually drawn to the same scale as the floor plan. Since the scale is small, it is not possible to show all the material symbols and connection details. These are shown on the detail drawings which are made

to a larger scale. Longitudinal and transverse sections do not replace typical wall sections or details. They do help people using the plans to have a better understanding of the entire structure.

Longitudinal and transverse sections are usually drawn with no portions broken out. The entire length or width of the building is drawn.

How successfully these may be used depends upon where the sections are made in the building. The designer needs to study the building and see where the sections will best help relate the various floor and ceiling levels. The location of stairs must also be considered since stairs are part of the system of moving people from one level to the other.

Typical longitudinal and transverse sections for residential buildings are shown in Fig. 6-13. Detailed sections for commercial buildings are shown in Chapter 16, Working Drawings and Special Design Considerations.

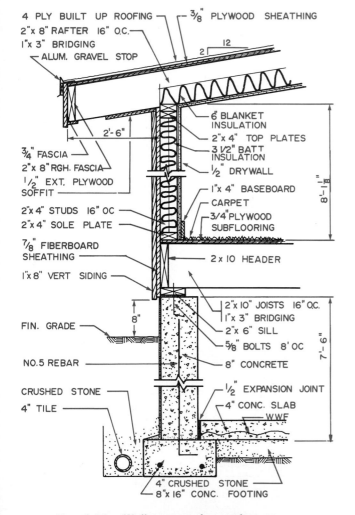

Fig. 6-11.   Wall section of a residence.

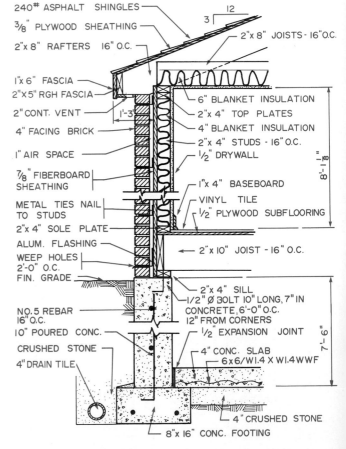

Fig. 6-12.   Typical wall section.

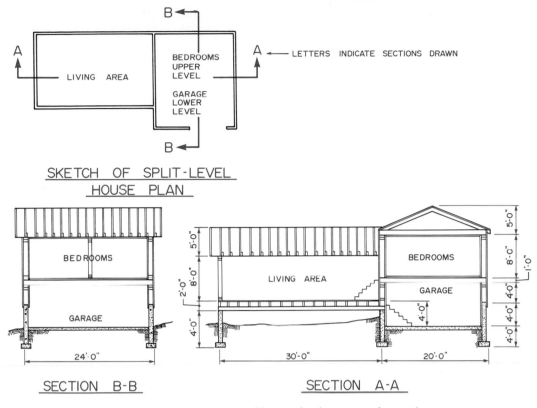

SKETCH OF SPLIT-LEVEL
HOUSE PLAN

SECTION B-B

SECTION A-A

Fig. 6-13. Typical transverse and longitudinal sections of a residence.

Drawing the Details

Many parts of a house cannot be shown clearly on the floor plan, foundation plan, and elevations because of the small scale at which they are drawn. Anything needing clarification is drawn as a separate drawing at a greatly enlarged scale. These special drawings are called **details.**

The items most frequently needing detailing are stairs, fireplaces, and cabinets and other built-ins. Also, some parts of the structure may be assembled in an unconventional manner and need special drawings. For example, details of cornice construction seem to be needed frequently. Typical construction details for residential buildings can be found in Chapter 7, Methods of House Construction. Details for commercial building design can be found in Chapter 16.

Stair Details

A simple, straight stairway can be indicated on the floor plan and dimensioned with a note. See Fig. 6-14. However, many designers prefer to draw a simple section through the stair. See Fig. 6-15. This clarifies most points and definitely checks to

see if sufficient room is available to accommodate the stair. Headroom can also be checked with such a detail.

If a stair is open and has ornate balusters and newel posts, a more elaborate detail is necessary. This is especially true if the stair is to be a custom job. (A **custom job** is designed and built especially for one house.) If the stair is a manufactured stock stair, a detail is unnecessary.

The dimensions required on the typical stair detail are the rise, run, tread and riser sizes, and headroom clearance. If balusters and newel posts are used, these sizes must also be indicated. A detailed discussion of stairs is in Chapter 2.

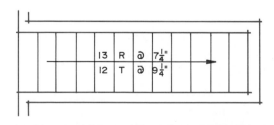

Fig. 6-14. Stair data can be recorded on a plan with a note.

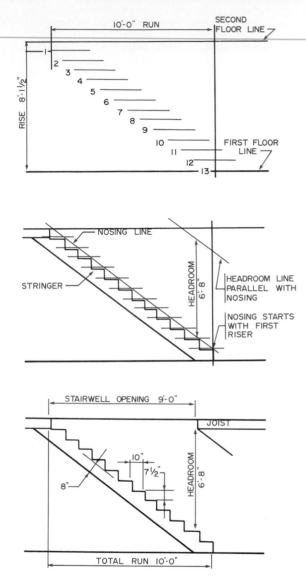

Fig. 6-15. Suggested procedure for drawing a stair detail.

Fireplace Details

A fireplace is only as good as its design. If improperly built or proportioned, it won't draw properly and will produce smoke inside the house. Detailing the fireplace assures that it will be built to proper dimensions, since this gives the mason a guide to follow.

Three views are needed in most cases. A **horizontal section** — showing the hearth, opening size, and flue linings — is indicated on the floor plan. Under most conditions, it can be dimensioned there and need not be redrawn as a separate detail.

A **vertical section** is necessary. This is drawn as a special detail on one of the sheets. Usually, it is tied in with the third fireplace detail, a **front elevation.**

A discussion of conventional fireplace design and construction is in Chapter 12. How to draw and dimension the required front elevation and horizontal and vertical sections is also shown.

There is increased use of factory manufactured fireplaces. These are indicated on the floor plan in the usual manner. It is not necessary to draw sections since the unit is simply set in place as the house is built. The only detail needed is an elevation to show the design of the mantel and wall area surrounding the unit. The brand wanted and any identifying numbers can be lettered on the floor plan or elevation and can be recorded in the specifications.

Proven specifications for fireplace design can be found in standards publications, such as **Architectural Graphic Standards** or **Time Saver Standards.**

A good detail will indicate the size of the fireplace opening, the depth, the splay, and the required masonry thicknesses. Also, it is wise to specify the flue size desired. On the elevation detail, materials are indicated by symbols and notes.

Cabinet Details

The various kitchen cabinets and built-in appliances are indicated on the floor plan. This shows the base cabinets, bars, and major appliances. Since the wall-hung cabinets were cut away with the upper half of the house, they are shown with dotted lines. See the kitchen on the working drawings in Fig. 6-29, pp. 169-178.

It is necessary to show an elevation of all walls with cabinets and built-in appliances. The cabinet heights and the spaces for appliances are dimensioned. In some cases, it helps if the length of a cabinet unit is dimensioned. See Fig. 6-16. Special features, such as built-in lighting or ventilating fans, are noted.

If the cabinets are to be custom built, a typical section showing construction details is necessary. See Fig. 6-17. The exact length does not need to be known because the cabinetmaker will measure the finished room and build the cabinets to fit. If the cabinets are ordered from a factory, the kitchen can be planned around the units available from the manufacturer's catalogue. The specifications may state the brand and style of cabinet. The elevations show the number and size required. These may be

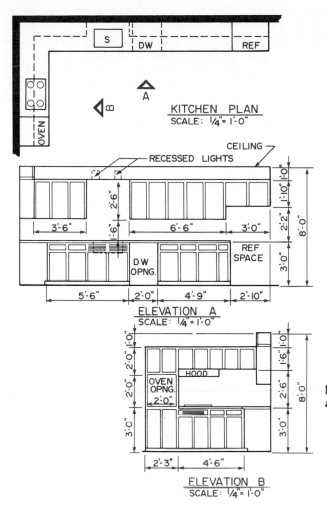

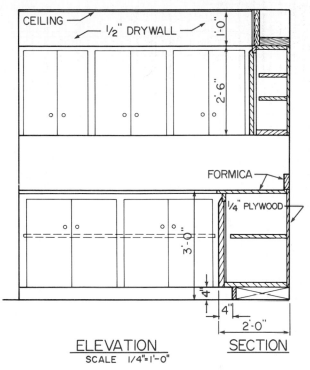

Fig. 6-17. Typical detail for custom-built shelving and storage unit.

Fig. 6-16. Typical kitchen cabinet details.

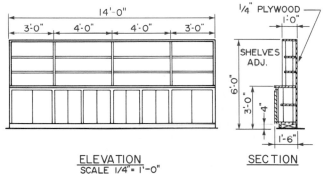

Fig. 6-18. Typical cabinet elevation with a section.

identified on the elevation with a manufacturer's model number.

Other cabinets, such as bathroom, den, or storage walls, are treated in the same manner as discussed for kitchen cabinets. If they are factory manufactured, the name and unit identification number are used. If they are custom built, dimensioned elevations and construction sections are needed. See Fig. 6-18.

Drawing the Roof-Framing Plan

A typical roof plan is in Fig. 6-19. Such a plan shows the location of each part of the roof-framing system. Usually, such plans are not drawn unless the roof has some unusual or complex feature. The scale $1/4'' = 1'-0''$ is commonly used.

Dimensioning a Working Drawing

Since the working drawings for a house are drawn to a very small scale, the actual sizes of the building and its many parts can be indicated accurately only by recording the actual size of the drawing. This process is called **dimensioning**.

A good way to study dimensioning is to examine carefully drawings made by trained architectural drafting technicians. You will notice that there are some slight differences in the dimensioning techniques, all of which are acceptable. When dimensioning, the primary consideration should be to

157

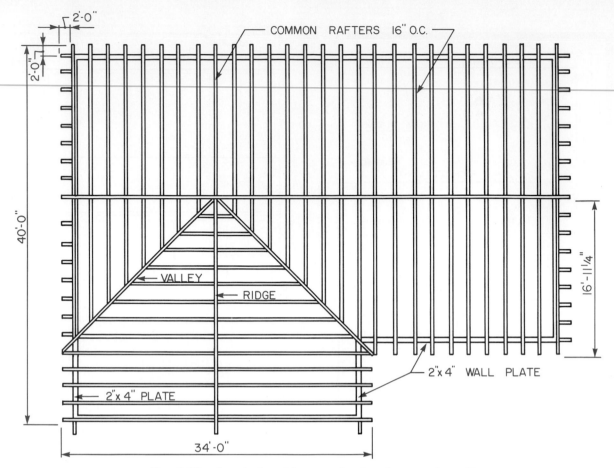

Fig. 6-19. A typical roof-framing plan for a house with an ell.

show the information in a way most helpful to those building the house. A knowledge of how a house is built helps the drafting technician decide where and how to place dimensions. Each drawing should be completely dimensioned. If it will help those using the drawings, dimensions can be repeated. At no time should a builder have to add or subtract to find the size of a portion of a house.

The following pages describe dimensioning practices that are in general use.

General Dimensioning Techniques

Following is a list of general dimensioning techniques. These are illustrated in Fig. 6-20. The numbers in color correspond to numbers on this list. Metric dimensioning information is presented later in this chapter.

1. Everything is dimensioned the actual size, even though it is drawn to scale.
2. Dimensions should not be crowded. Keep dimensions at least 1 to 1½ inches from the drawing and ½ to ¾ inches from each other.
3. An unbroken dimension line should be used; the dimension should be placed above it.
4. Distances over 12 inches are recorded as feet and inches. If the distance is an even number of feet, record the inches with a zero, i. e., 12′-0″.
5. Distances less than one foot are given in inches, i. e., 6″ not 0′-6″.
6. Dimensions are placed on a drawing so they can be read from the bottom or right side of the drawing.
7. The scale should be indicated on each drawing or detail, even if items are on one page and are drawn to the same scale.
8. Anything unusual should be explained with a note. Notes are lettered horizontally so they can be read from the bottom of the page. The letters are usually 1/8 inch high.
9. Dimension lines are drawn thinner than object lines used to draw the views.
10. Extension lines should not cross dimension lines.
11. Extension lines may cross each other.

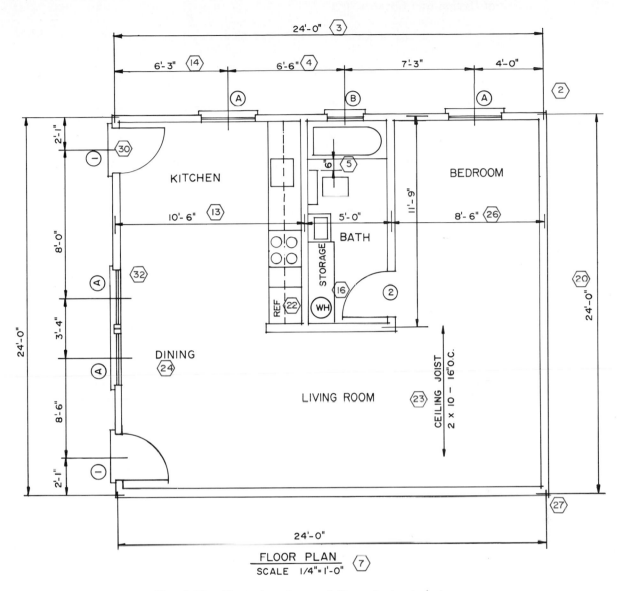

Fig. 6-20. Examples of general dimensioning techniques.

12. Do not dimension to hidden lines or object lines unless there is no other way.
13. Whenever possible, line up a series of dimensions.
14. Place the shortest dimensions near the object. The overall dimension should be on the outside.
15. When a space to be dimensioned is too small to receive the figures, place them outside of it.
16. Architectural abbreviations help to clarify items on a drawing.
17. Several types of symbols are used to indicate the end of a dimension line. Most frequently the arrowhead is used. The dot is used for tight

places where there is no room for an arrowhead. See Fig. 6-21.

Dimensioning the Floor Plan

Some of the following techniques are illustrated in Fig. 6-20.

18. Dimensions are placed on all sides of the plan.
19. All ells or projections are dimensioned.
20. Overall dimensions are placed on all sides of the drawing. The builder should not have to add or use a scale to find a dimension.
21. Intermediate dimensions should be added to see that their total equals the overall dimension.

159

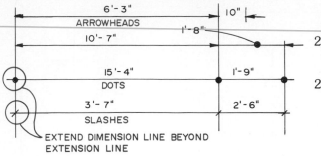

Fig. 6-21. **These symbols all indicate the end of a dimension line. The arrowhead is used most often.**

22. Identify fixtures and built-in items which cannot be clearly drawn, such as a clothes chute or dishwasher.
23. The size and direction of ceiling joists can be shown by a special symbol.
24. Rooms are identified by name, such as living room. The bedrooms should be numbered to assist the builder when decisions are made concerning the color of paint in the room and other related decisions. For example, the builder would record: bedroom #1 — walls

light blue, oak floors, light fixture No. 1324 Westinghouse.
25. The size and depth of the fireplace opening are commonly shown. Hearth material is indicated.
26. The width and length of each room is given. Interior walls are located by dimensioning to their centerline. (**O.C.** is often used to indicate a centerline measurement. It stands for "on center.") An alternate way is to dimension to the face of the interior stud. The size of the stud must be given. See Fig. 6-22.
27. When dimensioning an exterior frame wall, the dimension extends over the stud. It does not include the sheathing or siding. When the sheathing is flush with the foundation, the stud is set in the thickness of the sheathing. When the sheathing overhangs the foundation, the stud is flush with the edge of the foundation. Again see Fig. 6-22.
28. When dimensioning an exterior masonry wall, the dimension extends over the masonry unit. The masonry unit is set flush with the foundation. See Fig. 6-23.
29. When dimensioning an exterior masonry veneer wall, 6 inches is allowed beyond the face of the stud. The dimension is to the face

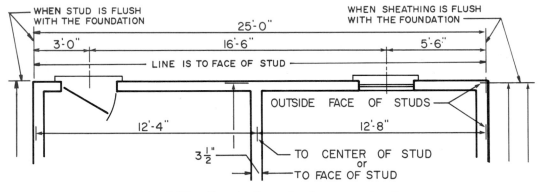

Fig. 6-22. **How to dimension frame construction.**

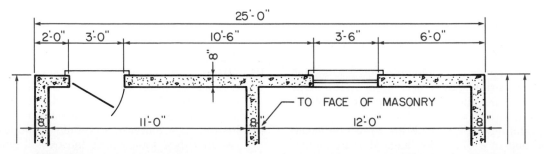

Fig. 6-23. **How to dimension solid masonry construction (poured concrete, concrete block, brick, stone).**

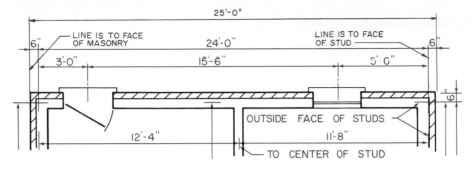

Fig. 6-24. How to dimension masonry veneer construction. Notice that the dimension is from the center of the inside stud to the face of the outside stud.

of the stud with the 6 inches also noted. The 6-inch allowance is for the masonry unit, sheathing, and air space. See Fig. 6-24.

30. Doors and windows in frame walls are dimensioned to their centerlines. These dimensions are located off the floor plan. Look again at Figs. 6-20, 6-22, and 6-24.

31. Doors and windows in solid masonry walls are usually dimensioned to the sides of the opening. These dimensions are located off the floor plan. See again Fig. 6-23.

32. Door and window sizes are coded on the floor plan, and detailed information is given in door and window schedules. These codes or symbols are usually placed inside a circle.

33. A stair has the number of risers shown, and Up or Down is indicated. Actual tread and riser dimensions are shown on the stair detail. Sometimes they are shown on the floor plan. Refer back to Fig. 6-8.

Schedules

Door and window data and interior finish identification are shown with **schedules**. Doors and windows are identified by the code previously recorded on the floor plan. All units that are the same are given the same code. Several code systems are commonly found. One uses numbers for doors and letters for windows. The letters and numbers are inside a circle. This clearly sets them apart from the dimensions. Another system uses a **W** before a number for windows and a **D** before a number for doors. See Fig. 6-25.

Dimensioning the Foundation and Basement Plan

1. All techniques mentioned for the floor plan apply to the foundation or basement plan. See Fig. 6-9.
2. Beams are located by their centerlines.
3. Piers are located by dimensioning to their centerlines.
4. If an area is filled, this should be noted.

5. The size and direction of floor joists should be noted.

Dimensioning the Elevations

1. The ridge, ceiling, floor and top of footing are dimensioned as shown in Fig. 6-10.
2. Roof slope is indicated by a symbol. The example in the elevation of the sample set of drawings is $4\ \overline{}^{12}$. This indicates 4 inches of rise in the roof for every 12 inches of run.
3. Exterior materials are noted as brick, asphalt shingles, or board and batten siding.
4. The overhang of the roof at the eaves is frequently dimensioned.

DOOR SCHEDULE						
MARK	NUMBER	SIZE	ROUGH OPENING	MATERIAL	TYPE	REMARKS
D1	3	3'-0" X 7'-0"	3'-4" X 7'-2"	BIRCH	FLUSH	SOLID CORE
D2	8	2'-0" X 6'-8"	3'-0" X 6'-10"	BIRCH	FLUSH	HOLLOW CORE

WINDOW SCHEDULE						
MARK	NUMBER	UNIT SIZE	ROUGH OPENING	MATERIAL	TYPE	REMARKS
W1	6	4'-0$\frac{1}{2}$" X 4'-1$\frac{1}{2}$"	4'-4$\frac{1}{2}$" X 4'-3$\frac{1}{2}$"	PINE	DOUBLE HUNG	

INTERIOR FINISH						
ROOM	FLOOR	BASE	WALLS	CEILING	TRIM	REMARKS
LIV. RM.	NYLON CARPET W/FOAM PAD	NONE	YELLOW LATEX ON GYPSUM BOARD	YELLOW LATEX ON GYPSUM BOARD	PINE BIRCH STAIN CLEAR LACQUER	

Fig. 6-25. Detailed information about doors, windows, and interior finishes is given in schedules. Complete schedules are included in Fig. 6-29.

5. The finished grade is indicated with a note.
6. Window sizes are sometimes lettered on top of the window elevation, but door and window schedules are preferred.

Dimensioning the Details

1. Cabinets and other built-ins require only overall dimensions. Notes indicating other information, such as material and finish, are recorded in the specifications. If cabinets are stock units, the manufacturer's name and stock number can be recorded on the cabinet elevation.
2. Fireplace details should be completely dimensioned. Face material and hearth material should be indicated with a note.
3. Stair details should have tread and riser size indicated. Total stair rise and run should be noted.
4. Other built-ins are treated the same as the kitchen details. Anything unusual, such as a special planter, should be detailed to a larger scale and completely dimensioned. This would include dimensions on joints and other construction details.

Dimensioning Wall Sections

The **wall section** shows the construction to be used for the wall of the house. On it are recorded the sizes and kinds of materials to be used. In addition, the footing and foundation are dimensioned. Other details, such as ceiling height and location of grade, are shown. Every detail of construction should be identified. Typical dimensioned wall sections for frame and masonry veneer walls are shown in Figs. 6-11 and 6-12.

Metric Dimensioning

The metric system and scales are explained in Chapter 5. Review this before attempting to draw and dimension metric plans.

The two measures used in drawing plans are the meter (m) and the millimeter (mm). **The centimeter is not used in building design applications.** The site plan is dimensioned in meters to three decimal places. This permits easy conversion to millimeters if it is needed. For example, 2.134 meters equal 2134 millimeters. Notice that it is converted by simply moving the decimal point three places to the right. Likewise, millimeters can be converted to meters by moving the decimal point three places to the left. For example, 975 millimeters equals 0.975 meters.

The floor plan, the foundation plan, and small detail drawings are dimensioned using millimeters. This eliminates the need for the use of a decimal point. See Figs. 6-26 and 6-27.

Following are some metric dimensioning techniques to be observed.

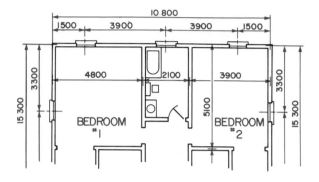

Fig. 6-26. A floor plan for a building designed using metric measure.

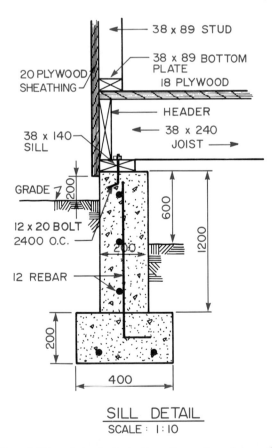

Fig. 6-27. A typical section dimensioned in millimeters.

1. Measurements in meters are always taken to three decimal places — 3.145.
2. Measurements in millimeters are always given in whole numbers — 870.
3. Since the difference between meters (m) and millimeters (mm) is so great and meters are always given to three decimal places, it is not necessary to label them **m** or **mm** on a drawing. To aid in reading the drawing, a note such as "all dimensions shown in millimeters" may be used.
4. In numbers containing many digits, group the numbers in threes from the decimal point and leave a space between the sets of three — 4 150 692.260 55. The space replaces the comma used in the U.S. Customary system. When a number has four digits before the decimal, a space is **not** used — 4500.
5. Common fractions are never used. Fractional parts are expressed in decimals. For example, 1/2 kg is wrong. It should be 0.5 kg.
6. If the number is smaller than 1.0, put a 0 before the decimal point — 0.457 mm.
7. Leave a space between the numerical value and the metric symbol — 100 mm.
8. When using non-SI expressions of plane angles — degrees (°), minutes ('), and seconds (") — do not leave a space between the numerical value and the symbol. Express them as follows: 60°25'.
9. Leave a space on each side of signs for multiplication, division, addition, and subtraction — 100 mm × 500 mm.
10. When writing sizes in sequence, such as length × width × height, the metric symbol is given only after the last number. For example, 975 × 450 × 52 mm.
11. Sizes are generally recorded in the following order: length × width × height. If so recorded, no identification after each number is needed. If written in any other sequence, each number should be followed by an identifying symbol and word — 250 mm high × 100 mm wide.
12. When the size of an item is shown, all dimensions should be of the same unit. For example, 2.100 m × 100 mm is incorrect. The item should be dimensioned 2100 mm × 100 mm.
13. Write out metric ton. The symbol **t** for metric ton can be confused with the symbol for customary ton.
14. For unit symbols such as squared or cubed, show the superscript immediately after the symbol — m^2 (square meter) and m^3 (cubic meter).
15. If possible, all the dimensions on a drawing should be the same unit. For example, all dimensions on a site plan will generally be meters, while on a detail drawing, they will all be millimeters.
16. Volume, such as tank capacity, is expressed in cubic meters (m^3).
17. Measurement of liquid volume is expressed in liters (L) and milliliters (mL).

Border and Title Block

The size of the border can vary. Generally, the top, bottom, and right side border are 1/2 to 3/4 inch (13 to 19 mm). The left border can be from 1 to 1 1/2 inches (25 to 38 mm). The left border is wider because the drawings are stapled together into a set on that side.

Title blocks also vary. A typical example is shown in Fig. 6-28.

Specifications

A house is made from many parts and many materials. The contractor purchases and assembles these into a finished house.

In order to prevent misunderstanding between the contractor and the owner, it is important to indicate on the working drawings what is to be built, how it is to be built, and what materials are to be used. Items that cannot be clearly shown on the drawings are described in writing in a document called the **specifications.**

Fig. 6-28. Title block.

The kinds of materials used and the quality of these materials must be specified. For example, finished floors could be made of oak, maple, ceramic tile, asphalt tile, or other materials. Flashing could be galvanized iron, aluminum, or copper. Water pipes could be galvanized iron, copper, or brass. The concrete mix desired should be indicated. These and many more material selections must be clearly specified.

Since the quality of materials is not uniform from product to product, it should be indicated. There are many grades of construction lumber. Roofing materials, hardware, and lighting fixtures vary in quality. The high-quality materials are more expensive, but the quality of materials directly influences the life of the house parts as well as the total price.

Another variable that cannot be shown on working drawings is the desired quality of the work. For example, after concrete is poured, it should be kept moist for seven days, to secure the most satisfactory curing job. This should be specified. It protects the owners against the possibility of shoddy work.

Certain parts of the construction job are the responsibility of the contractor, such as providing electricity for construction purposes or hiring labor and paying the employment taxes. These and other responsibilities should be clearly indicated.

If the owner is to assume certain responsibilities, these should be recorded. Frequently, the owner chooses to do interior and exterior painting. This becomes a part of the specifications.

All accessories to be supplied by the contractor should be specified. Do the windows have screens and storm sash? Are sidewalks and driveways included? What shrubbery, if any, is in the contract? Are the stove, refrigerator, and dishwasher included?

The architect occasionally will call for certain articles by their trade name. For example, perhaps a Westinghouse Range Platform Model PLB is required. This means that the bidders must supply this exact range.

While such a procedure guarantees the exact item the owner desires, it does tend to increase the cost, because it removes all other range manufacturers from bidding. To overcome this disadvantage, architects frequently specify (for example) Westinghouse Range Platform Model PLB **or equal.** This permits any other company with a range of equal quality to submit a bid. Such competitive bidding will lower the cost of the unit.

Most houses are built on a **bid basis.** This means that the owner or architect contacts several contractors, and each indicates what the charge will be for building the house. In order for a contractor to make a bid, every detail must be understood. This requires a complete set of working drawings, accompanied by detailed specifications.

If an item is not clearly recorded, disagreements can occur. Some such disputes lead to costly law suits. The contractor is not responsible for supplying any item not called for on the working drawings or in the specifications. Verbal promises are no stronger than the character of those making them.

Occasionally the working drawings will disagree with something recorded in the specifications. In these cases, the "General Conditions" of the specifications provide that the specifications will be held correct and have precedence over the working drawings.

When the owner decides to accept a bid from a contractor, a legal agreement binding on both parties is written and signed. This is called a **contract.** The working drawings and specifications become a part of this contract. The contractor, then, has agreed to build the house indicated by the drawings and specifications for a specific sum of money, and the owner has agreed to pay the contractor the sum indicated in the contract.

Writing the Specifications

The first portion of the specifications is called "General Conditions." These conditions record the relationships existing between the contractor and the owner. They also include items that are not part of any of the trades covered in other divisions of the specifications.

A very complete set of General Conditions has been published by the American Institute of Architects, and, usually, these are made a part of the General Conditions by referring to them with a statement such as "the latest edition of the 'General Conditions of the Contract' published by the American Institute of Architects shall be understood to be a part of this specification and shall be adhered to by this contractor." Doing this is equivalent to copying the General Conditions verbatim and attaching them to the specifications. The remainder of the specifications is arranged by trades, as nearly as possible, in the order in which the work is to be performed. An examination of a set of specifications will reveal that the first division is that involving excavating, filling, and grading. Of necessity, the excavation must come first, followed by concrete work in the footings and foundation, on through the order of work.

The exact form of the specifications varies somewhat from architect to architect, and the degree of detail also varies according to the architect's decisions. In general, the specifications for a residence will include the following divisions: General Conditions; Special Conditions; General Scope of Work; Excavating, Filling, and Grading; Concrete; Masonry; Carpentry and Millwork; Lathing and Plastering; Sheet Metal and Roofing; Glass and Glazing; Painting; Finish Hardware; Heating and Hot Water; Plumbing; and Electrical.

The following set of specifications is a simple set. These specifications were written for use with the working drawings in this Fig. 6-29, pp. 169-178.

General Conditions

The latest edition of the standard form of "General Conditions of the Contract" published by the American Institute of Architects shall be understood to be a part of this specification and shall be adhered to by this contractor.

Special Conditions

Sec. 1. Examination of Site It is understood that the contractor has examined the site and is familiar with all conditions which might affect the execution of this contract and has made provision therefor in his bid.

Sec. 2. Time of Completion. The work shall be completed within 90 calendar days after written Notice to Proceed is issued to the contractor.

Sec. 3. Existing Trees. Existing trees within 15 feet of the foundation line for the new structure shall be carefully protected by the general contractor from injury which might result from any operation connected with the execution of this contract.

Sec. 4. Cleaning. The general contractor shall periodically remove from the building and premises all rubbish and debris, and at the completion of the work shall leave the entire building and premises "broom clean," all glass washed clean and all wood and plaster work of walls and ceilings brushed clean and free from stains or discoloration.

Sec. 5. The acceptance of this contract carries with it a guarantee on the part of the general contractor to make good any defects in the work of the building arising or discovered within one year after completion and acceptance of same by the architects, whether from shrinkage, settlement or faults of labor or materials.

Sec. 6. House shall receive standard termite protection by reliable dealer.

General Scope of Work

Sec. 1. The work consists of the General Construction including mechanical and electrical work for a one-and-one-half-story, single-family residence as shown on the drawings.

Sec. 2. Responsibilities of Contractor. Except as otherwise specifically stated in the Contract, the contractor shall provide and pay for all materials, labor, tools, equipment, water, light, heat, power, transportation, superintendence, temporary construction of every nature, taxes legally collected because of the work, and all other services and facilities of every nature whatsoever necessary to execute the work to be done under this contract and deliver it complete in every respect within the specified time, all in accordance with the drawings and specifications.

Division 1: Excavating, Filling, and Grading

Sec. 1. Scope. This division includes excavating, removal of all obstructions, filling, grading, and related items required to complete this work as indicated on the drawings and/or specified.

Sec. 2. Finished Grades. The words **finished grades**, as used herein, mean the required final grade elevations indicated on the drawings.

Sec. 3. Excavation.
a. **Dimensions.** Excavate to elevations and dimensions indicated, plus ample space for construction operations and inspection of foundation.
b. **Drainage.** Keep excavation free from water. Do not conduct water from excavations onto private property.
c. **Footing trenches** may be excavated to the exact dimensions of the concrete, and side forms may be omitted if concrete is poured in clean-cut trenches without caving. Place footing and foundations upon undisturbed and firm bottoms. The contractor shall not pour foundation footings until the architect inspects and approves the bearing.

Sec. 4. Grading.
a. **Grading.** Do all cutting, backfilling, filling, and grading necessary to bring all areas within property lines to the following subgrade levels:
   (1) **For paving, walks and other paved areas** to the underside of the respective installation as fixed by the finished grades therefor.
   (2) **For lawns and planted areas** to four (4) inches below finished grades.
b. **Materials for backfill and fill.** All material used for backfill and fill shall be free from deleterious materials subject to termite attack, rot, decay or corrosion, frozen lumps, or objects which would prevent solid compaction.
   Materials for backfill and fill in various locations shall be as follows:
   (1) **For interior of building.** Use sand or an approved, properly graded mixture of sand and gravel. Foundry sand shall not be used.
   (2) **For exteriors, under paving.** Use excavated materials free from top soil, or other materials approved by the architect.
   (3) **For use under lawns and planted areas.** Use, after architect's approval, excavated materials with admixture of top soil or earth. Heavy clay shall not be used.
c. **Backfill against foundation walls** shall:
   Be deposited in six (6) inch layers, each to be solidly compacted by tamping and puddling.
d. **Subgrades for lawn and planted areas.** Slope the subgrade evenly to provide drainage away from building in all directions at a grade of at least 1/4 inch per foot.
e. **Settlement of fills.** Fill to required subgrade levels any areas where settlement occurs.

Division 2: Concrete

Sec. 1. Scope. This division includes all concrete and related items required to complete the work indicated on the drawings and/or specified.

Sec. 2. Materials.
a. Portland cement shall meet the requirements for Type I.
b. Aggregates.
   (1) **Fine aggregates** shall be natural sand, or sand prepared from inert materials having similar characteristics, if approved by the architect.
   (2) **Coarse aggregates** shall be crushed stone, ground clean and free from foreign matter. Size range for walls shall be designated, as "No. 4 to 3/4."
c. **Water.** Water used for concrete work shall be clean and free from injurious amounts of oils, acids, alkalies, organic, or other deleterious substances.
d. **Metal reinforcement.** Wire mesh reinforcing shall conform to A.S.T.M. Designation A82-34 and shall be free from excessive scale, rust, or coatings which will reduce bond to the concrete.

Sec. 3. Depositing Concrete.
a. Deposit concrete as nearly as practicable in its final position to avoid segregation due to rehandling or flowing.

b. Retempering. No concrete that has partially hardened or has been retempered shall be used.

c. Compaction. Concrete shall be thoroughly compacted by vibrating during placement.

Sec. 4. Curing. All concrete shall be covered with a polyethylene plastic, airtight cover for three days at 70° F or five days at 50° F.

Sec. 5. Cleaning. Clean all exposed concrete surfaces and all adjoining work which has been stained by the leakage of concrete.

Division 3: Masonry

Sec. 1. Scope. This division includes masonry and related items required to complete this work as indicated on the drawings and/or specified.

Sec. 2. Materials.

a. **Water** shall be clean and free from injurious amount of acids, alkalies, organic materials and other deleterious substances.

b. **Mortar** for masonry shall be one-to-two lime mortar.

c. **Brick** to be selected by owner.

Sec. 3. Installation.

a. **All masonry work** shall be laid plumb, true to line, and with level courses in common bond.

b. **Joints in masonry** shall be nominally 3/8-inch wide, shall be cut flush and as the mortar takes its initial set shall be tooled with a 1/2-inch diameter round tool 6 inches longer than the masonry unit.

Sec. 4. Cleaning. Upon completion, clean all masonry from top down with solution of non-staining soap powder and clean water using stiff fiber brushes or a solution of one part muriatic acid in ten parts water. Wet thoroughly before applying acid solution and protect other work. Rinse surfaces with clean water immediately after cleaning.

Division 4: Carpentry and Millwork

Sec. 1. Scope. This division includes the furnishing and installation of all carpentry, millwork, and related items required to complete this work as indicated on the drawings and/or hereinafter specified.

Sec. 2. Materials.

a. **Rough lumber** shall be Framing Lumber — No. 3 Grade Southern Pine.

b. **Exterior millwork.** Horizontal siding shall be 6″ bevel redwood.

c. **Sheathing** shall be 5/8″ aluminum-foil-backed foam as made by Dow Chemical.

d. **Attic insulation** shall be 12″ fiberglass batts installed in accordance with manufacturer's instructions.

e. **Interior trim** shall be of pine. Pattern selected by owner.

f. **Interior doors** shall be as indicated on door schedule.

g. **Exterior door frames** shall be 1 3/4″ thick, rabbeted with 1 1/8″ outside casings of white pine.

h. **Interior door frames** shall be 7/8″ thick pine.

i. **Double-hung sash** shall be Andersen or equal.

j. **Kitchen cabinets** shall be birch. Countertops Formica or equal.

k. **Closets and wardrobes** shall have 7/8″ by 12″ shelving and one clothes pole running length of space.

l. **Living room, hall, bedroom floors** to be straight line oak; select grade, finished natural; kitchen vinyl composition tile over plywood underlayment; bathroom, ceramic tile set in rubber into squares, cemented to plywood sub-floor; entrance hall imitation slate.

Division 5: Interior Walls

Sec. 1. Ceilings and walls to be 1/2″ dry wall, all joints taped and coated according to manufacturer's specifications.

Sec. 2. Stair walls to be 1/2″ dry wall.

Sec. 3. Baths to be ceramic tile 4′-0″ everywhere except at tub where it is ceiling height.

Division 6: Sheet Metal and Roofing

Sec. 1. Scope. This division includes furnishing and setting of all sheet metal, roofing and flashing and related items required to complete this work as indicated on the drawings and/or specified.

Sec. 2. Materials.

a. **All flashings, valleys,** and miscellaneous items of sheet metal work shall be 26-gauge galvanized iron extending 10 inches from valley or ridge.

b. **Asphalt shingles, 240-pound** installed according to manufacturer's specifications.

c. **Nails** for fastening shingles shall be zinc coated.

d. **Bituminous-saturated roofing felt** shall be 15-lb. felt.

Sec. 3. Application.

a. Roofing shall not be laid until all sheet metal, valleys, or work which extends under roofing material has been installed. Coordinate roofing work with flashing of intersections around vertical surfaces.

b. Roof and flashings shall be weathertight, free from leaks and other defects.

c. Cover roof surfaces with 15-lb. bituminous-saturated roofing felt before laying shingles. Double felt not less than 36 inches wide at valleys, hips and ridges. Extend up vertical surfaces 6 inches. Lap joints 3 inches. Nail sheathing 12 inches on center in both directions through metal discs.

d. Application of shingles shall be according to manufacturer's specifications.

Sec. 4. Guarantee. Roofing shall be weathertight and watertight. The roof shall be guaranteed by the roofing contractor against leaks due to defects in material and workmanship for a period of five years.

Division 7: Painting

Sec. 1. Scope. This division includes all the painting, both exterior and interior for the structure. The intent is to paint walls, ceilings, wood doors and bucks, and all other items painted in usual practice.

Sec. 2. Materials.

a. The term **paint** used herein includes emulsions, latex paint, enamels, oil paints, sealers, stains, varnishes, shellacs and any other allied coating.

b. **Latex base paint** will be "Ludens Smooth-Spread" or equal.

c. **Alkyd flat enamel** shall be "Ludens Alkyd Flat Enamel" or equal.

d. **Varnish** shall be "Durox-Seal" or equal.

e. **Exterior primer** shall be synthetic by "Ludens" or equal.

f. **Exterior paint** shall be "Ludens" exterior oil paint or equal.

g. **Stain** shall be by G. Dulac Co. or equal.

Sec. 3. Preparation of Surfaces. Surfaces to be painted shall be clean, dry, and free from dirt and frost. Cover knots and pitch streaks with orange shellac, aluminum paint or a resin sealer. Fill nail holes and minor imperfections with putty between first and second coats. All millwork and trim shall be back primed.

Sec. 4. Exterior Painting. Wood sash and doors shall be given one coat of exterior primer and two coats of exterior latex.

Sec. 5. Interior Painting.

a. **Wood trim** shall be stained to match birch doors.

b. **Wood doors and plywood ceiling** in living room shall be given two coats "Durox-Seal" or equal sanded between coats.

c. **All walls and ceilings** shall be given two coats latex base paint.

Division 8: Finish Hardware

Sec. 1. Scope. This division includes builder's Finish Hardware and related items required to complete the work specified and/or indicated on the drawings.

Sec. 2. Materials.

a. **Exterior swing doors** shall have 1 1/2 pair bronze butts 4 1/2″ × 4 1/2″.

b. **Interior swing doors** shall have 1 pair bronze butts 3 1/2″ × 3 1/2″.

c. **All knobs and spindles** shall be brass or bronze type NM-I by Lock-Tite Co. or equal.

d. **All exterior swing doors** shall have bronze threshold No. 861 by Chalmers Co. or equal.

e. **Cylinder locks and latches** shall be by Lock-Tite Co. or equal.

f. **Folding closet doors** shall be Simpson or equal.

Sec. 3. Installation. All hardware items listed here and others required to complete the work shall be installed in a skillful manner.

Division 9: Heating and Hot Water

Sec. 1. Scope. This division includes furnishing and installing space heating equipment and equipment for generating and storing hot water.

Sec. 2. Description of system — gas forced air duct system.

Sec. 3. Furnace should be Lennox forced air, 3-ton cooling, 120,000 BTU heating or equal. Installed with evaporative cooler, necessary duct work and thermostat. Will operate on natural gas.

Sec. 4. Installation — as recommended by manufacturer. Ducts to be installed according to furnace supplier recommendations.

Sec. 5. Tests and adjustments — furnace to be thoroughly tested and regulated.

Sec. 6. Hot water supply. Water heater to be 30 gallon unit, glass lined, with safety valve. To be electric. To be American Standard or equal.

Division 10: Plumbing

Sec. 1. Scope. This division includes the furnishing and installation of all plumbing and related items as required to complete this work as indicated on the drawings and/or hereinafter specified.

a. Lavatories, 20″ × 18″, one brown, one yellow, cast iron, acid resisting enamel, built-in stopper, with stainless steel rim, to be built into Formica top, trap below sink.

b. Water closet bowls — one yellow, one brown, vitreous china.

c. Seats for water closets — color of water closet, one brown, one yellow, closed front, with cover and self-sustaining hinge.

d. Bath tubs, one brown, one yellow, exact shade to be selected by owner; cast iron, acid resisting enamel, 5′-0″, stopper built-in, properly trapped.

e. Kitchen sink, white double compartment, cast iron with acid resisting enamel, swing type spout, trap below sink, 21″ × 32″ twin bowl.

f. Water pipe — copper.

g. Pipe cleanouts — cast iron with brass plugs or screwed fittings with brass plugs.

h. Waste and vent piping — galvanized wrought iron.

i. Soil pipe — cast iron with coated fittings.

j. Septic tank to be 1000 gallon, installed with sufficient drain field to insure proper operation. Drains set in gravel beds. To meet county health standards.

Sec. 2. Installation.

a. Drainage and vent piping shall be installed extending through roof not less than one foot with 4# lead flashing at least 24″ square. Piping shall be assembled and installed without undue strains or stresses and provisions shall be made for expansion and contraction.

b. Erect soil, waste, and vent stacks of sizes shown and extend above roof.

c. No change of direction in drainage piping shall be greater than 90 degrees.

d. Prohibited fittings: No double hub, or double tee branch shall be used on soil or waste lines. The drilling and tapping of building drains, soil, waste, or vent pipes and the use of saddle hubs or bands are prohibited.

e. Prohibited connections: No fixture or device shall be installed which will provide a backflow connection between a distribution system of water for drinking and domestic purposes and a drainage system, soil or waste pipe so as to permit or make possible the backflow of sewage or waste into water supply system.

f. All joints and connections shall be made gas- and water-tight. All exposed threads on ferrous pipe shall be given a coat of acid-resisting paint.

g. All caulked joints shall be firmly packed with oakum and shall be secured with caulking lead, not less than 1″ deep.

h. Cast iron pipe joints shall be caulked.

Sec. 3. Tests. Tests shall be made in accordance with City Authority or Utility company having jurisdiction.

Division 11: Electrical

Sec. 1. Scope. This division includes all interior electrical wiring, fixture installations, and related items required to complete the work indicated on the drawings and specified.

a. The work under this division shall commence at the point of attachment at the service entrance equipment.

b. Electrical service supplied to the structure will be 208 volts, 150-amp, 3-phase, 60 cycle, 3 wire.

c. Allowance of $1000 for selection of light fixtures by owner.

d. Post light to be electric.

Sec. 2. General Requirements.

a. Electrical system layouts indicated on drawings are diagrammatic and locations of outlets and equipment are approximate; exact routing of raceways, location of outlets and equipment shall be governed by structural conditions.

b. The right is reserved to make any reasonable change in location of outlets and equipment prior to roughing-in, without additional expense to the owner.

Sec. 3. Materials and Appliances.

a. Materials and appliances of the types for which there are Underwriter's Laboratories Standard requirements, listings, and labels, shall have listing of Underwriter's Laboratories and be so labeled, or shall conform to their requirements, in which case certified statements to the effect shall be furnished, if requested.

b. Materials other than those listed herein shall be the size, type, and capacity indicated by the drawings and the specifications. Insofar as possible use one type and quality.

c. Hood over surface unit to be Nutone 1600, 30 inch, or equal.

d. Surface unit and oven to be General Electric or equal.

e. Dishwasher to be Hotpoint H47 or equal.

Sec. 4. Final Inspection and Tests. Test system free from short circuits and grounds with insulation resistances, not less than outlined in Section 1119, 1985 National Electrical Code.

Build Your Vocabulary

Following are terms that you should understand and use as part of your working vocabulary. Write a brief explanation of each term.

details
elevations
floor plan
foundation plan
plot plan
scale

schedules
specifications
symbols
vellum
working drawings

Reading a Working Drawing

The following questions are designed to see how well you can read working drawings. Read the question, find the answer on the set of drawings in this chapter, and write the answer on a sheet of lined paper. Upon completion, your instructor will check these to see how well you have done.

Work rapidly. Remember that extra time spent looking for an answer on a set of working drawings is time and money wasted.

1. How many pairs of shutters are on the house?
2. How many lineal feet of closet space are in the upstairs bedrooms?
3. Where is the furnace located?
4. Where is the water heater located?
5. How many duplex outlets are in the house?
6. What kitchen appliances are specified?
7. What are the stair tread and riser sizes?
8. What size is the front door?

9. What size roof rafters are specified?
10. What size ceiling joists are required for the second floor ceiling?
11. How wide is the hall leading to the bedrooms?
12. What size and kind of windows are used in the kitchen?
13. What size beam is used to support the first floor joists?
14. How many toilets are in the house?
15. What size joists are used for the floor of the second level?
16. How is the concrete block foundation capped?
17. What is the maximum length of the house?
18. How is water in the crawl spaced removed?
19. How much insulation is required in the ceiling?
20. What size footing is specified for the exterior wall foundation?
21. What is the size of the piers and their footings?
22. What material is to be used on the ceilings of all rooms?
23. What size and type of sheathing is specified?
24. What size reinforcing bars are specified in the footing?
25. What is the spacing of the joists on the first floor?

The following questions will see how familiar you are with the specifications.

1. What provision is made for termite protection?
2. Bathroom lavatories are to be what colors?
3. How long should poured concrete be allowed to cure?
4. What width masonry joint is specified?
5. How high above the floor is the ceramic tile around the tub to be placed?
6. What type of roofing felt is required?
7. What is the capacity of the water heater?
8. Where does the responsibility of the electrical contractor begin?
9. Who is responsible for cleaning up the building site after the job is completed?
10. Explain how the foundation should be backfilled.
11. What size kitchen sink is required?
12. What type of water pipe is specified?
13. What restrictions are specified for the water to be used for concrete work?
14. How long must the contractor guarantee the work of the building?
15. What restrictions are placed on backfill materials?
16. What pattern of interior trim is specified?
17. What type of shingle is specified?
18. What type of flooring is to be used in the bedrooms?
19. What type of flashing material should be used?
20. How many coats of paint are to be applied to the exterior?
21. What kind of hinge is required on the interior doors?
22. What kind of guarantee is specified for the roof?

Class Activities

1. Design a small storage building. It is to be used to store a riding lawn mower, garden tools, and lawn furniture. It should contain about 100 square feet or 10 m² of floor space. Each exterior wall should have a window. The door opening should be at least 5'-0" (1524 mm) wide. The floor is a concrete slab. Electric lights and outlets are necessary. The inside surface of the walls is not finished. Assume it is to be built on a flat site. Use any building materials you desire. Keep the cost as low as possible. Use simple construction methods. Draw a complete set of plans for this building. Write what specifications are necessary to ensure the use of proper materials.

2. Design a summer cottage that will comfortably sleep four people. It should have a kitchen with stove, sink, refrigerator, and some storage. A bath is required but use minimum space. Materials used should require little maintenance. The maximum size is 750 square feet or 70 m². The lot upon which the cottage will be built slopes toward the lake with a one foot drop for every six feet or 0.300 m drop for every 2 m of horizontal distance. It is level in the other direction. Draw a complete set of plans for the cottage. Write a set of specifications to ensure the use of proper materials and construction techniques.

3. List the things you would like in a residence. For example, a workshop or family room may be important to you. Sketch a floor plan and preliminary elevations for a residence having all the things you included in this list. Calculate the total area needed for the plan. Secure an estimate of the local cost per square unit to build this residence. Do you think you will ever be able to afford to build this building? If not, reduce the size of the plan until it is such that you believe someday you might be able to build it. Draw a complete set of plans assuming the building is to be built on a level site. Write a complete set of specifications.

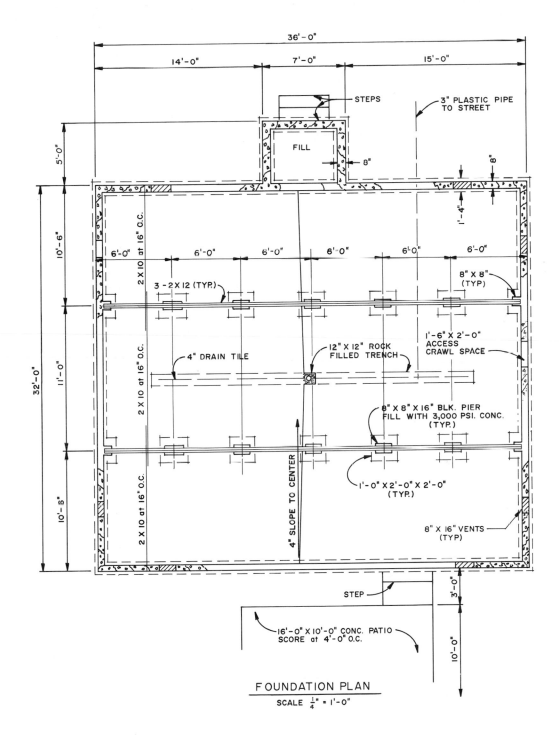

Fig. 6-29. A set of working drawings.

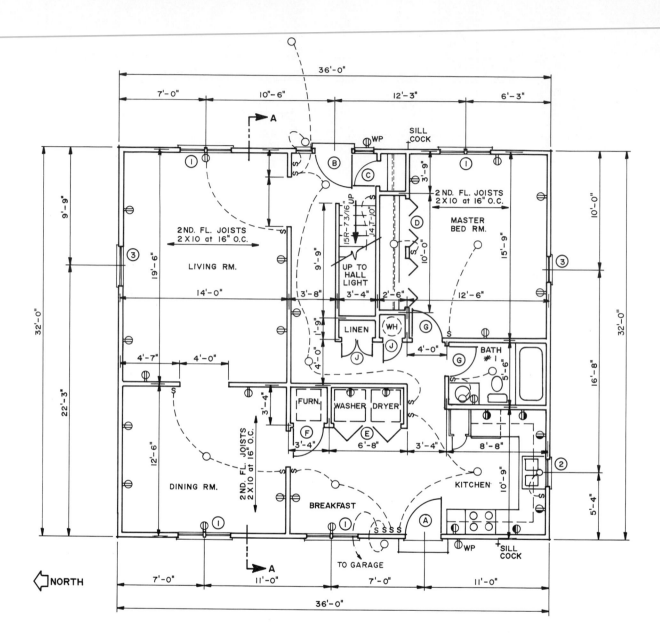

FIRST FLOOR PLAN

SCALE $\frac{1}{4}" = 1'-0"$

Fig. 6-29. (Continued)

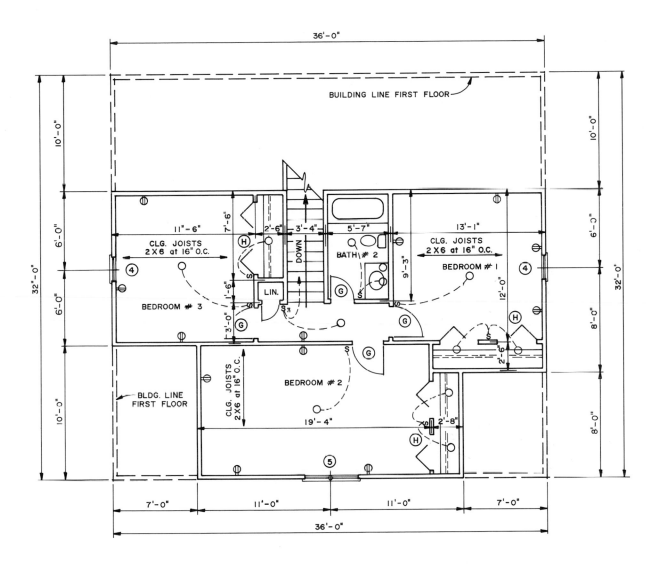

SECOND FLOOR PLAN

SCALE $\frac{1}{4}$" = 1'-0"

Fig. 6-29. (Continued)

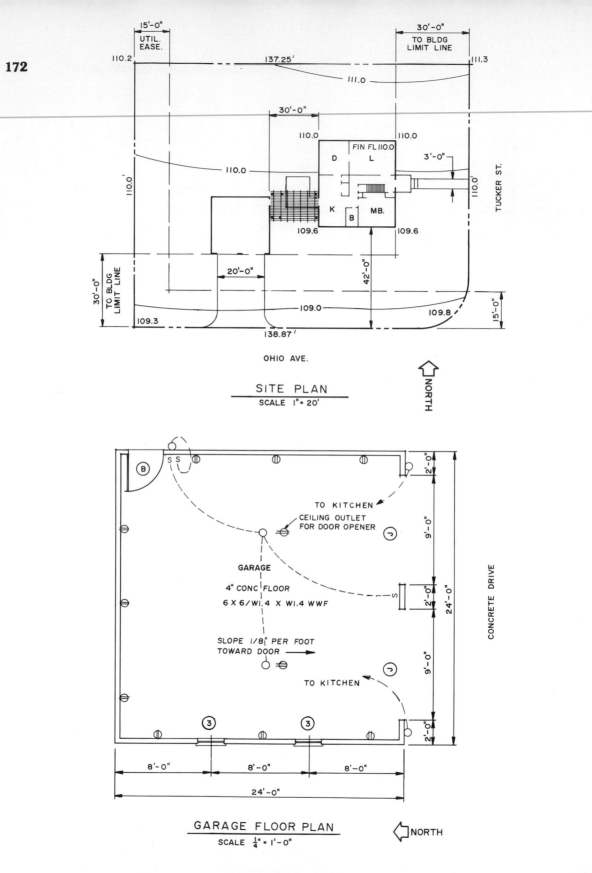

SITE PLAN
SCALE 1" = 20'

GARAGE FLOOR PLAN
SCALE ¼" = 1'-0"

Fig. 6-29. (Continued)

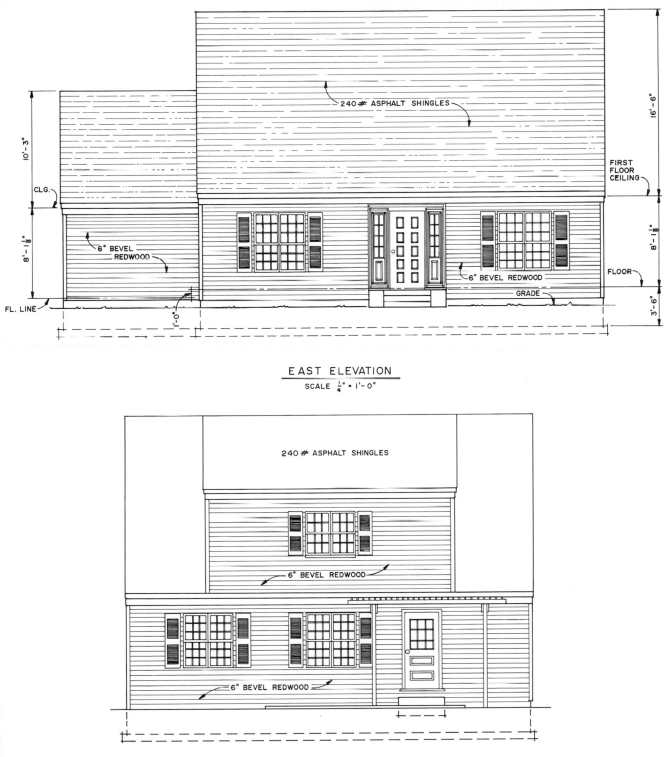

EAST ELEVATION
SCALE $\frac{1}{4}$" = 1'-0"

WEST ELEVATION
SCALE $\frac{1}{4}$" = 1'-0"

Fig. 6-29. (Continued)

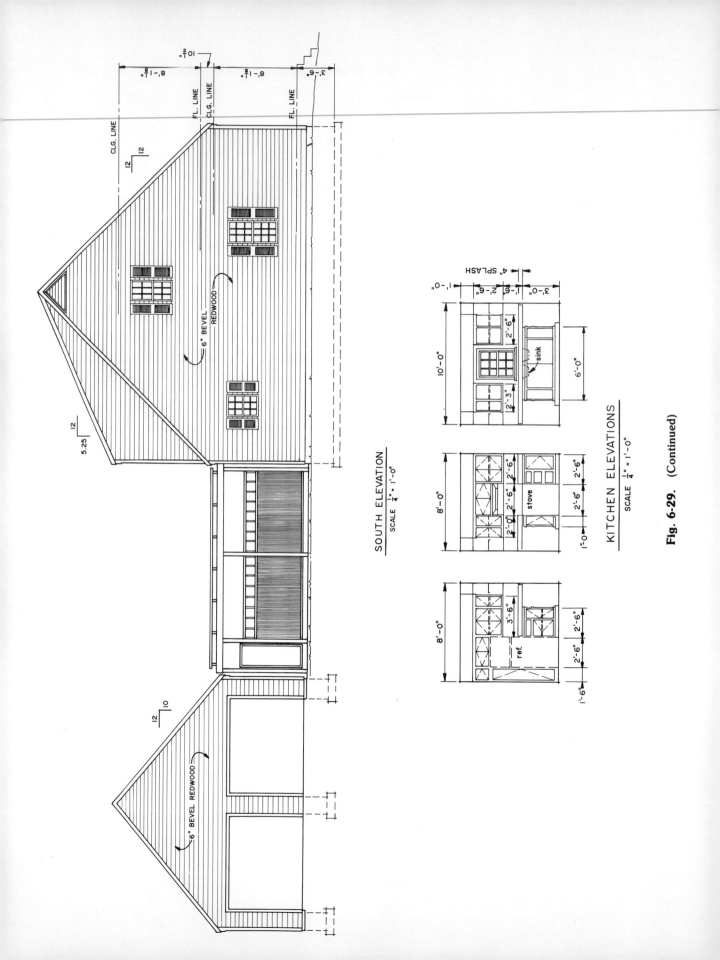

SOUTH ELEVATION
SCALE $\frac{1}{4}$" = 1'-0"

KITCHEN ELEVATIONS
SCALE $\frac{1}{4}$" = 1'-0"

Fig. 6-29. (Continued)

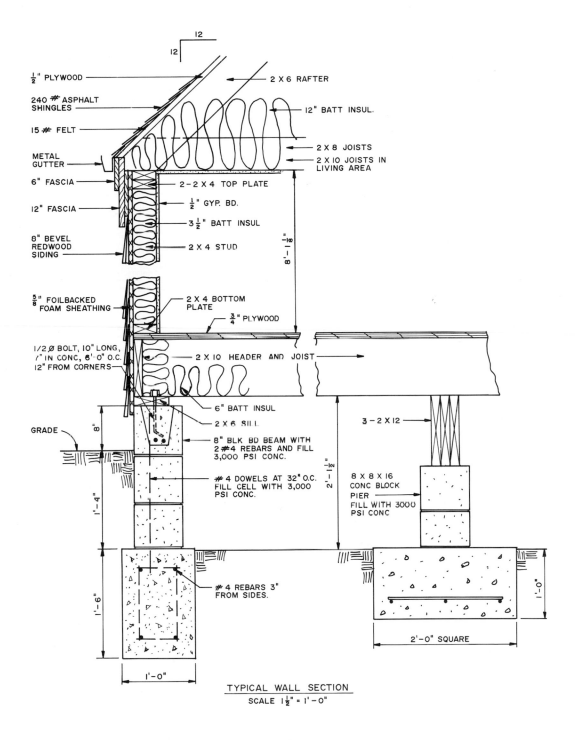

$\frac{1}{2}$" PLYWOOD

240 # ASPHALT SHINGLES

15 # FELT

METAL GUTTER

6" FASCIA

12" FASCIA

8" BEVEL REDWOOD SIDING

$\frac{5}{8}$" FOILBACKED FOAM SHEATHING

1/2 Ø BOLT, 10" LONG, 7" IN CONC, 8'-0" O.C. 12" FROM CORNERS

GRADE

12
12

2 X 6 RAFTER

12" BATT INSUL.

2 X 8 JOISTS
2 X 10 JOISTS IN LIVING AREA

2 - 2 X 4 TOP PLATE

$\frac{1}{2}$" GYP. BD.

$3\frac{1}{2}$" BATT INSUL

2 X 4 STUD

8'-1$\frac{1}{8}$"

2 X 4 BOTTOM PLATE

$\frac{3}{4}$" PLYWOOD

2 X 10 HEADER AND JOIST

6" BATT INSUL

2 X 6 SILL

8"

8" BLK BD BEAM WITH 2 # 4 REBARS AND FILL 3,000 PSI CONC.

4 DOWELS AT 32" O.C. FILL CELL WITH 3,000 PSI CONC.

1'-4"

1'-6"

4 REBARS 3" FROM SIDES.

1'-0"

3 - 2 X 12

8 X 8 X 16 CONC BLOCK PIER FILL WITH 3000 PSI CONC

2'-1$\frac{1}{2}$"

1'-0"

2'-0" SQUARE

TYPICAL WALL SECTION
SCALE 1$\frac{1}{2}$" = 1'-0"

Fig. 6-29. (Continued)

DOOR	SCHEDULE				
MARK	NO.	SIZE	TYPE	MATERIAL	REMARKS
A	1	1 3/4" X 3'-0" X 6'-8"	PANEL	PINE	9 LIGHT
B	1	1 3/4" X 3'-0" X 6'-8"	FLUSH	BIRCH	SOLID CORE
C	1	1 3/8" X 2'-0" X 6'-8"	LOUVERED	PINE	
D	2	1 3/8" X 4'-0" X 6'-8"	LOUVERED	PINE	BI-FOLD
E	1	1 3/8" X 5'-0" X 6'-8"	LOUVERED	PINE	BI-FOLD
F	1	1 3/8" X 2'-8" X 6'-8"	LOUVERED	PINE	
G	5	1 3/8" X 2'-8" X 6'-8"	FLUSH	BIRCH	HOLLOW CORE
H	3	1 3/8" X 6'-0" X 6'-8"	FLUSH	BIRCH	BI-FOLD
J	3	1 3/8" X 1'-6" X 6'-8"	FLUSH	BIRCH	HOLLOW CORE

WINDOW	SCHEDULE				
MARK	NO.	UNIT SIZE	ROUGH OPENING	TYPE	MATERIAL	REMARKS
1	4	2'-5 5/8" X 4'-9 1/4"	2'-6 1/8" X 4'-9 1/8"	D.H.	PINE	DBL UNIT 9 LIGHT
2	1	2'-1 5/8" X 3'-1 1/4"	2'-2 1/8" X 3'-1 1/8"	D.H.	PINE	6 LIGHT
3	2	2'-1 5/8" X 4'-9 1/4"	2'-2 1/8" X 4'-9 1/8"	D.H.	PINE	6 LIGHT
4	2	2'-1 5/8" X 4'-1 1/4"	2'-2 1/8" X 4'-1 1/8"	D.H.	PINE	6 LIGHT
5	1	2'-1 5/8" X 3'-5 1/4"	2'-2 1/8" X 3'-5 1/8"	D.H.	PINE	DBL UNIT 6 LIGHT

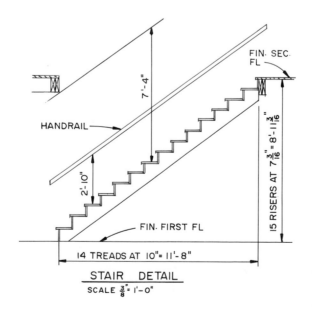

STAIR DETAIL
SCALE $\frac{3}{8}$" = 1'-0"

Fig. 6-29. (Continued)

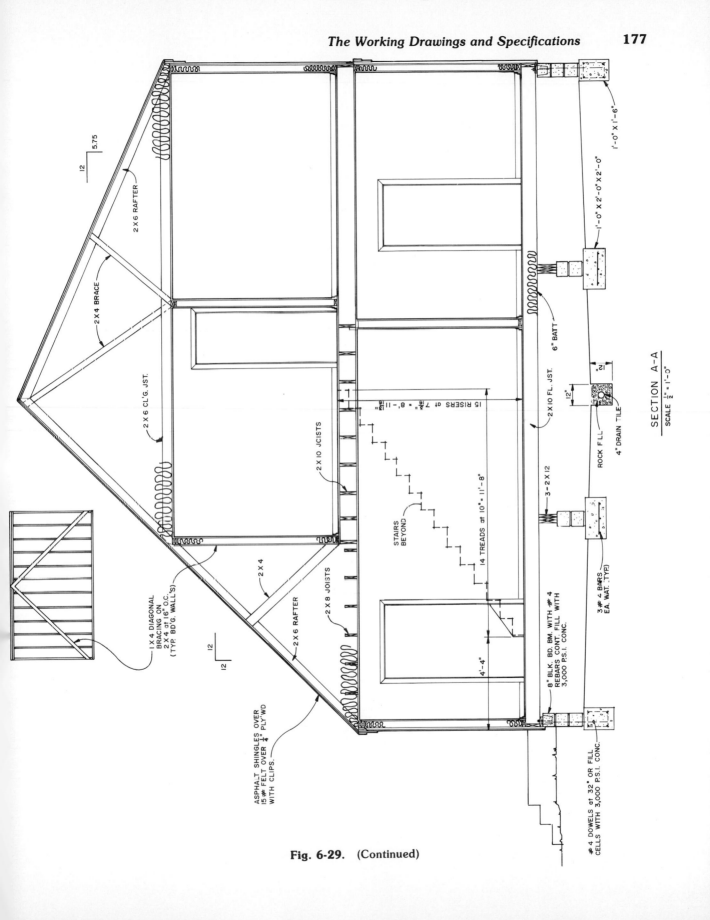

Fig. 6-29. (Continued)

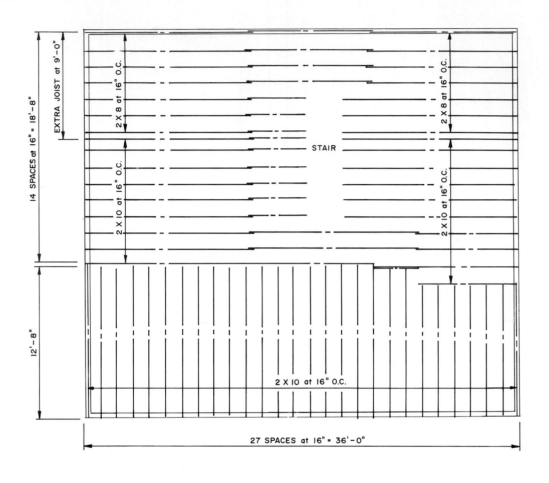

SECOND STORY FLOOR
FRAMING PLAN

SCALE $\frac{1}{4}$" = 1'-0"

Bill Wilson, Architect, Pittsburg, Kansas

Fig. 6-29. (Continued)

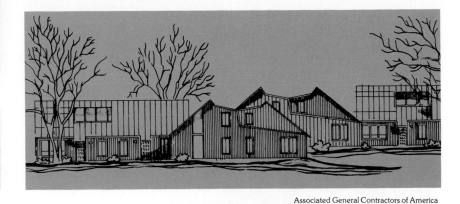

Chapter 7

Methods of House Construction

Certain basic techniques of house construction are used across the nation as standard practice. In some areas, unique or special practices have developed. Also, progressive builders, professional organizations, and companies supplying materials are constantly developing new and better ways to build houses. The commonly accepted methods as well as some of the new approaches to house construction are presented in this chapter.

How a Conventional House is Built

After the architect has considered the many factors involved in relating a house to the building site, a plot plan is drawn. This carefully locates the house on the site. The builder is responsible for laying out the exact location and for digging the foundation.

This layout is begun by measuring with a steel tape from the corner stakes of the site. At the corners of the foundation, batter boards are erected, with their posts set wider than the proposed foundation. See Fig. 7-1. By carefully measuring, each side is located and marked on the batter boards.

Fig. 7-2. The basement wall is prepared for pouring using metal forms.

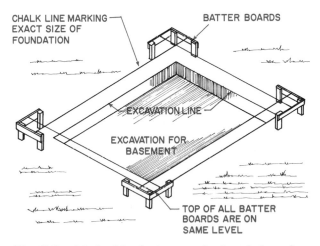

CHALK LINE MARKING
EXACT SIZE OF
FOUNDATION

BATTER BOARDS

EXCAVATION LINE

EXCAVATION FOR
BASEMENT

TOP OF ALL BATTER
BOARDS ARE ON
SAME LEVEL

Fig. 7-1. Method for laying out the foundation of a building.

Chalk lines are run to outline the foundation. These are sighted, the corners are squared, and the chalk lines are leveled. Usually, a surveying instrument called a transit is used for this job.

As soon as the foundation is laid out, the digging begins. This is most often done with a small bulldozer or a backhoe. If the house is to have a basement, the entire area is excavated until it is a little larger than the finished basement. This allows room for a footing and the basement walls. See Fig. 7-2. If the house is to be built without a basement, the trench for the footing is dug wide enough to permit erection of the foundation wall.

After the foundation trench has been dug, footings poured, and foundation wall constructed, the sills, girders, and floor joists are placed and the subfloor is applied. The sills are bolted in place.

Fig. 7-3. Workers raise the walls into position.

Once the subfloor is completed, the exterior walls are laid out on the subfloor; pieces are cut to size and nailed together to form a wall or part of a wall. This is then placed in a vertical position and nailed to the joists. Bracing is necessary to hold it in place. See Fig. 7-3.

Fig. 7-4. Concrete is machine-troweled to a smooth and uniform finish.

Next, the interior partitions are assembled and erected, followed by installation of the ceiling joists. This considerably stiffens the structure. The exterior sheathing is applied, while other workers cut and erect the rafters. The roof is completed by the application of roof sheathing, builder's paper, and shingles.

Then, windows are set in place and flashed. This is followed by the siding, which can be either wood, plastic, aluminum, or masonry. The house is now weathertight.

Miscellaneous pieces of trim, such as facia, frieze, and soffit, are installed. The prime coat of paint is applied to all exterior wood.

Inside the house, others are at work. The electrician installs the wiring; the plumber, the pipes and fixtures; and the heating contractor, the furnace and air-conditioning unit.

Cabinetmakers build and install the kitchen counters, cabinets, and other required items, such as room dividers, built-in bedroom drawers, and bathroom lavatory cabinets.

After all this is finished, the insulation is installed, and the interior walls are plastered, covered with drywall, or panelled.

The flooring contractor then installs the hardwood floors. Before these floors are sanded and

Fig. 7-5.

Standard Softwood Lumber Sizes*

THICKNESS (Inches)			FACE WIDTH (Inches)		
			Minimum Dressed		
Nominal	Dry	Green (Unseasoned)	Nominal	Dry	Green (Unseasoned)
1	3/4	25/32	2	1-1/2	1-9/16
1-1/4	1	1-1/32	3	2-1/2	2-9/16
1-1/2	1-1/4	1-9/32	4	3-1/2	3-9/16
2	1-1/2	1-9/16	5	4-1/2	4-5/8
2-1/2	2	2-1/16	6	5-1/2	5-5/8
3	2-1/2	2-9/16	7	6-1/2	6-5/8
3-1/2	3	3-1/16	8	7-1/4	7-1/2
4	3-1/2	3-9/16	9	8-1/4	8-1/2
			10	9-1/4	9-1/2
			11	10-1/4	10-1/2
			12	11-1/4	11-1/2
			14	13-1/4	13-1/2
			16	15-1/4	15-1/2

*Available in 2-foot increments from 8'-0" to 20'-0" long.

finished, painters and paperhangers decorate the interior. Then the electrician hangs the light fixtures. Finally, the flooring contractor sands and finishes the hardwood floors, installs tile in the bath and kitchen, and lays the carpet. This completes the basic house.

Before leaving the job, the builder is responsible for cleaning away all debris and for landscaping the house. Any driveways and sidewalks are installed at this time. See Fig. 7-4.

Frame House Construction

Frame Construction Materials

Typical sizes of frame construction materials are given in Figs. 7-5, 7-6, and 7-7. At the time of this writing, metric sizes have not been standardized; therefore, those shown are the best available estimate as to their sizes. Notice that metric standard sizes are not likely to be **exact** equivalents of customary standard sizes. For this reason, conversions to metric of present **standards** and related measurements could be misleading and will not be presented. If conversions are desired, however, use the tables in Chapter 5. The U.S. Customary sizes are given as nominal and actual. The **nominal size** is the dimension before the member is planed smooth. It is common to refer to members by their nominal size. The metric series shows the nominal size from which 4 mm is removed by planing.

Design information on wood, sheathing, and other materials is in Chapter 9.

Fig. 7-6.

Nominal Metric Sizes* for Softwood Lumber in Millimeters

Thickness	Width								
	75	100	125	150	175	200	225	250	300
16	x	x	x	x					
19	x	x	x	x					
22	x	x	x	x					
25	x	x	x	x	x	x	x	x	x
32	x	x	x	x	x	x	x	x	x
36	x	x	x	x					
38	x	x	x	x	x	x	x		
40	x	x	x	x	x	x	x		
44	x	x	x	x	x	x	x	x	x
50	x	x	x	x	x	x	x	x	x
63		x	x	x	x	x	x		
75		x	x	x	x	x	x	x	x
100		x		x		x		x	x
150				x		x			x
200						x			
250								x	
300									x

*Lengths: 1.8 to 6.3 m × 0.3 m
Allow: 4 mm for planing

Fig 7-7

Plywood and Other Sheathing Panels

U.S. Customary Sheet	4'-0" × 8'-0"
Metric Sheets	1200 mm × 2400 mm

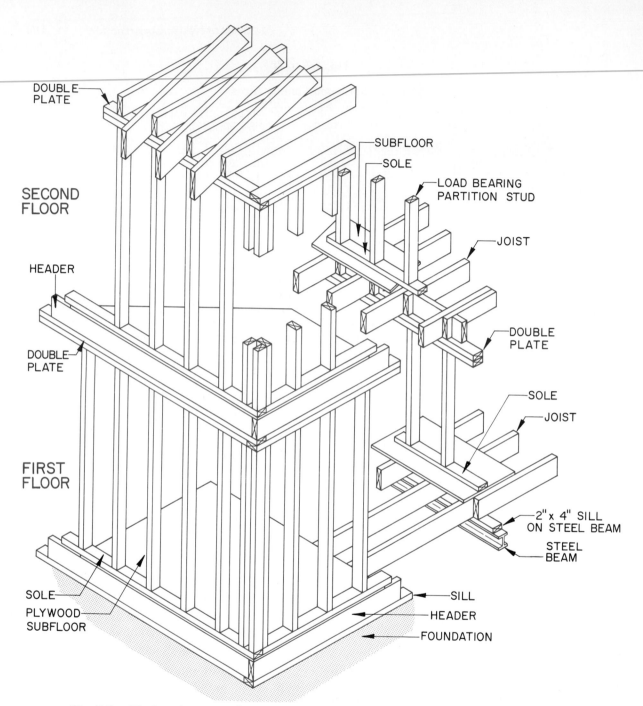

DOUBLE PLATE

SECOND FLOOR

SUBFLOOR

SOLE

LOAD BEARING PARTITION STUD

JOIST

HEADER

DOUBLE PLATE

DOUBLE PLATE

SOLE

JOIST

FIRST FLOOR

2" x 4" SILL ON STEEL BEAM

STEEL BEAM

SOLE

PLYWOOD SUBFLOOR

SILL

HEADER

FOUNDATION

Fig. 7-8. Platform framing is the most commonly used framing system for house construction.

Frame Construction Details

Two systems are commonly used for conventional house framing. These are the platform system and the balloon framing system. See Figs. 7-8 and 7-10.

The **platform framing system** is the most widely used. The floor joists are assembled and covered with the subfloor. The walls are then assembled and placed on top of the subfloor. In a two-story house, the second floor is built in the same manner as the first floor and then those walls are built and erected.

A typical section through the wall of a house using platform framing is in Fig. 7-9. A completely dimensioned section is in Chapter 6. This type of section is on the drawings to guide the carpenters.

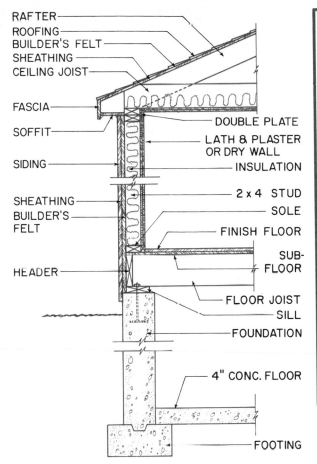

RAFTER
ROOFING
BUILDER'S FELT
SHEATHING
CEILING JOIST
FASCIA
SOFFIT
SIDING
SHEATHING
BUILDER'S FELT
HEADER

DOUBLE PLATE
LATH & PLASTER OR DRY WALL
INSULATION
2 x 4 STUD
SOLE
FINISH FLOOR
SUB-FLOOR
FLOOR JOIST
SILL
FOUNDATION
4" CONC. FLOOR
FOOTING

Fig. 7-9. A typical wall section through a frame house using platform framing.

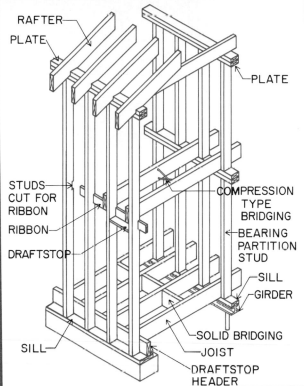

RAFTER
PLATE
STUDS CUT FOR RIBBON
RIBBON
DRAFTSTOP
SILL
PLATE
COMPRESSION TYPE BRIDGING
BEARING PARTITION STUD
SILL
GIRDER
SOLID BRIDGING
JOIST
DRAFTSTOP HEADER

Fig. 7-10. Balloon framing has studs running from the plate to the sill.

The **balloon framing system** uses a continuous stud for the full two floors. The exterior and interior supporting walls rest on the sill. The platform system has the advantage of having a floor available for use when framing the walls.

Sills

The sill, usually a 2 × 6 or 2 × 8, is bolted to the foundation. Between the sill and the foundation, a strip of sealer is installed. This sealer is usually about 1/2-inch (13 mm) thick and is composed of an insulation material held between two layers of heavy, asphalt-impregnated paper. When the sill is bolted down, the sealer is squeezed tightly and fills any openings between the foundation and the sill. Sometimes a thin layer of fresh mortar is used both as a sealer and as a leveling device for the sill. See Fig. 7-11.

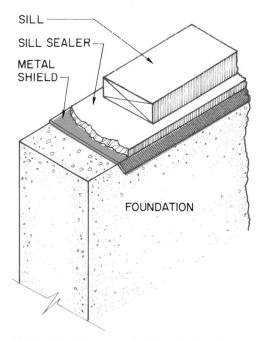

SILL
SILL SEALER
METAL SHIELD
FOUNDATION

Fig. 7-11. Metal termite shield and sill sealer.

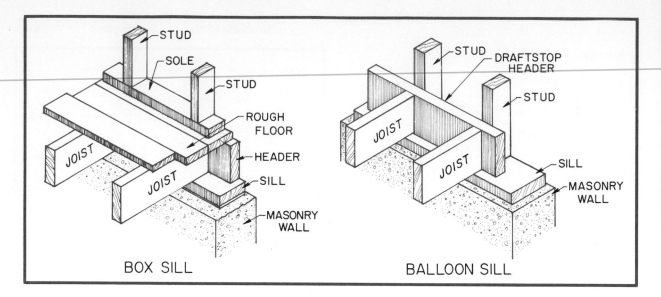

Fig. 7-12. Typical types of frame sill construction with wood floors.

Figure 7-12 illustrates common types of sill construction used on houses with wood floors. A typical detail for a frame house with concrete floors is shown in Fig. 7-13.

Sills are bolted to the foundation with 1/2-inch anchor bolts 16 inches long. They are threaded on one end and bent 90° on the other. This prevents the bolt from slipping. The bolts are placed 4'-0" on center along the foundation wall. There should be at least two bolts in every sill board.

Floor joists are toenailed to the sill (the nails are driven in at an angle), and the header is nailed to the ends of the joists.

Termites are found in almost every section of the United States. If they are prevalent, the sill should be protected against attack. The lumber can be treated with chemicals that repel the termite, or a termite shield can be used. The most common shield is a layer of metal placed under the sill and projecting on each side of the foundation. This prevents termites from traveling up the foundation and attacking the wood sill. Termites can travel inside hollow masonry foundations and can penetrate poor mortar and porous concrete. Refer again to Fig. 7-11 which illustrates a typical termite shield.

Framing Joists

The typical method for framing joists onto the foundation was shown in Fig. 7-12. A number of different methods are used when framing joists to a beam. The simplest method is shown in Fig. 7-14. The joist rests on top of the girder. If it is necessary to have the top of the joist flush with the top of the

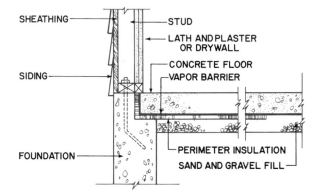

Fig. 7-13. Typical section through a frame house with a concrete slab on grade.

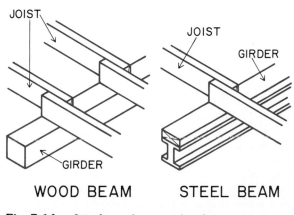

WOOD BEAM   **STEEL BEAM**

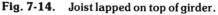

Fig. 7-14. Joist lapped on top of girder.

184

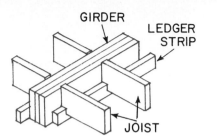

Fig. 7-15. Joist notched over ribband nailed to girder.

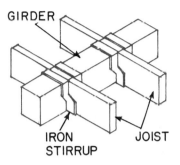

Fig. 7-16. Joists hung on girder with iron stirrups.

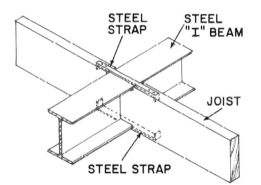

Fig. 7-17. Joist framed into steel I beam.

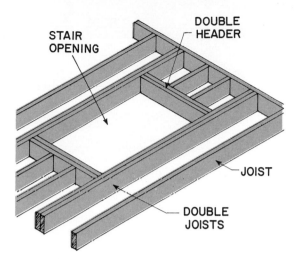

Fig. 7-18. An opening framed to receive a stairway. Notice the double joists and double headers.

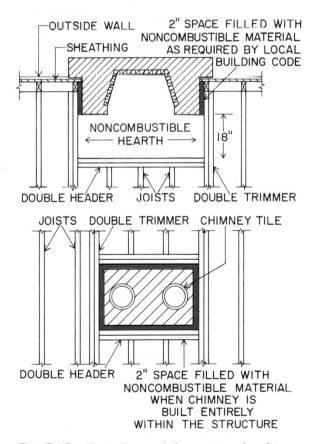

Fig. 7-19. Typical ways to frame around a chimney and a fireplace.

beam, the joist can be installed as shown in Figs. 7-15, 7-16, or 7-17. The system in Figs. 7-16 and 7-17 also permits a flush ceiling to be installed in the area below.

An opening in the floor, such as a stairway, requires that some joists be cut. Since this weakens the floor, a header is installed to support the cut joists. This header is doubled and sometimes tripled if the distance is great. See Fig. 7-18. Details for framing around a chimney or fireplace are shown in Fig. 7-19.

If an interior partition runs parallel with the joists, the joist supporting the partition must be doubled to carry the weight.

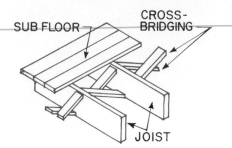

Fig. 7-20. Cross-bridging in place between floor joists.

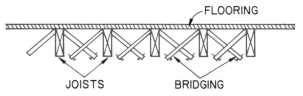

Fig. 7-21. Lower end of cross-bridging is not nailed in place until subfloor is nailed to joists.

The use of wood floor trusses in floor framing is increasing. Trusses are used instead of solid wood floor joists. Details are in Chapter 27.

Bridging

Bridging consists of rows of small diagonal braces nailed between the joists. See Fig. 7-20. These distribute any load on the floor above over a wider area, thus putting the strain on many joists, rather than permitting one or two to carry the load. Besides this, bridging stiffens the floor and tends to keep the joists in line by preventing warping.

Bridging should be installed in rows from one side of the house to the other. Presently, these rows are spaced not over 10′ apart. Typical bridging members are cut from 1 × 4 stock and are available as preformed metal units.

The subfloor should be installed before the bridging is nailed tightly. Usually, the top ends of the bridging are nailed in place before the subfloor is laid, but the lower ends are not fastened until the joists have adjusted to the subfloor. See Fig. 7-21.

Framing the Walls

Most builders choose either to cut all the studs to size or to buy precut ones. The exterior wall is laid out on the subfloor, and the studs are spaced and nailed to the sole and to a single 2 × 4 plate. The

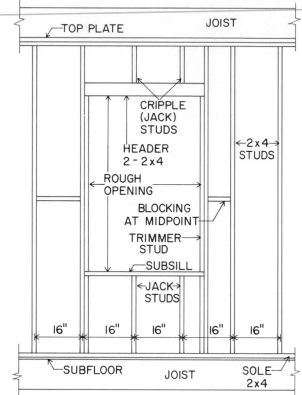

Fig. 7-22. Window openings have double studs on the sides and a header on top.

normal spacing of studs is 16 inches on center. As sizes of members change to metric, however, spacing will probably also change.

The placement of framing members to form rough openings for windows and doors is shown in Figs. 7-22 and 7-23. For most normal openings, the header can be made from two 2 × 4 members placed on edge. Studs are doubled on each side of the opening. The 16-inch spacing of studs is maintained by the cripple studs. The studs forming the opening are often not on the 16-inch module.

The corners of the exterior wall are most often built up from three 2 × 4 studs. One method is illustrated in Fig. 7-24. The small blocks at the sole are used for securing baseboards.

Frame walls can be strengthened by nailing steel straps diagonally to the sole, plate, and each stud. See Fig. 7-25. A wood 1 × 4 can be set diagonally into the studs and nailed to each member. See Fig. 7-26. Diagonal bracing is not needed if plywood sheathing is used. One technique is to use a sheet

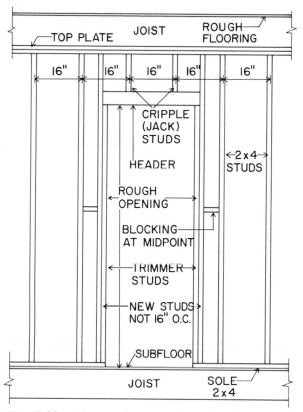

Fig. 7-23. The rough opening for a door frame needs double studs on the sides and a header at the top.

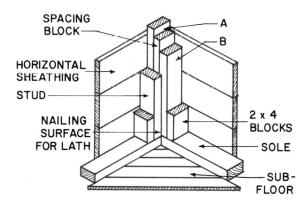

Fig. 7-24. A method for framing a corner. Studs **A** and **B** are spaced by nailing short lengths of 2 × 4 between them at 3′-0″ intervals.

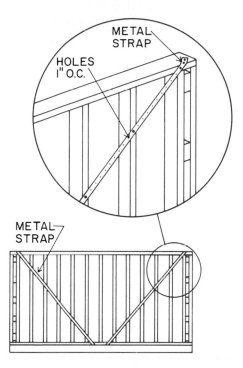

Fig. 7-25. Metal straps are used to brace frame exterior walls.

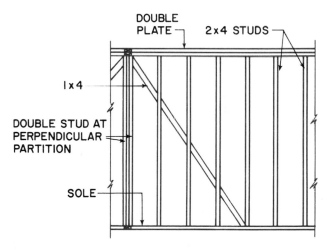

Fig. 7-26. A frame wall strengthened by a 1 × 4 set into studs.

of plywood on each corner of the building and finish sheathing with fiberboard. For greatest strength, the entire building should be sheathed with plywood.

After all exterior framing is up, the second 2 × 4 is added to the plate. This is used to increase the

strength of the plate, since it has to help support the roof.

Interior partitions are framed in the same way as the exterior structure. A special problem occurs where an interior partition meets an exterior wall. A method for framing when the partition meets a stud is illustrated in Fig. 7-27. This requires that an extra stud be inserted in the exterior wall to provide nailing space for lath or drywall sheets. If a partition does not meet a stud in the exterior wall, two extra studs must be inserted in the exterior wall.

Three ways to frame partitions to reduce sound transmission are shown in Fig. 7-28. These are useful in any area where noise could be a problem. Typical applications are the bathroom, the recreation room, and bedroom walls.

Sheathing the Walls

Wall sheathing is placed next to the stud framing. It strengthens the wall by resisting lateral movement. It also has insulation values. The siding is placed over the sheathing.

The types of sheathing include solid wood, plywood, composite, waferboard, oriented strand board, particleboard, fiberboard, gypsum sheets, and rigid foamed plastic sheets. Lumber sheathing can be applied perpendicular to the studs or on a 45° diagonal. The diagonal application makes the wall more rigid. If diagonal sheathing is used, the let-in brace can usually be omitted. It is covered with building paper before the siding is applied.

Sheathing is made in large sheets. The most common size sheet is 48″ × 96″. No builder's paper is needed between sheathing and siding. Plywood-sheathed walls have greater strength than lumber- or fiberboard-sheathed walls. Plywood also gives a solid base for nailing the siding.

Fiberboard sheathing has a bituminous coating which gives it a degree of moisture resistance. It will not hold nails. Wood siding must be nailed through it into the studs. Information on sheathing spans and thicknesses is given in Chapter 9.

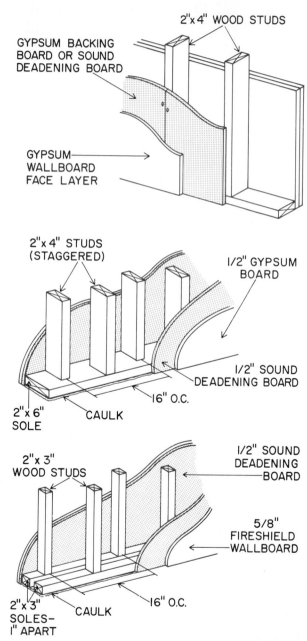

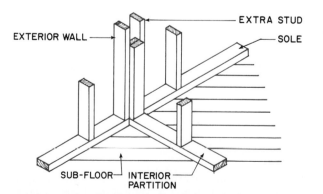

Fig. 7-27. An interior partition framed into the exterior wall.

Fig. 7-28. Double-stud walls and sound-deadening material help reduce the transmission of sound.

Fig. 7-29.

Typical Frame Siding Materials

Cedar Shakes	LENGTH	18″	24″	32″			
	EXPOSURE	8-1/2″	11-1/2″	15″			
Cedar Shingles	LENGTH	16″	18″				
	EXPOSURE	12″	14″				
Cement-Asbestos Shingles	SIZE	8-3/4″ × 48″	12″ × 24″				
	EXPOSURE	7-3/4″ × 48″	11″ × 24″				
Bevel Siding	ACTUAL WIDTH	3-1/2″	4-1/2″	5-1/2″	7-1/2″	9-1/2″	11-1/2″
	EXPOSURE	2-1/2″	3-1/2″	4-1/2″	6-1/4″	8-1/4″	10-1/4″
Shiplap Siding	ACTUAL WIDTH	5-1/4″	7-1/4″				
	EXPOSURE	5-1/4″	7-1/4″				
Plywood Panels	WIDTH	4′-0″	4′-0″	4′-0″			
	LENGTH	8′-0″	9′-0″	10′-0″			
Horizontal Plywood Siding	WIDTH	10″	12″				
	EXPOSURE	8-1/2″	10-1/2″				
Hardboard Panels	WIDTH	4′-0″	4′-0″	4′-0″			
	LENGTH	8′-0″	9′-0″	10′-0″			
Horizontal Hardboard Siding	WIDTH	10″	12″	16″	24″		
	EXPOSURE	8-7/8″	10-3/8″	14-3/8″	22-3/8″		

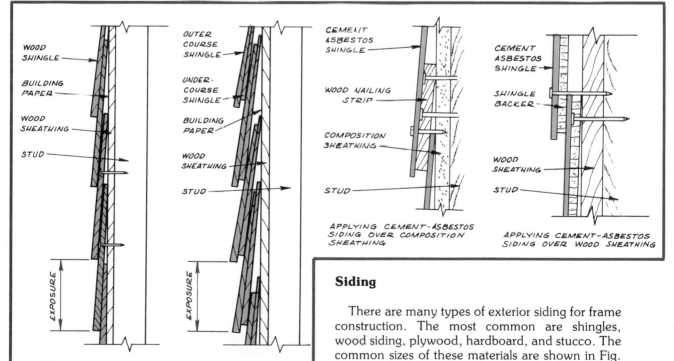

Fig. 7-30. Construction details for wood shingle and cement-asbestos siding.

Siding

There are many types of exterior siding for frame construction. The most common are shingles, wood siding, plywood, hardboard, and stucco. The common sizes of these materials are shown in Fig. 7-29.

Shingles are available in wood and cement-asbestos. **Wood shingles,** Fig. 7-30, are applied in single and double course. The single course wall has two thicknesses of wood at the thinnest place.

189

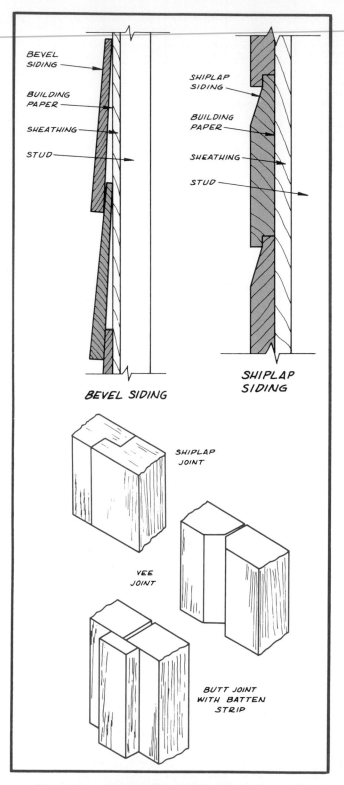

BEVEL SIDING
- BEVEL SIDING
- BUILDING PAPER
- SHEATHING
- STUD

SHIPLAP SIDING
- SHIPLAP SIDING
- BUILDING PAPER
- SHEATHING
- STUD

SHIPLAP JOINT

VEE JOINT

BUTT JOINT WITH BATTEN STRIP

Fig. 7-31. Construction with horizontal wood siding and plywood panel joints.

Much of the wall has three thicknesses. A double course wall uses a low grade shingle such as No. 3 for the undercourse. A No. 1 grade shingle is used for the outer course. Double coursing provides better wall coverage. It also gives the rows of shingles a deeper shadow line due to the extra thickness. A wider exposure is used with double coursed shingles. **Exposure** is the area of the shingle not covered by another shingle.

Cement-asbestos shingles, Fig. 7-30, have a baked-on factory finish. They are made in many colors. The surface is made in a variety of textures. When placed over wood sheathing, a fiber shingle backer is used. The unit is nailed directly to the wood sheathing. When placed over fiberboard sheathing, wood nailing strips are needed. Since the fiberboard sheathing will not hold nails, the shingle is nailed directly to the nailing strip.

There are many kinds of **horizontal wood siding,** Fig. 7-31. Two of the most common are bevel and shiplap. There are several other patterns of shiplap siding in addition to the one shown.

Plywood siding is available in sheets 48″ wide and 8′, 9′, and 10′ long. It is made with a variety of surface textures. This siding is nailed directly to the studs. The panels are placed vertically so there are no horizontal joints. A variety of joints are used to join these panels. See Fig. 7-31. The joints should be sealed with caulking compound.

Horizontal plywood siding is applied in the same manner as bevel siding. The pieces are usually lapped 1 1/2″ (38 mm).

Hardboard siding is available in sheets 48″ wide and 8′, 9′, and 10′ long. It is also made to resemble horizontal bevel siding. A variety of surface textures are available. Most panels have a factory applied prime coat. Hardboard panels are applied directly to studs or sheathing. The horizontal siding can be applied with a plain lap just like bevel siding. See Fig. 7-31. For additional shadow, it can have a wood nailing strip as shown for cement-asbestos siding in Fig. 7-30.

Stucco Construction

An exterior stucco finish is used a great deal in some parts of the United States. It is used in the warmer, dryer climates. In Fig. 7-32 are three commonly used types of wall construction utilizing stucco as the final exterior finish.

Wire lath is fastened to masonry walls with concrete nails and to frame walls with a large-headed galvanized nail. The nail has a space ring on it. This keeps the head about 3/8″ (10 mm) above the surface of the sheathing or masonry. The head holds the wire lath to the wall. See Figs. 7-33 and 7-34.

The exterior stucco is made of a portland cement base. Usually, three coats are applied. See Fig. 7-34.

The first coat is the scratch coat. It is worked into the wire mesh. The second coat is the brown coat. It builds up the thickness of the finish and smooths out many irregularities left in the scratch coat. Commonly these are made of one part portland cement to three parts sand. Some lime is also added. The final coat is the finish coat. The actual finished appearance of the house depends on this coat. It can be trowled smooth or brushed for a rougher texture. Coloring may be added to tint it. Stucco can also be painted with a paint suitable for this purpose.

A metal molding is applied on all edges and around all openings in a wall to be finished with stucco.

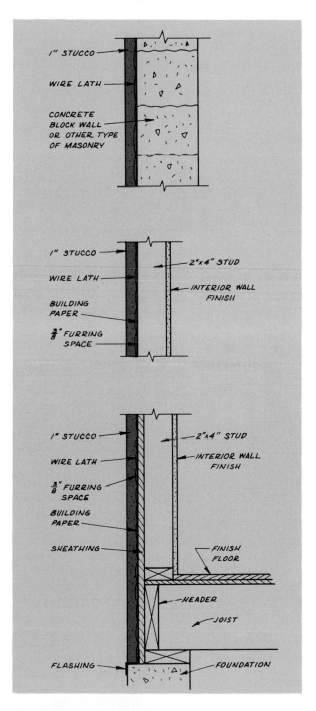

Fig. 7-32.   Stucco finishes.

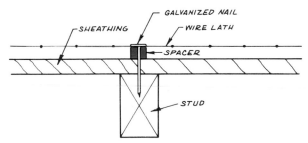

Fig. 7-33.   Wire lath is spaced out from sheathing so stucco can flow behind the lath.

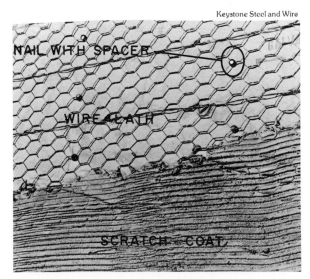

Keystone Steel and Wire

Fig. 7-34.   The basic preparation for a stucco finish includes wire lath and a scratch coat of mortar.

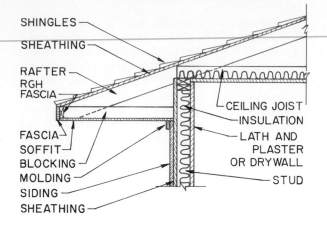

Fig. 7-35. Typical cornice detail for wide overhang.

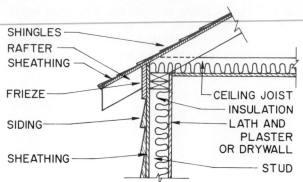

Fig. 7-37. Typical detail of cornice with exposed rafter ends.

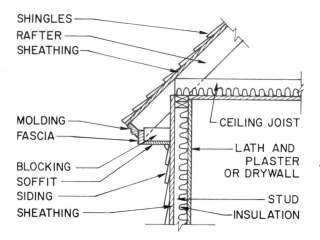

Fig. 7-36. Typical cornice detail for Colonial house.

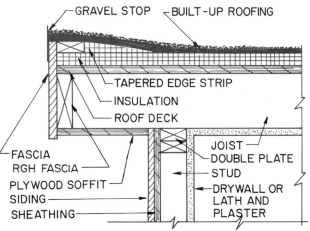

Fig. 7-38. Typical cornice detail for flat roof.

Cornice Details

The **cornice** is that portion of the rafters which extends beyond the exterior wall. There are many ways to frame a cornice. Some are typical of a particular style or period of house; for example, the eave on a Cape Cod Cottage, is built only one way. On a contemporary house, the cornice is designed for a desired appearance or for a special function, such as shielding windows from the sun's rays.

Some typical cornice designs are detailed in Figs. 7-35 through 7-38.

A 24-Inch On Center Framing System

This system utilizes the same basic construction techniques as the conventional 16-inch on center

system. It is based on spacing joists, studs, and rafters on 24-inch centers. The grade and species of wood used must meet the structural requirements. The structural members are kept in alignment. Studs, joists, and rafters are placed directly above each other to help transfer the loads. The system permits the use of a 24-inch module. Plywood sheets will always end on a structural member. To accomplish this, the floor joists must meet end to end. See Fig. 7-39.

The walls can use conventional 3/8-inch plywood sheathing plus siding or a single thickness exterior plywood siding which serves both purposes.

The floors can use a single 3/4-inch plywood sheet designed for spanning the 24-inch on center joists. Floors can use the conventional two-layer construction, if desired.

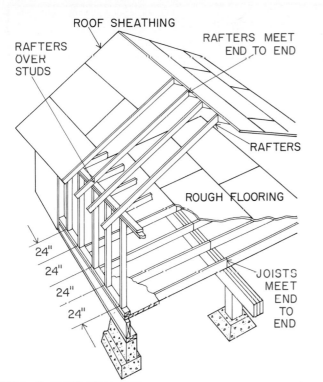

RAFTERS OVER STUDS

ROOF SHEATHING

RAFTERS MEET END TO END

RAFTERS

ROUGH FLOORING

24"
24"
24"
24"

JOISTS MEET END TO END

Fig. 7-39. All structural units are spaced 24 inches O.C. and are placed in alignment.

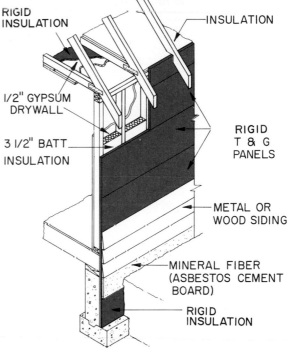

RIGID INSULATION

INSULATION

1/2" GYPSUM DRYWALL

3 1/2" BATT INSULATION

RIGID T & G PANELS

METAL OR WOOD SIDING

MINERAL FIBER (ASBESTOS CEMENT BOARD)

RIGID INSULATION

Fig. 7-40. Rigid foamed plastic panels seal the wall from the plate to the footing.

Attention should be given by designers to the new 2 × 5 member. It can be used in wall construction at 16 and 24 inches on center. It allows increased insulation in the wall and will make the wall cost less than one which uses 2 × 6 members.

Energy Saving Ideas

There are many variations to construction designed to save energy. One of these is to use 2 × 5 or 2 × 6 studs in place of the usual 2 × 4 studs. This permits wider spacing of studs and allows the use of 6 inches (152 mm) of insulation in the wall.

Another technique is to use tongue-and-groove rigid foamed plastic panels as sheathing and run them into the ground over the foundation. See Fig. 7-40. This seals the wall from the roof to the footing.

Glued Floor System

The American Plywood Association has developed a floor system using single-thickness tongue-and-groove plywood sheets, an adhesive, and nails. See Fig. 7-41. The plywood sheets are joined to the floor with the adhesive and then nailed before the adhesive dries. The ends of the sheets must meet over a joist. Blocking is not needed

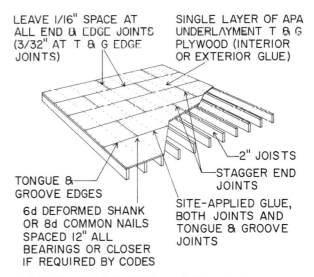

LEAVE 1/16" SPACE AT ALL END & EDGE JOINTS (3/32" AT T & G EDGE JOINTS)

SINGLE LAYER OF APA UNDERLAYMENT T & G PLYWOOD (INTERIOR OR EXTERIOR GLUE)

2" JOISTS

STAGGER END JOINTS

TONGUE & GROOVE EDGES

SITE-APPLIED GLUE, BOTH JOINTS AND TONGUE & GROOVE JOINTS

6d DEFORMED SHANK OR 8d COMMON NAILS SPACED 12" ALL BEARINGS OR CLOSER IF REQUIRED BY CODES

Fig. 7-41. The glued plywood floor system uses a single-layer, tongue-and-groove plywood sheet.

under the side joints of the plywood sheets. When glued, the tongue-and-groove joint provides the needed strength. This produces a floor system that is stronger than the conventionally built nailed floor. It also eliminates squeaks from loose nails.

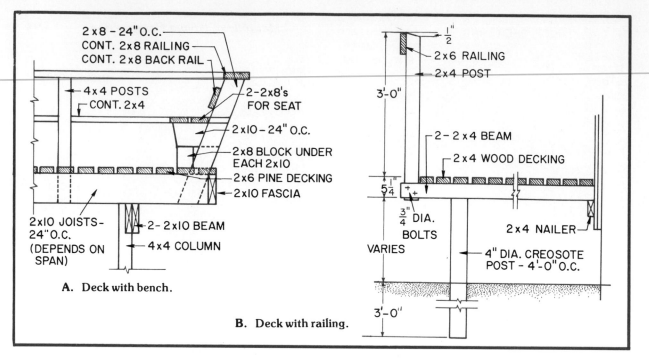

A. Deck with bench.

B. Deck with railing.

Fig. 7-42. Typical deck structural designs.

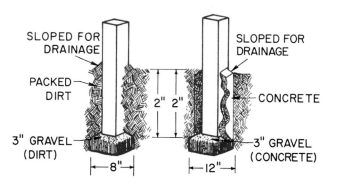

Fig. 7-43. Recommended ways to set posts for wood deck construction.

Framing Wood Decks

A number of wood decks are shown in Chapter 2. Their framing systems are rather simple. Two typical details are in Fig. 7-42. One shows a simple deck with a railing. The other deck has a bench. The actual size of the structural members might vary because of the span used. The sizes shown are typical of those in use.

The wood posts should be set on a 3-inch gravel bed. The post is steadier if cased in concrete. See Fig. 7-43. The posts should be pressure treated to resist rot.

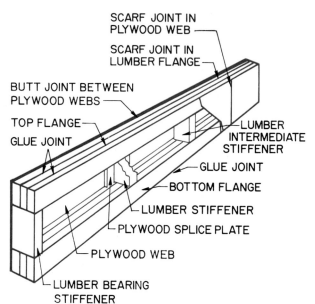

Fig. 7-44. Typical plywood box beam.

Box Beams

Box beams are simple in construction. They consist of one or more vertical plywood webs laminated to seasoned lumber flanges that are separated at intervals along the beam's length by vertical spacers. The spacers distribute concentrated loads and prevent web-buckling. Box beams can be designed to span distances up to 120 feet. Since they are hollow units, they have a high strength-to-weight ratio. This reduces the overall weight of the structure to be supported. Figure 7-44 illustrates a typical box beam.

These structural units can be used as garage-door lintels, ridge beams, flat-roof girders, or floor beams as in Fig. 7-45. They must be fabricated in accordance with carefully developed specifications. Figure 7-46 illustrates the common uses for box beams in residential construction.

Structural data for box beams can be found in Chapter 9.

Stressed-Skin Panels

Stressed-skin panels are prefabricated units used in floors, walls, or roofs. They are made from sheets of plywood glued to longitudinal framing members or other core materials. See Fig. 7-47.

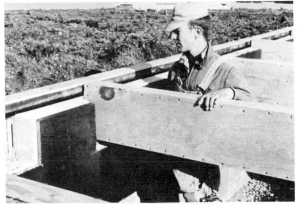

Lumber Dealers Research Council

Fig. 7-45. A typical box beam installation.

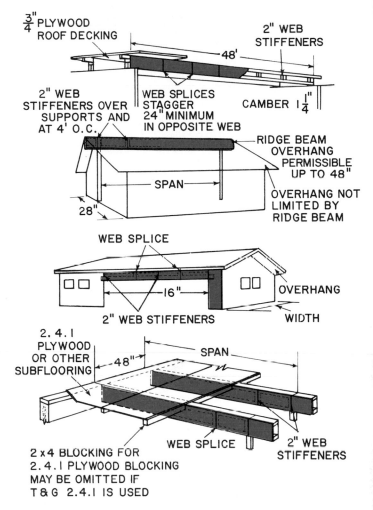

Fig. 7-46. Uses of box beams. Top to bottom: As flat-roof girders. As garage lintel. As floor beam.

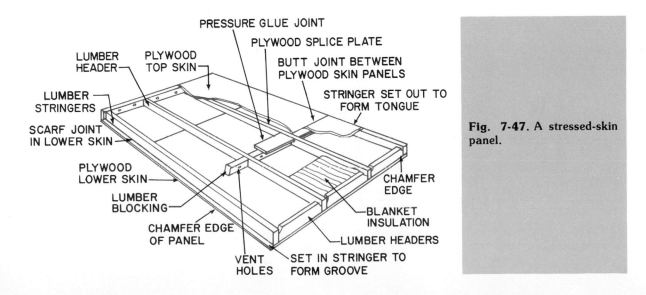

Fig. 7-47. A stressed-skin panel.

American Plywood Association

The glue joint between the plywood cover skins and framing members enables the panel elements to act as one unit in resisting loads. The action is similar to a series of adjoining, built-up, wooden I beams. See Fig. 7-48.

Since these panels are fabricated and can be quickly installed, they save on-the-site labor costs and enable a builder to get a house "under roof" rapidly. They save additional time because required insulation can be installed as they are manufactured. These panels are dimensionally stable, light in weight, and have high strength. Flat panels are shown in Fig. 7-49, and vaulted panels are shown in Fig. 7-50.

Structural design data for stressed-skin panels can be found in Chapter 9.

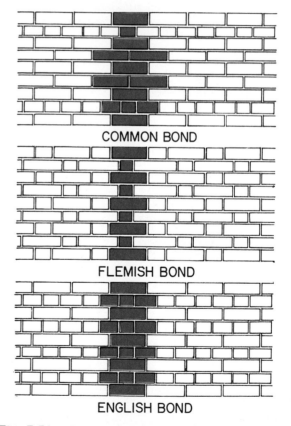

Fig. 7-50.   A stressed-skin panel vault. Vaults have honeycomb core with plywood skins.

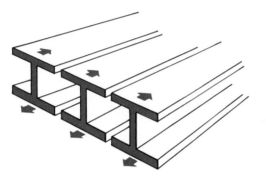

Fig. 7-48.   Stressed-skin panels produce an action similar to a series of wood I beams.

Lumber Dealers Research Council

Fig. 7-49.   A typical installation of stressed-skin panels on a roof.

COMMON BOND

FLEMISH BOND

ENGLISH BOND

Fig. 7-51.   Frequently used bond patterns for brick walls.

Common bond - every sixth row is turned to form a tie between the two rows of brick in a solid masonry wall.

Flemish bond - every other brick is turned to form a tie.

English bond - every other row is turned to form a tie.

Solid Masonry Construction

While many methods and materials are used to construct solid masonry walls, only those commonly used are discussed. Refer to a standards book for additional information.

Solid masonry walls are built entirely from one material — such as brick, stone, or concrete block — or are constructed with a brick or stone veneer covering concrete blocks or tile backing.

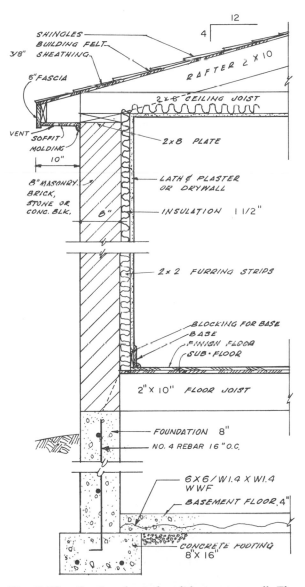

Fig. 7-52. Section through solid masonry wall. The hidden line shows the beveled cut on the end of the floor joists.

Solid Brick or Stone

Exterior walls for one- and two-story houses may be constructed of solid brick or stone. Careful selection of the bricks and the pattern of laying them is essential to the appearance of the house. See Fig. 7-51 for frequently used bond patterns for brick walls.

Details of solid masonry wall construction are shown in Fig. 7-52. Notice the beveled cut on the end of the floor joists. Since the masonry wall is heavy, any movement in the ends of the floor joists tends to crack the wall. The bevel cut reduces this danger. A solid masonry house with a concrete floor is illustrated in Fig. 7-53.

Furring strips are necessary for high-quality work. A **turring strip** is a 2 × 2 wood member fastened on the interior of a masonry wall to insure that an air pocket is left between the wall and the interior of the house. Solid masonry is a poor insulator, and the interior of the house would be very hot in summer and cold in winter without the air space provided by the furring strips. This space also permits installation of insulation, to further decrease heat loss through the wall.

If a solid masonry wall is not furred out, moisture will condense on the walls, due to warm, moist interior air contacting the cold, exterior wall.

The solid masonry wall needs to be "tied together" into a solid mass. This is most often done by turning an occasional brick or an entire layer of bricks sideways to form a tie. See Figs. 7-51 and 7-54.

Another variety of solid masonry wall has two thicknesses of brick, spaced 1 inch (25 mm) apart.

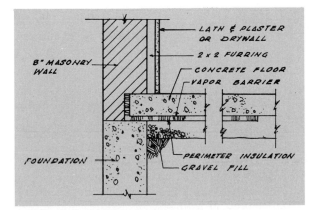

Fig. 7-53. Section through a solid masonry wall with concrete slab floor on grade.

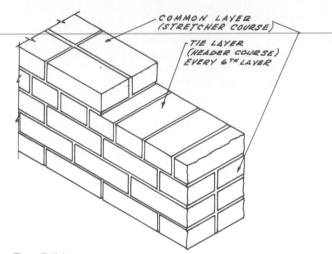

Fig. 7-54. Brick wall in common bond showing tie made with a header course.

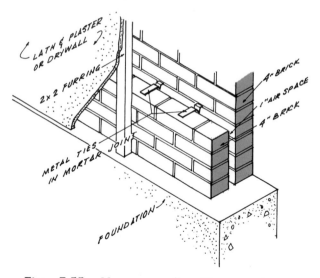

Fig. 7-55. Masonry wall with an air space separating each unit; this is frequently called a **cavity wall.**

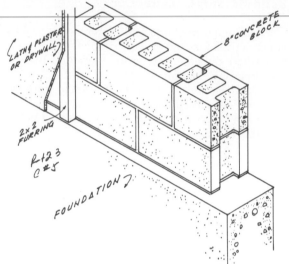

Fig. 7-56. Typical construction of a concrete block wall.

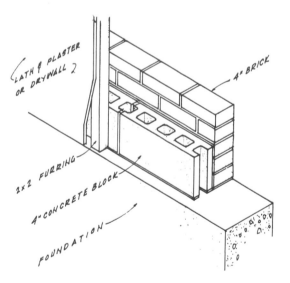

Fig. 7-57. Brick veneer over concrete block wall.

This provides a dead air space and reduces some of the difficulties of the solid masonry wall. However, even with this type wall, it is still best to furr out the interior. See Fig. 7-55.

A low-cost, solid masonry wall can be constructed from 8-inch concrete blocks. The wall can be erected quickly, and the holes in the block provide dead air space. Furring the interior surface is recommended. See Fig. 7-56. Concrete blocks are manufactured in a variety of textures and faces. This breaks the monotony of the large block and makes an attractive wall.

Brick Veneer over Concrete Block or Tile

To have a faster-built, less-expensive wall, a 4-inch brick facing can be used over 4-inch concrete block or over terra-cotta tile. This method now is widely used instead of building walls of solid brick or stone. See Fig. 7-57. These walls should be furred out for insulation purposes. Frequently, a 1-inch (25 mm) air space is left between the brick and the concrete block or tile.

Masonry Veneer over Frame Construction

A frame house with a masonry veneer has all the advantages of a frame house; yet, it has the low maintenance qualities of a masonry house. An examination of Fig. 7-58 shows that the conventional frame wall is used. This provides the fine insulation offered by a frame house. The high exterior-maintenance cost of the frame house is reduced by the use of the brick or stone veneer instead of wood siding. A masonry-veneer house with a concrete floor is illustrated in Fig. 7-59.

The masonry veneer is placed 1 inch (25 mm) away from the sheathing. This dead air space serves as an insulator, but more importantly, as a space to carry away condensation that forms on the inside of the masonry wall.

A house constructed with a masonry veneer over a frame wall must have a larger foundation than usual. This is not because of the extra weight, but because this construction is thicker than solid masonry walls and typical frame walls.

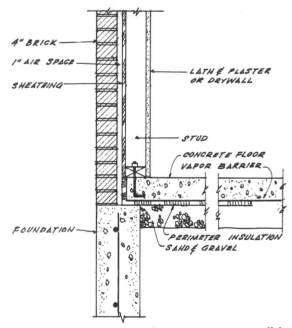

Fig. 7-59. Section through masonry veneer wall for a house with a concrete slab floor on grade.

Framing the Roof

A conventionally framed roof is composed of rafters and a ridge board. The ridge board is a horizontal board upon which the rafters rest at their upper ends. It adds little to the strength of the structure, but aids in the erection and alignment of the rafters. Usually it is lumber about the same width as the rafters. The **rafters** support the sheathing, builder's felt, and roofing material. Since there are various styles of conventionally built roofs, different kinds of rafters are necessary. In Fig. 7-60, the common kinds of rafters used in roof construction are indicated.

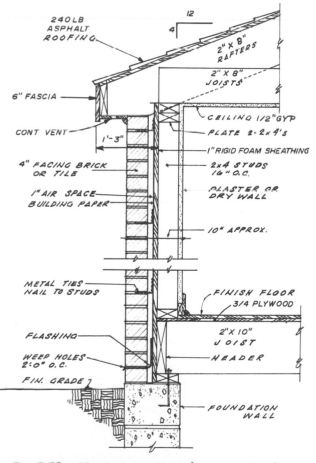

Fig. 7-58. Masonry veneer over frame construction.

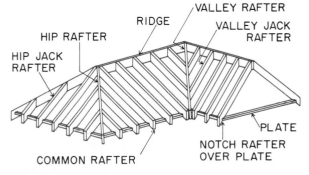

Fig. 7-60. Types of rafters used in roof framing. Compare the sizes of the various types.

Common rafters are straight-run rafters, extending without interruption from the ridge to the eave.

Valley rafters run from the ridge to the eave, but are located at the point where two sections of a roof come together forming a trough or valley.

Valley jack rafters are short members, running from the ridge to the valley rafter.

Hip rafters run from the ridge to the eave, but are located where two sections of a roof come together to form a peak or ridge. Since hip rafters and valley rafters span longer distances than common rafters, frequently they must be larger to carry the additional load caused by the greater distance.

Hip jack rafters run from a hip rafter to the eave. They are shorter than common rafters.

When a roof is framed by conventional methods, the rafters are cut on the ground. (The method for ascertaining the proper size rafters to use is explained in Chapter 9.) They are nailed into place one by one. Then the sheathing is applied followed by a covering of builder's felt. The roofing material, such as asphalt shingles, is nailed over the builder's felt.

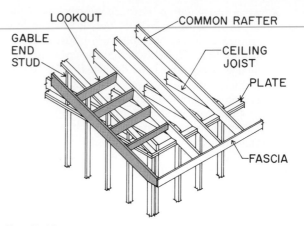

Fig. 7-63. The gable roof overhang at the gable end is formed with lookouts.

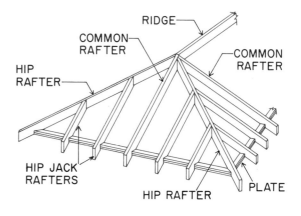

Fig. 7-64. Framing for a hip roof.

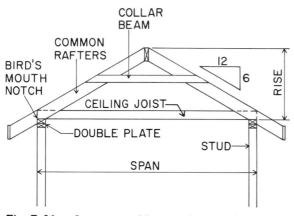

Fig. 7-61. Conventional framing for a gable roof.

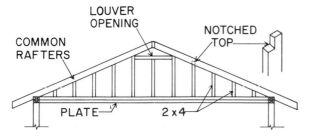

Fig. 7-62. The framing of a gable end. The louver opening is for attic ventilation.

The members for a conventional gable roof are shown in Fig. 7-61. Notice that the rafter is notched to fit on the plate. This notch is often called a bird's mouth. The collar beam may be added to help reduce stress on the rafters. At the end of the building, the gable is framed to receive the sheathing. See Fig. 7-62. Also, the roof usually has some type of lookout over the gable end. See Fig. 7-63.

Framing for a hip roof is shown in Fig. 7-64. This type of roof eliminates the gable end. Ventilation for the attic is accomplished by metal vents placed on the roof and screened openings in the soffit.

A typical shed roof is shown in Fig. 7-65. It slants in one direction. The shed rafter is much like the common rafter except it has a bird's mouth notch on each end.

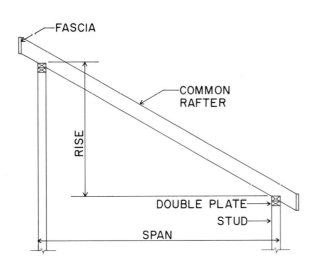

Fig. 7-65. Typical framing for a shed roof.

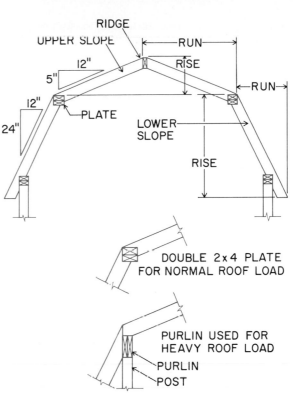

Fig. 7-67. Framing for a gambrel roof.

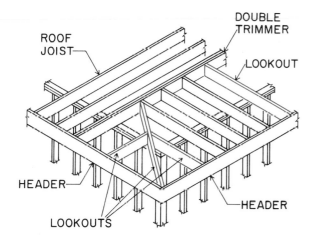

Fig. 7-66. Cantilever lookouts are used to give overhang to a flat roof.

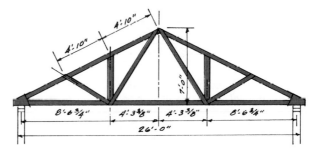

Fig. 7-68. Typical roof truss for a small house.

A flat roof has little or no slope. It requires the least materials and is the easiest to build. The rafters serve to hold any exterior ceiling which may be needed. If an overhang is wanted, the rafters can extend beyond the wall on two sides. Lookouts must be built on the other two sides. See Fig. 7-66.

A gambrel roof is one of the more difficult to build. It is much like the gable roof except it has two slopes. This style requires four rafters to span a house. See Fig. 7-67.

Frequently, a builder will use wood roof trusses, instead of conventional rafters. A **truss** is a preassembled unit, consisting of two rafters, a ceiling joist, and all necessary members to strengthen the roof. See Fig. 7-68. A definite savings in materials and labor results through the use of trusses.

Small wood members can be used in truss construction, because of the truss design. It is braced similar to the framing on a bridge, enabling small members to span long distances to support the necessary weight.

Another savings is in labor. Since trusses are assembled on the ground, they can be easily and quickly made. The time spent to install them is reduced considerably from that required for conventional roof framing.

Roof trusses also offer the advantage of increased flexibility for interior planning, since they span the entire width of a building and require no supporting wall inside the house. Design data for truss construction are in Chapter 9.

Post, Plank, and Beam Construction

Post-and-beam construction lends itself well to the design and construction of contemporary houses. It permits an open interior area, free from supporting walls. It also frees the exterior walls from supporting the roof, thus enabling the use of large areas of glass and lightweight materials in these walls. The need for large, long, heavy footings is eliminated.

Post-and-beam construction is a simple framing system that can be rapidly erected. There are several ways it can be used. Figure 7-69 shows a system using beams as rafters and joists. The roof and floor decking are structural planks. Figure 7-70 shows a system using beams the length of the structure. The roof decking is applied transversely over the beams. The top of the beam is shaped to match the roof slope.

The spacing of these members varies with their size and the species of wood used. Tables for selecting these structural members are in Chapter 9.

A frequently-used roof decking is wood planking. See Fig. 7-71. This decking becomes the finished interior ceiling and utilizes the natural color and grain of the wood. See Fig. 7-72. Composition decking is also manufactured for this use. See Figs. 7-73 and 7-74. Tables for selecting decking are in Chapter 9. Stressed-skin panels can be used. A description of these is also in Chapter 9.

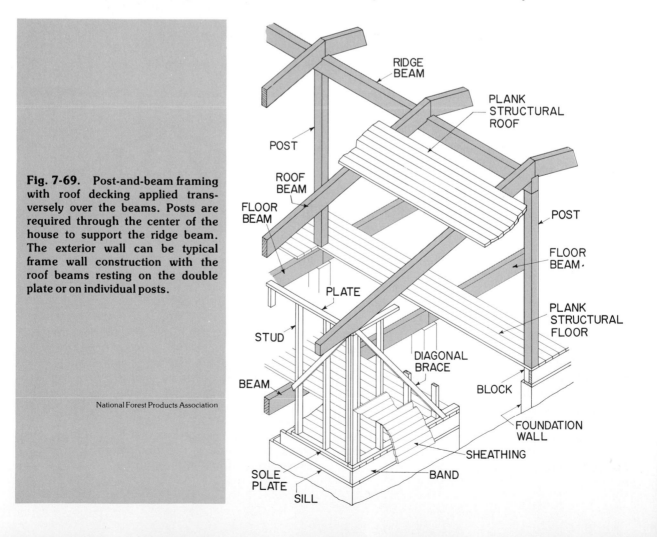

Fig. 7-69. Post-and-beam framing with roof decking applied transversely over the beams. Posts are required through the center of the house to support the ridge beam. The exterior wall can be typical frame wall construction with the roof beams resting on the double plate or on individual posts.

National Forest Products Association

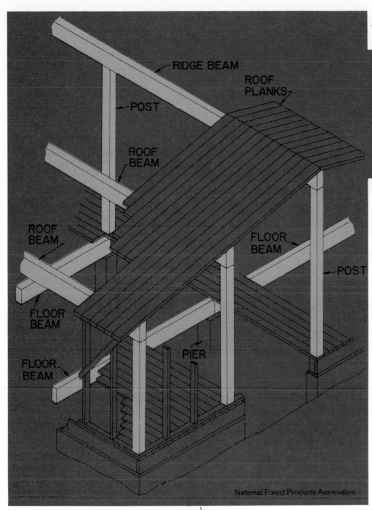

Fig. 7-70. Post-and-beam framing with roof decking applied transversely over the beams. Posts are required through the center of the house to support the ridge beam. The exterior wall can be typical frame wall construction with roof beams resting on the double plate.

Fig. 7-72. An interior view of a framed house with exterior walls in place. The ceiling is wood roof decking.

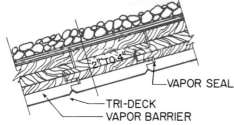

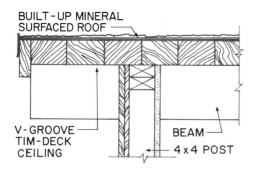

Fig. 7-71. Sections through composition roof decking (top) and wood roof decking.

Fig. 7-73. Composition roof decking being installed on a house with an open-beam ceiling.

Fig. 7-74. Interior view of a ceiling built with composition roof decking.

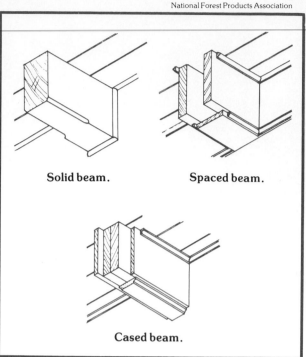

Solid beam.   Spaced beam.

Cased beam.

Fig. 7-76. Typical beams used in post, plank, and beam framing systems.

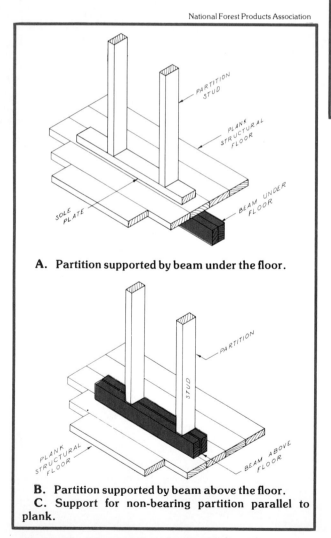

A. Partition supported by beam under the floor.

B. Partition supported by beam above the floor.
C. Support for non-bearing partition parallel to plank.

Fig. 7-75. Unusual floor loads require extra structural support.

The two-inch roof and floor decking provides adequate insulation in moderate climates. In cold climates, additonal insulation is often needed. Sheets of rigid insulation can be applied to the bottom of the planks. However, this hides the beauty of the natural wood interior ceiling. If this is undesirable, a rigid insulation can be applied on top of the roof decking and the roofing material over this. See Chapter 12 for further information.

The finished flooring is applied over the floor decking in the same manner as conventional construction. Typical construction details are shown in Figs. 7-75 through 7-85. The plank floors will carry normal uniform loads. If a load is to be concentrated in one area, as a bearing partition or bathtub, additional framing is needed. See Fig. 7-75. The extra support can be above or below the floor decking. Regardless of where they are placed, they must transmit the load to the main structural members.

Non-bearing partitions that run parallel with the floor planking also need extra support, Fig. 7-75.

Three types of beams are commonly used. These are solid, built-up cased, and spaced beams. See Fig. 7-76. The spaced beam permits electrical wiring and plumbing to run inside the beam, Fig. 7-77.

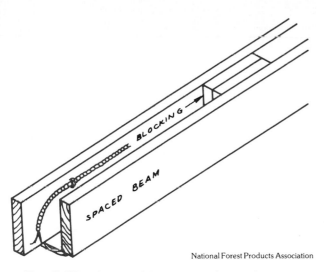

Fig. 7-77. A spaced beam provides a place to run electrical wiring and plumbing.

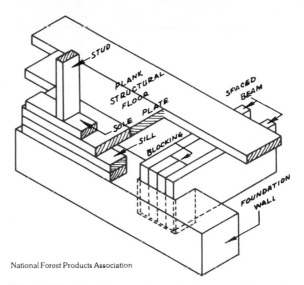

Fig. 7-79. Another way to set a beam and frame the building at the sill.

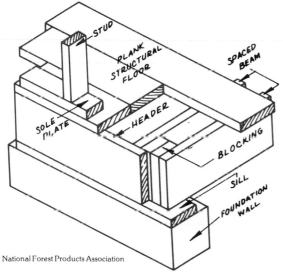

Fig. 7-78. Typical framing at the sill. A spaced beam is shown but a solid or built-up beam could be used.

Construction details at the sill are shown in Figs. 7-78 and 7-79. These will permit the house to have a crawl space or basement. If a slab floor is to be used, construction could be as shown in Fig. 7-80.

The floor beams will be supported in the center of the house by a column or pier. If solid beams are used, they would be joined as in Fig. 7-81. The column can be made wider by adding bearing blocks. This permits the bearing surface of the column to be increased. Normally, at least a 6-inch width parallel

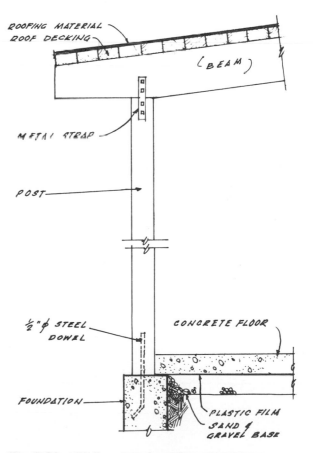

Fig. 7-80. Wall section through post-and-beam construction. What do you think is the purpose of the steel dowel?

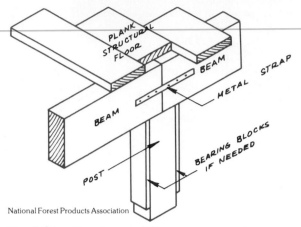

National Forest Products Association

Fig. 7-81. Framing a solid beam over a post.

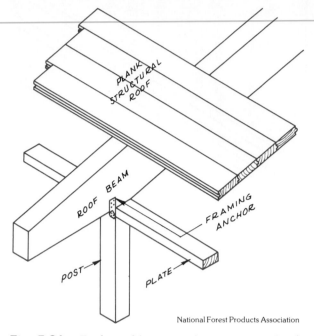

National Forest Products Association

Fig. 7-84. Each roof beam is above a post. Metal framing anchors join them together.

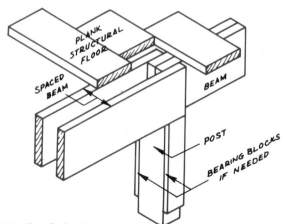

National Forest Products Association

Fig. 7-82. Framing a spaced beam over a post.

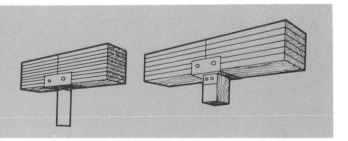

Fig. 7-83. The same method may be used to connect a pipe column to a beam or a wood post to a beam.

to the beam is needed. If solid or built-up and spaced beams are used, the construction would be as shown in Fig. 7-82. Another method of connecting columns and beams is shown in Fig. 7-83.

Roof beams serving as rafters should be directly above the posts. Figure 7-84 shows how a roof beam could be joined to the exterior wall. Metal framing anchors or angle clips are used. There are several methods for joining the roof beam to the ridge beam. These use metal connectors. See Fig. 7-85.

The areas between the posts on exterior walls are filled with windows, doors, or solid panels. The solid panel areas provide a place to frame in lateral bracing. These walls are usually framed with 2 × 4 studs as in conventional construction. Lateral bracing is accomplished by letting in a diagonal brace or using sheathing, such as plywood or fiberboard. Refer back to Fig. 7-69.

Posts may be solid or built-up from 2-inch material. They must be of a size to carry the design load. Tables for this are in Chapter 9. Usually, no posts smaller than 4″ × 4″ are used.

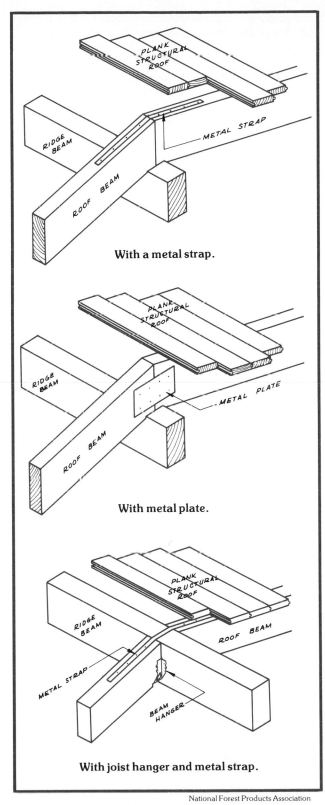

With a metal strap.

With metal plate.

With joist hanger and metal strap.

National Forest Products Association

Fig. 7-85. Typical ways to support the roof beam on the ridge beam. This absorbs horizontal thrust.

Steel House Framing

The steel-framed house can be framed with few members and can have long spans uninterrupted by columns. Spans of 30 feet (9144 mm) and longer are common; this allows big open areas. See Fig. 7-86. All structural members can be slender, and these are pleasing to the eye when left exposed. Construction of overhangs, balconies, and covered walkways is easily accomplished. See Fig. 7-87.

Since the steel frame carries all loads, the choice of wall panel material is unlimited. Metal, wood, plastic, or glass panels may be used with equal

Bethlehem Steel Company

Fig. 7-86. Steel framing allows large open areas. Heat and light ducts are carried in an interior soffit. Notice balcony at treetop height.

Bethlehem Steel Company

Fig. 7-87. Patio of steel-framed house. Notice wide overhang. Stairs lead to sun deck on roof.

Bethlehem Steel Company

Fig. 7-88. Entire floor area of house is supported by wide flange steel sections on four concrete pylons embedded in rock mass.

Bethlehem Steel Company

Fig. 7-89. Steel enables house to cantilever large sections.

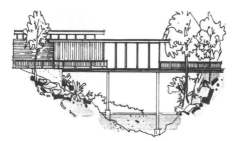

Fig. 7-90. A steel-framed house spanning a ravine.

ease. Interior planning is freed from the restrictions of supporting walls.

With steel framing, a house can be built on the side of a steep hill or on top of a rock formation. The house can be cantilevered over a cliff or can span a ravine. See Figs. 7-88 through 7-90.

Steel framing is not affected by termites. It will not sag, burn, shrink, or rot. Usually fabricated in a shop, the steel framing for an average residence can be erected on the site in less than one day.

The Factory-Manufactured House

Throughout the centuries, the search for a better and easier way to build dwellings has continued. Although the prefabricated structure was a natural development, it is not a current development. The pyramids built in 3000 B.C. were prefabricated. The limestone blocks from which they were con-

structed were preformed and then hauled to the building site where they were placed in predetermined locations.

Roman armies carried prefabricated shelters. Prefabricated houses were shipped from the East to California during the Gold Rush. The Union Army used prefabricated structures throughout the Civil War. Mark Twain's boyhood home in Hannibal, Missouri, is a prefabricated structure.

Before World War II, the biggest step in prefabrication in the United States was in the precut house. The factory-cut parts for the house were shipped to the site and most homeowners assembled the parts themselves. Builders began to realize the value of prefabricating houses in larger parts. The first break-throughs were the factory-assembled window and the prefitted door and door frame. Kitchen cabinets are also factory made.

Since World War II, modern construction methods and new design approaches have made possible the construction of houses in factories. These are transported to the building site for final erection and finishing.

Special-purpose machines, such as power nailers, are used. Adhesives have been developed that speed construction. Large jigs are used to assemble sections of the house. They can also turn sections over so both sides can be finished.

A large company can take advantage of mass purchases of material, thus lowering the cost of the house. Quality can be more closely controlled in a factory than on the site. The total working time required to build a house can be reduced. Bad weather does not stop construction.

Currently, there are two approaches being used to construct houses in factories. One uses **modular components.** These are panels or sections of walls, roofs, and floors. The other approach is **modular units.** A modular unit is a room size or larger portion of a house that is completely assembled in a factory. It is moved to the site much like a mobile home. These approaches are discussed in detail in the following pages.

Buildings from Modular Components

A **modular component** is a factory-assembled part of a building. Typical examples are wall panels, floor panels, roof panels, and trusses.

The use of standardized building components enables a builder to construct a house in the fastest possible time with the least waste of materials. This helps reduce the cost of the house. It is applying the principles of mass production of standardized parts as used in the auto industry to building construction.

The system of modular components described in the following pages was developed by The National Forest Products Association.

Standardization

Companies manufacturing building materials are making an effort to change their product sizes to fit a modular plan. All types of materials, such as floor tile, plywood paneling, and sheathing panels are sized to fit the modular plan.

The standard module for the U.S. Customary system of measurement is a 4-inch cube. This unit is used to build larger modules. A 4'-0" cube is a major module. It is made of 4-inch modules. Minor modules are 16-inch and 24-inch cubes. See Fig. 7-91.

The standard module for a metric modular system will probably be 100 mm. The major module will probably be 1200 mm, and the minor modules 400 mm and 600 mm. See again Fig. 7-91.

The modular system includes length, width, and height. All three must be considered for maximum effectiveness.

In Fig. 7-92, the modular coordination of house elements is shown. The roof trusses are designed to be spaced 24-inch on center. This allows the use of

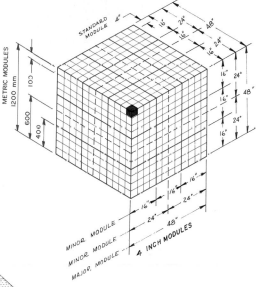

Fig. 7-91. The modular system is based on a 4" (U.S. Customary) or 100 mm (metric) standard module. Length, width, and height are all taken into consideration.

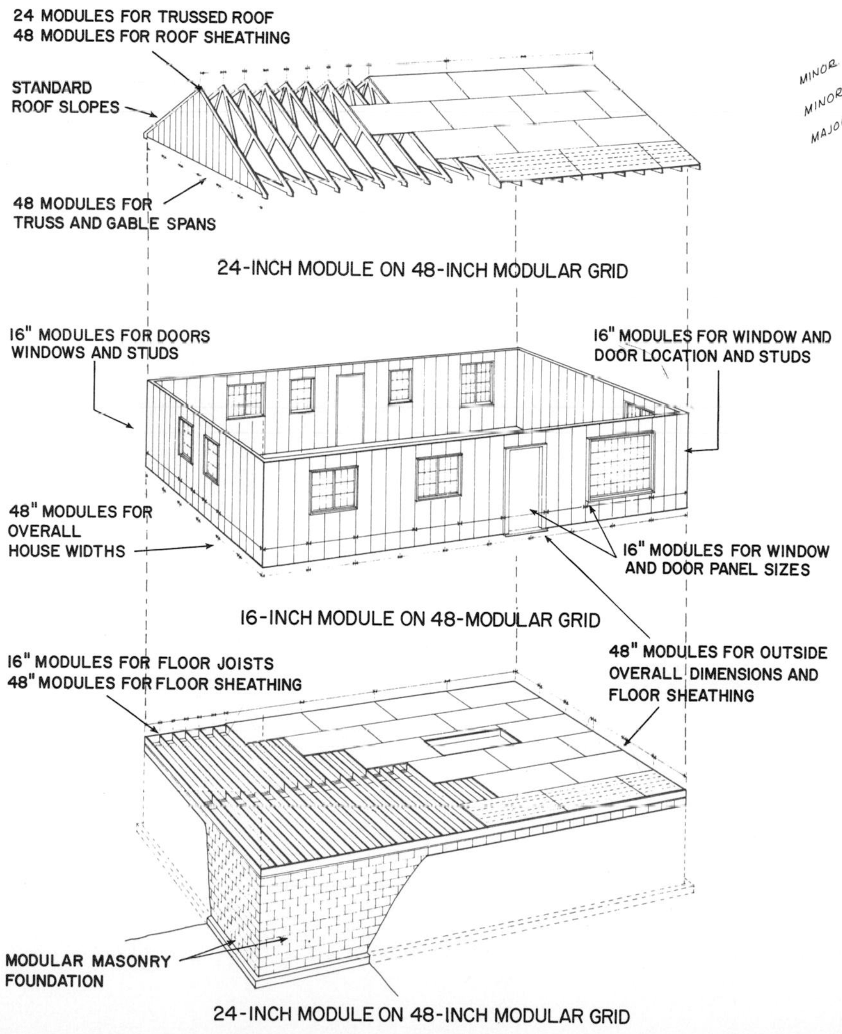

24 MODULES FOR TRUSSED ROOF
48 MODULES FOR ROOF SHEATHING

STANDARD
ROOF SLOPES

48 MODULES FOR
TRUSS AND GABLE SPANS

24-INCH MODULE ON 48-INCH MODULAR GRID

16" MODULES FOR DOORS
WINDOWS AND STUDS

16" MODULES FOR WINDOW AND
DOOR LOCATION AND STUDS

48" MODULES FOR
OVERALL
HOUSE WIDTHS

16" MODULES FOR WINDOW
AND DOOR PANEL SIZES

16-INCH MODULE ON 48-MODULAR GRID

16" MODULES FOR FLOOR JOISTS
48" MODULES FOR FLOOR SHEATHING

48" MODULES FOR OUTSIDE
OVERALL DIMENSIONS AND
FLOOR SHEATHING

MODULAR MASONRY
FOUNDATION

24-INCH MODULE ON 48-INCH MODULAR GRID

Fig. 7-92. Modular coordination of house elements.

All structural and aesthetic elements of a house are related. This example house is designed on the 48-inch module. The coordinated modular increments of its structural elements are shown in the drawings. Standard sizes of various existing materials will easily fit the design.

National Forest Products Association

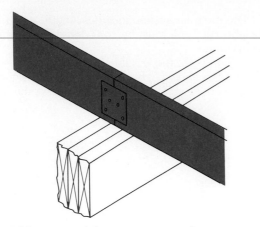

Fig. 7-93. In modular construction, floor joists are butted end to end.

plywood roof sheathing which is 48″ × 96″. The sheathing has no waste. If a 16-inch on center spacing was used, the same sheathing would be used with no waste.

The exterior wall components are built on a 16-inch module. This provides normal structural support. It also allows the use of a wide variety of doors and windows. The common panel widths in inches are 16, 32, 48, 64, 80, 96, and 144. The overall house width and length is based on the 48-inch module. This allows the most effective use of floor sheathing and joists.

When using the modular component design, the floor joists must be butted. See Fig. 7-93. This allows each joist to stay on the 16-inch module.

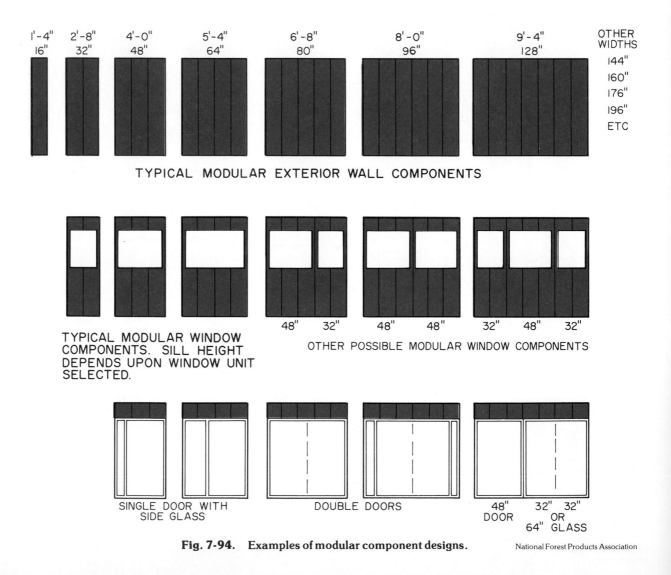

Fig. 7-94. Examples of modular component designs.

National Forest Products Association

Examples of typical door and window components are illustrated in Fig. 7-94. There are many other possible designs than those shown. Any type of door or window can be used. However, the component must be framed to receive them. The sill height of the window will vary with the window selected. A company could manufacture several window components having the same widths but different window sill heights.

An example of how a typical component is framed is shown in Fig. 7-95. It is framed using the same structural standards as conventional construction. The width is based on modular size. This example is 64 inches wide. This is equal to four 16-inch minor modules. The window opening is the rough opening size. The window unit is installed in this opening. Windows selected are those that fit the module. The finished unit has the window installed and sheathing, siding, and sometimes interior finished wall material in place.

Examples of how outside and inside corners are framed where exterior walls meet are in Figs. 7-96 and 7-97. An outside corner requires that one component have the end stud set-in the thickness of the frame wall it will meet. The sheathing overhangs and is nailed to the second component forming the corner.

Two exterior walls forming an inside corner are shown in Fig. 7-97. This requires an extra stud in the component. The wall is perpendicular to the stud.

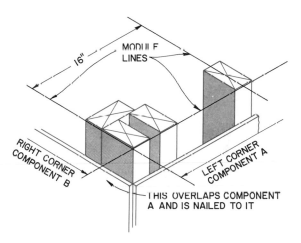

Fig. 7-96. The size of the component is from stud to stud. This is typical framing at an outside corner of exterior wall.

National Forest Products Association

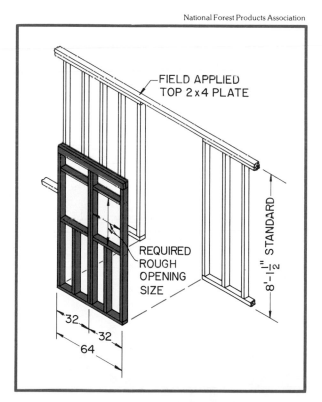

Fig. 7-95. Typical framing for window component. This 64″ unit is framed for two windows to fit in the rough opening between studs. Notice that the studs are doubled where the components join. The factory-finished component will have the window installed and the sheathing and siding in place. Often, the interior wall is also finished.

National Forest Products Association

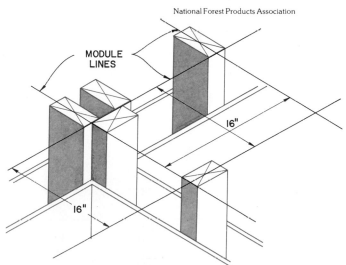

Fig. 7-97. Typical framing at inside corner formed when two exterior walls meet.

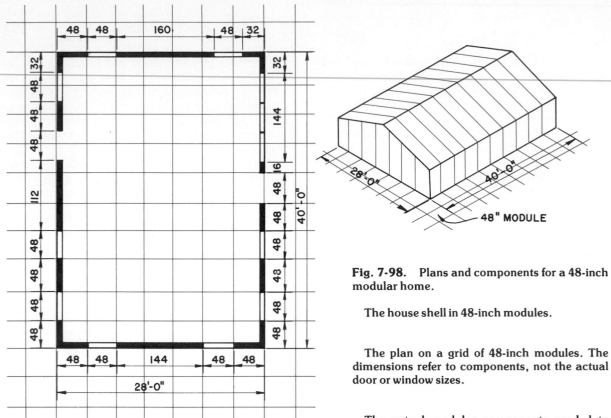

Fig. 7-98. Plans and components for a 48-inch modular home.

The house shell in 48-inch modules.

The plan on a grid of 48-inch modules. The dimensions refer to components, not the actual door or window sizes.

The actual modular components needed to form the exterior of the house plan below.

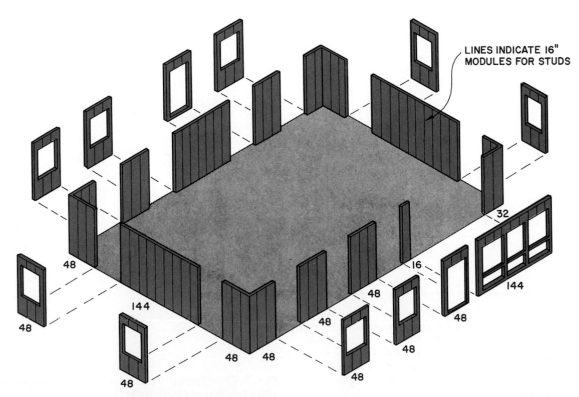

LINES INDICATE 16"
MODULES FOR STUDS

National Forest Products Association

Application of Modular Components

A plan for the exterior walls of a small house is shown in Fig. 7-98. It is based on a 48-inch major module grid. The walls are made up of modular components. Some of the components contain the needed doors and windows. Others are simply solid wall sections. The individual components can be as small as 16 inches. They can be of any width that permits the use of the 16-inch module. Reexamine Fig. 7-94 for examples.

Assembling a House with Modular Components

The foundation is located and constructed in the same manner as that for the conventional house. The floor joists, cut to the proper length and notched in the factory, are installed first. See Fig. 7-99. The subflooring of tongue-and-groove, plywood sheets covers a great area rapidly. The tongue-and-groove construction helps bind the subfloor together into a well supported, integral unit. See Fig. 7-100.

After the subfloor is installed, the wall panels are unloaded and placed. Usually, the first panels to be erected are those forming a corner. These panels consist of the studs, sole, plate, sheathing, siding, insulation, and the sheetrock on the interior of the panel.

Other exterior wall panels are unloaded and are joined to the corner panels. These panels may contain door openings and windows. Usually the windows are already installed. Doors are generally installed after the panel is erected.

After the exterior wall panels are placed, the gable ends are unloaded and installed. See Figs. 7-101 and 7-102.

National Homes Corporation

Fig. 7-99. Assembling factory-cut-and-notched floor joists. Notice the joists are numbered to speed installation. The notches are used to speed location of the joists.

National Homes Corporation

Fig. 7-100. Installing tongue-and-groove plywood subfloor.

National Homes Corporation

Fig. 7-101. A factory-assembled gable end being unloaded. Notice that the last wall panel is left out to enable workers to carry in the gable ends and trusses.

National Homes Corporation

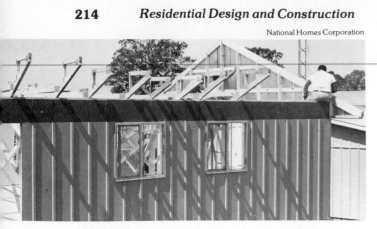

Fig. 7-102.   The gable end raised into place. Trusses are positioned and ready to be raised.

National Homes Corporation

Fig. 7-103.   Nearly completed exterior.

Next, factory-assembled roof trusses are raised in place. Many times they span the entire dwelling and require no interior supporting wall. Plywood sheathing is then applied.

On the interior, ceiling insulation is applied and sheetrock is installed in the conventional manner. The other trades, such as plumbing and heating, coordinate their work. Frequently, the plumbing is in factory-assembled sections, which are set into the house as a unit.

In Fig. 7-103, workers are putting the finishing touches on a factory-manufactured house. This particular house was built on a concrete slab which contained the plumbing and heating ducts. The shell for such a house can be erected in a few days by a very small crew.

Building from Modular Units

A **modular unit** is a factory-built, finished, room size or larger section of a building. See Fig. 7-104. It usually contains in assembled form all the required parts of that section of the building. Commonly, these include interior and exterior wall finish, roof, insulation, floor, wiring, heating, plumbing, and painting. The unit is ready for oc-

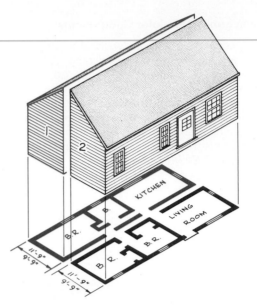

Fig. 7-104.   A two-module building. Modules are commonly 9′-9″ or 11′-9″ wide.

cupancy within a few days after it is set on the foundation. Generally, the only on-site work is the joining of the modules, applying a few trim pieces, and connecting the utilities. The foundation is prepared before the module is delivered to the building site.

There is no standard modular unit design. The designs in use vary considerably. They are the result of research and experimentation of many individuals and companies. The following discussion shows how some companies have designed and put modular units into production.

Modular Units in Residential Design

Following are some general factors for consideration when designing a residence to be constructed with modular units. Various approaches are being used and new designs are being developed regularly.

A preliminary consideration is the overall width and length of the module. Since modules are factory-built, they must be moved to the building site. State laws control the size of units that can be moved on highways. In this case, they would actually be moved like a trailer. Laws vary from state to state. In general, trailers over 10′-0″ (3048 mm) wide require a special permit. Some states allow moving up to a 12′-0″ (3658 mm) width with a special permit.

It is recommended that the 12′-0″ width be used whenever possible. The actual width of the module will vary depending on the eave overhang. A

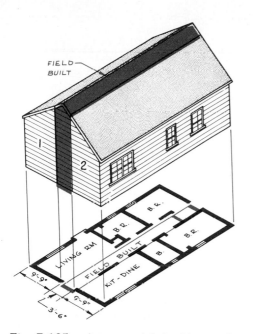

Fig. 7-105. A two-module building with a field-built section to add width.

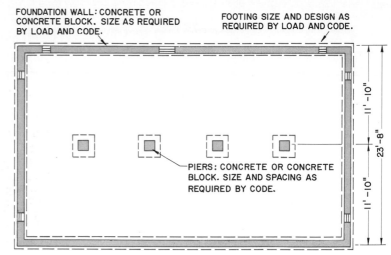

Small Homes Council, University of Illinois

Fig. 7-106. Typical foundation plan for two-module building.

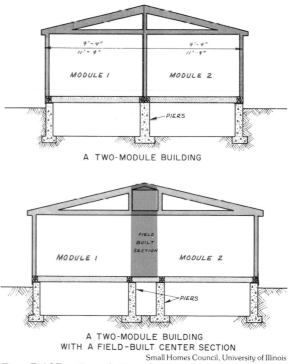

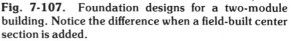

Small Homes Council, University of Illinois

Fig. 7-107. Foundation designs for a two-module building. Notice the difference when a field-built center section is added.

typical width is 11'-9" to the outside of the framing. The 3" remaining are used by roof overhang and trim. The 11'-9" width will normally give an interior room width of 11'-3".

When 12'-0" sections are used, the house can be divided into two equal modules. The division is made along the central bearing wall. Refer again to Fig. 7-104.

If the width must be held to 10'-0" and a 3" eave is used, the module size will be 9'-9". If the overall 20'-0" width is inadequate, a field-built center section can be added. See Fig. 7-105. Here two 10'-0" modules were built with the central hall section added on the job in the conventional manner.

The length of the module is also regulated by state law. The permissible length includes the module and the tractor pulling it. When length is a factor, the cab-over-engine tractor will permit longer modules to be moved.

Height is a third consideration. State laws govern the height of the truck and its load. A normal maximum height is 14'-0". To meet this requirement, some modules are built with ceiling heights of 7'-6". Low sloped or flat roofs can be used. Low-bed trailers or special-built trailers can reduce the height of the shipment.

Height can be saved by using a thicker but lower floor joist. For example, a 3 × 8 floor joist might be used instead of a 2 × 10. The same adjustments could be made on ceiling joists and rafters.

Foundation Design

A typical foundation design for modular residential design is shown in Fig. 7-106. It is built in a conventional manner before the modules are delivered to the building site. The two modules join over the piers. The foundation is designed to carry the live and dead loads exactly the same as a conventional house. In Fig. 7-107 are sectional views of typical foundation designs.

215

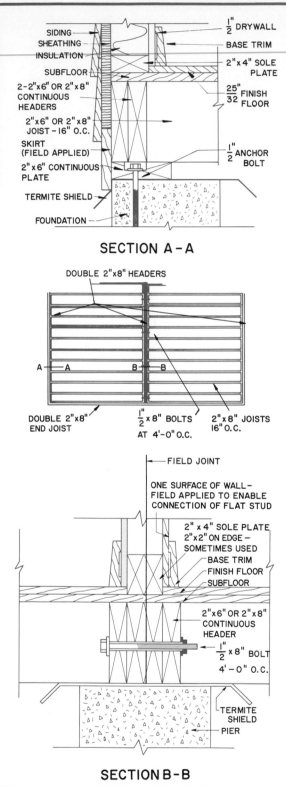

Fig. 7-108. Typical floor-framing plan and construction details.

Floor Construction

Conventional wood framed floor construction is commonly used. The wood headers are doubled around the entire modular unit. See Fig. 7-108. If the span is great enough, the headers must be tripled. A 2 × 8 floor joist spaced 16 inches on center would be satisfactory for most designs. Plywood subflooring is generally used. Any type of finished floor can be laid over it.

Wall Construction

The conventional 2 × 4 stud wall system is frequently used. The studs are placed conventionally except on the wall where the modules join. Here the studs were placed flat 16 inches on center. See Fig. 7-109. Any type of sheathing or siding can be used. If masonry exterior walls are desired, the wall is sheathed in the factory. The brick is applied on the job. The wall is insulated in the factory in the conventional manner. Door frames in the wall where modules join have split frames.

Roof Construction

Typical roof construction for a two-module building is shown in Fig. 7-110. Each module has half the roof. Conventional rafters and ceiling joists are used. The ridge is supported by 2 × 4 studs resting on the interior supporting wall. See Fig. 7-111. The ridge boards are bolted together. Conventional roof sheathing and shingles are factory-applied.

Framing for a two-module building with a field-built section is shown in Fig. 7-112.

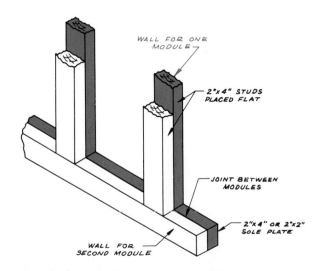

Fig. 7-109. Typical design for interior wall joining modules.

216

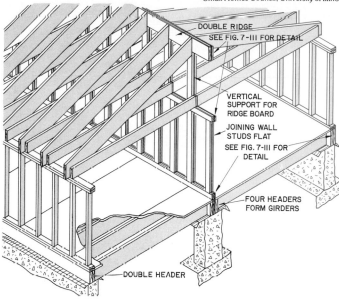

Small Homes Council, University of Illinois

DOUBLE RIDGE
SEE FIG. 7-III FOR DETAIL

VERTICAL
SUPPORT FOR
RIDGE BOARD

JOINING WALL
STUDS FLAT
SEE FIG. 7-III FOR
DETAIL

FOUR HEADERS
FORM GIRDERS

DOUBLE HEADER

Fig. 7-110. Typical framing plan for residence made from two modules.

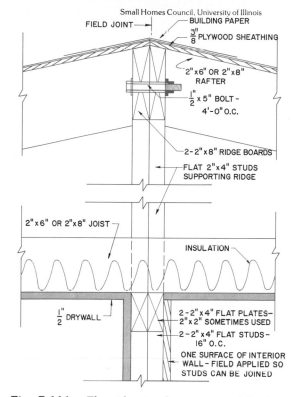

Small Homes Council, University of Illinois

FIELD JOINT — BUILDING PAPER
⅜" PLYWOOD SHEATHING

2"x6" OR 2"x8" RAFTER

½" x 5" BOLT –
4'-0" O.C.

2-2"x8" RIDGE BOARDS

FLAT 2"x4" STUDS
SUPPORTING RIDGE

2"x6" OR 2"x8" JOIST

INSULATION

½" DRYWALL

2-2"x4" FLAT PLATES–
2"x2" SOMETIMES USED
2-2"x4" FLAT STUDS–
16" O.C.
ONE SURFACE OF INTERIOR
WALL – FIELD APPLIED SO
STUDS CAN BE JOINED

Fig. 7-111. The ridge is often supported by 2 × 4 studs above the interior supporting wall.

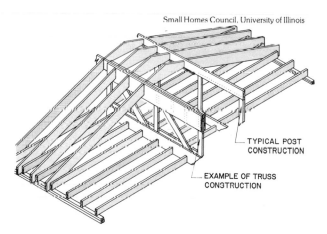

Small Homes Council, University of Illinois

TYPICAL POST
CONSTRUCTION

EXAMPLE OF TRUSS
CONSTRUCTION

Fig. 7-112. Roof construction detail for two-module building with a field-constructed center section.

Instead of using rafters, half-section roof trusses could be used. A half-section roof truss is designed like a full truss except a vertical member is placed at the end over the interior supporting wall. See Fig. 7-113.

Still another approach to roof design is given in Fig. 7-114. Here the roof is pivoted on bolts where the rafters meet the ceiling joists. The module is shipped with the roof in an almost flat position. On the site, the roof is lifted and a knee wall is used to support it. The ridge opening is filled by a small section set in place. This is supported by the knee wall.

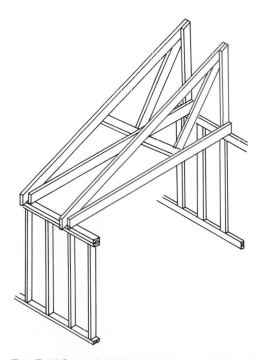

Fig. 7-113. A half truss used on a modular unit.

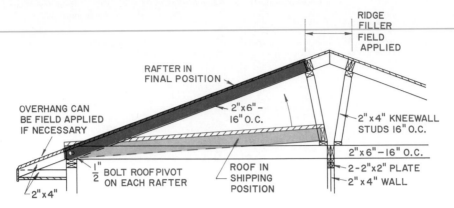

Fig. 7-114. Another roof-framing plan. The roof is shipped in an almost flat position. This reduces the overall height during shipping. On the site, the roof is lifted and supported in place by the knee wall. The ridge filler is set in place. The overhang can be field-applied if the extra width causes shipping problems.

The overhang at the eaves can be fastened to the roof in the factory if the extra width does not hinder shipping. It can be field-applied if necessary.

Preparation for Lifting

Included in the design is some means of lifting the completed module. The module is designed to be supported from below the floor. Therefore, it must be lifted from the same place. A common lifting system is to use a cable sling from a crane. These are attached to some type of ring in the top of the module. The ring must be part of a rod that carries through to the bottom of the module. The sling usually goes to the four corners. However, additional cables can be attached to other rings in the walls. The lifting rods are usually removed after the module is set in place.

Mechanical Considerations

On most modular units, all wiring, plumbing, heating, and air-conditioning requirements are

Fig. 7-115. A two-modular residence. This building is designed for a long, narrow lot.

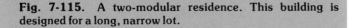

Scholz Homes, Inc.

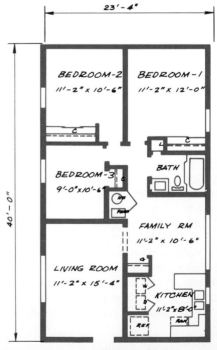

FLOOR PLAN

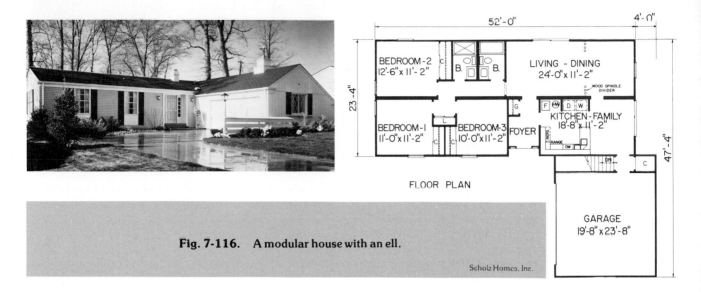

Fig. 7-116. A modular house with an ell.

Scholz Homes, Inc.

factory-installed. Whatever systems are used, the module must be carefully designed to accommodate them. The problems presented by having to move the unit must be carefully studied. The heating system that is easiest to use is electric. It requires no ducts or central heat generating plant. If central heat is used, it appears easiest to run the ducts overhead.

No parts of the plumbing system should extend below the floor joists. Use lightweight plastic drain, waste, and vent piping where accepted. This reduces the weight of the module.

Finishing

Plan for as little on-the-site finishing as possible. Typically, a house can be set and finished in three to seven days. The actual time will depend upon the extent of factory finishing. Some modules are completely finished in the factory. Even outside painting is complete. These require some exterior work to conceal the joint between the modules. A board and batten exterior is excellent for this purpose. If horizontal siding is used, a vertical strip can be used. A better appearance results if the siding is field-applied over the joint. Masonry veneers are field-applied.

The joint in the floor usually needs field-finishing. If it has tile, the tile over the joint is field-applied. It is easier to cover the joint with carpeting.

Some Solutions

The two-module residence shown in Fig. 7-115 is designed for a narrow city lot. Nothing in the ex-

terior appearance indicates it is a factory-built modular house.

A larger el-shaped residence is shown in Fig. 7-116. The house is made of two modules. The garage is field-built from factory-made wall panels.

Modules for Commercial Buildings

A modular building concept using a welded steel frame is illustrated in Fig. 7-117. The steel framing

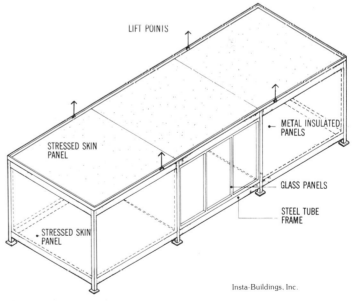

Insta-Buildings, Inc.

Fig. 7-117. This modular system uses a steel tube frame and stressed skin panels. Notice the lift points built into the structure. They are located so the unit is balanced when it is lifted.

provides the structural support for both live and dead loads. The siding, roofing, and flooring panels are attached to it.

Shown in Fig. 7-118 is a bank building constructed from three modules, each 12'-0" wide by 40'-0" long. The use of a standard module permitted the designer to plan a small drive-in bank with one module. See Fig. 7-119. With the use of additional modules, the size of the bank was expanded considerably. Figure 7-119 shows plans for bank buildings using single and multiple modules.

The interior can have large open spaces, as seen in Fig. 7-120. The only evidence of the line of union of the modules is the double post.

This module is constructed on an assembly line. First the steel frame is welded together. The frame has the roof, siding, windows, doors, utilities, partitions, interior siding, carpeting, and other required

Fig. 7-120. Interior showing the union of two modules. Notice the double steel posts at the line of union.

features installed. The unit is completely finished when it reaches the end of the assembly line.

The units are loaded on trucks for delivery to the building site. See Fig. 7-121. At the site, the units are removed by a crane and set on the foundation.

Fig. 7-118.   This is a three-module building.

Fig. 7-121.   The finished units are moved to the site.

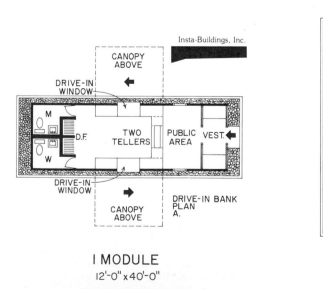

I MODULE
12'-0" x 40'-0"

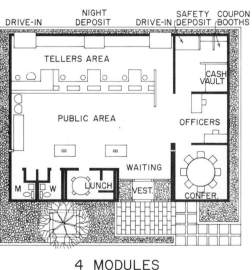

4 MODULES
48'-0" x 40'-0"

Fig. 7-119.   Notice how the addition of modules enables the designer to expand the activities housed in this bank building.

Following are the design specifications for this modular system. See Fig. 7-122.

Exterior walls:	Cedar boards or brick veneer over plastic foam insulation.
Frame:	Steel columns and beams.
Exterior trim:	Anodized aluminum, sealed and caulked.
Floor purlins:	6″ junior 4.4 I-beams 4′-0″ O.C.
Subfloor:	2″ insulated panel with 1/2″ plywood underlayment.
Ceiling purlins:	4″ I-beam 17.7 4′-0″ O.C.
Partition studs:	2 × 4 construction grade Douglas fir, 16″ O.C.
Roofing:	2″ insulated panel, 5-ply asphalt with 15# felt layer plus vapor barrier.
Interior walls:	Drywall (glued) or wood paneling.
Ceilings:	Drywall (glued) or acoustical.
Floors:	Vinyl asbestos tile, ceramic tile, carpeting.
Interior doors:	1 3/4″ flush type, birch.
Windows:	Anodized aluminum, awning type or floor-to-ceiling glass.
Insulation:	Walls 1 1/2″ rigid plastic foam, ceiling and floor 2″ rigid plastic foam.
Heating:	Electric baseboard perimeter.
Air conditioning:	Thru-wall units.

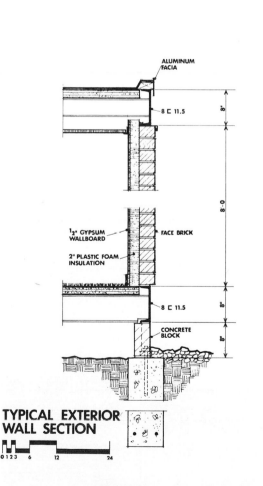

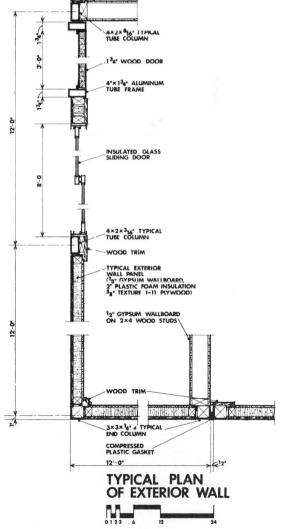

Fig. 7-122.   Construction details for steel-framed modular system.

A Modular System for Apartments

The system to be described was developed by the Magnolia Homes Division of Guerdon Industries, Inc., as experimental housing under the Federal Housing Administration's Experimental Housing Program.

The basic technique used was to design the apartments in units that could be factory-built and stacked on top of each other on the building site. The final design had two modules — an upper module and a lower module — which made up one apartment. See Fig. 7-123.

The lower floor module was 12'-0" (3658 mm) wide and 32'-0" (9754 mm) long. It contained a living room, a kitchen-dining room, a stairwell, and a pantry.

The upper floor was a module 12'-0" (3658 mm) and 34'-0" (10 363 mm) long. It contained a linen closet, a complete bath, a stairwell, and two bedrooms with wardrobes.

The modules were designed using wood frame construction. The interior walls and ceilings were covered with fire-rated sheetrock. These were sprayed with a plaster veneer finish. When dry, this plaster develops a hardness approaching that of concrete. The floor surfaces were 1/8-inch vinyl asbestos black tile.

A totally electrical system was used. The heating system was a baseboard convection radiant heat.

Magnolia Homes Division of Guerdon Industries, Inc.

1ST FLOOR

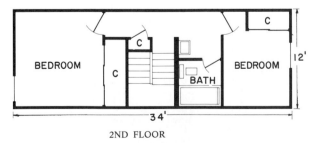

2ND FLOOR

Fig. 7-123. Floor plan for two-story apartment designed for modular construction.

Guerdon Industries, Inc.

Fig. 7-124. The unit is built on an assembly line. Here the floor system is shown. It is mounted on a cart. The floor system has 2 × 6 joists, 16" O.C. with 5/8" plywood subfloor.

Fig. 7-125. The walls are assembled on jigs and moved to the floor system by overhead conveyors. The ceiling is also constructed on jigs and carried to the assembly area with an overhead crane. The wall and ceiling units are covered with drywall and insulated before moving to the floor system. The walls are constructed of 2 × 4 studs, 16" O.C. The ceiling has 2 × 6 joists, 16" O.C.

Guerdon Industries, Inc.

There was no forced air system, no blower, no filter, and no register or heat ducts. The radiant heat system was placed on the outside walls. The water heater and cooking stove were electric. Air conditioning was not planned.

All kitchen cabinets had plastic veneer facing. In the bath area, a fiberglass surface pressure laminated to exterior grade plywood was used instead of tile. The exterior surfaces were exterior grade cedar plywood.

The process of constructing and erecting the apartments is shown in Figs. 7-124 through 7-132.

Guerdon Industries, Inc.

Fig. 7-128. Plaster being sprayed over interior drywall material. This eliminates the need to tape the seams between the drywall sheets. Windows, floors, and appliances are covered with plastic sheets to protect them from the spray.

Guerdon Industries, Inc.

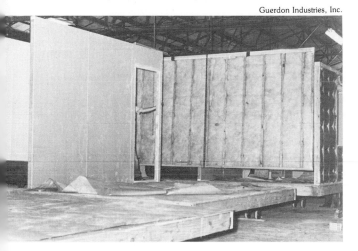

Fig. 7-126. Here the walls are being set in place on the floor. Notice the insulation in the walls in the background.

Fig. 7-127. This is a second floor unit on the production line. Notice the waste vent plumbing tree has been set in place. Factory-installed plumbing greatly speeds the erection time on the site. The bath fixtures are all installed in the factory. On the site, sewer and water lines are connected to the public utilities.

Fig. 7-129. After the units are assembled, the exterior siding is attached, and the aluminum windows are installed. Above and below each window are painted panels. These are painted before they are installed.

Guerdon Industries, Inc.

Guerdon Industries, Inc.

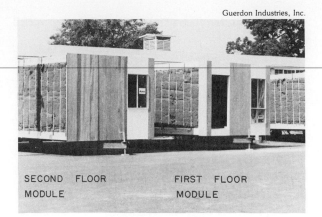

SECOND FLOOR MODULE　　　FIRST FLOOR MODULE

Fig. 7-130. Completed units are placed on trailers for movement to the building site.

Fig. 7-133. The floor plan of modular-designed townhouse.

Fig. 7-131. On the site, a first floor module is lifted from its trailer. It will be set on a foundation. Notice the completed units in the background.

Fig. 7-132. The first floor units are set on a masonry foundation. The second floor units are placed on top. The units are joined together with lag bolts.

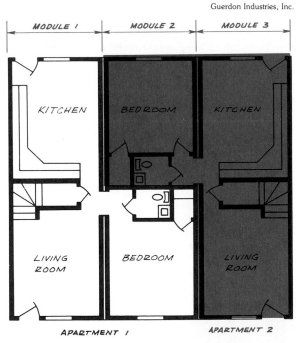

MODULE 1　MODULE 2　MODULE 3

KITCHEN　BEDROOM　KITCHEN

LIVING ROOM　BEDROOM　LIVING ROOM

APARTMENT 1　　APARTMENT 2

Fig. 7-134. First floor, two townhouse units.

There are many approaches to modular construction. The floor plan for a townhouse is shown in Fig. 7-133. It was designed especially for mass-production modular construction. The basic plan is flipped over as units are put together to form a long building. See Fig. 7-134. A brick veneer was installed on this modular-designed townhouse. In this case,

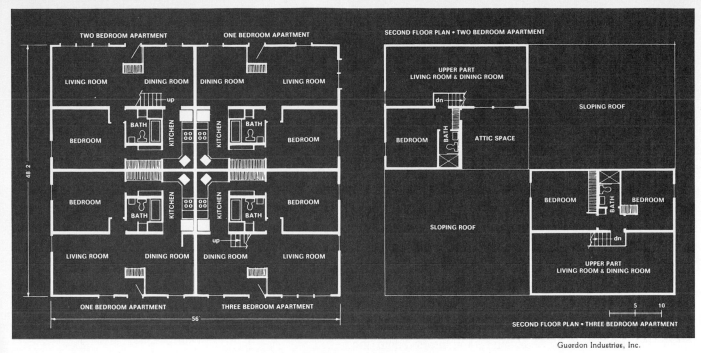

Guerdon Industries, Inc.

Fig. 7-135. Plans for modular apartments.

the lower module was sheathed in the factory. The foundation was designed to support the building and provide a footing for the brick veneer. The brick was installed on the site after the modules were erected.

Another modular design is shown in Fig. 7-135. It has four apartments on the first floor and two on the second. The five modules used are shown in Fig. 7-136. Some trim and panels were installed on the site.

The factory-produced modules are 12'-0" (3658 mm) wide, 56'-0" (17 069 mm) long, and 11'-0" (3353 mm) high. The air conditioning, furnace, kitchen equipment, and bath fixtures were installed at the factory. The modules are wood frame construction. The exterior siding is exterior grade cedar plywood. A metal roof was used. All win-

dows are double-insulating glass. Interior walls and ceilings are sprayed-on plaster and grooved cedar plywood.

Steel and concrete is finding use in production line, factory-produced modules. An example is shown in Fig. 7-137. This is an office building constructed from 16 steel and concrete modules. Eight modules were used for each level. Any exterior siding can be used. In this example, the exterior is a brick veneer.

The structural load is carried in wall columns of structural steel. See Fig. 7-138. The steel framed walls may be sheathed and finished with any exterior siding. Notice in Fig. 7-138 that wood blocking is used around window openings. The wall is cross-braced with steel bars. The entire steel assembly is welded together. The use of steel col-

Guerdon Industries, Inc.

Fig. 7-136. Plans showing the five module units for the apartments in Fig. 7-135.

Fig. 7-137. This office building was built from 16 rectangular steel and concrete modules. They were 12′ wide and 40′ long. Eight modules formed each floor. The building was erected in one day. The exterior was bricked after erection.

umns in the wall gives sufficient strength to stack units five stories high.

The floor of the module has a steel perimeter frame. Inside this floor frame is a subflooring of corrugated metal. Steel studs are inserted into the corrugated metal to stretch the steel concrete reinforcing mesh. The studs also serve as a depth gage for pouring the concrete floor. A special lightweight concrete is used. See Fig. 7-139. In total weight, the modules are lighter than wood frame modules; they are also stronger and more durable. In cost, the two systems are about the same.

Fig. 7-138. Steel-framed walls are welded together. Wood blocking is used around window opening.

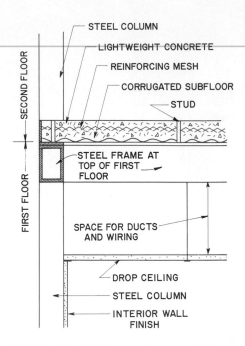

Fig. 7-139. Detail of union of two modules.

Another advantage to the steel and concrete module is the great reduction of combustible materials used.

In modules that use suspended ceilings, the space between the corrugated steel floor of the second floor unit and the suspended ceiling of the first floor unit is used for ducts and electrical wiring. See again Fig. 7-139.

Mobile Homes

A mobile home is a factory-built unit. See Fig. 7-140. The design of a floor plan for a mobile home follows basically the same principles as for house planning. However, space is at a premium and its use requires more careful consideration.

Some typical floor plans for 12′-0″ wide homes are in Fig. 7-141. A 24′-0″ wide unit is called a double wide since it is made from two 12′-0″ wide units placed together. In some areas a 14′-0″ wide unit is built. Some as long as 70′-0″ are available. State highway regulations must be observed in the selection of size, since the unit must be moved on public highways.

In Fig. 7-141, notice that one plan has an expandable area to enlarge the living room. This type of home is called an expandable mobile home.

Fig. 7-140. A mobile home on a landscaped lot.

Fig. 7-142. A kitchen-dining area with a Spanish influence.

The plans in Fig. 7-141 show representative room sizes. The length shown is the actual length of the living area. The length listed by the manufacturer includes the 3' 0" hitch extending beyond the end wall of the mobile home.

Storage is an important aspect of planning. It is necessary to locate storage closets and cabinets carefully. They must be near the area in which the items to be stored are to be used. Mobile homes must be designed to use space economically. However, with careful planning and tasteful decorating, a mobile home can be a pleasant and attractive residence.

A combination kitchen-dining area is used a great deal. See Fig. 7-142. Standard size ranges and

Fig. 7-141. Typical floor plans of a mobile home.

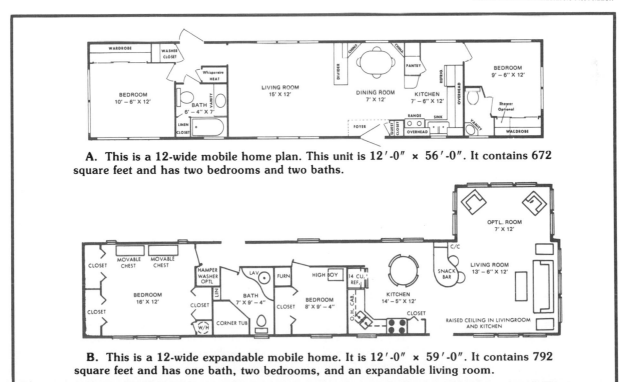

A. This is a 12-wide mobile home plan. This unit is 12'-0" × 56'-0". It contains 672 square feet and has two bedrooms and two baths.

B. This is a 12-wide expandable mobile home. It is 12'-0" × 59'-0". It contains 792 square feet and has one bath, two bedrooms, and an expandable living room.

Champion Home Builders Company

Fig. 7-143. This bath illustrates the principles of good planning.

Boise Cascade

Fig. 7-145. A living area with contemporary furnishings.

Boise Cascade

Fig. 7-144. This bath is located off the bedroom.

Mobile Home Parks

A great deal of planning is required to design a functional mobile home park. Basically, a successful park involves finding a good location, careful site planning, and competent management. The Mobile Homes Manufacturers Association gives these details in their publication, **How to Build and Operate a Mobile Home Park.**

Some major factors involving location include:
1. Proper zoning for mobile homes.
2. Will the park fit into the community?
3. Proximity to community facilities, such as shopping, churches.
4. Cost of land and improvements.
5. Availability of utilities.
6. Availability of public transportation.
7. Competition by other parks.
8. Anticipated growth of area.

Some major factors to consider when planning the site include:
1. Each home needs a degree of privacy.
2. Mobile homes should form neighborhood clusters. This breaks a large group of homes into small, neighborly communities.
3. Landscaping is essential for privacy and appearance.
4. Paved roads to each home are needed. It should be easy to move about the subdivision.
5. Parking for two cars should be provided for each mobile home. Off the street parking is most attractive.

refrigerators are used. The bath is held to the minimum acceptable size, Fig. 7-143. Small shower units or corner bathtubs are used. A bath adjoining a bedroom is shown in Fig. 7-144. The living area usually is larger than the other areas. See Fig. 7-145.

6. Mobile homes have to be moved occasionally.
7. Utilities must be located so homes can be easily connected and disconnected.
8. Surface water must be drained away.
9. Most parks provide a concrete entry patio on each lot.
10. Sidewalks are needed to move from the home to the parking area.
11. A refuse disposal system must be planned.
12. The community building should be located so it can be easily reached by everyone living in the development.
13. Street lights are very beneficial.
14. Laundry facilities are needed.
15. Recreational facilities are valuable. Swimming pools and playgrounds are often built.

There are many ways to plan the mobile home lots. The most commonly used plans are shown in Figs. 7-146 through 7-154.

Angled lots are shown in Fig. 7-146. They are positioned so that the front side of the mobile home faces toward the street. Angle lots are usually 40'-0" to 45'-0" (12.190 m to 13.720 m) wide and 85' to 100' (25.910 m to 30.480 m) deep.

Rectangular lots are shown in Figs. 7-147 through 7-150. They can be made perpendicular to the street, as in Fig. 7-147, or parallel with the street, as in Fig. 7-149. The perpendicular plan is most frequently used because it permits more lots to have street frontage. Notice how the mobile home is near the edge of the lot in Fig. 7-148, Part A. This has the disadvantage that the rear door opens onto the neighbor's lot. In Fig. 7-148, Part B, the home is moved to leave a service area on this side. Notice how the auto parking area in Fig. 7-148, Part B, forms a part of the service area. This permits the rear side door to be used as a service entrance.

The mobile homes on the rectangular lots in Fig. 7-149 are near the back of the lot. This is a poor practice because it leaves no room for a patio or access to the rear door. Privacy is at a minimum since the full outdoor area is exposed to the street. This is improved in Fig. 7-150. Here the home was moved to the center of the lot.

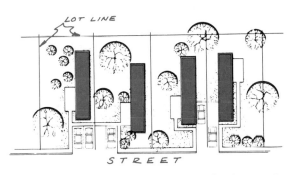

Fig. 7-147. Rectangular lots perpendicular to the street.

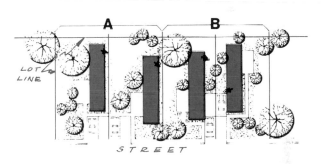

Fig. 7-148. Mobile homes in A are near the edge of the lot. This is not a good practice. In B, they are located to leave a service area.

Fig. 7-146. Angled mobile home lots.

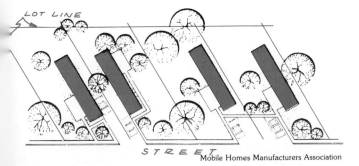

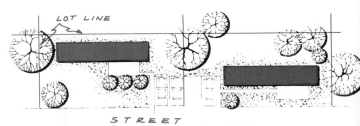

Fig. 7-149. This is a rectangular lot parallel to the street. Putting the home near the rear of the lot is a poor practice.

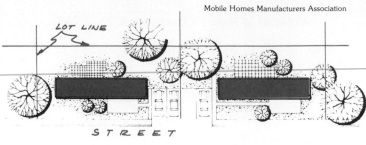

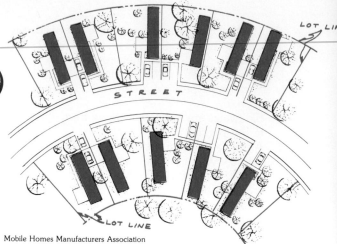

Fig. 7-150. On lots parallel with the street, the mobile home should be in the center of the lot.

Radial and cul-de-sac lots are shown in Figs. 7-151 and 7-152. They show design possibilities from curved streets or a looped street end. These lots have the same planning characteristics as the rectangular lots perpendicular to the street. The angular relationship of one home to the other gives a feeling of greater lot size. They can help create small communities within a larger development. A plan for a total development is shown in Fig. 7-153.

The community center building can serve many purposes. It can include such things as a lounge, recreation facilities such as pool and table tennis, game and meeting rooms, a small kitchen, an office for the park manager, and laundry facilities. A typical building is shown in Fig. 7-154.

Fig. 7-151. These are cul-de-sac lots.

Fig. 7-152. These are curved street lots.

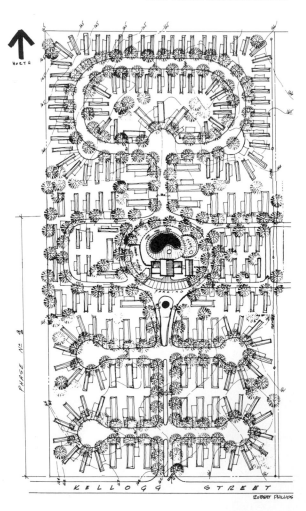

Fig. 7-153. A high-density mobile home community.

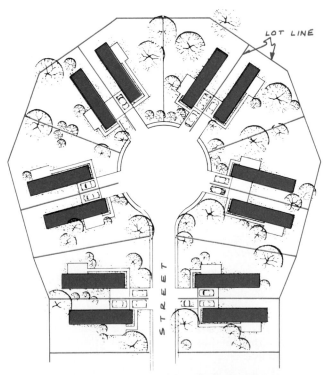

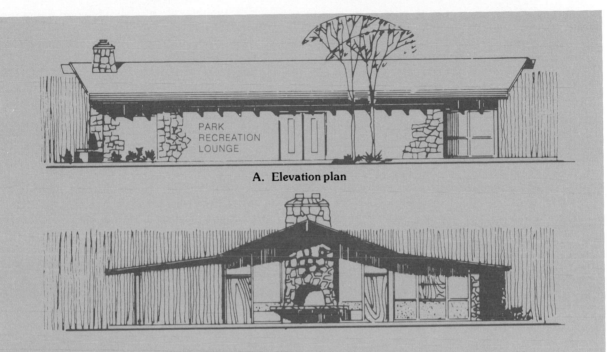

Fig. 7-154. Elevations and floor plan for a mobile home park community center building.

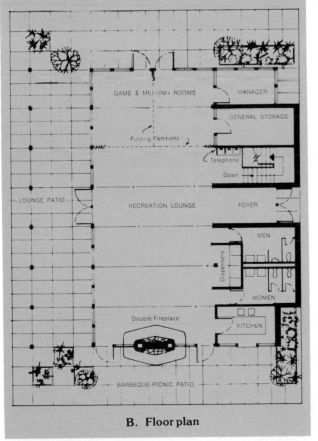

B. Floor plan

Mobile Homes Manufacturers Association

Build Your Vocabulary

Following are terms that you should understand and use as part of your working vocabulary. Write a brief explanation of each term.

batter boards	post-and-beam
box beam	construction
brick veneer	prefabrication
bridging	sheathing
fascia	sill
flashing	soffit
footing	sole
foundation	steel framing
frame construction	studs
furring	subfloor
header	termite shield
joists	trusses
partition	valley, valley jack, hip
plate	jack, and common
	rafters

Class Activities

1. Locate a house being built by conventional construction methods. Visit it regularly and photograph the various stages of construction.

A suggested sequence of photos is given below.

For a frame house:
A. Building located on site with batter boards.
B. Foundation excavated.
C. Foundation completed.
D. Floor joists in place.
E. Exterior walls in place.
F. Roof framing in place.
G. All sheathing installed.
H. Windows installed.
I. Roofing and siding completed.
J. Yard landscaped and walks and driveway finished.

For a concrete-slab house:
A. Building located on site with batter boards.
B. Foundation ready to pour.
C. Floor with plumbing, heating, and electrical wiring installed, before concrete is poured.
D. Floor after concrete is poured.
E. Walls erected.
F. Roof framing in place.
G. All sheathing in place.
H. Windows installed.
I. Roofing and siding completed.
J. Yard landscaped and walks and driveway finished.

2. Build, to scale, the framing of a conventional house. Cut wood framing materials to a scale of $1'' = 1'-0''$ or 1:20 or 1:10 if using a metric scale. Tack or glue these together, using the proper house-framing methods. Include interior partitions. The completed project should represent the skeleton framing of a house. Several class members can work together on one house.

3. Using full-size framing members, build sections of a house frame. These should be nailed together. Suggested projects are listed below.
A. Illustrate box sill construction for platform framing. Include sill, header, joists, rough flooring, sole and studs.
B. Illustrate a typical cornice construction for a frame house with an overhang of $2'-0''$ or 600 mm. Include stud, plate, rafter, ceiling joist, facia, blocking, soffit, wall sheathing and wood siding.
C. Illustrate one method of framing an interior partition into a frame exterior wall. Include subfloor, sole, and studs.
D. Illustrate a common method of framing an exterior corner of a frame house. Include the sole, studs and sheathing.

4. Construct a scale model of a wood frame roof truss.

5. Using clay and wood, construct a scale model of a typical wall section through the sill of a frame house with a concrete slab floor. Use any scale desired. Include gravel fill, perimeter insulation, concrete slab floor, foundation, studs, sheathing, siding and sole.

6. Build, to scale, the framing for a house, using the post-and-beam method. Observe how this skeleton differs from the conventional house framing.

7. Bring samples of building-construction materials to class for examination. These could become a part of a permanent collection for the school drafting area.

8. Using your ingenuity, develop ways to test building materials. Some examples follow:
A. Support a wood member ($2 \times 4 \times 12'-0''$) on each end, and apply a load to the center. This load could be bricks, concrete blocks or metal. Continue to load until the member fractures. Record the number of pounds needed to break the member. Repeat this with several members, and average the breaking load. Compare this average with that reported in a standard load table.
B. Expose 1-inch wood sheathing and 1/2-inch sheetrock to the same source of heat. With a thermometer attached to the outer surface, keep a record of temperature changes every minute. After the temperature on the thermometer begins to change, what conclusion can you draw about the insulation qualities of these two materials?
C. Measure the thickness of a piece of wood siding and a piece of masonite siding. Allow these materials to soak in water several days. Remove and observe the condition of the surface. Measure the thickness. What conclusion can you draw concerning the resistance of these materials to moisture?

9. Find a floor plan for a residence in a magazine or newspaper. Try to plan modular wall components for the exterior walls. If the design does not permit this, change it until modular components can be used. Show the size of each component on the revised floor plan.

10. Develop a design for a small one story residence that can be built as modular units. Draw the floor plan. Draw a section through the unit to show construction details.

Chapter 8

Footing, Foundation, and Basement Construction

If a building has an inadequate foundation and footing, it will settle; the plaster and bathroom tile will crack; doors and windows will be forced out-of-square and not operate properly. It is, therefore, extremely important that the foundation and footings be properly designed.

Soil Investigation

A key to the success of foundation or basement construction is the soil upon which it is to be built. Unsuitable soil can cause flooding or permit the house to settle. In some cases, the hillside near a house may, under proper conditions, begin to slide and bury everything in its way. Excess settling can completely destroy a house. Before you buy a lot, consider the following:

1. Will the soil support the weight of the house?
2. Will a septic tank work if one is needed?
3. Will the water table get high enough to flood the basement or crawl space?
4. Will water cause the house to settle regardless of the footing used?
5. Is the site undisturbed soil or has it been filled? If it has been filled, the foundation may have to rest on pilings.
6. Is the site in an area that will flood from heavy rain?

To avoid these problems, carefully investigate the neighborhood before building. Have a soil specialist perform tests to see if the normal footing and foundation design will be adequate. If problems exist and are known, a foundation can be designed to overcome many of them.

Footings

A **footing** is a concrete pad upon which the foundation wall is built. It must provide support for the building without excessive settlement or movement. A footing is to a house as snow shoes are to a person walking on snow. The snow shoes distribute the weight over a large surface, thus preventing the person from sinking.

Conventional Rectangular Footings

The following factors should be considered when designing and constructing footings.
1. Footings should be made of poured concrete.
2. The concrete should be poured continuously. No load should be placed on a footing until the concrete has thoroughly set up.
3. If a footing trench is dug too deeply, the excess space should be filled with concrete.
4. Footings can be poured in earth trenches without side forms if the soil is firm enough to retain its shape.
5. Footings should be protected from freezing.
6. Footing width should be designed to support with safety the load to be placed upon it. This involves (1) bearing value of the soil, (2) stability of the soil, (3) earth pressure on the foundation, and (4) the weight of the house on each lineal foot of footing.
7. In residential construction, a footing should never be less than 6 inches (152 mm) thick. The thickness should never be less than one and one-half times the projection of the footing from the foundation. A much used standard is

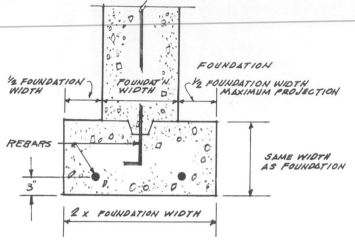

Fig. 8-1. A rectangular footing. The footing is often the same thickness as the foundation wall.

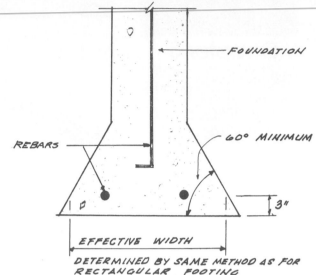

Fig. 8-2. A flared footing. The minimum slope of the flare is 60°.

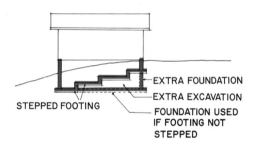

Fig. 8-3. Stepped footings save labor and materials.

to make the footing the same thickness as the foundation wall. See Fig. 8-1.

8. If it is necessary to increase the width of the footing to support the determined load, and if the footing projection exceeds one-half the width of the foundation thickness, it is necessary to increase the thickness of the footing. The thickness-to-width ratio of a footing should be kept close to 2 to 1. For example, if a footing is 8″ (203 mm) thick, it should be 1′-4″ (406 mm) wide. If this width is not enough to support the load and must be increased to 1′-8″ (508 mm), then the footing thickness must be increased to one-half of this, or 10″ (254 mm).

9. Reinforce footings when the projection exceeds two-thirds of the thickness. Under normal conditions this will require two No. 4 rebars placed 3 inches (76 mm) from the bottom of the footing.

Flared Footings

10. If flared footings are used, they should be poured at the same time that the foundation wall is poured.
11. The effective bearing area is figured in the same way as that for rectangular footings.
12. The slope of the flare should not be less than 60° from horizontal. See Fig. 8-2.

Stepped Footings

Stepped footings are a means of cutting the cost of basementless houses built on sloping lots. Step-

ping the footings down the hill removes the necessity of digging very deep into the hillside, thus eliminating the cost of extra foundation material. See Fig. 8-3.

13. Stepped footings have horizontal steps and vertical steps. The vertical step should not be **higher** than three-quarters of the horizontal distance between steps. The horizontal distance between steps should not be **less** than 2′-0″ (610 mm). See Fig. 8-4.
14. The horizontal and vertical steps should be poured monolithically (at the same time). Their widths should be the same.

Pier and Column Footings

15. Pier footings and column footings support considerably more load per square foot than do

Fig. 8-4. Stepped footing design details.

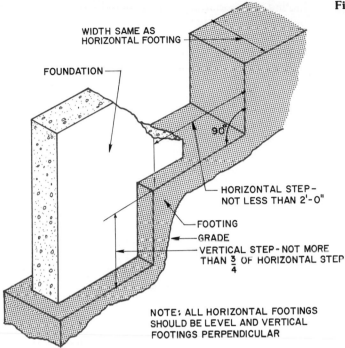

Fig. 8-5. Typical column footing proportions.

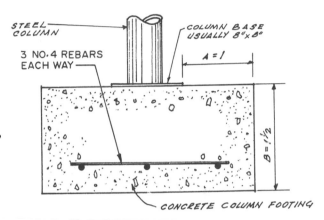

footings under foundation walls. The load on each pier or column should be figured, and footings should be designed that are large enough to carry this load.

The load per square foot calculated for exterior wall footings cannot be used for pier and column footings.

16. The minimum thickness for a pier or column footing is 8 inches (203 mm). The thickness should be one and one-half times the projection. See Fig. 8-5. The usual minimum size of a footing for typical residential construction is 1'-0" × 2'-0" × 2'-0" (304 mm × 610 mm × 610 mm).

17. Reinforcing bars are added to strengthen the footing. Figure 8-5 shows a typical design.

How to Ascertain Foundation Footing Sizes

For the average single-family residence, footing sizes can be ascertained using the standard sizes indicated in Fig. 8-6. Another satisfactory, simple method was shown in Fig. 8-1.

If a house design may have foundation loads greater than normal, the footing sizes should be calculated. To figure footing loads, it is necessary to ascertain the total load that will be placed upon one

Fig. 8-6.

Standard Footing Sizes for Residential Construction

	FRAME HOUSE		MASONRY OR MASONRY VENEER HOUSE	
	Minimum Thickness	Projection Each Side of Foundation	Minimum Thickness	Projection Each Side of Foundation
One-Story				
No Basement	6"	2"	6"	3"
Basement	6"	3"	6"	4"
Two-Story				
No Basement	6"	3"	6"	4"
Basement	6"	4"	8"	5"

lineal foot of footing. Such a load is comprised of two forces — dead load and live load.

Dead and Live Loads

Dead load is the weight of the materials used to construct the building. This includes the foundation, exterior walls, floors, roof, and partitions.

Live load is the weight of, or force exerted by, items that are not a part of the building itself. Furniture is an example of a live load. Live and dead loads are listed in Fig. 8-7.

A roof is subject to two live load forces — wind and snow. See Fig. 8-8. The actual loads vary widely even within a state or region. Examine local codes to find the loads used. Recommended minimum live roof loads are in Fig. 8-9. In areas where the snow load is greater than these, use the snow load as the live load figure. Complete design details can be found in ANSI A58.1-1982, **Minimum Design Loads in Buildings and Other Structures**.

The average weights of conventional materials are fairly standardized. Some of these are listed at the end of this chapter, in Fig. 8-61.

Footing Design Problems

The following problem illustrates the calculation of weight on a footing and the size of the required footing. The same procedure would be followed if metric measures were used.

The weight to be placed on one lineal foot of footing is calculated by adding the weights of a strip of the house. This strip should be one foot wide, should run from the top of the footing to the ridge, and should include all live- and dead-load design figures.

Assume we are calculating the footing requirements for a one-story frame house built by conventional methods and with standard materials. See Fig. 8-10. The bearing soil is clay. The roofing material is asphalt shingles.

Figure 8-11 shows the tabulation of both the live and dead weights of this house.

The foundation section is 7'-6" high and 8 inches thick. A 1'-0" section is cut from the house. This gives a foundation surface of 7.5 square feet. An 8-inch concrete wall weighs 100 pounds per square foot, Fig. 8-9; therefore, this 7'-6" section weighs 750 pounds.

The first-floor area of the one-foot-wide slice runs from the foundation to the steel girder in the center of the house. The girder is supporting half of the weight of the first floor; the foundation wall supports the other half. Therefore, the area supported by the foundation is 1'-0" wide and 7'-6" long. This gives an area of 7.5 square feet. The dead weight of the first floor is 10 pounds per square foot, and 40 pounds per square foot is the live weight, giving a total of 50 pounds per square foot. Since the footage held by the foundation is 7.5 square feet, the total weight held is 375 pounds.

The ceiling joist load is computed in the same manner. It is assumed, for this problem, that the attic will have limited storage use.

The roof weight is found by figuring on a one-foot-wide slice of the roof. The roof area, then, is 1'-0" × 18'-0" (rafter length) or 18 square feet. Since the roof has asphalt shingles, the dead weight is 10 pounds per square foot. The average live load is 30 pounds per square foot, giving a total roof

Fig. 8-7.

Design Figures for Live and Dead Loads

Design Factor	Live Load PSF*	Dead Load PSF*
Floors of rooms used for sleeping area	30	10
Floors of rooms other than sleeping	40	10
Floors with ceiling attached below	—	10
Ceiling joists with limited attic storage	20	10
Ceiling joists with no attic storage	10	10
Ceiling joists if attic rooms are used	30	10

*Pounds per square foot

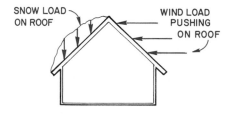

Fig. 8-8.   Live loads bearing on roof.

Fig. 8-9.

Minimum Live Roof Loads*

	PSF
Slope 3 in 12 or less	20
Roof used as deck	40
Slope over 3 in 12	15

*Higher in areas with snow

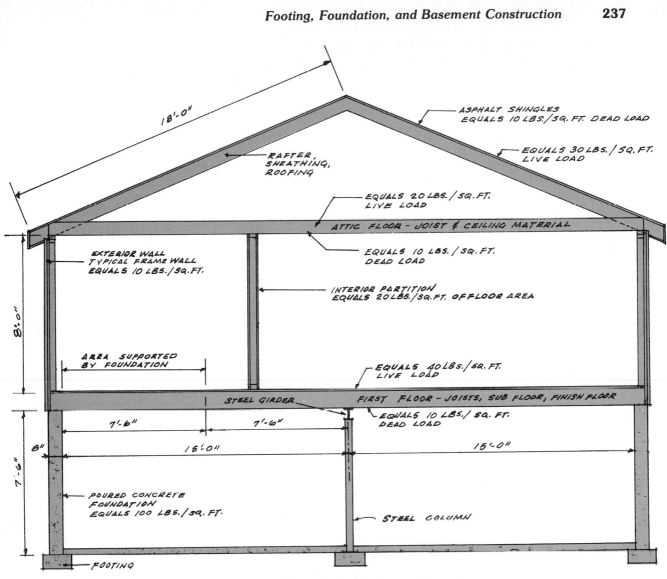

Fig. 8-10. Typical live and dead loads for a frame house.

Fig. 8-11.

Example of Load Calculation for a Residence

Part	Size	Square Feet	Weight PSF	Total Weight in Pounds
Foundation	8″ x 1′-0″ x 7′-6″	7.5	100	750
First Floor	1′-0″ x 7′-6″	7.5	50	375
Attic Floor	1′-0″ x 7′-6″	7.5	30	225
Roof	1′-0″ x 18′-0″	18.0	40	720
Exterior Wall	1′-0″ x 8′-0″	8.0	10	80
Partitions	1′-0″ x 7′-6″	7.5	20	150
Total Load Without Footing				2300
Footing	9″ x 12″ x 18′	Cubic Feet	Weight PCF	168
		1.12	150	
Total Load				2468

load of 40 pounds per square foot. The weight of the roof is then 18 square feet times 40 pounds per square foot, or 720 pounds.

The entire length of the rafter is bearing on the foundation, so the entire weight of the roof is supported by the foundation. The steel girder usually does not assist in supporting the roof, so no roof load is figured into the load on the steel beam.

If the roof used trusses, the roof and ceiling loads for half the width of the house would bear on the exterior wall.

The exterior wall section is 1'-0" × 8'-0", or 8 square feet. A typical frame exterior wall weighs 10 pounds per square foot, so 80 pounds is added to the weight on the footing.

The actual load of the partition is bearing more on the steel beam than on the exterior foundation, but since these computations are rough averages it can be assumed to be bearing evenly. A one-foot strip of floor area equals 15 square feet; half of this amount is assumed to be bearing on the exterior foundation. The dead load for the partition is 20 pounds per square foot of floor area (7.5 square feet). This adds 150 pounds to the total load.

These figures show that approximately 2300 pounds are bearing on each lineal foot of footing. The house is built upon clay which, at a depth of 8', offers a 2400-pound bearing capacity to each square foot of footing bearing upon it. See Fig. 8-12. The normal footing required for this house is 1'-4" wide. A one-foot slice of this offers 1.3 square feet of surface bearing on the soil. This will offer support to 2400 pounds per square foot times

1.3 square feet or 3120 pounds. Since support for only 2300 pounds per square foot is needed, the standard footings are more than ample even when the weight of the footing is figured and added.

If the foundation extended only 2' into the soil, the clay would support 1400 pounds per square foot. This times 1.3 square feet would equal 1820 pounds which is inadequate. To find the required footing width, divide 2300 by 1400. This equals 1.6 feet or 1'-8". Thus the footing width must be 1'-8". The footing thickness will have to be increased to 10" to maintain the approximate 2 to 1 proportion.

Now find the weight of the footing. The footing contains 2400 cubic inches of concrete. Structural concrete weighs 150 pounds per cubic foot (1728 cubic inches). The footing contains 1.38 cubic feet of concrete. It weighs 150 times 1.38 or 207 pounds per lineal foot. Since the building weighs 2300 pounds per lineal foot, the total weight is 2507 pounds per lineal foot. Because the 1'-8" footing will support only 2300 pounds per lineal foot, the footing is inadequate. It must be widened to 1.8 feet or 1'-10". This size footing will carry 2520 pounds.

Footing for Piers and Columns

The weight placed upon a pier or column is concentrated on a small area. It is important that this footing be designed properly, to prevent the house from settling and sagging. (The term **column** refers to a wood or metal post used to support a beam. A **pier** is a masonry unit, usually 12 to 16 inches square, that supports a beam.)

An examination of Fig. 8-13 indicates why a pier or column footing carries a greater, more concentrated load than does a foundation footing. The column supports all the floor load extending halfway from the column to the foundation wall or to another column. In Fig. 8-13 the distance from the column to the foundation wall on the width of the house is 15'-0". Half of this is supported by the column. The same is true for the opposite direction; so a total width of 15'-0" of floor is supported by the column.

The column in Fig. 8-13 is located 10'-0" from the foundation wall on the length of the house. It supports half of this or 5'-0". It is also 10'-0" from the next column. Since the second column supports half this load, the first column supports only an additional 5'-0". The **total area** supported by the column is 15'-0" × 10'-0", or 150 square feet.

Fig. 8-12.

Allowable Foundation Pressure (Load) on Various Soils[1]

Class of Materials	Allowable Foundation Pressure (lb. per sq. ft.)[2]	
	Footing 1' deep	Footing 8' deep
Massive crystalline bedrock	4000	9600
Sedimentary and foliated rock	2000	4800
Sandy gravel and/or gravel	2000	4800
Sand, silty sand, clayey sand, silty gravel, and clayey gravel	1500	3600
Clay, sandy clay, silty clay, and clayey silt	1000	2400

[1]Soil samples must be tested to determine class and bearing capacity. Bearing pressure must not exceed those in this table.
[2]Add 20% to one foot depth for each additional foot of footing depth to a maximum of three times the one-foot depth.

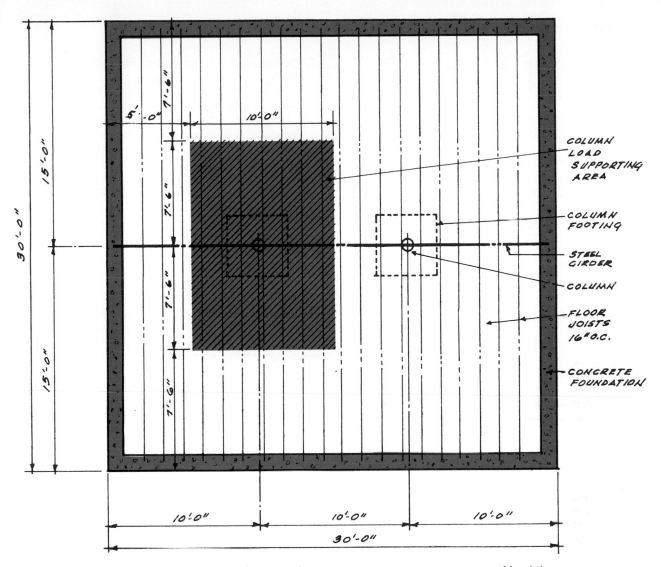

Fig. 8-13. A pier load example. A pier footing must carry a more concentrated load than a foundation footing.

Assuming this column is the one indicated in Fig. 8-10, the total load on the footing would be as shown in Fig. 8-14.

If this house is built upon sandy gravel and the column footing is 8' below ground, the soil will support 4800 pounds per square foot. The area of the footing is found by dividing 15,000 pounds by 4800 pounds per square foot which equals 3.1 square feet. By taking the square root of the area, the length of the sides of the footing are found. The sides are 1.8 feet or 1'-10". Figure 8-15 indicates the solution to this problem. As the projection beyond the steel base lengthens, the footing must be made thicker. The thickness should be approximately 1 ½ times the size of the projection.

Fig. 8-14.

Example of Load Calculation on Pier Footing

Part	Square Feet	Weight PSF	Total Weight (Pounds)
Attic Floor	150	30	4,500
First Floor	150	50	7,500
Partition Load	150	20	3,000
Total Load on Column			15,000

239

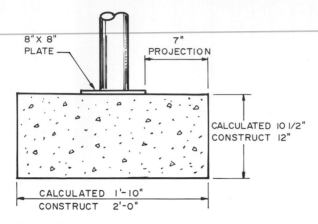

Fig. 8-15.   Solution to typical column footing problem.

In this instance, the width of the footing was calculated to be 1'-10" square. Ordinarily, column footings are made no smaller than 2'-0" square; therefore, a large safety margin exists. The same is true for thickness. It was calculated to be 10 ½", but footings are normally cast 12" thick minimum.

Fireplace Footings

A chimney is very heavy. It cannot be supported by the foundation footing, but must have a special footing poured for it. The procedure for calculating the proper footing size is the same as that for a pier or column, except the weight to be considered is the actual weight of the materials in the chimney.

The weight of the brick and concrete masonry units used in the fireplace must be ascertained.

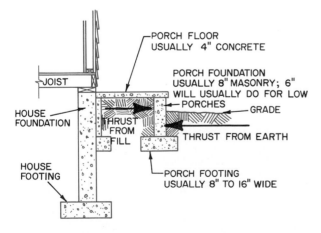

Fig. 8-16.   Typical porch construction.

Standard weights are given in Fig. 8-61, later in this chapter.

Since most fireplaces have a tile flue lining, the weight of this must also be considered.

The base of most fireplaces is quite large. Generally, it is as large as the size required for the footing. If this is the case, the footing is made approximately 6 inches (152 mm) larger than the fireplace base, to increase the stability of the unit. A fireplace footing must be at least 12 inches (305 mm) thick.

Porch and Stair Footings

The usual porches and stairs built are light enough that they require no footings. The end surface of a porch wall is sufficient to support the weight of the wall and the porch floor. However, it is a common practice to pour a footing about twice as wide as the foundation wall to help stabilize the wall and to help overcome the thrust of earth. See Fig. 8-16.

Depth of Footings

Footings must be below the frost line of the area in which the building is to be built. The **frost line** is the depth to which the soil freezes. Freezing and thawing cause the soil to expand and contract. This generates great force. If the soil beneath a footing freezes, it causes the foundation wall to move. This often causes cracks.

Figure 8-17 gives general minimum footing depths for the United States based on the frost line of each area. Specific depths must be obtained from local building codes. Footings are often placed 12" below the frost line to give a margin of safety.

Footings must rest on natural, undisturbed soil which will provide adequate bearing. Footings on filled land must meet local codes for such construction. When the bearing capacity or stability of the soil is questionable, soil analysis, bearing tests, or special footing designs may be required.

Foundations

A **foundation** serves as the base upon which a house is built. It holds the house above the grade level and keeps the area under the house dry.

There are many kinds of foundations that can be used successfully. The type chosen depends upon many factors. If a basement is desired, the foundation forms the basement wall. If a house is to be built over a crawl space, the foundation may be a masonry wall or piers. A curtain wall may become part of a foundation. In areas where the soil does not offer sufficient support for usual foundations, pilings must be used to support the foundation wall.

Basement Construction

If a house is to have a basement, the foundation must be designed to serve also as the basement wall. It must withstand the thrust of the soil and repel surface and subsurface water. It also must resist the action of freezing and thawing, which can cause a building to rise and fall slightly and break in places. Figure 8-18 illustrates a house with a basement wall as a foundation, as compared to the same house with a crawl space.

Poured concrete foundation walls are considered best for residential basement construction. The entire foundation is poured at one time, thus becoming a single, integral wall. It tends to crack less and leak less than other types. Unskilled labor can be used to place the forms and pour the walls, while walls built of concrete blocks and other masonry units require highly paid masons to build them.

The top course must be capped with 4 inches (102 mm) of solid masonry or concrete or have the cells in the top course filled with concrete grout.

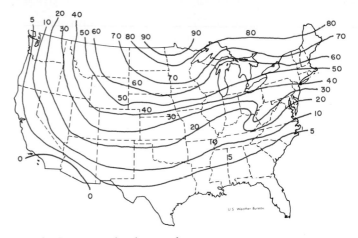

Fig. 8-17. Frost depth in inches.

The size and placement of reinforcing bars can vary with design requirements. Those in Fig. 8-18 are typical.

Foundations for Crawl Spaces

If a house is to be built over a crawl space, the foundation is made exactly the same as explained for the basement. The same materials and sizes are

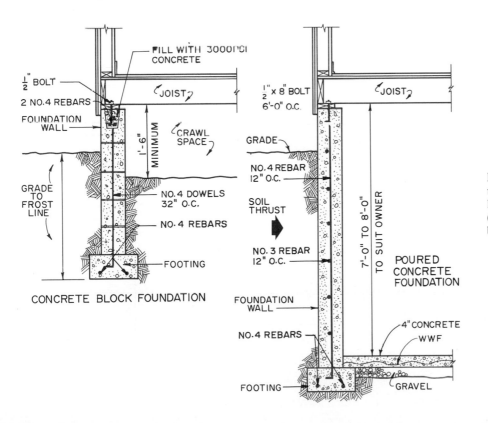

CONCRETE BLOCK FOUNDATION

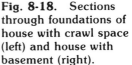

Fig. 8-18. Sections through foundations of house with crawl space (left) and house with basement (right).

used. The only difference is that the foundation for the house with the crawl space is dug only to the frost line or to a solid soil, whichever is the deeper. See again Fig. 8-18.

The ground level should be at least 18 inches (457 mm) below the bottom of the floor joists and 12 inches (305 mm) below the bottom of the girders. Usually, it is best to try to get 2'-0" (610 mm) or more. This permits easier access to plumbing, wiring, and furnace connections that are below the floor. The larger space also allows more air to circulate below the joists, thus reducing moisture damage.

If moisture is a problem, the soil beneath the joists should be covered with a vapor barrier, with 2 inches (51 mm) of sand on top to hold it in place.

Foundations for Basements and Crawl Spaces

The thickness of the foundation wall should not be less than that of the exterior wall of the house supported, except that exterior walls of masonry veneer over frame construction may be corbelled one inch maximum. See Fig. 8-19.

Standard foundation wall thicknesses for residential construction are given in Fig. 8-20. This table may be used for normal conditions.

The upper portion of the foundation can be reduced in thickness to 4 inches, to permit placement of brick veneer or to permit the edge of a concrete floor slab to rest on the foundation. This reduced 4-inch portion should not exceed 4 inches in height. If it does, reinforcing rods are required. Seldom should this height exceed 12 inches, even if reinforced. See Fig. 8-21.

Waterproofing Foundations

In order to perform its total function, the foundation should repel all water that runs against it. If a house does not have a basement, the foundation still should be expected to shed water and to keep the area under the house dry.

A number of things can be done to prevent water from penetrating a foundation. An easy thing to do

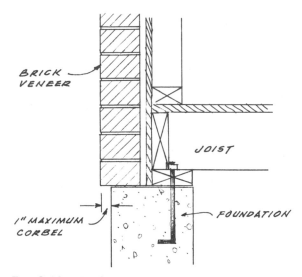

Fig. 8-19. Brick veneer corbelled over foundation.

Fig. 8-20.

Standard Thicknesses of Residence Foundation Walls

TYPE OF FOUNDATION WALL CONSTRUCTION	MAXIMUM HEIGHT OF UNBALANCED FILL[1]	MINIMUM WALL THICKNESS (in Inches)	
		Frame	Masonry or Masonry Veneer
Hollow Masonry, such as concrete block	3	8	8
	5	8	8
	7	12	10
Solid Masonry	3	6	8
	5	8	8
	7	10	8
Poured Concrete	3	6[2]	8
	5	6[2]	8
	7	8	8

[1]Height in feet of finish grade above basement floor.
[2]Provided forms are used on both sides for full height of wall.

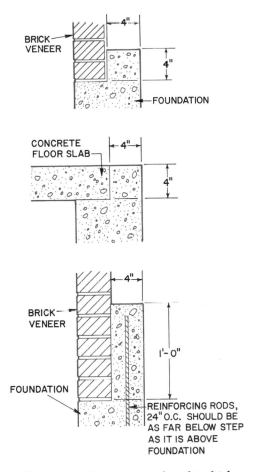

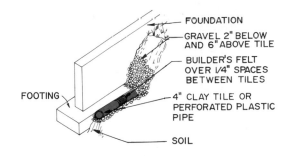

Fig. 8-22. Drain tiles used to carry away subsurface water.

Fig. 8-21. Top of foundation reduced in thickness for special application.

Fig. 8-23. Applying a coat of bituminous material to a masonry foundation to make it waterproof.

is to bank the earth up against the foundation so the soil slopes away from it. This provides a natural drain away from the house for surface water.

Drain tiles are of great value in helping to carry away water found below the surface. These tiles are placed in a 2-inch-thick (51 mm) gravel bed just above the footing. They are placed ¼ inch (6 mm) apart, and this space is covered with building paper. The tiles are sloped so they drain to a storm sewer or dry well. See Fig. 8-22. About 6 inches (152 mm) of coarse gravel or crushed rock is placed over the tile.

Poured concrete foundations can be waterproofed by applying at least one coat of bituminous material to the wall. For severe conditions, several layers of bituminous material can be applied, or builder's felt can be adhered to the foundation. Polyethylene film (a plastic) is widely used as a waterproofing material. It comes in large sheets and is adhered to the exterior of the foundation wall. See Fig. 8-23.

Other waterproofing materials include synthetic rubber sheet membranes, liquid-applied membranes, cementitious materials, and bentonite, a natural clay that swells when wet and resists penetration.

Termite Protection

In almost every part of the United States, a small, destructive insect called the termite exists. It lives in large colonies beneath the surface of the earth and thrives on the cellulose of wood.

Actually, two classes of termites exist — the subterranean and the nonsubterranean. The subterranean termites, who need moisture to live, do the greatest damage to buildings. They tunnel through the earth and consume any wood in their path.

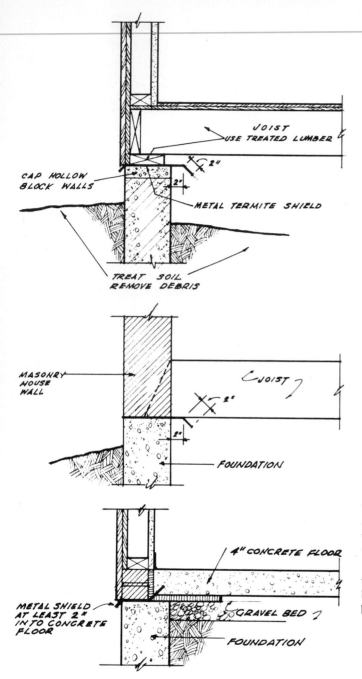

CAP HOLLOW
BLOCK WALLS

JOIST
USE TREATED LUMBER

2"

2"

METAL TERMITE SHIELD

TREAT SOIL
REMOVE DEBRIS

MASONRY
HOUSE
WALL

JOIST

2"

2"

FOUNDATION

4" CONCRETE FLOOR

METAL SHIELD
AT LEAST 2"
INTO CONCRETE
FLOOR

GRAVEL BED

FOUNDATION

Fig. 8-24. Methods for installing metal termite shields.

These insects build mud tunnels up foundation walls and infest the wood portions of a house. They actually consume the wood, leaving only a thin shell to protect themselves. Any wood members thus damaged must be replaced as they have little structural value.

The nonsubterranean termites are able to live in dry wood and without moisture or a colony in the ground. They are fewer in number and do less damage than the subterranean termite.

Several things can be done to reduce the danger of termite infestation. After a house is completed, all wood scraps should be removed. Any scraps buried in backfilling or grading become possible pockets for termite growth. From here they can tunnel into a house.

Poured concrete walls best resist penetration, since termites can get through the smallest crack in a foundation. If a foundation wall is built of hollow concrete block, these insects can crawl unobserved up the inside of the wall and attack the sill. To help prevent this, hollow walls should be capped with a solid masonry block, or they should have a metal termite shield installed, Fig. 8-24. This tends to repel the insect, but is far from perfect protection. Termites can penetrate the masonry cap if it has a tiny crack, and they have been known to build their clay tunnels around metal shields.

The wood members likely to be infested can be treated with a chemical that repels termites. The chemical can be brushed or sprayed on; however, it is much more effective to use lumber that has been treated under high pressure. By this method, the chemical is forced **into** the wood.

Another very effective preventative measure is to treat the soil with a termite-repellent chemical. Soil beneath concrete floors should be treated before they are poured and also the soil around footings and foundation walls. Termites cannot live in the treated soil around the house; yet plants and grass are not affected. The chemical usually is applied by trained exterminators who guarantee their work for many years.

As the house is being designed and built, some of these termite-repellent methods should be used. Usually, no one method will suffice, and a combination of several will be necessary.

Piers, Columns, and Curtain Walls

The purpose of a pier or column is to provide a foundation to support a load placed upon it. Usually, piers are freestanding masonry units, while columns are usually steel or wood members. See Figs. 8-25 and 8-26. A **curtain wall** is a masonry wall built between piers. It is not as thick as a pier, but does help carry some load. However, the pier carries the main load. See Fig. 8-27.

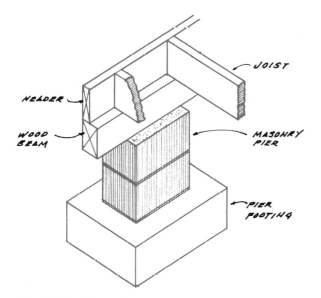

Fig. 8-25. Typical masonry pier.

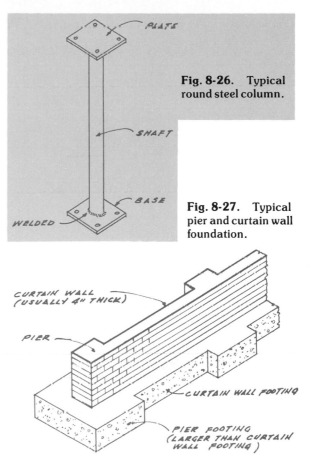

Fig. 8-26. Typical round steel column.

Fig. 8-27. Typical pier and curtain wall foundation.

Pier Design

Freestanding piers are divided into two classes — exterior piers and interior piers. **Exterior piers** are subject to a load from above, as well as a horizontal load due to the wind. Such piers must be either poured concrete, solid masonry, or hollow masonry with the cells filled with concrete to make a solid unit.

The freestanding exterior pier should not be built over three times its least dimension above grade, unless special reinforcement is added.

Interior piers usually are not subject to wind load. They can be poured concrete, solid masonry, or hollow masonry. If hollow masonry, they should be topped with 4 inches of solid masonry, or the top masonry course should be filled with concrete.

An interior pier can be built above grade ten times its least dimension if it is concrete or solid masonry, or four times its least dimension if it is hollow masonry. See Fig. 8-28.

Piers supporting frame construction should extend at least 12 inches above finished grade.

The actual size and spacing of piers should be determined by the load they must support. Figure 8-29 gives pier design data usable for single-story residences built on average soil.

Pier with Curtain Wall

The use of a curtain wall with pier construction is limited to houses without basements. A basement requires a full foundation wall.

Lumber Dealers Research Council

Fig. 8-28. Round, poured, interior concrete piers.

Fig. 8-29.

Standard Pier Design Data for Single-Story Residences on Average Soil

PIER MATERIAL	MINIMUM PIER SIZE (in Inches)	MINIMUM FOOTING SIZE (in Inches)	PIER SPACING	
			Right Angle to Joists	Parallel to Joists
Solid or Grouted Masonry	8 × 12	12 × 24 × 8	8'-0" O.C.	12'-0" O.C.
Hollow Masonry, Interior Pier	8 × 16	16 × 24 × 8	8'-0" O.C.	12'-0" O.C.
Plain Concrete	10 × 10 or 12" dia.	20 × 20 × 8	8'-0" O.C.	12'-0" O.C.

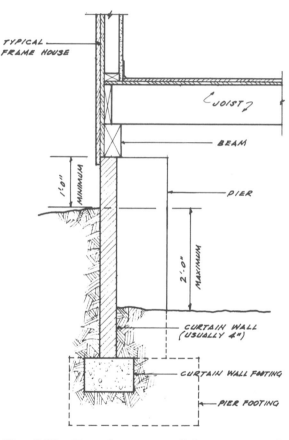

Fig. 8-30. Typical curtain wall between piers for frame construction.

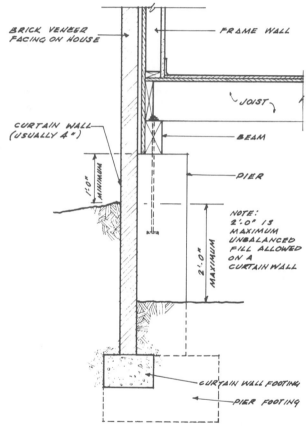

Fig. 8-31. Typical curtain wall between piers for brick veneer over frame construction.

Since the curtain wall helps carry some of the load, braces the pier, and helps resist the wind, an exterior pier can be built higher if it is combined with a curtain wall. See Fig. 8-30. A pier with a curtain wall can be built ten times its least dimension above the footing, if it is concrete or solid masonry; it can be built four times its least dimension, if it is hollow masonry. Figure 8-29 can be used as a guide for

spacing piers with curtain walls. A curtain wall can be safely built above the footing to a height fourteen times the thickness of the curtain wall, if it is concrete or solid masonry; and ten times the thickness of the curtain wall, if it is hollow masonry.

Most curtain walls are made from one, 4-inch course of brick. The thrust of the earth against such a wall is great. Therefore, no 4-inch curtain wall should have more than 2'-0" unbalanced fill placed against it. See Fig. 8-31.

The curtain wall must be bonded or anchored to the pier so the two become a single unit. The pier and curtain-wall footing should also be a single, integral unit.

It is possible to support a brick veneer facing on a building up to one and one-half stories high on a curtain wall with sufficient footing to carry the weight. Refer to Fig. 8-31.

Pilasters

A **pilaster** is a masonry unit used to support the ends of beams or to stiffen a long foundation wall. See Fig. 8-32. The pilaster must be bonded to the foundation wall so the two become an integral unit.

If a foundation wall is very high or long, pilasters can be added about every 20 feet to help brace the wall against the soil pressure. In the typical residence, this case seldom occurs.

More frequently, pilasters are used to provide the necessary support for the ends of beams on the foundation. A beam requires at least 4 inches of solid bearing surface. A pilaster can increase the thickness of the foundation at one place for adequate support.

If a foundation wall is only 6 inches thick, or if it is an 8-inch hollow masonry wall, pilasters are required for adequate beam support. If the pilaster is poured concrete, it must be at least 2" × 12", and if solid masonry, it must be 4" × 12".

Grade Beam on Piling Foundation

Upon occasion, a residence may be constructed upon pilings. A **piling,** as used in residential construction, is usually a concrete pier sunk into the ground. Frequently, it is cast in a hole dug in the ground or cast in a steel shell in the ground. The shell is then removed.

This type of construction can be used for normal residential construction and, generally, is used in areas where the soil cannot adequately support footings at the usual depths. The pilings are spaced in a manner similar to that discussed under piers. See Chapter 23 for more details.

On top of these pilings, a poured concrete grade beam is placed. This beam forms the foundation wall upon which the house is built. A house built this way does not have a basement. It can have a crawl space or a concrete slab on grade. The examples in Figs. 8-33 and 8-34 illustrate some common methods of construction of this type of foundation.

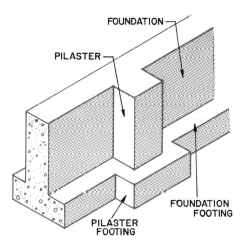

Fig. 8-32. Pilaster on a foundation wall.

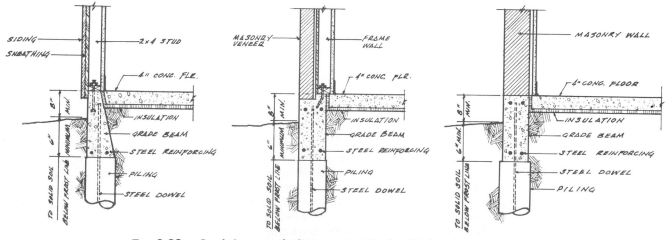

Fig. 8-33. Grade beam and piling construction for slab floor on grade.

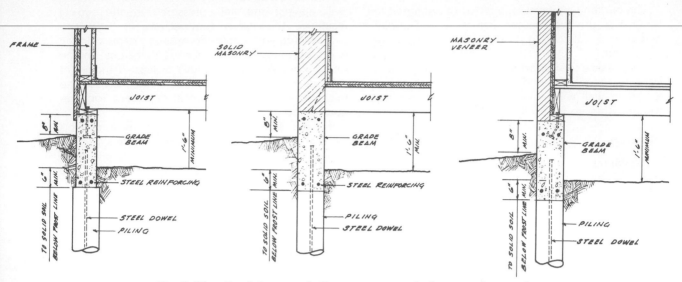

Fig. 8-34. Grade beam and piling construction for house with a crawl space.

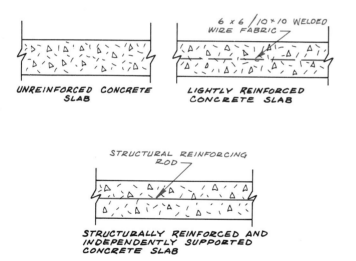

Fig. 8-35. Common concrete slabs.

The following design factors are satisfactory for single-story residences built under normal conditions. If soil is of such a nature that it has unusually poor bearing qualities, pilings of larger diameters must be used.

1. Piling should be spaced 8'-0" on center maximum.
2. Piling should be at least 10 inches in diameter.
3. Piling should extend below frost line and have a bearing area of 2 square feet, for average soil.
4. Piling should have one No. 5 bar running full length of piling and into grade beam.

5. Grade beam for frame construction should be at least 6 inches wide by 14 inches high.
6. Grade beam for masonry and masonry veneer construction should be 8 inches wide by 14 inches high.
7. Grade beam should have four No. 5 steel bars for masonry and masonry veneer construction or four No. 4 steel bars for frame construction.
8. The bottom of the grade beam should be below the frost line, unless provision is made so that moisture does not collect under the beam. Coarse rock or gravel should replace the soil under the grade beam to enable moisture to drain away. If this is not done, freezing will cause the house to heave.

Slab Foundation and Floor

The design of the foundation and floor of a house to be built with a concrete slab requires consideration of the soil properties and size of the slab. There are four basic slab designs. The first is an **unreinforced** slab. See Fig. 8-35. It depends upon support from the fill below. This slab will crack when subjected to tension or warping. It may crack while drying due to shrinkage. The unreinforced slab must be in an area providing excellent drainage. It should be at least 4 inches (102 mm) thick. Seldom are slabs larger than 32'-0" (9754 mm) in length poured. If a larger slab is needed, control joints should be provided. If a nonrectangular slab is to be poured, it must be divided into rectangular elements. Control joints are used between elements. Heating coils and pipes cannot be used in unreinforced slabs. The stresses generated by the heat will crack the slab.

A second slab is the **lightly reinforced** slab. It is reinforced over its entire area with a welded wire fabric reinforcement. See Fig. 8-35. The most common wire fabric used is 6 × 6/W1.4 × W1.4 WWF. It has 6-inch (152 mm) squares. The W numbers refer to the wire gauge (diameter). The wire fabric resists the stresses of drying shrinkage. It also helps resist cracking due to temperature change. This is required when heating pipes are embedded in the slab. The wire fabric helps the slab resist some warpage and slab movement. This type of slab should be at least 4" (102 mm) thick. It requires the support of the fill beneath the slab.

The lightly reinforced slab can have perimeter dimensions up to 75'-0" (22 860 mm) without a control joint.

A third type of slab is the **structurally reinforced** slab. The slab is reinforced with steel reinforcing rods. While it still depends upon the fill beneath the slab for support, the reinforcing enables the load to be distributed over a larger portion of the slab. See Fig. 8-35.

The **independently supported** slab is used where soil is very poor and will not support a ground-supported slab. The slab is reinforced to carry all the load to the foundation and piers. It has structural reinforcing rods. See Fig. 8-35. It does not depend on the soil to carry any of the load on the slab. It is designed much like a concrete beam or other

structural member. The distance it must span and the load it must carry are used to calculate the reinforcing and thickness of the slab.

The three commonly used slab and foundation designs are shown in Figs. 8-36, 8-37, and 8-38. If

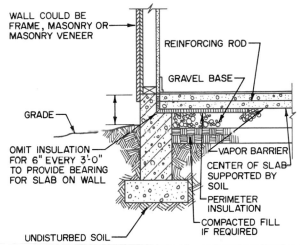

Fig. 8-37. This slab is supported on the edges by the foundation. The compacted soil supports the remainder of the slab. This design is most often used for structurally reinforced slabs.

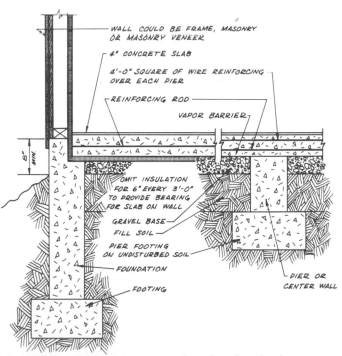

Fig. 8-38. This slab is supported on the edges by the foundation. The center of the slab is supported by piers or a center wall.

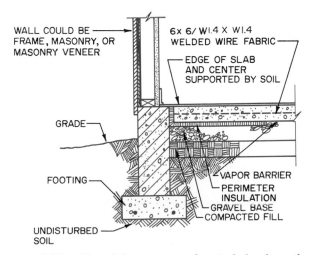

Fig. 8-36. This slab is supported entirely by the soil underneath. This design is most often used with unreinforced and lightly reinforced slabs.

the slab is to be totally supported by the ground, it would be constructed as in Fig. 8-36. This places no load on the foundation wall or footing. They carry the load of the walls and roof but not the floor. Usually, the unreinforced and lightly reinforced slabs are constructed this way.

The construction of a foundation in which the foundation carries part of the slab load is shown in Fig. 8-37. This construction is used with structurally reinforced and independently supported slabs. The structurally reinforced slab carries most of the load.

The soil carries some, but the load is spread over a wide area due to the strength of the reinforcing rods.

If the soil is poor, an independently reinforced slab is used. See Fig. 8-38. Here the slab is supported by piers or supporting foundation walls. See Fig. 8-39. The area beneath the slab is filled to serve as support for the concrete when it is poured. Since this soil is not stable, it will eventually drop below the bottom of the slab. Figure 8-40 gives design data for an independently supported structural slab.

In Fig. 8-41 is a monolithically-cast slab and grade beam. **Monolithically cast** means the slab and grade beam are cast at the same time. They are one solid concrete unit. This type of construction

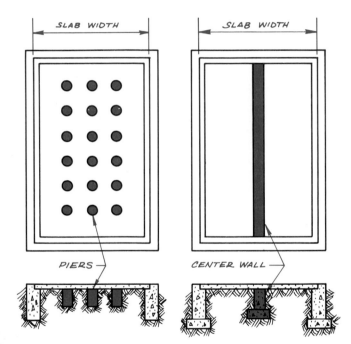

Fig. 8-39. Pier and center wall support for independently supported slab.

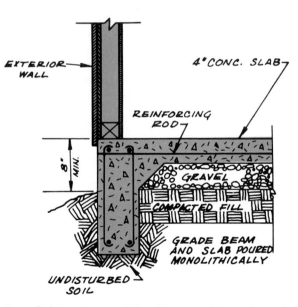

Fig. 8-41. A monolithically-cast slab and grade beam is a solid concrete unit.

Fig. 8-40.

Design Data for Structural Slab on the Ground[1]

Structural Unit	Maximum Spacing Between Piers		
	6'-0" O.C.	7'-0" O.C.	8'-0" O.C.
Piers (Minimum)[2]			
Poured Concrete (Round)	10"	12"	14"
Concrete or Masonry (Square)	8" x 8"	10" x 10"	12" x 12"
Footing Area (Minimum)	115 sq. in.	130 sq. in.	175 sq. in
Welded Wire Fabric Required	6 x 6 / W1.4 x W1.4		

[1]Slab supported on edges by foundation and in center by piers.
[2]Piers are designed to support slab only; bearing walls must have separate footings designed to carry their load.

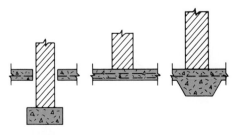

Fig. 8-42. Designs for carrying extra load of heavy interior partitions.

generally uses a structurally reinforced slab or an independently supported slab.

The means of designing load bearing partitions in slab construction are shown in Fig. 8-42. The unreinforced and lightly reinforced slab cannot support loads in excess of 500 pounds per lineal foot. If excess loads are to be placed on the slab, one of the designs shown in Fig. 8-42 must be used. In each case, the reinforcement and concrete thickness must be determined by engineering analysis of the loads to be carried.

The exterior wall load should always be on a footing independent of the slab. In structurally reinforced and independently supported slabs, exterior wall loads are to be to a footing or grade beam. Interior load-bearing partitions may be supported by the slab if they do not exceed the structural design capacity of the slab. Special reinforcing is needed if slab load-carrying capacities will be exceeded.

Following are some additional factors related to concrete slab construction.

1. Slab on ground construction should not be attempted in areas having ground water or a hydrostatic pressure condition near the ground surface.
2. The site should be graded so water cannot collect beneath the slab.
3. Perimeter insulation should be used, except in areas where freezing weather seldom occurs. See Figs. 8-36, 8-37, 8-38, and 8-43. Figure 8-44 gives perimeter design data.
4. The slab should be poured continuously if possible. If a construction joint occurs, reinforcement should be provided for transfer of stress. See Fig. 8-45.

5. Concrete should be allowed to develop strength before being subjected to a load. It should be protected from freezing for at least two days after pouring.
6. The minimum floor slab should be 4 inches (102 mm) thick.
7. Metal heating coils and reinforcement should be covered with at least 1 inch of concrete.
8. Hot air ducts in a slab should be completely encased in at least 2 inches (51 mm) of concrete, unless they are crush-resistant, noncorrodible, nonabsorbent, and have tight-fitting

Fig. 8-44.

Perimeter Insulation Design Data for Heating Requirements[1]

Outside Design Temperature (°F.)	Distance Insulation Should Extend Under Slab (in Inches)	Thickness of Insulation (in Inches)
+ 31 and above	none; 12[2]	none; 1[2]
+ 21 to + 30	Vertical edge of slab only	1
+ 11 to + 20	12	1
+ 1 to + 10	18	1
+ 0 to − 9	24	1
10 to 19	24	1 ½
− 20 to − 29	24	1 ½
− 30 and lower	24	2

[1]These figures are for an unheated slab. If slab is heated, insulation thickness should be increased 50 percent.

[2]If house is air conditioned, install 12 inches of 1-inch thick insulation.

Wire Reinforcement Institute

Dow Chemical Company

Fig. 8-43. Perimeter insulation being installed on slab floor edges. Plastic vapor barrier is also being installed.

Fig. 8-45. Reinforcement is used in some concrete slab construction.

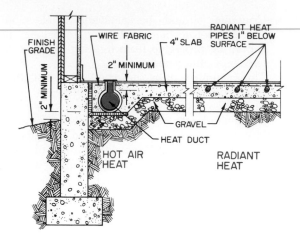

Fig. 8-46. Construction details for hot-air heating duct installation in concrete-slab floor.

Union Carbide Corporation

Fig. 8-47. Forms ready for pouring of concrete-floor slab. Notice the polyethylene vapor barrier, wire reinforcing, heat ducts and plumbing.

joints. A two-inch coverage over the duct is required in all cases. See Fig. 8-46.

9. Slabs supported by compacted fill should not have more than 12 inches (305 mm) of earth fill or 24 inches (610 mm) of sand or gravel fill.

10. Slabs supported by foundation at the edges and piers in the center allow a maximum fill of 3 feet (914 mm) of earth or 6 feet (1829 mm) of sand or gravel.

11. A vapor barrier should be placed below all slabs forming floors for habitable spaces. See Fig. 8-47.

12. The top of the floor slab should be at least 8 inches (203 mm) above finished grade.

13. The bottom of heat ducts in a slab should be at least 2 inches (51 mm) above finished grade.

14. Any wood sills should be 8 inches (203 mm) above finished grade.

15. All interior bearing partitions and non-bearing masonry partitions must be supported by footings resting on natural soil independent of the slab. It is possible to thicken the slab at such a point and provide special reinforcement to distribute this load. The 4-inch (102 mm) slab by itself can support only nonload-bearing, frame, interior partitions.

16. Structural slabs should be reinforced according to the design data in Fig. 8-40.

17. All piers should extend to natural, undisturbed soil.

18. Basement floor slabs and garage floor slabs should be designed according to the requirements set forth for house floor slabs.

19. Porch floor slabs should be reinforced if they exceed 3'-6" (1067 mm) in width.

All-Weather Wood Foundation

The **All-Weather Wood Foundation** is a load-bearing wood-frame wall system designed for below grade use as a foundation for light-frame construction. It is basically the same as above-ground frame construction except for several factors. The lumber and plywood used in framing is stress graded to withstand the lateral soil pressures. It also carries the usual live and dead loads. Vertical loads on the foundation are distributed to the supporting soil by a special footing made of a wood footing plate and a structural gravel layer.

Following are some design specifications:

1. All lumber and plywood in contact with or close to the soil are pressure treated with wood preservatives. This protects against decay and insects. The preservative treatment must meet the requirements set by The American Wood Preservers Bureau. If lumber is cut after treatment, the cut surface should be brush coated with preservative.

2. Extensive moisture control measures are used. Moisture reaching the upper part of the wall is deflected by polyethylene sheeting or the plywood sheathing. This moisture flows into a porous gravel layer built around the lower part of the basement. The moisture flows through the gravel to a positively-drained sump which removes the water.

3. Lumber should be of a species and grade for which allowable unit stresses are set forth in "The National Design Specification for Stress Grade Lumber and Its Fastenings" by the National Forest Products Association.

4. Plywood must be bonded with exterior glue and be grade marked as meeting U.S. Department of Commerce Product Standard PS 1-74, Construction and Industrial Plywood.
5. The fasteners used should be silicon, bronze, copper, or stainless steel types 304 or 316 as defined by The American Iron and Steel Institute Classification. They may also be hot-dipped zinc-coated steel meeting requirements of The American Society for Testing and Materials, Standard A153.
6. Framing anchors are zinc-coated sheet steel which conform to Grade A, set forth by The American Society for Testing and Materials, Standard A446.
7. The gravel used should be washed and graded. The largest size stone acceptable is 3/4 inch (19 mm). It must be free from organic, clay, or silt types of soils.
8. Sand should be coarse and have grains not smaller than 1/16 inch (1.6 mm). It should be free from organic, clay, or silt types of soils.
9. Crushed rock should have a maximum size of 1/2 inch (13 mm).

Some Unique Features

The All-Weather Wood Foundation does not require the customary concrete footing. This reduces the overall weight and cost of construction. Since it can be assembled in a shop and erected on the site, it allows builders to install foundations in any kind of weather. Installation is fast and does not require the usual crew of concrete workers.

The basement tends to be warmer. The wooden wall provides good insulation. In addition, it can be insulated on the inside in the same manner as above-grade frame construction. This tends to reduce heating costs.

Since considerable attention is given to water-proofing and handling below-surface water, the basement produced is dry. This reduces mildew often associated with basements.

The basement area is easier to finish into living area. Normal procedures are used to apply finish over the insulation. Electrical outlets can be easily installed around the basement walls.

Installation Procedure

1. Excavate for the footings, pipes, conduit, and sump system. The footings should go below the frost line.
2. Fill the footings with the required gravel or coarse sand and level them. Fill the floor area to the proper depth needed for pouring the concrete floor.
3. Set the preassembled foundation wall panels on the footings. Plumb and square them, and fasten them together. Install the upper top plates to tie the units together.
4. Next, install the floor joists and subfloor for the first floor of the house.
5. Caulk the joints between the panels and the nailing strip and the plywood panels where they are above the ground.
6. Glue a 6-mil polyethylene film to the exterior of the panels where they will be below grade.
7. Pour the concrete floor before backfilling. This braces the bottom of the panels against the lateral thrust of the earth.
8. Backfill the foundation by first filling in at least one foot gravel above the footings. Then backfill with earth, and the job is finished.

Typical Construction Details

A typical foundation panel is shown in Fig. 8-48. The thickness of the plywood depends upon the grain direction, height of soil, and soil pressure. Recommended design data are in Fig. 8-49.

If the wall has openings for doors or windows, support members around them should be doubled.

In Fig. 8-50, the details for forming a corner are shown. Notice that after the panels are joined, a top plate is nailed in place. The method for joining panels in the wall is shown in Fig. 8-51.

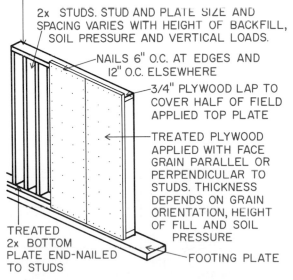

2x TOP PLATE END-NAILED TO STUDS

2x STUDS. STUD AND PLATE SIZE AND SPACING VARIES WITH HEIGHT OF BACKFILL, SOIL PRESSURE AND VERTICAL LOADS.

NAILS 6" O.C. AT EDGES AND 12" O.C. ELSEWHERE

3/4" PLYWOOD LAP TO COVER HALF OF FIELD APPLIED TOP PLATE

TREATED PLYWOOD APPLIED WITH FACE GRAIN PARALLEL OR PERPENDICULAR TO STUDS. THICKNESS DEPENDS ON GRAIN ORIENTATION, HEIGHT OF FILL AND SOIL PRESSURE

TREATED 2x BOTTOM PLATE END-NAILED TO STUDS

FOOTING PLATE

Fig. 8-48. A frame foundation wall panel.

Fig. 8-49.

Minimum APA Plywood Grade and Thickness for Exterior Foundation Wall Sheathing[1,6]
(30 PCF Equivalent Fluid Density)

Height of Fill (Inches)	Stud Spacing (Inches)	FACE GRAIN ACROSS STUDS[2]			FACE GRAIN PARALLEL TO STUDS		
		Grade[3]	Minimum Thickness (Inches)	Identification Index	Grade[3]	Minimum Thickness (Inches)	Identification Index
24	12	B	1/2	32/16	B	1/2	32/16
	16	B	1/2	32/16	B	1/2 (4,5 ply)	32/16
48	12	B	1/2	32/16	B A	1/2 ^{5}(5 ply) 1/2	32/16 32/16
	16	B	1/2	32/16	A B	5/8 3/4	42/20 48/24
72	12	B	1/2	32/16	B	5/8 ^{5}(5 ply)	42/20
	16	A^4	1/2 ^{5}	32/16	A	3/4 ^{5}	48/24
86	12	B	1/2 ^{5}	32/16	A B	5/8 ^{5} 3/4 ^{5}	42/20 48/24

American Plywood Association

[1] For crawl space construction, use 3/8-inch minimum thickness.

[2] Blocking between studs required at all horizontal panel joints more than 4 feet below adjacent ground level.

[3] Minimum grade: A. STRUCTURAL I C-D; B. C-D (exterior glue). If a major portion of the wall is exposed above ground, a better appearance may be desired. In this case, the following Exterior grades would be suitable: A. STRUCTURAL I A-C, STRUCTURAL I B-C, or STRUCTURAL I C-C (Plugged); B. A-C Exterior Group 1, B-C Exterior Group 1, C-C (Plugged) Exterior Group 1, or MDO Exterior Group 1. All plywood to carry the grade-trademarks of the American Plywood Association.

[4] Only STRUCTURAL I A-C may be substituted.

[5] For this combination of fill height and minimum panel thickness, panels that are continuous over less than three spans (across less than three stud spacings) require blocking 2 feet above bottomplate. Offset adjacent blocks and fasten through studs with two 16d corrosion-resistant nails at each end.

[6] More information on plywood sheathing is included in Chapter 9.

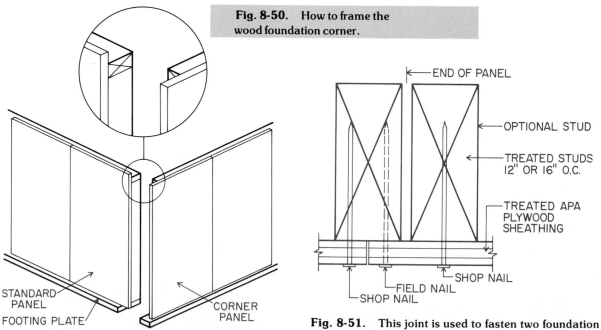

Fig. 8-50. How to frame the wood foundation corner.

STANDARD PANEL

FOOTING PLATE

CORNER PANEL

NOTE: PLYWOOD LAP ON CORNER PANEL IS EQUAL TO STUD DEPTH PLUS PLYWOOD THICKNESS ON STANDARD PANEL

END OF PANEL

OPTIONAL STUD

TREATED STUDS 12" OR 16" O.C.

TREATED APA PLYWOOD SHEATHING

SHOP NAIL

FIELD NAIL

SHOP NAIL

Fig. 8-51. This joint is used to fasten two foundation panels together.

Fig. 8-52.

Minimum Structural Requirements for Exterior Foundation Wall Framing[1,2]
(2,000 Lbs. Per Sq. Ft. Allowable Soil Bearing Pressure; 30 Lbs. Per Cu. Ft. Soil Equivalent Fluid Density)

House Width (Feet)	Number of Stories	Height of Fill (Inches)	Roof—40 PSF Live, Ceiling—10 PSF Live / 1st Floor—50 PSF Live and Dead / 2nd Floor—50 PSF Live and Dead				Roof—30 PSF Live, Ceiling—10 PSF Live / 1st Floor—50 PSF Live and Dead / 2nd Floor—50 PSF Live and Dead			
			Lumber Species and Grade[3]	Stud and Plate Size (Nominal)	Stud Spacing (Inches)	Size of Footing[4] (Nominal)	Lumber Species and Grade[3]	Stud and Plate Size (Nominal)	Stud Spacing (Inches)	Size of Footing[4] (Nominal)
24 to 28	1	24	C / B	2×4 / 2×4	12 / 16	2×8 / 2×8	C	2×4	16	2×6
		48	C	2×6	16	2×8	B / C	2×4 / 2×6	12 / 16	2×6 / 2×8
		72	B / A	2×6 / 2×6	12 / 16	2×8 / 2×8	B / A	2×6 / 2×6	12 / 16	2×8 / 2×8
		86	A	2×6	12	2×8	A	2×6	12	2×8
29 to 32	1	24	C / B	2×4 / 2×4	12 / 16	2×8 / 2×8	C / B	2×4 / 2×4	12 / 16	2×8 / 2×8
		48	C	2×6	16	2×8	C	2×6	16	2×8
		72	B / A	2×6 / 2×6	12 / 16	2×8 / 2×8	B / A	2×6 / 2×6	12 / 16	2×8 / 2×8
		86	A	2×6	12	2×8	A	2×6	12	2×8
24 to 32	2	24	C	2×6	16	2×10	C	2×6	16	2×10
		48	C	2×6	16	2×10	C	2×6	16	2×10
		72	B / A	2×6 / 2×6	12 / 16	2×10 / 2×10	B / A	2×6 / 2×6	12 / 16	2×10 / 2×10
		86	A	2×6	12	2×10	A	2×6	12	2×10
24 to 28	1	Max. 2-ft. difference in outside and inside fill height	B / C	2×4 / 2×6	16 / 16	2×8 / 2×8	B / C	2×4 / 2×6	16 / 16	2×6 / 2×6
29 to 32	1		B / C / B	2×4 / 2×6 / 2×6	12 / 12 / 16	2×8 / 2×8 / 2×8	B / C	2×4 / 2×6	16 / 16	2×8 / 2×8
24	2		B / C	2×6 / 2×6	16 / 12	2×8 / 2×8	B / C	2×6 / 2×6	16 / 12	2×8 / 2×8
25 to 32	2		B / C	2×6 / 2×6	16 / 12	2×10 / 2×10	B / C	2×6 / 2×6	16 / 12	2×10 / 2×10

[1]Loading and framing conditions are based on the assumption that roof framing spans between exterior walls and floor framing spans from interior bearing walls located midway between exterior walls.

[2]Interior bearing walls support floor loads only and must be at least grade C studs on grade C foundation plates 2 inches wider than studs. Studs shall be 2 inches by 4 inches at 16 inches on center where supporting one floor and 2 inches by 6 inches at 16 inches on center where supporting two floors.

[3]Species groups and grades having the following minimum properties (surfaced dry or surfaced green).

		A	B	C
F_b (repetitive member) psi:	2×6	1750	1450	1150
	2×4	—	1650	1300
F_c psi:	2×6	1250	1050	850
	2×4	—	1000	800
$F_{c\perp}$ psi:		385	385	245
F_u psi:		90†	90	75
E psi:		1,800,000	1,600,000	1,400,000

†Length of end splits or checks at lower end of studs not to exceed width of piece.

Examples of wood grades are: A. No. 1 grade Douglas fir-larch or southern pine; B. No. 2 grade Douglas fir-larch or No. 2 medium grain southern pine; and C. No. 2 grade Western hemlock, No. 2 grade hem-fir, No. 1 grade Douglas fir south, or No. 2 grade southern pine.

[4]Where width of footing plate is 4 inches or more wider than that of stud plate, use 3/4-inch-thick continuous treated plywood strips under footing; minimum grade C-D with exterior glue. Use plywood of same width as footing with face grain perpendicular to wall run. Fasten to footing with two 16d nails spaced 16 inches on center.

National Forest Products Association

Fig. 9-1.

Floor Joists[1]

Size and Spacing In.	Grade O.C.	30 PSF LIVE LOAD[2]				40 PSF LIVE LOAD[3]			
		Dense Sel Str KD and No. 1 Dense KD	Dense Sel Str, Sel Str KD, No. 1 Dense and No. 1 KD	Sel Str, No. 1 and No. 2 Dense KD	No. 2 Dense, No. 2 KD and No. 2	Dense Sel Str KD and No. 1 Dense KD	Dense Sel Str, Sel Str KD, No. 1 Dense and No. 1 KD	Sel Str, No. 1 and No. 2 Dense KD	No. 2 Dense, No. 2 KD and No. 2
2 × 5	12.0″	10′-3″	10′-0″	9′-10″	9′-8″	9′-3″	9′-1″	8′-11″	8′-9″
	13.7″	9′-9″	9′-7″	9′-5″	9′-3″	8′-11″	8′-9″	8′-7″	8′-5″
	16.0″	9′-3″	9′-1″	8′-11″	8′-9″	8′-5″	8′-3″	8′-2″	8′-0″
	19.2″	8′-9″	8′-7″	8′-5″	8′-3″	7′-11″	7′-10″	7′-8″	7′-6″
	24.0″	8′-1″	8′-0″	7′-10″	7′-8″	7′-4″	7′-3″	7′-1″	7′-0″[4]
2 × 6	12.0″	12′-6″	12′-3″	12′-0″	11′-10″	11′-4″	11′-2″	10′-11″	10′-9″
	13.7″	11′-11″	11′-9″	11′-6″	11′-3″	10′-10″	10′-8″	10′-6″	10′-3″
	16.0″	11′-4″	11′-2″	10′-11″	10′-9″	10′-4″	10′-2″	9′-11″	9′-9″
	19.2″	10′-8″	10′-6″	10′-4″	10′-1″	9′-8″	9′-6″	9′-4″	9′-2″
	24.0″	9′-11″	9′-9″	9′-7″	9′-4″	9′-0″	8′-10″	8′-8″	8′-6″[4]
2 × 8	12.0″	16′-6″	16′-2″	15′-10″	15′-7″	15′-0″	14′-8″	14′-5″	14′-2″
	13.7″	15′-9″	15′-6″	15′-2″	14′-11″	14′-4″	14′-1″	13′-10″	13′-6″
	16.0″	15′-0″	14′-8″	14′-5″	14′-2″	13′-7″	13′-4″	13′-1″	12′-10″
	19.2″	14′-1″	13′-10″	13′-7″	13′-4″	12′-10″	12′-7″	12′-4″	12′-1″
	24.0″	13′-1″	12′-10″	12′-7″	12′-4″	11′-11″	11′-8″	11′-5″	11′-3″[4]
2 × 10	12.0″	21′-0″	20′-8″	20′-3″	19′-10″	19′-1″	18′-9″	18′-5″	18′-0″
	13.7″	20′-1″	19′-9″	19′-4″	19′-0″	18′-3″	17′-11″	17′-7″	17′-3″
	16.0″	19′-1″	18′-9″	18′-5″	18′-0″	17′-4″	17′-0″	16′-9″	16′-5″
	19.2″	18′-0″	17′-8″	17′-4″	17′-0″	16′-4″	16′-0″	15′-9″	15′-5″
	24.0″	16′-8″	16′-5″	16′-1″	15′-9″[4]	15′-2″	14′-11″	14′-7″	14′-4″[4]
2 × 12	12.0″	25′-7″	25′-1″	24′-8″	24′-2″	23′-3″	22′-10″	22′-5″	21′-11″
	13.7″	24′-5″	24′-0″	23′-7″	23′-1″	22′-3″	21′-10″	21′-5″	21′-0″
	16.0″	23′-3″	22′-10″	22′-5″	21′-11″	21′-1″	20′-9″	20′-4″	19′-11″
	19.2″	21′-10″	21′-6″	21′-1″	20′-8″	19′-10″	19′-6″	19′-2″	18′-9″
	24.0″	20′-3″	19′-11″	19′-7″	19′-2″[4]	18′-5″	18′-1″	17′-9″	17′-5″[4]

[1]Data for southern pine structural members
[2]Sleeping rooms and attic floors.
[3]All rooms except sleeping rooms and attic floors.
[4]Deflection limitation of *l*/360.

Southern Forest Products Association

Selecting Rafters

A rafter must be of such capabilities that it can carry the normal live loads imposed upon it, as well as the dead load of its own weight. As the length of the rafter increases, the total design load has more effect upon the member. The length of a rafter is the clear distance between the two points support-ing it. See Fig. 9-3. For an ordinary rafter, this is from the exterior wall plate to the ridge board.

Roof Pitch

The selection of rafters is also influenced by the pitch of the roof. The pitch influences the length of the rafter.

Fig. 9-2.

Ceiling Joists With Drywall Ceiling[1]

Size and Spacing In.	Grade O.C.	10 PSF Live Load[2]				20 PSF Live Load[3]			
		Dense Sel Str KD and No. 1 Dense KD	Dense Sel Str, Sel Str KD, No. 1 Dense and No. 1 KD	Sel Str, No. 1 and No. 2 Dense KD	No. 2 Dense, No. 2 KD and No. 2	Dense Sel Str KD and No. 1 Dense KD	Dense Sel Str, Sel Str KD, No. 1 Dense and No. 1 KD	Sel Str, No. 1 and No. 2 Dense KD	No. 2 Dense
2 × 4	12.0″	13′-2″	12′-11″	12′-8″	12′-5″	10′-5″	10′-3″	10′-0″	9′-10″
	13.7″	12′-7″	12′-4″	12′-1″	11′-10″	10′-0″	9′-9″	9′-7″	9′-5″
	16.0″	11′-11″	11′-9″	11′-6″	11′-3″	9′-6″	9′-4″	9′-1″	8′-11″
	19.2″	11′-3″	11′-0″	10′-10″	10′-7″	8′-11″	8′-9″	8′-7″	8′-5″
	24.0″	10′-5″	10′-3″	10′-0″	9′-10″	8′-3″	8′-1″	8′-0″	7′-10″
2 × 5	12.0″	16′-11″	16′-7″	16′-3″	15′-11″	13′-5″	13′-2″	12′-11″	12′-8″
	13.7″	16′-2″	15′-10″	15′-7″	15′-3″	12′-10″	12′-7″	12′-4″	12′-1″
	16.0″	15′-4″	15′-1″	14′-9″	14′-6″	12′-2″	11′-11″	11′-9″	11′-6″
	19.2″	14′-5″	14′-2″	13′-11″	13′-8″	11′-5″	11′-3″	11′-0″	
	24.0″	13′-5″	13′-2″	12′-11″	12′-8″[4]	10′-8″	10′-5″[4]	10′-3″	
2 × 6	12.0″	20′-8″	20′-3″	19′-11″	19′-6″	16′-4″	16′-1″	15′-9″	15′-6″
	13.7″	19′-9″	19′-5″	19′-0″	18′-8″	15′-8″	15′-5″	15′-1″	14′-9″
	16.0″	18′-9″	18′-5″	18′-1″	17′-8″	14′-11″	14′-7″	14′-4″	14′-1″
	19.2″	17′-8″	17′-4″	17′-0″	16′-8″	14′-0″	13′-9″	13′-6″	
	24.0″	16′-4″	16′-1″	15′-9″	15′-6″[4]	13′-0″	12′-9″[4]	12′-6″	
2 × 8	12.0″	27′-2″	26′-9″	26′-2″	25′-8″	21′-7″	21′-2″	20′-10″	20′-5″
	13.7″	26′-0″	25′-7″	25′-1″	24′-7″	20′-8″	20′-3″	19′-11″	19′-6″
	16.0″	24′-8″	24′-3″	23′-10″	23′-4″	19′-7″	19′-3″	18′-11″	18′-6″
	19.2″	23′-3″	22′-10″	22′-5″	21′-11″	18′-5″	18′-2″	17′-9″	
	24.0″	21′-7″	21′-2″	20′-10″	20′-5″[4]	17′-2″	16′-10″[4]	16′-6″	
2 × 10	12.0″	34′-8″	34′-1″	33′-5″	32′-9″	27′-6″	27′-1″	26′-6″	26′-0″
	13.7″	33′-2″	32′-7″	32′-0″	31′-4″	26′-4″	25′-10″	25′-5″	24′-11″
	16.0″	31′-6″	31′-0″	30′-5″	29′-9″	25′-0″	24′-7″	24′-1″	23′-8″
	19.2″	29′-8″	29′-2″	28′-7″	28′-0″	23′-7″	23′-2″	22′-8″	
	24.0″	27′-6″	27′-1″	26′-6″	26′-0″[4]	21′-10″	21′-6″[4]	21′-1″	

Southern Forest Products Association

[1]Data for southern pine structural members
[2]With no future sleeping rooms and no attic storage.
[3]With no future sleeping rooms and limited attic storage.
[4]Deflection limitation of $l/240$.

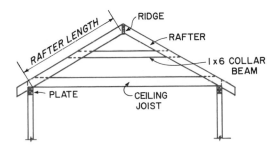

Fig. 9-3. Calculation of rafter length.

Pitch is the angle of the roof from the ridge board to the plate. It is usually expressed in the form of a ratio of the total rise to the total span of a building.

$$\text{Pitch} = \frac{\text{total rise}}{\text{total span}}.$$ See Figs. 9-4 and 9-5.

The angle of the roof is expressed on working drawings in terms of slope. **Slope** is a ratio between rise and run. The run is half the span. Refer again

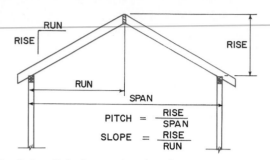

Fig. 9-4.   Calculation of roof pitch.

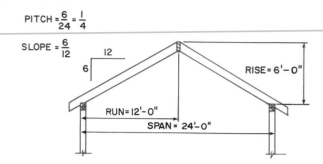

Fig. 9-5.   Calculation of pitch and slope.

Fig. 9-6.

Rafters — Low Slope[1,2]

Size and Spacing In.	O.C.	20 PSF LIVE LOAD			30 PSF LIVE LOAD				40 PSF LIVE LOAD			
Grade		Dense Sel Str KD and No. 1 Dense KD	Dense Sel Str, Sel Str KD, No. 1 Dense and No. 1 KD	Sel Str	Dense Sel Str KD and No. 1 Dense KD	Dense Sel Str and Sel Str KD	No. 1 Dense	Sel Str	Dense Sel Str KD and No. 1 Dense KD	Dense Sel Str and Sel Str KD	No. 1 Dense	No. 1 KD
2 × 5	12.0″	13′-5″	13′-2″	12′-11″	11′-8″	11′-6″	11′-6″	11′-13″	10′-8″	10′-5″	10′-5″	10′-5″
	13.7″	12′-10″	12′-7″	12′-4″	11′-2″	11′-0″	11′-0″	10′-9″	10′-2″	10′-0″	10′-0″	10′-0″
	16.0″	12′-2″	11′-11″	11′-9″	10′-8″	10′-5″	10′-5″	10′-3″	9′-8″	9′-6″	9′-6″	9′-6″
	19.2″	11′-5″	11′-3″	11′-0″	10′-0″	9′-10″	9′-10″	9′-8″	9′-1″	8′-11″	8′-11″	
	24.0″	10′-8″	10′-5″[3]	10′-3″	9′-3″	9′-1″	9′-1″	8′-11″	8′-5″	8′-3″		
2 × 6	12.0″	16′-4″	16′-1″	15′-9″	14′-4″	14′-1″	14′-1″	13′-9″	13′-0″	12′-9″	12′-9″	12′-9″
	13.7″	15′-8″	15′-5″	15′-1″	13′-8″	13′-5″	13′-5″	13′-2″	12′-5″	12′-3″	12′-3″	12′-3″
	16.0″	14′-11″	14′-7″	14′-4″	13′-0″	12′-9″	12′-9″	12′-6″	11′-10″	11′-7″	11′-7″	11′-7″
	19.2″	14′-0″	13′-9″	13′-6″	12′-3″	12′-0″	12′-0″	11′-9″	11′-1″	10′-11″	10′-11″	
	24.0″	13′-0″	12′-9″[3]	12′-6″	11′-4″	11′-2″	11′-1″	10′-11″	10′-4″	10′-2″		
2 × 8	12.0″	21′-7″	21′-2″	20′-10″	18′-10″	18′-6″	18′-6″	18′-2″	17′-2″	16′-10″	16′-10″	16′-10″
	13.7″	20′-8″	20′-3″	19′-11″	18′-0″	17′-9″	17′-9″	17′-5″	16′-5″	16′-1″	16′-1″	16′-1″
	16.0″	19′-7″	19′-3″	18′-11″	17′-2″	16′-10″	16′-10″	16′-6″	15′-7″	15′-3″	15′-3″	15′-3″
	19.2″	18′-5″	18′-2″	17′-9″	16′-1″	15′-10″	15′-10″	15′-6″	14′-8″	14′-5″	14′-5″	
	24.0″	17′-2″	16′-10″[3]	16′-6″	15′-0″	14′-8″	14′-8″	14′-5″	13′-7″	13′-4″		
2 × 10	12.0″	27′-6″	27′-1″	26′-6″	24′-1″	23′-8″	23′-8″	23′-2″	21′-10″	21′-6″	21′-6″	21′-6″
	13.7″	26′-4″	25′-10″	25′-5″	23′-0″	22′-7″	22′-7″	22′-2″	20′-11″	20′-6″	20′-6″	20′-6″
	16.0″	25′-0″	24′-7″	24′-1″	21′-10″	21′-6″	21′-6″	21′-1″	19′-10″	19′-6″	19′-6″	19′-6″
	19.2″	23′-7″	23′-2″	22′-8″	20′-7″	20′-2″	20′-2″	19′-10″	18′-8″	18′-4″	18′-4″	
	24.0″	21′-10″	21′-6″[3]	21′-1″	19′-1″	18′-9″	18′-8″	18′-5″	17′-4″	17′-0″		
2 × 12	12.0″	33′-6″	32′-11″	32′-3″	29′-3″	28′-9″	28′-9″	28′-2″	26′-7″	26′-1″	26′-1″	26′-1″
	13.7″	32′-0″	31′-6″	30′-10″	28′-0″	27′-6″	27′-6″	27′-0″	25′-5″	25′-0″	25′-0″	25′-0″
	16.0″	30′-5″	29′-11″	29′-4″	26′-7″	26′-1″	26′-1″	25′-7″	24′-2″	23′-9″	23′-9″	23′-9″
	19.2″	28′-8″	28′-2″	27′-7″	25′-0″	24′-7″	24′-7″	24′-1″	22′-9″	22′-4″	22′-4″	
	24.0″	26′-7″	26′-1″[3]	25′-7″	23′-3″	22′-10″	22′-8″	22′-5″	21′-1″	20′-9″		

[1]Data for southern pine structural members
[2]3 in 12 or less; no finished ceiling.
[3]Deflection limitation of $l/240$.

Southern Forest Products Association

Fig. 9-7.
Rafters — High Slope[1,2,5]

Size and Spacing In.	Grade O.C.	20 PSF LIVE LOAD[3]				30 PSF LIVE LOAD[3]				40 PSF LIVE LOAD[3]			
		Dense Sel Str KD	Sel Str KD and Dense Sel Str	No. 1 Dense KD	Sel Str	Dense Sel Str KD	Sel Str KD and Dense Sel Str	No. 1 Dense KD	Sel Str	Dense Sel Str KD	Sel Str KD and Dense Sel Str	No. 1 Dense KD	Sel Str
2 × 4	12.0"	11'-6"	11'-3"	11'-6"	11'-1"	10'-0"	9'-10"	10'-0"	9'-8"	9'-1"	8'-11"	9'-1"	8'-9"
	13.7"	11'-0"	10'-9"	11'-0"	10'-7"	9'-7"	9'-5"	9'-7"	9'-3"	8'-8"	8'-7"	8'-8"	8'-5"
	16.0"	10'-5"	10'-3"	10'-5"	10'-0"	9'-1"	8'-11"	9'-1"	8'-9"	8'-3"	8'-1"	8'-3"	8'-0"
	19.2"	9'-10"	9'-8"	9'-10"	9'-5"	8'-7"	8'-5"	8'-7"	8'-3"	7'-9"	7'-8"	7'-9"	7'-6"
	24.0"	9'-1"	8'-11"	9'-1"	8'-9"	7'-11"	7'-10"	7'-11"	7'-8"	7'-3"	7'-1"	7'-3"	7'-0"
2 × 5	12.0"	14'-9"	14'-6"	14'-9"	14'-3"	12'-11"	12'-8"	12'-11"	12'5"	11'-8"	11'-6"	11'-8"	11'-3"
	13.7"	14'-1"	13'-10"	14'-1"	13'-7"	12'-4"	12'-1"	12'-4"	11'-10"	11'-2"	11'-0"	11'-2"	10'-9"
	16.0"	13'-5"	13'-2"	13'-5"	12'-11"	11'-8"	11'-6"	11'-8"	11'-3"	10'-8"	10'-5"	10'-8"	10'-3"
	19.2"	12'-7"	12'-5"	12'-7"	12'-2"	11'-0"	10'-10"		10'-7"	10'-0"	9'-10"[4]		
	24.0"	11'-8"	11'-6"			10'-3"	10'-1"[4]			9'-3"	9'-1"[1]		
2 × 6	12.0"	18'-0"	17'-8"	18'-0"	17'-4"	15'-9"	15'-6"	15'-9"	15'2"	14'-4"	14'-1"	14'-4"	13'-9"
	13.7"	17'-3"	16'-11"	17'-3"	16'-7"	15'-1"	14'-9"	15'-1"	14'-6"	13'8"	13'-5"	13'-8"	13'-2"
	16.0"	16'-4"	16'-1"	16'-4"	15'-9"	14'-4"	14'-1"	14'-4"	13'-9"	13'-0"	12'-9"	13'-0"	12'-6"
	19.2"	15'-5"	15'-2"	15'-5"	14'-10"	13'-6"	13'-3"		13'0"	12'-3"	12'-0"[4]		
	24.0"	14'-4"	14'-1"			12'-6"	12'-3"[4]			11'-4"	11'-2"[4]		
2 × 8	12.0"	23'-9"	23'-4"	23'-9"	22'-11"	20'-9"	20'-5"	20'-9"	20'-0"	18'-10"	18'-6"	18'-10"	18'-2"
	13.7"	22'-9"	22'-4"	22'-9"	21'-11"	19'-10"	19'-6"	19'-10"	19'-2"	18'-0"	17'-9"	18'-0"	17'-5"
	16.0"	21'-7"	21'-2"	21'-7"	20'-10"	18'-10"	18'-6"	18'-10"	18'-2"	17'-2"	16'-10"	17'-2"	16'-6"
	19.2"	20'-4"	19'-11"	20'-4"	19'-7"	17'-9"	17'-5"		17'-1"	16'-1"	15'-10"[4]		
	24.0"	18'-10"	18'-6"			16'-6"	16'-2"[4]			15'-0"	14'-8"[4]		
2 × 10	12.0"	30'-4"	29'-9"	30'-4"	29'-2"	26'-6"	26'-0"	26'-6"	25'-6"	24'-1"	23'-8"	24'-1"	23'-2"
	13.7"	29'-0"	28'-6"	29'-0"	27'-11"	25'-4"	24'-11"	25'-4"	24'-5"	23'-0"	22'-7"	23'-0"	22'-2"
	16.0"	27'-6"	27'-1"	27'-6"	26'-6"	24'-1"	23'-8"	24'-1"	23'-2"	21'-10"	21'-6"	21'-10"	21'-1"
	19.2"	25'-11"	25'-5"	25'-11"	25'-0"	22'-8"	22'-3"		21'-10"	20'-7"	20'-2"[4]		
	24.0"	24'-1"	23'-8"			21'-0"	20'-8"[4]			19'-1"	18'-9"[4]		

[1]Data for southern pine structural members.
[2]Over 3 in 12; no finished ceiling.
[3]Plus 7 PSF deadload.
[4]Deflection limitation $l/180$.
[5]Data indicate rafter run.

Southern Forest Products Association

to Fig. 9-5. The symbol $6\overline{)12}$ found on working drawings is an expression of slope. The numbers placed upon this symbol mean there are 6 inches of rise for every 12 inches of run. The pitch of such a roof is $\frac{rise}{span}$ or $\frac{6}{24}$ or $\frac{1}{4}$ pitch. The carpenter uses expression of slope when laying out the cuts on a rafter.

Rafter Size Selection

Data for selecting rafters of four grades of southern pine are given in Figs. 9-6 and 9-7. These data are typical of design data available by species and grade of lumber.

The figures show data for low slope roofs — 3 in 12 or below — and high slope roofs — 3 in 12 and over. They are for rafters which support no interior ceiling. The live loads used are 20, 30, and 40 pounds per square foot.

The grades of lumber are indicated across the top of the table. The figures in the columns below are the maximum clear run of the rafters. The size and spacing of the member is indicated in the column at the left. For example, a 2 × 6 rafter of select structural southern pine spaced 16 inches on center, used to build a roof sloping over 3 in 12 with a 20-pound live load will have a run of 15'-9". If the live load is 40 pounds per square foot, it will span 12'-6". This member will span 14'-4" for a roof that slopes less than 3 in 12 with a 20-pound per square foot live load.

The actual length of a rafter is determined by the horizontal rafter run and roof slope. The diagram in Fig. 9-8 explains how to find rafter length.

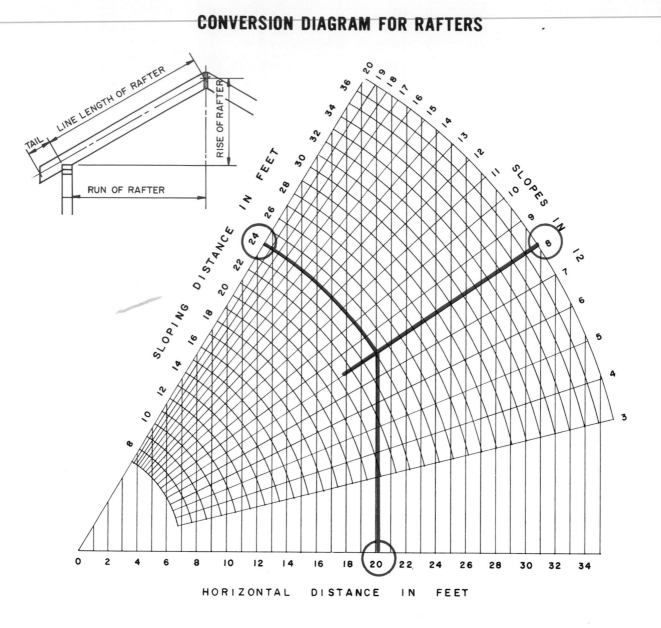

CONVERSION DIAGRAM FOR RAFTERS

To use the diagram select the known horizontal distance and follow the vertical line to its intersection with the radial line of the specified slope, then proceed along the arc to read the sloping distance. In some cases it may be desirable to interpolate between the one foot separations. The diagram also may be used to find the horizontal distance corresponding to a given sloping distance or to find the slope when the horizontal and sloping distances are known.

Example: With a roof slope of 8 in 12 and a horizontal distance of 20 feet the sloping distance may be read as 24 feet.

Fig. 9-8.

Southern Forest Products Association

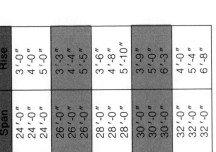

Slope	Span	Rise
3/12	24'-0"	3'-0"
4/12	24'-0"	4'-0"
5/12	24'-0"	5'-0"
3/12	26'-0"	3'-3"
4/12	26'-0"	4'-4"
5/12	26'-0"	5'-5"
3/12	28'-0"	3'-6"
4/12	28'-0"	4'-8"
5/12	28'-0"	5'-10"
3/12	30'-0"	3'-9"
4/12	30'-0"	5'-0"
5/12	30'-0"	6'-3"
3/12	32'-0"	4'-0"
4/12	32'-0"	5'-4"
5/12	32'-0"	6'-8"

Midwest Plan Service

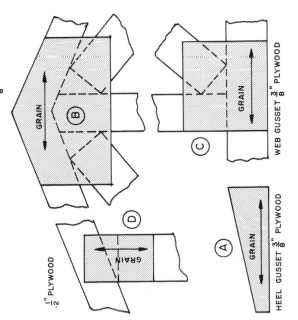

RIDGE GUSSET $\frac{3}{8}$" PLYWOOD

GRAIN

Ⓑ

WEB GUSSET $\frac{3}{8}$" PLYWOOD

Ⓒ

GRAIN

$\frac{1}{2}$" PLYWOOD

Ⓓ

GRAIN

HEEL GUSSET $\frac{3}{8}$" PLYWOOD

Ⓐ

GRAIN

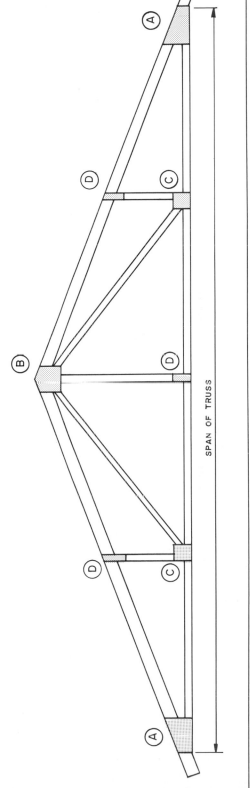

SPAN OF TRUSS

Fig. 9-9. A design for a wood roof truss using glued and nailed plywood gussets.

Fig. 9-10. A split ring as used in truss construction.

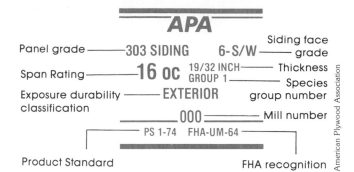

Fig. 9-11. Typical plywood grade marks.

Collar Beams

Collar beams frequently are used to stiffen a roof. A collar beam is a 1 × 6 tie placed between rafters on opposite sides of a roof. Refer again to Fig. 9-3. This member is usually placed on every second or third rafter and is most effective if placed midway between the ridge and the ceiling joist.

Some believe that the use of a collar beam reduces the required span of the rafter, but, in general practice, this is not accepted.

Roof Trusses

Designing a truss is a job for someone trained in structural design, such as an architect or engineer. It is beyond the scope of this text to present such information.

The trusses generally used in residential construction have plywood or metal gussets. A **gusset** is a plate used to reinforce a joint. Trusses are mass-produced in factories, but may sometimes be assembled in the field. The design requirements must be followed carefully.

A typical two-web truss design is shown in Fig. 9-9. Complete design data for this and other designs are available from the Midwest Plan Service, Agricultural Engineering, Iowa State University, Ames, Iowa 50010.

When plywood gussets are used, their thickness and the direction of the face grain is critical to producing a strong truss. The structural members are cut square on the ends and are joined by the gussets.

Large, heavy-duty trusses, especially those used in commerical construction, are joined using a special bolt and split-ring connector. The **split ring** is installed in grooves cut into each member. The bolt pulls the members together, closing them over the ring. See Fig. 9-10.

Plywood and Other Structural Wood Panels

Panels used in construction are available in several forms — plywood, composites, waferboard, oriented-strand board, and structural particleboard. Plywood is made of an odd number of cross-laminated veneers. Composites have a core of reconstituted wood bonded between a veneer face and the back plies. Waferboard is made from wafer-like wood chips compressed and bonded together. Oriented-strand board is made from strand-like wood particles arranged in layers with the strands in each layer at right angles to those in the layers above and below. Structural particleboard is made of small wood particles that are compressed and bonded into panels.

Structural wood panels are divided into two types — interior and exterior. Interior panels are marked "EXPOSURE 1" or "EXPOSURE 2." EXPOSURE 1 is an interior-type plywood made with exterior glue. It is used where long exposure to weather during construction is expected. EXPOSURE 2 panels use an intermediate glue and are used where only a moderate amount of exposure is likely.

Exterior panels are made with a waterproof glue and are used where permanent exposure to the weather is expected. These are marked "EXTERIOR."

Plywood veneers are graded for appearance. Grades are identified as N, A, B, C, and D. The N and A veneers are the best in appearance.

APA (American Plywood Association) panels have identifying data on the **trademark stamp**. See Fig. 9-11. This stamp on a panel verifies that the panel has been manufactured to meet APA standards.

Construction plywood panel grades are identified in one of two ways. They may be identified in terms of the veneer grade on the face and back of the panel, such as C-D. Or they may have a name that suggests the panels' intended use, such as APA Rated Sheathing or APA Rated Sturd-I-Floor. See Fig. 9-12.

The **span rating** is shown numerically. Refer again to Fig. 9-11. An example is 32/16. The first number is the maximum span if the panel is used as roof sheathing. The second number is the maximum span if it is used as subflooring.

The **group number**, when used, gives an indication of the strength of the wood used to make the panel. Woods are divided into 5 groups — 1, 2, 3, 4, and 5. The woods in group 1 are the strongest. Those in group 5 are the weakest.

Fig. 9-12.

Guide to APA Performance-Rated Panels[1,2]

Interior Exposure 1 and 2 and Exterior	Interior Exposure 1 and Exterior
APA RATED SHEATHING	**APA STRUCTURAL I AND II RATED SHEATHING[3]**

APA RATED SHEATHING

TYPICAL TRADEMARK

APA
RATED SHEATHING
32/16 15/32 INCH
SIZED FOR SPACING
EXPOSURE 1
000
NRB-108

Specially designed for subflooring and wall and roof sheathing. Also good for broad range of other construction and industrial applications. Can be manufactured as conventional veneered plywood, as a composite, or as a nonveneered panel. For special engineered applications, veneered panels conforming to PS 1 may be required. EXPOSURE DURABILITY CLASSIFICATIONS: Exterior, Exposure 1, Exposure 2. COMMON THICKNESSES: 5/16, 3/8, 7/16, 15/32, 1/2, 19/32, 5/8, 23/32, 3/4.

APA STRUCTURAL I AND II RATED SHEATHING[3]

TYPICAL TRADEMARK

APA
RATED SHEATHING
STRUCTURAL I
48/24 23/32 INCH
SIZED FOR SPACING
EXTERIOR
000
PS 1-83 C-C   NRB-108

Unsanded all-veneer plywood grades for use where strength properties are of maximum importance, such as box beams, gusset plates, stressed-skin panels, containers, pallet bins. Structural I more commonly available. EXPOSURE DURABILITY CLASSIFICATIONS: Exterior, Exposure 1. COMMON THICKNESS: 5/16, 3/8, 15/32, 1/2, 19/32, 5/8, 23/32, 3/4.

APA RATED STURD-I-FLOOR	Interior Exposure I
	APA RATED STURD-I-FLOOR 48 O.C. (2-4-1)

APA RATED STURD-I-FLOOR

TYPICAL TRADEMARK

APA
RATED STURD-I-FLOOR
20 OC 19/32 INCH
SIZED FOR SPACING
EXTERIOR
000
NRB-108

Specially designed as combination subfloor-underlayment. Provides smooth surface for application of carpet and possesses high concentrated and impact load resistance. Can be manufactured as conventional veneered plywood, as a composite, or as a nonveneered panel. Available square edge or tongue-and-groove. EXPOSURE DURABILITY CLASSIFICATIONS: Exterior, Exposure 1, Exposure 2. COMMON THICKNESSES: 19/32, 5/8, 23/32, 3/4.

APA RATED STURD-I-FLOOR 48 O.C. (2-4-1)

TYPICAL TRADEMARK

APA
RATED STURD-I-FLOOR
48 OC 1-1/8 INCH
2-4-1
SIZED FOR SPACING
EXPOSURE 1
T&G 000
PS 1-83   UNDERLAYMENT
NRB-108   FHA-UM-66

For combination subfloor-underlayment on 32- and 48-inch spans and for heavy timber roof construction. Manufactured only as conventional veneered plywood. Available square edge or tongue-and-groove. EXPOSURE DURABILITY CLASSIFICATIONS: Exposure 1. THICKNESS: 1-1/8.

[1]Specific grades, thicknesses and exposure durability classifications may be in limited supply in some areas. Check with your supplier before specifying.

[2]Specify Performance-Rated Panels by thickness and Span Rating. Span Ratings are based on panel strength and stiffness. Since these properties are a function of panel composition and configuration as well as thickness, the same Span Rating may appear on panels of different thickness. Conversely, panels of the same thickness may be marked with different Span Ratings.

[3]All plies in Structural I plywood panels are special improved grades and limited to Group 1 species. All plies in Structural II plywood panels are special improved grades and limited to Group 1, 2, or 3 species.

American Plywood Association

Fig. 9-13.

Roof Decking[1]

Panel Span Rating	Panel Thickness (in.)	MAXIMUM SPAN (IN.)		ALLOWABLE LIVE LOADS (PSF)[4]							
		With Edge Support[2]	Without Edge Support	Spacing of Supports Center-to-Center (in.)							
				12	16	20	24	32	40	48	60
12/0	5/16	12	12	30							
16/0	5/16, 3/8	16	16	55	30						
20/0	5/16, 3/8	20	20	70	50	30					
24/0	3/8, 7/16, 1/2	24	20[3]	90	65	55	30				
24/16	7/16, 1/2	24	24	135	100	75	40				
32/16	15/32, 1/2, 5/8	32	28	135	100	75	55	30			
40/20	19/32, 5/8, 3/4, 7/8	40	32	165	120	100	75	55	30		
48/24	23/32, 3/4, 7/8	48	36	210	155	130	100	65	50	35	
48 O.C.[5]	1 1/8	60	48				375	205	100	65	40

[1]APA Structural Rated Sheathing and APA Structural I and II Rated Sheathing.

[2]Tongue-and-groove edges, panel edge clips (one between each support, except two between supports 48 inches on center), lumber blocking, or other.

[3]24 inches for 1/2" panels.

[4]10 PSF dead load assumed.

[5]Span Rating applies to APA Rated Sturd-I-Floor "2-4-1".

American Plywood Association

Data for selecting panels for roof decking, sheathing, subfloors, and underlayment are in Figs. 9-13 to 9-15. Figure 9-16 gives data for selecting panels that are a combination subfloor-underlayment. These panels are identified by the name Sturd-I-Floor and are available for interior and exterior use depending on thickness. The interior type is 1 1/8-inch thick.

Fig. 9-14.

Panel Subflooring

Panel Span Rating (or Group Number)	Panel Thickness (in.)	Maximum Span (in.)
24/16[3]	7/16	16
32/16	15/32, 1/2, 5/8	16[1]
40/20	19/32, 5/8, 3/4, 7/8	20[2]
48/24	23/32, 3/4, 7/8	24
1 1/8″ Groups 1 & 2[3]	1 1/8	48
1 1/4″ Groups 3 & 4[3]	1 1/4	48

[1]Span may be 24 inches if 25/32 inch wood strip flooring is installed at right angles to joists.
[2]Span may be 24 inches if 25/32 inch wood strip flooring is installed at right angles to joists, or if a minimum 1 1/2 inches of lightweight concrete (or 1 inch of some gypsum concrete products) is applied over panels.
[3]Check dealer for availability. Note that a Span Rating of 32/16 is the minimum recommended on joists spaced 16 inches on center when topped with lightweight or gypsum concrete.

American Plywood Association

Fig. 9-15A.

Plywood Underlayment

Plywood Grades* and Species Group	Application	Minimum Plywood Thickness (in.)
Groups 1, 2, 3, 4, 5 APA UNDERLAYMENT INT (with interior or exterior glue) APA UNDERLAYMENT EXT APA C-C Plugged EXT	Over smooth subfloor	1/4
	Over lumber subfloor or other uneven surfaces.	11/32
Same grades as above, but Group 1 only.	Over lumber floor up to 4″ wide. Face grain must be perpendicular to boards.	1/4

*When 19/32 inch or thicker underlayment is desired, APA Rated Sturd-I-Floor may be specified. In areas to be finished with thin floor coverings such as tile, linoleum, or vinyl, specify Underlayment, C-C Plugged or Sturd-I-Floor with fully sanded face.

American Plywood Association

Fig. 9-15B.

Plywood Wall Sheathing[1]

Panel Span Rating	Panel Thickness (in.)	Stud Spacing (in.)
16/0	3/8	16 O.C.
20/0	1/2	16 O.C.
24/0	15/32	24 O.C.
24/16	1/2	24 O.C.
32/16	15/32	24 O.C.

[1]Panel continuous over two or more studs.

American Plywood Association

Fig. 9-16.

Subfloor-Underlayment[1]

Span Rating (Maximum Joist Spacing) (in.)	Panel Thickness (in.)
16	19/32, 5/8, 21/32
20	19/32, 5/8, 23/32, 3/4
24	11/16, 23/32, 3/4
	7/8, 1
48 (2-4-1)	1 1/8

[1]APA Rated Sturd-I-Floor

American Plywood Association

Fig. 9-17.

Rigid Foamed Plastic Sheathing

Thickness	1/2″, 3/4″, 1″, 1.1/2″, 2″
Panel Length and Width	24 ″ x 96″

Dow Chemical Company

Other Sheathing Materials

Solid wood can be used for sheathing. One-inch (3/4″ actual) solid wood sheathing can span up to 24 inches on roofs and 16 inches on wall construction. If applied diagonally on a wall, it can be used over studs spaced 24 inches on center.

Fiberboard sheathing is available in 1/2-inch and 25/32-inch thicknesses. It is used for wall sheathing and can span studs 16 inches on center. If the wall studs are diagonally braced, it can span 24 inches on center.

Gypsum sheathing is 1/2-inch thick. Maximum stud-spacing is 16 inches on center. If the wall is diagonally braced, the studs can be 24 inches on center.

Rigid foamed plastic sheathing panels are available in tongue-and-groove and square-edge panels. Sizes are given in Fig. 9-17. Some of these panels are covered with a thin layer of aluminum. Normal spacing of the studs is 16 inches on center.

Heavy Decking for Plank-and-Beam Roofs

Roof decking customarily is surfaced two sides, is tongue and groove, and, usually, is run to a V-joint pattern. The 2-inch decking has a single tongue and groove; 3- and 4-inch decking have a double tongue and groove. See Fig. 9-18. This decking is installed by spiking each course to the preceding course at regular intervals through factory-drilled holes and toenailing to each beam.

Where there is a combination of warmth and high humidity inside a building and low temperature outside, the moisture vapor inside tends to move outward through walls and ceilings. In extreme cases, a vapor barrier and 1-inch rigid insulation should be placed above the deck to prevent condensation on top of the decking. This is purely a provision for moisture control, since wood is a good insulating material. Decking 3 and 4 inches thick provides adequate thermal insulation.

Maximum spans for decking used on roofs are given in Fig. 9-19. The span is influenced not only by the species and grade of lumber, but by the method of installing the decking. Three common methods are the simple span, random length, and combination simple and two-span continuous. Figure 9-19 illustrates data for random length decking.

Other types of decking are available. For example, a decking composed of long, wood fibers bonded with an inorganic, hydraulic cement is available. The wood-fiber layer forms the base upon which a urethane-foam layer is bonded. The thicknesses and panel sizes are shown in Fig. 9-20.

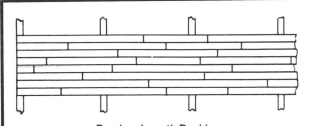

Random Length Decking

Economical, random lengths of plank having well scattered joints, and each plank bearing on at least one support, is the basic arrangement for this type. The pieces should be end matched or end splined when of 2″ nominal thickness.

WEST COAST DOUGLAS FIR

Nominal Thickness	Grade	Live Load (lbs. per sq. ft.)			
		20	30	40	50
2″	Construction	10′-3″	9′-0″	8′-2″	7′-7″
2″	Standard	10′-3″	9′-0″	8′-2″	7′-7″
2″	Utility	7′-2″	6′-3″	5′-8″	5′-3″
3″	Select Dex	16′-9″	14′-6″	13′-3″	12′-3″
3″	Coml. Dex	16′-9″	14′-6″	13′-3″	12′-3″
4″	Select Dex	22′-0″	19′-3″	17′-6″	16′-3″
4″	Coml. Dex	22′-0″	19′-3″	17′-6″	16′-3″

WESTERN RED CEDAR

Nominal Thickness	Grade	Live Load (lbs. per sq. ft.)			
		20	30	40	50
2″	Select Merch.	8′-10″	7′-9″	7′-0″	6′-6″
2″	Construction	8′-10″	7′-9″	7′-0″	6′-6″
2″	Standard	6′-9″	6′-0″	5′-5″	5′-0″
3″	Select Dex	14′-3″	12′-3″	11′-3″	10′-6″
3″	Coml. Dex	13′-6″	12′-0″	10′-9″	10′-0″
4″	Select Dex	19′-0″	16′-3″	15′-0″	14′-0″
4″	Coml. Dex	18′-0″	16′-0″	14′-9″	13′-3″

West Coast Lumbermen's Association

Fig. 9-19. Span data for heavy wood decking laid in random lengths.

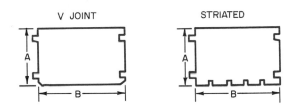

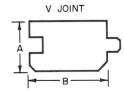

Heavy Wood Roof Decking

SIZES	
Nominal	Actual
2 x 6	1 5/8 x 5 face
2 x 8	1 5/8 x 7 face
3 x 6	2 5/8 x 5 1/4 face
4 x 6	3 1/2 x 5 1/4 face

Fig. 9-18. Typical heavy wood decking.

Fig. 9-20.

Tectum® Roof Decking

Thickness	R-Value	Span	Load (PSF)	Weight (PSF)
2″ with 2″ Foam	19	48″	200	3.8
2 1/2″ with 2″ Foam	20	60″	165	4.7
3″	21	72″	200	5.5

National Gypsum Company

Columns

Steel columns used in residential construction are manufactured in standard sizes. The load-bearing qualities of these members have been ascertained through actual load tests.

These columns are made from a steel shell filled with a special concrete mix that is carefully vibrated in place to fill all voids. A steel plate is welded to the top of the column to provide a bearing surface for the beam, and another steel plate is welded to the bottom of the column to form a base. Refer to Fig. 9-21.

These structural members are available in many sizes. Figure 9-22 records design data for those commonly used in residences. W-shapes for columns are on page 546. Heavy steel round and square columns are on page 547.

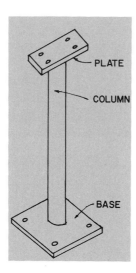

Fig. 9-21. Typical steel column.

Fig. 9-22.
Maximum Loads for Steel Columns

	Outside Diameter of Column (Inches)	Weight Per Foot of Column (Pounds)	Maximum Load for Length of Column (Feet)*					
			6	7	8	9	10	12
Lightweight Steel Columns	3½	12	23	21	19	17		
	4	15	30	28	26	24		
Standard Weight Steel Columns	3½	15	38		32		27	
	4	20	49		43		37	31

*Safe concentric load in thousands of pounds.   The Lally Company

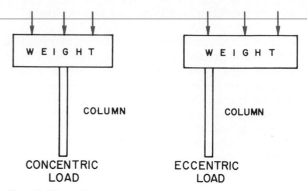

Fig. 9-23. Concentric and eccentric loading.

Fig. 9-24.
Steel Column Base Dimensions

Column Diameter	Base in Inches
3½	8 × 8
4	9 × 9

The Lally Company

Columns are subject to two forces — **concentric loading** and **eccentric loading.** A column is considered concentrically loaded if the vertical load bearing upon it is directly on top of the column. This is the case in the usual residential construction.

Eccentric loading is a load bearing on a column that is unbalanced by another load opposite it. See Fig. 9-23. If such a case exists, the equivalent concentric load must be calculated. This is beyond the scope of this study.

Figure 9-24 indicates the common sizes of the bases of steel columns. This information is useful in designing column footings.

Wood columns may be used in residential construction. They may be part of a post-and-beam structure or columns to hold the weight of the floor in conventional construction. They must be strong enough to carry the loads placed on them. They also have to provide a flat surface upon which beams can rest. Wood columns are seldom smaller than 4″ × 4″. They can be solid stock or made from glued, laminated layers. These layers are usually 2 inches thick.

The longer the column, the thicker it should be. Adequate thickness is needed to prevent buckling. The loads columns can support vary with the species of wood. A timber construction manual should be consulted for design data.

The length is determined by a slenderness ratio. This is a proportion of length in inches to width in inches. For example, typical residential construction would use a 4″ × 4″ column 8 feet long. This

/>
Fig. 9-25.

Design Data for Solid Wood Columns 8′-0″ in Length [1]

Nominal Size (In.)	Actual Size (In.)	Area Sq. In. A	Slenderness Ratio for 8′-0″ Length (Length/Least Dimension) (L/d)	Eastern Hemlock Select Str No. 1 E = 1,200,000 F_c = 875 (Lb.)	Calif. Red-wood, Select Str. No. 1 E = 1,300,000 F_c = 1030 (Lb.)	Douglas Fir Select Str No. 1 E = 1,600,000 F_c = 1000 (Lb.)	Eastern Spruce Select, No. 1 E = 1,200,000 F_c = 725 (Lb.)	Southern Pine Select Str No. 1 E = 1,800,000 F_c = 1250 (Lb.)	Oak Select Str No. 1 E = 1,650,000 F_c = 1300 (Lb.)
4 × 4 [2]	3½ × 3½	12.25	27.4	5,870	6,360	7,830	5,870	8,800	8,070
4 × 6 [2]	3½ × 5½	19.25	27.4	9,230	10,000	12,300	9,230	13,830	12,690
6 × 6	5½ × 5½	30.25	17.5	26,460	31,150	30,250	21,930	37,800	39,300
6 × 8	5½ × 7½	41.25	17.5	36,090	42,400	41,250	29,900	51,550	53,600
8 × 8	7½ × 7½	56.25	12.8	49,200	57,900	56,250	40,750	70,300	73,100
8 × 10	7½ × 9½	71.25	12.8	62,340	73,350	71,250	51,650	89,050	92,600
10 × 10	9½ × 9½	90.25	10.1	78,960	92,950	90,250	65,430	112,800	117,300

[1]Figured for 19% moisture content and surfaced four sides.
[2]Allowable load controlled by slenderness ratio.

$$\text{Allowable Load} = P = A \left[\frac{0.3\,E}{(l/d)^2} \right] \text{ or } P = A\,F_c \text{ whichever is smaller}$$

E = Modulus of Elasticity in PSI
A = Cross-Sectional Area in In²
L = Column Length in Inches
D = Least Dimension of Column
F_c = Allowable Unit Stress in PSI

would give 96 inches divided by 3.5 inches (actual size). The slenderness ratio would be 27.4. Recommended ratios are found in local building codes.

As a general rule, the column should be the same width as the beam. A column for a 6″ × 8″ beam would be 6 inches in width. Thus the **wood** column should be the same width as the **wood** beam.

Load-carrying capacities for selected sizes of solid wood columns made from several species of wood are given in Fig. 9-25. The formula and design data used to compute the load are included. The capacities for other lengths can be computed with this formula.

Stressed-Skin Panels

Stressed-skin panels are flat panels with stressed plywood skins and spaced lumber stringers. Panels may have a skin on one or both sides. A typical one-sided stressed-skin panel is shown in Fig. 9-26. Notice the use of a plywood member as a splice plate at the butt joint between the plywood skin panels. One-skin panels may require bridging between stringers. Design data for selected roof panels are in Fig. 9-27, page 274. These have the stringers spaced 12, 16, and 24 inches on center. Many other designs are possible.

A typical two-skin panel is shown in Fig. 9-28, page 274. Bridging between stringers is not required. A two-skin panel usually is fully insulated. The bottom panel can serve as the finished interior ceiling.

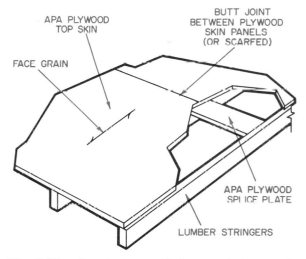

Fig. 9-26. A typical one-sided stressed-skin panel. Panels may also have skins on both sides.

Girders

A **girder** is a large, horizontal beam. Several computations must be made before girder size can be selected. First, the distance the girder must span must be ascertained. Then, the girder load area must be computed; this gives the area of the building the girder is to support. Next, the weight of one square foot of the house bearing on the girder must be ascertained. The total load bearing on the girder can then be found by multiplying the girder load area by the weight per square foot of the house.

Fig. 9-27.

Recommended Spans for PS 1 Plywood Stressed-Skin Roof Panels

Top Skin[1]	Stringer Size[2,3]	No. of Stringers (spacing)	Allowable Single-Span Uniform Live Load (PSF) (10 PSF dead load assumed)						
			20	25	30	35	40	45	50
3/8″ APA STRUCTURAL I RATED SHEATHING EXP 1	2x4	4 (12″ O.C.)	12′-9″	11′-10″	11′-1″	10′-6″			
	2x6		18′-10″	17′-5″	16′-5″	15′-7″			
	2x8		24′-0″	22′-3″	20′-11″	19′-10″			
	2x10		29′-11″	27′-9″	26′-1″	24′-9″			
	2x12		35′-8″	33′-1″	31′-1″	29′-3″			
15/32″ APA RATED SHEATHING EXP 1 32/16 5-ply	2x4	3 (16″ O.C.)	12′-3″	11′-4″	10′-1″	8′-11″	8′-1″	7′-4″	6′-8′
	2x6		17′-11″	16′-7″	15′-7″	14′-8″	13″-11″	13′-3″	12′-4″
	2x8		22′-11″	21′-3″	19′-10″	18′-9″	17′-9″	16′-11″	16′-3″
	2x10		28′-4″	26′-3″	24′-8″	23′-3″	22′-1″	21′-1″	20′-2″
	2x12		33′-9″	31′-3″	29′-5″	27′-9″	26′-4″	25′-2″	24′-1″
15/32″ APA STRUCTURAL I RATED SHEATHING EXP 1 5-ply	2x4	2 (24″ O.C.)	9′-5″	8′-1″	7′-1″	6′-3″	5′-8″		
	2x6		15′-3″	14′-1″	12′-5″	11′-0″	9′-11″		
	2x8		19′-6″	18′-0″	16′-10″	15′-8″	14′-1″		
	2x10		23′-8″	21′-11″	20′-6″	19′-4″	18′-4″		
	2x12		27′-5″	25′-4″	23′-9″	22′-4″	21′-3″		
19/32″ APA RATED SHEATHING EXP 1 40/20 5-ply	2x4	4 (12″ O.C.)	13′-6″	12′-6″	11′-9″	11′-2″	10′-8″	10′-3″	9′-7′
	2x6[4]		19′-9″	18′-3″	17′-2″	16′-4″	15″-7″	15′-0″	14′-5″
	2x8[4]		25′-0″	23′-2″	21′-10″	20′-8″	19′-9″	19′-0″	18′-4″
	2x10[4]		30′-11″	28′-8″	27′-0″	25′-7″	24′-6″	23′-6″	22′-8″
	2x12		36′-10″	34′-2″	32′-1″	30′-5″	29′-1″	23′-0″	27′-0″

[1] Skin may be spliced or scarfed.
[2] Stringers may be No. 2 Douglas fir, larch, or No. 2 KD15 southern pine.
[3] Net size is less 1/16-inch for resurfacing glued surface.
[4] Qualifies as one-hour roof-ceiling assembly when specified ceiling is applied.

American Plywood Association

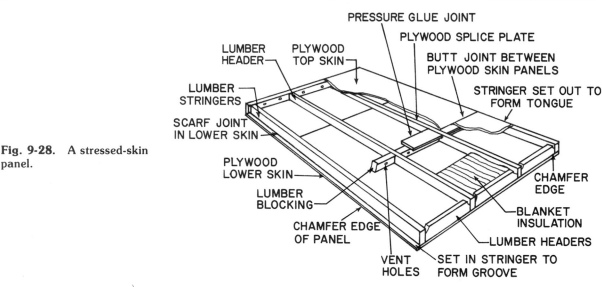

Fig. 9-28. A stressed-skin panel.

Once the total load is known, the material from which the girder is to be made must be decided. Finally, from safe load tables, a girder is chosen which will span the specified distance and carry the determined load.

Girder Length

The actual length of the girder includes the unsupported length plus that needed to rest on the supports. The determination of this length is a decision of the designer. The longer the span, the heavier the girder must be; the shorter the span, the greater the number of columns that must be used. The designer must reach a compromise between excessive girder weight and an overabundance of columns. If the span is excessive, the girder must be so large that it reduces the headroom in a basement. By using several columns, one can select a shallower beam, thus increasing the headroom.

How to Compute Girder Load Area

The girder in a typical house must support itself, the live and dead loads of the first floor, any partitions, and the live and dead loads of the ceiling. For a two-story house, the weight of the second floor must be considered. The roof is usually supported by the exterior walls.

A girder in residential construction is assumed to carry the load imposed upon it uniformly along its length. This principle is considered as the total load is computed.

An examination of Fig. 9-29 reveals how the load imposed by a house is distributed on a girder. Assume that a 50-pound load exists on each joist. Half of this load is supported by the foundation and half by the girder. Since the girder supports the ends of two joists, each with a 50-pound load,

it supports a total of 50 pounds. Each foundation supports only one end of the joist and, therefore, supports 25 pounds.

To compute the load on a girder, the number of square feet of building supported by the girder must be multiplied by the weight of the house per square foot. The plan view in Fig. 9-30 shows a girder running through the center of a house. This girder helps support the load of half of the force applied downward by the joists and other load forces. In other words, half of the distance between the girder and the foundation is load upon the girder. The width of the girder load area is, therefore, 7 1/2 feet on each side or a total width of 15 feet. The remainder of the load is supported by the foundation wall. The girder runs the entire length of the building, so the load area is 15 feet × 40 feet or 600 square feet. If the load were 90 pounds per square foot, the girder would have to support a total of 54,000 pounds, uniformly distributed along its length.

The weight per square foot can be ascertained by the method given in Chapter 8.

If a column is inserted in the plan view in Fig. 9-30, the situation changes. See Fig. 9-31. In this illustration, the girder span is reduced. The span of each piece is 20'-0", and the width of the girder load area is still 15'-0". The total girder load area that each part of the girder must support is 300 square feet. If the load is 90 pounds per square

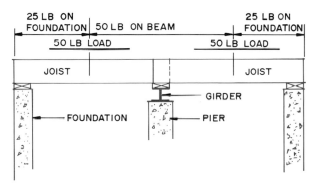

Fig. 9-29. Calculation of load on a girder.

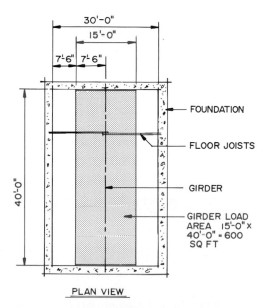

Fig. 9-30. Visualization of load zone on a girder.

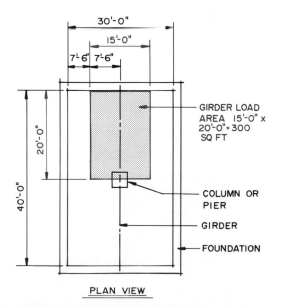

Fig. 9-31. Visualization of girder load with supporting column.

foot, the total load is 27,000 pounds or just half of that in Fig. 9-30. The column reduces the girder span, and this reduces the load the girder must carry. Since the load is reduced, the girder can be a smaller structural member.

Sometimes a girder is not spaced in the center of a structure. See Fig. 9-32. Here the girder is 10'-0" from one foundation and 25'-0" from the other. The method of computing the girder load area is exactly the same as that just described. The girder supports half the load each way and has a total width of 5'-0" plus 12'-6" or 17'-6". The length of the girder is 15'-0". The total girder load area is 262.5 square feet.

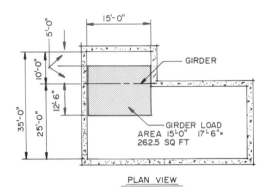

PLAN VIEW

Fig. 9-32.   Load on girder placed off center.

Solid and Nailed Wood Girders

Wood girders can be made from one solid piece of material or can be assembled from several 2-inch pieces, carefully nailed together so that the separate pieces act as a single member.

If the span, the total load on the girder, and the strength or allowable working stress of the lumber to be used are known, it is possible to compute the necessary size of the girder. This information, however, is available in tables that show the maximum allowable loads on beams of different sizes and for various spans. The size needed will vary with the lumber species and grade used, since some lumber is stronger than others.

The size of the beam to be used varies with the load it must carry and the distance to be spanned. Actual design data should be secured from a timber construction manual.

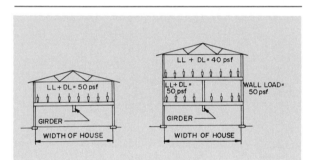

Fig. 9-33.

Design Data for Built-Up Wood Girders   National Association of
Home Builders Research Foundation, Inc.

Nominal Lumber Sizes	One-Story House Spans (in Feet and Inches)								
	20	22	24	26	28	30	32	34	36
2 - 2x6	5'-3"*	4'-10"	4'-5"	4'-1"	—	—	—	—	—
3 - 2x6	6'-9"	6'-5"	6'-2"	5'-11"	5'-8"	5'-3"	4'-11"	4'-8"	4'-5"
2 - 2x8	7'-0"	6'-4"	5'-10"	5'-4"	5'-0"	4'-8"	4'-4"	4'-1"	—
3 - 2x8	8'-11"	8'-6"	8'-1"	7'-9"	7'-5"	7'-0"	6'-6"	6'-2"	5'-10"
2 - 2x10	8'-11"	8'-1"	7'-5"	6'-10"	6'-4"	5'-11"	5'-7"	5'-3"	4'-11"
3 - 2x10	11'-4"	10'-10"	10'-4"	9'-11"	9'-6"	8'-11"	8'-4"	7'-10"	7'-5"
2 - 2x12	10'-10"	9'-10"	9'-0"	8'-4"	7'-9"	7'-2"	6'-9"	6'-4"	6'-0"
3 - 2x12	13'-9"	13'-2"	12'-7"	12'-1"	11'-7"	10'-10"	10'-2"	9'-6"	9'-0"

Nominal Lumber Sizes	Two-Story House Spans (in Feet and Inches)								
	20	22	24	26	28	30	32	34	36
3 - 2x6	4'-2"*	—	—	—	—	—	—	—	—
2 - 2x10	4'-8"	4'-3"	—	—	—	—	—	—	—
3 - 2x8	5'-6"	5'-0"	4'-7"	4'-3"	—	—	—	—	—
2 - 2x12	5'-8"	5'-2"	4'-9"	4'-5"	4'-1"	—	—	—	—
3 - 2x10	7'-0"	6'-5"	6'-0"	5'-6"	5'-1"	4'-9"	4'-6"	4'-3"	4'-0"
3 - 2x12	8'-6"	7'-9"	7'-2"	6'-8"	6'-2"	5'-9"	5'-5"	5'-1"	4'-10"

*Maximum spans for girders supporting floor loads using lumber that has an allowable bending stress of not less than 1500 psi

Some species and grades of lumber that can be used with this table:

Douglas Fir-Larch No. 1   Douglas Fir South Select Structural   Mountain Hemlock Select Structural   Southern Pine No. 1

Figure 9-33 shows span data for built-up girders for one- and two-story houses. These are limited to lumber having an allowable bending stress of not less than 1500 psi. They assume a uniformly loaded floor.

Steel Girders

The most common structural steel shapes used in residential construction as girders are the **S** (standard) beam and the **W** (wide-flange) beam. See Fig. 9-34. The method for ascertaining the size of steel girders follows the same principle as that for wood girders. As wood girders vary in width for a given depth, so steel girders vary in weight, depth, and thickness of web and flanges. The **W** beam has the advantage of presenting a wide bearing surface to rest upon the foundation wall. Frequently, the standard **S** beam must have a flange welded to it to distribute the load to the foundation.

Tables presenting the allowable loads on **W** and American Standard **S** beams can be found in Chapter 23. They assume the total load is uniformly distributed. The members are laterally supported. The loads include the weight of the beam, which should be added to the computed beam load to arrive at the total load upon the member.

Bethlehem Steel Company

Fig. 9-34. An American Standard S beam (left) and a W beam (right).

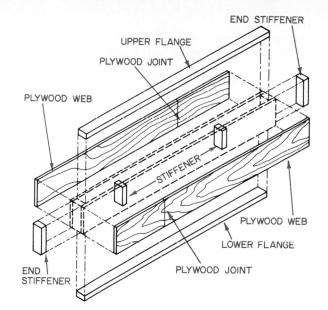

Fig. 9-36. The parts forming a nailed box beam.

Open-Web Steel Joists

A steel-framed house may use steel **S** beams for roof framing or, if long spans are necessary, may use open-web joists, Fig. 9-35. This is especially true of houses with flat roofs and open planning, where interior partitions are not used for support and/or do not reach ceiling height.

Allowable loads and spans for selected lightweight, open-web joists may be found in Chapter 23. Many other stock sizes and types are commercially available.

Nailed Plywood and Lumber Box Beams

A box beam is made from plywood and lumber. It is used to span long distances. A typical use is as a header above the opening for a garage door. A box beam is lightweight, resists shrinking and twisting, and is easy to make and install.

The parts of a typical box beam are shown in Fig. 9-36. The lumber flanges carry most of the bending stress. The plywood webs transmit the shear stress.

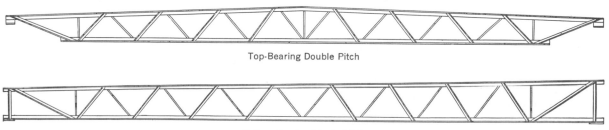

Bethlehem Steel Company

Top-Bearing Double Pitch

Bottom-Bearing Single Pitch

Fig. 9-35. Two types of open-web steel joists.

Allowable Load* for 12"-Deep Roof Beam or Header (Lb/Lin. Ft.)

Plywood	Cross Section	Approx. Wt. per Ft.(Lb.)	Span (Ft.) 10	12	14	16	18	20	22	24
½" 32/16	A	6	304	253	189	145	114	92	76	64
½" 32/16	B	8	332	276	237	207	173	140	116	97
¾" 48/24	B	10	536	391	287	220	173	140	116	97
¾" 48/24	C	12	—	410	323	247	195	158	130	110

Allowable Load* for 16"-Deep Roof Beam or Header (Lb/Lin. Ft.)

Plywood	Cross Section	Approx. Wt. per Ft.(Lb.)	Span (Ft.) 10	12	14	16	18	20	22	24
½" 32/16	A	7	436	364	270	207	163	132	109	92
½" 32/16	B	9	465	387	332	290	258	216	178	150
¾" 48/24	B	11	749	600	441	338	267	216	178	150
¾" 48/24	C	14	—	—	499	409	323	262	216	182

Allowable Load* for 20"-Deep Roof Beam or Header (Lb/Lin. Ft.)

Plywood	Cross Section	Approx. Wt. per Ft.(Lb.)	Span (Ft.) 10	12	14	16	18	20	22	24
½" 32/16	A	8	579	479	352	269	212	172	142	119
½" 32/16	B	10	597	498	426	373	332	293	242	204
¾" 48/24	B	13	957	798	599	459	362	293	242	204
¾" 48/24	C	15	—	—	650	569	460	372	308	258

Allowable Load* for 24"-Deep Roof Beam or Header (Lb/Lin. Ft.)

Plywood	Cross Section	Approx. Wt. per Ft.(Lb.)	Span (Ft.) 10	12	14	16	18	20	22	24
½" 32/16	A	9	732	590	433	332	262	212	175	147
½" 32/16	B	11	—	606	519	454	404	363	307	258
¾" 48/24	B	14	1164	970	759	581	459	372	307	258
¾" 48/24	C	17	—	—	802	701	600	486	402	337

*Includes 15% snow loading increase.

CROSS SECTIONS

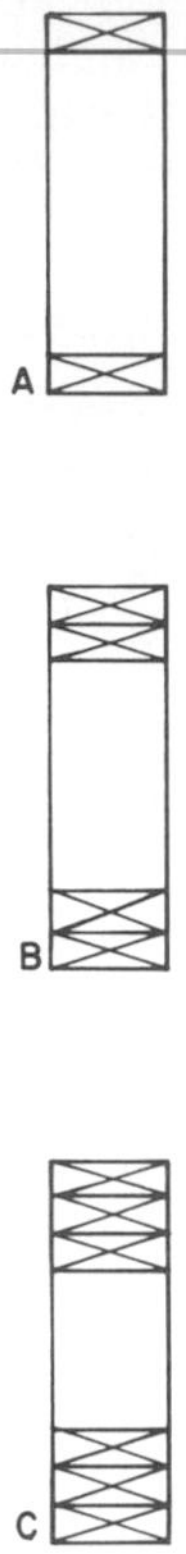

A

B

C

American Plywood Association

Fig. 9-37. Selected designs for nailed plywood and lumber box beams.

The vertical stiffeners distribute concentrated loads and end reactions. The nails transfer the stresses between the lumber and plywood webs.

Selecting a Box Beam

To select a box beam, first ascertain the load to be on the beam. Then, select the proper beam design.

Once the load is known, the box beam can be selected from pre-engineered designs. See Fig. 9-37. The loadings for beams with various thicknesses and sizes of plywood webs are given in these tables. The data assume the plywood is C-D INT-APA with exterior glue. The lumber members are assumed to be 2 × 4 No. 1 Douglas fir or southern pine. The cross section of the beam is also shown in Fig. 9-37.

To read the tables, follow this example. A beam is to span 18'-0" and carry a load of 385 pounds per lineal foot. Check the 18-foot span column in the tables for one that will carry 385 pounds. The closest is a 24-inch deep beam with 1/2-inch

plywood webs which uses cross section **B.** This section has double 2 × 4 members at the top and bottom. See Fig. 9-37.

To make this member, the number and location of the stiffeners is needed. This is shown in Fig. 9-38. This illustration also shows where the joints for the plywood webs should occur. For the beam just discussed, find the joist and stiffener layout for the 18-foot member. It has double 2 × 4 members on each end and four single stiffeners in the center section.

In order to make the box beam most effective, a nail pattern is recommended. This is shown in Fig. 9-39. The nails recommended are 8d common. The plywood web is installed with the face grain in the direction of the flanges.

Glued Laminated Wood Beams

Glued, laminated structural members are manufactured in stock lengths and shapes. These are engineered to carry specified loads of various

278

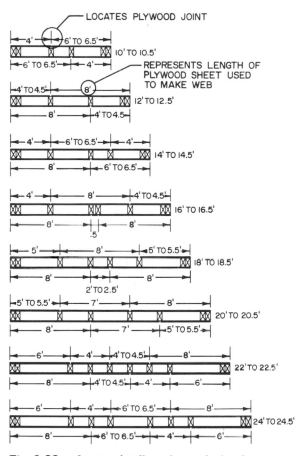

Fig. 9-38. Joint and stiffener layout for box beams.

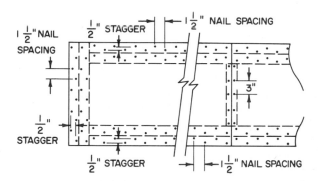

Fig. 9-39. Nailing layout for box beam.

spans. These manufactured members are about one-third stronger than a sawn timber of equal section of seasoned material.

For interior applications, the laminations are adhered with water-resistant, casein-type glues. For exposure to weather or excessive moisture, special

Fig. 9-40.

Typical Sizes of Glued Laminated Beams and Purlins

(Wide spacing. 1/240 Deflection.
f = 2200 + 15% for Short Term Duration of Live Load)

BEAM OR PURLIN		TOTAL LOAD (PSF)		
Span	Spacing	30	40	50
16′	6′	$3^{1}/_{4} \times 8^{*}$	$3^{1}/_{4} \times 9^{5}/_{8}$	$3^{1}/_{4} \times 9^{5}/_{8}$
	8′	$3^{1}/_{4} \times 9^{5}/_{8}$	$3^{1}/_{4} \times 11^{1}/_{4}$	$3^{1}/_{4} \times 11^{1}/_{4}$
	10′	$3^{1}/_{4} \times 9^{5}/_{8}$	$3^{1}/_{4} \times 11^{1}/_{4}$	$3^{1}/_{4} \times 12^{7}/_{8}$
	12′	$3^{1}/_{4} \times 11^{1}/_{4}$	$3^{1}/_{4} \times 12^{7}/_{8}$	$3^{1}/_{4} \times 14^{1}/_{2}$
	14′	$3^{1}/_{4} \times 11^{1}/_{4}$	$3^{1}/_{4} \times 12^{7}/_{8}$	$3^{1}/_{4} \times 14^{1}/_{2}$
	16′	$3^{1}/_{4} \times 12^{7}/_{8}$	$3^{1}/_{4} \times 14^{1}/_{2}$	$5^{1}/_{4} \times 12^{7}/_{8}$
20′	8′	$3^{1}/_{4} \times 11^{1}/_{4}$	$3^{1}/_{4} \times 12^{7}/_{8}$	$3^{1}/_{4} \times 14^{1}/_{2}$
	10′	$3^{1}/_{4} \times 12^{7}/_{8}$	$3^{1}/_{4} \times 14^{1}/_{2}$	$5^{1}/_{4} \times 12^{7}/_{8}$
	12′	$3^{1}/_{4} \times 12^{7}/_{8}$	$5^{1}/_{4} \times 12^{7}/_{8}$	$5^{1}/_{4} \times 14^{1}/_{2}$
	14′	$3^{1}/_{4} \times 14^{1}/_{2}$	$5^{1}/_{4} \times 12^{7}/_{8}$	$5^{1}/_{4} \times 14^{1}/_{2}$
	16′	$3^{1}/_{4} \times 14^{1}/_{2}$	$5^{1}/_{4} \times 14^{1}/_{2}$	$5^{1}/_{4} \times 16^{1}/_{8}$
	18′	$5^{1}/_{4} \times 12^{7}/_{8}$	$5^{1}/_{4} \times 14^{1}/_{2}$	$5^{1}/_{4} \times 16^{1}/_{8}$
24′	8′	$3^{1}/_{4} \times 14^{1}/_{2}$	$5^{1}/_{4} \times 12^{7}/_{8}$	$5^{1}/_{4} \times 14^{1}/_{2}$
	10′	$3^{1}/_{4} \times 14^{1}/_{2}$	$5^{1}/_{4} \times 14^{1}/_{2}$	$5^{1}/_{4} \times 14^{1}/_{2}$
	12′	$5^{1}/_{4} \times 14^{1}/_{2}$	$5^{1}/_{4} \times 14^{1}/_{2}$	$5^{1}/_{4} \times 16^{1}/_{8}$
	14′	$5^{1}/_{4} \times 14^{1}/_{2}$	$5^{1}/_{4} \times 16^{1}/_{8}$	$5^{1}/_{4} \times 17^{3}/_{4}$
	16′	$5^{1}/_{4} \times 14^{1}/_{2}$	$5^{1}/_{4} \times 16^{1}/_{8}$	$5^{1}/_{4} \times 17^{3}/_{4}$
	18′	$5^{1}/_{4} \times 16^{1}/_{8}$	$5^{1}/_{4} \times 17^{3}/_{4}$	$5^{1}/_{4} \times 19^{3}/_{8}$
32′	8′	$5^{1}/_{4} \times 16^{1}/_{8}$	$5^{1}/_{4} \times 17^{3}/_{4}$	$5^{1}/_{4} \times 19^{3}/_{8}$
	10′	$5^{1}/_{4} \times 16^{1}/_{8}$	$5^{1}/_{4} \times 19^{3}/_{8}$	$5^{1}/_{4} \times 19^{3}/_{8}$
	12′	$5^{1}/_{4} \times 17^{3}/_{4}$	$5^{1}/_{4} \times 19^{3}/_{8}$	$5^{1}/_{4} \times 21$
	14′	$5^{1}/_{4} \times 19^{3}/_{8}$	$5^{1}/_{4} \times 21$	$5^{1}/_{4} \times 22^{5}/_{8}$
	16′	$5^{1}/_{4} \times 19^{3}/_{8}$	$5^{1}/_{4} \times 21$	7×21
	18′	$5^{1}/_{4} \times 21$	$5^{1}/_{4} \times 22^{5}/_{8}$	$7 \times 22^{5}/_{8}$
40′	8′	$5^{1}/_{4} \times 19^{3}/_{8}$	$5^{1}/_{4} \times 21$	7×21
	10′	$5^{1}/_{4} \times 21$	$5^{1}/_{4} \times 22^{5}/_{8}$	$7 \times 22^{5}/_{8}$
	12′	$5^{1}/_{4} \times 22^{5}/_{8}$	$7 \times 22^{5}/_{8}$	$7 \times 24^{1}/_{4}$
	14′	7×21	$7 \times 22^{5}/_{8}$	$7 \times 25^{7}/_{8}$
	16′	$7 \times 22^{5}/_{8}$	$7 \times 24^{1}/_{4}$	$7 \times 25^{7}/_{8}$
	18′	$7 \times 22^{5}/_{8}$	$7 \times 25^{7}/_{8}$	$7 \times 27^{1}/_{2}$
50′	8′	$7 \times 22^{5}/_{8}$	$7 \times 24^{1}/_{4}$	$7 \times 25^{7}/_{8}$
	10′	$7 \times 24^{1}/_{4}$	$7 \times 25^{7}/_{8}$	$7 \times 27^{1}/_{2}$
	12′	$7 \times 24^{1}/_{4}$	$7 \times 27^{1}/_{2}$	$7 \times 29^{1}/_{8}$
	14′	$7 \times 25^{7}/_{8}$	$7 \times 29^{1}/_{8}$	$7 \times 30^{3}/_{4}$
	16′	$7 \times 27^{1}/_{2}$	$7 \times 30^{3}/_{4}$	$9 \times 29^{1}/_{8}$
	18′	$7 \times 29^{1}/_{8}$	$9 \times 29^{1}/_{8}$	$9 \times 30^{3}/_{4}$

*Cross section of beam.　　Timber Structures, Inc.

glues are used. These are made of waterproof, exterior-type phenol, resorcinol, or melamine resin.

The structural members shown in Fig. 9-40 are but a few of the many stock types available. Manufacturers' catalogs should be consulted for a more complete listing. These tables are for butterfly, peaked, peaked and cambered, slightly tapered, and simple straight beams, as shown in Fig. 9-41.

To use the tables in Fig. 9-40, the span and desired spacing must be decided. The total roof load must be ascertained. Then the correct beam size can be chosen from the table.

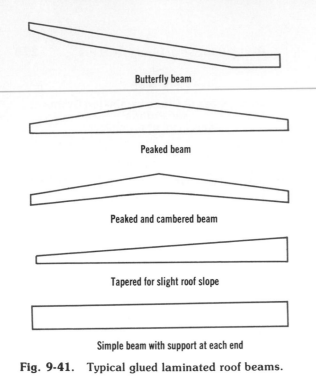

Butterfly beam

Peaked beam

Peaked and cambered beam

Tapered for slight roof slope

Simple beam with support at each end

Fig. 9-41. Typical glued laminated roof beams.

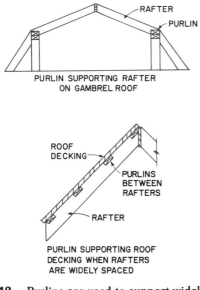

RAFTER

PURLIN

PURLIN SUPPORTING RAFTER
ON GAMBREL ROOF

ROOF
DECKING

PURLINS
BETWEEN
RAFTERS

RAFTER

PURLIN SUPPORTING ROOF
DECKING WHEN RAFTERS
ARE WIDELY SPACED

Fig. 9-42. Purlins are used to support widely spaced rafters.

These members can also serve as purlins. A **purlin** is a small beam laid at right angles to the rafters and used to support the rafters or roof decking when rafters are widely spaced. See Fig. 9-42.

Lintels

A **lintel** is a horizontal structural member which spans window and door openings and supports the

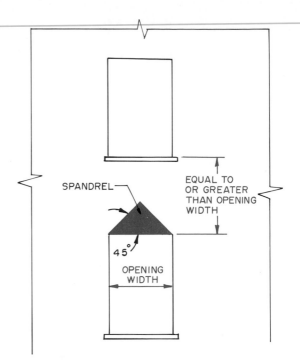

SPANDREL

EQUAL TO
OR GREATER
THAN OPENING
WIDTH

45°

OPENING
WIDTH

Fig. 9-43. When the vertical distance between the two openings is equal to or greater than the width of the opening, the load on the lintel is the weight of the materials in a spandrel formed by an isosceles triangle.

wall above the opening. The wall area supported by a lintel is called a **spandrel.** See Figs. 9-43 and 9-44. Lintels may be made from concrete or steel.

Computing Lintel Load

In order to select the proper lintel size, the weight of the material and other loads supported by the lintel must be computed. This usually involves computing the number of square feet in the spandrel and multiplying this by the weight of one square foot of the material used (generally brick). This gives the total load on the lintel. The span of the lintel is the width of the window opening. The proper lintel can then be selected from a table of safe loads.

Under most conditions, the lintel carries only the weight of the masonry units above it. Since these units tend to form an arch and support each other, the weight on the lintel is computed as the weight of the materials in an isosceles triangle above the window. See Fig. 9-43. However, this is true only if the vertical distance between the openings is equal to or greater than the width of the opening.

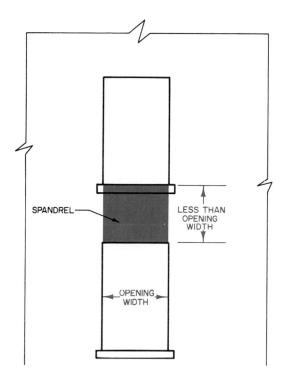

Fig. 9-44. When the distance between two openings is less than the width of the opening, the load on the lintel is the weight of materials in a rectangular spandrel.

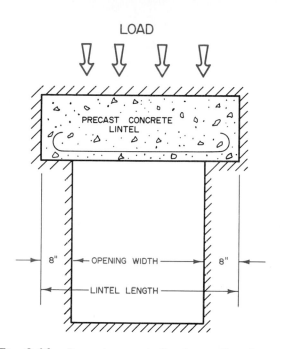

Fig. 9-46. Precast concrete lintels rest 8 inches on each side of the masonry opening. The lintel length is the opening width plus 1′-4″.

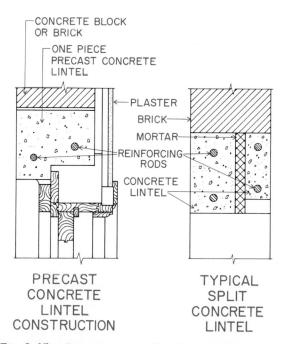

PRECAST CONCRETE LINTEL CONSTRUCTION

TYPICAL SPLIT CONCRETE LINTEL

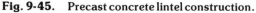

Fig. 9-45. Precast concrete lintel construction.

If the distance between openings is less than the width of the lower opening, the weight is figured for the rectangular area between the two openings. See Fig. 9-44.

If the lintel is to carry any other load, such as several floor joists, this load must be added to the weight of the masonry. The lintel selected would be expected to carry both loads.

Concrete Lintels

Precast concrete lintels are available in a variety of shapes and sizes. They are cast in standard shapes and reinforced with steel bars. Basically, they are made in two ways — as a single cast unit or as two separate pieces called a **split lintel**. See Fig. 9-45.

A typical precast concrete lintel diagram is in Fig. 9-46. Notice that the lintel rests 8 inches on each side of the masonry opening. The typical lintel is, therefore, the width of the opening plus 1′-4″.

If the load is uniform, such as in a rectangular spandrel, uniformly loaded lintel designs are used. Data for one size are in Fig. 9-47.

Steel Lintels

Steel lintels are made from steel angles. Since steel is stronger than concrete of the same thickness, it is most frequently used for lintel construction. A steel lintel can be much shallower than one of concrete and can support as much or more load.

Fig. 9-47.

Reinforced Concrete Lintels
(Capacity in Pounds)

8 x 8 In. Nominal Lintels (7 ⁵/₈ x 7 ⁵/₈)							
Clear Opening L	Reinforcement¹					Lintel Length	Lintel Weight (lb.)
	2-#3	#3 + #4	2-#4	#4 + #5	2-#5		
4'-0"	697	1010	1320	1545	1870	5'-4"	340
5'-0"	452	662	870	1130	1386	6'-4"	405
6'-0"	309	460	612	800	980	7'-4"	469
7'-0"	218	330	446	586	730	8'-4"	533
8'-0"	157	247	336	446	556	9'-4"	597
9'-0"	93	186	258	346	436	10'-4"	661
10'-0"	78	136	192	264	338	11'-4"	725
11'-0"	58	108	157	219	280	12'-4"	790
12'-0"	40	82	124	176	228	13'-4"	854

Concrete Reinforcing Steel Institute

¹No. and size of rebar

The typical use of steel lintels is illustrated for masonry veneer over frame and solid masonry walls in Figs. 9-48 and 9-49. Notice that a separate lintel is used for each 4 inches of masonry.

The size of steel lintels is recorded on drawings by giving the length of the two legs first and then the thickness of the metal from which the lintel is made; for example — 4" × 3 1/2" × 1/4".

The safe load-carrying potential for the commonly used angles is indicated in Fig. 9-50. The heading of the table indicates that these angles are considered beams, since they are horizontal supporting members. The allowable uniform loads are in kips (1000 pounds of dead load).

The proper angle size to use for a lintel is selected by first computing the total load on the lintel. Assume a load of 3,000 pounds is carried by a lintel spanning 5'-0". An examination of Fig. 9-50 reveals that a 4 × 3 1/2 × 5/16 angle will support 3.5 kips over a distance of 5'-0". This meets the demands of load.

Notice that angles are available with both legs the same length as well as with legs of unequal length. Usually, the longer leg is placed in a vertical position when the angle is used as a lintel. Frequently on a blueprint, the abbreviation **L.L.V.** (long leg vertical) is used. An angle will carry a heavier load when the long leg is in a vertical position.

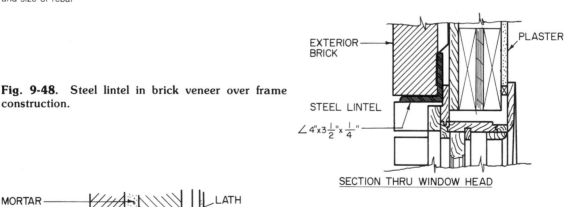

SECTION THRU WINDOW HEAD

Fig. 9-48. Steel lintel in brick veneer over frame construction.

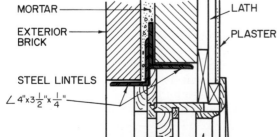

Fig. 9-49. Steel lintel in 8″ solid brick construction.

Fig. 9-50.   **283**

Loads for Regular Series Angle Beams*

Angle Size	Wt. per Ft.	Span in Feet								
		2	3	4	5	6	7	8	9	10
5 × 3½ × ½	13.6	20.0	13.3	10.0	8.0	6.7	5.7	5.0	4.4	4.0
³/₈	10.4	15.3	10.2	7.7	6.1	5.1	4.4	3.8	3.4	3.1
⁵/₁₆	8.7	12.7	8.4	6.3	5.1	4.2	3.6	3.2	2.8	2.5
4 × 3½ × ⁵/₁₆	7.7	8.7	5.8	4.3	3.5	2.9	2.5	2.2	1.9	
3½ × 3½ × ⁵/₁₆	7.2	6.5	4.4	3.3	2.6	2.2	1.9	1.6		
¼	5.8	5.3	3.5	2.6	2.1	1.8	1.5	1.3		

*Allowable uniform loads in kips for angles laterally supported. Neutral axis is parallel to horizontal leg.

American Institute of Steel Construction

Build Your Vocabulary

Following are terms that you should understand and use as part of your working vocabulary. Write a brief explanation of each term.

allowable unit
  working stress
American standard
  beams
box beams
collar beam
concentric loading
eccentric loading
girder load area
glued laminated beams
grading factors

heavy roof decking
lintel
lumber grade
open-web joists
roof pitch
roof slope
span
split ring
stressed-skin panels
trusses
wide-flange beams

Class Activities

1. What is the maximum span for a southern pine floor joist with a 40-pound live load, spaced 16 inches on center, if the joist is 2 × 8? . . . If it's 2 × 10? . . . 2 × 12?
2. What is the maximum span for a southern pine ceiling joist with a 20-pound attic load, spaced 16 inches on center, if the joist is 2 × 6? . . . If it's 2 × 8? . . . 2 × 10?
3. What is the required size of a joist for a southern pine, low-slope roof, if the joists are spaced 24 inches on center and must span 21'-6"?
4. A roof rafter must span 19'-7". What size joist must be used to span this distance, if the rafter is made of southern pine and is spaced 16 inches on center?
5. A roof has a total rise of 6'-0" and a total span of 24'-0". What are the pitch and slope of the roof?
6. If floor joists are spaced 16 inches on center, what thickness of plywood sheathing should be used?
7. A roof having a span of 24 inches between rafters is to be decked with plywood. What is the minimum thickness that can be used?
8. A roof is to be decked with 2-inch Douglas fir, using a random span decking plan. What is the maximum spacing for the rafters in this roof, if the live load is 40 pounds per square foot?
9. A round concrete-filled steel column must support 35,000 pounds with an unbraced column length of 8'-0". What size steel base is on this column? What size wood column would be necessary to support this load?
10. The roof of a post-and-beam house is to be decked with stressed-skin panels. The panels must span 14'-10". What is the allowable single-span live load?
11. Compute the girder load area for your house, and ascertain load on the girder. Select the girder that would safely carry this load. Measure the girder installed in your house and see if it is the proper size.
12. The roof of a post-and-beam house is to be framed with a glued, laminated, peaked beam. If this beam is spaced 10 feet on center, must span 24'-0", and must support a total load of 40 pounds per square foot, what are the dimensions of the stock beam?
13. If the load on a box beam is to be 600 pounds per lineal foot, what size beam should be used? What should be the design of the beam?
14. A window in a brick house is 3'-6" wide. The lintel supports a 4-inch brick veneer. What size concrete lintel would be required? What size steel lintel would be required?

architecture
Design Engineering Drawing

Electrical and Mechanical Systems

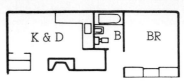

Chapter 10
Electrical Features

Chapter 11
Plumbing

Chapter 12
Heating and Air Conditioning

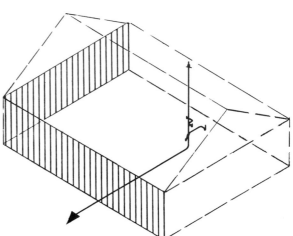

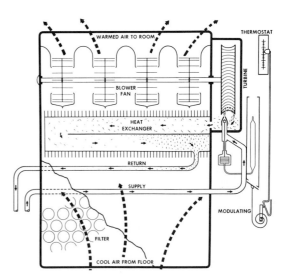

Chapter 10

Electrical Features

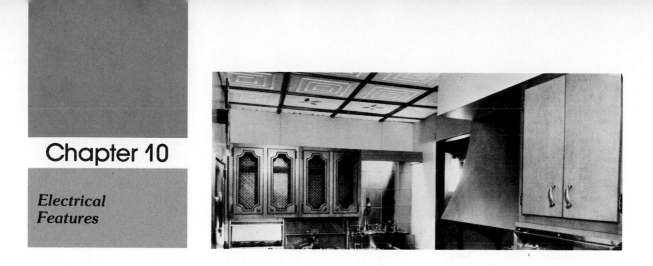

In today's home, electricity is used to perform so many tasks that a good wiring system is vital. The power company can make a full supply of electricity available, but the extent of its use in the home is determined by the house wiring. Every part of the system, from the service entrance to each circuit, should be planned to serve a specific purpose.

The expanded use of appliances, the rapid increase in air-conditioning installations, and the revolution in house lighting — all increase the demand for electrical power. If the load requirements exceed the supply, operating efficiency suffers and the voltage drops. Lights operate poorly, and appliances are inefficient. A 5 percent voltage loss produces a 10 percent loss of heat in the heating appliance or a 17 percent loss of light from an incandescent lamp.

While designing a house, try to anticipate future needs and allow extra circuits for growth. Bring in enough current to handle the requirements of this potential demand.

Wiring should be done by a reputable electrical contractor. In most cities, an inspector checks the job while it is in progress and certifies that it meets the minimum requirements for residential wiring.

Wiring Characteristics and Service Requirements

The planning of the electrical features of a home involves an understanding of voltage, wattage, and amperes. These are defined as follows:
1. A **volt** is the unit used in measuring electrical pressure (similar to pounds of pressure in a water system).
2. An **ampere** is the unit used in measuring the electrical rate of flow (similar to gallons per minute in a water system).
3. A **watt** is the unit of electrical power, which is composed of both voltage and amperage. For example, 1 ampere at a pressure of 1 volt equals 1 watt. If 1 watt is used for 1 hour, it is 1 watt-hour. One thousand watt-hours equals 1 kilowatt-hour.

The standards for electrical wiring and associated equipment are based on the National Electrical Code, prepared by the National Fire Protection Association. The **National Electrical Code** is a basic minimum standard to safeguard persons and property from the hazards of misused electricity. It is not a design or specification manual. The National Electrical Code and the local electrical code should always be observed in engineering a home wiring system, to assure safety in operation. Maximum convenience and safety with load growth can be designed into a wiring system only by observance of sound and realistic standards.

Wiring System Characteristics

The major characteristics of a wiring system designed to carry the load demanded are:

1. **Accessibility** — Convenience outlets and switches should be numerous, conveniently located, and of the proper types.
2. **Capacity** — All parts of the electrical system should be capable of operating at rated voltage and of supplying current in sufficient quantities to handle the full load. On a general-purpose circuit, the normal load of lights and appliances should not exceed 50 percent of the rated circuit capacity.
3. **Isolation** — To assure efficiency of automatic operation, most automatic appliances should be served by individual circuits. Lights, televisions, and other devices sensitive to voltage fluctuations should not be served by circuits to which motor-driven or automatic appliances are connected.
4. **Safety** — The system should comply with the provisions of the National Electrical Code.
5. **Control** — The system should provide maximum operating convenience. Switches and other controls should be carefully positioned and should be in sufficient number to be suitable for family living habits.

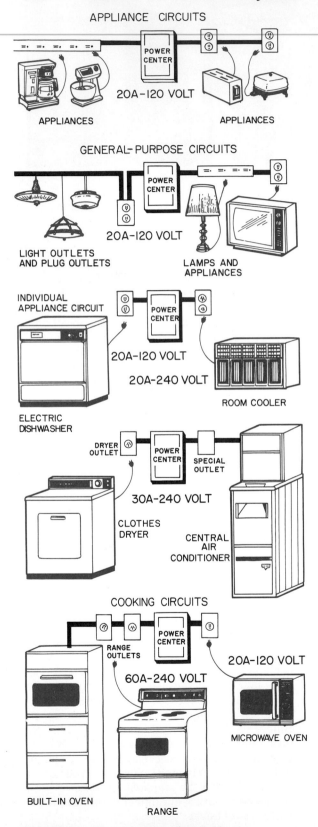

APPLIANCE CIRCUITS

POWER CENTER

20A–120 VOLT

APPLIANCES

APPLIANCES

GENERAL-PURPOSE CIRCUITS

POWER CENTER

20A–120 VOLT

LIGHT OUTLETS AND PLUG OUTLETS

LAMPS AND APPLIANCES

INDIVIDUAL APPLIANCE CIRCUIT

POWER CENTER

20A–120 VOLT

20A–240 VOLT

ROOM COOLER

ELECTRIC DISHWASHER

DRYER OUTLET

POWER CENTER

SPECIAL OUTLET

30A–240 VOLT

CLOTHES DRYER

CENTRAL AIR CONDITIONER

COOKING CIRCUITS

POWER CENTER

RANGE OUTLETS

60A–240 VOLT

20A–120 VOLT

MICROWAVE OVEN

BUILT-IN OVEN

RANGE

Fig. 10-1. Circuit types and typical uses.

Circuit Requirements

A sound approach to the design and layout of an electrical system is important. Probable loads of lighting, heating, and appliances should be determined; then circuits and services to handle these loads should be installed. This prevents overloads, and allows unrestricted use of electrical appliances. It also ensures the attainment of the five major characteristics of a good wiring system — (1) accessibility, (2) capacity, (3) isolation, (4) safety, and (5) control.

The designer should remember that a typical 20-ampere, 120-volt circuit carries 2400 watts. This can operate a 1000-watt iron and 1200-watt dishwasher, but if a 600-watt percolator were also connected, the circuit would be overloaded.

Types of Residential Circuits

To provide all the requirements of functional wiring, several types of circuits are essential. The first type of 120-volt circuit serves two or more convenience outlets for appliance connection in the kitchen, pantry, dining, and laundry areas. This is classified as an **appliance circuit.** It should be of No. 12 wire and should be protected by a 20-ampere fuse or circuit breaker. Generally, two such circuits are needed in today's kitchens because of the large number of electrical appliances in use.

The second type of 120-volt circuit, called a **general-purpose circuit,** is also a 20-ampere circuit, but is used for lighting, convenience outlets, and small appliances, such as lamps and radios. It serves all convenience outlets except those wired for appliances. This circuit will carry 2400 watts.

The third type of 120-volt circuit is also a 20-ampere circuit, but operates a single outlet. It is called a **special-purpose circuit.** This circuit would operate a heavy-duty appliance, such as a dishwasher or air conditioner. This unit operates on 120 volts, but has a high wattage rating, so each unit needs to be on a separate circuit.

Another type of special-purpose circuit is used for such items as ranges, dryers, or central air conditioners which operate on 240 volts. The differences in the various 240-volt circuits are the size of wire used, the protective device incorporated, and the resultant capacity. The types of circuits used are a No. 6 wire with a 60-ampere rating, 14,400 watts capacity; a No. 8 wire with a 40-ampere rating, 9600 watts capacity; a No. 10 wire with a 30-ampere rating, 7200 watts capacity; and a No. 12 wire with a 20-ampere rating, 4800 watts capacity. (The smaller the wire number, the larger the diameter of the wire.) These various types of circuits are illustrated in Fig. 10-1.

The Number of Circuits Required

In a completely electrified, combination kitchen-laundry area, it would not be unusual to find five single-outlet circuits, one each to serve the range, broiler, water heater, refrigerator-freezer, and dryer. A sixth circuit would serve fixed lights and an exhaust fan, while two appliance circuits would provide sufficient capacity to serve a dishwasher, clothes washer, toaster, radio, food mixer, iron, coffee maker, and similar 120-volt appliances. So, with the appliances and services mentioned, eight circuits would be needed for this area.

In the utility area, an additional three circuits could be used to advantage: the first one could serve a blower and central heating unit; the second one, a summer cooling unit; and the third, such loads as workshop motors or a soldering iron. General lighting could be connected to a lighting circuit serving other areas.

Three more circuits might also be used advantageously in the living-dining area, front entry and terrace, and for exterior lighting. For example, one circuit could serve both fixed and portable lights, and a second could serve the radio, television set, circulating fan, etc. Then, if central cooling were not installed or were insufficient, local comfort could be obtained by operating a portable cooling unit from the third circuit in this area.

Finally, three circuits could be used to advantage in the bedroom-bathroom section of the home; two circuits could serve all lights, radios, fans, clocks, sun lamps, electric blankets, etc., while the third circuit could be used to serve built-in bathroom heaters or a room air-conditioning unit.

In the house just described, the wiring system calls for seventeen circuits. This is a satisfactory electrical service for an average home.

Service Entrance

The **service entrance** is that part of the wiring system that brings power from the pole to the house. The service entrance is the focal point of electrical adequacy in the home, for it is the ultimate limit on the total energy which may be used. The size of this entrance should be decided only after a careful analysis of such factors as number and types of electrical appliances and devices to be in the home, and possible future additions. Codes require a service entrance to have a minimum capacity of 20.0 kw and rated not less than 100 amperes. See Fig. 10-2.

As a guide to the selection of the proper size for a service entrance, the following sizes are outlined:

125 ampere — General-purpose circuits, electric cooking, water heater, electric laundry.

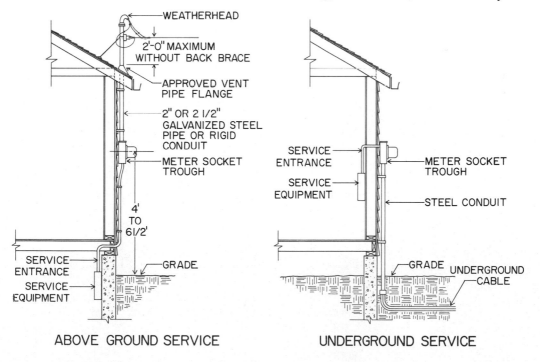

ABOVE GROUND SERVICE   UNDERGROUND SERVICE

Fig. 10-2. Service entrance installations for residential construction.

150 ampere — General-purpose circuits, electric cooking, water heater, electric laundry, air conditioning, and electric heating (small homes). If the house has total-resistance heating, amperage will have to be higher and should be computed to suit the load of the heating unit.

200 ampere — General-purpose circuits, electric cooking, water heater, electric laundry, air conditioning and electric home heating. If the house has total-resistance heating, amperage will have to be higher and should be computed to suit the load of the heating unit.

The Service Panel

The service entrance brings electrical power into the house and to a **service panel**. From this panel, 120- and 240-volt circuits are run to the various parts of the house. Each circuit is controlled by a **circuit breaker**. If there is a short (a redirection of power flow), the circuit breaker opens, much like a light switch, cutting off power to the troubled circuit. A typical array of circuits from a service panel is shown in Fig. 10-3. The main service panel should be located where it is easy to reach, so that it can be found quickly in an emergency or in the dark. See Fig. 10-4. This service panel is usually located near where the service entrance enters the house. It is also possible to locate a subpanel in another part of the house and run circuits from it.

Electrical codes permit up to six subpanels to be serviced directly from the service entrance conductor. These can be used to serve heavy-duty appliances and as feeders to subpanels serving as distribution centers for light and outlet circuits. A typical installation might include the following circuits: (1) 50-amp range, (2) 30-amp dryer, (3) 20-amp water heater, (4) 40-amp air conditioner, (5) 40-amp kitchen panel for kitchen circuits, and (6) 40-amp lighting panel for entire house.

At the initial installation, each service panel should be larger than needed. This provides for future expansion. The service panel shown in Fig. 10-4 will handle twelve circuits and would be suitable for a kitchen subpanel.

To calculate the load to each service panel, the demands upon it must be estimated. For a kitchen, the following loads would be typical:

Small appliance load	1500	watts
Dishwasher	1200	watts
Disposal	300	watts
Refrigerator	300	watts
Freezer	300	watts
Range and oven	8000	watts
	11,600	watts

This load totals 11,600 watts. Divide this by 240 volts, and you will find the current capacity of feeder needed to be 48 amps. Select wire diameters large enough to carry this current. Figure 10-5 indicates the accepted ampere rating of selected standard wire sizes. This situation would require a No. 6 copper wire or a No. 4 aluminum wire.

Approximate wattage requirements of various appliances are given in Appendix B. These can be used as a guide in estimating total load.

Calculating the Required Service Entrance Load

To ascertain the size of service entrance required, the electrical demands must be calculated. Most service entrances now are three-wire, single-phase, 120/240 volts. The calculation procedure follows.

1. Calculate the **lighting load** by multiplying the square feet of floor area by 4 watts per square foot. If the actual electrical layout is known, the number of circuits can be ascertained by examining the floor plan.

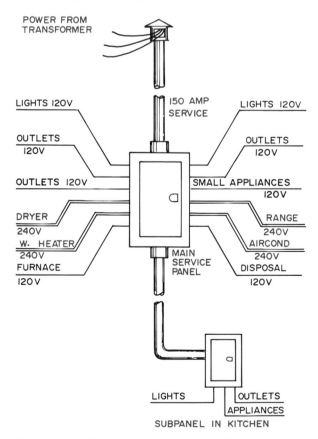

Fig. 10-3. **Circuits are run from a service panel to various parts of the house.**

Frank Adam Electric Company

Fig. 10-4. A typical enclosed panel for indoor power centers. This is a twelve-circuit center with circuit breakers.

Fig. 10-5.

Allowable Current-Carrying Capacities of Selected Wires

WIRE SIZE	AMPERES	
	Copper	Aluminum or Copper-Clad Aluminum
14	15	—
12	20	15
10	30	25
8	40	30
6	55	40
4	70	55
3	80	65
2	95	75
1	110	85
0	125	100

2. Add the total circuit capacity (in watts) allowed for small appliances in the kitchen, dining room, pantry, laundry, and utility areas. To get this total, decide how many 20-ampere appliance circuits are needed, and multiply this by 2000 watts.

3. Add the lighting load and small appliance load. Of this total, take 3000 watts at 100 percent. Take 35 percent of the total load remaining after removing the 3000 watts. Add this to the 3000-watt load. This gives needed load for lighting and small appliance circuits. The entire

load was not taken at 100 percent because all lights and appliances are never operated at the same time.

Note: These calculations are for residential wiring only; consult Code for information on other types.

4. Add the wattage loads of the range and other fixed appliances, such as dryer, water heater, and washer. Add 100 percent of this to the adjusted lighting and small appliance load (from Step 3).

5. Divide this total wattage by 240 volts to find the required amperage for the service entrance.

From these calculations the required size of entrance wire can be found. See Fig. 10-6 for a sample problem in calculating the size of the service entrance.

For a 1500 sq. ft. home with electric range, electric water heater and electric home laundry

LIGHTING AND GENERAL PURPOSE LOAD:

1500 sq. ft. x 4 watts/sq. ft.....6000 watts

SMALL APPLIANCE LOAD:

Number of appliance circuits included in branch circuit layout — 4

4 x 2000 watts/circ.........8000 watts

total 14000 watts

take 3000 at 100%.................3000 watts

and the remainder of 11000 at 35%....3850 watts

RANGE DEMAND LOAD:.................................8000 watts

FIXED APPLIANCE LOAD:

Water heater ....3000 watts
Clothes dryer ....4500 watts
Clothes washer ....500 watts
Ironer ....1650 watts

9650 watts at 100%...9650 watts

Total watts of service capacity: 24500 watts

Required current carrying capacity of service entrance conductors (at 120/240 volts, 3 wire, single-phase):

$$\frac{24500}{240} = 102.8 \text{ amperes}$$

Service entrance conductors must be: No. 1's Type R in 1½" Conduit
or No. 1's Type TW in 1½" Conduit
or No. 2's Type RH in 1¼" Conduit

This service entrance can be described as a 25 kw service, the value obtained by multiplying the current rating of the service entrance by the voltage between the two ungrounded conductors (240 volts). A wide range of possible service entrance arrangements could be used to carry out the design set forth in this typical calculation. If a main disconnect switch is not required by local regulations, six or less circuit breakers can be used as protective subdivisions as set forth in Section 2351 a. of the National Electrical Code. If a main disconnect switch were required in this particular service entrance, either a 200-amp switch fused at 110-amps or, because the calculation contains spare capacity, a 100-amp switch fused at 100-amp or a 100-amp circuit breaker may be used as the main disconnect.

Fig. 10-6. Sample service entrance calculation.

Safety Devices

Electricity has the potential to injure or kill. Safety devices are used to try to prevent this when unexpected accidents occur. Overcurrent devices are in every circuit. They interrupt the flow of electricity when it exceeds the capacity of the conductor. **Fuses** and **circuit breakers** are types commonly used. A **ground-fault interrupter** opens the circuit when it detects a leakage of electrical current to the ground. The electrical fault will not be great enough to trip a circuit breaker, but without the interrupter a person could receive an electrical shock strong enough to cause injury. This is especially likely in a wet area such as a patio or swimming pool. Ground-fault interrupters can be installed on a circuit at the service panel, or each outlet can have an interrupter attached.

Planning the Electrical System

Illumination

The choice of **illuminating fixtures** should be based upon the various visual tasks carried on in the home. Lighting must be planned for recreation, entertainment, reading, cooking, eating, and the other activities in the home. In most cases, best results are achieved by a blend of lighting from both stationary fixtures and portable lamps. Some fixtures provide wide, general illumination, while others focus light in only one spot. The selection of the best fixture, and the best location for this fixture to perform properly, requires study and planning. The use of electrical energy can be reduced by using fluorescent lighting and dimmer switches on incandescent lights.

Convenience Outlets

Convenience outlets should be located near the ends of wall space, rather than near the center, thus reducing the likelihood of being concealed behind large pieces of furniture. They should be placed so that no point along any usable wall space (in any room except a bathroom or utility room) is more than 6 feet (1829 mm) from an outlet.

If a section of wall between two doors is 2 feet (610 mm) wide, a convenience outlet is needed.

In some rooms, it is desirable to install split-wired convenience outlets. A **split-wired outlet** has the top outlet on one circuit and the bottom outlet on a different circuit. For example, in a living room, the top outlet can be controlled by a switch at the door. The bottom outlet, being on a separate circuit, can be hot all the time.

Under normal conditions, convenience outlets should be placed 12 inches (305 mm) above the floor line. Exceptions to this might be found in baths and in kitchens with outlets above the countertop and with outlets above or built into fireplaces. There should be one convenience outlet for every 150 square feet (13.95 m²) (or major fraction thereof) in a room.

All convenience outlets both inside and outside must be grounded. Those outside must have a ground fault interrupter. A kitchen must have at least two grounded convenience outlets over the counter. A bathroom must have one grounded outlet near the mirror.

Switches

Wall **switches** should be located on the latch side of doors and on the traffic side of arches. They should be placed within the room for which the switch controls the light. Exceptions to this would be a switch controlling a yard light or stair lights for an adjoining area. Wall switches are normally located 48 inches (1219 mm) above the floor line.

Frequently, a room will have two or more entrances, and it is necessary to control a light or a switch to a convenience outlet from these locations. Multiple switch control will accomplish this. If a light is controlled by two switches, it has a three-way switch. If controlled from three locations, it has a four-way switch.

It is especially important to have switch locations so placed that people can enter the house from any door and progress to any other part of the house, having the way lighted before them. Also, it should be possible to turn off the lights in the area just vacated without retracing one's steps.

The various electrical features of a plan are indicated on a drawing by symbols. These have been standardized and are common language in the construction industry. Figure 10-7 lists these symbols.

Living Rooms

The types of lighting controls, and outlets which should be installed in the living room are:

1. General illumination throughout the entire room. This may be provided by ceiling or wall fixtures or by lighting in valances, cornices, or portable lamps.
2. Local lighting at each furniture grouping or at each area where some specific activity (sewing, reading, etc.) is carried on. See Fig. 10-8.
3. Decorative lighting, such as picture illumination and bookcase lighting, to create focal points in the room. See Fig. 10-9.

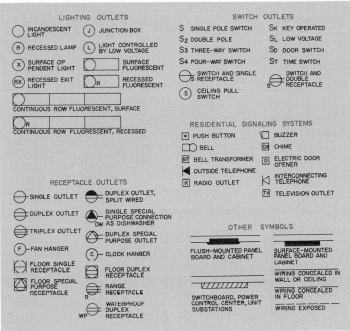

LIGHTING OUTLETS

◯	INCANDESCENT LIGHT	Ⓙ	JUNCTION BOX
Ⓡ	RECESSED LAMP	Ⓛ	LIGHT CONTROLLED BY LOW VOLTAGE
Ⓧ	SURFACE OP PENDENT LIGHT		SURFACE FLUORESCENT
Ⓡ⬡	RECESSED EXIT LIGHT		RECESSED FLUORESCENT

CONTINUOUS ROW FLUORESCENT, SURFACE

CONTINUOUS ROW FLUORESCENT, RECESSED

SWITCH OUTLETS

S	SINGLE POLE SWITCH	Sₖ	KEY OPERATED
S₂	DOUBLE POLE	Sₗ	LOW VOLTAGE
S₃	THREE-WAY SWITCH	Sᴅ	DOOR SWITCH
S₄	FOUR-WAY SWITCH	Sₜ	TIME SWITCH
S	SWITCH AND SINGLE RECEPTACLE	S	SWITCH AND DOUBLE RECEPTACLE
S	CEILING PULL SWITCH		

RESIDENTIAL SIGNALING SYSTEMS

▣	PUSH BUTTON	▢	BUZZER
◖▢	BELL	CH	CHIME
BT	BELL TRANSFORMER	D	ELECTRIC DOOR OPENER
◀	OUTSIDE TELEPHONE	◣	INTERCONNECTING TELEPHONE
R	RADIO OUTLET	TV	TELEVISION OUTLET

RECEPTACLE OUTLETS

⊖	SINGLE OUTLET	⊜	DUPLEX OUTLET, SPLIT WIRED
⊖	DUPLEX OUTLET	▲	SINGLE SPECIAL PURPOSE CONNECTION DW AS DISHWASHER
⊖	TRIPLEX OUTLET	⬚	DUPLEX SPECIAL PURPOSE OUTLET
Ⓕ	FAN HANGER	Ⓒ	CLOCK HANGER
	FLOOR SINGLE RECEPTACLE		FLOOR DUPLEX RECEPTACLE
	FLOOR SPECIAL PURPOSE RECEPTACLE	R	RANGE RECEPTACLE
		WP	WATERPROOF DUPLEX RECEPTACLE

OTHER SYMBOLS

FLUSH-MOUNTED PANEL BOARD AND CABINET

SURFACE-MOUNTED PANEL BOARD AND CABINET

WIRING CONCEALED IN WALL OR CEILING

WIRING CONCEALED IN FLOOR

SWITCHBOARD, POWER CONTROL CENTER, UNIT SUBSTATIONS

WIRING EXPOSED

Fig. 10-7. Graphical electrical symbols for residential wiring plans. These are shown approximately half the size they appear on drawings.

Lightolier

Fig. 10-9. Decorative or accent lighting provides visual appeal.

Lightolier

Fig. 10-8. Local lighting is provided for reading, sewing, and other activities.

4. Multiple switching and dimmer controls. Maximum use of these enables the lighting effects to be adjusted to suit requirements.
5. Convenience outlets located to provide flexibility in arranging furniture. An outlet might be placed in the fireplace mantel. Often, one is placed with each wall switch for use when using a vacuum cleaner.
6. A special-purpose outlet for a window air conditioner. This is needed if central air conditioning is not planned.

Dining Areas

For a combination living-dining room, the lighting system and fixtures in the dining area should harmonize with those in the living area. The types of lighting and outlets needed are:

1. In the dining room, a permanent lighting fixture for general illumination. A dimmer switch helps regulate the intensity of the light.
2. Accent lighting on the table. This can be provided by spot lights built flush into the ceiling. Some hanging fixtures have accent spots built into them.
3. Supplementary and decorative lighting, including any of a variety of perimeter lighting techniques. See Fig. 10-10.

Lightolier

Fig. 10-10. A hanging light fixture provides general illumination in the dining room. The amount of light is varied by a dimmer switch.

4. Convenience outlets should be placed in the room. Split-wired outlets located at table height are helpful when portable appliances are to be used. See Fig. 10-11.

Bedrooms

Lighting and outlet needs for bedrooms are:

1. General illumination, which can be provided by a ceiling fixture or cove or valance lighting. It should be controlled by a switch near the door to the bedroom. There are many styles of lights from flush fixtures to hanging fixtures that provide satisfactory general illumination.
2. Local lighting, such as bed lights for reading in bed, vanity lights, or lights on full-length mirrors, should be considered.
3. Convenience outlets placed with special attention to furniture location. Outlets near the bed are needed to handle bedside table lamps, clocks, radios, and electric blankets. See Fig. 10-12. A special-purpose outlet for a window air conditioner is needed if central air conditioning is not planned.

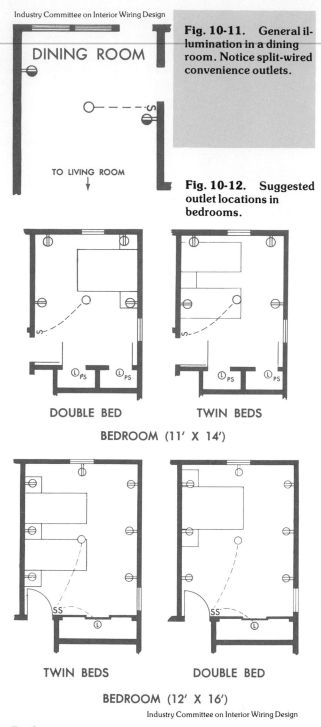

Industry Committee on Interior Wiring Design

DINING ROOM

TO LIVING ROOM

Fig. 10-11. General illumination in a dining room. Notice split-wired convenience outlets.

Fig. 10-12. Suggested outlet locations in bedrooms.

DOUBLE BED   TWIN BEDS

BEDROOM (11′ X 14′)

TWIN BEDS   DOUBLE BED

BEDROOM (12′ X 16′)

Industry Committee on Interior Wiring Design

Bathrooms

Bathroom electrical needs are somewhat more specialized.

1. The average bath should have a ceiling light for general illumination and wall lamps on each side of the mirror as supplemental lighting. If the

mirror is large, side lighting is inadequate and overhead lighting must be used. A large bath should have a vapor-proof light installed over the tub or shower. A small bath can be satisfactorily lighted by using wall lamps on each side of the mirror. The use of large, luminous plastic panels provides an excellent means of general illumination.

2. If a bath has no exterior window, it is necessary to install a ventilating fan in the ceiling. These are available as a compact unit combined with a light and an electric heater. See Fig. 10-13. This should be operated by a wall switch.

3. Convenience outlets are generally part of the wall lights installed on the sides of the mirror. However, an outlet should be installed at each mirror or vanity to accommodate electric razors and hair dryers. The receptacle which is a part of a lighting fixture will not satisfy requirements, unless it is rated at 15 amperes and wired with at least 15-ampere-rated wires. See Fig. 10-14.

Kitchen

Two types of lighting are needed in the kitchen. Outlets require careful consideration.

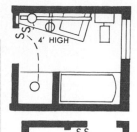

Fig. 10-14. Typical bath outlet layout. Notice the convenience outlet separate from the lights.

Industry Committee on Interior Wiring Design

1. General illumination can be provided by a large ceiling fixture centered in the room. It may be mounted flush with the ceiling, or it may be a pendant type. An excellent means of general illumination is provided by using luminous plastic ceiling panels that are lighted from above by fluorescent lamps. They create a skylight appearance. See Fig. 10-15.

2. Local lighting for work areas, such as the sink, surface unit, counter tops, or inside cabinets, requires individual lamps. These should be controlled by a switch located on or near the area to be lighted. Fixtures are frequently built into the bottom of cabinets or hoods.

3. Special attention to the location of convenience outlets is necessary. The refrigerator, washer, freezer, and other major appliances must be located. The countertop needs one outlet for every 4 lineal feet (1219 mm) of work space. These are usually located 44 inches (1118 mm) above the floor. An outlet at table height is

NuTone, Inc

Fig. 10-13. This unit serves as a rapid electric heater, light, and exhaust fan for bathroom use.

Diffusa-Lite Company

Fig. 10-15. An entire lighted ceiling of plastic panels over fluorescent lights.

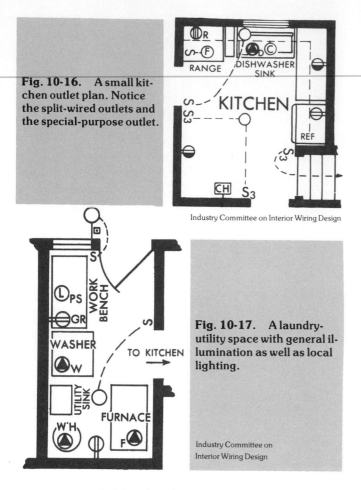

Fig. 10-16. A small kitchen outlet plan. Notice the split-wired outlets and the special-purpose outlet.

Fig. 10-17. A laundry-utility space with general illumination as well as local lighting.

needed by the dining table. All convenience outlets should be split-wired to prevent overloads. See Fig. 10-16.

4. Attention is needed to locate appliances needing special-purpose outlets, such as the electric dryer. Electric ovens, ranges, and surface units are wired directly into the circuit and do not require outlets.

Laundry and Utility Areas

Considerations for lighting and outlets in the laundry and utility areas are:

1. Lighting fixtures should be placed to illuminate the work areas, such as laundry tubs, sorting area, and ironing area. All lighting, especially in the laundry area, should be controlled by a wall switch. At least one large light, centered in the room, is needed. If the room is large, local lighting of work areas may be needed.

2. Convenience outlets are needed for activities such as ironing, or electrical tools if a workshop is planned. These should be split-wired outlets. See Fig. 10-17.

3. Special-purpose outlets are needed for appliances such as an electric dryer. The electric water heater and furnace are wired directly into their circuits and do not require outlets.

Closets

Each closet should have one light. Wall switches or door switches which automatically operate as the door is opened are recommended. A pull-chain light is satisfactory, but inconvenient to use.

Halls and Stairs

Halls and stairs are traffic areas, and lighting must be planned accordingly.

1. Sufficient wall-controlled lights should be provided to illuminate the entire hall and stair areas. These may be ceiling or wall fixtures. Lights should have multiple-switch control at each end of the hall or the head and foot of the stair. Switches should be grouped together and never located so close to steps that a fall might result from a misstep while reaching for a switch.

2. One convenience outlet should be provided for each 15 lineal feet (4572 mm) of hallway. If a hall contains over 25 square feet (2.325 m²), it should contain one convenience outlet. Stairways require one outlet conveniently located for a vacuum cleaner or night light. This could be on an intermediate landing in a long stair.

Basement Area

The lighting demands depend upon the use of the basement area. If it is simply a storage area, all that is needed are enough ceiling lights to provide general illumination and several convenience outlets located around the walls. Most often, this area is to be used in the future for a recreation room, workshop, laundry, or other activity. The considerations mentioned for these areas have been covered.

Garages and Carports

1. At least one, 100-watt fixture should be provided in a one-car garage. However, it is preferred that two be used. These should be placed on each side of the car and 6 feet (1829 mm) back from the front bumper. In a two-car garage, two lights will provide sufficient illumination. All fixtures should be controlled by wall switches located near the outside door of the house and on a convenient wall near the outside garage door.

If the garage is to contain a workshop or is to be used for other activities, these should receive additional attention and should be illuminated by local means.

2. At least one convenience outlet should be in a one- or two-car garage. However, one outlet in each wall of the garage is much better.

Exterior Lighting and Outlets

1. All exterior entrances should have one or more light fixtures. See Fig. 10-18. Where a single light is attached to a house at the door, it should be on the latch side. Post lights set in the yard and lights along the driveway are of much value. See Fig. 10-19. They should be controlled by switches from within the house. Terraces and patios can be tastefully lighted by lights built into fences or walls or by freestanding luminaries. Large areas can be illuminated by floodlights. These are usually mounted under the eaves of the house and should be controlled by switches within the house.

2. A weatherproof convenience outlet near the front entrance is essential. This outlet should be 18 inches (457 mm) above the grade. Many post lamps contain a weatherproof outlet built into the post. Such outlets are of great value on a patio or terrace for radios and electric cookers.

Porches

The lighting and outlet needs for a porch are as follows:

1. Each porch, breezeway, or other similar roofed area of more than 75 square feet (6.975 m²) in floor area should have one ceiling light controlled by a wall switch within the house. If the porch is large or irregular in shape, additional lights may be needed. If it has more than one entrance, the light should be multiple switched to all entrances.

2. One weatherproof convenience outlet should be installed for each 15 lineal feet (4572 mm) of porch or breezeway. It is recommended that such outlets be controlled by an inside wall switch.

Kinds of Light Fixtures

As lighting is planned, the kinds and styles of fixtures must be decided. The common kinds of fixtures are floor lamps, table lamps, tree lamps, lighted valances, interior spotlights, flush and cone lights, hanging fixtures, flush-with-the-ceiling fixtures, surface-mounted fixtures, cove lighting, and luminous ceilings.

The examples that follow illustrate some of the units that are on the market. See Figs. 10-20 through 10-26. A careful study of manufacturers' catalogs is of great value in planning lighting and in selecting fixtures.

Signal and Communication Systems

Signaling systems range from the minimum — a front doorbell — to elaborate chimes with multi-point control, or even to audible and visual annunciators in homes with servant accommodations.

The recommended minimum system is a chime with front and rear (trade) door controls which sound different tones or notes to identify each point of operation. If separate bells or buzzers are used, each unit should have a different tone to distinguish the front, rear, or side door.

Flush-mounted loudspeaking telephone equipment installed at each commonly used exterior entrance, with an interior telephone station in the basement, kitchen, and second-floor hall (or master bedroom) saves steps.

An annunciator may be installed in the kitchen with push-button stations in each bedroom, living room, family room, porch, etc., and is especially

Lightolier

Ryther-Purdy Lumber Co., Inc.

Fig. 10-18. Outdoor lighting provides safety as well as an attractive appearance.

Fig. 10-19. Outdoor lights are often freestanding.

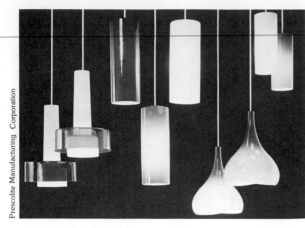

Prescolite Manufacturing Corporation

Fig. 10-20.   Pendant luminaries.

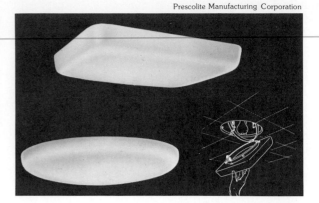

Prescolite Manufacturing Corporation

Fig. 10-24.   Shallow, surface-mounted fixtures with light source and box above ceiling.

Prescolite Manufacturing Corporation

Fig. 10-21.   A hanging, metal cone spotlight used to highlight a special area.

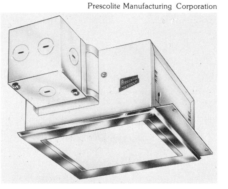

Prescolite Manufacturing Corporation

Fig. 10-22.   A flush-with-the-ceiling fixture for general illumination.

Prescolite Manufacturing Corporation

Fig. 10-25.   A tree lamp, for dramatic lighting.

Prescolite Manufacturing Corporation

Fig. 10-23.   An adjustable spotlight for highlighting an object of special importance, such as a fireplace. It is flush with the ceiling, and the box is above the ceiling.

Prescolite Manufacturing Corporation

Fig. 10-26.   A floor-style tree lamp.

convenient in large, two-story houses. Or, inter-communicating telephones may be used to serve the same purpose.

A fire-alarm system is recommended in every home. A system may use ionization-chamber or photoelectric smoke detectors in critical places. Detectors should be placed in the furnace area, storage areas, garage, workshop, and other critical places. A more elaborate central alarm system can be installed. There are systems available that operate off a transformer. These have rechargeable batteries that supply power in case of a power failure. No switches are permitted in the circuit. See Chapter 13.

Wiring channels for telephone conductors should be planned when a new home is built. Service conduit, interior raceways, outlet boxes, and telephone niches should be installed while the home is under construction. This allows a telephone installation to have a better appearance. Conventional outlets may be used for permanently connected units or the jack-plug receptacle for portable telephones. The local telephone company provides technical assistance, and should be consulted regarding local service facilities, regulations, etc.

An intercommunicating telephone system is a desirable convenience for the larger home. Manufacturers of the various systems should be consulted for specific features of equipment, installation details, and required wiring.

Multistation television antenna, cable, or satellite systems provide outlets for operation of one or more TV sets, or for moving one set from room to room as the occasion requires.

Undercarpet Wiring System

When an area such as an office is rearranged, it occasionally becomes necessary to move electrical outlets. The undercarpet wiring system permits easy and rapid rewiring. The conductor has wires formed flat and embedded in a plastic coating. A three-conductor cable is .031 inches (0.787 mm)

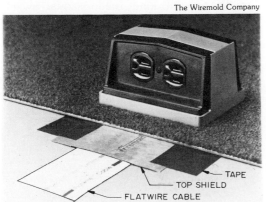

The Wiremold Company

TAPE
TOP SHIELD
FLATWIRE CABLE

Fig. 10-27. An outlet box and the flat cable for an undercarpet wiring system.

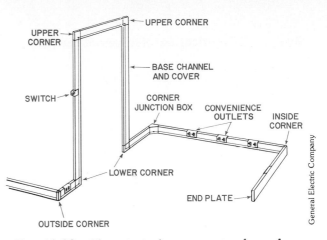

UPPER CORNER
UPPER CORNER
BASE CHANNEL AND COVER
SWITCH
CORNER JUNCTION BOX
CONVENIENCE OUTLETS
INSIDE CORNER
LOWER CORNER
END PLATE
OUTSIDE CORNER

General Electric Company

Fig. 10-28. The principal components of a surface-mounted wiring system. Notice that the raceway forms the baseboard and door trim. The wires carry around the door opening inside the metal channel.

thick and 2 1/2 inches (63.5 mm) wide. The conductor is rolled out on the floor, a top shield is taped over it, and the carpet is installed. Outlets are placed along the cable where needed. See Fig. 10-27. To turn a corner, the flat cable is folded like a piece of paper.

Surface-Mounted Wiring

Many new homes are built of prefabricated sandwich or solid-core panels. This limits the use of wall cavities for wiring. The surface-mounted wiring system is an economical solution to this problem. It is also a good system to use when rewiring an old house. Look again at Fig. 10-27.

The main part of the system is a metal raceway. It serves as a baseboard, door trim, or chair rail, and provides a space for carrying wires. See Fig. 10-28. Specially designed outlets can be mounted where needed on the raceway, and switches are available for mounting on the raceway to control any of the outlets desired.

After the system is installed, the raceway cover can be easily removed for rearranging the outlets or for adding more outlets.

In Fig. 10-29, the actual dimensions of the raceway for one such wiring system are given.

Low-Voltage, Remote-Control Wiring

Modern living has created many demands on the electrical wiring system of the home. One big problem is that of flexibility of control. The low-voltage, remote-control system* was designed to give the homeowner wider control of the electrical system. For example, all the lights in a house can be turned on or off from one point, such as the bedroom.

This system uses relays to perform the actual switching of the 120-volt line current. These relays

*Material on low-voltage, remote-control systems supplied by General Electric Company

General Electric Company

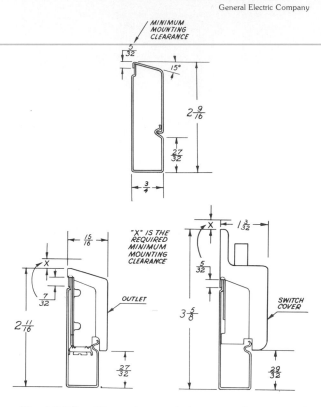

Fig. 10-29. Dimensions and clearances for a typical surface-mounted raceway wiring system.

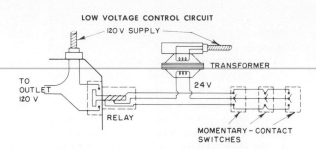

Fig. 10-30. A typical low-voltage circuit.

The 120-volt outlet is fed 120 volts and has a set of relay contacts connected in series with one side of the supply line. The relay coil is connected to three momentary-contact switches. These are operated by a transformer connected to a 120-volt line that steps the voltage down to 24 volts to operate the switches and relay. If a switch is touched, 24 volts flow to the relay. The relay actuates the outlet (which could be a light). This example is a 4-way switch, but special switches are not required as in line voltage switching. Any number of other switches could be connected in to control this outlet.

are controlled by small switches operating on 24 volts. This permits the use of inexpensive wiring similar to that used for door chimes. In addition, because these relays only require a momentary impulse to change from off to on (or on to off), the control switches are momentary-contact switches. This means that the low voltage causes a small current to flow only for the time the switch is pressed. See Fig. 10-30.

Because all switches used in this system are momentary-contact switches, as many as desired can be wired in parallel. They are not actually in the circuit, except during the short period that they are pressed. One type of switch performs the functions of a single-pole switch, a three-way switch, or a four-way switch, with no complicated wiring. Even after installation, switches can be added to the system without changing the wiring.

Another outstanding feature of this system is that the heavy-gauge, current-carrying wires required for 120-volt service do not have to loop down to the switches or make several runs across a long distance to permit several switches to control one outlet. Instead, the current-carrying wires go di-

rectly to the outlet where the low-voltage relay is located. The control switches are looped to the relay with low-cost, small-gauge wire. This saves in the cost of expensive wire and prevents a voltage drop in the long runs of heavy-gauge wire, thus increasing the voltage at the outlet.

By arranging switch points close together, a dial-type switch can rotate and make many contacts in a fraction of a second. In this way, master control of many circuits is possible. Switches of this type are available in nine-circuit and twelve-circuit controls. With this master control located in the bedroom, a person can turn on the coffee maker or turn on the electric heaters in the driveway to melt the snow, all before arising from bed. See Fig. 10-31.

Planning a Low-Voltage Remote System

To realize the full advantage of this modern wiring system, proper planning is essential.

1. Decide on the kind of individual switches desired. Those available are the dual, push-button switch (lighted or non-illuminated), the locking-style switch, the trigger-style switch, and the pilot-type switch. The push-button type is shown in Fig. 10-32. The locking style requires a key to operate, while the pilot-light style has a red light that glows to indicate that an unseen light is on.

2. Decide on the controls needed. It is necessary to study the electrical layout and to decide from how many locations each light or outlet should be controlled. Multiswitch control can greatly increase the convenience of living in a house.

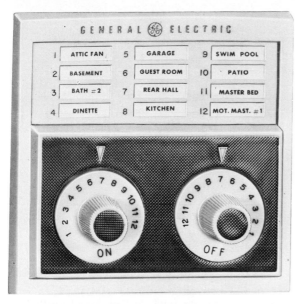

Fig. 10-31. A master selector switch for low-voltage wiring. This one panel controls twelve lights.

Fig. 10-32. Push-button switch used in low-voltage wiring system.

The location of master selector switches is important. Essentially, a **master selector switch** is a console of individual switches mounted on a single plate for convenient selection of any of the individual circuits. Usually, the best locations for these master switches are beside the bed, garage door, and other frequently used entrances. They should be located wherever it would be convenient to be able to operate all the circuits at once from a single location.

3. Decide on the type of installation needed. Manufacturers of low-voltage, remote-control wiring systems recommend the use of the relay-center method of installation. This method has a master relay-center box (a large steel box as

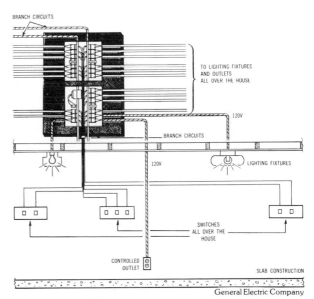

Fig. 10-33. A master relay-center box for a low-voltage wiring system. Each relay is located here and labeled for easy identification of the circuit.

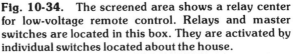

Fig. 10-34. The screened area shows a relay center for low-voltage remote control. Relays and master switches are located in this box. They are activated by individual switches located about the house.

used for fuses or circuit breakers) in which are located the relays and master selector switches. See Figs. 10-33 and 10-34. Any relays or switches that need to be added later are easily

General Electic Company

Symbol	Description
—‖—T—	Remote-Control Low-Voltage Wire
T	Low-Voltage Transformer
►‖—	Rectifier for Remote Control
B R	Box for Relays and Motor Master Controls
R	Remote-Control Relay
R P	Remote-Control Pilot-Light Relay
P 11	Separate Pilot-Light, R.C. Plate
P 10	Separate Pilot-Light, Inter. Plate
MS	Master-Selector Switch
MM R	Motor Master Control for ON
MM B	Motor Master Control for OFF
S M	Switch for Motor Master
S F6	R.C. Flush Switch
S F7	R.C. Locator-Light Switch
S F8	R.C. Pilot-Light Switch
S K6	R.C. Key Switch
S K7	R.C. Locator Light Key Switch
S K8	R.C. Pilot-Light Key Switch
S T6	R.C. Trigger Switch
S T7	R.C. Locator-Light Trigger Switch
S T8	R.C. Pilot-Light Trigger Switch
S T4	Interchangeable Trigger Switch, Brown
S T5	Interchangeable Trigger Switch, Ivory
⊖ RO	Remote-Control Outlet for Extension Switch

Fig. 10-35. Wiring symbols for low-voltage remote-control wiring.

inserted in the box. It has the advantage of being easy to wire, and improper connections can be readily corrected. It also eliminates the slight buzz and click made by the relays as they operate.

The symbols in common use for low-voltage wiring are shown in Fig. 10-35.

Build Your Vocabulary

Following are terms that you should understand and use as part of your working vocabulary. Write a brief explanation of each term.

accent lighting
ampere
circuit
circuit breakers
convenience outlets
electrical load
fuses
general illumination
light fixture
low-voltage, remote-
control wiring

National Electrical
Code
service entrance
service panel
special-purpose outlets
surface-mounted wiring
switch
volt
watt

Class Activities

1. Trace the electrical circuits in your home. List the electrical units (such as refrigerator) connected to each circuit, as well as the number of lights and convenience outlets. Ascertain the number of amperes each circuit can carry. Are any of the circuits overloaded?
2. Compute the size of the electrical service entrance needed for your home, or for the home economics room or industrial arts shop in your school. In your report, list each major appliance or machine and give its amperage rating.
3. Make a freehand sketch of the home economics room or industrial arts shop in your school and indicate the locations of all machines, all 120-volt and 240-volt electrical outlets, and all lights. Ascertain the number of circuits needed, and list the equipment to be on each circuit.
4. With the teacher's permission, invite a speaker from the electric company to talk to the class about the all-electric home.
5. Examine the location of convenience outlets in your home. Cite areas that have inadequate outlets.
6. Compile a list of all the special-purpose outlets in your home.
7. Make a freehand sketch of the plot plan of your home. Locate any sidewalks and driveways. Plan an adequate system of exterior lighting, and show it on the plot plan with conventional symbols.
8. Design a modern telephone system for your home. Make a freehand sketch of the floor plan and locate the principally used telephone. Then indicate where other telephones or telephone jacks should be located.
9. Plan a communication system for your home. Make a freehand sketch of the floor plan. Locate the master control unit and speakers on the plan.

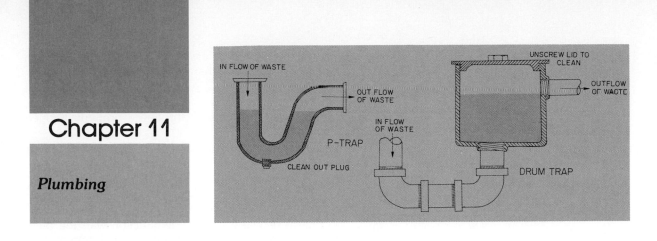

Chapter 11

Plumbing

A satisfactory plumbing system requires foresight on the part of the person planning the building. Also, it requires the services of a competent plumber to install the system correctly and to overcome unforeseen difficulties that always seem to arise after a building is under construction.

Plumbing design and installation requirements are available in the **Uniform Plumbing Code.** It is published by the International Association of Plumbing and Mechanical Officials.

The plumbing system contains two major parts — the waste disposal system and the fresh water distribution system.

Waste Disposal System

The main parts of a waste disposal system are the soil pipes, waste pipes, house or building drain, and house or building sewer. The **soil pipe** is the part of the drainage system which receives the waste from water closets and carries it to the house or building drain. It may have other fixtures emptying into it. The **waste pipe** receives the waste from sinks, lavatories, and bathtubs or other fixtures not receiving human excreta. It carries the waste to a soil pipe or to the house or building drain. The **house** or **building drain** receives the discharge from the soil pipe and waste pipes within the building and carries it to the house or building sewer. The **house** or **building sewer** begins just outside the foundation of the building and carries the waste to the sewer in the street or to the septic tank. The parts of a typical residential waste disposal system are illustrated in Fig. 11-1. In Fig. 11-2, a multistory commercial system is shown.

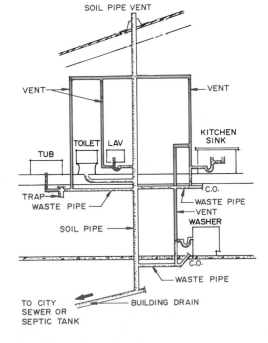

Fig. 11-1.   Parts of a typical residential waste disposal system.

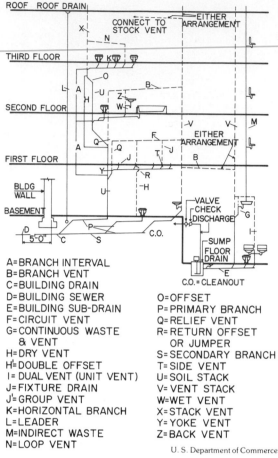

A=BRANCH INTERVAL
B=BRANCH VENT
C=BUILDING DRAIN
D=BUILDING SEWER
E=BUILDING SUB-DRAIN
F=CIRCUIT VENT
G=CONTINUOUS WASTE
    & VENT
H=DRY VENT
H¹=DOUBLE OFFSET
I=DUAL VENT (UNIT VENT)
J=FIXTURE DRAIN
J¹=GROUP VENT
K=HORIZONTAL BRANCH
L=LEADER
M=INDIRECT WASTE
N=LOOP VENT

O=OFFSET
P=PRIMARY BRANCH
Q=RELIEF VENT
R=RETURN OFFSET
    OR JUMPER
S=SECONDARY BRANCH
T=SIDE VENT
U=SOIL STACK
V=VENT STACK
W=WET VENT
X=STACK VENT
Y=YOKE VENT
Z=BACK VENT

U. S. Department of Commerce

Fig. 11-2. A typical multistory waste disposal system.

Parts of the Waste Disposal System

The following parts can be located in Fig. 11-2:

Branch interval — A section of soil pipe at least 8 feet long into which the waste from fixtures on the floor enters.

Branch vent — Any vent pipe connecting a branch of the drainage system with the vent stack.

Circuit vent — A group vent extending from in front of the last fixture connection of a horizontal branch to the vent stack.

Continuous waste and vent — A vent that is a continuation of and in straight line with the drain to which it connects.

Dry vent — A vent that does not carry water or water-borne wastes.

Dual vent — A group vent connecting at the junction of two fixture branches and serving as a back vent for both branches.

Group vent — A branch vent that performs its function for two or more traps.

Leader — A pipe draining water from the roof to a storm sewer or other means of disposal.

Indirect waste — A waste pipe which does not connect directly with the building drainage system, but discharges into it through a properly trapped fixture.

Loop vent — Same as a circuit vent except it loops back and connects with a soil or waste stack vent instead of the vent stack.

Primary branch — A part of the drainage system that is the single sloping drain from the base of a soil stack or waste stack to its junction with the main building drain or with another branch thereof.

Relief vent — A branch from a vent stack, connected to a horizontal branch between the first fixture branch and the soil or waste stack. Its function is to provide for circulation of air between the vent stack and the soil or waste stack.

Secondary branch — Any branch of a building or house drain other than the primary branch.

Side vent — A vent connecting to the drain pipe through a 45° fitting.

Stack — A general term referring to any vertical pipe, such as a soil pipe, waste pipe, or vent pipe.

Vent stack — The main vertical vent pipe installed primarily to provide circulation of air to or from any part of the building drainage system.

Wet vent — A soil or waste pipe that serves also as a vent.

Yoke vent — A vertical or 45° relief vent of the continuous waste and vent type formed by the extension of an upright **Y** branch or 45° **Y** branch inlet of the horizontal branch to the stack.

Back vent — A branch vent installed primarily for the purpose of protecting fixture traps from self-siphonage.

General Design Considerations

The typical cast-iron soil pipe used for residential work is usually 4 inches in diameter. It has a large bell shape on one end of each length, and the largest outside dimension approaches 5½ inches. This bell shape is necessary for the joining of two pieces of pipe. When it must be hidden inside an interior wall, the wall must be framed with 2 × 6 studs to conceal the pipe. The designer must plan for this when designing room layouts and partitions. If a second-floor partition is not directly above a first-floor partition, with both concealing a soil pipe, great difficulties are encountered in running the soil pipe through the second floor to the roof. Because of the nature of the waste carried by a soil pipe and because of the large-diameter pipe required, it cannot turn sharp corners or make many

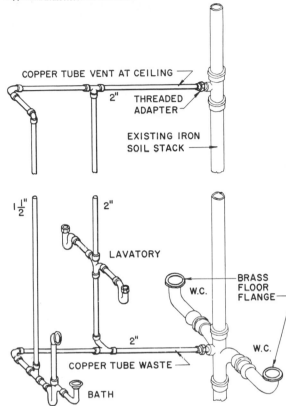

Fig. 11-3. Copper waste disposal system for two baths built back to back.

Horizontal branches of soil pipe are run between floor joists from the vertical stack to the water closets. These pipes have a slight slope from the water closets to the vertical soil pipe. Since the slope is small, they tend to clog easier than the vertical stack. The longer the horizontal pipe is, the more likely it is to clog. Pipes 4 inches in diameter are most often used in residential work, but for large projects with many fixtures, the size is ascertained by the unit method. This is explained in the next section of this chapter.

The waste pipe is usually smaller in diameter than the soil pipe, because it carries a different type of waste. Waste pipes may be cast iron, galvanized steel, wrought iron, copper, or ABS, PVC, and polybutylene plastic. A typical waste-pipe layout is shown in Fig. 11-3. If a lavatory is to discharge unusual materials, such as chemicals, special consideration should be given to the material from which the pipe is made. A plan view of a bath showing a suggested layout for the waste pipe, vent pipe, and hot and cold water lines is in Fig. 11-4.

Waste disposal systems should have clean-out plugs located at convenient places. It is inevitable that a system will become clogged and will need rodding out. However, clogging can be decreased by making horizontal waste lines as short as possible and by avoiding sharp offsets. The standard plumb-

turns. The designer should plan the waste disposal system so gradual bends and long straight runs of soil pipe are used.

Copper soil pipes have thin walls and small joints and can be placed inside the typical 4-inch interior partition. See Fig. 11-3.

Steel or wrought-iron soil pipe is used for industrial buildings where heavy work is done or where temperatures vary. These materials are also used for multistory buildings.

Another major consideration is to design the waste disposal system so it can be installed with a minimum of cutting of structural members. Roof and floor joists generally should not be notched more than one-sixth of the joist depth. The exception is when the notch is in the top only and no further from the face of support than the depth of the joist. There, the notch can be a maximum of one-third of the joist depth.

The soil pipe consists of the vertical stack which forms the central core of the waste disposal system. It runs from the house drain through the roof.

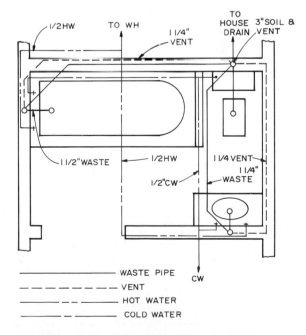

Fig. 11-4. Typical piping layout for a bathroom.

ing symbols used on architectural drawings are shown in Fig. 11-5.

The house drain may be located below the basement floor if the sewer or septic tank is below this level. Then waste will drain into it. If this condition does not exist, the house drain must be hung below the floor joists. The ideal situation, of course, is to locate it below the basement floor, so headroom is not reduced and unsightly pipes are hidden. This arrangement also permits satisfactory placement of fixtures in the basement. If a house drain is hung from the floor joists, all waste in the basement needs to be lifted up to it with a sump pump. This is a rather unsatisfactory arrangement.

The house drain is cast-iron pipe. The average house drain carries the waste from the fixtures in the house only. For residential work, a 4-inch cast-iron pipe serves as a house drain. See Fig. 11-6.

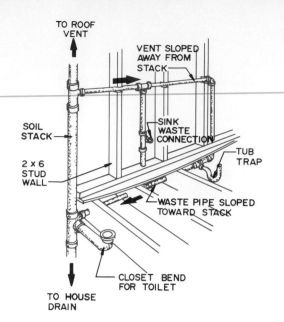

Fig. 11-6. A waste system installation must penetrate studs, floor, and ceiling joists.

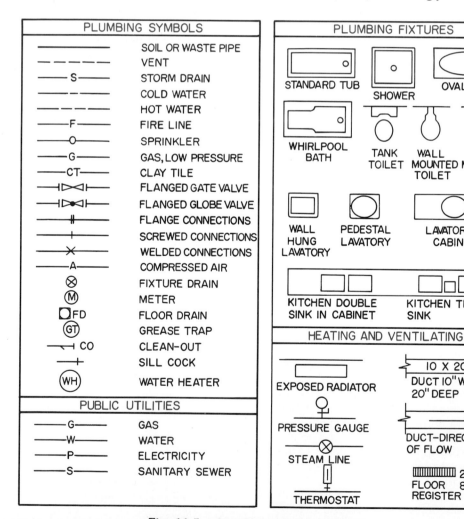

Fig. 11-5. Standard plumbing symbols.

The house sewer most frequently used is made of cast-iron pipe. It is suitable for installation on unstable soil, it resists clogging by tree roots, and it withstands vibrations and heavy loads (such as trucks) passing over it. Vitrified clay-tile pipe is satisfactory if the above conditions do not exist. A cast-iron pipe buried in cinders or ashes will deteriorate due to chemical action, and clay pipe must be used under such conditions.

In Chapter 2, consideration was given to planning bathrooms. The advantages of back-to-back building of a kitchen and a bath or other areas requiring water and sewer were discussed. The plumbing required to make this installation is illustrated in Fig. 11-7. This principle is violated in Fig. 11-8. Notice the extra pipe (thus, extra expense) involved. The decision to be made is whether the owner considers the value of the split location to be worth the cost of the additional plumbing.

It is most economical to locate second-floor bathrooms directly above a kitchen or first-floor bath, so all fixtures can discharge into a common stack. In Fig. 11-9, the drain and vent piping needed for such an installation is shown. Remember that the common wall also reduces the cost of water piping.

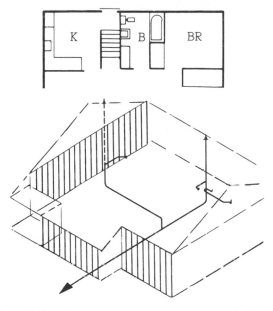

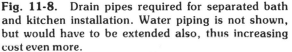

Fig. 11-8. Drain pipes required for separated bath and kitchen installation. Water piping is not shown, but would have to be extended also, thus increasing cost even more.

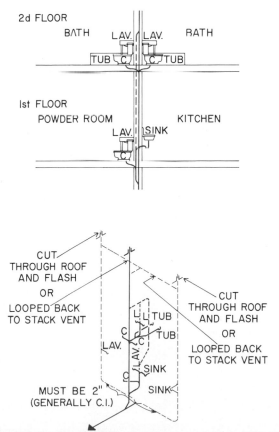

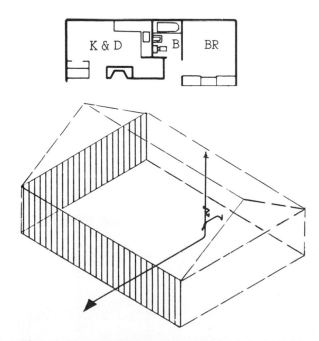

Fig. 11-7. Drain pipes required for back-to-back kitchen and bath installation.

Fig. 11-9. Two bathrooms using a common wall and soil pipe, placed directly above a kitchen and powder room.

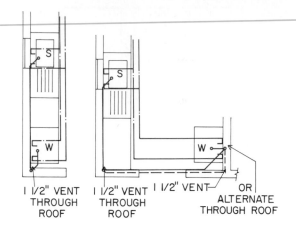

I 1/2" VENT THROUGH ROOF   I 1/2" VENT THROUGH ROOF   I 1/2" VENT   OR   ALTERNATE THROUGH ROOF

Fig. 11-10. Compare the differences in plumbing requirements for these two washer and sink locations.

Clothes washers and dishwashers should be located as near to the other plumbing in the house as is possible. If a washer is poorly placed, the cost of plumbing a kitchen can easily be doubled. Figure 11-10 illustrates such planning.

Sizing Drains, Soil Pipes, and Waste Pipes

Once the number and locations of fixtures have been decided, it is necessary to ascertain the size of the pipe used for the house or building drain, soil pipes, and waste pipes. The size of pipe depends upon the number and kinds of fixtures connected to each section. Each fixture discharges different volumes of waste. This flow of waste is measured in **fixture unit values.**

A study of the volume of water discharged by various fixtures disclosed that a lavatory discharges

Fig. 11-11.

Unit Values of Plumbing Fixtures

Fixture	Units	
	Private Use	Public Use
Lavatory or washbasin	1	2
Lavatory — public, barber, beauty parlor		2
Kitchen sink	2	4
Kitchen sink with disposal	3	
Bathtub, with or without overhead shower	2	4
Hose Bib	3	5
Laundry tub	2	4
Drinking fountain	1/2	1
Mobile Home	6	6
Urinal wall or stall, public		5
Urinal, pedestal, public; syphon jet, blowout		10
Shower bath	2	4
Dishwasher, domestic	2	
Water closet, tank operated	3	5
Water closet, valve operated	6	10

International Association of Plumbing and Mechanical Officials, Copyright, 1976.

about 7½ gallons of liquid per minute. This equals almost one cubic foot of water and represents "one fixture unit" of discharge.

Discharge capacity per unit is as follows:
Up to 7½ gallons per minute = 1 unit
  (Up to 30 liters) *
8 to 15 gallons per minute = 2 units
  (30 to 60 liters)
16 to 30 gallons per minute = 4 units
  (60 to 115 liters)
31 to 50 gallons per minute = 6 units
  (115 to 190 liters)

*Liter measures are approximate.

Fig. 11-12.

Capacities of Building Drains and Sewers

DIAMETER OF PIPE (Inches)	MAXIMUM NUMBER OF FIXTURE UNITS THAT MAY BE CONNECTED TO ANY PORTION OF THE BUILDING DRAIN OR THE BUILDING SEWER[1]			
	Fall per Foot			
	1/16-Inch	1/8-Inch	1/4-Inch	1/2-Inch
2			21	26
2½			24	31
3		20[2]	27[2]	36[2]
4		180	216	250
5		390	480	575
6		700	840	1000
8	1400	1600	1920	2300
10	2500	2900	3500	4200
12	3900	4600	5600	6700
15	7000	8300	10,000	12,000

From National Plumbing Code,
American Society of Mechanical Engineers

[1]Includes branches of the building drain.
[2]Not over two water closets.

The established unit values for selected fixtures are given in Fig. 11-11. For example, a bathtub in a residence has a unit value of 2. It adds 2 units or 15 gallons per minute to the needed waste-carrying capacity of the drain pipe.

The pipe diameters required for building or house drains, horizontal branch pipes, or vertical soil pipes are given in Figs. 11-12 and 11-13.

Computing Size of Building Drain or Sewer

Problem: Assume a motel contains 30 water closets (tank type), 30 lavatories, 30 bathtubs, 6 urinals, 5 slop sinks, and 4 laundry tubs. What size drain pipe will be required to handle the entire flow, if the slope of the drain pipe is to be 1/8 inch to the foot?

Solution: The first step is to find the total fixture units that the pipe will carry. Examine Fig. 11-11 for the fixture units for each type of fixture in the motel. These are listed as follows:

30 bathroom groups × 6 units	180 units
6 wall urinals × 4 units	24 units
5 slop sinks × 3 units	15 units
4 laundry tubs × 2 units	8 units
	227 fixture units total flow

It can be seen that the building drain will have to carry 227 fixture units of flow. An examination of Fig. 11-12 reveals that a 4-inch pipe sloped 1/8 inch to the foot carries 180 units, while a 5-inch pipe carries 390 units. A 5-inch pipe, therefore, is required to carry the 227 units of waste from the building to the sewer in the street or to a septic tank.

A similar problem exists when the size of a soil pipe must be ascertained. Standards have been established on the unit flow basis; these are given in Fig. 11-13.

No soil or waste stack should be smaller than the largest horizontal branch connected to it. A 4 × 3 water closet connection is not considered a reduction in pipe size.

A stack vent should be carried full size through the roof; the vent should not be less than 3 inches in diameter or the size of the building drain, whichever is the smaller.

When provision is made for the future installation of fixtures, those to be added should be provided for in the calculation of required drain pipe sizes. Construction should provide for such fixture installation by terminating and plugging fittings at the stack where they eventually will be connected.

Computing Size of Horizontal Fixture Branch

Problem: If the motel were a one-story building and two baths were on a single, horizontal fixture branch, what diameter pipe would be required? (Assume that each bath contains a water closet, lavatory, and bathtub with overhead shower.)

Solution: It can be seen in Fig. 11-11 that such a combination of fixtures has a unit value of 6. Since two baths feed into a single, horizontal branch, it must carry a total flow of 12 fixture units. An examination of Fig. 11-13 reveals that of "any horizontal fixture branch," a 2½-inch pipe carries 12 units; therefore, this would be a satisfactory pipe size for this fixture branch.

Fig. 11-13.

Capacities of Horizontal Fixture Branches and Stacks

DIAMETER OF PIPE (Inches)	MAXIMUM NUMBER OF FIXTURE UNITS			
			More Than Three Stories in Height	
	Any Horizontal Fixture Branch[1]	One Stack Three Stories in Height or Not Over Three Branch Intervals	Total at One Story or Branch Interval	Total for Vertical Soil Pipe or Stack
1 ¼	1	2	1	2
1 ½	3	4	2	8
2	6	10	6	24
2 ½	12	20	9	42
3	20[2]	30[3]	16[3]	60[3]
4	160	240	90	500
5	360	540	200	1100
6	620	960	350	1900
8	1400	2200	660	3600
10	2500	3800	1000	5600
12	3900	6000	1500	8400
15	7000	....	....	....

From National Plumbing Code.
American Society of Mechanical Engineers.

[1]Does not include branches of the building drain.
[2]Not over two water closets.
[3]Not over six water closets.

Uniform Plumbing Code, The International Association of Plumbing and Mechanical Officials, Copyright, 1976.

Computing Size of Soil Pipe or Stack

Problem: What size soil pipe or stack would be required for the motel, if the total flow were fed into the soil pipe by all horizontal branches?

Solution: The total flow of the motel was found to be 227 fixture units. Figure 11-13 reveals that one stack, 3 stories (or less) in height and 3 inches in diameter, carries 30 fixture units; if 4 inches in diameter, it carries 240 units. Therefore, the 4-inch stack would serve adequately.

Traps

An important function of waste disposal is performed by the trapping devices. They must permit the passage of waste into the system; yet they must prevent offensive sewer gases from backing up in the waste pipes and entering the building. Each fixture should be separately trapped. The trap should be located as near the fixture as possible, and each trap should be provided with a clean-out plug.

Fig. 11-15.

Commonly Used Trap Sizes

Fixture	Size of Trap and Fixture Drain (Inches)	Fixture Units
Bathtubs	1 ½	2
Clothes washer	2	2
Drinking fountains	1 ¼	1
Floor drains	2	2
Laundry	1 ½	2
Lavatories	1 ¼	1
Mobile home park, per trailor	3	6
Shower stalls	2	2
Sinks, residence kitchen	1 ½	2
Sinks, commercial, school	1 ½	3
Sinks, small (pantry or bar)	1 ½	1
Sinks, dishwasher	1 ½	2
Sinks, service	2	3
Urinals, wall	1 ½	2
Water closet, tank	3	4
Water closet, flush valve	3	6

Fig. 11-16.

Maximum Unit Loadings on Traps

Fixture Drain or Trap Size (in Inches)	Fixture Unit Value
1 ¼ and smaller	1
1 ½	3
2	4
3	6
4	8

Uniform Plumbing Code, International Association of Plumbing and Mechanical Officials, Copyright, 1976.

There are many types of traps in use. Two common types are illustrated in Fig. 11-14. The water enters the trap and pushes the waste down the waste pipe. The last of the water entering remains in the trap, thus preventing sewer gas from backing up through the pipe. Trap sizes commonly used are given in Fig. 11-15. Usually, the trap should be the same diameter as the drain pipe to which it is connected. The maximum unit loadings that can be placed on various size traps are given in Fig. 11-16.

Vents

It is necessary to vent a waste disposal system properly so that atmospheric pressure is maintained. Proper venting also prevents the loss of trap seals, enables the waste to flow normally, and prevents deterioration of materials. Refer again to Figs. 11-1 and 11-6 for examples of vents. Figure 11-2 illustrates and defines vents commonly found in residential and commercial plumbing systems.

If a waste disposal system is not vented properly, the atmospheric pressure inside the waste pipe may become less than the pressure in the room (a partial

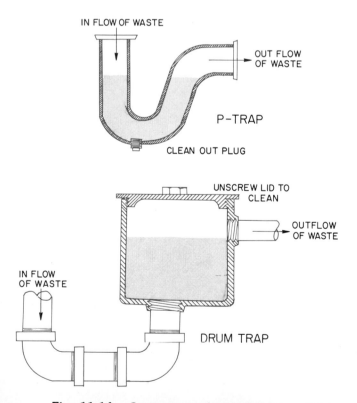

Fig. 11-14. **Comparison of typical lavatory P-trap and drum trap used on bathtubs and showers.**

vacuum), and the water in the trap will be forced into the waste pipe, leaving the trap empty. The foul gases from the sewer system can then enter the building. Occasionally, the waste pipe will build up excessive pressure if not vented properly, and the waste in the trap will be sprayed into the room. Or, the flow of waste may compress the air in the pipe and seriously retard the flow of liquid. This is the same as having a waste pipe blocked by a physical obstruction causing the waste to drain out slowly.

The gas in a waste system frequently contains acid-forming elements. If not vented, these will attack the pipe and cause it to deteriorate and, eventually, to fail.

There are two main classifications of vents — (1) the pipes used to vent the soil and waste pipes and (2) individual vents to each fixture, designed to maintain atmospheric pressure in the disposal system.

The soil and waste pipe vent is obtained by extending the soil pipe up through the roof. It is usually the same diameter as the portion of the pipe that carries waste. Attached to this is the main vent pipe, which serves as the ventilation stack for the smaller fixture traps. This vent begins in the basement, at the bottom of the soil pipe, and runs the full height of the soil stack, connecting again in the attic at least 3 feet (914 mm) above the highest fixture branch. See Figs. 11-1 and 11-5. The main vent must run straight up, with no offsets to slow the venting process.

The individual vents run from single traps and are connected into the main vent pipe above the overflow of the fixture vented. They are connected as near to the trap as possible, below and behind the fixture vented. If several fixtures are closely grouped, one individual vent can serve the group. The size of the vent pipe and the number of fixtures that can be vented with a pipe of a particular size can be found in Fig. 11-17.

It should be pointed out again that all residential piping must be hidden inside the walls of the building. In many industrial installations, the piping problems are very complex, and it is not possible or necessary to hide the piping.

Sizing Vent Piping. The nominal size of vent piping is determined by its necessary length and by the total of fixture units connected with it. Design data are reported in Fig. 11-17.

Note: Only 20 percent of the total length of vent piping can be installed in a horizontal position.

Fig. 11-17.

Sizes and Lengths of Vents

Size of Soil or Waste Stack (Inches)	Fixture Units Connected	Diameter of Vent Required (Inches)								
		1¼	1½	2	2½	3	4	5	6	8
		Maximum Length of Vent (Feet)								
1¼	2	30								
1½	8	50	150							
1½	10	30	100							
2	12	30	75	200						
2	20	26	50	150						
2½	42	. . . .	30	100	300					
3	10	. . . .	30	100	200	600				
3	30	. . . .	. . . .	60	200	500				
3	60	. . . .	. . . .	50	80	400				
4	100	. . . .	. . . .	35	100	260	1000			
4	200	. . . .	. . . .	30	90	250	900			
4	500	. . . .	. . . .	20	70	180	700			
5	200	. . . .	. . . .	. . . .	35	80	350	1000		
5	500	. . . .	. . . .	. . . .	30	70	300	900		
5	1100	. . . .	. . . .	. . . .	20	50	200	700		
6	350	. . . .	. . . .	. . . .	25	50	200	400	1300	
6	620	. . . .	. . . .	. . . .	15	30	125	300	1100	
6	960	. . . .	. . . .	. . . .	. . . .	24	100	250	1000	
6	1900	. . . .	. . . .	. . . .	. . . .	20	70	200	700	
8	600	. . . .	. . . .	. . . .	. . . .	. . . .	50	150	500	1300
8	1400	. . . .	. . . .	. . . .	. . . .	. . . .	40	100	400	1200
8	2200	. . . .	. . . .	. . . .	. . . .	. . . .	30	80	350	1100
8	3600	. . . .	. . . .	. . . .	. . . .	. . . .	25	60	250	800
10	1000	. . . .	. . . .	. . . .	. . . .	. . . .	. . . .	75	125	1000
10	2500	. . . .	. . . .	. . . .	. . . .	. . . .	. . . .	50	100	500
10	3800	. . . .	. . . .	. . . .	. . . .	. . . .	. . . .	30	80	350
10	5600	. . . .	. . . .	. . . .	. . . .	. . . .	. . . .	25	60	250

From National Plumbing Code, American Society of Mechanical Engineers

Fig. 11-18.

Usual Residential Vent Sizes

Fixture	Vent Pipe Diameter
Water closet	2″
Lavatory	1¼″
Bathtub	1¼″
Sink, kitchen	1¼″
Laundry tray	1¼″
Sink and tray combination	1¼″
Shower	1¼″
Dishwasher	1¼″

From Capacities of Stacks in Sanitary Drainage Systems for Buildings, Federal Housing Administration

Fig. 11-19.

Recommended Distances from Fixture Traps to Vents

Size of Fixture Drain	Distance Trap to Vent
1¼″	2 ft. 6 in.
1½″	3 ft. 6 in.
2″	5 ft. 0 in.
3″	6 ft. 0 in.
4″	10 ft. 0 in.

From National Plumbing Code, American Society of Mechanical Engineers

The length of the vent stack or main vent should be its developed length from the lowest connection of the vent system with the soil stack, waste stack, or building drain to the vent terminal. It can terminate in the open air through the roof, or it can connect to the stack vent. In the latter case, the length includes the added length of the stack vent to its termination in open air.

Individual vents should be not less than one-half the diameter of the drain, and never less than 1¼ inches (32 mm) in diameter. For residential use, the usual vent sizes are indicated in Fig. 11-18.

It is necessary to place vents away from traps, so the water in the trap will not be siphoned away, thus permitting sewer gas to enter the building. Figure 11-19 lists the recommended distances that vents should be located from fixture traps to prevent seal loss.

The Septic Tank

In areas beyond the city sewer lines, a septic tank must be installed to treat and filter waste material. Factors to be considered in planning for such a unit include the type of soil and the size of the tank.

Liquid waste is purified in the tank and is then dispersed into drain tile which permits the liquid to drain into the soil. If the soil is sandy or if it contains gravel, this leaching into the soil is aided. However, soil that contains an abundance of clay does not absorb the liquid waste as well.

The contour of the land should be sufficient to permit a gravity flow of waste from the septic tank

Fig. 11-20.

Minimum Capacities for Septic Tanks

Number of Bedrooms	Minimum Liquid Capacity (Gallons)
2 or less	750
3	900
4	1000
Each additional bedroom, add	250

HUD Minimum Property Standards

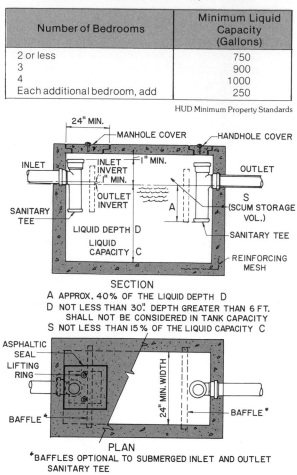

SECTION

A APPROX. 40% OF THE LIQUID DEPTH D

D NOT LESS THAN 30″. DEPTH GREATER THAN 6 FT. SHALL NOT BE CONSIDERED IN TANK CAPACITY

S NOT LESS THAN 15% OF THE LIQUID CAPACITY C

PLAN

*BAFFLES OPTIONAL TO SUBMERGED INLET AND OUTLET SANITARY TEE

Fig. 11-21.   Single-compartment septic tank.

to the drain tile and thus to the soil. The slope also should permit the tank and drain lines to be installed close to the surface of the soil. The bacteria that act to purify the waste cannot exist deep in the ground.

The more fixtures that discharge waste into the tank, the larger the tank must be. Satisfactory tank sizes for residential use are given in Fig. 11-20.

If a house draws its water supply from a well, the septic tank must be at least 50 feet (15 240 mm) from the well, and the drain field must be 100 feet (30 480 mm) away.

A typical septic tank is illustrated in Fig. 11-21. The raw sewage enters through the inlet. It consists of heavy, solid matter that settles to the bottom, as well as liquid and materials that remain in solution.

The heavy materials that settle to the bottom are called sludge. The lighter organic materials rise to the surface to form scum. It is believed that the bacteria that thrive in the tank aid in the decomposition of the organic waste materials. A gas is produced which must be discharged to the atmosphere. This can be accomplished by equipping the tank with a vent above the ground or by inserting an inverted inlet pipe in the tank to permit the gas to be expelled through the soil pipe in the roof of the house. Figure 11-21 illustrates the latter method.

Certain solids settle out and do not decompose. These must be removed periodically by pumping or dipping.

The liquid in the tank overflows into the outlet pipe and into the drain field pipe where it is percolated (filtered) into the soil.

If large volumes of water, such as rain from the roof, are run into the tank, solid matter cannot settle and is washed into the drain field pipes. Here the solids cannot be liquefied and purified, and, consequently, they clog the pipes. This reduces the effectiveness of the system and, frequently, necessitates the replacement of the drain field.

The size of the drain field is best ascertained by the local health department. They will make percolation tests to see how well the soil will absorb water. The faster the water is absorbed, the smaller the drain field needs to be.

The Water System

The purpose of the water system is to provide a continuous supply of clean, pure water to the points of use within a building.

The water supply line from the water main to the house should be buried below the frost line, to protect it from freezing. This depth will vary from one section of the country to another.

The diameter of the pipe also will vary according to local conditions. If the water pressure on the city main is low, a large water service pipe must be run from the main to the house. This service pipe should be as short as possible, and turns and offsets should be avoided, since they retard the flow of water and reduce the amount of water flowing to the building. Generally, two piping systems are used within a building — one for the cold water supply and one for the hot water supply.

Cold Water System

Usually, the cold water system consists of a supply main suspended below the first floor. From this main, supply pipes are run to each fixture. The house service pipe should be ¾-inch diameter. The minimum pipe sizes to supply fixtures are indicated in Fig. 11-22.

All riser connections to the water main within a building should be made at a 45° angle to the main, and the riser connections to this branch from the main should be on a 90° angle. See Fig. 11-23.

The water system should be installed on a slight grade and should drain to one low point, if possible. This allows the entire system to be drained by one valve located at the low end.

Valves should be located in the system so that service to various parts of the building can be shut off without turning off the entire system. Frequently, each fixture will have a separate shut-off valve, while other times, one valve may serve a branch and control several fixtures. These valves should be easily accessible. A simple water system for a small house is illustrated in Fig. 11-24.

Fig. 11-22.

Minimum Sizes of Water Supply Pipes

Fixture or Device	Pipe Size (Inches)
Bathtubs	1/2
Drinking fountain	3/8
Dishwasher (domestic)	1/2
Kitchen sink, residential	1/2
Kitchen sink, commercial	3/4
Lavatory	3/8
Laundry	1/2
Shower (single head)	1/2
Urinal (direct flush valve)	3/4
Water Closet (tank type)	3/8
Water closet (flush valve type)	1
Hose bibbs	1/2
Wall hydrant	1/2

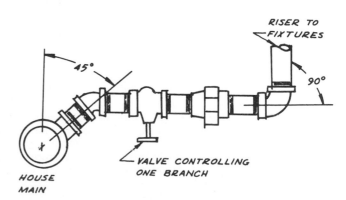

Fig. 11-23. Typical house main, branch, and riser pipe in a water system.

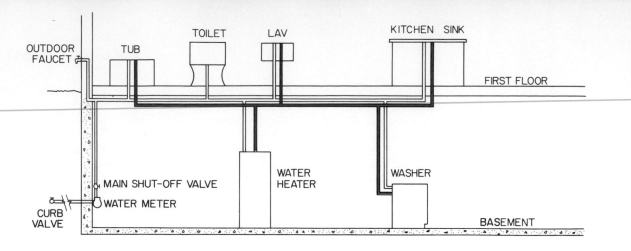

Fig. 11-24. A simple water-piping system.

Hot Water System

The hot water system carries the water from the water heater to the various fixtures. One system of hot water distribution is a single pipe to each fixture as indicated in Fig. 11-24.

In residential building, water is usually heated by gas or electricity. Units heat the water to a predetermined temperature and automatically shut off. When hot water is drained off and cold replaces it, they automatically turn on and again heat water to the desired temperature. All water heaters must have a pressure **relief valve** to prevent explosions. If the heater should malfunction and continue to heat water beyond a safe level, great steam pressure is built up. This is bled off by the relief

Lennox Industries, Inc.

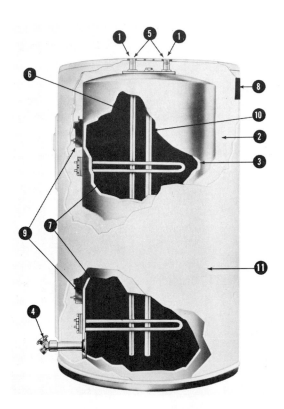

Fig. 11-25. Details of an electric water heater.
1. Hot and cold water connections. 2. Fiber insulation.
3. Glass-lined, heavy-gauge steel tank. 4. Drain valve.
5. Pipe connections threaded with ¾-inch pipe thread.
6. Diffuse tube to stratify incoming cold water at bottom of tank. 7. High-intensity heating units. 8. Outlet box for electrical connections. 9. Automatic thermostats. 10. Magnesium rod to increase tank life in all kinds of water. 11. Metal exterior shell.

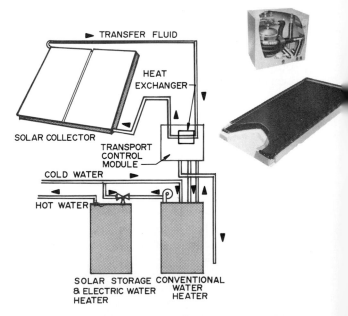

Fig. 11-26. A domestic hot water system utilizing solar collectors as the energy source. A conventional water heater serves as storage and a standby heater.

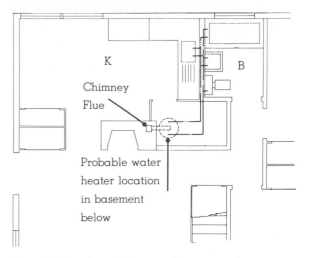

K   B

Chimney
Flue

Probable water
heater location
in basement
below

Fig. 11-27. A good location for a water heater, providing a vent and requiring only a short run to fixtures.

valve before the tank explodes. When this occurs, the user knows that something is wrong with the temperature control on the heater. An electric water heater is illustrated in Fig. 11-25.

Solar systems for heating hot water for domestic use are available. See Fig. 11-26. The storage tank should hold 30 gallons (114 liters) of water for each person who lives in the house. The system shown in Fig. 11-26 uses solar collectors to heat the water before it enters an electric water heater. When there is sufficient solar heat to keep the water at the desired temperature, the heater only serves as a storage tank. When the solar heat is insufficient, the electric heater heats the water.

A complete explanation of solar energy utilization is in Chapter 13.

It is important to locate water heaters as close to the fixtures using hot water as possible. However, gas- and oil-fired units need a chimney to vent the gas from the burning of the fuel. This makes it more difficult to locate these types than it is to locate an electric water heater since the electric heater does not need venting. Refer to Fig. 11-27. This drawing illustrates a satisfactory installation for a water heater requiring venting.

Water pipe can be galvanized steel or iron, or copper. Plastic pipe is approved in some cities. Outside faucets should be of the frost-proof type or should have a shut-off valve so they can be drained in freezing weather.

It is a good practice to insulate both hot and cold water pipes. Insulation on hot pipes retains the heat and thus saves money, while on cold pipes it keeps warm, moist air from causing them to sweat.

Build Your Vocabulary

Following are terms that you should understand and use as part of your working vocabulary. Write a brief explanation of each term.

back vent
branch interval
branch vent
circuit vent
cold water supply main
continuous waste and
  vent
dry vent
dual vent
group vent
house drain
house sewer
indirect waste
leader
loop vent

plumbing fixtures
primary branch
relief valve
relief vent
secondary branch
septic tank
side vent
soil pipe
trap
valves
vent stack
waste pipe
water risers
yoke vent

Class Activities

1. Trace the plumbing system in your home. On graph paper, make a freehand sketch of the floor plan and locate all fixtures, hot and cold water pipes, and sewer lines. Be certain to indicate where the water and sewer lines leave the house.

2. Visit the local building inspector. Write a report giving the major requirements for the installation of a sewage system in the typical residence and small commercial building. Especially note any differences between these two types of buildings.

3. Using wooden dowel rods, construct a three-dimensional scale model of a complete plumbing system for a residence. Be certain to include hot and cold water lines, sewer lines, and vents. Fixtures and water heater can be molded from clay. Color hot water lines red, cold water lines green, and sewer lines yellow.

4. Visit a house under construction when the plumbing is being installed. Notice how the pipes are placed through the walls and under the floor. Observe how the joints are sealed. Check the venting system. See if the pipes used meet minimum size standards.

5. Assume that the fixtures in one of the school rest rooms all flow into one sewer line. Compute the number of fixture units the line would have to carry. Ascertain the correct size sewer pipe, vent pipe, and trap size for this job.

Chapter 12

Heating and Air Conditioning

The all-year air-conditioning system supplies heat when the outside temperature drops, and cool air when the temperature rises. This makes living more healthful (especially for people suffering from hay fever or heart conditions), decreases the amount of housework since air is filtered the year around, and reduces family fatigue and irritation. It also reduces paint peeling, warped doors, and furniture damage by controlling the **humidity** (moisture content) of the air.

If air is too dry (low humidity), it drains moisture from wallpaper, furniture, and woodwork; nose and throat membranes become dry and sore; lips become chapped; colds are frequent and bronchial ailments result. If air is too moist, the result is mildew, musty odors, and warped doors and furniture. Humidity must be controlled all year.

A good all-year air-conditioning system is one that is able to maintain an even, 75°F (24°C) temperature the year around. Such a system should give perfect circulation of air, with no feeling of air movement. During the winter, room temperatures should not vary more than 2°F or 1°C throughout the day. From room to room, the temperature should not vary more than 4°F or 2°C. When the outside temperature is 30°F (-1°C), there should be no more than 3°F or 1.5°C difference in temperature from the floor to sitting level or 4.5°F or 2.5°C from the floor to the ceiling. At 0°F (−18°C), the difference between the floor and sitting level should be no more than 5.5°F or 3°C; between floor and ceiling, a maximum of 8°F or 4.5°C.

Technical publications relating to the design, installation, and maintenance of heating and air-conditioning systems are published by the American Society of Heating, Refrigerating, and Air-Conditioning Engineers.

Placement of Heating and Cooling Units

Heat always flows from a warm object to a cold one. As a house loses heat to the colder outdoors, the human body feels cold because it is radiating its heat toward the cold walls, ceiling, and floors. The heating system must supply heat as fast as it is lost. However, even if a furnace does this, a person can still feel cold. Comfort depends upon how heat is supplied, as well as how much heat is supplied. Older homes often have heat registers placed on inside walls. The warm air rises to the ceiling and gradually descends to the floor as it cools. As the air passes cold windows, it is chilled rapidly and thus becomes heavier and descends faster. This causes a draft and a layer of cold air on the floor. The temperature inside the house may be 70°F (21°C) on the thermometer, yet the occupants feel cold, because the exterior walls (being colder than the interior walls) draw away body heat. See Fig. 12-1.

This situation can be corrected by placing the heat outlets on the outside walls under windows. The heat enters the house on the outside perimeter and is, therefore, called **perimeter heating.** Perimeter heating sends a curtain of warm air up the outside wall; this reduces the loss of body heat to this wall. It also eliminates drafts and enables the heating system to maintain even heat. See Fig. 12-2.

Heat-emitting units should be low in height, but should be as large in area as possible. This is done in order to maintain a low output per square foot and to avoid a concentration of heat.

Fig. 12-1. Heat is drawn to the cold outside wall, is chilled, and settles to the floor.

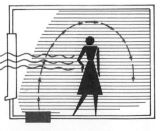

Fig. 12-2. Heat forms a protective curtain over the outside wall.

314

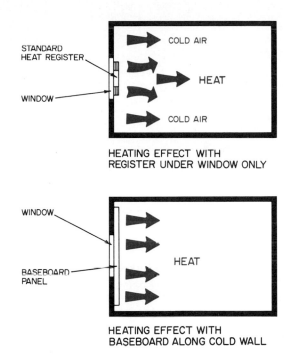

Fig. 12-3. Heating effects with standard heat registers and baseboard panels.

The **standard heat register** distributes heated air from an opening about 18 inches (457 mm) long. Some of the heated air is cooled by the cold walls on each side. See Fig. 12-3. While this unit provides satisfactory heat, it does not provide the ultimate in correct heating effect.

The **baseboard panel** has the advantage of emitting heat along the entire cold wall. If baseboard panels are used, the cold wall and the cold air descending from a window are warmed, providing

Honeywell

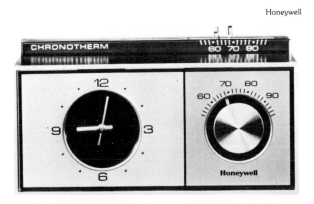

Fig. 12-4. This thermostat regulates the temperature according to the time of day.

better distribution of radiant heat. Figure 12-3 illustrates the difference between the two types of heat emitters.

In the summer, cool air is provided from forced-air cooling units. The cool air flows up the warm outside walls with a protective curtain. This prevents the wall from radiating heat and absorbs heat from people in the house, thus cooling them. The all-year air-conditioning system, then, provides heat in winter and cool air in summer, to the places in the house where these are needed most.

There should be an air return to the furnace from every room except the bathroom and garage. A large hall should also have an air-return register. This aids in the circulation of air in the house and prevents "starving" the system for lack of air. It greatly increases the efficiency of the installation. Air returns located in crawl spaces and attics should be insulated to prevent excess loss of heat or coolness as the air returns. These returns are placed on inside walls in a perimeter system.

Fuel can be saved using a **thermostat** with a timer. See Fig. 12-4. This thermostat holds the temperature at a low temperature, such as 60°F (16°C), until the timer activates it to raise the temperature to a higher preset level, such as 70°F (21°C).

A wide range of devices is available for all-year air conditioning. Devices in common use are forced-air, electric, hot-water, steam, and solar. (Refer to Chapter 13 for information about solar air-conditioning systems.)

Forced-Air Heating Systems

In a forced-air heating system, air is heated by oil, natural gas, LP gas, or electricity. In gas and oil systems, the fuel is burned in the furnace, and the heated air is forced through pipes to the outlets in the house. In electric systems, heating elements are located in the ducts or in a firebox. The blower forces air past the heating elements, thus warming the air.

There are three types of gas- and oil-fired forced-air systems in common use: the upflow, the downflow, and the horizontal furnaces. In the **upflow system,** the furnace is installed in the basement, and the ductwork is placed between or below the floor joists. See Figs. 12-5 and 12-6. The **downflow furnace** forces air into ducts in the floor below it. It is ideal for perimeter heat in houses that have no basements, because the ducts can be placed in a concrete slab or in the crawl space beneath wood floors. See Fig. 12-7.

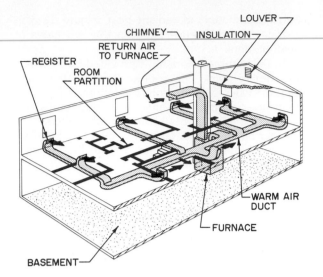

Fig. 12-5. An upflow hot-air heating system.

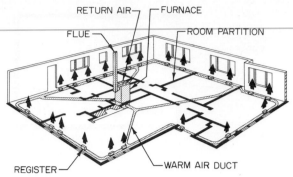

Fig. 12-7. A downflow hot-air heating system.

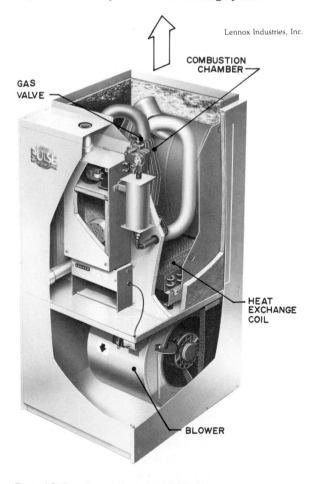

Lennox Industries, Inc.

Fig. 12-6. An upflow gas-fired furnace. Air-conditioning coils are placed on top of this unit.

The **horizontal forced-air furnace** is designed to hang in the crawl space below the floor in a basementless house or in the attic of a house with concrete floors. If installed below the floor, the ducts are hung below the floor joists; if in the attic, the ducts run above the ceiling joists. See Fig. 12-8.

There are two types of duct distribution systems in use for forced-air furnaces in houses with basements or crawl spaces. Both systems originate from the **plenum,** a metal box located at the top, side, or bottom of the furnace. In the radial-distribution system, all the ducts radiate from the plenum to the outlets in the various rooms. In the extended-plenum distribution system, a long duct extends the plenum the length of the house, and smaller ducts lead from it to the various rooms. See Fig. 12-9.

Another plenum system is shown in Fig. 12-10. The entire crawl space is the plenum. Normally, the

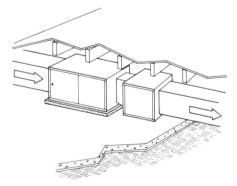

Fig. 12-8. A horizontal furnace can be installed below the floor in the crawl space.

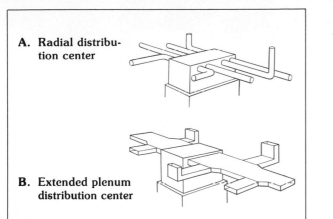

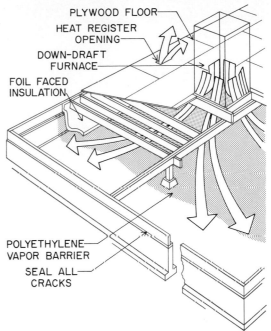

Fig. 12-9. Types of distribution systems.

Fig. 12-10. This hot-air system uses a downflow furnace. The entire crawl space becomes a plenum.

ground level is 12 to 24 inches (305 mm to 610 mm) below the floor. If the joists and plywood subfloor are pressure treated, the ground can be as little as 3 inches (76 mm) below them.

The soil is covered with a 0.6 mil polyethylene film vapor barrier. This vapor barrier is extended up the inside of the foundation wall. The crawl space must be dry. The ground outside the building must be sloped to provide good drainage. A positive drain which connects to a storm sewer or other drain is required in the crawl space. Plumbing can be located in the area of the positive drain. The concrete foundation wall and wood headers are covered with foil-faced insulation having a minimum rating of R-11. It is also extended from the wall out over the edge of the vapor barrier on the ground. The foil faces the inside. Additional insulation is placed between the floor joists.

All possible leaks should be sealed. Cracks between the floor and foundation should be sealed with caulking or sill sealer. The subfloor is glued and nailed to the joists.

The furnace used is a conventional downflow heating and air-conditioning unit. Holes for standard floor heat registers are cut through the plywood floor wherever a heat source is wanted.

The duct distribution system for a house built on a concrete slab can be either a loop perimeter or a radial perimeter. See Fig. 12-11. The ducts are usually transite or fiber that is lined with aluminum foil and wrapped with asphalt-filled duplex kraft paper. They are used for supply and return lines in hot air furnaces. The series of pictures in Fig. 12-12 illustrates the installation of this system.

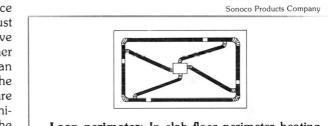

Loop perimeter: In slab floor perimeter heating, the loop system is generally preferred and should be used wherever practical because it produces more uniform temperatures. The duct extends completely around the perimeter of the building, and is supplied by feeder ducts.

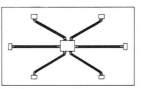

Radial perimeter: In the radial system, air distribution consists of individual ducts running spoke-wise from the plenum chamber of a centrally located furnace to each perimeter warm air outlet.

Fig. 12-11. Types of perimeter distribution systems for concrete slab construction.

Sonoco Products Company

Fig. 12-12. Installation of duct system in slab.

1. The foundation has been poured and the ditch dug for the duct.
2. Gravel fill has been made. The moisture barrier is in place with joints sealed. Heat ducts go on top of vapor barrier.
3. Metal is being fitted to duct to form T-joint.
4. Duct is fitted to plenum chamber.
5. Mitered sections are in place to form elbows and T's.
6. Joints are covered with mastic tape.
7. Registers are fitted to perimeter section of duct.
8. Duct is installed, ready for concrete workers.
9. Reinforcing wire mesh is in place, ready for pouring slab.

Forced-Air Cooling Units

There are many types of cooling units available for forced-air systems. Some furnaces are built with the heating and cooling system in two separate units, while other furnaces have the heating and cooling systems built into a single unit. Both types of cooling systems use the same ducts, blower, and filters as the heating unit.

Electric Cooling Units

When a liquid changes to a gas, it absorbs heat. This principle is utilized by the electric cooling unit. The refrigerant frequently used in electric cooling units is **freon**, which boils at $-22°F$ ($-30°C$). The freon is pumped through coils placed in the furnace, and the blower forces air around these pipes. The freon absorbs large amounts of heat from the air in the furnace, thus cooling the air. The blower forces the cooled air through the ducts into the house. The freon, because it has absorbed heat, becomes a gas and is circulated to the compressor. This unit compresses the gas, thus concentrating the heat and raising the gas to an even higher temperature. The freon is then cooled by being pumped through an outside air unit called a condenser; it is circulated through tubes which are cooled with air or water. This causes the freon to return to a liquid.

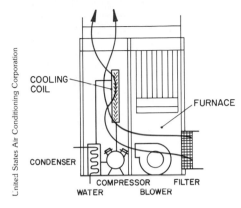

United States Air Conditioning Corporation

Fig. 12-14. Year-round combination furnace and air conditioner.

Year-round combination: This is a compact twin unit consisting of a warm-air furnace and matching summer air conditioner which will cool, dehumidify, heat, filter, and circulate air the year round. Occupying little more space than the average furnace, it can be easily installed in a closet, basement, or utility room, using the same ducts to distribute both warm and cool air.

The liquid freon is circulated back to the furnace cooling coil where it absorbs more heat from the air and repeats the cycle. The only electricity consumed is that needed to run the blower and the compressor. Figure 12-13 illustrates this process for an air-cooled unit.

Water-cooled units must handle large quantities of water. Because of the great expense of water in many areas, they find limited use. Some units salvage some of the water and recirculate it, but the operating cost can still be high where water is scarce.

Air-cooled systems have the cooling coil in the furnace and the air-cooled condenser outside the building. The condenser can be located in the yard or on a rooftop. This system can be used on upflow, downflow, and horizontal-flow furnaces, as was illustrated in Figs. 12-5, 12-7, and 12-8.

The typical heating and air-conditioning system utilizes a furnace that contains the heating and cooling units in a single cabinet. See Fig. 12-14.

The self-contained, electric air-cooling system is frequently used in homes in which the existing heating system cannot be easily converted to cooling. The self-contained system is also useful in homes with hot water or steam heat. The self-contained air conditioner contains all elements — refrigerator circuit, evaporator, compressor, and condenser — in one sealed cabinet.

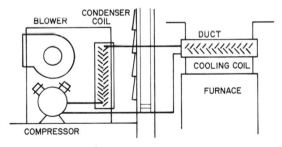

Fig. 12-13. Air-cooled systems.

For areas where water is scarce or expensive and where water disposal is a problem, an "add-on" is available which uses outside air rather than water to cool the refrigerant in the condenser. This unit is also perfectly suited for homes where indoor space is very limited, because the refrigeration section can be located in almost any area where there is sufficient air circulation, indoors or out. Copper tubing carries the refrigerant from the refrigeration section to the housed cooling coil, usually installed on the outlet side of the air supply system.

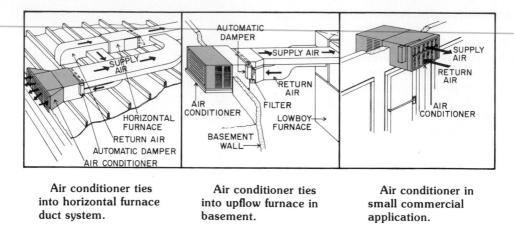

Air conditioner ties into horizontal furnace duct system.

Air conditioner ties into upflow furnace in basement.

Air conditioner in small commercial application.

Fig. 12-15.   Installation of self-contained air conditioner.

This type of air conditioner can be installed in the attic, with its own ducts leading to ceiling outlets. Or, it can be connected to the duct system of a horizontal furnace installed in the attic. It can be placed in a basement, crawl space, utility room, closet, or attached garage. It can be installed through the exterior wall or can be located entirely outside the building on a roof, in a carport, or on the ground. In these positions, it can tie into the existing duct system, or it can have a system of its own. In small commercial buildings, it can be installed through the wall and cool a large, open area. Several applications of this system as it ties into existing systems are shown in Fig. 12-15.

The electric, air-cooled air conditioner is also useful in cooling buildings having wet heat (hot water or steam). Such a building can be cooled in the summer by installing a chiller unit to send cold water through the pipe system or by installing a separate, forced-air cooling system using ducts and an electric air-conditioning unit.

A room air conditioner can be installed in a window or through the wall. This enables a house to be easily zoned for cooling, permitting some sections to be cooler than others or not cooled at all. Only electrical service to the unit is required. Since the unit blows air directly into the room, no ductwork is necessary. This type of installation is illustrated in Fig. 12-16. It is frequently found in motels and hotels, where room occupancy is irregular. Units are available that can heat in the winter as well as cool in the summer. These units can be either small heat pumps or an electric air conditioner with built-in resistance-heating elements.

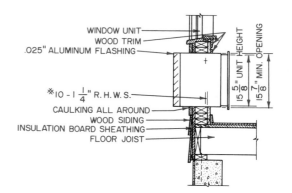

Fig. 12-16.   A single-room air conditioner installed in a frame wall.

Electric Heat

Electric heat is quiet, convenient, clean, and safe. It provides even heat, requires little maintenance, saves floor space, and has a low installation cost. It enables a building to have **zoned** heat. Each room can be controlled by a separate thermostat, thus offering complete flexibility.

An important factor in electric heat is proper insulation of the building. Engineers recommend that the ceiling have an R-value of 38, walls R-19, and floors R-19 or better. (Refer to Chapter 13 for information on determining R-values.) Doors and windows should be weatherstripped. Storm windows and doors are essential.

Before the decision to use electric heat is made, the cost of electricity in the area must be ascertained. Many power companies offer reduced rates to those using this type of heat.

Many kinds of electric heating systems are available. Those in common use are recessed heaters, baseboard panels, concealed cables, perimeter convectors, central furnaces, and heat pumps.

Types of Electric Heating Systems

Recessed electric wall heaters are available in either the radiant or convection systems. The radiant heater delivers a high percentage of its heat in the form of direct radiation to a limited area. A bathroom heater is an example of this type. A convection-type heater is shown in Fig. 12-17. This type distributes heat to a larger area, without any appreciable radiant heat. It usually has a small fan which assists in distributing the heat but does not increase the amount of heat generated. In Fig. 12-18, the difference between radiated heat and convected heat is illustrated.

Fig. 12-17. A convection-type recessed heater.

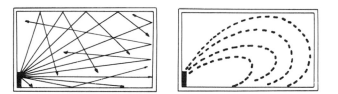

Fig. 12-18. Radiated heat (left) and convected heat.
Radiated heat: All surfaces and objects in the rooms are reached and warmed by radiant heat rays. Some are absorbed and some reflected to other cooler surfaces.

Convected heat: Convected heat is also provided by all radiant-heat distributors. Air coming in contact with the radiant surfaces is warmed and then circulates about the room with natural, gentle motion, without chilling drafts.

The electric baseboard panels use both radiant and convection principles. Heated air is gently circulated by convection throughout the room, thus assuring even heat distribution. Heat is also radiated in sufficient amount to allow body comfort, while the overall room temperature remains at a healthy level. A typical electric baseboard heater is shown in Fig. 12-19. See also Fig. 12-20.

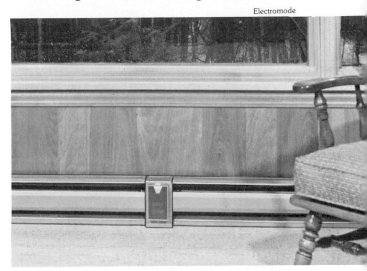

Fig. 12-19. An electric baseboard heater.

Fig. 12-20. A section through a typical baseboard convector.

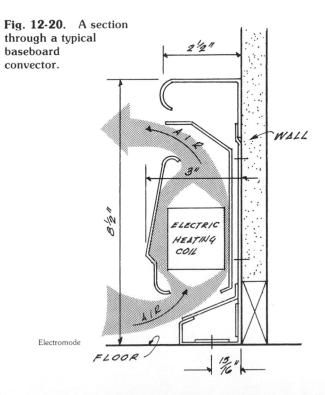

Electromode

Fig. 12-21. The electric heating cable is fastened to the ceiling and covered with plaster.

Electromode

Fig. 12-22. A typical floor-type perimeter convector.

A third method of heating with electricity is to use concealed cables in the ceiling, walls, or floor. The heating cables are buried in the plaster ceiling or wall or in the concrete slab floor. This system tends to have slow heating response, because it must heat the wall or floor before it can heat the room. The ceiling heating cable should have at least a $^3/_8$-inch (10 mm) plaster coating over it. In concrete floors, the cable should have 4 inches (102 mm) of insulating concrete below and $1^1/_2$ inches (38 mm) of noninsulating concrete above. The recessed heater or baseboard heater tends to be more effective and efficient. Figure 12-21 illustrates the installation of this cable. Also available are heating units suitable for installation in sidewalks and driveways to melt ice and snow.

Another effective means of electric heat is the floor-type perimeter convector. This type of heater is recessed in the floor, with a grill flush with the floor. Each unit is a separate heater. It can be located wherever a heat source is needed, such as under a picture window. In Figs. 12-22 and 12-23, such a unit in a typical installation is shown.

Central electric furnaces are available which operate exactly like gas- or oil-fired furnaces, except that the source of heat is from resistance-type electric elements. The heat is distributed by a blower through ducts.

Many other types of electric heaters are available. Unit heaters, such as those shown in Fig. 12-24, are used in industrial applications.

Infrared heaters are available for spot heating, indoors and outdoors. This heater produces a heat similar to that of the sun and warms the person who steps under its rays. Industrial applications can be found at drive-in bank windows, at bus stops, and at display windows of stores. Residentially, these heaters are used in baths, on patios, and near swimming pools. An infrared heater is shown in Fig. 12-25.

Another means of heating or cooling a building is to use a heat pump. See Figs. 12-26 and 12-27. The most commonly used heat pumps employ either air or water as their source of heat. This device literally pumps heat out of the air or water. Basically, it operates on much the same principle as

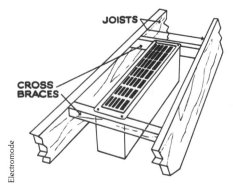

Electromode

Fig. 12-23. Installation of floor-type, individual-unit perimeter convectors.

a refrigerator. In the summer, the heat pump removes heat from inside the building and discharges it outside. In the winter, this cycle is reversed. Heat is removed from the outside air and discharged inside the building. Small heat pumps actually reverse their cycle, while large commercial units follow a slightly different system of reversing. The small heat pump is a single machine that uses the same components to provide heat or cooling. It

Lennox Industries, Inc.

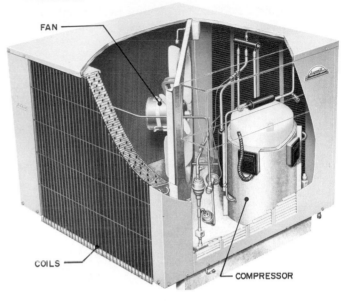

Fig. 12-26. The outdoor unit of a heat pump.

Electromode

Fig. 12-24. Industrial application of electric unit heaters.

Fig. 12-25. Infrared electric heaters on a patio.

Electromode

Fig. 12-27. A large heat pump for commercial construction. This unit provides 140 tons of cooling capacity in summer, and in winter will transfer heat generated by people, lights, and business machines in inner office spaces to offices on outside walls needing heat.

requires no flue. Single-unit heat pumps are installed outside of buildings as shown in Fig. 12-28. Two-unit heat pumps have the compressor unit outside and cooling unit inside. Figures 12-29 and 13-30 illustrate several applications of this type.

Water provides a greater amount of heat than air, in proportion to the electrical energy put into the system. However, water is becoming less satisfactory in quantity and quality. In many areas, its scarcity makes use in large quantities prohibitive. Water also creates scale and causes corrosion. Another problem is created when the large quantities of used water must be discharged.

The earth is also used as a source of heat for small heat pumps. Coils buried below the frost line extract heat from the earth. The initial cost is high, and good performance depends upon the quantity of moisture in the soil.

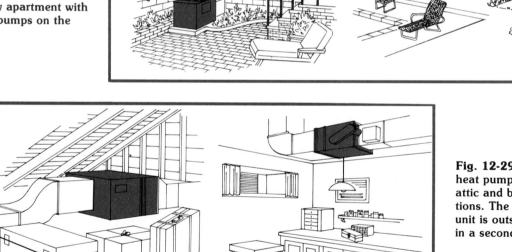

Fig. 12-28. A single-unit heat pump can be installed outside a home on the ground or on the roof. This illustration shows a four-family apartment with heat pumps on the roof.

Fig. 12-29. Two-unit heat pumps installed in attic and basement locations. The compressor unit is outside the house in a second cabinet.

Fig. 12-30. A two-unit heat pump in commercial installation. The ceiling-hung unit uses no floor space. Floor-mounted units provide zone control without ducts. A compressor unit is located outside, usually on the roof.

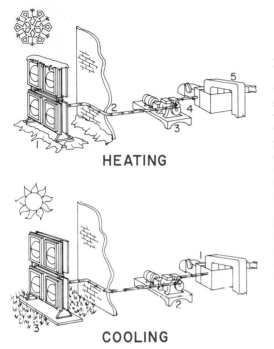

HEATING

COOLING

Fig. 12-31. Air-to-air heat pump operation.

Heating cycle operation: When passed through an outdoor unit (1), outside air gives up low-temperature heat to the system's refrigerant. Through a line (2), the refrigerant carries this low-temperature heat to a compressor (3) which increases the heat to a desirable high-temperature level. From the compressor, the refrigerant moves this high-temperature heat to a coil in an indoor unit (4). Here the indoor unit moves air for conditioning through the coil, and heat from the refrigerant is transferred to the air. Ductwork (5) conducts the heated inside air to the conditioned areas where it is desired.

Cooling cycle operation: Circulating inside air carries heat from conditioned areas to the indoor unit (1) which transfers this heat to the system's refrigerant. Carried to the compressor (2) by the refrigerant, the heat is elevated to a high-temperature level. The refrigerant then moves this high-temperature heat to the outdoor unit (3) where it is discharged into outside air.

Fig. 12-32. Air-to-water heat pump operation.

Heating cycle operation: Low-temperature heat from outside air is transferred to the system's refrigerant by an outdoor unit (1). The refrigerant carries this low-temperature heat to a compressor (2) which elevates the heat to a desirable high temperature. From the compressor, this high-temperature heat is moved by refrigerant to a condenser-chiller unit (3) in which the refrigerant gives up the heat to circulating water. Through the building's pipes (4), the circulating water carries this heat to room units (5) or central-station indoor units, which take the heat from circulating water and transfer it to the air in the building for comfort.

Cooling cycle operation: Air in the conditioned spaces gives up heat to room units (1) which transfer it to recirculating water. Through piping (2), the water carries this heat to the condenser-chiller (3) where the heat is transferred from the water to the system's refrigerant. Moved on to the compressor (4) by refrigerant, the heat is elevated to a high-temperature level. From the compressor, the refrigerant carries this high-temperature heat to the outdoor unit (5) which releases it to outside air.

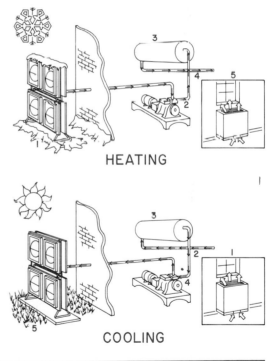

HEATING

COOLING

There is also a fourth source of heat: In areas of some large buildings, heat is produced by certain activities, such as the operation of heavy machinery. A heat pump can remove the excess heat and pump it to parts of the building in which heat is needed. See Fig. 12-27.

Heat pumps for residences and small commercial buildings usually supply heated or cooled air to the building by blowing it through a duct system. Larger commercial units, of the water-to-water or air-to-water type, send heated or chilled water to indoor room units, which then heat or cool the air in the room. Two systems are shown in Figs. 12-31 and 12-32.

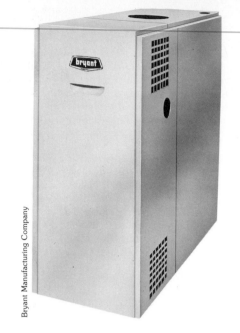

Bryant Manufacturing Company

Fig. 12-33. Residential boiler.

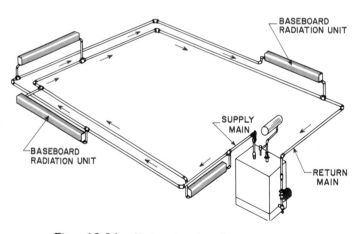

Fig. 12-34. Hot-water heating system, reverse-return type.

Hot-Water Heat

The typical hot-water heating system consists of a boiler, a burner (usually gas- or oil-fired), a flue to carry away waste fumes from the burner, a pump, a system of pipes to carry the hot water to the various rooms, and heat-emitting devices such as radiators. Also, a large number of control valves are used.

The typical boiler is very small and is housed in an attractive shell. It can be exposed to view and

not detract from the room in which it is located. See Fig. 12-33. The boiler can be installed in a basement, a utility room, or a garage.

The types of heat-emitting devices used in hot-water heating systems include radiators, convectors, unit heaters, baseboards, and panels. These units transfer heat from the circulating water to the space to be heated by radiation and convection. The design of the heat-emitting unit determines how much heat is supplied each way.

Radiators, radiant baseboards, and panels supply a major proportion of their heat by radiation; while convectors, unit heaters, and finned-tube baseboards supply their greatest amount by convection.

Hot-water systems have the advantage of requiring smaller distribution mains and riser pipe sizes than atmospheric-pressure gravity systems. They also permit greater flexibility in piping design. Better heat control is provided, since the system can vary the average water temperature to meet changes in outdoor temperature. This control holds fuel consumption to a minimum, since overheating of the water is prevented.

A hot-water system has two pipes. One supplies hot water to each radiation unit, which then discharges the water passed through it into a second pipe. This pipe returns the cooled water to the boiler. See Fig. 12-34.

Tall buildings, long buildings, and groups of buildings are more easily heated with the forced-circulation system. If the proper pressure-temperature relationship is maintained in the system, water design temperatures approaching 400°F (204°C) can be used in the transmission circuits, and high temperatures, usually associated with steam, can be employed in the space-heating circuits. The temperature of the forced-circulation hot-water system can be modulated between values of 80°F (27°C) and 400°F (204°C).

Types of Heat Emitters

The typical small-tube **radiator** is made of cast iron. Generally, the small-tube radiator is installed in a wall recess.

Radiant baseboards are long, low units resembling the usual wood baseboards. See Fig. 12-35. They are installed along outside walls in place of wood baseboards. Radiant baseboards are finned tubing installed behind a metal enclosure.

Widely used in industrial construction are wall-hung convectors, consisting of a finned heat emitter covered with a sheet-metal shield. The fins provide

Fig. 12-37. Forced hot-water unit heaters.

Fig. 12-35. Hot-water baseboard convectors.

Edwards Engineering Corporation

Fig. 12-36. Finned-type wall convectors.

Young Radiator Company

Fig. 12-38. Freestanding, convector hot-water radiator, also available in semirecessed, recessed, wall-hung, and long, low units.

Fig. 12-39. Hot-water heating panel formed of 3/8″ copper tubing.

additional surface area for the release of heat to the air passing over the pipe. They may be recessed or surface mounted. See Fig. 12-36.

Unit heaters, such as that shown in Fig. 12-37 are also widely used in industrial heating. The hot water is forced through a coil in the unit, and an electric fan forces air through the coil, thus creating air circulation in the room and heating it by convection.

A **convector** is a space-heating device composed of a sheet-metal casing, with an outlet grill and a fin-tube heating element. The casing usually contains a damper to control air circulation. The air enters the casing near the floor line below the heating element, is heated in passing through the element, and is delivered to the room through the outlet grill located near the top of the enclosure. Convectors usually have electric fans to increase air flow. See Fig. 12-38.

Radiant panels are built either by securing copper tubing to the ceiling and plastering over it or by burying tubing in the concrete floor slab. Shown in Fig. 12-39 is the installation for a ceiling before plaster was applied, and a floor heating panel is

327

Fig. 12-40. Hot-water floor heating panel.

shown in Fig. 12-40. These panels heat by radiant heat. While there are a number of ways to install these panels in the ceiling and floor, Fig. 12-41 illustrates sound and proven installations.

Hot-water systems can provide zoned heating by regulating the flow of hot water to each room. Each room has a thermostat that controls a flow valve at the boiler. This controls the flow of hot water to each room. A zoned heating system is illustrated in Fig. 12-42.

Copper coils for circulating hot water can be buried in driveways and sidewalks to melt ice and snow.

Cold-Water Air-Cooling Systems

Some water chillers are designed to work in combination with a forced hot-water heating system. However, they can be installed as a separate, space-cooling system if another type of heat is in use. The chiller is a unit consisting of a compressor, condenser, and evaporator tank. Basically, this unit cools and circulates chilled water to the same units that emitted heat from hot water. There the cold is emitted into the air, and the water returns to the cooling unit to be rechilled and recirculated.

Another way to air condition with a chilled-water system is to use an absorptive-type water chiller. Refrigeration is produced by the absorptive principle, utilizing steam as the energy source. The cooling unit is a sealed refrigeration package; the unit operates under a vacuum at all times.

There are many systems designed for producing chilled water for cooling purposes. Figure 12-43 illustrates one type of water chiller. This type of unit cools and does not heat. Such a chiller can be used

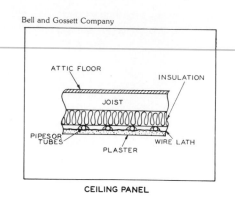

CEILING PANEL

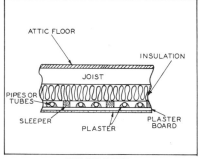

CEILING PANEL

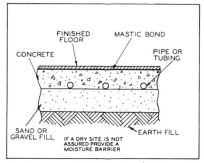

FLOOR SLAB ON GROUND

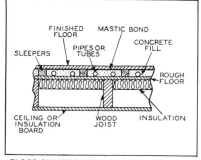

FLOOR ON WOOD JOIST, AIR SPACE BELOW

Fig. 12-41. Typical installations for floor and ceiling hot-water panels.

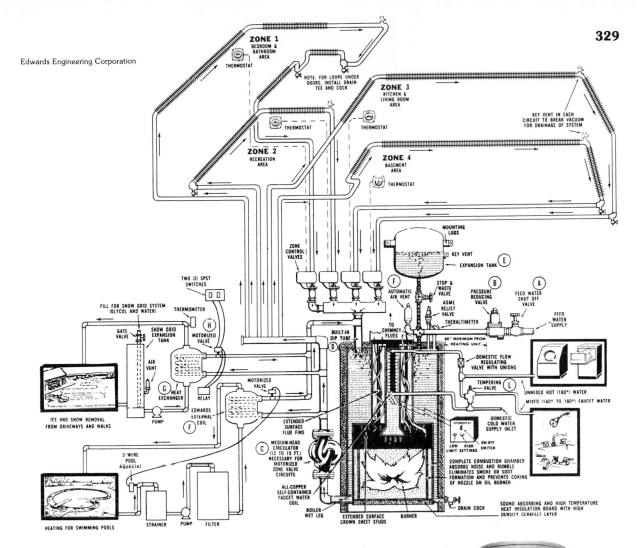

Edwards Engineering Corporation

Fig. 12-42. Zone-controlled hot-water heating system.

When a zone thermostat calls for heat, its zone control-valve motor begins to run, opening the zone control valve slowly. When the valve is fully opened, the motor stops. The burner, either gas- or oil-fired, will now light and burn until the water in the boiler reaches the high limit (usually 220°F).

When the high limit is reached, the burner shuts off, but the water circulation to the zone calling for heat continues until the zone reaches the temperature called for by the thermostat. If the zone removes enough heat to lower the water temperature to a preset low temperature (usually 180°F), the burner ignites and heats the water up again. When the zone reaches the desired temperature, as called for by the thermostat, the zone control-valve motor closes the valve, thus stopping the circulation of hot water to that area.

The boiler will keep the water between 180°F and 220°F, even if the various zones do not call for heat.

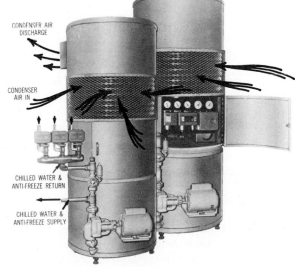

Fig. 12-43. A zone-controlled water-chiller unit with zone valves. Unit can be installed indoors or outdoors.

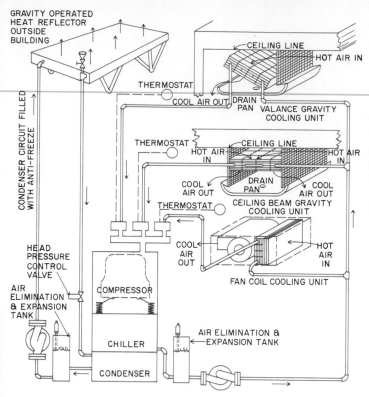

Fig. 12-44. Zone-controlled, chilled-water cooling system.

The heat removed from the space by the refrigerant is rejected to the outside air through a gravity heat rejector located on a roof, under an overhang, or other inconspicuous location. The gravity cooling units are hung from the ceiling. The hot air in the space rises, is cooled by the coils, and descends, causing a gentle circulation of air. Each space can have a separate thermostat. Each thermostat controls a valve that regulates the flow of chilled water to that space, thus providing zoned cooling.

for cooling commercial buildings; it can also provide chilled water for industrial applications, such as the manufacture of plastics.

A cold-water space-cooling system that is not a part of a hot-water heating system, but is installed as a separate system, utilizes two types of units to emit the cold into the space. It uses cooling coils with air pulled through by fans and cooling coils with air pulled through by gravity. In Fig. 12-44, a typical system with both types of units is illustrated.

Steam Heat

Steam heat is generally used for large buildings. The system operates in much the same way as the hot-water system. The fuels commonly used are gas, oil, and coal. The design of such systems requires the services of a heating specialist. In considering a steam system, it is important to select the type of

fuel carefully. Considerations that are among the most important for commercial installations are:

1. The comparative cost of gas, oil, and coal heating equipment,
2. Cost of labor and supervision to operate the system,
3. Fuel cost, based on heating value in BTU's,
4. Maintenance,
5. Efficiency of equipment,
6. Cost of fuel storage and handling,
7. Insurance, and
8. Availability of fuel.

Steam heat can provide thermostatic control of heat in each room by using individual room heaters. An explanation of how this steam heat unit works is given in Fig. 12-45.

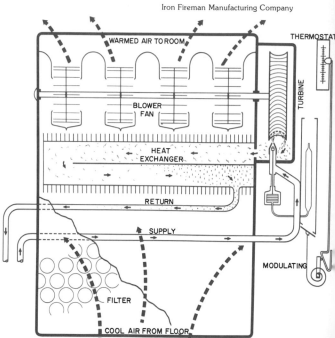

Iron Fireman Manufacturing Company

Fig. 12-45. A steam heat unit with individual thermostatic control.

The steam enters the supply line and reaches the valve controlled by the thermostat. When the thermostat calls for heat, the valve opens, permitting steam to rush into the turbine. The steam turns the turbine which activates the blower. The steam passes into the heat exchanger and heats it. The blower pulls cool air from the floor over the heat exchanger, warming it, and blows it out the grille on the top of the unit. As the steam cools, it condenses and returns to the boiler through the return line.

Fig. 12-46. A fire and smoke detector.

Square D Company

Fire and Smoke Detectors

In many areas, fire and smoke detectors are mandatory in new construction. The fire and smoke detector senses combustion at the earliest stage of the fire, often before smoke becomes visible or flames break out. Plug-in and battery-operated units are available, as well as permanently wired installations. See Fig. 12-46.

Units should be located in the sleeping area. In most houses, one detector in the hall is adequate. Others should be located at the top of each stair. Place one each in the areas near the furnace, utility room, and garage. It is necessary to make a study of the house to determine where fires are most likely to start.

Heat Loss and Heat Gain Calculations

Heat loss and heat gain calculations are used to determine the proper size of heating and cooling units. Following are terms used in connection with these calculations.

Heat Flow and Heat Transfer — The movement of heat from one substance to another. Heat flows to the material that has the lower temperature.

BTUH or British Thermal Units per Hour — The unit of heat flow, used to express heat loss or heat gain.

HTF or Heat Transfer Factor — A value used to express heat flow (BTUH) through a square foot or linear foot of material at a given design condition.

R-Value — A number which indicates a material's resistance to heat flow. The larger the number, the greater the resistance to heat flow.

Infiltration — The flow of heat with the air into or out of a structure, as through cracks around doors and windows.

Heat Loss — The amount of heat flow for BTUH from the inside of a structure to the outside at a given design temperature.

Sensible Heat Gain — The calculated heat gain of a structure, in BTUH.

Latent Heat Gain — Heat gain introduced by moisture in the air. This moisture must be removed by the cooling system. The latent heat gain is commonly calculated as 30 percent of the sensible heat gain.

Duct Heat Loss or Duct Heat Gain — The heat flow that occurs when the air surrounding ducts is a different temperature from that of the air traveling inside ducts.

Design Temperature — The temperature that is assumed when designing a heating or cooling system and when calculating heat loss or heat gain. The inside design temperature is the temperature to be maintained inside the building. The outside design temperature is the assumed highest or lowest average temperature of outside air for a given geographic location.

Design Temperature Difference — The difference between the inside design temperature and the outside design temperature.

Design Heat Loss — The heat loss of a structure when design conditions are met.

Design Heat Gain — The heat gain of a structure when design conditions are met.

Unit Heat Loss — An expression of the total BTUH heat loss of a structure in relation to its square foot floor area. The lower the unit heat loss, the more energy-efficient the structure.

Calculating Heat Loss and Heat Gain

The total heat loss and heat gain will be the sum of the losses or gains of each of the components of the building. These include exposed walls, windows, doors, ceilings, and floors. Following is the procedure for determining heat loss or gain. The data should be recorded on a form such as that shown in Fig. 12-47.

1. Determine the outside design temperature for your area. The local electric power company can usually give this information.

Fig. 12-47.

Heat Loss and Heat Gain Calculation Form

DESIGN TEMPERATURES

Winter: outdoor _10°_ indoor _70°_ temperature differential _60°_
Summer: outdoor _95°_ indoor _75°_ temperature differential _20°_
Floor area of building: _1800_ square feet

SECTIONS OF BUILDING		AREA[1]	HEATING HTF	HEATING LOSS	COOLING HTF	COOLING GAIN
GLASS North (1) Single Pane and		85	37	3145	22	1870
(2) Storm Sash						
South (1) Single Pane and		65	37	2405	34	2210
(2) Storm Sash						
East & (1)						
West (2)						
DOORS (1) Hollow Core and		84	39	3276	15	1260
(2) Storm Door						
WALLS (1) Frame R-11		1440	5	7200	2	2880
(2)						
INFIL-TRATION[2] Windows Storm Sash		220	21	4620	10	2200
Doors Storm Doors		80	21	1680	7	560
CEILING (1) R-19		1800	3.7	6660	2.2	3960
(2)						
FLOOR (1) Double Wood R-19		1800	3.1	5580	.8	1440
(2)						
PEOPLE & APPLIANCE GAIN						3000
COOLING SUBTOTAL						19,380
LATENT GAIN 30% OF SUBTOTAL						5814
SUBTOTAL				34566		25194
DUCT LOSS & GAIN				5185		3779
TOTAL HEAT LOSS & GAIN IN BTUH				39751		28973

[1]Square Feet
[2]Lineal Feet

UNIT HEAT LOSS _22.0_ BTUH/SQ FT

Mississippi Power and Light Company

2. Decide on the desired inside design temperature. Commonly used temperatures are 70°F (21°C) in the winter and 75°F (24°C) in the summer.
3. Calculate the design temperature difference. This is the difference between the outside and the inside design temperatures.
4. Calculate the area in square feet of the windows on each wall. Note the direction each window faces.
5. Calculate the area in square feet of exterior doors.
6. Calculate the crackage in lineal feet for the doors and windows. Crackage is the width plus length of each unit. These measures are used to figure infiltration.
7. Calculate in square feet the net exposed wall area. This is the inside length of exposed walls multiplied by their inside height. Subtract the area of doors and windows.
8. Calculate in square feet the area of the ceiling.
9. Calculate in square feet the area of the floor for houses not built on a concrete slab. For slab floors, use the length of the perimeter of the house.
10. Calculate the heat loss and heat gain for each component by multiplying the area by the HTF factor. The value calculated is BTUH. Selected HTF factors are given in Fig. 12-48.
11. Calculate the heat gain caused by people and appliances. For a typical house, figure a minimum of six people, each generating 300 BTUH.

HEAT TRANSFER FACTORS (HTF)
(HTFs are in BTUH/Sq Ft or BTUH/Lin Ft)

BUILDING COMPONENTS	HEAT LOSS		HEAT GAIN	
	Design Temperature Differential (TD)			
	60°	55°	20°	
GLASS - Weather stripped. Multiply cooling factor by .85 if drapes will be closed. Multiply heating factor by .9 if wood windows are used.			Unshaded	Shaded
NORTH				
Single Pane	68	62	31	27
Single Pane & Storm Sash	37	34	24	20
Double Pane with ¼ " air gap	43	40	24	20
SOUTH				
Single Pane	68	62	48	37
Single Pane & Storm Sash	37	34	38	29
Double Pane with ¼ " air gap	43	40	38	29
EAST & WEST				
Single Pane	68	62	89	66
Single Pane & Storm Sash	37	34	73	55
Double Pane with ¼ " air gap	43	40	73	55
DOORS				
Hollow Core	56	48	22	
Hollow Core & Storm Door	39	36	15	
Solid Wood	26	24	10	
Solid Wood & Storm Door	18	16	6.8	
Metal with Urethane Core	4.5	4.1	1.7	
WALLS				
Frame — No Insulation	13	12	5.1	
Brick — No Insulation	14	13	5.3	
Frame or Brick — 2" of Ins. (R-7)	6.5	5.8	2.5	
Frame or Brick — 3½" Ins. (R-11)	5	4.6	2	
Frame or Brick — R-5 Sht. & R-11 Ins.	3.7	3.4	1.5	
Frame or Brick — R-5 Sht. & R-13 Ins.	3.5	3.2	1.4	
Frame or Brick — 6" Ins. (R-19)	3.2	2.9	1.3	
INFILTRATION (Doors & Windows)				
Without Storm Sash or Door	30	28	10	
With Storm Sash or Door	21	19	7	
CEILING - Dark roof. Multiply cooling by .85 for light roof.			Nat. Vent.	Power Vent.
No Insulation	22	20	8.4	4.2
4" Batt or R-13 or equivalent	4.6	4.2	2.6	2
6" Batt or R-19 or equivalent	3.7	3.4	2.2	1.7
12" Blown Fiberglass R-26 or equiv.	2.1	2.0	1.4	1.2
2 — 6" Batts or R-38 or equiv.	1.6	1.5	1.1	1
FLOOR				
Slab with No Insulation	49	45	—	
Slab with 1" of Styrofoam	31	31	—	
Slab with 1½ " of Urethane	21	21	—	
Double Wood Floor — No Insulation	15	13	3.7	
Double Wood Floor — 2" Ins. (R-7)	5.6	5.2	1.4	
Double Wood Floor — 3½" Ins. (R-11)	4.3	4	1	
Double Wood Floor — 6" Ins. (R-19)	3.1	2.9	.8	

American Society of Heating, Refrigerating, and Air Conditioning Engineers

Fig. 12-48. Heat loss and heat gain factors.

Appliances in a typical home are usually figured together at 1200 BTUH. Record this on the cooling heat gain column.

12. Add up the heat gain and heat loss subtotals. Add 30 percent of the total gain to the heat gain in the cooling column for latent gain.

13. Calculate duct heat loss and heat gain. For normal construction with insulated ducts, figure 15 percent of the subtotal for heat loss and heat gain. If ducts are located inside the area to be heated and cooled, figure a 3 percent loss and gain. Add these to the subtotals to find the total BTUH heat loss and heat gain values.

14. The Unit Heat Loss Factor is now found by dividing the total BTUH heat loss by the floor area.

Fireplaces

A fireplace adds much to the attractiveness of a room and to the joy of living in a house. Besides providing the pleasure of an open fire, a fireplace can serve as a source of heat when normal means of heating fail.

The design of fireplace elevations has changed with the times. Many materials are in common use for facing the elevation, such as brick, marble, slate, stone, tile, and glass. Wood trim and paneling usually complete the elevation design.

The styling of the elevation should be in keeping with the house type. The appearance of a fireplace designed for a formal living room should be entirely different from the appearance of one designed for a family room. See Fig. 12-49.

The size of the fireplace opening should be in proportion to the room in which it is to be located. A large fireplace in a small family room loses much of its appeal, because it is out-of-scale with its surroundings. A large fireplace, if it is to project heat properly, must have a large fire. If too large a fireplace is built, it will produce too much heat for comfort in an average room. If a small fire is built in a large fireplace, the heat goes up the chimney.

Another consideration in planning a fireplace is its location. An attempt should be made to create an area of comfort and repose in the room, subject to the least disturbance from those moving about. Consideration should be given to the location of comfortable chairs, books, and other fireside comforts. Such roominess is most likely to be found at the side or end of a room. An end position is very

Fig. 12-49.   This fireplace is an example of excellent design.

likely to be preferable because of its seclusion. However, a side-wall position may allow more room for furniture to be grouped about the fireplace. A fireplace should not have doorways on either side, since this detracts from the sense of security and repose necessary for full enjoyment. Windows located very near a fireplace also intrude upon the atmosphere.

The fireplace may be built flush with the interior wall. Or, it may project into the room, allowing a mantel to be installed or allowing bookshelves and cabinets to be built on either side of the fireplace.

It is considered best to locate a fireplace on an inside wall. This location reduces heat loss to the outside and keeps the chimney warm, thus helping the fireplace to draw better.

Fireplace Design

Through experience, the proportions necessary to enable a fireplace to draw and burn properly have been established. If a fireplace is to be built conventionally (built brick-by-brick on the job), it is important that a proven design be followed.

Designing the Conventional Fireplace

Architectural details for a proven, single-opening, conventional design for a fireplace are illustrated in

Fig. 12-50. The dimensions for fireplaces of different sizes based on this design are given in Fig. 12-51.

The total thickness of the firebox wall, if it is lined with at least 2 inches (51 mm) of firebrick, should be a minimum of 8 inches (203 mm).

The hearth should extend at least 16 inches (406 mm) in front of the fireplace opening and at least 8 inches (203 mm) on each side of the opening.

Combustible material should not be placed within 3½ inches (89 mm) of the edges of the fireplace

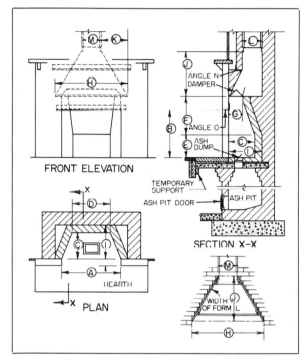

Fig. 12-50. A proven design for a single-opening, conventionally built fireplace. The letters refer to the dimensions given in Fig. 12-51.

opening. A distance of 8 inches (203 mm) is strongly recommended.

An ash dump can be built so that ashes can be cleaned out from a door in the basement, rather than from the hearth in the house. The dump is a metal door built in the floor of the firebox. This door pivots, permitting the ashes to drop inside the foundation walls of the fireplace to a cleanout door below in the basement. Figure 12-50 illustrates this convenience.

The generally accepted method of constructing the conventional fireplace is to complete the rough brickwork of the fireplace from the footings to the chimney top before undertaking the installation of the finished interior and front. See Fig. 12-52.

The entire hearth, including that portion outside the fire area, should be supported by the chimney footings and foundation. Notice in Fig. 12-50 that the hearth cantilevers over to meet the wood floor. This requires that steel reinforcing bars be placed in the 3½-inch (89 mm) concrete slab under the exposed hearth material. The floor joist does not support the end of the hearth.

Fig. 12-51.
Dimensions for Conventionally Built, Single-Opening Fireplace

A	B	C	D	E	F	G	H	I	J	K	L × M
24	24	16	11	14	18	8¾	32	20	19	10	8 × 12
28	24	16	15	14	18	8¾	36	20	21	12	8 × 12
30	29	16	17	14	23	8¾	38	20	24	13	12 × 12
32	29	16	19	14	23	8¾	40	20	24	14	12 × 12
36	29	16	23	14	23	8¾	44	20	27	16	12 × 12
40	29	16	27	14	23	8¾	48	20	29	16	12 × 16
42	32	16	29	14	26	8¾	50	20	32	17	16 × 16
48	32	18	33	14	26	8¾	56	22	37	20	16 × 16

Donley Brothers Company

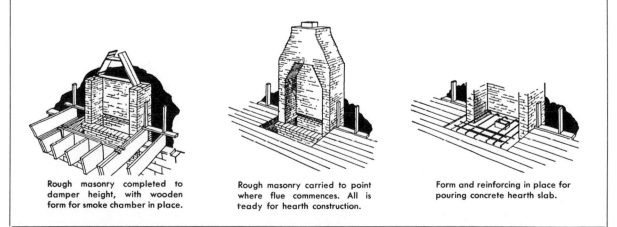

Rough masonry completed to damper height, with wooden form for smoke chamber in place.

Rough masonry carried to point where flue commences. All is ready for hearth construction.

Form and reinforcing in place for pouring concrete hearth slab.

Donley Brothers Company

Fig. 12-52.   Conventional fireplace construction.

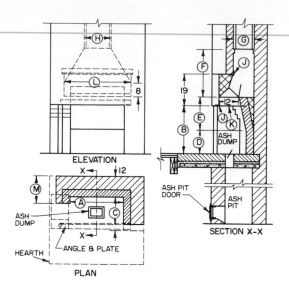

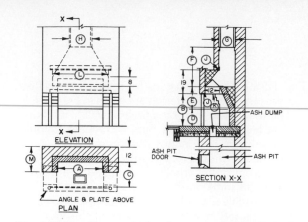

Fig. 12-55. Proven design for a conventionally built, three-way fireplace. Angles are indicated by **J**, and **K** indicates a plate lintel.

Fig. 12-53. A proven design for a conventionally built, projecting-corner fireplace. Angles are indicated by **J**, and **K** indicates a plate lintel.

Fig. 12-54.

Dimensions for Conventionally Built, Projecting-Corner Fireplace

A	B	C	D	E	F	G	H	L	M	Corner Post Height
28	26$\frac{1}{2}$	16	14	20	29$\frac{1}{3}$	12	12	36	16	26$\frac{1}{2}$
32	26$\frac{1}{2}$	16	14	20	32	12	16	40	16	26$\frac{1}{2}$
36	26$\frac{1}{2}$	16	14	20	35	12	16	44	16	26$\frac{1}{2}$
40	29	16	14	20	35	16	16	48	16	29
48	29	20	14	24	43	16	16	56	20	29

Donley Brothers Company

Fig. 12-56.

Dimensions for Conventionally Built, Three-Way Fireplace

A	B	C	D	E	F	G	H	L	M	Corner Post Height 2 req'd
28	26$\frac{1}{2}$	20	14	18	27	12	16	36	20	26$\frac{1}{2}$
32	26$\frac{1}{2}$	20	14	18	32	16	16	40	20	26$\frac{1}{2}$
36	26$\frac{1}{2}$	20	14	18	32	16	16	44	20	26$\frac{1}{2}$
40	29	20	14	21	35	16	16	48	20	29
48	29	20	14	21	40	16	20	56	20	29

Donley Brothers Company

Fig. 12-58.

Dimensions for Conventionally Built, Double-Opening Fireplace

A	B	E	F	G	H	L	Tee
28	24	30	19	12	16	36	35
32	29	35	21	16	16	40	39
36	29	35	21	16	20	44	43
40	29	35	27	16	20	48	47
48	32	37	32	20	20	56	55

Donley Brothers Company

Fig. 12-57. Proven design for a conventionally built, double-opening fireplace. An angle is shown by **J**.

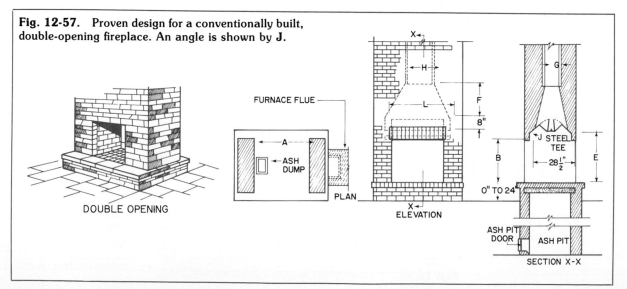

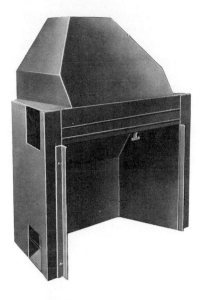

Fig. 12-59. A steel, prefabricated, heat-circulating fireplace unit.

Figure 12-53 illustrates a proven design for a projecting-corner fireplace; dimensions are given in Fig. 12-54. This type of fireplace has two open faces. It can be built facing left or right.

A third type of conventionally built fireplace is the three-way fireplace. It affords the novelty of giving a profile view of the fire in rooms where there is no projecting corner for the projecting-corner fireplace. Everyone within the room can see the fire, since the fireplace projects from the wall into the room. The design of such a fireplace is illustrated in Fig. 12-55. The dimensions are given in Fig. 12-56.

The double-opening fireplace is a fourth type in common use. Double-opening fireplaces open from one room to another and establish a sort of spiritual bond between persons in the two rooms. While there are varieties in design, the proven design shown in Fig. 12-57 uses two dampers mounted back to back. Dimensions for the double-opening fireplace are given in Fig. 12-58.

Welded Steel Prefabricated Fireplaces

There are a number of different types of welded steel fireplaces available. The steel heat-circulating unit increases the heat delivery of the fireplace. It also has the advantage of being carefully designed. All the mason has to do is to brick around the steel unit, build the chimney, and install a finished face on the room side of the unit. See Figs. 12-59 and 12-60. This unit heats the air in the space between the metal liners and ejects it into the rooms through

grilles. See Fig. 12-61. In Fig. 12-62 is a list of some common sizes available for this unit.

Another type of steel fireplace unit is a self-contained, all-metal unit. This unit requires no masonry. Because it has adequate insulation below the hearth and behind the fire area, it can be installed by being placed on a wood floor and backed to a

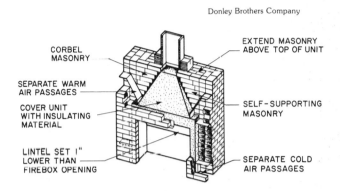

Fig. 12-60. A steel, prefabricated, heat-circulating fireplace with unit partially bricked up. Note cold-air intakes at the floor and the hot-air ejection registers above them.

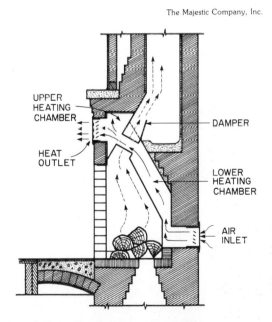

Fig. 12-61. Section through welded-steel, prefabricated fireplace. The fire warms the air in the lower heating chamber; the warm air rises, drawing in cool air at the air inlet and discharging warm air at heat outlet. Electric fans can be used to increase the circulation of warm air.

Fig. 12-62.

Common Sizes of Prefabricated Steel Heat-Circulating Fireplaces

Finished Opening		Overall Dimensions			Approximate
Wide	High	Wide	High	Deep	Flue Size
28	22	34¾	44	18	8½ × 13
32	24	38½	48	19	8½ × 13
36	25	42½	51	20	13 × 13
40	27	47¾	55	21	13 × 13
46	29	55	60	22	13 × 13
54	31	63	65	23	13 × 18

The Majestic Company, Inc.

frame wall. The elevation can be faced with brick, glass, or marble in the same manner as a conventional fireplace or a heat-circulating fireplace. In Fig. 12-63, a projecting-corner fireplace of this type is shown. This unit uses a prefabricated chimney built of corrosion-resistant alloys and stainless steel. It is comprised of three pipes, one inside the other, with Tyrex insulation between each set. A single-opening welded steel fireplace with a prefabricated chimney is illustrated in Fig. 12-64. In Fig. 12-65, a finished installation is shown.

Fig. 12-63. A self-contained steel fireplace.

The Majestic Company, Inc.

The Majestic Company, Inc.

Fig. 12-65. Welded steel fireplace with facing in place and plywood paneling above.

Fig. 12-64. A self-contained, steel single-opening fireplace with prefabricated chimney.

The Majestic Company, Inc.

Vega Industries, Inc.

Fig. 12-66. Wall-hung, metal fireplace.

The Majestic Company, Inc.

Fig. 12-67. Freestanding, welded steel fireplace.

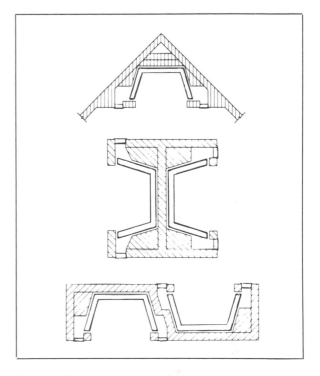

Fig. 12-68. Solutions to back to back and corner fireplace requirements.
Top: Foundation plan for corner fireplace showing location of intake grilles.
Center: Foundation plan for back to back fireplaces with grilles in sides or face.
Bottom: Foundation plan for two fireplaces with intake grilles in face, installed side by side.

Two examples of yet another type of all-metal fireplace are illustrated in Figs. 12-66 and 12-67. These units can be added easily after a house is built. They are lightweight and require no footing.

Placement of Fireplaces

Frequently, it is necessary to place two fireplaces back to back or in a corner. While the exact solution may vary, some commonly accepted solutions are illustrated in Fig. 12-68. These are heat-circulating fireplaces. Notice the location of the cool-air inlets and warm-air outlets.

Chimneys

The following discussion refers only to the masonry chimney. Prefabricated metal chimneys are also available.

The chimney should have a fireclay flue lining. When more than two flues are located in the same chimney, a 4-inch (102 mm) masonry division is necessary. The chimney wall should be at least 4 inches thick; it should be separated from the framing members of the building by a 2-inch (51 mm) air space and from flooring and sheathing by a ¾-inch (19 mm) air space. The top of the chimney should be 2'-0" (610 mm) above the ridge of the roof or any other obstruction nearby. If more than one flue

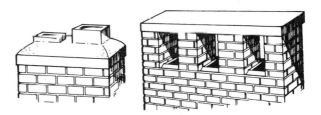

Fig. 12-69. Solutions to chimney problems.
Left: Vary height of flue tile to prevent smoke of one flue from coming down other flue.
Right: Hood flues to help discharge smoke and shed rain.

is in a chimney, the flue tile should be projected unequally, as a safeguard against smoke pouring out of one flue and down the other.

If a high building or trees interfere with the free discharge of smoke, the chimney can be hooded. See Fig. 12-69. A hood also protects the flue from rain.

A fireplace chimney is a good place to locate other flues for furnaces and for other units needing venting. Figure 12-70 illustrates a section through a fireplace above the hearth, showing a second flue for a furnace; it also illustrates the elevation, showing how these flues are located. The furnace flue should have a fireclay lining and be surrounded by 4 inches (102 mm) of masonry.

Mantels

Period-style wood mantels are fairly standardized. Certain types produced by millwork companies are available as stock items. These are usually attractive and well designed. For contemporary homes, the mantel as it is commonly conceived is frequently omitted, and the design is produced through the use of the facing materials. Simplicity is an important factor.

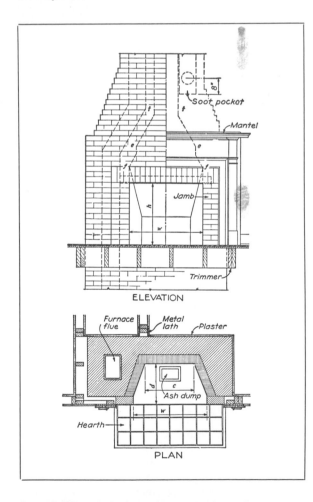

Fig. 12-70. Elevation and section through a conventionally built fireplace, illustrating flue location.

Build Your Vocabulary

Following are terms that you should understand and use as part of your working vocabulary. Write a brief explanation of each term.

boiler	hot-air heat
BTUH	hot-water heating
chilled-water cooling	system
system	humidity
chimney	latent heat gain
compressor	mantel
convected heat	perimeter heating
convectors	plenum
cooling coil	plenum heat distribution
damper	radial heat distribution
design temperature	radiant baseboard
downflow furnace	radiant heat
duct	radiator
hearth	R-value
heat flow	steam heat
heat gain	thermostat
heat loss	unit heat loss
heat pump	upflow furnace
heat transfer factor	zoned heat control
horizontal-flow furnace	

Class Activities

1. Make a study of the effectiveness of the heating plant in your home or classroom. Set the thermostat on 72° F (22° C). Place a series of thermometers on a wall, starting with one at the ceiling and then 2 feet (610 mm) apart to the floor. Record the temperatures indicated by each, every hour for eight hours. This will give the floor to ceiling difference. Record the outside temperature. Repeat this on days when the outside temperature is 30° F, 20° F, 10° F, and 0° F (-1° C, -7° C, -12° C, and -18° C). Does the heating system keep the room within acceptable limits?
2. Compute the heat loss for one room in your home. Explain what could be done to reduce this heat loss.
3. Using an architectural standards book, ascertain the insulating values for a frame exterior wall of a house with no insulation. Compare this with the insulating value of this wall when 2 inches (51 mm) of blanket insulation is added.
4. From local gas, electric, and oil companies ascertain the relative costs for the use of each fuel for heating. Prepare a report giving the advantages and disadvantages for using gas, electricity, and oil for heating purposes in your community.

UNIT FOUR

Architectural Design for Energy Conservation

Chapter 13
Energy Efficient Design

Chapter 13

Energy Efficient Design

The conservation of energy in building design is the result of a combination of common sense and creative thinking. Planning for proper insulation and ventilation within a structure is very important. It includes carefully selecting doors and windows and controlling condensation and sound. Taking such steps can make life more pleasant.

The type of energy used within a structure must also be considered. Traditional energy sources such as oil, gas, and coal are **nonrenewable resources.** That is, they are limited in quantity and can be used up. It is necessary to begin to rely more heavily on resources that are, for all practical purposes, inexhaustible. These resources reoccur or "renew themselves." **Renewable resources** include sun, wind, and climate. The most practical of these to use at this time is radiant heat energy from the sun — **solar energy**.

A special type of building design that can incorporate the above-mentioned energy saving design factors especially well is **earth sheltered design**. Earth sheltered buildings will be discussed in detail later in this chapter.

Energy Saving Techniques

Following are some general suggestions for designing buildings to conserve energy. These are explored in greater detail later in this chapter.

1. **Protect the house from extreme exposure to the sun in the summer.** Use a white roof and insulated exterior doors; double- or triple-glaze the windows; and use trees, shrubs, awnings, and overhangs to shade the roof, walls, and windows. See Fig. 13-1. Also use adequate insulation in the walls, ceiling, and floor. Consider using 6 inches (152 mm) of insulation in the walls and 12 to 24 inches (305 mm to 610 mm) in the ceiling.
2. **Provide air movement inside the house.** A person feels cool when air passes over the skin, causing evaporation. Therefore, use electric fans to move the air, place windows to take advantage of summer breezes, and do not let plants, such as trees, block natural summer breezes. See Fig. 13-2.
3. **Use the sun for winter heating.** Orient the house so it faces south and can utilize solar heat. Face windows and glass walls south. Remove or design sun shields so they do not block the sun in the winter. Also, a small solar greenhouse is a good addition. See Fig. 13-3. (Review Chapter 4.)
4. **Block winter storms.** Have few or no windows on the north side. Use the garage, trees, and fences to block winter winds, and use an entrance foyer with doors on each side to reduce air infiltration. (Review Chapter 4.)
5. **Use earth sheltering principles.** If the house is not completely earth sheltered, build berms (mounds of earth) next to the walls receiving the extremes of heat, cold, and storms.
6. **Provide adequate ventilation.** The attic should be ventilated in both summer and winter. If the crawl space is ventilated, insulate the floor. The humidity inside the house must be controlled. In summer, use a dehumidifier to remove moisture from the air; an air conditioner will help do this as well. In winter, use a humidifier to increase the humidity. Excess moisture can be a problem, especially in the kitchen and bathrooms. To remove this moisture, ventilate the interior of the house. See Fig. 13-4.

Insulation

Proper use of insulation is a major factor in conserving energy. The placement, the kind, and the amount of insulation used are all important considerations.

Where to Insulate

Any heated area should be surrounded by insulation. This normally places insulation in the ceiling,

Use trees and shrubs to shelter the house.

Orient the house to have windows facing south. Concrete or tile floors will absorb and hold warmth from the sun.

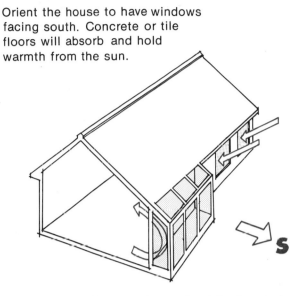

Use solar greenhouse to collect and store heat. Fans can be used to move the heat into the house.

Fig. 13-3. Techniques to use the winter sun to help heat the house.

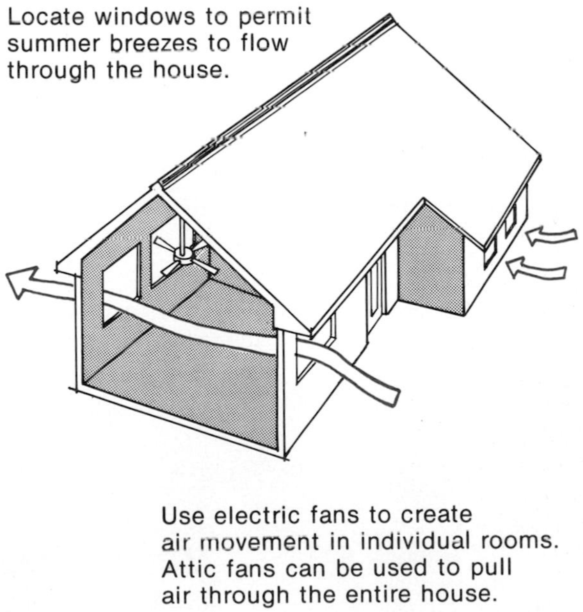

Protect windows with awnings or a large roof overhang.

Fig. 13-1. Techniques to protect the house from the hot summer sun.

Locate windows to permit summer breezes to flow through the house.

Use electric fans to create air movement in individual rooms. Attic fans can be used to pull air through the entire house.

Fig. 13-2. Techniques to provide good air circulation in the house.

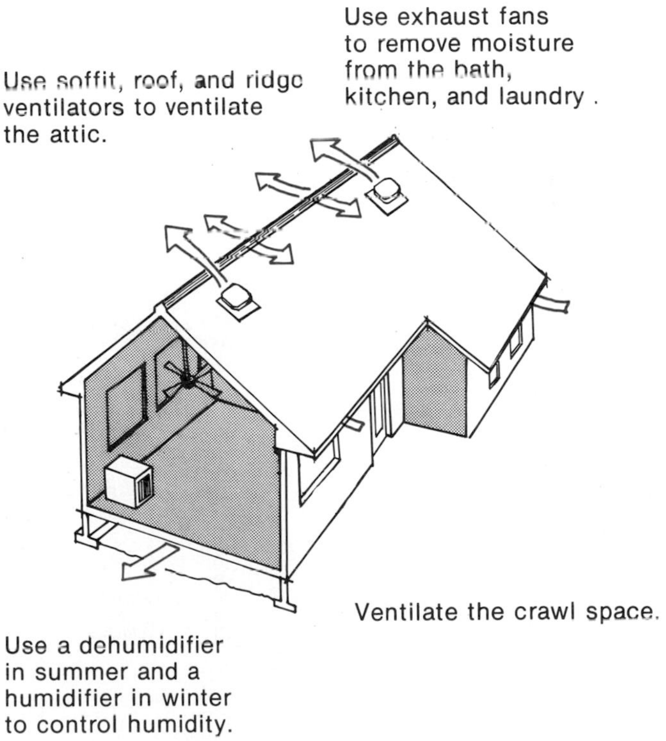

Use soffit, roof, and ridge ventilators to ventilate the attic.

Use exhaust fans to remove moisture from the bath, kitchen, and laundry .

Ventilate the crawl space.

Use a dehumidifier in summer and a humidifier in winter to control humidity.

Fig. 13-4. Ventilation techniques.

walls, and floor. Exterior walls of a building should always be insulated. If an attic is not used, the insulation should be in the attic floor rather than in the roof. Any rooms built in an attic should have the walls and ceiling insulated.

Floors are a major source of heat loss. Insulation is needed to control this. The floors of rooms over porches or unheated garages should be insulated. Water pipes in these floors also need insulation to prevent their freezing and breaking. Floors over a heated basement require no insulation.

Homes built over a crawl space or on a slab have special insulation needs. The proper placement of insulation in a home is illustrated in Fig. 13-5.

Kinds of Insulation

Several kinds of insulation are being used today. **Insulation batts** are made of glass fiber or rock wool and come in 15- and 23-inch widths to fit between floor joists and wall studs. They are available in thicknesses from 1 to 7 inches and in lengths of 4 feet and 8 feet. They come with or without a vapor barrier. **Insulation blankets** are identical to batts except they are available in rolls 50 to 60 feet long. See Fig. 13-6.

Foamed-in-place insulation is a sprayed urethane foam. It is used in frame walls, between ceiling joists, and over roof decks. See Figs. 13-7 and 13-8. The system has a two-coat process. The urethane foam is sprayed over the clean roof decking and covered with two coats of liquid-applied silicon rubber. The second coat has a surface, mineral granual embedment. Typical construction details are shown in Fig. 13-9.

Rigid board insulation is available in extruded polystyrene, urethane, expanded polystyrene (bead board), and glass fiber. The extruded

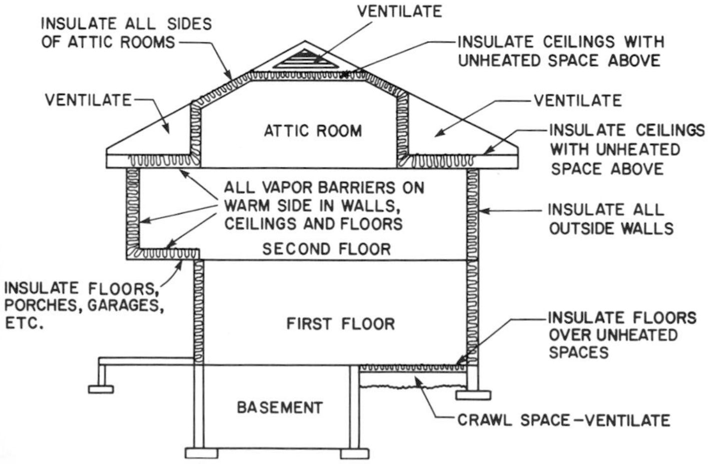
INSULATE ALL SIDES
OF ATTIC ROOMS
VENTILATE
INSULATE CEILINGS WITH
UNHEATED SPACE ABOVE
VENTILATE
VENTILATE
ATTIC ROOM
INSULATE CEILINGS
WITH UNHEATED
SPACE ABOVE
ALL VAPOR BARRIERS ON
WARM SIDE IN WALLS,
CEILINGS AND FLOORS
INSULATE ALL
OUTSIDE WALLS
SECOND FLOOR
INSULATE FLOORS,
PORCHES, GARAGES,
ETC.
FIRST FLOOR
INSULATE FLOORS
OVER UNHEATED
SPACES
BASEMENT
CRAWL SPACE—VENTILATE

Fig. 13-5. Proper placement of insulation in a home.

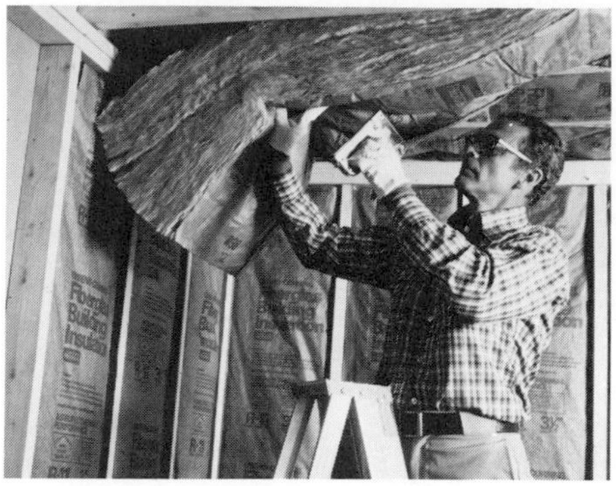

Owens-Corning Fiberglas Corporation

Fig. 13-6. Paper-backed insulation blankets being stapled in place.

Sprayfoam Southwest Inc.

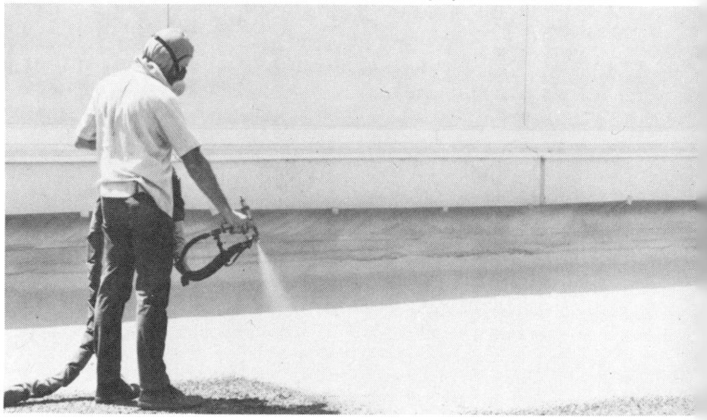

Fig. 13-7. The urethane foam insulation material is sprayed over the roof deck.

Fig. 13-8. The liquid silicone rubber layers are sprayed over the hardened urethane foam layer.

Sprayfoam Southwest Inc.

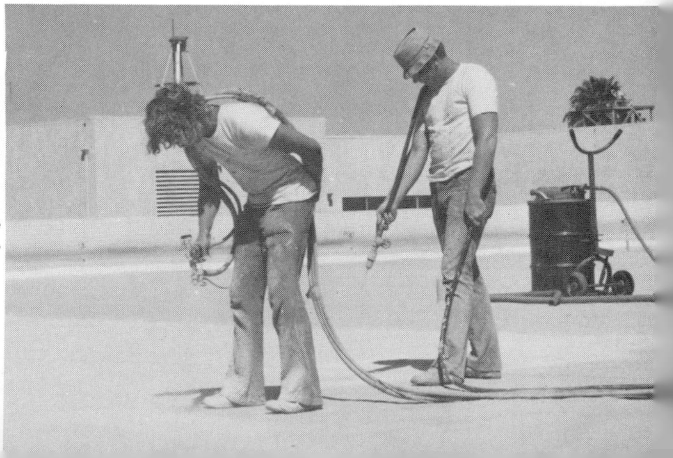

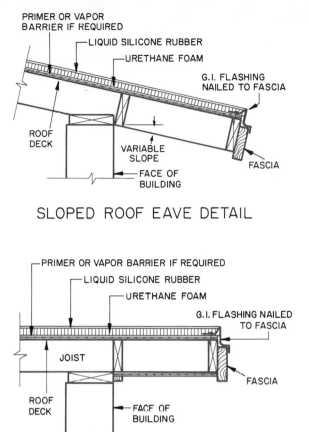

SLOPED ROOF EAVE DETAIL

FLAT ROOF EAVE DETAIL

Fig. 13-9. Typical construction details for using urethane foam as a roof insulation.

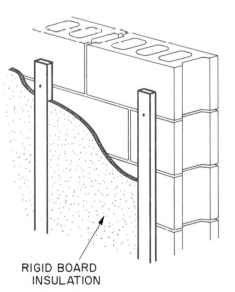

Fig. 13-10. Rigid board insulation can be glued to masonry walls.

polystyrene and urethane boards also serve as vapor barriers. They must be covered with ½-inch gypsum wallboard to meet fire safety requirements. The sheets are from ¾ to 4 inches thick, 24 or 48 inches wide, and 8 feet long. In some cases, these sheets can be used as sheathing in frame wall construction. They are excellent for insulating masonry walls. See Fig. 13-10.

Insulation is available in **loose form.** It can be installed by being poured from bags. Insulation contractors blow it into place. See Fig. 13-11. The following types are available: glass fiber, rock wool, cellulosic fiber, vermiculite, and perlite.

When insulation is installed, it should have no air space around it. In walls, for example, it should touch the sheathing. See Fig. 13-12.

Fig. 13-11. Loose insulation may be blown into place.

Fig. 13-12. There should be no air space between the insulation and the sheathing.

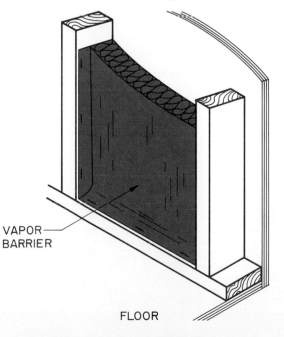

Insulation Values

The effectiveness of an insulation is specified by its R-value. **R-value** is a measure of the resistance a material has to the flow of heat. The higher the R-value the greater the resistance to heat. The R-value is the total resistance value of a particular material. It is calculated from the k-value. The **k-value** represents the amount of heat that will pass in one hour through one square foot of one-inch-thick material per one degree Fahrenheit of temper-

Fig. 13-13.

Resistance Values of Structural and Finish Materials, Air Spaces, and Surface Films

Structural and Finish Materials	R-Values
Wood bevel siding, 1/2 × 8, lapped	R-0.81
Wood siding shingles, 16″, 7½″ exposure	R-0.87
Asbestos-cement shingles	R-0.03
Stucco, per inch	R-0.20
Building paper	R-0.06
1/2″ nail-base insul. board sheathing	R-1.14
1/2″ insul. board sheathing, regular density	R-1.32
25/32″ insul. board sheathing, regular density	R-2.04
1/4″ plywood	R-0.31
3/8″ plywood	R-0.47
1/2″ plywood	R-0.62
5/8″ plywood	R-0.78
1/4″ hardboard	R-0.18
Softwood, per inch	R-1.25
Softwood board, 3/4″ thick	R-0.94
Concrete blocks, three oval cores	
Cinder aggregate, 4″ thick	R-1.11
Cinder aggregate, 12″ thick	R-1.89
Cinder aggregate, 8″ thick	R-1.72
Sand and gravel aggregate, 8″ thick	R-1.11
Lightweight aggregate (expanded clay, shale, slag, pumice, etc.), 8″ thick	R-2.00
Concrete blocks, two rectangular cores	
Sand and gravel aggregate, 8″ thick	R-1.04
Lightweight aggregate, 8″ thick	R-2.18
Common brick, per inch	R-0.20
Face brick, per inch	R-0.11
Sand-and-gravel concrete, per inch	R-0.08
Sand-and-gravel concrete, 8 inches thick	R-0.64
1/2″ gypsum board	R-0.45
5/8″ gypsum board	R-0.56
1/2″ lightweight-aggregate gypsum plaster	R-0.32
25/32″ hardwood finish flooring	R-0.68
Asphalt, linoleum, vinyl, or rubber floor tile	R-0.05
Carpet and fibrous pad	R-2.08
Carpet and foam rubber pad	R-1.23
Asphalt roof shingles	R-0.44
Wood roof shingles	R-0.94
3/8″ built-up roof	R-0.33

Glass	
Single glass (winter)	U = 1.13
Single glass (summer)	U = 1.06
Insulating glass (double)	
1/4″ air space (winter)	U = 0.65
1/4″ air space (summer)	U = 0.61
1/2″ air space (winter)	U = 0.58
1/2″ air space (summer)	U = 0.56
Storm windows	
1″ to 4″ air space (winter)	U = 0.56
1″ to 4″ air space (summer)	U = 0.54

Air Spaces and Surface Films	R-Values
Air Spaces (3/4″)	
Heat flow UP	
Non-reflective	R-0.87
Reflective, one surface	R-2.23
Heat flow DOWN	
Non-reflective	R-1.02
Reflective, one surface	R-3.55
Heat flow HORIZONTAL	
Non-reflective (also same for 4″ thickness)	R-1.01
Reflective, one surface	R-3.48
Note: The addition of a second reflective surface facing the first reflective surface increases thermal resistance values of an air space only 4 to 7 per cent.	
Surface Air Films	
INSIDE (still air)	
Heat flow UP (through horizontal surface)	
Non-reflective	R-0.61
Reflective	R-1.32
Heat flow DOWN (through horizontal surface)	
Non-reflective	R-0.92
Reflective	R-4.55
Heat flow HORIZONTAL (through vertical surface)	
Non-reflective	R-0.68
OUTSIDE	
Heat flow any direction, surface any position	
15 mph wind (winter)	R-0.17
7.5 mph wind (summer)	R-0.25

American Society of Heating, Refrigerating, and Air Conditioning Engineers

R-Values for Insulation

Material	Thickness	R-Value
Glass Fiber	2″ Batt	R-7
	4″ Batt	R-11
	6″ Batt	R-19
	6″ Blown	R-13
	8½″ Blown	R-19
	12″ Blown	R-26
	18″ Blown	R-38
	2-6″ Batts	R-38
Rock Wool	4″ Blown	R-11
	6½″ Blown	R-19
	13″ Blown	R-38
Cellulose	4″ Blown	R-11
	6½″ Blown	R-19
	13″ Blown	R-38
Styrofoam®	1″ Board	R-5
	1½″ Board	R-7.5
Urethane Foam	1½″ Board	R-9.3
	4″ Injected	R-25

ature difference between the two faces of the material. The k-value is measured in **British Thermal Units (BTU).** The **lower** the k-value the better the insulating value; the **higher** the R-value the better the insulating value. R-value is equal to $\frac{1}{k}$.

The combined thermal value of a combination of materials, as in the wall of a house, is expressed in terms of **U** — the **overall coefficient of heat transmission.** This includes the materials, air spaces, and surface air films. **U** is the reciprocal of **R** or $U = \frac{1}{R}$. If the R-value of a wall is 14.43, the **U** of the wall is 1/14.43 or 0.07. The smaller the **U** the higher the insulating value. Following is an example for a typical frame wall. R-values are given in Fig. 13-13.

Example problem for calculating the coefficient of heat transmission:

Material	R-value
Outside surface film	0.17
Wood bevel siding	0.81
½-inch Insulation board sheathing	1.32
4" glass fiber insulation batts	11.00
½" gypsum wallboard	0.45
Inside surface film	0.68
R-value	**14.43**

$$U = \frac{1}{R} = \frac{1}{14.43} = 0.07$$

Construction Suggestions

The building should be designed to utilize the recommended thicknesses of insulation. Frame walls should be made from 2 × 6 members spaced 24 inches on center. (Refer to Chapter 7.) This will allow the use of 6 inches (152 mm) of insulation with a rating of R-19. An effective vapor barrier is needed. Recommended vapor barriers are polyethylene sheets or foil-backed gypsum board.

Ceilings should have an insulation rating of R-26 to R-30. To achieve this, 12 to 13 inches (305 mm to 330 mm) of insulation is usually needed. The roof must be framed to accommodate this thickness. If the roof framing is a truss, the design can be altered as shown in Fig. 13-14. When a truss is not

used for roof support, a vertical member similar to that used for the truss is needed to raise the rafter to allow for the increased thickness of insulation. A typical solution is given in Fig. 13-15.

A home on a slab foundation should have a foamed plastic sheet or urethane insulating board bonded to the edge of the slab around the entire exterior of the house. See Fig. 13-16. A home with a

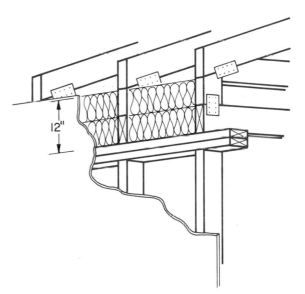

Fig. 13-14. Truss design can be altered so 12 inches of insulation can be used in the ceiling.

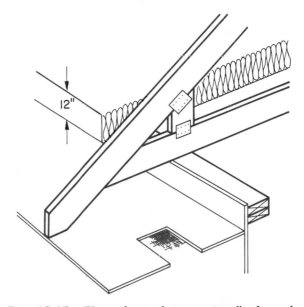

Fig. 13-15. The rafters of conventionally framed roofs can be raised to allow the use of 12 inches of insulation.

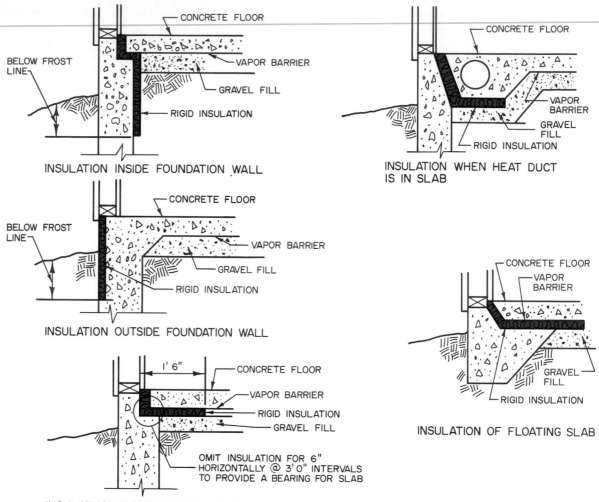

Fig. 13-16. Some typical methods of insulating concrete floors.

two-piece or floating slab needs a foam sheet 2 feet wide placed over the foundation and under the edge of the slab. Refer again to Fig. 13-16. The concrete floor is poured over the insulation. See Fig. 13-17.

A house with a vented crawl space should have batt insulation with an R-19 or better rating placed between the joists as shown in Fig. 13-18. The vapor barrier is next to the floor. This installation requires supports to hold it in place. Usually, wire or wire mesh is nailed below it.

If the crawl space is unvented, insulation should be placed on the header and foundation wall. A polyethylene film should be placed over the soil for a vapor barrier. See Fig. 13-19. This type of construction can make the crawl space a plenum for heating. (See Chapter 12.)

Fig. 13-17. Vapor barrier being installed before a concrete floor is poured. It is important that no open places be left in the barrier.

Portland Cement Association

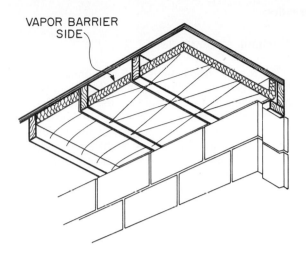

Fig. 13-18. Floors over vented crawl spaces need to be insulated.

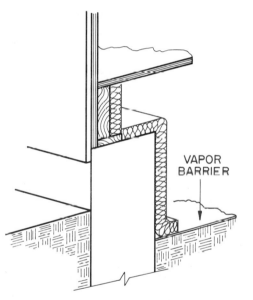

Fig. 13-19. The walls of an unvented crawl space should be insulated, and the ground underneath should be covered by a vapor barrier.

Ventilation

Attic Ventilation

In addition to removing moisture, attic ventilation can reduce energy consumption. A large percentage of the air-conditioning load is due to solar heat. Attic temperatures can reach in excess of 150°F (66°C). This solar heat is transmitted through the roof, and all materials in the attic become very hot. This heat passes through the ceiling to the living area below. While ceiling insulation helps reduce heat transfer, adequate ventilation is essential.

For a ventilating system to be effective, air must be moved over the heated surfaces and removed from the building. Since the roof is the hottest area, the air must flow up along the surface of the roof and be expelled at the ridge. This can be done with natural ventilation or exhaust fans.

The most effective natural means is to provide vents in the soffits and a continuous vent at the ridge. This causes air to flow along the inside of the roof, thus cooling the hot surface. Soffit vents with a power attic ventilator are also effective. See Fig. 13-20. Soffit vents alone, however, are not sufficient, because the air flows across the attic floor. The hottest area, at the ridge, is not cooled. The metal roof ventilator without an electric fan is also insufficient. This type requires wind outside the building to cause a draft and pull out the hot air. In the hottest conditions, there is often little or no wind.

Following are some suggested guidelines for attic ventilation:

1. There should be at least ten air changes per hour of attic air, or an air flow of 1.0 to 1.5 cubic feet per minute per square foot of floor space.

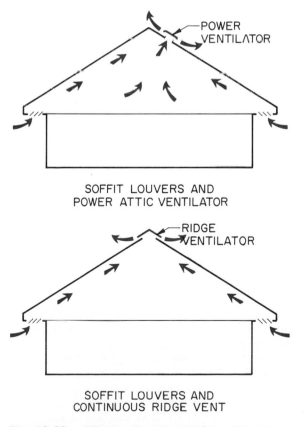

Fig. 13-20. Effective ways to ventilate attic spaces.

Fig. 13-21.

Vent Area Suggested For Natural Ventilation of Attic

Type of Vent	Vent Area in Square Inches Per Square Foot of Attic Floor
Roof Louvers only	3.3
Soffit Louvers only	2.9
Gable Louvers only	2.4
Roof Louvers and Soffit Vents	3.1
Gable Louvers and Soffit Vents	2.7
Continuous Ridge and Continuous Soffit	1.5

Fig. 13-22.

Power Attic Fan Ventilation Capacities

Capacity of Fan cfm	m³m	Attic Floor Area ft²	m²	Vent Area in.²	m²
1500	42	1000	93	1250	0.807
2250	63	1500	139.5	1750	1.129
3000	84	2000	186	2500	1.613
3750	105	2500	232.5	3000	1.936
4500	126	3000	279	3500	2.258
5000	140	3500	325.5	4000	2.581

2. There should be at least 80 square inches of net free air flow through attic vents for every 100 cubic feet per minute of power ventilator fan capacity.

Suggested guidelines for sizing vents for natural ventilation are given in Fig. 13-21. In Fig. 13-22 are suggested guidelines for selecting the proper attic power ventilation fan.

Crawl Space Ventilation

The crawl space between the floor and the earth must be ventilated. The minimum net area of ventilating openings should not be less than 1 square foot to every 150 square feet (approximately 0.09 m² to every 14 m²) of crawl space. **Net area** is the actual open area free from screen wire. The screen wire cuts down the amount of free open space in a ventilator. One ventilator should be within 3 feet (914 mm) of each corner of the building. Each wall should have at least one ventilator. Typical ventilator sizes are 5 × 8 1/8 inches, 5 × 16 1/2 inches, and 7 3/4 × 16 1/2 inches (the size of one concrete block).

Mechanical Ventilation

All rooms should have natural ventilation equal to 5 percent of the floor area. If this is not possible in a kitchen, a mechanical exhaust system that provides fifteen air changes per hour should be installed. It must exhaust to the outside of the building. A bath system should provide eight air changes per hour, and may exhaust into an attic that has cross ventilation and a net free area 1/150 of the capacity of the exhaust fan. Furnace rooms should be ventilated according to building codes.

Windows and Doors

Heat is lost and gained through windows and doors. This is due to direct transmissions through the glass and filtering of outside air through loosely fitted or poorly installed doors and windows. Air infiltration can be reduced by using quality-built doors and windows. After they are installed, the space between the window frame and the studs should be filled with insulation. Storm windows greatly reduce air infiltration, because they reduce the flow of outside air across the surface of the window. Refer back to Fig. 13-13. Notice the low U-values for storm windows. Heat loss is greatly reduced by storm windows.

Transmission through the glass can be reduced by using double-pane or insulated glass. The space between the two panes adds to the insulation qualities. A single pane of glass has a U-value of 1.13; the value for 1/4 inch double-pane glass is 0.65. Triple-glazed glass with 1/4 inch air spaces has a U-value of .44.

Doors have great differences in R-value. See Fig. 13-23. The metal door with a urethane core is by far the most effective in reducing heat gain and loss. Infiltration can be reduced by using a storm door. If a door has glass, this should be double panes. The use of weather stripping around doors greatly reduces heat loss.

Ductwork

Forced-air heating and cooling systems require that ducts run from the furnace to each room. A great deal of energy can be lost in the process. Ducts should be tight so no loss occurs due to leaks in the pipe.

There is a loss of heat or cooling through the walls of the duct. Whenever possible, ducts should be located inside the area to be conditioned. In this

Fig. 13-23.

R-Values of Common Doors

Type of Door	R-Value
Hollow core wood	R-1.0
Hollow core wood with storm door	R-1.5
Solid core wood	R-2.3
Solid core wood with storm door	R-3.5
Metal with urethane core	R-13.5

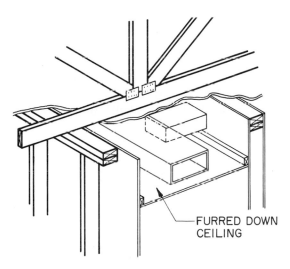

FURRED DOWN CEILING

Fig. 13-24. Locate ducts inside the space to be conditioned.

way, the loss is used to heat or cool the intended area. Ducts can be run down a hall or above closets, where the ceiling can be dropped. See Fig. 13-24. If located outside the conditioned area, ducts must be insulated. Even then, some loss of heat or cooling will occur. Most frequently, the ducts run below the floor in a crawl space or in the attic.

Other Planning Considerations

1. The more exterior walls a house has, the greater the heat loss. A square house is the most efficient shape.
2. A two-story house has a lot of wall space, but only half the ceiling area that the same house would have if it were on one level. The ceiling is the major source of heat loss.
3. In warm climates, a light-colored roof will be helpful because it will reflect solar heat. This keeps attic temperatures lower.

4. A roof with a high pitch will have a lower heat gain and can be ventilated more easily.
5. When glass areas are properly located, solar energy can be used to heat the house. Reduce the amount of glass in wall areas where it would increase heating or cooling costs. See the section on orientation in Chapter 4.
6. Energy can be saved by keeping hot water pipes as short as possible.
7. All places on the exterior of the house which could permit air to infiltrate should be caulked. Some typical places are around windows and doors where they meet the siding, along sills, and where siding meets in corners.
8. The use of sun shielding screens or draperies controls heat loss and gain through windows. Special reflecting glass can reduce solar heat through windows by 40 to 70 percent.
9. Using heating and cooling units that are the proper size conserves energy.

Condensation Control

Moisture damage in buildings can be caused by leaks from the outside, such as around improperly installed flashing or poorly fitted windows. It can also be caused from the condensation of water vapor produced inside the building by such activities as cooking and bathing or by manufacturing processes giving off moisture to the air. Condensation can occur in summer or winter. It can be seen when windows sweat, but it also occurs inside the walls, unseen but damaging. Excess moisture inside a building can penetrate the interior wall. When it strikes the cold sheathing, it condenses and runs down inside the wall. If paint peels off a frame wall, condensation inside the wall can be suspected.

Other common sources of condensation are the ground under crawl-space houses, the concrete slab in this type building, unvented clothes dryers, and fuel-burning devices. Summer condensation is usually the result of excess humidity in the air or in the soil.

Condensation can be controlled by providing exhaust fans in areas of high humidity, such as baths and kitchens, and by venting such devices as clothes dryers and heaters. A vapor barrier should be applied behind the interior wall finish, on the warm side of the insulation. This is usually adhered to the insulation which already has been installed in the wall. Mechanical devices such as dehumidifiers help remove moisture from the air.

Crawl spaces require at least four ventilators in the foundation wall (one located close to each corner of the space). The aggregate ventilating area (net free) should be not less than 1/150 of the area of the space. As previously mentioned, a vapor barrier placed upon the ground is of considerable value in preventing moisture from entering the crawl space. It may be covered with several inches of sand to hold it in place. The wall ventilators should be placed as high in the foundation wall as possible and should remain open all year. To prevent cold floors, insulation should be installed below the floor.

For a concrete-slab house, a vapor barrier should be placed below the slab before the concrete is poured. If this is not done, the floors will sweat whenever the outside temperature changes considerably. As a result, carpets will be wet and much moisture will be released to the air. The permeance of the vapor barrier should not exceed 1 perm (1 grain of vapor transmission per square foot, per hour, per inch of mercury vapor pressure difference).

A basement should have windows or other means of ventilation totaling not less than 1 percent of the basement floor area. Ventilation of unheated attic spaces helps remove water vapor before it can cause damage.

Sound Control

Architectural designers must consider **acoustics**. That is, they must utilize techniques of sound control to create pleasant living environments. Sound control is a specialty of acoustical engineers. One thing these engineers do is calculate the reverberation characteristics of a room. **Reverberation characteristics** relate to the amount of sound absorption in a room. Established formulas are used for these calculations. A standard measure called **reverberation time** is used. This is the time required for a specified sound to die away to one-thousandth of its initial pressure. If the calculated reverberation time is too long, the engineer must increase the amount of acoustical treatment. Figure 13-25 shows how acoustical treatment influences reverberation time.

In residential design, it is especially important that the noise penetrating the bedrooms be reduced and that bathrooms be soundproofed. In commercial work, the applications are more numerous and more difficult because of the range of activities performed in commercial buildings.

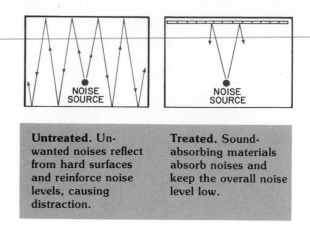

| Untreated. Unwanted noises reflect from hard surfaces and reinforce noise levels, causing distraction. | Treated. Sound-absorbing materials absorb noises and keep the overall noise level low. |

Fig. 13-25.   Hard, untreated surfaces reflect sound.

Sound travels in waves that radiate from the source through the air until they strike walls, floors, ceilings, furniture, curtains, or people. Some of these surfaces absorb or dampen the sound. Other surfaces are set into vibration by the pressure of the sound waves. As a result, some of the sound is reflected back into the room. At the same time, some of the sound is conducted through the material to the other side. The resistance of a material to the passage of airborne sound is its **sound transmission class (STC).** The higher the STC of a material the better it is as a sound barrier.

Several STC sound levels are listed below to give you some understanding of the scale used.

STC No.	
25	Normal speech can be understood easily
35	Loud speech is audible but not intelligible
45	One must strain to hear loud speech
50	Loud speech is not audible

Ideally the STC rating for walls of bedrooms, baths, kitchens, and walls between private apartments should be about 45 to 50.

Three types of noise or sound must be considered when attempting to provide a quieter acoustical environment—direct noise, sound transmission, and vibration. Direct noise is sound reaching the ear without being reflected, such as traffic noise through an open window. Acoustical materials can help reduce the reflections of sound reaching the room, but they will have no effect on direct noise until it strikes the absorbing surface. This noise is one type of airborne sound. Outside airborne noises penetrate through open doors and windows,

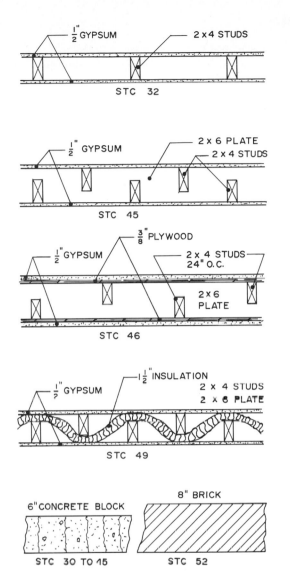

Fig. 13-26. Staggered stud wall construction and the use of sound-deadening materials reduce noise transmission.

Airborne noise control also involves reducing noise flow from one space to another. Noise flow is often reduced by separating the spaces with a double-stud wall, with no connections between the independent wall faces. It can be further reduced by adding sound-deadening material, flexible or rigid, between the studs. See Fig. 13-26.

Airborne sound is also transmitted through floors to the living spaces below. The same techniques discussed for walls apply. In addition floors must resist impact sounds, such as the sound of someone walking. Impact noise is expressed as **impact isolation class (IIC)**. The greater the IIC value the more resistant a material is to impact noise.

Some suggested floor and ceiling construction details are in Fig. 13-27. Both the STC and IIC ratings are given.

The floor plan arrangement may control transmission of airborne sound. For example, a room such as a bathroom can be surrounded with closets. The extra wall, plus the clothing in the closet, greatly reduces sound transmission.

The most effective barrier to sound transmission is a wall of solid mass, such as a solid-masonry wall. The thicker the wall, the more effective the barrier. Dead air space, as found in the typical interior partition, is rather ineffective. It takes solid mass to stop sound transmission.

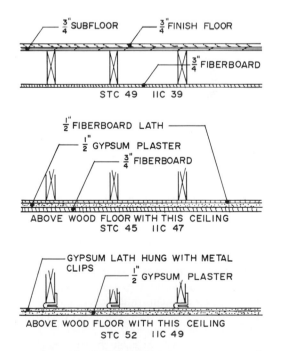

Fig. 13-27. Methods of reducing transmission of airborne noise and conducted noise through floors.

as well as through cracks around these openings and through glass and thin doors. Storm windows, double-glazed sash, and weather stripping help reduce this problem. Also, objects outside the building, such as trees, fences, and balconies, can deflect the sound, thus reducing the amount permitted to penetrate the building.

Airborne noises developed within the building can be reduced by treating ceilings with acoustical materials, by laying carpeting, or by hanging curtains. (Wall surfaces are treated only in extreme cases.) However, the muffling of sound within a space does not prevent it from being transmitted to other spaces.

Fig. 13-28. A suspended ceiling with removable acoustical panels provides access to mechanical systems above the ceiling.

There are a number of stock ceiling systems that provide for a dropped ceiling, with sound-deadening panels as the finished surface. See Fig. 13-28. These usually are suspended by wires from the floor above, so that little sound is transferred by conduction. However, sound can easily penetrate acoustical materials. These materials deaden sound within a space, but do little to stop it from passing through them to another space; therefore, sound could enter another space by going through an acoustical ceiling, bouncing off the floor overhead, and entering the next space. The only way to stop this is to put a solid barrier, such as masonry, through the ceiling to the floor overhead, or to put a sound-deadening mass directly above the acoustical ceiling. See Fig. 13-29.

Sound can be conducted through the framing members of a building. Suspended ceilings and double floors help reduce this somewhat. The most effective corrective procedure is to isolate the sound-producing device, so that it is not in direct contact with the structure. This can be done by placing machinery on sound-deadening pads or on a foundation that is entirely independent of the building foundation. Other noises are conducted by water, sewer, and heating systems. If the plumbing system can be hung with hangers lined with rubber or other sound-deadening material, sound transmission will be reduced materially.

A significant amount of noise can be reduced if the mechanical equipment is carefully selected. Some types of such devices, such as furnaces and attic fans, run more quietly than others. Quiet operation should be one criterion for selecting such units.

Good hearing or listening conditions are important in auditoriums, music rooms, and theaters. The major factors in design for good hearing and listening include the cubic capacity of the space, the basic proportions of the space, and the shape of the surfaces of ceilings and walls in relation to a stage or platform. The proper design of such facilities requires the services of an acoustical engineer. Figure 13-30 shows what happens to sound waves when an auditorium is acoustically treated and when it is not.

Acoustical Materials

There are many acoustical materials commonly used to dampen sound within a space. Curtains and carpeting are two in general use. Many types of

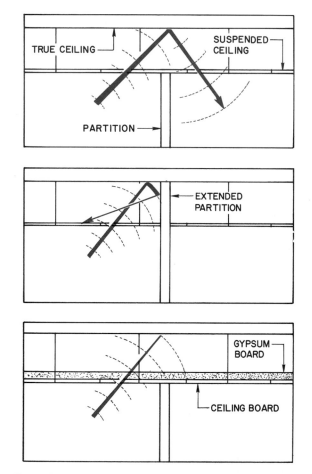

Fig. 13-29. Factors to consider when using acoustical materials.

Top: Sound can enter another space through ceiling.

Center: Partition blocks sound transmission.

Bottom: A solid mass, such as gypsum board, in a ceiling prevents sound transmission.

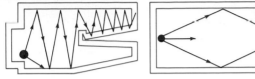

Owens-Corning Fiberglas Corporation

Fig. 13-30. Sound patterns in auditoriums with and without acoustical treatment.
Left: Without treatment, sound bounces around auditorium and garbles the performance.
Right: Treated auditorium reduces reverberation times to give improved hearing conditions.

wall panels and ceiling tiles are available. These are made from fiber glass, wood fibers, mineral wool, and aluminum. Sound-absorptive plaster is available. It requires careful installation if it is to be effective.

These acoustical materials typically have many small openings, some of which are interconnected while others are not connected. The sound waves strike these openings and "get lost," thus dissipating their energy. Some surfaces absorb sound by vibrating, similar to a diaphragm when struck, and use up the sound energy in that manner.

When selecting acoustical materials, consider the following factors:

1. Appearance — A wide selection of textures, shapes, and colors is available.
2. Performance — The coefficient of noise reduction should be ascertained.
3. Thermal insulation value — Most acoustical materials offer some insulation value.
4. Permanence — The product should be dimensionally stable and free from rot and vermin; it should resist moisture and withstand normal impact.
5. Maintenance cost — The material should be easily cleaned.
6. Fire resistance — The product should meet building code requirements for fire resistance.
7. Sanitary requirements — The material should not absorb or give off odors and should offer no sustenance to vermin.
8. Cost — Consider the cost of the material plus the cost of installation.

Using Solar Energy

The use of solar energy is becoming more attractive because conventional fuels are increasing in cost and have uncertain availability. Also, solar energy does not present the environmental prob-

lems associated with other fuels. Considerable research and development activities are in process to make the use of solar energy more economical.

Every day the sun showers the earth with several thousand times as much energy as is consumed. The solar energy reaching the earth every three days is greater than the estimated total of all the fossil fuels on earth. The solar energy annually striking the roof of a typical residence is ten times as great as its annual heat demand. Present solar technology permits the utilization of only a small percentage of this energy.

Solar heating and cooling systems can provide a large percentage of the total needs. Using present technology, the systems cannot provide 100 percent of the needs; therefore, a conventional backup heating and cooling system is needed.

Solar Design Factors

The design of a solar heating system involves many factors too lengthy to present in this book. However, the major factors which must be considered when designing a solar heating system for a building are climate, comfort, building characteristics, available solar systems, and site conditions.

A study of climate includes sun, wind, temperature, and humidity. The designer needs to know the following:

1. Temperature averages, changes, extremes, and ranges day to night,
2. Humidity in various seasons,
3. Amount of solar radiation,
4. Wind movements,
5. Snowfall, and
6. Special conditions, such as hurricanes, hail, and thunderstorms.

The day-night temperature ranges greatly influence solar cooling. A wide range is needed for the system to be effective. The storage capacity is also dependent on temperature ranges.

The heat content of air having high humidity with high temperature is greater than that having low humidity with high temperature. Solar cooling is most effective if conditions of high humidity and high temperature exist most of the year.

Winds can provide natural ventilation, thus reducing the amount of cooling needed from a cooling system. Winter winds will increase the heat loss from solar collectors as well as from buildings. Special structural considerations are necessary to protect the collectors from the wind and keep them free from snow. (Winds are discussed in Chapter 4.)

The designer needs an understanding of the relationship of the sun to the building. The most effective situation is when the sun's rays are perpendicular to the collector. Since the earth tilts on its axis, the angle at various latitudes differs. It also changes during the seasons. The solar collector must be tilted at the angle determined to be most effective locally. Some collectors are adjustable so that the tilt can be changed as the seasons change.

The angle of tilt may be determined as follows: For heating — degrees of latitude plus 15 degrees; for cooling — the degrees of latitude plus 5 degrees. The angle of tilt influences the size of the solar collector. The more the angle differs from the most effective angle, the larger the surface area of the collector must be. The solar collector is usually faced 10 to 20 degrees either side of true south.

The building needs to be designed to maintain bodily comfort. The heat loss and heat gain of the human body must be considered. Plans will vary depending upon the activities occurring in the building. Conditions required for sleeping are quite different from those for an exercise room.

The designer must have information about the characteristics of the building. The heat loss through walls, ceiling, floor, and windows must be considered as well as heat generated in the building by electric lights or various devices in use. The total volume of space to be heated or cooled is significant. Also, the orientation of the building to utilize the natural weather conditions can influence the amount of solar heat needed. (Refer again to Chapter 4.)

Another factor is the effectiveness of available solar systems. This varies considerably, but will improve in the years ahead. Cost will be reduced as systems become accepted.

Solar Energy and Building Design

The use of solar heating and cooling imposes certain design considerations. Systems can be planned which are consistent with traditional or contemporary design. One major factor is the effect on the exterior appearance. Often the collectors are placed on the roof, as in Fig. 13-31. See how easily they are made a part of contemporary house design.

Solar collectors can be integrated into the design of condominiums and townhouses. See Fig. 13-32. They can be placed on any part of a building which gives them the proper exposure. In Fig. 13-33, they

Fig. 13-31. Solar collectors are often placed on the roof. If the back of the building faces south, the collectors will not be seen from the front.

Fig. 13-32. Solar collectors are used on apartments and condominiums.

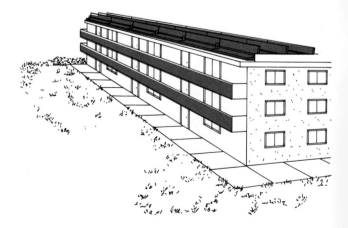

Fig. 13-33. Solar collectors can be placed on flat roofs, balconies, and other parts of buildings.

Fig. 13-34. Solar collectors can form a sun shade.

Fig. 13-35. Solar collectors can be mounted separately from the building.

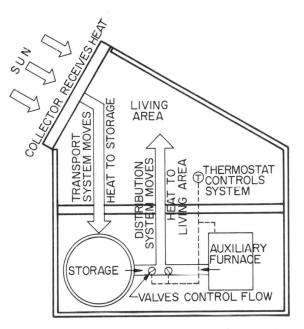

Fig. 13-36. The major components of an active solar heating system.

are on the balconies and the flat roof. In Fig. 13-34, the collector forms a sun shade. Solar collectors can be mounted separately from the building. See Fig. 13-35.

Solar Systems

There are two major types of solar systems: active and passive. **Active solar systems** require the use of pumps, motors, and blowers. They use collectors, valves, pipes, electrical controls, and some form of heat storage. Together, these form a system that collects, stores, and distributes solar energy.

Passive solar systems have few, if any, moving parts. They permit solar energy to enter through windows and glass-covered walls. Heat is stored in floors and walls that are heavy concrete or brick masses. Cylinders of water are also used for heat storage. Sometimes small fans are used to circulate the air.

Active Solar Systems

An active solar system has three major components: collectors, storage, and distribution. Radiation is absorbed by the **collector**, placed in **storage**, and **distributed** to the living space when needed. A system must have **automatic or manual controls,** such as a thermostat. Since existing solar systems cannot provide all the heat needed, an **auxiliary heating source** is needed. For example, a water system might have a gas-fired burner to heat the water when solar energy is insufficient. See Fig. 13-36.

The Collector

The collector converts solar radiation into usable thermal energy. **Thermal energy** is energy produced by heat. A collector is really a heat trap. It transfers the heat to a heat transfer medium within the collector. The medium is usually a gas or liquid. Collectors are classified as focusing or nonfocusing. A **focusing system** concentrates the sun's energy on a small area. This is much like using a magnifying glass to concentrate the sun's rays. The **nonfocusing system** absorbs the rays as they are received on the earth's surface.

Flat-plate collectors. These are nonfocusing collectors. A section through a typical flat-plate collector is shown in Fig. 13-37. It consists of a cover

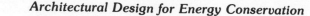

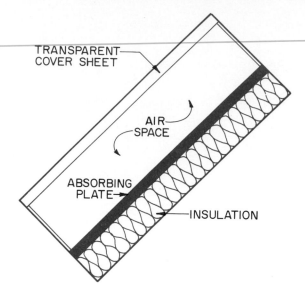

Fig. 13-37.   A section through a typical flat-plate collector.

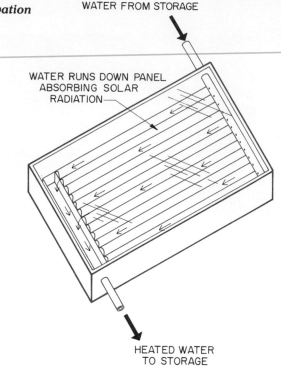

Fig. 13-38.   An open water collector.

sheet, an absorbing plate, and heavy insulation on the back.

Collectors generally use a transparent cover sheet to trap the heat in the collector. Glass or plastic is commonly used to form the cover sheet.

The absorber portion of the collector receives and retains the heat energy. The material used to make the absorber must absorb solar radiation and not reradiate it. If the absorber is used to transfer heat to a liquid or gas, it must be a good thermal conductor. Absorbers are usually coated with a black paint or chemical coating.

An **open water collector** has open troughs for trickling water to flow down. The water comes from a water storage tank and is heated by the air trapped in the collector. It passes from the collector back to the storage tank, increasing the temperature of the stored water. See Fig. 13-38.

An **air-cooled collector** receives air from the building. The air is heated as it passes through the collector. The heated air is transported through ducts back into the building. See Fig. 13-39.

A **liquid-cooled collector** passes a liquid through the collector in pipes. The liquid is usually water or an antifreeze solution. The liquid flows from a storage tank, is heated as it passes through the collector, and returns to the storage tank. See Fig. 13-40.

Flat-plate collectors are mounted facing south to southwest. The tilt varies with the latitude. A rule of thumb is to tilt the collector at an angle which equals the latitude plus 15 degrees. It is more efficient if the tilt can be adjusted every few weeks to place the

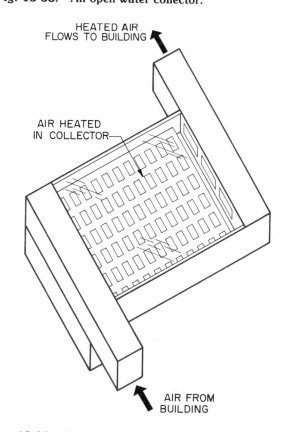

Fig. 13-39.   An air-cooled collector.

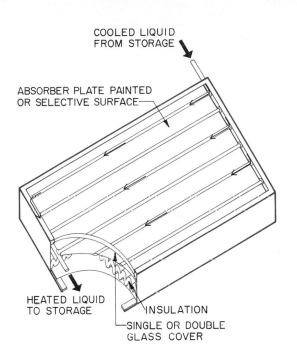

Fig. 13-40. A liquid-cooled collector.

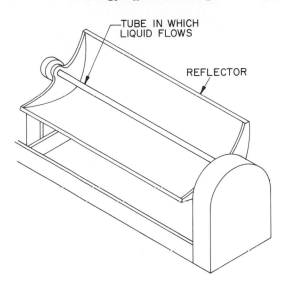

Fig. 13-41. A linear concentrating collector. The reflector focuses the sun's rays on the tube, heating the water inside.

collector in the most favorable position with the sun. If the collector can rotate on an axis each day to follow the sun, even greater efficiency can be achieved. This requires an automatic tracking mechanism.

Concentrating collectors. These collectors are a focusing system. They use curved reflectors to increase radiation on a small target area, either a tube or a point absorber. See Fig. 13-41. While they are more expensive than flat collectors, in sunny climates they more than double the temperature generated by flat-plate collectors. Concentrating collectors are most effective in areas which have clear skies most of the time.

A **linear concentrating collector**, such as the one shown in Fig. 13-41, focuses radiation on a tube which carries the liquid. The heated liquid returns to the storage tank. The liquid used should have a boiling point above the temperatures generated in the collector. It should also resist freezing. Linear concentrating collectors can be designed with the long axis horizontal in an east-west direction or at the most efficient angle (tilt) while facing in a north-south direction.

A **circular concentrating collector** uses a dish-shaped reflector. The reflector focuses the solar radiation on a point absorber. The liquid flowing

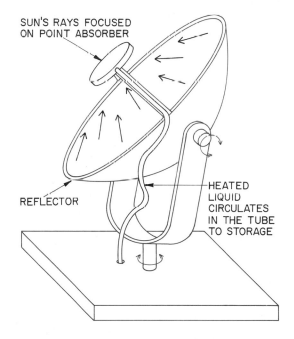

Fig. 13-42. A circular concentrating collector. The reflector concentrates the energy on the point absorber.

through the absorber is heated and flows to a storage tank. These units are adjusted in one of two ways: the collector may remain fixed while the absorber follows the sun, or the collector and absorber may follow the sun as a unit. See Fig. 13-42.

Solar Heat Storage

Since solar heat is intermittent, the heat given off while radiation is intense must be stored for use at a time when the amount of solar radiation is insufficient. Heat may be stored by raising the temperature of an inert material (rock, masonry, water, adobe) or by reversible chemical or physical-chemical reactions, such as the dehydration of salts. The type, cost, operation, and size of solar storage are determined by the method of collection, the amount of heating and cooling required, and the efficiency of the heat transfer from storage to the building.

Some heat can be stored in the **air** in a room. For example, the solar heat which enters a room through a glass wall heats the air. The air will store the heat for a while, often becoming uncomfortably warm. During cool periods, the glass must be insulated so the air is not cooled.

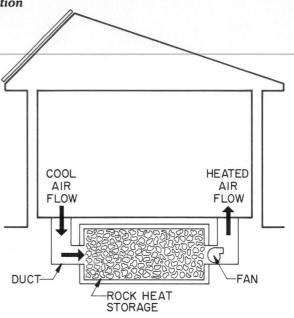

Fig. 13-44. Heat can be stored in insulated bins full of rocks. A fan forces air through the bin and warm rocks heat the air.

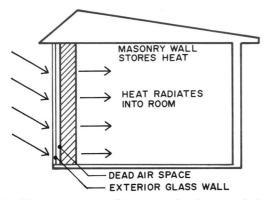

A. The masonry wall stores solar heat and then radiates the heat into the room.

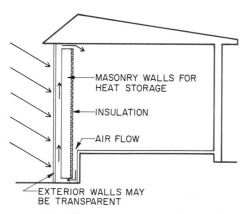

B. This Trombe wall uses the thermosiphoning principle to move heated air from between the glass and masonry wall into the room.

Fig. 13-43. A Trombe wall can supply solar heat.

Heat can be stored in exposed surfaces in a room. The **Trombe wall** is a typical example of such a surface. There are two varieties of Trombe walls. See Fig. 13-43. One consists of a thick solid masonry exterior wall covered by a glass wall. The solar heat is intensified by the glass, and warms the masonry. The masonry stores the heat and radiates it into the room. The other type of Trombe wall uses thermosiphoning principles. The masonry wall has vents at the floor and the top of the wall. Refer again to Fig. 13-43. The vents permit a natural circulation of air from the room through the air space between the glass and the masonry wall. Sometimes fans are placed in the vents.

Another method of heat storage is the use of **pebble beds or rock piles.** See Fig. 13-44. The pebbles or rocks are contained in an insulated storage unit. The rock storage is heated as air from a collector is forced through the rock container by a blower. Cool air can then be heated by flowing through the storage area. Generally, rock storage will require about two and one-half times the volume of water storage.

Water can be used for heat storage. It has the highest heat storage capacity per pound of all common materials, is inexpensive, and serves well as a heat storage material. The water is stored in an insulated storage tank which is often underground. See Fig. 13-45. Such a system has problems of freez-

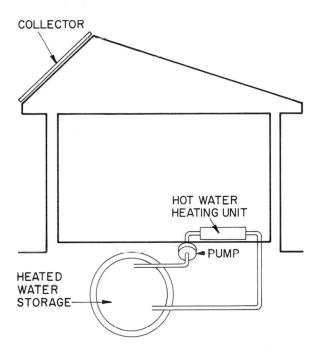

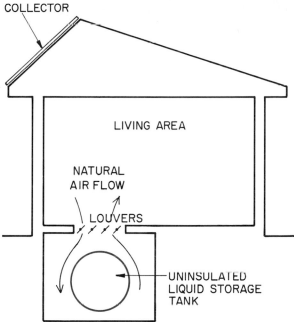

Fig. 13-45. Heat can be stored in liquid held in insulated tanks. A pump circulates the heated liquid to heating units in the living area.

Fig. 13-46. Heat can be distributed by the natural air flow caused by temperature differences.

ing and corrosion. The heated water is circulated from storage to heat-emitting units in the building by an electric pump.

The use of **heat of fusion** or **heat of vaporization** offers possibilities of heat storage. These have yet to be developed to a reliable stage. One possibility is the use of salt hydrates. This heat storage process involves a chemical change — from solid to liquid and back to solid. The temperature of the salts is raised by solar heat. The salts absorb the heat and release water of crystallization which melts the salts. When the temperature drops below the crystallization temperature, the stored heat is released, and the solution returns to solid salts.

Energy Distribution

In an active system, there are three primary ways to distribute solar heat: gas flow, liquid flow, and radiation. A variety of techniques may be used for each of these. Some distribution methods require the use of mechanical and electrical equipment. Others rely upon natural convection and radiation.

One gas flow system is by **natural convection**. It is the circulation of air caused by temperature differences. The hot air rises and the cool air falls, thus creating a circulation system. This is illustrated in Fig. 13-46. The thermal storage tank heats the air

around it. The air rises, flowing into the living space. The cooled air descends to the floor and flows into the area around the thermal storage tank. It is heated and begins to rise again.

Forced-air is another gas flow system. See Fig. 13-47. This system uses mechanical equipment, such as a fan, to distribute the solar heat. The fan forces air over or through some type of thermal heat storage. The heated air is carried by ducts to the living space. Return ducts carry the cooled air back to the thermal storage area. This is much like a conventional forced-air heating system. The ducts for solar systems are larger than those for conventional systems.

A **forced-radiation** system moves heated water through pipes to living spaces. See Fig. 13-48. The heat is transferred to the air by radiation and convection. Usually, fin-type units are used to radiate the heat. They must be larger than those used in conventional hot-water heating systems. The piping can be in the ceiling, floor, or walls. The pipes used for heating can also be used for cooling. A mechanical water-chilling unit is used to lower the water temperature and pump it through the pipes.

Natural radiation involves the transfer of heat to the air with no mechanical device assisting. The heat from the thermal source is absorbed by sur-

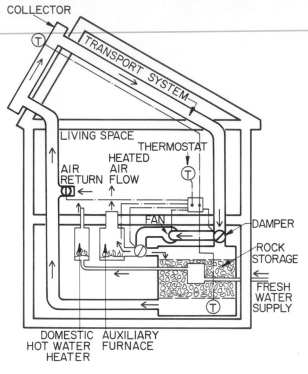

Fig. 13-47. A warm-air flat-plate solar system.

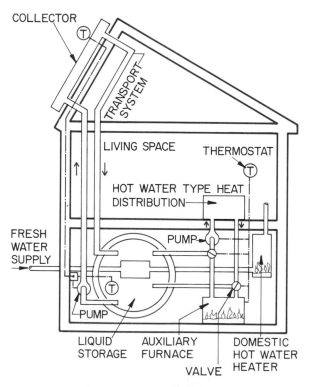

Fig. 13-48. A warm-water flat-plate solar system.

faces of the room, such as walls and furniture. The cooler the absorbing surfaces, the more rapid the transfer of heat.

Typical Active Solar Systems

Two examples of active solar systems are the warm-water flat-plate system and the warm-air flat-plate system.

A **warm-water flat-plate system** is illustrated in Fig. 13-48. The collector is a liquid-cooled flat plate. Energy is removed from the collector by a liquid which flows through tubes in the collector plate. The liquid is usually water or a water-antifreeze mixture. Ordinarily, the liquid is moved through the collector only when the collector surface is hotter than the liquid in storage.

The storage area is a large tank near or beneath a building. It must be well insulated, able to stand high temperatures and to resist corrosion.

The distribution system uses an electric pump to move the heated water through pipes. Most systems use heating coils with electric fans as heat-emitting units in the rooms. A thermostat controls the water flow and the use of the fan coil in each room.

A conventional boiler, either oil- or gas-fired, is needed. It heats the liquid when solar heat is not sufficient to maintain proper temperatures. The liquid is directed through this boiler by valves which change the direction of flow.

The piping for hot water used for domestic purposes, such as washing, is run through the storage tank. Here, the water in the pipe is preheated by the liquid in the tank. The water then flows to a conventional water heater where it is brought to the desired temperature. The tank of the water heater serves as storage.

A solar system for heating water that is only for domestic use is shown in Chapter 11. This system uses a storage tank which holds 30 gallons of water for each person living in the house. It uses the collector shown in Fig. 13-40.

A **warm-air flat-plate system** is shown in Fig. 13-47. The air-cooled flat-plate collector has a solid absorbing surface. It collects direct and diffuse radiation. The heated air is moved from the collector by air flowing in ducts beneath the absorber plate.

The system consists of four different operations:

1. Moving heat to storage and collecting it there,
2. Moving heat from the collector directly to the living area,
3. Drawing heat from storage to heat the living area, and
4. Starting the auxiliary heating plant to heat the living area.

These operations are controlled either automatically or manually by dampeners.

The rock pile method of heat storage is most often used. Rocks that are about 2 inches (51 mm) in diameter are kept in a concrete bin in the basement or beneath the building. The bin must be heavily insulated. Heated air flows into the top of the bin and out to the collector from the bottom. Refer again to Fig. 13-47.

The auxiliary heating system may be integrated with the solar system or be entirely separate. A separate system is generally more expensive. When the systems are integrated, the heated air from the solar collector is usually run through the auxiliary heating unit and into the living area.

This system provides for domestic hot water needs in the same way as the warm-water system.

Passive Solar Systems

Unlike active systems, which are added to a building and operate independently, passive solar systems are an integral part of the overall design of a building. They influence room location and exterior styling.

A passive collector may be as simple as a window or a glass wall. It may involve heat storage and a simple distribution system. Following are examples of basic types of passive systems.

The simplest example is a large **glass wall** facing south. See Fig. 13-49. This is especially effective in northern climates because a wide overhang can be used to protect the window from the sun's rays in

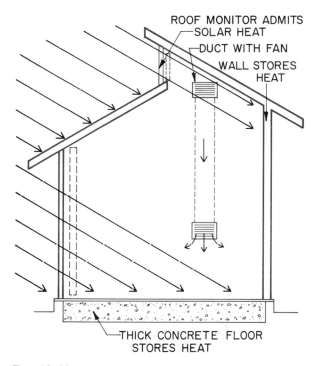

Fig. 13-50. A roof monitor admits solar heat.

the summer. (See the section on orientation in Chapter 4.) At night, some form of insulation over the window will help reduce heat loss. Heavy draperies are most commonly used.

The disadvantage of using windows for solar heat is that the room may become overheated. To avoid this, **masonry units**, such as a heavy concrete floor or masonry interior wall, are used to absorb heat during the day. At night, they will radiate the absorbed heat to the room. The storage effect of a masonry wall or floor can be calculated.

A **roof monitor** is another type of collector. See Fig. 13-50. A monitor is a skylight, cupola, or clerestory shed roof window. All will admit heat into the room. Since heat rises and monitors are at the ceiling, the heat remains next to the ceiling. A fan arrangement is needed to move heat to the floor.

Heat collects in the walls of buildings and in the roofs. This trapped heat can be moved into a building. Since heat rises, a natural flow through a wall or roof can be developed. See Fig. 13-51. This heat can be moved to other parts of a building using **ducts and fans**. This method is rather marginal, but can be improved by making the exterior wall or roof transparent. This will increase the temperature of the air trapped in a wall or roof.

Passive solar heating can also use a **thermosiphoning wall**. This was shown earlier in Fig. 13-43.

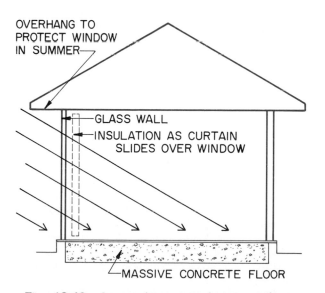

Fig. 13-49. Large glass areas facing south are a simple form of solar heating.

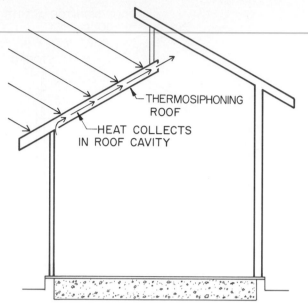

Fig. 13-51. Heat can be collected in a roof or wall cavity and moved into the living area.

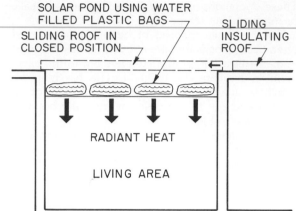

Fig. 13-53. A solar pond can be used for heating and cooling.

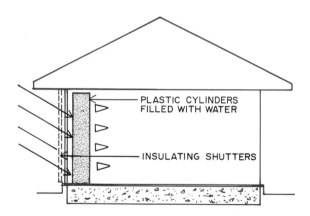

Fig. 13-52. Water-filled cylinders within a thermosiphoning wall absorb and store heat from the sun.

Heat can also be stored in cylinders filled with water. These can be placed inside a greenhouse or other glass wall. Water is an excellent heat storage material. See Fig. 13-52.

Some experimentation has been done using solar ponds for heating and cooling. A **solar pond** is a body of water. It is often located on the roof of a building as shown in Fig. 13-53, but it could also be on ground level. In the winter, it is exposed to solar radiation. It heats up and radiates this heat into the building, usually through a metal ceiling. At night, an insulated roof closes over the pond to help hold in the heat.

In the summer, the insulated roof is kept closed over the pond during the day. The cool water draws heat from the building. At night the roof is opened, and the heat absorbed by the water during the day is lost to the cooler night air. In the morning, the roof is closed again.

One type of energy efficient home that uses a passive solar system is sometimes called a **double envelope house**. It resembles a house within a house. First the basic energy efficient house is built. Then on one or more sides, a second exterior wall is built, forming an enclosed interior courtyard (sometimes called a solarium). This provides a thermal break between the weather and the interior wall of the house. When glassed in, it becomes a huge passive solar collector. The floor is massive masonry and the exterior wall of the basic house is 8 to 12 inches (203 mm to 305 mm) of masonry. These serve as heat storage units releasing the heat in the night. Fans are used to move the heated air into the house. The cool air settles to the floor of the rooms and is moved into the solar space by convection or with small fans. See Fig. 13-54. Windows on the interior masonry wall are used to circulate air between the solarium and the interior of the house. In the summer the glassed area must be shielded from the sun, and wall and roof ventilators are required. These could use small fans or rely on natural convection.

The solarium provides a pleasant space to grow plants, even trees. It can be furnished like a patio or perhaps like an interior room. It serves not only as a solar heater, but also as a bright and cheerful living space.

An attached solar greenhouse economically adds direct gain solar heat to a building. See Chapter 27.

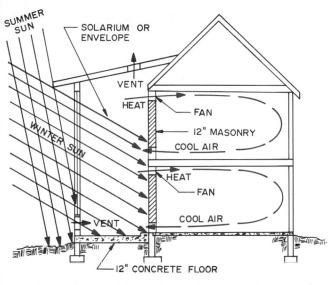

Fig. 13-54. A basic design for an envelope or solarium type house.

Oklahoma State University Architectural Extension

Fig. 13-55. An earth sheltered house.

Earth Sheltered Housing

Earth sheltered housing refers to those buildings that use earth to cover some parts of the walls and very often the roof. This provides a protective barrier, moderates extremes of temperature in the building, and has aesthetic and environmental benefits as well. See Fig. 13-55.

Earth sheltered buildings have the potential to produce substantial energy savings. This is due to the reduction of heat loss and gain through the walls and roof. The enclosed space does not become extremely cold or extremely hot, regardless of outside temperatures. Thus a passive solar heating system can be used effectively.

The earth covering acts as a barrier to the effects of storms and airborne exterior noises, permitting the use of land near noisy areas such as busy highways. The earth covering also supports vegetation, making it possible to blend living areas naturally into the surroundings. The presence of vegetation also helps the ecology.

The structure itself is generally concrete. Therefore it is fire-resistant and not subject to decay or to attacks by insects. Exterior maintenance is minimized.

There are some disadvantages to earth sheltered buildings. Because of the weight of the earth on the roof, heavy loads must be carried. This limits the size of clear interior spans and reduces the number of alternatives for arranging rooms. Also, since some rooms will have no natural light or ventilation, special design solutions are needed.

The concept of an earth sheltered house is in violation of parts of most building codes and zoning regulations. This limits where such housing can be built.

Since the construction is somewhat different from that of conventional houses, many contractors are unfamiliar with it. As a result, the initial cost of the building tends to be greater than that of a conventionally constructed house of the same size. Also, wall or roof repairs are difficult to make and are expensive. (Still, over a period of ten to fifteen years, the energy savings will make up for increased costs.)

Suitable building sites are hard to find: A suitable site should have proper slope and a low water table, and the terrain should allow an exposed wall to be built facing south. Most earth sheltered houses are designed long to increase the area exposed to light. Therefore, the typical residential lot is too small.

Some people do not want to live in earth sheltered housing because they do not like rooms without windows. Also, since this type of house is unconventional, some people feel they might have difficulty selling it. Some banks even hesitate to grant loans for earth sheltered housing. These disadvantages will be of less significance, however, as more earth sheltered houses are built.

Site Planning

Many factors must be considered in the selection of a site. **Positioning** of the house in relation to the sun, wind, and view is very important.

Sun. The house should be placed to take advantage of the sun as a source of heat. Either a passive or an active solar system can utilize heat from the sun effectively. If an active system is planned, the collectors must face south. But a passive system traps the radiant energy that enters the house through windows; therefore, the exposed wall of the house must face south.

An efficient plan is to place the exposed window wall south and have the other three sides covered with earth. Just as the orientation to the sun is considered, the angle the sun's rays make with the earth at that location must be considered. In winter, the sun's rays are at a lower angle. With southern orientation and proper overhang, the winter sun can enter and warm the house. In summer, the sun makes a steeper angle and can be blocked with an overhang or sun screen of some type. See Fig. 13-56. Also see Chapter 4 for more information.

When planning a house with an atrium, keep in mind that the extent of light penetration is limited by the width of the court. See Fig. 13-57.

Wind. It is also very important to position the house to be protected from cold winter winds and to take advantage of cooling summer winds. The effect of the **prevailing winds** has a great deal of influence on the energy used to heat the house. Find

out the usual direction of the cold winter winds and place the earth sheltered sides of the house toward them. If possible, have no door or window openings facing that exposure. See Fig. 13-58. Such a design barrier is a major feature of the earth sheltered home. If a design has an atrium or courtyard, the barrier also shelters the exposed glass from the winds; however, some turbulence may occur. This turbulence can be reduced by building a protective wall or by planting shrubs. See Fig. 13-59.

Natural **ventilation** is important to a successful design, and the summer winds can often be used to provide it. Keeping in mind that a southern orientation is necessary for solar heating, locate windows to take the greatest possible advantage of the prevailing summer winds. Sometimes a skylight or clerestory windows can be used to help ventilation. But these often lose considerable heat in cold weather and, if used, need to be heavily insulated in the winter. The placement of interior walls and halls must be considered to help support natural ventilation. If the earth cover is not broken for ventilation, some means of mechanical ventilation will be needed.

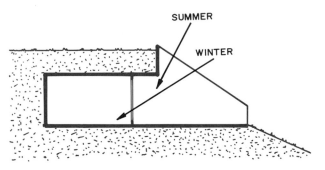

Fig. 13-56. Earth sheltered houses can be oriented to take advantage of the winter sun as a source of heat.

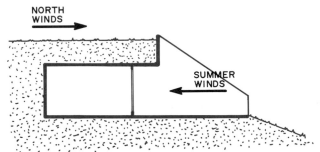

Fig. 13-58. Place the earth sheltered side of the house toward the cold winter winds.

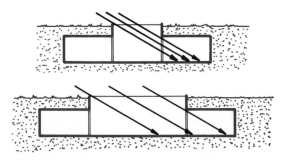

Fig. 13-57. The larger atrium permits greater light penetration of the building and of the atrium itself.

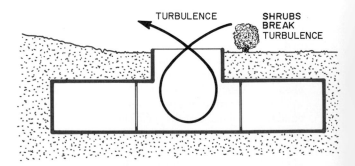

Fig. 13-59. Turbulence in an atrium can be reduced by barriers such as shrubs or a wall.

View. Positioning with regard to view is important, since a good view adds much to the enjoyment of a home. An ideal southern exposure will provide a pleasant view, but since the ideal is not always the case, some compromises may need to be made. Remember, though, that windows placed to the east or west will increase air infiltration, thus increasing heating and cooling costs. If the house is built on a flat site, the view is limited unless a two-story structure is built with the top story above grade.

Other Site Considerations

The **topography** of the building site affects not only the view but the direction of surface water run-off and the basic design of the building. On a nicely

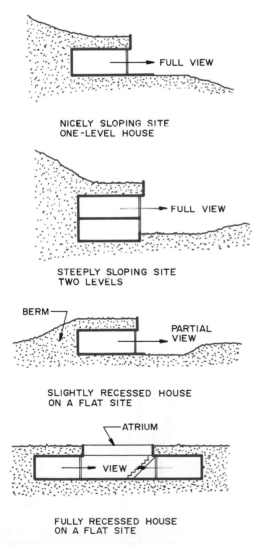

NICELY SLOPING SITE
ONE-LEVEL HOUSE

STEEPLY SLOPING SITE
TWO LEVELS

SLIGHTLY RECESSED HOUSE
ON A FLAT SITE

FULLY RECESSED HOUSE
ON A FLAT SITE

Fig. 13-60. The slope of the site helps determine the design and view.

sloping site, a one-level design is easily accommodated. See Fig. 13-60. If the site is steeply sloped, a two-level design is practical. On relatively flat sites the building can be slightly recessed, with a mound of earth called a **berm** built around it. Or the building can be fully recessed and have an atrium. Steps can lead from grade to the living level. Recessed design possibilities are usually limited to one-level buildings. These designs also need special provision for a system of water removal.

Vegetation on the site helps control the interior temperature and noise. Plant growth on the roof is especially important. It is not only attractive and helpful to the ecology, but it also reflects the summer heat, reducing the solar gain in the house. In the summer, it is important to have shading on the south side of the house. Deciduous trees and shrubs can effectively shield the house from the sun. In the winter, deciduous trees lose their leaves and the sun filters through, warming the house. If there are windows exposed to the east or west, shrubs will serve as windbreaks, reducing air infiltration.

The **lot size** in most residential developments is too small to accommodate an earth sheltered house. The typical earth sheltered house is either very long or has a large square shape and an atrium. The large area is necessary to get as much natural light and ventilation as possible. In addition, the earth built up around the house as a berm requires extra space. See Fig. 13-61.

In most developments, the **zoning ordinances** require that the house be set back a specified distance from the street and property lines. This further limits the area available for a house. Attempts have been made to permit the portion of the house that is below grade to extend into the set-back area, because above ground the roof looks like a normal yard area. See Fig. 13-62. Zoning is discussed in greater detail later in this chapter.

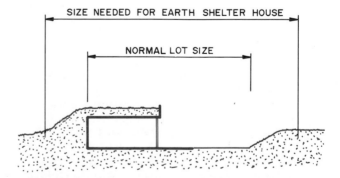

SIZE NEEDED FOR EARTH SHELTER HOUSE

NORMAL LOT SIZE

Fig. 13-61. An earth sheltered house requires a larger lot than a conventionally built house.

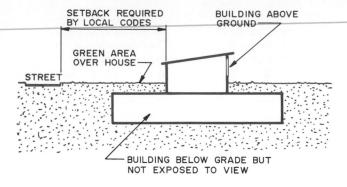

Fig. 13-62. Setbacks, as required by local zoning ordinances, limit the space available for earth sheltered homes.

In established developments, **adjacent houses** may block the earth sheltered house's sunlight, or view of the southern exposure. A good solution is to establish separate developments for earth sheltered housing, with setbacks and height requirements suitable for this type of dwelling.

When selecting a site, investigate the type of **soil and groundwater conditions.** Since the typical earth sheltered house is heavier than a conventional house, it is important to determine whether the soil will support the load without excessive footings. Some soils expand when wet. This can exert great pressure (hydrostatic pressure) on the walls and cause them to crack.

Ideally the water table should be several feet below the floor of the house. An accurate water level should be determined, because it influences the waterproofing and floor design. Water below the floor can exert great pressure that can lift or buckle the floor.

Building Codes

The national codes in common use include the Uniform Building Code (UBC), the Basic Building Code (BBC), the National Building Code (NBC), and the Southern Building Code (SBC). The regulations related to housing are published as the "One and Two Family Dwelling Code" (FDC). In addition, the Hud Minimum Property Standards influence housing design.

Before building an earth sheltered house, it is necessary to make a thorough study of the codes to identify those that would be violated, then to try to alter the design or get a variance by showing how your design meets the **intent** of the regulation. Most codes have an "alternate methods and mate-

rials" section that allows for any new material or method to be tested to prove it is equivalent to the code regulation. If the tests are rejected, the codes have an appeal process.

Following are some of the general areas in which difficulties occur.

Fire Safety and Egress

> Sleeping rooms should have one operable window or exterior door for emergency exit. Windows, if used, must have a sill not more than 44 inches above the floor. Windows must have a net area of 5.7 square feet with a minimum opening height of 24 inches and width of 20 inches (FDC Sec. R-211).

The intent of this requirement is to provide two separate escape routes. This requirement can be met by using a floor plan such as that shown in Fig. 13-63. This does cause the design to be very long. An alternate solution is to provide two separate ways to leave the bedroom but no direct exterior exit. See Fig. 13-64. This plan does not meet the

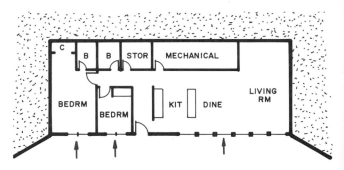

Fig. 13-63. All habitable rooms have natural light, adequate ventilation, and emergency exits.

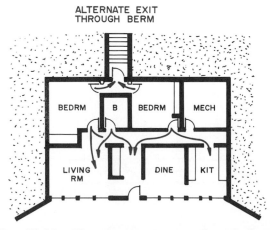

Fig. 13-64. This plan gives two paths out of each room without windows, but it does not meet existing codes.

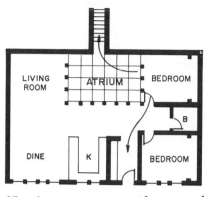

Fig. 13-65. An atrium can provide a second exit but this plan usually does not meet codes.

code and requires special approval. Another possibility is to provide a door or window to an atrium and give the atrium an outside exit. See Fig. 13-65. Again, this plan would require approval.

Sprinklers and smoke detectors are other ways to try to meet this requirement. The use of a sky light with a ladder provides exterior escape, but it violates the sill requirement. Many people are not physically able to climb a ladder, and the sky light would tend to fill with smoke.

Natural Light and Ventilation

All habitable rooms shall be provided with an aggregate glazing area of not less than 10 square feet or 1/10 of the floor area of such rooms. One-half the required area of glazing shall be openable (FDC Sec R-204)

Habitable rooms include those used for sleeping, living, cooking, and dining. Areas such as closets, pantries, bathrooms, halls, laundries, and storage and utility rooms are not considered habitable rooms.

The ventilation problem for earth sheltered homes can be met by mechanical means. Glazed areas need not open where there is an approved mechanical ventilation system capable of producing a change of air every thirty minutes. The opening for the ventilation system need not be one that is required for fire egress.

There are several ways to provide natural light to living areas that have no exterior view. One is to use a skylight. This provides light but results in a large heat loss. Another way is to put glass between rooms borrowing natural light from an outside room. See Fig. 13-66. An atrium can supply natural light but no view or direct exterior access. See Fig. 13-67. One solution for living areas where privacy

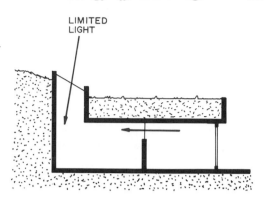

Fig. 13-66. Natural light can be admitted through a skylight or through glass in an interior wall.

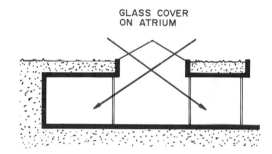

Fig. 13-67. An atrium admits natural light into the rooms facing it.

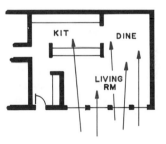

Fig. 13-68. Natural light reaches rooms that have no interior walls.

is not important is to combine several rooms into one without walls. In Fig. 13-68, the living room, kitchen, and dining room are combined and have natural light and an exterior view, even though two of them have no windows. This plan meets most codes for light and ventilation.

Prohibition of Below-Grade Space

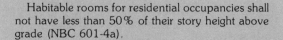

Habitable rooms for residential occupancies shall not have less than 50% of their story height above grade (NBC 601-4a).

While not all codes specifically prohibit below-grade residential space, they all have regulations concerning basements and cellars, and these must be handled in some way.

Guardrails

> All enclosed floor and roof openings, open and glazed sides of landings and ramps, balconies or porches which are more than 30 inches above grade or floor below and roofs used for other than service of the building shall be protected by guardrails. Guardrails shall not be less than 42 inches in height (UBC Sec. 1716).

A guardrail placed at the edge of the roof is not usually an aesthetically pleasing design feature. Shrubs could be used to mark the danger point, but these do not prohibit someone from walking on through. One possible solution is to build a barrier that keeps everyone off the roof. See Fig. 13-69.

There are also a variety of provisions for guardrails on retaining walls having a drop of 24 inches (610 mm) or more.

Structural Design

The building codes have sections on masonry, steel, wood, concrete, and general design requirements such as for foundation loads, stresses, materials, and tests. Because these codes include all types of commercial construction, they adequately cover structural design features needed for earth sheltered houses. Since the structural design is different from a conventional house and the design loads are many times greater, the services of an engineer are required.

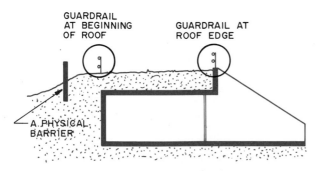

GUARDRAIL AT BEGINNING OF ROOF

GUARDRAIL AT ROOF EDGE

A PHYSICAL BARRIER

Fig. 13-69. Unmarked, the edge of the roof of an earth sheltered house may not be obvious. Barriers are needed to prevent accidents.

Waterproofing

At the time of this writing, there are no specific code requirements for waterproofing below-grade earth covered roofs. Codes do specify acceptable roofing materials for above-ground space. They have sections, addressing the construction of basements, in which specifications are given for the waterproofing of earth covered walls. But standards for waterproofing below-grade habitable space are yet to be developed. Codes do specify vapor barriers under concrete slab floors. The waterproofing of the floor is very important in earth sheltered houses.

Waterproofing is affected by the water table. Some codes require that floors of habitable spaces that are partially below grade be at least 2 feet (610 mm) above the highest known groundwater table.

Energy Use

An earth sheltered house must comply with the same energy use codes as a conventional house. The house must meet minimum standards for energy use. This creates problems in the planning of an earth sheltered house, because the standards are based on above-grade construction. Codes do include provisions for alternative building systems and equipment design, but it must be shown that the energy consumption is not greater than that of a similar building with similar forms of energy requirements. The builder must provide the comparative energy use analysis.

Zoning

Zoning ordinances are developed locally and vary from city to city. There are no national zoning ordinances as there are building codes. Zoning ordinances are set up to regulate the size, height, bulk, location, and use of buildings. Since these are written for conventional housing, builders of earth sheltered houses may experience some zoning difficulties. Exceptions to zoning ordinances can be made by filing an appeal with the local zoning board. Following are examples of difficulties that can arise:

1. **Prohibition of underground spaces** — This prohibits living in basements or other below-ground space. Although a well-designed earth sheltered house presents a pleasant and healthful environment, this ordinance makes building an earth sheltered house difficult.

2. **Minimum height requirement** — Minimum height requirements are designed to ensure the

relative uniformity of house heights within a residential area.

3. **Minimum floor area** — This requirement is designed to protect property values by regulating the minimum size of the house that can be built. These exclude basements in their calculations, and this exclusion is often interpreted to apply to earth sheltered housing.

4. **Maximum lot coverage** — This specifies the maximum lot area that can be covered by buildings. It is intended to provide a certain amount of yard area and space between adjacent houses. This also controls the amount of hard surface area, which can create water runoff that overloads storm sewers and is ecologically undesirable. Since most earth sheltered houses are on one level, they often exceed these maximum requirements. However, they do meet the intent of the regulation because the earth-covered roof provides open space and the soil reduces runoff by absorbing the water.

5. **Setbacks** — Setback requirements specify how far a house must be built from the street and from the sides of the property. Because of the nature of the design of earth sheltered houses, they often cannot be built within these restrictions. If the regulations would consider that portion of the house that is above grade, the regulations could be met. Refer back to Fig. 13-62.

Relationship to Grade

When deciding the best way to position the house in relation to the grade, consider the slope of the site, the water table, the view, the exterior appearance, the entry, and the garage. The final solution will sometimes be a compromise. For example, placing part of the house above grade to obtain a good view may reduce energy efficiency. Examples are shown in Fig. 13-10.

On a flat lot, the house can be semi-recessed or fully recessed. A **semi-recessed** design permits a limited view and helps get the floor above the water table. Three sides of the house are bermed. The berm aids in providing surface water runoff. The roof may be earth covered or a conventional roof.

A **fully recessed** design has a view usually limited to an excavated area. It is fully earth sheltered and requires a stair or ramp for access down to the front entrance. The garage may be above grade and bermed to blend into the overall design. An alternate plan is to design the house around an atrium. This provides no view except the plantings in the atrium. It provides a great deal of privacy but less of a relationship with the surrounding community.

A sloping site provides the best opportunity for a view, and the house can be semi-recessed or fully recessed. Steeply sloping sites enable two-story houses to be built. The second level can either be bermed or fully exposed. It is usually a single room where a view is important. Too much exposure will severely limit the energy efficiency of the house.

Basic Layouts

The three basic approaches to developing a floor plan are the single elevation type, the atrium type, and the penetration or perimeter type. See Fig. 13-70.

The **elevation type** is earth covered on three sides. The fourth side is exposed. This design permits maximum earth cover and, if the windows face south, provides a source of solar heat, natural light, and ventilation. All habitable spaces are required to face along this south wall. Other spaces, such as a bath or utility room, must be along the rear wall. These requirements produce a long narrow plan with a lengthy interior traffic pattern. In a two-story design, halls are shorter.

The **atrium type** is fully recessed, and the habitable rooms open onto the atrium. Rooms can be on all four sides of the atrium, or on three sides with one side left open for access.

The atrium may or may not be covered with a glass roof. In either case, the plan presents difficulties in the development of a traffic pattern. In most climates, it is necessary to be able to go from room to room without going outside. If the atrium is covered, people are protected as they cross it, but this violates codes requiring direct openings to the outdoors. The traffic can be directed along the inside of the exposed wall, but this creates a long traffic pattern. It is usually necessary to permit traffic to flow through rooms such as a living room or dining room even though this restricts the use of the room somewhat. Private rooms, such as bedrooms, must not be used as part of the traffic system.

An atrium type of plan often requires exceptions to the codes because of the problem of getting windows in each room. Other design possibilities include using two atriums, building a two-story house, or cutting windows through the berm (as is done in the penetration type).

The **penetration or perimeter type** plan cuts through the berm to place windows and doors in various locations. This allows light, ventilation, and

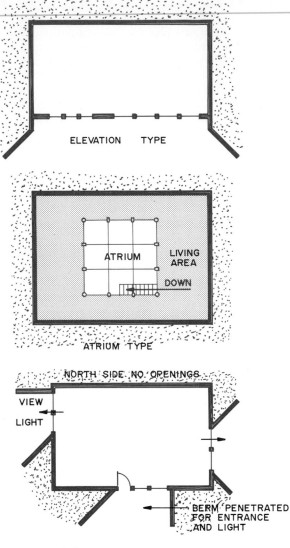

ELEVATION   TYPE

ATRIUM   TYPE

PENETRATION OR PERIMETER TYPE

Fig. 13-70. The basic layouts for earth sheltered houses.

a view from any direction. It also allows a floor plan that is much like that of a conventional house and that aids in the planning of circulation through the house.

The doors and windows do break the earth shelter; therefore, locations should be carefully planned. Windows to the north would be especially suspect. Such windows could be covered with insulated shutters during periods of extreme heat or cold. It is best to face the house south and place most of the windows in this direction, while keeping perimeter windows and doors to a minimum.

Design Considerations

The following design considerations relate to earth sheltered homes. Basic planning principles and minimum size and space data can be found in detail in Chapter 4.

1. **Entry** — The entry should be easily recognizable and provide a smooth transition from the outside to the interior of the house. It is the "face" of the house and deserves careful attention both from an aesthetic viewpoint and with respect to its location on the floor plan. Consider either building an overhang to protect visitors or building a patio entrance with trellises and trees.

   The entry should be light and spacious, especially since people may mistakenly feel they are going into a dark cavern when they enter an earth sheltered house. The entry should be near the main living level and should not require visitors to descend a long flight of steps. Steps should be outside the house, leading up to the entry.

   Locating an entry in relation to an atrium presents special problems. The entry could be through the atrium, but this violates the privacy of the space. Or, an entry could be cut through the berm. Another possibility is to build a second, smaller atrium especially for the entry.

2. **Garage** — The placement of the entry into the house is planned in direct relationship to the garage location. Quick access from the garage to the entry is necessary. The garage could be built as part of the earth sheltered building, but this is very expensive. It could be above grade and bermed to blend in with the landscape. Care must be taken that the garage does not block the view, the sun, or the summer breezes. See Fig. 13-71.

3. **Landscaping** — The landscaping of an earth sheltered house is an important part of the exterior design. It should be planned as the house is designed.

   Since grass and ground covers such as ivy reflect solar heat and help keep the house cool in the summer, the roof should have good coverage. Trees and shrubs can be used as solar breaks on the south wall.

   The amount of soil needed for various plants to grow must be known. Some small shrubs and grass require a minimum of 18 inches (457 mm) of soil and must be watered frequently. Small shrubs require 30 to 36 inches (762 mm to 914 mm) of soil covering. It is recommended that

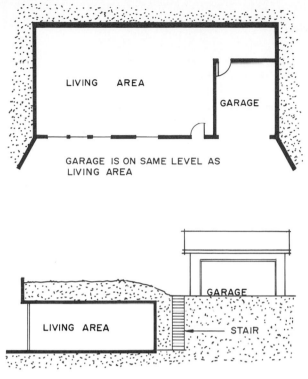

LIVING AREA

GARAGE

GARAGE IS ON SAME LEVEL AS LIVING AREA

GARAGE

LIVING AREA

STAIR

GARAGE IS ON GRADE

Fig. 13-71. Suggested garage locations.

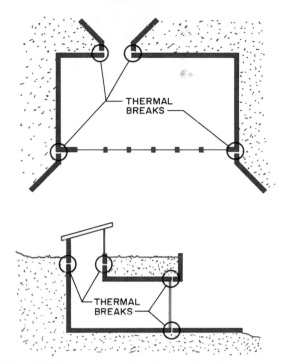

THERMAL BREAKS

THERMAL BREAKS

Fig. 13-73. Locations for thermal breaks.

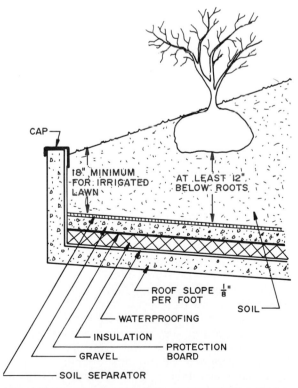

CAP

18" MINIMUM FOR IRRIGATED LAWN

AT LEAST 12" BELOW ROOTS

ROOF SLOPE $\frac{1}{8}$" PER FOOT

SOIL

WATERPROOFING

INSULATION

GRAVEL

PROTECTION BOARD

SOIL SEPARATOR

Fig. 13-72. Recommended minimum soil coverage for lawn and small shrubs.

trees be planted around the house but not on the roof. See Fig. 13-72.

The soil on the roof must have adequate drainage so that plants can grow properly. Usually a 4-to 6-inch (100 to 150 mm) layer of gravel and some slope to the roof is adequate. The maximum slope of the earth for growing grass and ground cover is 3 to 1. After the house is built, the ground should be covered with sod to control erosion. Sod is the best choice because it provides ground cover quickly.

4. **Thermal break** — Earth sheltered houses are usually built of concrete, and concrete conducts heat and cold. Wherever the wall or roof is continuous and runs from interior to exterior, a thermal break is needed. See Fig. 13-73. These occur at retaining walls, floors, roofs, and skylights. A layer of rigid foamed plastic insulation makes a good thermal break, though expansion joint material or wood could be used.

Structural Design

Since the weight of the roof and walls of an earth sheltered home is much greater than the comparable weight on a conventional home, the design of the structure must be considered as the floor plan is developed. It is especially important to provide support for interior walls, since these carry the roof loads over the spans required by the rooms. Developing such designs requires the services of an engineer. The following paragraphs describe a few structural considerations that might have to be made for an earth sheltered home. They are simply examples, and are **not** to be used as design recommendations.

Loads. The load considerations include dead and live loads as well as special problems related to earth sheltered houses. **Dead loads** include the structural system, roof, interior finish, and waterproofing. **Live loads** include, for example, wind, snow, furniture, people, and surface water from rains. If trees are to be planted, they must be considered as live loads because they will increase in weight over the years. Normally the roof is not designed to hold trucks, tractors, or autos. Providing support for such live loads would greatly increase both the structural requirements and the cost.

Vertical and horizontal loads must be considered as well. See Fig. 13-74. Vertical loads, such as soil on the roof, are figured at 100 to 120 pounds per square foot for each foot of earth (488 to 585 kg/m²). Horizontal loads are created by lateral earth pressures. While these loads vary with the type of soil, 30 pounds per square foot for each foot of depth (146 kg/m²) is a typical load.

Groundwater causes an extra load on the building. Water in the soil adds to the weight of the soil. If the soil is saturated and water is standing, **hydrostatic pressure** is produced. This must be considered, since it can cause walls to collapse and floors to lift.

Footings. Footings are reinforced cast-in-place concrete. Footings for an earth sheltered house are generally larger than those for a conventional house because they must support heavier loads. It is

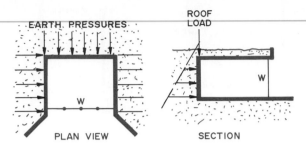

Fig. 13-75.   Walls carry roof loads and resist earth pressures.

therefore critical to the eventual house design to do an analysis of the soil to learn its bearing qualities.

Walls. Exterior walls carry part of the roof load and resist horizontal earth pressure. In Fig. 13-75, exposed wall **W** carries a roof load, but does not have to resist horizontal earth pressures. The exposed wall may or may not have to carry a roof load. The roof could be self-supporting, in which case the wall would simply close out the elements. The design of a wall depends upon the loads, the soil conditions, and the height of the wall. Openings must have a steel beam across them to carry the load.

Exterior walls are usually made from cast-in-place concrete or concrete masonry units. In general they range from 10 to 18 inches (254 to 457 mm) thick. The placement and amount of reinforcement vary with the design. The masonry wall must be reinforced with steel reinforcing bars. The use of bond beam blocks every other course is a possible design solution. See Fig. 13-76.

In an earth sheltered design, interior walls are also supporting walls. They carry roof loads and vertical loads. Their thickness and design depend upon the loads to be placed upon them. They are usually cast-in-place concrete or are concrete masonry units. Openings in these interior walls require a steel beam across the opening.

Floor. The ground floor is usually a 4-inch (102 mm) cast-in-place concrete slab. Slabs subjected to underfloor water pressures must be designed to withstand these pressures. That is, they must be thicker and more heavily reinforced. When two-story construction is used, the roof, intermediate floor, and ground slab provide reactions to the lateral pressures on the wall. The second floor should be designed with no large openings next to any exterior wall that is resisting earth pressures. The floor actually serves as a beam laid on its side. See Fig. 13-77. Wood is usually not suitable for

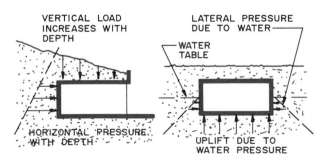

Fig. 13-74.   Vertical and horizontal loads.

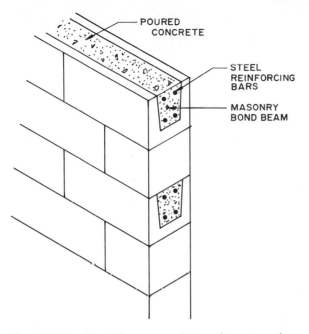

Fig. 13-76. Bond beams can be used to strengthen masonry walls.

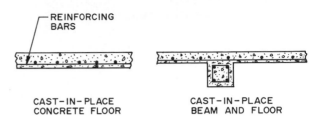

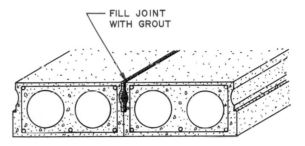

Fig. 13-78. Cast-in-place reinforced concrete floors.

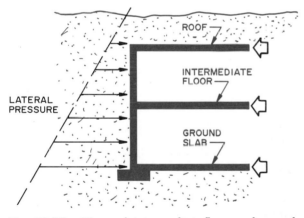

Fig. 13-77. The roof, intermediate floor, and ground slab provide reactions to lateral earth pressures.

Fig. 13-79. A typical precast reinforced concrete plank.

second-floor construction because it cannot handle the shear stresses.

Roofs. The common structural designs for roofs include cast-in-place concrete, precast concrete planks, heavy timber planking and beams, and open-web steel joists with concrete or precast plank decking. See Chapter 23 for details.

Reinforced cast-in-place concrete roofs having 18 to 24 inches (457 mm to 610 mm) of earth usually span 15 to 20 feet (4572 mm to 6096 mm).

Beyond this the cost increases and they are less economical. A cast-in-place beam and slab roof can be used to span longer distances. Typical roof thicknesses range from 10 to 12 inches (254 mm to 305 mm). See Fig. 13-78.

Precast concrete planks span a roof in one direction. With 18 to 24 inches (457 mm to 610 mm) of earth they can span 20 to 25 feet (6096 mm to 7620 mm) economically. See Fig. 13-79. Often a cast-in-place concrete slab is poured over the planks.

Heavy timber roof systems span the roof in one direction. The planking transfers the load to the beams. Practical spans depend upon the load, size, and species of wood, but usually approach those for cast-in-place concrete.

Open-web steel joists have the advantage of being light in weight. They are covered with a corrugated metal decking and 3 or more inches (76 mm or more) or reinforced concrete. They are available in a wide range of load-carrying capabilities and can span distances equal to precast concrete planks. See Fig. 13-80.

To prevent subsurface water from forming a pond, a roof should either be designed to have slight slope, such as 1/8 inch to the foot (approximately 3 mm per 305 mm), or should be crowned in the center. This reduces the load on the roof and helps prevent leaks. The slope should not be too great. If

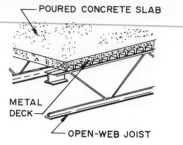

Fig. 13-80. A typical deck using open-web joists and concrete slab over metal decking.

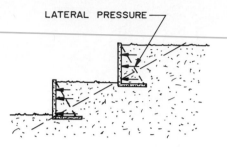

Fig. 13-82. A retaining wall can be stepped up the slope, reducing the structural requirements.

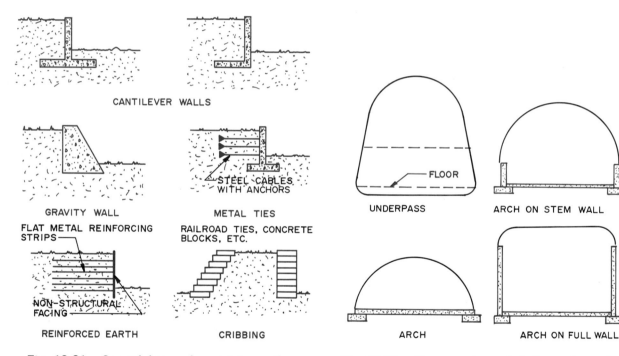

Fig. 13-81. Several designs for retaining walls.

Fig. 13-83. Corrugated metal shells can be used to form earth sheltered buildings.

too much subsurface water is drained away, grass and other plants may not live.

The roof also helps support the walls which are under horizontal loads.

Retaining Walls. Earth on the exposed wall on the outside of the house must be held back with a retaining wall. The most common retaining wall is cast-in-place concrete, but wood, stone, sheet steel, or other materials could be used. Various wall designs are shown in Fig. 13-81. Retaining walls can also be stepped. See Fig. 13-82. Stepping reduces the structural requirements of the retaining wall, so a lighter wall can be used.

Shell Structures. The use of curved shell structures is another way of forming an earth sheltered house. The geometry of the curve acts to support the loads without requiring the heavy mass of material that is required by concrete structures. The shell is either steel or precast concrete. Corrugated steel culvert sections provide a variety of shapes useful for this purpose. See Fig. 13-83.

Waterproofing

Keeping the inside of the structure dry is a major concern of those building earth sheltered houses. One of the first things to do is to avoid a site that has a high water table or is in a low area or flood plain. Next, plan to carry away surface water. This can be done by forming a swale, by grading a slope, or by using a gravel-filled trench with drain tile near the roof. See Fig. 13-84.

The backfill could contain gravel to help water to drain away from the house. The top layer can be

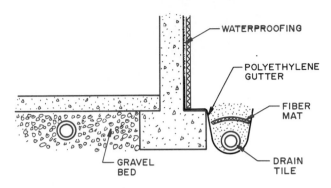

Fig. 13-85. Drain tile are used at the footing and under the floor if needed.

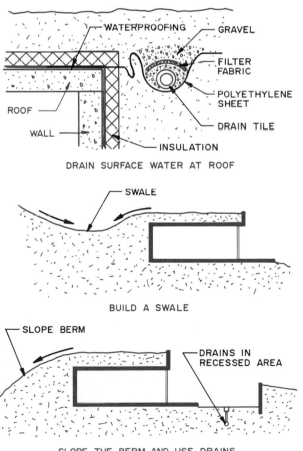

Fig. 13-84. Ways to carry away surface water.

soil to permit grass to grow. Filter fabrics should be placed over the drain tile to keep them from clogging with soil particles.

If the water table is always at or near the floor level, the floor should be on a gravel bed with additional drains below. See Fig. 13-85. The drains should run out of the side of a slope so that the water runs off on the surface. They need provisions for entry to permit cleaning, and should be capped with a screen to keep animals out of the system. Generally, running the drains into the sanitary sewer is prohibited.

Special attention to the concrete mix and the admixtures can produce a concrete that will reduce the permeability of the wall. Proper pouring, vibrating, and curing increases the wall's watertightness.

Waterproofing materials should be applied to the outside of the wall. This allows the water pressure to press the material against the wall. The wall itself will be dry, which protects the reinforcing steel. If it gets wet the steel rusts.

Following are the commonly used waterproofing techniques. Some are more successful than others, but in all cases proper installation by trained workers is essential to success.

Parging is the technique of coating the wall with a dense cement plaster. Since concrete walls tend to crack, the brittle cement plaster is also likely to crack.

Asphalt coatings can be hot or cold and are applied by troweling, brushing, or spraying. Hot asphalt coatings tend to become brittle as they age. Cold coatings remain more pliable and bridge small surface cracks but not larger structural cracks. Asphalt tends to slowly dissolve due to water action and will need to be replaced eventually. See Fig. 13-86.

Fig. 13-86. Applying insulation over asphalt waterproof coating.

Pitch is applied much like asphalt and is a more stable material. It tends to get brittle and will crack.

Polyethylene sheets can be wrapped around the walls. Underground it lasts a long time, but if exposed to sunlight it degrades. It is thin and easily punctured when the area is backfilled. It is most widely used on top of gravel layers under concrete slab floors.

Built-up membranes are made up of alternate layers of asphalt or pitch and layers of glass fiber fabric. At least three or four plies are usually applied. Since the walls are vertical, it is difficult to apply the hot asphalt or pitch and get it to saturate the fabric before it cools.

Bituthene is a membrane made up of polyethylene — coated rubberized asphalt. It is applied to the wall and firmly rolled. It is not used where the wall is under continuous water pressure.

Polyethylene sheets are applied to a wall that has been covered with mastic. Sheets should be overlapped and sealed with mastic. Polyethylene requires a foundation drain; it is not recommended for use on roofs.

Butyl rubber, epdm, and **neoprene** membranes can be used on the roof and walls. The sheets are glued to the roof or wall. They may be completely embedded in the adhesive, or the adhesive can be applied in strips. It is essential to take great care when applying these, and careful inspection during the application process is important.

Liquid polymer membranes are polyurethanes in liquid form and are applied with a trowel. They can be especially useful for waterproofing awkward areas such as around pipes. They are not recommended for use on precast roof systems.

Bentonite panels are made of a layer of sodium bentonite with a cardboard facing on each side. Bentonite is a clay that, when wet, swells up to fifteen times its dry volume. The clay absorbs moisture, creating a thick, impenetrable gel membrane that keeps water from getting through. It will expand to fill cracks in the concrete.

The sheets are nailed to the wall. After the area is backfilled, the biogradable cardboard rapidly decomposes.

Sprayed-on bentonite is a mixture of bentonite and a mastic binder that is sprayed onto the walls and roof to about a 3/8-inch (10 mm) thickness. Some types can be applied with a trowel. The bentonite coating should be covered with rigid insulation or a 6 mil sheet of polyethylene to protect it from damage during the backfilling operation. If the backfill is rocky, a rigid protection board should be used to cover the bentonite.

Waterstops are used to prevent leaks. The most likely place for a house to leak is in a joint. Various types of waterstops are available to reduce water leakage at expansion joints or cold joints in concrete. (A cold joint is an area where one concrete pour ends and another abuts it.) One type of joint is shown in Fig. 13-87. A waterstopping material is placed between cold joints in roofs, floors, footings, and walls. It is also applied directly to precast concrete decking, and the cast-in-place slab is poured

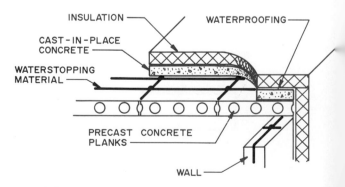

Fig. 13-87. Using waterstopping material on a concrete roof cast over a precast concrete deck.

over it to form a seal. Bentonite is then applied over the completed work.

Waterproofing Design Details

The designing of a waterproofing system should be done by those experienced and knowledgeable in this field. The following examples illustrate a system using bentonite.

A standard wall detail with a drain tile at the footing is shown in Fig. 13-85. The top of the drain tile should be below the bottom of the floor. A filter matt, such as 2- to 4-inch fiberglass insulation, is placed above the drain tile. The filter matt is covered with a coarse granular material. A polyethylene sheet is used to form a gutter below the drain tile.

A roof drain system is shown in Fig. 13-84. This system is recommended in addition to the footing drain. It keeps the surface water from passing down the walls.

A system for waterproofing a precast concrete roof is shown in Fig. 13-87. The precast concrete members usually have a cast-in-place concrete layer on top. A grid of waterstopping material is placed on the precast unit before the cast-in-place layer is poured.

Waterproofing at the parapet wall is shown in Fig. 13-88. Notice that the top of the parapet has cap flashing.

When a wall is exposed rather than sheltered, the sill must be waterproofed. Suggested details are in Fig. 13-89.

Details for waterproofing a skylight or other type of roof penetration are in Fig. 13-90. In this process,

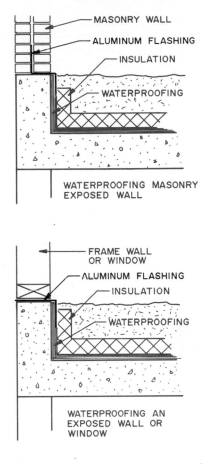

Fig. 13-89. Waterproofing at an exposed masonry or frame wall of window.

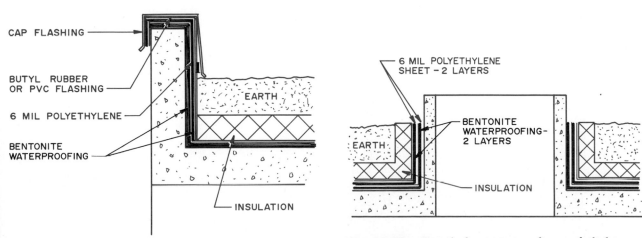

Fig. 13-88. A design for waterproofing at a parapet.

Fig. 13-90. Details for waterproofing a skylight or other penetration through a roof.

bentonite is placed on the roof deck, and a polyethylene sheet is cut in the center over the roof penetration and stuck to the vertical sides with bentonite. Next, bentonite is troweled over the polyethylene, and a second polyethylene sheet is wrapped around

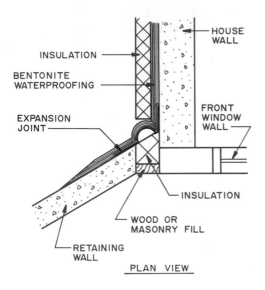

Fig. 13-91. A way to waterproof expansion joints and thermal breaks.

and adhered to the vertical walls. Insulation is then installed, and finally backfilling is done.

Another area needing waterproofing is the thermal break where a retaining wall meets the wall of the house or where an expansion joint occurs.

Build an expansion strip by gluing half of a piece of 1½-inch foam rubber pipe insulation to a strip of P.V.C. butyl rubber. Glue this to the wall and cover with bentonite. See Fig. 13-91. This type of joint will move with the wall yet will remain watertight.

Insulation

The insulation used on the building envelope should be placed on the outside of the building. This keeps the temperature of the mass relatively uniform and eliminates rapid temperature changes. This means that the heating and air-conditioning equipment can be sized for average conditions rather than requiring extra capacity to handle peak periods.

Insulation should be able to resist earth loads of 20 to 30 psi (138 kPa to 207 kPa), have low water absorption, and resist deterioration due to chemical properties in the soil. Tongue-and-grooved sheets resist water movement between the sheets. Extruded polystyrene boards are highly recommended.

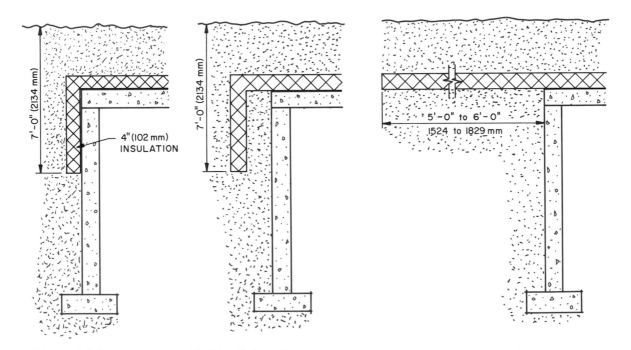

Fig. 13-92. Suggested ways to insulate walls.

The amount and placement of insulation should be determined by someone experienced in the use of insulation for earth sheltered housing. Following are several approaches. Walls may be insulated as shown in Fig. 13-92. One approach is to insulate the upper half of the wall, or to about 7 feet (2134 mm) below the surface of the earth. A common thickness is 4 inches (102 mm). This leaves the temperature of the lower half of the wall to fluctuate with the ground temperature, helping to stabilize the interior air temperature. Another approach is to extend the insulation past the wall and then drop it about half way down the wall. This allows an additional mass of earth to be exposed to the wall, controlling interior temperatures. The same idea is carried out when the insulation is extended horizontally from the roof, except that the mass of earth exposed to the wall is greater.

Roof insulation is typically placed on top of the waterproof covering. A thickness of 4 to 6 inches (102 mm to 152 mm) is commonly used.

The thermal conductivity of soil is about twenty-five times greater than that of rigid foam insulation. Thus the earth covering provides very little insulation. While one inch of extruded polystyrene has an R-value of about 5, the R-value of earth is approximately 0.16 per inch of thickness. The value of an earth covering, however, is that it tends to cause the concrete mass of the structure to maintain relatively stable temperatures. This produces a substantial energy savings.

Since the floors are located in an area where temperatures are relatively stable, they are not insulated. Any edges exposed to the weather should be insulated in the same manner as an above-grade concrete slab house.

Insulation can be used in other ways. Windows can have insulated shutters and draperies. The atrium could have a glass cover that would be put in place during periods of extreme cold. Double glazed windows could be used on the south wall. These permit about 70 percent solar gain (the degree of effectiveness of solar energy), but shutters or draperies will be needed for extreme times. Triple glazing cuts the solar gain to about 60 percent.

Heating, Ventilating, and Air Conditioning (HVAC)

Earth sheltered structures reduce energy needs because heat loss or gain is greatly reduced and inside temperatures vary less due to the earth mass. Heat losses due to ventilation are unavoidable because fresh air must be brought into the house, but these can be reduced by using some type of heat recovery unit. Another factor in the reduction of energy needs is that the building, being of concrete construction, has a substantial thermal mass. In winter, solar and other forms of heat are stored in the floor, walls, and roof. At night when the heat source is gone, low, or not operating, the mass slowly gives off heat into the air. In most areas, temperatures in a properly designed earth sheltered house will never get below freezing, even with no source of heat. It is in this controlled environment that the heating and cooling systems must operate.

Passive and active solar systems operate very well in earth sheltered structures. A wood-burning furnace makes a good back-up unit, but any conventional back-up heating system can be used. The important factor is that the HVAC design consider the limited need for heating and cooling as compared to the comparable need in a conventional house of the same size.

Humidity control is a very important part of the HVAC system. It includes both humidification (addition of water vapor) and dehumidification (removal of water vapor). Humidification is sometimes required in winter if conventionally fired furnaces are used, since these tend to dry the air. But humidification is not a common need in earth sheltered houses. In summer, dehumidification is needed, since the levels of humidity are high in many geographic areas.

Build Your Vocabulary

Following are terms that you should understand and use as a part of your working vocabulary. Write a brief explanation of each term.

acoustics	R-value
active solar system	shell structure
auxiliary heating source	solar collector
berm	solar energy
condensation	solar pond
dehumidification	solar transport system
double envelope house	STC
earth sheltered design	thermal break
hydrostatic pressure	thermal energy
insulation	thermosiphoning wall
k-value	vapor barrier
natural radiation	ventilation
passive solar systems	waterproofing
reverberation	water stops
characteristics	zoning
reverberation time	

Class Activities

1. Write a list of things you could do to improve the efficiency of the energy being used to heat and cool your place of residence. Ascertain the approximate cost of each improvement.
2. Calculate the R-value for the ceiling, walls, and floor of your place of residence or your school.
3. Invite a speaker from the local electric or gas company to come to your class and discuss energy conservation.
4. Examine your place of residence or your school and report on sound control problems. Then plan for a remodeling of the space to reduce this problem.
5. Prepare a report listing the devices in your school or place of residence which will help control moisture. List problem areas and make recommendations for correcting each problem.
6. For each of the following cases, calculate the size of attic vents needed for a 2000 square foot building with a gable roof:

    Roof louvers only
    Soffit louvers only
    Gable louvers and soffit vents
    Continuous ridge and continuous soffit

7. What capacity (cubic feet per minute) power attic ventilation fan and vent area is needed for your place of residence?
8. Invite to class a representative from a local company which sells solar heating systems. Ask the representative to explain how the systems work.
9. Prepare a list of items the class wants to discuss, such as the cost of a solar heating system, the cost to operate it, and how long it will take to save enough in energy costs to pay for the installation.
10. Make a simple solar panel from metal. Devise a way to let water run down it and collect heat. Keep a record of the changes in the temperature of the water over a period of days.
11. Visit a solar house under construction. Have the contractor trace the system. Make a sketch of the system and label all parts. Sketch the piping in red, the electrical parts in blue, and the remainder in black.
12. Visit an earth sheltered house under construction. Sketch the floor plan and describe the heating-cooling system that will be used and the measures being taken for good ventilation and for waterproofing.

Winston House, Lyme, NH
Don Metz, Architect

Fig. 13-93. Earth sheltering and passive solar heating design principles are effectively combined in this energy efficient home.

architecture
Design
Engineering
Drawing

Finance and Presentation

Chapter 14
Getting the House Built

Chapter 15
*Architectural Rendering
and Model Construction*

Chapter 14

Getting the House Built

This chapter discusses the factors involved with getting the house built. The important task of selecting the lot must be completed. As this is done, some consideration should be given to locating the house on the lot. An architect or designer must develop the plans, the cost of the house must be estimated, a contractor must be selected, and financing must be arranged. Certainly important is a consideration of the insurance needs.

The designing and building of a house is one of the largest business transactions which the average family ever undertakes. If properly handled, the building of a house can be a pleasant experience. If poorly handled, it can lead to displeasure, disappointment, and actual financial loss.

Basically, designing and building a house involves:

1. Obtaining a set of working drawings and specifications,
2. Getting bids and signing a contract,
3. Checking the work as the house is constructed, accepting it, and making final payment according to the contract.

Plans and Specifications

A reliable source of plans and specifications is a qualified architect. The architect will study your housing needs and your financial status and then design a house to meet these requirements. Before making a final decision to contract with an architect, talk to former clients to get their opinions concerning performance. Be certain the ideas and styles emphasized by the architect are in keeping with your own. Ask to see plans that have been completed. Remember that a fee is charged for designing the house and for drawing the plans, and the fee will be larger if the architect is employed to oversee the construction of the house. Be certain to ask the amount of the fee, before you contract to have a house designed.

Another source of plans and specifications is from companies specializing in this work. They have available hundreds of plans for houses of all sizes and styles. Usually, these are designed by qualified architects. If such plans are to be used, they need to be examined to see if any changes are needed to meet local building codes or the requirements of the agency lending the money. Before the final contract with the contractor is signed, the plans and specifications should be altered to reflect the exact agreement.

A large number of companies manufacture various types of prefabricated houses. Most companies have catalogs showing the houses available. The customer selects a house from the catalog. It is possible that the customer may never see a set of plans. Changes can be made in these plans, but if this is done, they should be clearly noted and be made a part of the final contract. Since these changes also change the price of the house, each expense should be noted and recorded.

Some companies have finished houses of various types available for examination. These exhibit houses give the prospective buyer a good chance to examine construction and quality.

Getting Bids and Selecting a Contractor

After the plans and specifications have been developed, they must be approved by the agency loaning the money to build the house. Now the plans can be put out to several reliable contractors for firm bids. The architect, if you have employed one, will solicit bids from reliable contractors and help select the best bid. If you have no architect, choose only contractors presenting recommendations from loaning agencies or who are bonded.

Usually, a date is set as a deadline for contractors to return the plans and specifications along with their bid. If all bidders are reliable and if their bids are based on materials of equal quality, the lowest bid is usually accepted. Occasionally, the lowest bid is not accepted because of a prolonged completion date. Bids should be examined for substitutions of materials for those in the plans and specifications. The low bid is not always the best bid. If all bids are too high, it is not unusual to confer with the low bidder to cut the plan and make substitutions until the price is acceptable.

Once the price is accepted, a contract must be drawn. Your architect will supply the information for contracts or without an architect, you must select the contractor and have the necessary contracts drawn up. Either way, you should have the help of an attorney.

A written agreement or contract between the owner and contractor is essential. Each party should have a signed copy of the contract.

A contract usually contains all points agreed upon by the owner and the contractor. Such items as method of payment and time of completion are included. It also includes a statement of general conditions and lists the duties of the contractor, the obligations of the owner to the contractor, and the duties of the architect if one is employed. The plans and specifications are also a legal part of the contract.

Before signing a contract, have your loaning agency make a routine check on the contractor's financial position. The contractor should be able to finance the materials for construction and should cause no delays in construction by taking time to raise money to continue operating.

Owner-Architect Relationships

If an architect is employed, the owner and architect should have a written agreement clearly detailing the duties and responsibilities of the architect, the method of payment, and the responsibilities of construction left to the owner.

Usually, the architect designs a house to suit the family's needs and budget. Help is given in locating the house on the lot and even in choosing a lot. The drafting of preliminary plans and estimating cost are part of the duties.

The plans are revised, working drawings are made, and specifications are written. In most cases, the architect selects contractors and collects and analyzes their bids for the owner. The architect also provides the information necessary for writing the contract between the owner and the selected contractor.

If the owner desires to supervise the construction personally, the architect's responsibility ends with the signing of the contract to build. On the other hand, if the owner desires the assistance of the architect during construction, the agreement between the architect and the owner should state that the architect will see that all work is carried out according to the plans and specifications, will examine proposals for substitutions, and will make recommendations to the owner concerning the desirability of these changes. The architect will issue orders pertaining to the changes. In addition, the architect should make a final inspection of the finished dwelling, along with a check to be certain the dwelling is free of liens. If all is according to plan, the architect gives orders for the owner or loaning agency to make the final payment to the general contractor.

Usually, the owner is responsible for finding out about such things as easements, lot boundaries, utilities, and streets.

Owner-Contractor Relationships

As stated earlier, the owner and contractor sign an agreement which states clearly what the contractor will do for the price of the bid. The contractor employs and supervises all subcontractors, orders and pays for building materials and labor, schedules delivery of materials to the job, and serves as general supervisor.

The owner usually has no direct responsibility for construction, but must select light fixtures, hardware, bath fixtures, and other such items, so that construction will not be delayed. The owner must also see that the contractor is paid according to the contract. Usually, this payment is in four parts: one-fourth when the foundation is laid; one-fourth when the building is enclosed and when plumbing, wiring, and heating are roughed in; one-fourth when plastering is completed; and the final one-fourth when the building is inspected and accepted.

Occasionally, a contract is drawn to permit the owner to do some of the work, such as painting. This reduces the cost, but must be clearly stated in the owner-contractor agreement.

Estimating the Cost of a House

Before a house is built or before it even goes out for bids, it is best to secure an estimate of costs to determine if the house is within your price range as a prospective owner.

Several schemes have been used through the years for quick estimates of costs. These are called **preliminary estimates.**

Preliminary Estimates

One such scheme is to compute the number of square feet in the building by multiplying the outside dimensions — width × length. Nothing is subtracted for wall thickness.

For example, a one-story house with a full basement measuring 24'-0" × 60'-0" having in addition a 20'-0" × 24'-0" two-car garage would cost about $79,680. This is figured as follows:

Basic house costs $40 per square foot	
24'-0" × 60'-0" = 1440 sq.ft.	
1440 sq.ft. × $40 =	$57,600
Basement costs $12 per square foot	
1440 sq.ft. × $12 =	17,280
Garage costs $10 per square foot	
20'-0" × 24'-0" = 480 sq. ft.	
480 sq.ft. × $10 =	4,800
Cost of house	$79,680

This cost does not include the cost of land and improvements, such as sidewalks, driveways, shrubs, and seeding the lawn.

Other builders prefer to estimate costs on a cubic foot basis. This is computed by multiplying the outside length by the outside width by the height from the basement floor to the eave line. The cubic footage of the attic is found by multiplying the length of the house by the width by the rise (from the eave to the top of the ridge) and dividing this figure by two. This takes into consideration the cubic footage lost by the sloping roof. Porches are usually figured at one-half their total cubic footage.

Securing an Accurate Cost By Regular Estimates

After a complete set of working drawings and specifications has been developed, the house is ready to be put out for bids. Interested contractors secure a set of the drawings and specifications and make a detailed analysis of materials and labor. If a contractor sets the price too high, someone whose bid is lower will get the job. If the bid is too low, however, the contractor could actually lose money.

Figuring the **actual cost** is an important task. It is done by a process builders call "taking off." Many large contractors employ an estimator whose sole job is to figure the cost of buildings. Small contractors may do this themselves or have the superintendent of the firm do it.

These estimates are made by carefully examining the plans and specifications, listing the amount of all materials needed and computing the cost of each. The labor involved in construction of the building is also figured. Standard printed forms for estimating are available.

The estimate is usually divided according to the types of work required. The estimator divides this in such a way as to simplify the estimating process.

The process of estimating is too complex to explain in detail. Those desiring to learn in detail about this process should secure a text on estimating and follow the procedure suggested. Yearly publications are available to contractors giving detailed, up-to-date figures on labor and other construction costs.

Fitting the House into Your Budget

An important question for the average family is: "How much can we afford to spend on a house?" Funds may be available in the form of cash savings, equity in a house, and/or money from a loan agency.

A "rule of thumb" to follow when considering how much one can spend for a house is to base the estimates upon income. Most people can afford a house costing between two to two and one-half times their annual income after taxes. The actual figure will vary according to the family's size and lifestyle. Two families with the same income cannot necessarily afford the same price house. One family may have conservative buying habits or only one or two children. Such a family could probably afford a more expensive house than a family that wants an active social life and the best of material things or than one in which there are many people to support. These and other factors influence the amount of money available for a house.

The typical house payment includes principal, interest, taxes, and insurance on the house. The total obligations of the family for all purposes, such as house and car payments, should not exceed 33 percent of the monthly income. These figures **do not** include the cost of living in the house. Living costs include the costs of utilities, repairs, and improvements.

The monthly payments on a house should not exceed 25 percent of the monthly income after taxes. As an example, a family with an income of $30,000 a year ($2500 a month) would have approximately $24,000 left after taxes. They could spend about $500 a month on house payments. Applying the previously mentioned "rule of thumb," the family should be able to afford a house in the $48,000 to $60,000 price range.

If this family secured an 80 percent conventional loan for a house costing $50,000, they could make a $10,000 cash down payment and assume a $40,000 loan at 12 percent interest for 30 years. In round figures, the monthly payment would be $412. After 30 years they would have paid approximately $158,320 for the $50,000 house. Of this, $108,320 would be interest. This is summarized as follows:

Loan $40,000 for 30 years at 12% interest	
Monthly payment (principal, interest)	$ 412
Cost per year	4,944
Cost of loan after 30 years	148,320
Cost of house (down payment, principal, and interest)	$158,320

Following is an approximation of the monthly costs the family would have to pay:

Loan $40,000 for 30 years at 12% interest	
Monthly payment (principal and interest)	$412
Mortgage insurance	25
Taxes	100
Insurance on house	15
Total monthly costs	$552

Based on the lifestyle of the average family, this family should be able to afford a $50,000 house if they have no other large, outstanding debts. Remember, however, that homeowners must pay for such things as utilities, repairs, furniture, lawn care, and any desired additions, such as a patio.

Insurance

Homeowners assume obligations and undertake risks not applicable to renters. Consider these risks and obligations when buying insurance.

First of all, be certain to choose a reliable contractor who will be able to finish the job. Occasionally, a general contractor goes broke before the house is finished. The house is left in an unfinished condition. Those who supplied the materials for its construction then file liens against it to try to get their money. As a prospective owner, you can protect yourself from such a loss by being certain that the contractor is bonded. This certifies that an insurance company is convinced that the contractor can and will complete the job. If this fails to happen, the insurance company will employ another contractor to finish the work. In this way, the owner is protected from loss.

The matter of liability for accidents occurring during construction is of great importance. Be certain the general contractor carries a manufacturer's and contractor liability policy. This will cover all types of accidents on the job, including mishaps involving curious children who like to explore buildings under construction.

If you are serving as your own general contractor and are employing workers in the various trades, liability insurance is absolutely essential. Without it, you risk losing everything if an accident occurs on your property.

It is also necessary to carry insurance on the house and materials, to cover loss by fire or storm during construction.

After the house is completed, the owner must reexamine insurance needs. Before a mortgage company will approve a loan, the house must be insured against loss by fire and lightning and, usually, against loss by windstorm, hail, explosion, smoke, vandalism, riot, falling aircraft, and vehicles.

An additional **extended coverage** can be bought to include damage by falling objects, glass breakage, collapse and landslide, theft and burglary, water or steam, rupture of hot-water appliances, ice and snow, and freezing of plumbing.

It is important to consider detached garages. Usually, the policy on the house will cover outbuildings up to 10 percent of the policy value. For example, a $36,000 policy will provide $3600 coverage on a detached garage. If the garage is worth more, additional coverage should be obtained.

Insurance on the contents of the house is necessary also. This will cover loss from fire and, occasionally, theft on such articles as furniture, clothing, sports and hobby equipment, and other personal property.

Another type of insurance needed by the property owner is liability insurance. This coverage protects the owner and family against claims for damages which may result from accidents that take place on the property.

The insurance coverages mentioned as available for the homeowner — house, personal property, theft, and liability — are provided in some states as a "package policy." One policy provides coverage for all four perils.

Mortgage insurance is another valuable protection. Briefly, this life insurance policy is taken out on the breadwinner of the family for an amount equal to that of the mortgage. If the breadwinner dies, the insurance company pays off the mortgage, and the dependents live in a house clear of debt.

Financing a House

As a house is being planned, the prospective owner should carefully consider how it will be financed. Usually, the purchase of a house is the largest and most important investment made by the average family. Much money can be lost through careless or unfortunate mortgage arrangements.

A **mortgage** is a contract. It requires that a loan be repaid, and it specifies the terms. If the homeowner fails to meet obligations, the mortgage holder can foreclose the mortgage and resell the house.

State laws set limits on the kinds of mortgages that can be issued and their terms. Mortgage practices are fairly well standardized, but a person planning to finance a home should investigate a number of agencies and compare their mortgage terms.

One method of financing is through government-insured loaning agencies, such as the Federal Housing Administration and the Veterans Administration.

The Federal Housing Administration is an element of the U.S. Department of Housing and Urban Development. This agency was set up by the National Housing Act for the purpose of "encouraging improvement in housing standards and conditions, to provide a system of mutual mortgage insurance, and for other purposes." It does not loan money, but insures the loan so the loaning agency is protected in case of a foreclosure.

The government limits the interest rates on these loans; it also sets up construction specifications which must be met before the agency will insure a loan on a house. The FHA will insure loans made to the general public, while the V.A. or G.I. Loan is restricted to veterans. This second type of loan usually has a lower down payment and a lower interest rate than the FHA loan.

A second source of money, the largest source, is through conventional loans. These are made by savings and loan associations, mortgage companies, banks, insurance companies, and individuals. These loans are not insured by the government. The interest rate and down payment on conventional loans are usually higher than on government-insured loans.

Traditionally, real estate loans were made at a fixed interest rate over many years, such as for twenty-five or thirty years. This practice is now used only in a limited way because the wide changes in interest rates have made it unprofitable to the lender. A more common practice is to fix interest rates for a short time, such as for one year. The rates may be raised or lowered after that for the next year and each following year of the loan. Under this plan the person getting the loan does not know what interest will be charged in the years ahead.

Another plan is the **graduated payment mortgage.** This plan has a reasonable down payment and low monthly payments in the early years of the loan. The monthly payments gradually get larger over a period of five to eight years until they reach a predetermined maximum. Then they remain at this level until the loan is paid.

Other Financial Considerations

When you begin to make arrangements for a loan, consider the following:

1. What additional fees or charges must be paid? This includes charges for such things as appraisals and credit reports.
2. What types of insurance are required? These may include property, life, and disability insurance. You should insist on using the insurance agency of your choice.
3. Is there a charge for late payments?
4. Can you repay the loan before its final due date without penalty? The faster a loan can be repaid, the less interest is paid. See Fig. 14-1.
5. Will the lender permit the loan to be assumed by someone else?

Fig. 14-1.

Effect of Repayment Period on the Cost Per $1000 of a Loan at 12%

Payment Period (Years)	Monthly Payment (Principal Plus Interest)	Total Interest
5	$22.25	$ 335.00
10	$14.35	$ 722.00
15	$12.01	$1,161.80
20	$11.02	$1,644.80
25	$10.54	$2,162.00
30	$10.29	$2,704.40

Fig. 14-2.

Effect of Interest Rate
on the Cost of a $10,000 Loan
Over a 25-Year Period

Interest Rate	Monthly Payment (Principal Plus Interest)	Total Interest Over 25-Year Period
10.0	$ 90.88	$17,264.00
10.5	$ 94.42	$18,326.00
11.0	$ 98.02	$19,406.00
11.5	$101.65	$20,495.00
12.0	$105.33	$21,599.00
12.5	$109.04	$22,712.00
13.0	$112.79	$23,837.00
13.5	$116.57	$24,971.00
14.0	$120.38	$26,114.00

6. If you sell the house and the purchaser assumes your loan, will an assumption fee be charged?
7. Can you borrow money against the part of the mortgage that has been paid?
8. Will you be required to place in escrow (a special account) funds to pay the taxes and insurance?
9. Are you getting the best overall terms? Do not take the first loan offered — shop around before deciding. A difference of 1.0%, or even less, over many years makes a big difference. See Fig. 14-2.
10. Can you include appliances and other extras as part of the loan?
11. What are your rights if you cannot make a payment? How long before you can be evicted?

Build Your Vocabulary

Following are terms that you should understand and use as a part of your working vocabulary. Write a brief explanation of each term.

actual cost	estimating
bids	insurance
bond	interest
contract	loans
down payment	mortgage

Class Activities

1. Visit the office of a local architect. Secure information concerning the services performed for the prospective homeowner. Write a paper detailing these services. Include information about the type of owner-contractor contract used when awarding a construction job.
2. Visit a local bank or home finance agency. Secure information about the requirements an owner must meet to secure a loan. Write a paper detailing these requirements.
3. From local architects, loan agencies, or contractors, ascertain the local cost per square foot and cost per cubic foot to build a brick veneer residence.
4. Invite a local contractor and/or an architect to visit the class. Ask them to explain how they estimate the cost of a house. Have them bring the forms they use when figuring material and labor costs.
5. A typical family with an income of $21,000 after taxes can build a house in what price range?
6. From a local bank or a home finance agency, ascertain the following information concerning a $35,000 house loan:
   A. Interest rate currently used.
   B. Total monthly payment if loan is for twenty years.
   C. Total interest for twenty-year period.
   D. The actual cost of the house after twenty years.
   E. Cost of insurance for twenty years.
   F. Estimated taxes for twenty years.
7. Visit an insurance agency. Prepare a report detailing the types of insurance coverage available on residences and small commercial buildings.

Chapter 15

Architectural Rendering and Model Construction

Frequently, it is advantageous to present a planned building in a way that relates it to its intended surroundings, so that others can visualize its finished appearance. Several methods of presentation are in common use. Presented in this chapter are perspectives, display drawings, models, and architectural renderings.

Pictorial drawings and models help the client visualize the finished building. Many people cannot look at a set of plans and picture the end product. Pictorial drawings help the architect "sell" the customer on the design of the structure. They are used with sales literature and other advertising campaigns. They help rent apartments before they are built, and they sell prefabricated houses. A rendering of a proposed church building will help in its fund-raising campaign. Models help the directors of corporations make decisions concerning new plants and retail outlets.

Architectural Perspective

The most commonly used perspective drawings are one- and two-point. The one-point perspective, sometimes called parallel perspective, is used primarily for interior design. Two-point perspective drawings are used for interior and exterior illustration. They can be rather simple line drawings, shaded in pencil, drawn and shaded in ink, or produced in color with the use of watercolors or other media.

One-Point Perspective

A one-point perspective always has one face of the object parallel to the viewer. All receding edges and surfaces extend to a single vanishing point. Following are the steps in drawing a one-point perspective. See Fig. 15-1.

1. Locate the **ground line.** It represents the location of the bottom of the object to be drawn.
2. Locate a line called the **picture plane.** This is the edge view of the plane (drawing paper) upon which the perspective will be drawn.

3. Draw the **plan view** of the room or item to be drawn, locating the front edge on the picture plane line.
4. Draw the **horizon line.** It can be any distance above the ground line desired. The distance between the ground and horizon lines represents the height of the viewpoint above the ground. Normally this is 6 to 8 feet (1829 mm to 2438 mm).
5. Locate the **station point.** Generally, it is located nearly perpendicular to the plan view at a distance equal to two or three times the longest side of the plan view. The station point can be moved around until the perspective meets the needs intended. See Fig. 15-2.
6. Project the **width** of the plan view to the ground line.
7. Draw the **front view** on the ground line. This is true size.
8. Project a line from the station point perpendicular to the horizon line to locate the **vanishing point.**
9. Draw lines from the corners of the front view to the vanishing point. This **"boxes in"** the object.
10. Project lines from the station point to the corners and intersections on the plan view. Mark the points at which the lines cross the picture plane. Project these points down to the front view perpendicular to the picture plane. The point at which each of these lines crosses a line projected from the perspective view to the vanishing point is the location of a corner. Locate all needed corners and connect them, forming the **perspective.**

Moving the horizon and vanishing points changes the perspective. Notice in Fig. 15-2 how the amount shown of wall, ceiling, or floor changes when the vanishing point and horizon are placed in different positions. This enables the designer to emphasize the important features.

A one-point perspective is in Fig. 15-3.

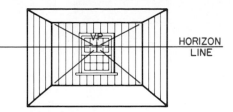

A. Horizon through center of room and vanishing point in center of object.

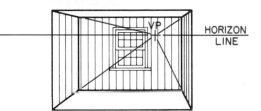

B. Horizon above center with vanishing point to right of center.

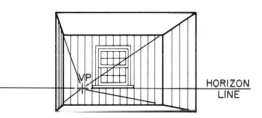

C. Horizon below center with vanishing point to left of center.

Fig. 15-2. The emphasis on the ceiling, floor, or walls can be changed by the location of the horizon and the vanishing point.

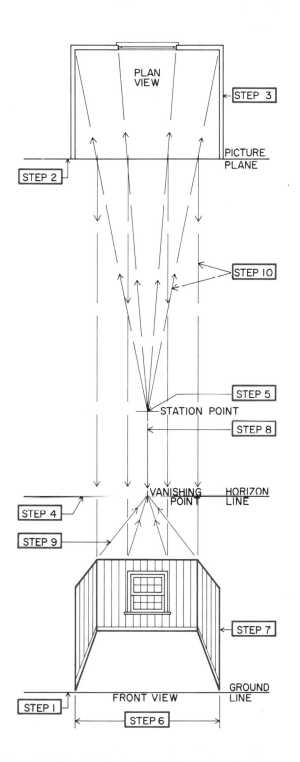

Fig. 15-1. Steps to draw a one-point perspective.

Home Planners, Inc.

Fig. 15-3. A one-point perspective.

Andersen Corporation

Fig. 15-4. The greater the distance an object is from the eye, the smaller it appears. This building is the same height at the rear, but it appears lower.

Two-Point Perspective

Perspective drawings illustrate the theory that the further away an object is from the eye, the smaller it appears. See Fig. 15-4. This is noticed when one drives an automobile; the road ahead appears to narrow and diminish to a point on the horizon. The same phenomenon occurs when one observes a series of telephone poles or a long building. The portion of the building nearest the viewer appears larger.

The steps to follow when drawing a two-point perspective are given in the following paragraphs. Remember that **any portion of a building touching the picture plane is true length,** while those portions behind the picture plane are smaller than normal. For purposes of clarification, the total top view is kept clear of the front view, showing the actual perspective. In practice, these are usually overlapped by permitting the **station point (S.P.)** to drop down over the front view, reducing the size of paper needed.

Layout for top view:

1. Near the top of the drawing paper, locate the **picture plane.** See Fig. 15-5.

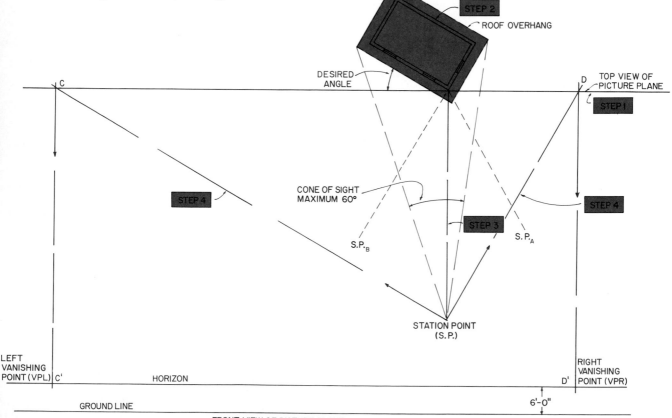

Fig. 15-5. Preliminary layout necessary before projecting points to draw perspective on picture plane.

2. Draw the outline of the floor plan, with door and window locations as shown in Fig. 15-5. It is not necessary to draw interior details or details of the back and the unseen end of the house. Time can be saved by making a print of the floor plan and taping it in position, thus removing the necessity to redraw it. Place one corner of the floor plan so that it touches the picture plane. Draw the house away from the plane on any angle desired; usually, an angle from 15 to 45 degrees is best. The angle chosen depends upon the details to be shown. The scale used to draw the floor plan is usually the same as that used on the working drawings; however, it may be changed if a smaller or larger perspective is desired.

3. Draw a line from the intersection of the floor plan and the picture plane to the **station point (S.P.)**. This is usually drawn perpendicular to the picture plane, but it can be on any angle desired. Consider how this angle will influence the perspective. If the station point is at the place such as **S.P.$_B$**, very little of the end of the building will be shown; while if it is at **S.P.$_A$**, very little of the front will be shown.

The distance the station point should be located from the building varies with the structure and the desired appearance of the rendering. If it is too close, the perspective will appear distorted. For most residences and small buildings, this distance is about 100 feet (30 480 mm). Most commonly, the station point is placed back far enough so the entire structure and its immediate surroundings can be included in a cone of 60 degrees. The cone is constructed with its tip at the station point and its central axis on a line of sight to the corner of the building touching the picture plane. Experience will help ascertain satisfactory distances. The beginner will probably have to try several distances before achieving the desired appearance.

4. From the station point, draw lines parallel to the sides of the floor plan, until they cross the picture plane at **C** and **D**. From these points, drop perpendiculars into the front view, until they cross the horizon at **C'** and **D'**. These points are the **vanishing points.** All horizontal lines run to these points. They are referred to as **vanishing point right (VPR)** and **vanishing point left (VPL).**

Layout for front view:

5. Figure 15-5 also shows the **front view** of the picture plane, upon which the perspective is to be drawn. First, locate the **ground line** upon which the building rests. This can be anywhere near the bottom of the page.

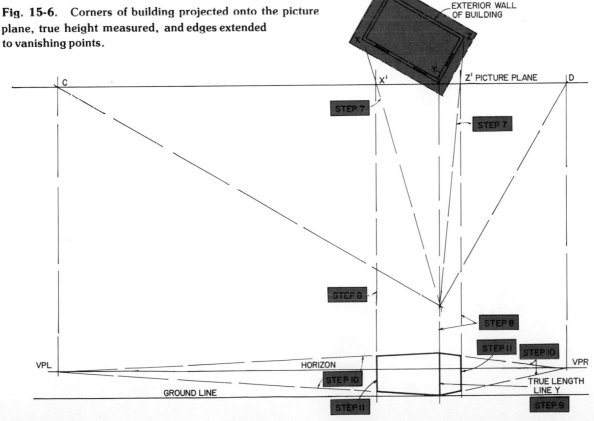

Fig. 15-6. Corners of building projected onto the picture plane, true height measured, and edges extended to vanishing points.

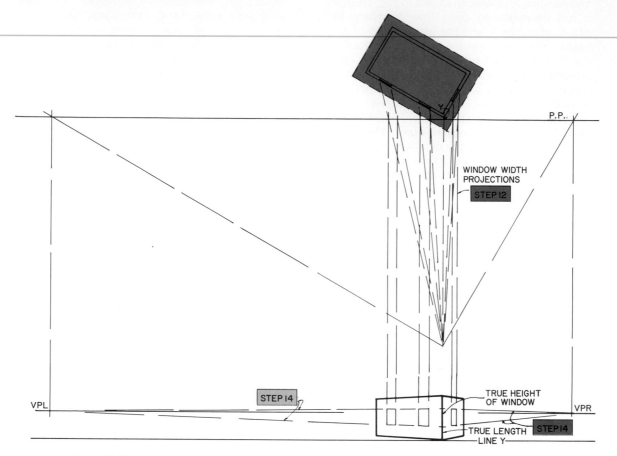

Fig. 15-7.   **Window widths projected onto picture plane, true height measured, and edges extended to vanishing points.**

6. Next, locate the **horizon.** As in one-point perspective drawing, this represents the height of the viewer's eyes above the ground line. If the horizon is very low, 1 to 2 feet (305 mm to 610 mm), little of the roof of the building is shown, and the walls loom large above the viewer. If the horizon is 6 to 8 feet (1829 mm to 2438 mm) above the ground line, the building appears as the viewer would normally see it. If the horizon is high, such as 15 to 20 feet (4572 mm to 6096 mm) or more, a bird's-eye view is given, and considerable roof and little wall are shown. A decision must be made as to the view desired.

Procedure for projecting into front view:

7. Draw projectors from the corners of the building, **(X, Y, Z)** to the station point. See Fig. 15-6.

8. Where these cut the edge view of the picture plane **(X′, Y, Z′)** in the top view, drop perpendiculars into the front view. All points

can be projected into the front view in this manner.

9. Corner **Y** is touching the picture plane and, therefore, will appear true length in the front view. All other parts of the building (except the roof overhang projecting beyond the picture plane) are behind the picture plane and are, therefore, foreshortened. **All vertical measurements must be made on a true-length line.** In the front view, measure the true height of corner **Y.**

10. Run projectors of the horizontal edges to **VPL** and **VPR.** The junctions of lines **X′** and **Y′** with these vanishing lines locate the corners of the building.

11. Repeat Steps 6 and 7 to locate the doors and windows. See Fig. 15-7.

12. Project the widths into the front view from the top view.

13. Measure the actual heights along the true-length line.

14. Project these heights to the proper vanishing point for whichever side of the house is being drawn.

    Locate any other parts in a similar manner.

    When a perspective is actually being drawn, all these points are located on one drawing. They are separated here for clarity. The number of lines of projection are many, and the drawing soon becomes crowded. Projectors should be very light lines that may be removed once they have served their purpose.

15. Details such as window mullions usually are not carefully measured, but are estimated and drawn rapidly.

Roofs in perspective:

16. The drawing of roofs other than flat roofs presents special problems. One problem is locating the height; another is allowing for the overhang. The procedure used is shown in Fig. 15-8.

17. Project the ridge in the top view until it cuts the picture plane at **Q**. Then project it into the front view. This becomes a **true-length line** in the front view.

18. The actual height of the ridge can be measured along this true-length line in the front view and projected to a vanishing point — in this case, **VPL.**

19. Repeat Steps 6 and 7 to find the length of the ridge **V′W′**. In Fig. 15-8, note projections of **V** and **W** to **V′** and **W′**.

20. The overhang **S²** projects on the other side of the picture plane. To locate it, project it back from the station point through the corner of the overhang to the picture plane at point **S**. This gives the point to use to project the corner of the overhang into the front view.

21. To find the true height of the overhang at the eave, project the point at which the overhang cuts the picture plane **R** into the front view. Measure the true height here, since this is a **true-length line.** Ascertain the true height of the eave by making a skeleton drawing of the end of the building to the scale used in the drawing. Project, horizontally, the true height of the eave to the true-height line on the perspective drawing. Then project this true-height point to the proper vanishing point. Draw the perspective where these cut the projections of the corners of the overhang, **T′** and **S′** front eave.

22. From corner **S′**, run a line to **VPR.** The point where the projection of corner **U** cuts this line (**U′**) is the corner of this rear eave. The rear eave is vanished to **VPL.**

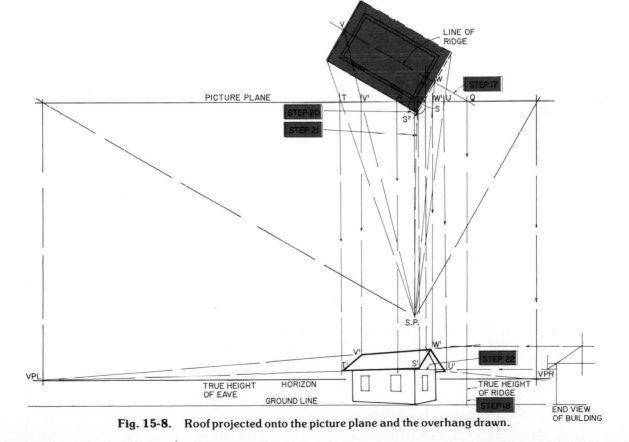

Fig. 15-8. Roof projected onto the picture plane and the overhang drawn.

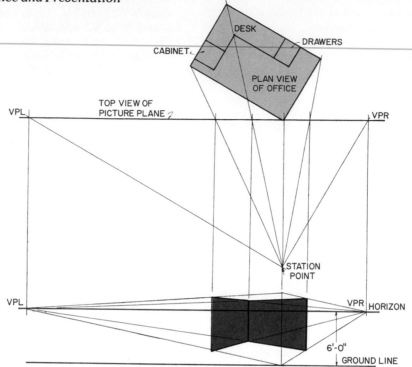

Fig. 15-9. The office walls are drawn in perspective as though the room were a box. The front walls are needed only for construction and are removed on the finished perspective.

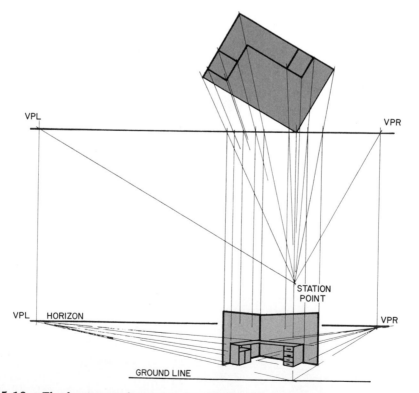

Fig. 15-10. The furniture is located in the room and drawn in perspective. Details can be as elaborate as desired. This drawing shows the office desk simply blocked in.

Two-Point Perspectives of Interiors

The principles of two-point perspective discussed for exterior views of buildings apply to interior perspectives. The top view of the room, with furniture located, should be drawn to the angle that seems to show the interior to best advantage. The horizon should usually be 6 feet (1829 mm) above the floor or ground line. This can be varied if a higher or lower view shows the room better.

The easiest beginning procedure is to draw the room in perspective as if it were a box. Then the furniture can be located. All furniture that is parallel with the sides of the room can use the same vanishing points as the walls of the room. If furniture is not square with the room, vanishing points must be established for each piece.

The walls of an office drawn in perspective prior to the placement of furniture is shown in Fig. 15-9. In Fig. 15-10, the office desk is blocked in and such details as the drawers and cabinet doors can now be located and drawn. Figure 15-11 shows the finished perspective with chairs in position. A complete room perspective is shown in Fig. 15-12.

Interior perspectives can be rendered in the same manner as exterior perspectives. An example of this type of drawing is shown in Fig. 15-13.

Perspective Drawing Boards

The system for drawing perspectives described on the preceding pages is accurate, but very slow. Industrial drafting technicians generally use some type of perspective drawing board. A perspective drawing board is a device that enables the user to develop a perspective in a wide range of scales. One such device is shown in Fig. 15-14. This is the KLOK Professional Perspective Drawing Board. It has a wide range of graduated scales and eight main vanishing points. The extreme left vanishing point is represented by the concave curve. A special T-square head slides along this curve. There are five vanishing points on the horizon line. The horizon line is in the exact center of the board. The points are actually holes in the board. A pin is inserted in the hole of the vanishing point to be used. This device can be used to rapidly draw one-point or two-point perspectives.

Better Kitchens Institute

Fig. 15-13. Rendered interior perspective of kitchen and family area.

John H. Klok and Bruning Division, Addressograph Multigraph Corporation

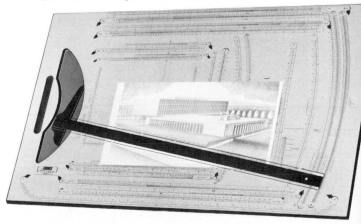

General Electric Company

Fig. 15-11. A detailed two-point perspective of the office.

Revco, Inc.

Fig. 15-12. Two-point perspective of kitchen and adjoining family room.

Fig. 15-14. A perspective drawing board.

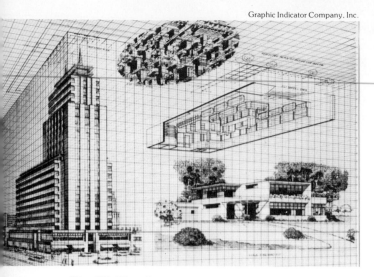

Fig. 15-15. A perspective chart with several perspective drawings.

Perspective Charts

Perspective charts are printed sheets with lines in true perspective. See Fig. 15-15. The chart is placed under a sheet of vellum. The printed lines show through, giving the drafting technician the direction to each vanishing point. The charts are available in various sheet sizes. They are ordered by giving the scale of the drawing, such as $1/4'' = 1'-0''$, and the distance of the station point from the object. Typical distances are $50'$, $75'$, $100'$, and $150'$. There are several different perspective chart and grid systems in use.

Presentation Drawings

Presentation drawings are made to give the basic information about a building in an attractive form. They omit almost all details, such as dimensions, heating, and plumbing. They usually consist of the floor plan, site planting details, and a front and side elevation drawn in great detail, with shrubs and walks artistically presented. The building could be presented in perspective. Occasionally, a section through the building is required. Presentation drawings may give some data in table form, such as the area of each room and the total area of the building. Some drawings show furniture on the floor plan.

The exterior and interior walls of the floor plan are usually drawn as solid lines. The lines for exterior walls are drawn thicker than those for interior walls. Other lines on the floor plan are thinner so they do not compete with the room layout.

The elevation or perspective is often rendered in color, and all the drawings are commonly placed on one sheet. The paper is usually of heavy, high-quality, white stock that will take pencil, ink, or watercolors. Neat lettering is vital to an attractive presentation drawing. Effective use can be made of templates of trees and other items. (Examples of these templates are shown in Chapter 5.)

Before the final drawing is made, careful consideration must be given to the placement of the various items on the page. The final arrangement should be balanced and attractive. The floor plan and perspective are the most important parts and should receive first consideration. Time should be spent deciding where to locate the lettering. Again, balance is necessary.

A common planning practice is to draw, in blocked-out form on separate sheets, the various parts to be included and then to arrange these in various ways on the display sheets. When a balanced, pleasing arrangement is reached, the final drawing can be done. A typical presentation drawing is shown in Fig. 15-16.

Architectural Models

Scale models are the best means for presenting a proposed building or group of buildings to those who must approve the project. Models are also very useful in advertising and fund-raising campaigns. The effectiveness of the exterior design can be better determined from a three-dimensional, colored model. Most models are built by companies specializing in this work. See page 472.

The following discussion presents some of the common techniques used in model construction. In actual practice, model builders are limited only by skill and imagination. Procedures for building a simple model are shown in Fig. 15-17, page 370.

Of first consideration is the base upon which the model is to be built. Plywood and hardboard sheets are good. If the building site slopes, sheets of insulation board can be glued together, and the contours can be cut with a power disc sander. This material is easily shaped and painted. Since it has a rough texture, it represents grass very nicely when painted green. Felt flocking and blotter material are also used for grass.

The models are made to scale. Special attention must be given to details, such as post lamps, cars, trees, and shrubs, so they are in the same scale as the building. Most residential models are built to the scale $1/4'' = 1'-0''$ or $1/8'' = 1'-0''$. Commercial buildings are built to the scale $1/8'' = 1'-0''$, though very large projects frequently use $1/16'' = 1'-0''$ or $1/32'' = 1'-0''$. The designer must ascertain the size. Obviously, the larger the scale,

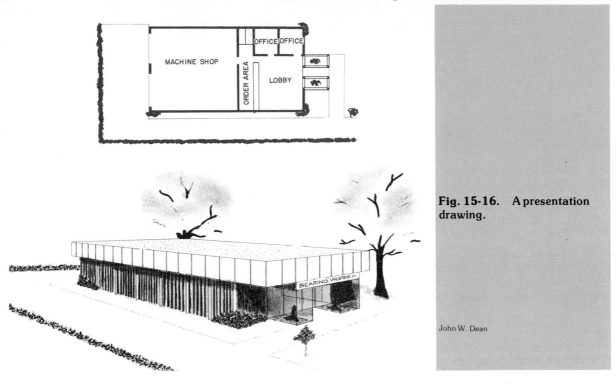

Fig. 15-16.   A presentation drawing.

John W. Dean

the greater the amount of detail that can be included.

The building shell can be constructed of almost any convenient material. Heavy cardboard and thin balsa sheets are most frequently used. These are glued together or taped on the inside.

Exterior details, such as siding and windows, can be represented in a number of ways. Paper sheets are available with bricks, stone, and other siding materials printed upon them. Also, printed paper or plastic sheets with windows, doors, and roof shingles are sold. These are simply cut out and glued on. They are available from model shops supplying materials to model-railroad enthusiasts.

The siding materials and windows can also be indicated by simply drawing them on cardboard walls in black ink, much the same as the building elevations are drawn. These can be colored with watercolors to add realism.

Balsa sheets can be lightly scored with a knife, to give a feeling of depth to siding. Small strips can be glued on to form batten strips, and shutters can be drawn and glued on to add shadow lines.

Thin, opaque plastic can be used to represent porcelain enamel or colored-glass panels, and clear plastic can be used for windows and plate-glass window walls.

Sheets of sandpaper make a good roofing material. For tar-and-gravel roofs, the entire sheet can be cemented in place. If a shingle roof is to be represented, the sandpaper sheet works well; however, some persons prefer to cut it into strips and glue it on, overlapping it to give the shadow-line appearance of shingles. Either balsa or wood veneer can be cut and glued on to represent wood shingles effectively, but it is a tedious task. Tar-and-gravel roofs can be further formed by gluing on finely crushed gravel or seeds, such as bird seed.

Wood can be stained with oil stain to represent redwood or other natural siding. Sidewalks and driveways can be painted on the base material. Gravel walks and patios can be made by gluing seeds or other such materials to the base.

Trees and shrubs can be purchased from model shops, or they can be handmade. Trees can be easily and economically made from dried twigs and weeds. The bare twigs can represent trees in winter, or they can be covered with sponge material or yarn to represent leaves. Small shrubs can be made from sponges. Thin wire stems with steel wool foliage painted green can also form attractive shrubs and small trees.

Wire can be used for railings and post lamps. Again, the thickness of the wire should be in scale with the entire building.

Plaster of paris is useful for walks, patios, stone walls, and pools.

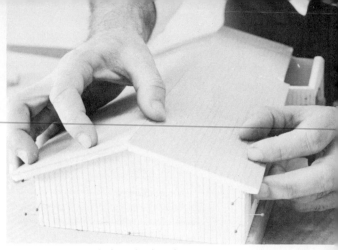

A. Wood is used for the foundation and floor. The exterior walls are grooved to simulate siding and glued in place.

D. The assembled roof is fitted on the house.

B. If the model is to show the interior arrangement of the rooms, the interior walls are glued in place.

E. The exterior is painted.

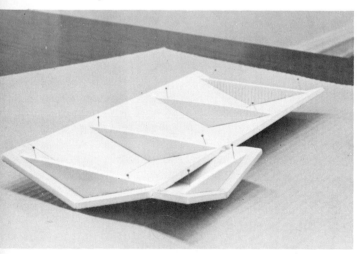

C. The roof is cut to size and assembled. In this model, the roof is made as a removable unit so that interior details can be seen.

F. The house is mounted on a large board. Grass, shrubs, trees, and driveway are added.

Fig. 15-17. Typical model construction techniques.

Fig. 15-18. A model of an urban renewal project for the Central Englewood District, Chicago, Illinois. Notice that most of the details of the buildings are omitted. A scale of 1″ = 50′ was used.

Fig. 15-19. A model of a one-story building designed to house an auto sales and service company. Notice the use of scale autos to enhance the overall scene. A scale of 1″ = 8′-0″ was used.

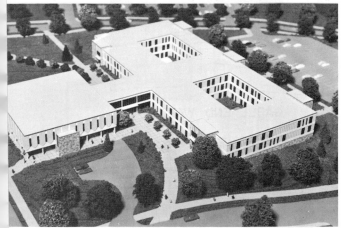

Fig. 15-20. A model of the Education Building, Western Michigan University, Kalamazoo, Michigan. Notice the use of trees and shrubs to place the building in its natural surroundings. A scale of 1″ = 32′-0″ was used.

The coloring of models should be carefully considered. Watercolors and tempera paints are frequently used. Care should be taken that the colors used are not so intense that they appear unnatural. Few buildings are fire-engine red or sky blue. Very light tints are more lifelike and do not obscure other details that may be drawn on the walls.

Models of projects of varying sizes are illustrated in Figs. 15-18 through 15-22.

Rendering Architectural Drawings

Architectural renderings are attempts to express the concepts of the architect in a form that people can understand and accept. Their purpose is to make clear to the viewer the ideas and plans that only the architect visualizes. They are more effective in imparting ideas than are models or black and white elevations, because they relate the building not only to the site but to the areas surrounding the site. Color is vital to expressing the complete experience.

Fig. 15-21. A model of the Riksha Inn, Chun King Corp. The outstanding feature is the great detail involved. A scale of 1″ = 4′-0″ was used.

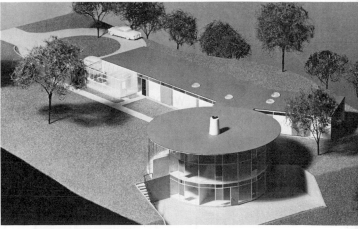

Fig. 15-22. A model of a contemporary home with fine detail. Notice the chimney, plastic bubble skylights, window mullions, and stair details. A scale of 1″ = 8′-0″ was used.

lines of various weights and colors. The degree of darkness or lightness (value) of the color determines the strength of the stimulus that attracts the eye.

The trained illustrator, when relating the elements of an architectural rendering, uses repetition, continuity, direction, rhythm, phrasing, balance, counterpoint, harmony, contrast, dynamic placement, tension, and transition to develop meaning in the rendering and to give emphasis or direction.

The effectiveness of a rendering can be increased by the use of shadows. This is a rather difficult subject, and entire books have been devoted to the topic. A satisfactory approach is to estimate the extent and location of the shadows. To do this, a decision must be made concerning the location and direction of the sun. Usually, the sun's rays are assumed to be coming from the right front of the building at a 45-degree angle; however, they can come from any direction. The suggested angle enables roof overhangs and porches to cast a shadow on the walls of the house. Practice in achieving a realistic appearance is necessary. Again, a study of finished renderings is helpful as a guide.

Colors must be used carefully. One frequent failing is to use colors that are too brilliant or that are abnormal for a building. The building should draw the attention of the viewer, and the colors on grass, trees, and sky should be such that they do not dominate the scene.

Trees and shrubs should be used only to provide a normal and proper setting for the building and should be secondary to the building itself. They should not be drawn in careful detail, nor should an attempt be made to show the thousands of branches and leaves. For best results, trees and shrubs should be grouped. Most landscape gardeners do this as they design the plantings. Examine finished renderings or specialized books for methods of representing trees and shrubs. Selected examples are shown in Fig. 15-25.

Preliminary Layout

While there are numerous ways to make architectural renderings, the most successful for the beginner is to make the perspective of just the building on tracing paper. When this is completed, a second sheet of tracing paper can be placed over the perspective, and the details of the foreground,

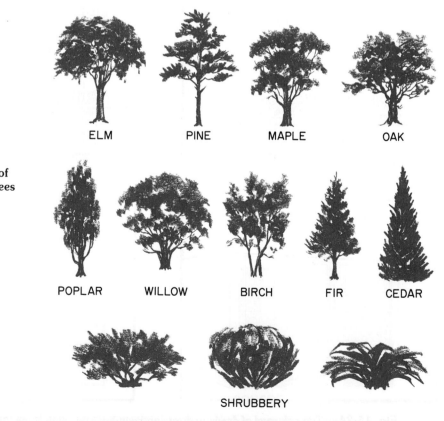

ELM   PINE   MAPLE   OAK

POPLAR   WILLOW   BIRCH   FIR   CEDAR

SHRUBBERY

Fig. 15-25. Methods of representing selected trees and shrubs.

middle ground, and background (such as trees, shrubs, parking areas, and mountains) can be drawn. In this way, the details can be erased and changed without damaging the perspective sketch. It is also possible to draw many settings for the building, each on a separate sheet, without disturbing the building perspective. The importance of making many preliminary sketches cannot be overemphasized. After satisfactory drawings have been sketched, they are traced upon illustration board. This is a heavy cardboard upon which the final rendering is drawn and colored.

When the perspective is drawn on the tracing paper, most details should be made freehand. Rendering should not have a mechanical appearance. This is especially important on details, such as windows and shutters.

Careful consideration must be given to the overall picture to be presented. Basic to this is the placement of the objects on the sheet. While most perspectives of small buildings are drawn from eye level, as in Fig. 15-26, some buildings have a good appearance from other positions. A rendering can be made from any position desired. Large buildings or groups of buildings, as in a shopping center, are frequently more attractive and descriptive if viewed from above and far away. This enables the artist to show the parking area, streets, and other items of importance. It also gives a better picture of the entire plan. The final decision is up to the artist. See Figs. 15-27 and 15-28.

Placement on the sheet involves consideration of three areas — the foreground, middle ground, and background. The building should be placed so each of these areas is pleasing and related to the others. If one area is more important to explaining the building, it should receive a larger proportion of the drawing. At all times, the building should be the center of importance. See Fig. 15-29.

The eye level should always be into the background. The sky should never be the eye level.

The building should be drawn large enough to fill the sheet, yet space should be allowed around it for the foreground, middle ground, and background. The building should not appear cramped on the sheet. Usually, more space at the sides is required by a low view than by a high view. The building should be placed above or below the center of the sheet. Most artists prefer to place large buildings below center and small ones above center.

Fig. 15-26. This rendering makes effective use of shadows including that of the flag pole cast on the building. Human figures assist in establishing the size of the building.

Fig. 15-28. An architectural rendering of an apartment complex. Notice how the high viewpoint enables the artist to relate the project to the surrounding area.

Fig. 15-27. A rendering of a school showing the relationship between buildings, parking areas, tennis courts, and athletic fields. Notice that the background shows no sky.

Fig. 15-29. In this rendering, the building is located below the center of the page and is framed by trees on each side. The street plus the trees directs the eye upon the building.

Remember the building is between the foreground and the background; therefore, these should seem to relate to the structure.

As the surroundings are rendered, they appear more spacious than is usually expected. It is suggested that the surroundings be fully rendered and their extent be reduced by matting the rendering exactly the same as a painting.

The rendering should give a feeling of balance, but should not be symmetrical. Seldom are building surroundings symmetrical, and such a rendering appears unnatural.

Fig. 15-30. Black and white illustrations. Notice the different visual reaction to each caused by varying the values of the black.

Matting a Rendering

Matting involves framing the rendering in a heavy cardboard frame. The material used is matt board. It is available in white and pastel colors. The matt frame should be kept a constant width at the top, left, and right sides of the rendering, but slightly wider at the bottom. This tends to give the rendering a base.

Black and White Illustrations

As a preliminary to the use of color in architectural renderings, work with black and white. These shades produce striking illustrations that can be varied considerably in effect. Black is used for all dark values and white for all light values. Any of several mediums can be used, including pencil, ink, charcoal, pastels, watercolors, and acrylic polymer emulsion paints. The grading of the values on the surfaces further increases the variety of effects possible. See Fig. 15-30 for selected examples.

Principles of Light and Shade

When an object is subjected to rays of light, that portion receiving the light is highlighted and tends to reflect the light, while the shaded portion tends to be darker. The farther the surface from the light, the darker it is because it reflects less light.

The shape of the surface influences the amount of light reflected and the shape of the reflected area. Figure 15-31 illustrates this for selected basic shapes. On the sphere, notice that the reflected area is round and the shade becomes darker as the edges are approached.

The size of the reflected area is influenced by the shape of the object. Notice the cone. The base reflects a greater amount of light than any other part because it is larger in diameter; on the cylinder, the reflected area is approximately the same for the entire length. These simple observations are of great value in increasing the effectiveness of architectural renderings. Details such as shrubs and windows should be highlighted to give a natural appearance.

Color Media

The most successful renderings are made by using a variety of media. The most used have been **watercolors.** Watercolors are easily diluted for varying value and intensity, and they can be intermixed to change the hue.

ARROWS INDICATE
SOURCE OF LIGHT

Fig. 15-31. The shape of an object and the direction of the light source influence the areas that are highlighted and shaded. As the light source (indicated by the arrow) moves, the area highlighted also moves.

Colored inks are also useful. Because they dry fast, they are more difficult to use for large areas of color. However, they are especially useful for small details and can be used under or over watercolors.

Pastel sticks (similar to chalk) are extremely useful for coloring large areas. The sticks can be sharpened to a point for drawing lines of various widths. To color large areas, rub the stick over sandpaper and pick up the powder on a piece of cotton. Rub the cotton over the area to be colored. Graduations in value are easily obtained in this way.

A rough-surfaced, heavy paper is best for pastels, but a surface such as common artist's tracing paper accepts the pastel very well. Interesting effects can be achieved on rough-textured materials.

When the rendering is colored with pastel sticks, it should be sprayed with a clear fixative after it is completed to prevent smearing. If watercolor and pastel sticks are to be used on the same rendering, apply the watercolor first, then the pastel, then the fixative.

If a grayed-color tone is desired, the pastel stick can be used on the reverse side of the tracing sheet. The tracing paper tends to gray the intensity of the color.

A kneaded eraser (eraser material that is similar to a putty) is an indispensible tool when using pastels. It can be shaped to a point, and small areas can be removed from the rendering to add "life" to the finished job. Examples of this are mortar joints, reflections, and highlights in a window. For straight lines, draw the eraser along the edge of a ruler.

Water-soluble paints of an **acrylic polymer emulsion** type are excellent for rendering. They dry fast and will not smudge and wash off with water. They can be diluted with water for variations in value and are applied with a brush, as are watercolors.

The beginner can probably achieve the most satisfactory results using pastel sticks. They are easily handled and are available in sets representing the entire spectrum of colors.

It is advisable to consult art instructors for assistance in handling these materials.

The Use of Values

One major problem in rendering is to make the building stand out from the surroundings so it can be seen and understood at a glance. The use of **value** (light and dark) or color is the means for doing this. In Fig. 15-32, the use of value is illustrated. At **A** the outline of the building is shown. Notice that this has little to attract attention. If the building is made of a light colored material, it should remain light, and the surroundings should be darkened, as shown at **B.** A dark building is usually contrasted against a light background, as shown at **C.** If the end of the building is in the shade, it will appear as at **D,** and if the front is in the shade, it will appear as at **E.** The shaded portion immediately stands out against the light background, and the other portion loses prominence. None of these is sufficient to stand alone and must be combined to give the desired attention. At **F** the light front is against a dark background. Immediately, the light front attracts attention. At **G** the light end becomes prominent.

The surfaces of the building to be darkened can be graded rather than simply dark or light. At **H** the end of the building is graded, and the front is light. There are unlimited variations of value, and the artist must decide upon the value that does the best job.

An expansion of the use of value is shown in Fig. 15-33. At **A** the building outline is shown. At **B** the roof is dark, thus becoming the center of attention.

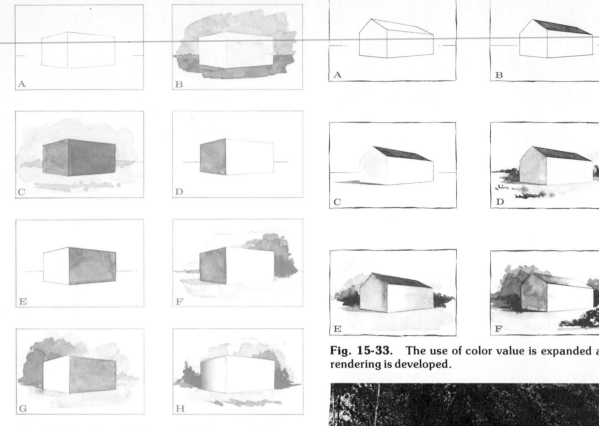

Fig. 15-32. Several of many possible uses of value.

At **C** an end has been graded; the roof is still prominent. At **D** a darkened foliage and foreground direct attention to the light front. The variations of this are many, but the principles of the use of value apply as seen at **E** and **F.** The artist must try to represent the building as clearly and as advantageously as possible.

As an excellent practice piece, obtain a photograph of a building, lay tracing paper over it, and sketch and major outlines. Then experiment by using light and dark tones and graded shading. See Fig. 15-34.

The Use of Hue and Intensity

Hue refers to the quality of a color that enables the viewer to distinguish it from other colors. It is the name we give to a color, such as blue hue or red hue. Hue can be changed by adding another color; for example, yellow can be added to red to produce an orange hue.

If one hue is brought against another, even if they are similar hues, **contrast** exists. The more dissimilar these hues are, the greater the contrast.

Fig. 15-33. The use of color value is expanded as a rendering is developed.

Fig. 15-34. Use of shading and value to vary building rendering. The photograph shows the actual appearance of the building. The drawing illustrates how the artist can vary the shading and value to present the building at a different time of day.

The farther apart the hues are on the color wheel in Fig. 15-35, the more dissimilar they appear. **Analogous colors** (colors near each other on the color wheel) are less useful to the artist making a rendering than **complementary colors** (colors opposite each other on the color wheel), because he is trying to build contrasts as a means of emphasizing a building. For example, a red brick building can be brought into prominence by making a background of green trees, shrubs, and lawn. The green is complementary to the red (opposite it on the color wheel). If the same building were surrounded with a yellow-orange sky and brown foreground, it would lose much of its prominence.

An architectural rendering can accommodate very little color of full intensity. (**Intensity** refers to the strength of the color). Intensity can be changed without changing value or hue by adding neutral gray of equal value. Intense color may be used on small portions of a rendering to call attention to a particular detail.

Complementary Color Schemes

While architectural renderings are sometimes done in analogous color schemes, the most used are the complementary schemes. They greatly aid in developing contrast and are more natural. Nature uses complements such as the red rose on the green bush. The proper use of complementary colors produces a harmonious picture. If contrast is uncontrolled, harmony is destroyed.

When a color scheme for a rendering is selected, one color should dominate. Never should a color scheme be based on complementary colors used in equal areas and full strengths. If an area is to be one color, such as red, then its complement, green should be used in a smaller area or should be neutralized so it is not the same strength as the red.

A simple, yet effective, use of complementary colors is to use two — one warm and one cool.

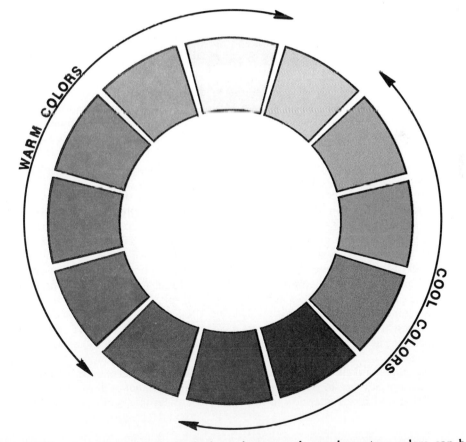

Fig. 15-35. A color wheel from which analogous and complementary colors can be chosen.

Red, yellow, and blue are the primary colors. A mixture of two primary colors is a secondary color, and a mixture of a primary color and a secondary color produces a tertiary color.

Usually, these are not used in full intensity. For example, portions could be done in orange with combinations of yellow and brown, while other areas could be in blue with green and violet indications. The cool colors complement the warm and are less assertive.

A person trained in the use of color can successfully use complementary color schemes in which the colors used are not exact opposites on the color wheel, but are near opposites. For example, green is the true opposite of red, but yellow-green and blue-green are near opposites and can be used effectively in complementary schemes.

Build Your Vocabulary

Following are terms that you should understand and use as a part of your working vocabulary. Write a brief explanation of each term.

analogous colors	matting
color value	model
color wheel	perspective
complementary colors	rendering
horizon	station point
hue	vanishing point
intensity	

Class Activities

1. Secure pictures of houses and small commercial buildings which show the front elevation. Catalogs of manufacturers of precut and factory-built homes are excellent sources. Draw several of these houses, in two-point perspective. Draw one with the station point 6'-0" (1829 mm) above the horizon and the house at an angle of 30 degrees to the picture plane. Lay a piece of paper over this perspective and draw another perspective, with the station point 30'-0" (9144 mm) above the ground line. Notice the difference in appearance.

2. Try shading one of these perspectives in black and white. Use ink or a black pastel stick.

3. On another perspective, try color. Use watercolors, followed by pastel sticks; then spray a fixative over this.

4. Practice drawing trees and shrubs. On another perspective drawing, attempt to place these to frame the building and direct attention to it.

5. Construct a model of a house or small commercial building. Include as much detail as you have time and skill to include. Since this is time-consuming, much of it could be done after school and at home.

6. Secure a photograph of a building. Place a sheet of tracing paper over it and copy the outline and details. With a soft pencil or pastel stick, render it to give an appearance different from the photo. Vary the amount of light on the building or the direction of the sun. Vary the degree of darkness of the surroundings (value). You could repeat this procedure on other sheets until you have several different sketches of the same house. Notice how the variance of light and dark changes the atmosphere and character of the house.

7. Make a color wheel by mixing watercolors and painting these in the proper places. Secure an art book with a color wheel in it and use it for a guide. Consult your art instructor or an art book for proportions. Perhaps the art instructor could demonstrate this process for your class.

8. Prepare a report on the use of color on buildings in your neighborhood. Cite examples of buildings making good use of color in relating the walls, roof, and trim. Explain why these are good. Cite examples of misuse of color and what could be done to improve the appearance of these buildings. Learn to look at your surroundings critically.

9. Make a display drawing of a building. Make the perspective in color and the remainder in black ink.

10. Mat your best rendering by framing it with heavyweight white or pastel matt board.

11. Try a group project. If your class is designing a shopping center or cluster of homes, prepare models of these and place them on a large base, showing roads, sidewalks, and trees. Model automobiles and figures enhance such a display. This could be a project for the entire class, with each person completing a portion of it. Arrange to have this displayed in your school lobby and in show windows of local stores.

12. With the permission of your instructor, invite an architect or illustrator to speak to your class concerning renderings and to show examples.

Commercial Design and Construction

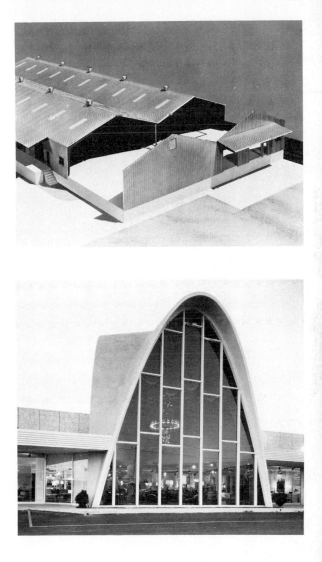

An architectural model of a large apartment building.

Cornerstone Development Co.

Joseph M. Seiler

An architectural model of a portion of the Chicago, Illinois lake front.

Joseph M. Seiler

An architectural model of a residence.

A manufacturing office/plant complex.

A steel-framed office building with native stone exterior.

Architectural sculpture used to identify a medical building.

A well-designed office building that fits well into its surroundings. Notice how it appears to extend from the hillside.

413

Burroughs Wellcome Co., photo by Greg Plachta

Reinforced concrete panels, strong yet relatively light in weight, fastened over an interesting architectural interpretation of a basically A-frame design.

Cold Springs Granite Co.

The granite exterior of this building projects a feeling of dignity and permanence.

Burroughs Wellcome Co., photo by Greg Plachta

Close-up of panels that also shows skylights. Skylights allow natural light and some solar heat into the building.

Slender columns with flared tops provide a graceful appearance.

Interesting texture is provided by exposed aggregate in precast concrete panels.

Use of identical precast concrete wall panels creates a unified appearance.

A panel system providing semi-private work stations. The rectangular units are easily adjusted to suit the size of the particular office area.

Herman Miller, Inc.

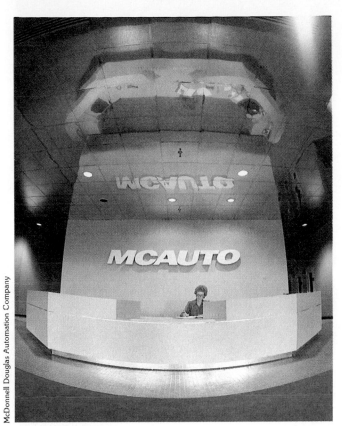

The essentially simple design of this reception area is made dramatic through the use of stainless steel panels on the ceiling.

McDonnell Douglas Automation Company

A warm atmosphere is created for this merchandising area of an interior design business.

417

Holiday Inns. Inc.

A motel lobby that is attractive and comfortable.

Holiday Inns. Inc.

A motel should have a large, eye-catching sign. An inn can be designed as a multistory structure.

Motel/Motor Inn Journal

A typical motel room.

Owens-Corning Fiberglas Corporation

Interior of a modern, attractive shopping mall. Notice the translucent fiber glass fabric roof that eliminates the need for artificial lighting during the day.

Owens-Corning Fiberglas Corporation

Exterior of the mall shown above. Notice how the trees enhance the appearance.

419

McDonnell Douglas Automation Company

Creative lighting makes this building design interesting and attractive after dark.

Columbia Lighting. Inc.

The interior lighting system of this corridor complements the natural light from the windows.

Columbia Lighting. Inc.

Lighting design can create interesting ceilings.

National Forest Products Association

The architectural design of this church complements the setting.

National Forest Products Association

The exterior is an architectural reflection of the function of this building.

National Forest Products Association

The interior was designed for large audiences.

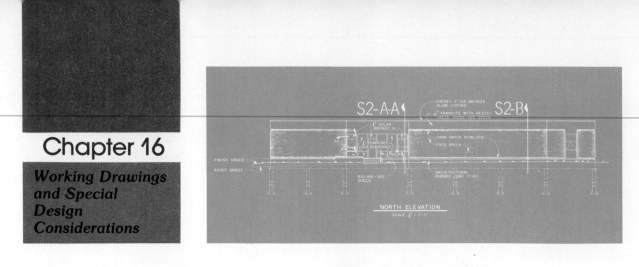

Chapter 16

Working Drawings and Special Design Considerations

Building designs are presented as a set of working drawings. These designs are the actual plans from which buildings are constructed. Working drawings plus written specifications provide a complete description of all aspects of a building. Environmental impact studies have become an important part of the planning of construction projects. And there is an ever-increasing awareness that building designs must accommodate the special needs of those with physical disabilities in order for buildings to be accessible to everyone and usable by all.

Working Drawings for Small Commercial Buildings

The complete set of drawings for a small commercial building represents the work of a team of specialists. The original planning and design are done by an architect who spends considerable time in conferences with clients. The architect must first ascertain exactly what functions are to be performed in the building, then must plan the interior arrangement, and finally must design a shell to house the required activities. For small structures, the architect may do the structural, mechanical, and electrical design. For larger buildings, engineers design these systems. These tasks are usually assigned to a firm of consulting engineers.

A detailed set of specifications is written to accompany the information given on the working drawings. Generally, this is written by specialists in this field. For small buildings, the architect may use a standard specification form and supplement it with specifics for the building under consideration.

Decisions concerning needed details must be made as the building is designed. Details usually needed are window and door heads, jambs and sills, cabinets, stairs, eaves, and a wide range of structural details. If some part of this structure can be clarified by a drawing, it should be drawn. Everything about the structure should be set forth in a drawing or in the specifications.

A Set of Working Drawings

Since the working drawings usually involve a large number of sheets and the specifications many typewritten pages, it is not possible to present a complete set in this text. Complete sets of plans and specifications for classroom use can often be obtained from architectural firms. Careful examination of these will reveal the many necessary layouts and details.

The drawings on pp. 427-440 were selected from a set of drawings for an office building for School District 250, Pittsburg, Kansas. The building was designed by Mr. Bill Wilson, Architect, Pittsburg, Kansas.

The original drawings were made on 22 × 34-inch vellum. They have been greatly reduced to fit the pages of this book. The scales indicated are those used on the original drawings. The scales commonly used on architectural drawings are shown in Fig. 16-1.

Drawing Techniques

Lines on drawings must meet or cross to form corners. There should never be a gap. See Fig. 16-2.

The main outline of each drawing is drawn with a thick line. For example, the walls on a floor plan, the outline of an elevation, and the main structural members on a wall section are drawn with thick lines. This makes them stand out at first glance. Other features are drawn with thin lines. On a floor plan, such features include stairs, bath fixtures, doors, cabinets, and windows. On elevations, details such as windows, doors, and material symbols are drawn with thin lines. On wall sections, material symbols are shown with thin lines. All dimension lines are thin.

Title Block and Border

The title block is placed in the lower right corner of each sheet. Although the exact layout of the

Fig. 16-1.

Scales Commonly Used on Commercial Drawings

Drawing	U.S. Customary	SI Metric (mm)
Floor plan	1/8″ = 1′-0″	1:100
Foundation plan	1/8″ = 1′-0″	1:100
Elevations	1/8″ = 1′-0″	1:100
Construction details	1/2″ to 1 1/2″ = 1′-0″	1:25 or 1:10
Interior details	1/2″ = 1′-0″	1:25
Building sections	1/4″ = 1′-0″	1:50
Framing details	1/8″ = 1′-0″	1:100
Lighting, electrical, heating, plumbing, air conditioning plans	1/8″ = 1′-0″	1:100
Site plan	1″ = 20′ or 40′	1:100, 1:200, or 1:500

The most frequently used sheet sizes are type **D**, 22 × 34 inches; type **E**, 34 × 44 inches; type **A1**, 594 × 841 mm; and type **A0**, 841 × 1189 mm.

Identification of Drawings

Commercial architectural drawings are identified with a letter symbol and a drawing number. The symbol used also tells the major purpose of the drawing. For example, a drawing containing electrical information is marked **E**. If there are three pages of electrical drawings, they are coded E-1, E-2, and E-3. All drawings except the title sheet are numbered. Following is a typical system:

Symbol	Drawing
A	Area design drawings such as floor plan, foundation plan, elevations, stair plan
E	Electrical drawings
M	Mechanical drawings
P	Plumbing drawings
S	Structural drawings

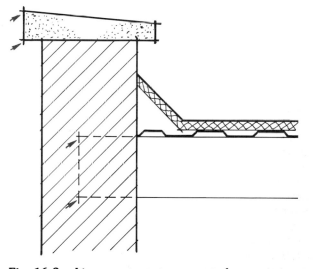

Fig. 16-2. Lines must meet or cross to form corners. Notice also how thick line widths are used to emphasize the important features on a drawing.

block will vary, the example in Fig. 16-3 is typical. Some companies place the title information in a narrow area across the bottom of the sheet or along the right side. See Fig. 16-4.

A 1/2-inch (13 mm) border is used on the top, bottom, and right sides, and a one-inch (25 mm) border on the left is common. The wider border on the left is needed because the set of drawings is bound on that side. Look again at Fig. 16-4.

Fig. 16-3. A typical title block giving the necessary information.

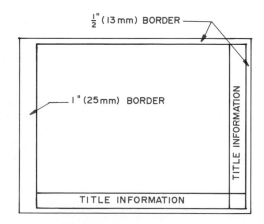

Fig. 16-4. Alternate locations for title blocks.

Placement of Drawings

While the placement of the various drawings and details on sheets varies, related drawings are placed together on a sheet or on successive sheets in numerical order. The following list shows a typical arrangement of drawings for a small commercial building.

Title Sheet — Name and location of building, names and locations of architect and consulting engineers, index to drawings, materials symbols.

Sheet A-1 — Site plan, related symbols, general notes.

Sheet A-2 — Floor plan, door schedule and details, window schedule and details, room finish schedule, interior door sections.

Sheet A-3 — Foundation plan, sections showing grade beams, control joints in floor, floor-foundation-exterior stair section at entrances, slab construction, thickened slab construction and footings at columns.

Sheet A-4 — Exterior elevations, interior elevation of rest rooms, cabinets.

Sheet E-1 — Locations of all electrical features, fixture schedules, electrical legend, distribution panel data, lighting.

Sheet M-1 — Plumbing diagram, heating and air conditioning plan, grille schedule, a site plan showing the entrances of utilities into the building.

Sheet S-1 — Roof-framing plan, column and footing schedule, grade beam schedule, general structural notes.

Sheet S-2 — Typical wall section, section through the building, other sections showing structural details.

Details of the Title Sheet

The **title sheet** identifies the project and gives the name of the city in which it is to be built. Lettering is done with templates or adhesive letters. The overall appearance should be balanced and attractive. The most important title is given first, in letters about twice as high as those used for other information. The names and locations of the architects and consulting engineers are lettered in the smaller letters, below the project name. See Fig. 16-5, p. 427.

The index to drawings gives the sheet numbers and the names of the major drawings on each sheet. Drawings are listed in alphabetical order beginning with the design drawings — the **A** series.

Material symbols are often located here because they relate to most of the drawings. Sometimes, if it fits without crowding the other information, the site plan is placed on this sheet.

Details of Sheet A-1

Sheet A-1 contains the **site plan.** See Fig. 16-6. The site plan shows the size and location of the property line, locates the building on the site, and gives the sizes and locations of sidewalks, driveways, parking areas, and planters. The elevations of the corners of the site are shown, as well as original and new contour lines. Elevations of selected points such as the finished floor, sidewalks, and parking lot are noted. Information about curbs and on-street parking is given. The sizes and locations of available utilities are recorded.

A **symbols legend** identifies the symbols on the site plan. See Fig. 16-7. This particular sheet also contains some general notes pertaining to the overall plan.

Details of Sheet A-2

Sheet A-2 contains the finished **floor plan.** See Fig. 16-8, p. 428. Locations of all doors, windows and partitions are shown. Rest room and janitorial facilities are also indicated. All interior and exterior sizes are dimensioned, including partition, door, and window locations. Codes that relate to the door and window schedules and the interior finish schedule are shown. Rooms are identified both by a descriptive name indicating their intended use and by a room number. Sections that appear on later drawings are noted, such as S2-B. This means this section can be found on sheet S-2 and is labeled Section **B**. Material symbols are drawn, such as those shown next to the partitions in Fig. 16-8. The title, floor plan, and the scale of the drawing are placed below it. Notice that the dimensions are placed well away from the outline of the building.

The **door and window schedules** show the elevations, sizes, and types of all windows and doors. The mark shown on the schedule is the code used on the floor plan. The type indicates a particular kind of door that may be used in several different ways. Details are listed in chart form. The data included on the chart will vary depending upon the situation. Door and window schedules are drawn separately and are identified with a proper title. See Fig. 16-9.

The **room finish schedule** is coded to the rooms on the floor plan. It details the finish for the floor, walls, ceiling, and the base, and also shows the finished ceiling height. See Fig. 16-10, p. 429.

Details of Sheet A-3

Sheet A-3 gives the details for the **foundation** of the building. See Fig. 16-11. It shows the exact location of each pier. Several sections were taken through the grade beam on the foundation plan. And under the interior partitions the slab is thickened to carry the extra weight. These are shown on the foundation plan and keyed to a typical section drawing. Codes are used to indicate where the large scale sections are drawn in the plans. For example, the code 2-A means the section is found on page A-2, labeled Section **A**.

The dimensions of all parts of the foundation are given, including several columns on the interior. Grade beams are identified by a code, such as GBI. Complete specifications for each grade beam are found on the **grade beam schedule** located on the first structural drawing, S-1. Information about the concrete slab floor — such as thickness, reinforcing, and fill — is given. Each control joint is located. A section through the control joint, 2-D, is noted. The section drawing will give full details of the joint. The title and scale are given for the foundation plan.

On this same sheet all **sections through the elements of the foundation** are drawn to a large scale and keyed to the foundation plan by a symbol. See Fig. 16-12, p. 430. One detail, **A**, shows the grade beam and piling for the exterior wall. It shows the sizes and placement of reinforcing steel, the location of insulation, and the relationship between the slab and the grade beam. Another detail, **B**, shows the details at the front entrance. In addition to the grade beam, floor, and insulation details, it shows how to step the slab down to the grade beam. Notice the use of a plastic vapor barrier under the floor slab in all details. Detail **C** is a section at the rear entrance. Of special interest is the relationship between the floor of the building and the floor of the rear porch — there is a one-inch step down. The porch floor and steps are cast monolithically (as one piece).

Section **D** shows a typical control joint. The location of control joints in the floor is shown on the foundation plan. Notice that the slab is thickened here to 1'-6" and the joint sets on a pier. One-half inch of expansion joint material is used to separate parts of the slab at this point. The joint is offset on the pier so the interior partition will not cover the joint. Required steel reinforcing is noted.

Section **E** details the thickened slab used under non-load-bearing partitions. The slab is poured 8 inches thick and has a base of 12 inches. Notice that a keyed joint is required to keep the slabs from rising above one another. Steel reinforcing bars are placed in the thickened portion.

The last detail, **F**, is a plan view of the interior columns. These are located on the foundation plan. This detail shows (enlarged) the steel column resting on a pier. It also shows the control joints and thickened slab at that point.

All of these details have individual descriptive titles, and the scale of each is given. The scale of details is given by each one even when they are all drawn to the same scale.

Details of Sheet A-4

Sheet A-4 shows the four **elevations**. These are drawn to the same scale as the floor plan. The foundation below grade is shown with dashed lines. All doors and windows are detailed, as are all features of each exterior wall. Railings, porches, and ramps are shown. See Fig. 16-13.

Several sections are cut through the building. These are shown with plane symbols and are coded. For example, Section S2-B is found on sheet S-2, labeled Section B.

Symbols for materials in elevations are used on various surfaces. Large surfaces need not be completely covered by the symbol. The symbol can be placed around the edges and next to doors and windows. Descriptive terms used to identify the types of materials are lettered on the elevations. The elevation of the finished floor and the bearing point for the roof joists are given. The total height between these is dimensioned. Often the height of the major elements from the footing to the roof is shown. Each elevation has a title and a scale.

Elevations of each wall of each rest room are shown. These elevations show all fixtures, stalls, and wall-hung items such as mirrors and towel dispensers. The elevations are dimensioned to show the various locations. See Fig. 16-14, p. 431. The individual fixtures are not dimensioned, just their locations. The required wall and floor coverings and the base are noted. Each wall has a title and a scale. Any built-in units, such as wall cabinets, wardrobes, and other storage units are also detailed. Elevations and sections are required.

Details of Sheet E-1

Electrical details are given on sheet E-1. A drawing of the floor plan is made with the dimensions. It shows the **power** and **signal data**, including receptacles, special-duty receptacles, motor controllers, power distribution panel, lighting panel, disconnecting switches, telephones, and other required features. See Fig. 16-15, p. 432. A series of outlets are connected to form a circuit with a triple arrow on the end indicating the lead to a power panel. Power data for mechanical units, such as air conditioning units, are shown.

A second drawing of the floor plan shows the **lighting**. See Fig. 16-16, p. 433. The fluorescent lights are located in the grid formed by the dropped ceiling. They are connected to junction boxes and into circuits. Switches for each light are located in such a way that it is possible to move through the building, lighting it as you proceed and turning off unneeded lights behind you. Some lights are connected to a time switch so they can be controlled in the absence of personnel.

The light fixtures are keyed by number to a **lighting fixture schedule**. The schedule gives identifying data for each type of light. It gives the code number, manufacturer name, catalog number, wattage, type, and use for each light, as well as the number of fixtures in the building and a place for remarks. See Fig. 16-17.

A **legend** to the electrical symbols used on E-1 is included. On it, each symbol is drawn and identified. See Fig. 16-18, p. 434.

The key to the electrical system is the **main distribution panel**. A schedule shows each circuit, what area it serves, its load, the frame size, the branch breaker size, the safety switch size, the size of the fuse, wire size, and conduit size. See Fig. 16-19.

A **riser diagram** is drawn to show the flow of power from the transformer to the main distribution panel. See Fig. 16-20, p. 435.

Details of Sheet M-1

Sheet M-1 is devoted to details of the plumbing and heating and air conditioning systems. The **plumbing diagram** shows the location of each fixture, the size of each water and sewer pipe, and the locations of the water heater, floor drains, and vents. See Fig. 16-21.

The **heating and air conditioning system** is developed on a drawing of the floor plan. See Fig. 16-22. Air conditioning condensers are located and tied into the system. Air handling units are located and connected to ducts to carry air to the various rooms. Each duct is sized, and the direction of air flow is shown with arrows. The grilles in each room are located and identified with a keyed letter. Data for grilles are given in a grille schedule. The loca-

tions of lights are shown with dashed lines to make certain the placement of the heating system will not conflict with them.

The **grille schedule** gives the size, a detailed description of each grille, and the required finish. See Fig. 16-23.

The **utility site plan** shows (on a very small scale) the water service, electrical service, and sanitary sewer connections from nearby lines to the building. See Fig. 16-24, p. 436. The size of existing and required new service is given.

Details of Sheet S-1

Sheet S-1 is the first of two sheets devoted to **structural details**. The locations of the piers, grade beams, and columns were given on the foundation plan. This sheet has a **roof framing plan** showing the size and spacing of the open web joists and the bridging. See Fig. 16-25. Several wide flange beams are required, and these are located and sizes given. The use of 3″ × 3″ angles to form shelves for the roof at exterior walls is dimensioned.

On the **column and footing schedule,** each column and footing is identified by a certain mark, which appears on the foundation plan. This mark shows how many of each, the size of the piers and footing, and the elevation of the bottom of the footing. See Fig. 16-26, p. 437.

The **grade beam schedule**, too, identifies each grade beam by a certain mark. The mark is found on the foundation plan. The schedule gives the size of each beam, the number of beams required, the size of reinforcing in each beam, the spacing of reinforcing, and the elevation of the top of each beam. See Fig. 16-27.

General structural notes are included. These provide specifications that cannot be shown on the drawings. See Fig. 16-28.

Details of Sheet S-2

Structural information is continued on sheet S-2. On this sheet, a **wall section** from the bottom of the grade beam to the top of the parapet has been drawn. All parts are identified by notes and material in section symbols. Insulation details, wall structure, setting of the open web joists, and roof and parapet construction are shown. See Fig. 16-29, p. 438.

The **building section** shows the relationship between the interior partitions, the dropped ceiling, and the exterior walls. Construction details too small to show are coded so the enlarged details can be found. For example, detail S2-F is detail **F** on sheet S-2. The thickened slab is shown, as are ceiling heights in various parts of the building. See Fig. 16-30, pp. 438-439.

The **structural details** show the construction at the head, jamb, sill, and mullion of each large glass entrance panel. The structure must be strong enough to span the opening since these units are not able to carry a roof load. Details for joining these units to the structure are obtained from the manufacturer of the units. Often these details are printed in bulletins published as part of the Sweet's Catalog Service. See Figs. 16-31 through 16-36, pp. 439-440.

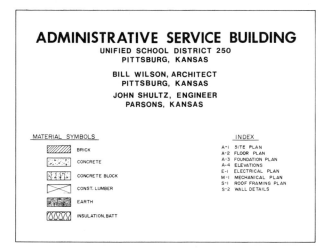

Fig. 16-5. The title sheet provides necessary identification of the project.

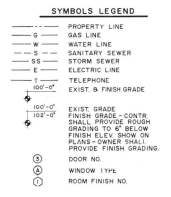

Fig. 16-7. The symbols legend for the site plan gives the meaning of each symbol used.

Fig. 16-6. The site plan shows the building location, lot contours, parking, sidewalks, and other site details.

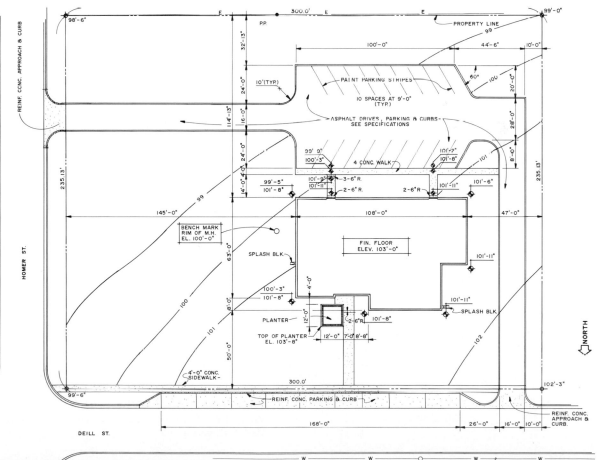

SITE PLAN

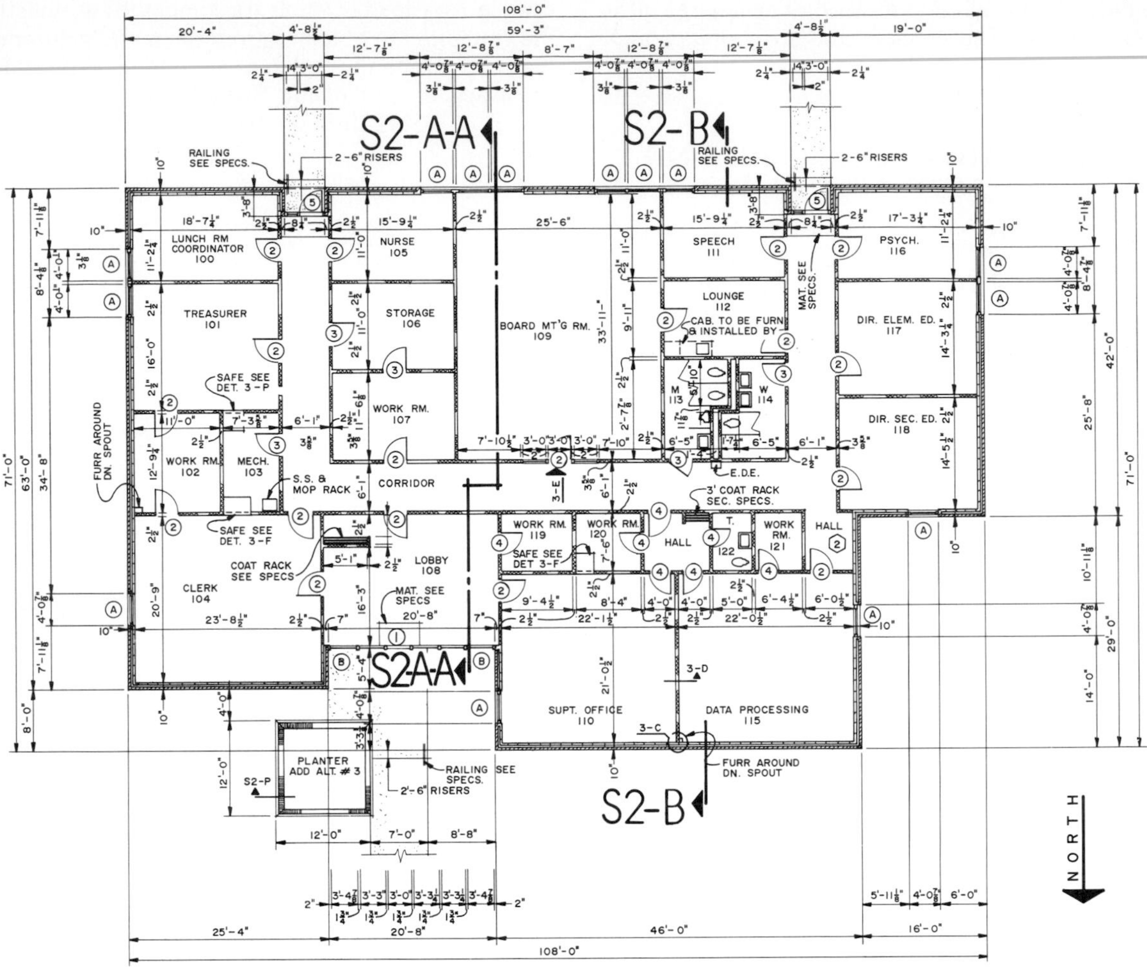

FLOOR PLAN

SCALE ⅛" = 1'-0"

Fig. 16-8. A floor plan shows room arrangement and sizes.

DOOR SCHEDULE					
MARK	SIZE	MATL.	FRAME NO.	FRAME MATL.	REMARKS
1	3'-0" X 6'-8" X 1¾"	ALUM.	1	ALUM.	¼" TEMP. SOLARBRONZE GLASS
2	3'-0" X 6'-8" X 1¾"	RED OAK	2	STEEL	¼" TEMP. SOLARBRONZE GLASS
3	3'-0" X 6'-8" X 1¾"	RED OAK	2	STEEL	¼" TEMP. SOLARBRONZE GLASS
4	2'-8" X 6'-8" X 1¾"	RED OAK	3	STEEL	¼" TEMP. SOLARBRONZE GLASS
5	3'-0" X 6'-8" X 1¾"	ALUM.	1	ALUM.	¼" TEMP. SOLARBRONZE GLASS

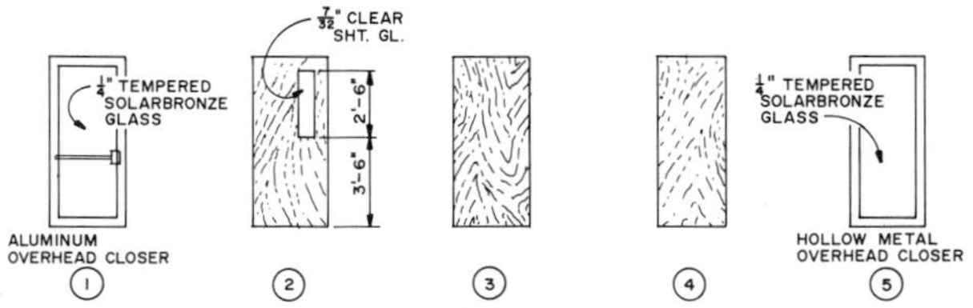

DOOR TYPES

DOOR SCHEDULE

SCALE ¼" = 1'-0"

WINDOW SCHEDULE			
MARK	SIZE	MATL.	REMARKS
A	4'-0⅞" X 8'-4"	ALUM.	¼" SOLARBRONZE GLASS
B	3'-4⅞" X 5'-5"	ALUM.	" " " "

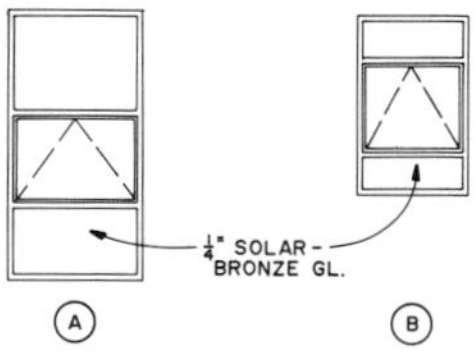

NOTE: WINDOWS TO BE COMPLETE WITH SCREENS & HARDWARE.

WINDOW SCHEDULE

SCALE ¼" = 1'-0"

Fig. 16-9. The door and window schedules give complete information for each type of door and window that will be used.

INTERIOR FINISH SCHEDULE										
ROOM	ROOM NO.	FLOOR		BASE		WALL		CEILING		CLG. HEIGHT
		MATL.	FIN.	MATL.	FIN.	MATL.	FIN.	MATL.	FIN.	
L. R. COORD	100	CARPET	—	CARPET	—	GYP. BD.	PAINT	ACOUS. T.	—	8'–4"
TREASURER	101	DO	—	DO	—	DO	DO	DO	—	DO
WORK ROOM	102	CONC.	TROWEL	NONE	—	C. BLOCK	DO	DO	—	DO
MECHANICAL	103	DO	DO	—	—	DO	DO	DO	—	8'–0"
CLERK	104	V. TILE	—	VINYL	—	GYP. BD.	DO	DO	—	8'–4"
NURSE	105	DO	—	DO	—	DO	DO	DO	—	DO
STORAGE	106	DO	—	DO	—	C. BLOCK	DO	NONE	—	—
WORK ROOM	107	DO	—	DO	—	DO	DO	ACOUS. T.	—	8'–4"
LOBBY	108	CARPET	—	CARPET	—	GYP. BD.	DO	DO	—	DO
BOARD RM.	109	DO	—	DO	—	WOOD PL.	NATURAL	DO	—	DO
SUPT. OFFICE	110	DO	—	DO	—	GYP. BD.	PAINT	DO	—	DO
SPEECH	111	DO	—	DO	—	DO	DO	DO	—	DO
LOUNGE	112	DO	—	DO	—	DO	DO	DO	—	DO
REST ROOMS	113	CER. T	—	CER. T	—	WATER RES. GYP. BD.	DO	DO	—	DO
DATA PROCESSING	115	CARPET	—	CARPET	—	GYP. BD.	DO	DO	—	DO
PSYCH.	116	DO	—	DO	—	DO	DO	DO	—	DO
ELEM. ED.	117	DO	—	DO	—	DO	DO	DO	—	DO
SEC. ED.	118	DO	—	DO	—	DO	DO	DO	—	DO
CORRIDOR	—	—	—	DO	—	DO	DO	DO	—	0'–0"
WORK ROOM	119 – 121	CONC.	TROWEL	—	—	C. BLOCK	DO	DO	—	7'–4"
REST ROOM	122	CER. T	—	CER. T	—	WATER RES. GYP. BD.	DO	DO	—	DO

Fig. 16-10. The interior finish schedule specifies the finish requirements for the floor, base, wall, and ceiling.

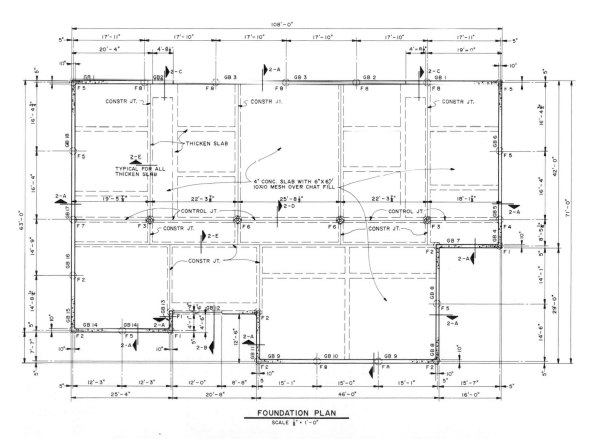

FOUNDATION PLAN
SCALE ⅛" = 1'–0"

Fig. 16-11. The foundation plan shows all footings, grade beams, columns, and control joints.

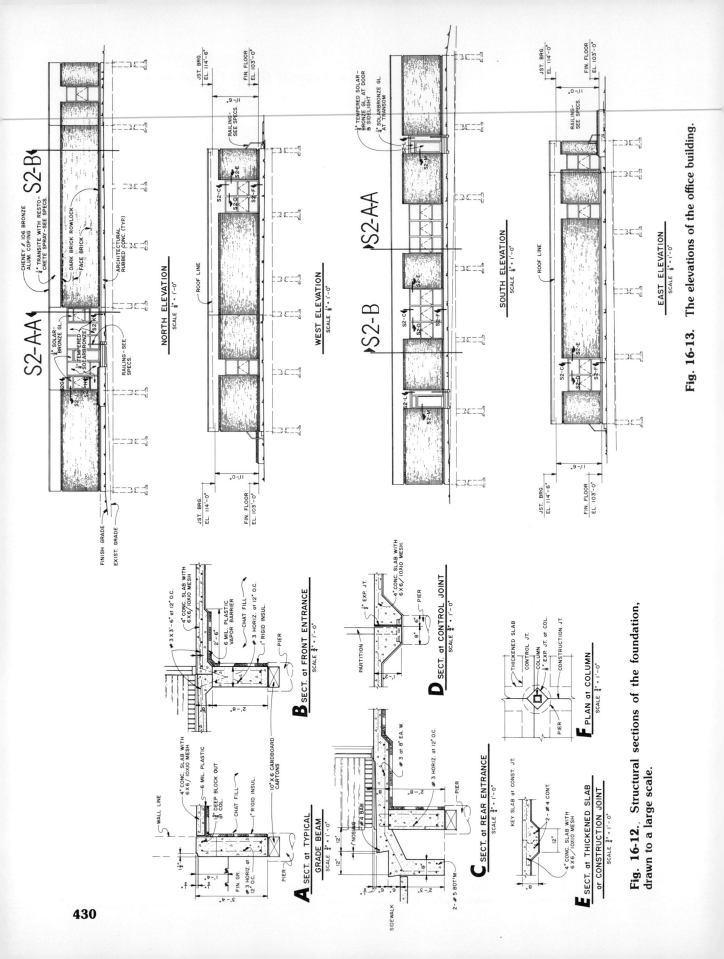

Fig. 16-13. The elevations of the office building.

Fig. 16-12. Structural sections of the foundation, drawn to a large scale.

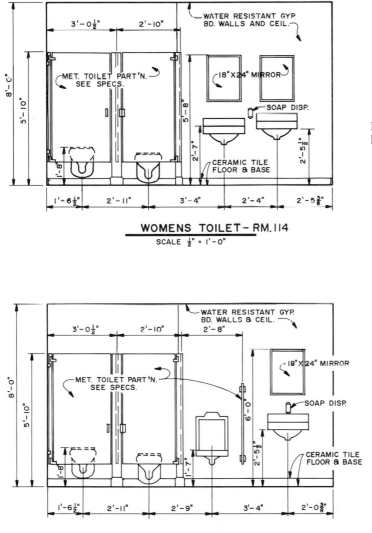

WOMENS TOILET – RM. 114
SCALE ½" = 1'-0"

Fig. 16-14.   These elevations show the locations of the fixtures in the rest rooms.

MEN'S TOILET – RM. 113
SCALE ½" = 1'-0"

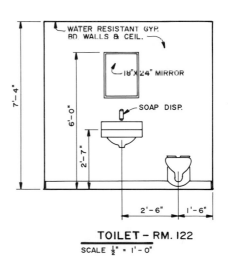

TOILET – RM. 122
SCALE ½" = 1'-0"

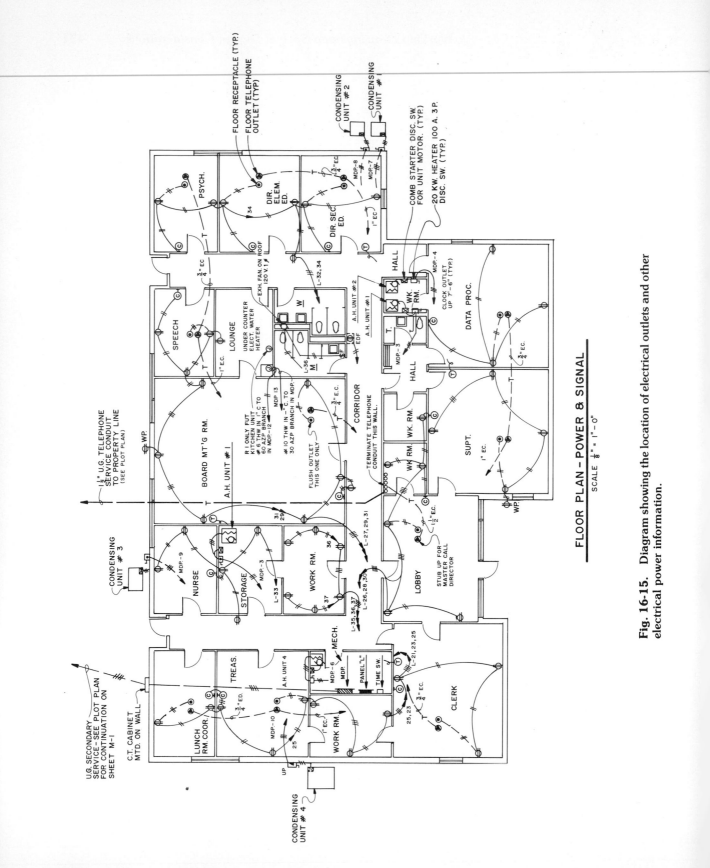

FLOOR PLAN – POWER & SIGNAL

SCALE $\frac{1}{8}$" = 1'-0"

Fig. 16-15. Diagram showing the location of electrical outlets and other electrical power information.

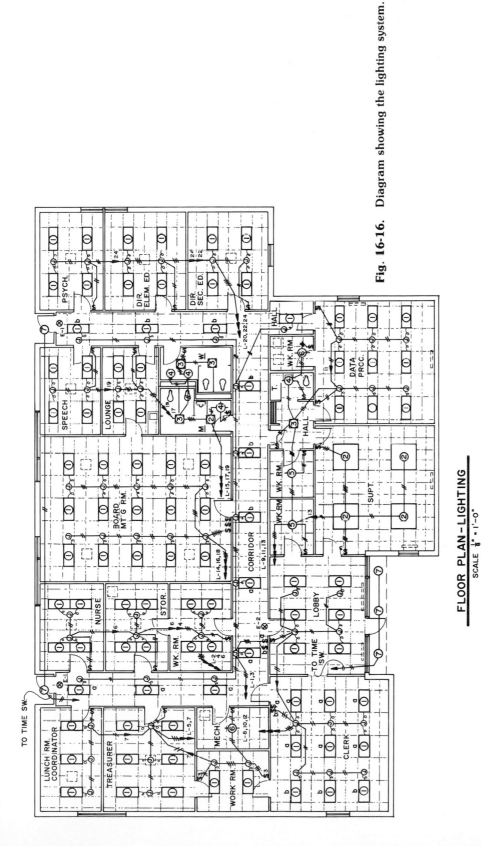

FLOOR PLAN – LIGHTING

SCALE $\frac{1}{8}$" = 1'-0"

Fig. 16-16. Diagram showing the lighting system.

LIGHTING		FIXTURE				SCHEDULE		
	FIXTURE		LAMPS					
TYPE	MFG.	CAT'G. NO.	NO.	WATTS	TYPE	MOUNTING	USE	REMARKS
1	WILLIAMS	5224-K12A-R	4	40 W.	F40-T12/CW	RECESSED	GEN. ILLUM.	FURNISHED & FINAL CONNECTED BY OWNER (THIS FIXT. ONLY)
2	LIGHTOLIER	13198	8	40 W.	F4C-T12/CW	SURFACE	GEN. ILLUM.	
3	PERFECLITE	KF-30-8	1	200 W.	TYPE A	RECESSED	DO	
4	LIGHTOLIER	6349	2	60 W.	TYPE A	WALL	DO	W / RECPT.
5	PERFECLITE	HH-42	2	100 W.	TYPE A	SURFACE	DO	
6	P & S	44	1	150 W.	A-21	SURFACE	OUTSIDE LT.	
7	PERFECLITE	FR-257	1	150 W. FL	FAR 38 FL	RECESSED	DO	
E-1	PERFECLITE	P-424	2	25 W.	T 10	CEIL'G. MTD.	EXIT LT.	
E-2	PERFECLITE	P-424-D	2	25 W.	T 10	DO	DO	

Fig. 16-17. This chart shows the types of lighting fixtures to be used and provides other technical information.

ELECTRICAL LEGEND

FLUORESCENT FIXTURE NO. DENOTES TYPE ON SCHEDULE
INCANDESCENT CEILING FIXTURE SAME FOR NO.5
RECESSED INCANDESCENT FIXTURE – SAME FOR NO. 5
INCANDESCENT WALL FIXTURE – SAME FOR NO.5
SINGLE POLE SWITCH
THREE WAY SWITCH
KEY OPERATED SWITCH
FUSED SWITCH
SWITCH & DUPLEX RECEPTACLE
DUPLEX RECEPTACLE
FLOOR RECEPTACLE
SPECIAL DUTY RECEPTACLE (AS NOTED)
TELEPHONE OUTLET
FLOOR TELEPHONE OUTLET
JUNCTION BOX
MOTOR – SIZE INDICATED
MOTOR – CONTROLLER
DISCONNECTING SWITCH
LIGHTING PANEL BOARD
POWER OR DISTRIBUTION PANEL BOARD
TELEPHONE CABINET
CONCEALED CONDUIT RUN IN WALLS OR CEILING
CONCEALED CONDUIT RUN IN OR UNDER FLOOR SLAB
EXPOSED CONDUIT
NO. OF MARKS ON PANEL INDICATES NO. OF WIRES
HOMERUN TO PANEL – ARROWS INDICATE NO. OF CIRCUITS
TELEPHONE CONDUIT
W.P.   WEATHERPROOF UNITS
THERMOSTAT
EXIT LIGHT

Fig. 16-18. The electrical legend identifies the symbols used on the electrical drawing.

MAIN		DISTRIBUTION					PANEL		
SERVICE: 120/208 V. 3 ∅ 4 W							LUGS: 400 AMP		
MAIN: W/Q.M.B. MAIN SWITCH FUSED 400 AMPS							FEED: BOTTOM		
BRANCHES: DISTRIBUTION TYPE BOLD IN BREAKERS							MOUNTING: SURFACE		
CIR. NO.	SERVING	LOAD	FRAME SIZE	BRANCH BREAKER SIZE	SFTY. SW. SIZE	SFTY. SW. FUSE ✳✳	WIRE SIZE		CONDUIT SIZE
MDP– 1	MAIN SWITCH	116 KW	——	400 ✳	——	——	——		——
MDP– 2	LIGHT PANEL "L"	31 KW	FH	150 A 3P	——	——	4 #2/0 THW		2½"
MDP– 3	A.H. UNIT #1	28 KW	FH	100 A 3P	100 A 3P	70 A	3 # 4 THW		1¼"
MDP– 4	" " #2	28 KW	FH	100 A 3P	100 A 3P	70 A	3 # 4 THW		1¼"
MDP– 5	" " #3	28 KW	FH	100 A 3P	100 A 3P	70 A	3 # 4 THW		1¼"
MDP– 6	" " #4	28 KW	FH	100 A 3P	100 A 3P	70 A	3 # 4 THW		1¼"
MDP– 7	CONDENSING UNIT #1	26.3 FLA	FA	60 A 3P	60 A 3P	35 A	3 # 6 THW		1"
MDP– 8	" " #2	26.3 FLA	FA	60 A 3P	60 A 3P	35 A	3 # 6 THW		1"
MDP– 9	" " #3	26.3 FLA	FA	60 A 3P	60 A 3P	35 A	3 # 6 THW		1"
MDP– 10	" " #4	21.7 FLA	FH	50 A 3P	60 A 3P	——	3 # 6 THW		¾"
MDP– 11	SPARE	——	FH	100 A 3P	——	——	——		——
MDP– 12	KITCHEN UNIT	8 KW	FH	60 A 3P	——	——	# 6 THW		1"
MDP– 13	WATER HEATER	5 KW	FH	30 A 3P	——	——	# 10 THW		——

✳   400 AMP 3 POLE Q.M.Q.B. MAIN SWITCH FUSED WITH BUS TYPE LPN FUSES.
✳✳ SFTY. SWITCHES FUSED WITH BUSS FUSETRON FUSES.

LIGHTING PANEL "L"   NLAB TYPE, SURFACE MOUNTED 150 AMP LUGS ONLY, TOP FEED 120/208 V. 3∅ 4 WIRE SERVICE, & WITH 42–20 A. 1 POLE BRANCH BREAKERS.

MOTOR STARTERS   FURNISH AND INSTALL FOR A.H. UNITS // 1, 2, 3, 4 ALLEN BRADLEY BUL, // 712 3 POOLE MAGNETIC ACROSS-THE-LINE COMBINATION STARTER AND DISCONNECT SWITCHES WITH START-STOP BUTTON IN COVER.

Fig. 16-19. This chart shows the electrical circuits leading from the main distribution panel. It also provides technical information.

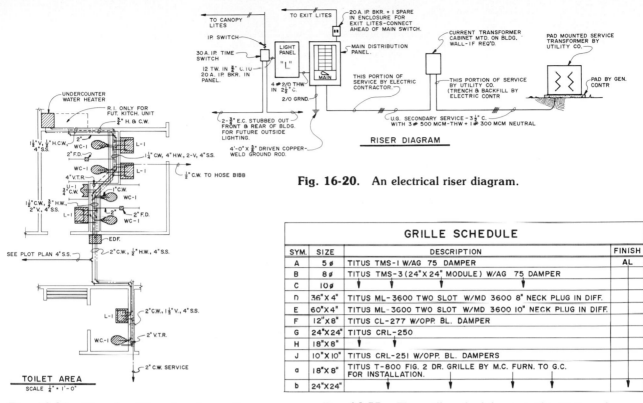

RISER DIAGRAM

Fig. 16-20. An electrical riser diagram.

TOILET AREA
SCALE ¼" = 1'-0"

Fig. 16-21. The plumbing plan shows the sizes of water lines and sewer connections.

SYM.	SIZE	DESCRIPTION	FINISH
\multicolumn{4}{c}{**GRILLE SCHEDULE**}			
A	5ø	TITUS TMS-1 W/AG 75 DAMPER	AL
B	8ø	TITUS TMS-3 (24"X 24" MODULE) W/AG 75 DAMPER	
C	10ø		
D	36"X 4"	TITUS ML-3600 TWO SLOT W/MD 3600 8" NECK PLUG IN DIFF.	
E	60"X4"	TITUS ML-3600 TWO SLOT W/MD 3600 10" NECK PLUG IN DIFF.	
F	12"X8"	TITUS CL-277 W/OPP. BL. DAMPER	
G	24"X24"	TITUS CRL-250	
H	18"X8"		
J	10"X10"	TITUS CRL-251 W/OPP. BL. DAMPERS	
a	18"X8"	TITUS T-800 FIG. 2 DR. GRILLE BY M.C. FURN. TO G.C. FOR INSTALLATION.	
b	24"X24"		

Fig. 16-23. The grille schedule gives the sizes and type of grilles used with the heating system.

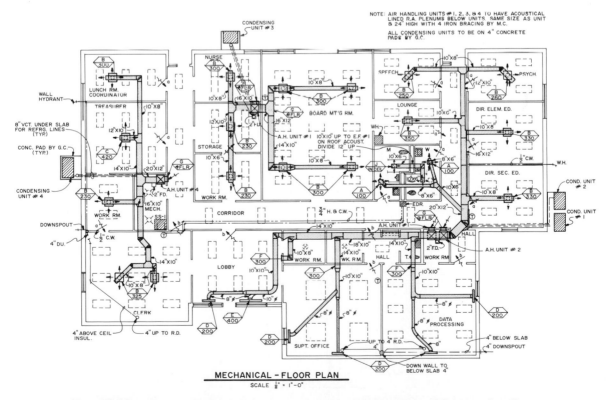

NOTE: AIR HANDLING UNITS #1, 2, 3, & 4 TO HAVE ACOUSTICAL LINED R.A. PLENUMS BELOW UNITS. SAME SIZE AS UNIT & 24" HIGH WITH 4 IRON BRACING BY M.C.

ALL CONDENSING UNITS TO BE ON 4" CONCRETE PADS BY G.C.

MECHANICAL - FLOOR PLAN
SCALE ⅛" = 1'-0"

Fig. 16-22. The mechanical plan shows the sizes and locations of ducts and grilles.

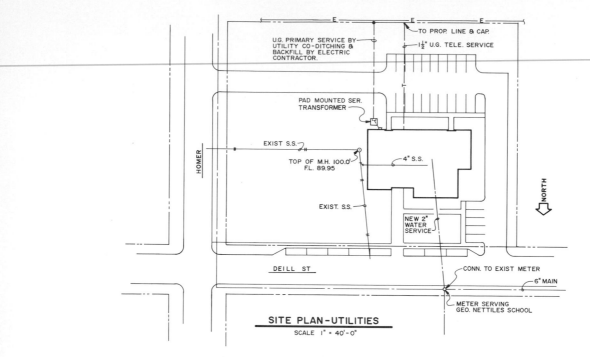

Fig. 16-24. An abbreviated version of the site plan is used to locate utilities and show required connections.

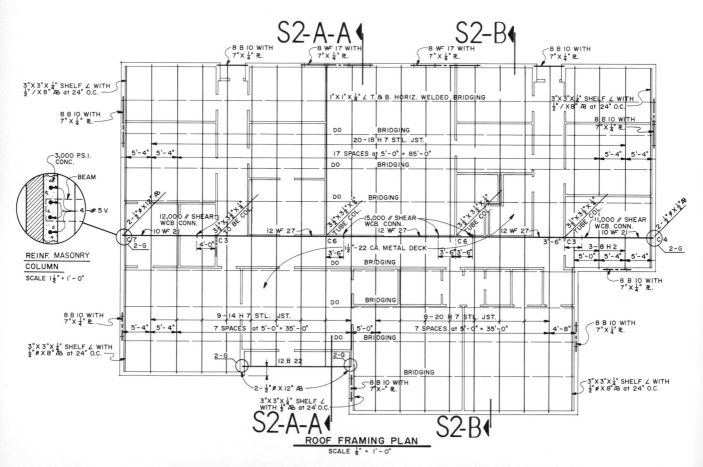

Fig. 16-25. A roof-framing plan shows the size and location of all structural members.

COLUMN & FOOTING SCHEDULE

MARK	F1	F2	C3 F3	C4 F4	F5	C6 F6	C7 F7	F8
TOTAL NO.	3	6	2	1	6	2	1	7
SUPPORTING ROOF	MASONRY	MASONRY	3½"□X¼" TUBE 8X½X1'-2" B.R. 4-¾"ØX18" AB	4-#5 V. IN CONC. FILLED MASONRY	MASONRY	3½"□X¼" TUBE 8X½X1'-2" B.R. 4-⅝"ØX18" AB	4-#5 V. IN CONC. FILLED MASONRY	MASONRY
PIERS	14"Ø 4-#5 V. #3T at 12"	14"Ø 4-#5 V. #3T at 12"	18"Ø 4-#5 V. #3T at 12"	14"Ø 4-#5 V. #3T at 12"	14"Ø 4-#5 V. #3T at 12"	14"Ø 4-#5 V. #3T at 12"	14"Ø 4-#5 V. #3T at 12"	14"Ø 4-#5 V. #3T at 12"
FOOTINGS	18"Ø	24"Ø	30"Ø	30"Ø	30"Ø	36"Ø	36"Ø	36"Ø
ELEV at BOT'M. OF FOOTINGS	94'-0"	94'-0"	94'-0"	94'-0"	94'-0"	94'-0"	94'-0"	94'-0"

Fig. 16-26. The column and footing schedule gives the complete design data.

GRADE BEAM SCHEDULE

MARK	SIZE	NO.	REINFORCING	HK (TOP BARS)	TOP BARS	HK	HK (BOTTOM BARS)	BOTTOM BARS	HK	NO.	SIZE	LENGTH	TYPE	SPACING	TOP OF GRADE BM. ELEV.
GB 1	10X40	2	2-#6 X 23'-4" 2-#6 X 18'-8"	8	22'-8"	—	—	18'-8"	—	15	#3	7'-5"	⊔	1 at 3", 3 at 12" O.C. BAL. at 18" O.C.	103'-0"
GB 2	10X 32/40	2	2-#6 X 26'-10" 2-#6 X 10'-10"	—	26'-10"	—	—	18'-10"	—	12	#3	7'-5"	⊔	DO	101'-6" at DR. 103'-0"
GB 3	10X40	2	2-#5 X 26'-10" 2-#5 X 18'-10"	—	26'-10"	—	—	18'-10"	—	12	#3	7'-5"	⊔	DO	103'-0"
GB 4	10X40	1	2-#5 X 11'-8" 2-#4 X 7'-4"	7	11'-1"	—	—	7'-4"	—	4	#3	7'-5"	⊔	DO	103'-0"
GB 5	10X40	1	2-#5 X 25'-8" 2-#5 X 18'-0"	—	25'-8"	—	—	18'-0"	—	12	#3	7'-5"	⊔	DO	103'-0"
GB 6	10X40	1	2-#5 X 22'-9" 2-#5 X 18'-3"	—	22'-9"	—	—	18'-3"	—	12	#3	7'-5"	⊔	DO	103'-0"
GB 7	10X40	1	2-#5 X 17'-8" 2-#6 X 16'-6"	7	16'-6"	7	—	16'-6"	—	12	#3	7'-5"	⊔	DO	103'-0"
GB 8	10X40	2	2-#5 X 19'-1" 2-#5 X 15'-4"	7	18'-6"	—	—	15'-4"	—	10	#3	7'-5"	⊔	DO	103'-0"
GB 9	10X40	2	2-#5 X 19'-8" 2-#6 X 15'-10"	7	19'-1"	—	—	15'-10"	—	11	#3	7'-5"	⊔	DO	103'-0"
GB 10	10X40	1	2-#5 X 22'-6" 2-#5 X 16'-0"	—	22'-6"	—	—	16'-0"	—	11	#3	7'-5"	⊔	DO	103'-0"
GB 11	10X40	1	2-#4 X 14'-0" 2-#6 X 13'-0"	6	13'-0"	6	—	13'-0"	—	9	#3	7'-5"	⊔	DO	103'-0"
GB 12	10X32	1	2-#4 X 23'-0" 2-#5 X 22'-0"	6	22'-0"	6"	—	22'-0"	—	14	#3		⊔	DO	103'-0"
GB 13	10X40	1	2-#4 X 6'-0" 2-#5 X 5'-0"	6	5'-0"	6	—	5'-0"	—	4	#3	7'-5"	⊔	DO	103'-0"
GB 14	10X40	2	2-#5 X 16'-3" 2-#5 X 13'-1"	7	15'-8"	—	—	13'-1"	—	9	#3	7'-5"	⊔	DO	103'-0"
GB 15	10X40	1	2-#5 X 18'-5" 2-#5 X 14'-0"	7	17'-0"	—	—	14'-10"	—	10	#3	7'-5"	⊔	DO	103'-0"
GB 16	10X40	1	2-#5 X 21'-10" 2-#5 X 15'-0"	—	21'-10"	—	—	15'-0"	—	10	#3	7'-5"	⊔	DO	103'-0"
GB 17	10X40	1	2-#5 X 25'-0" 2-#5 X 18'-3"	—	25'-11"	—	—	18'-3"	—	12	#3	7'-5"	⊔	DO	103'-0"
GB 18	10X40	1	2-#5 X 22'-6" 2-#5 X 18'-1"	—	21'-11"	7	—	18'-1"	—	12	#3	7'-5"	⊔	DO	103'-0"

Fig. 16-27. The grade beam schedule gives complete information about beam sizes and reinforcing.

GENERAL
STRUCTURAL NOTES

1. FOOTING DESIGNS ARE BASED UPON A SHALE BEARING VALUE OF 6,000 LBS./SQ. FT.

2. CONC. TO HAVE A MIN. COMPRESSIVE STRENGTH OF 3,000 LBS./SQ. FT. AT END OF 28 DAYS.

3. REINFORCING STEEL TO MEET A.S.T.M. SPECIFICATION A-615, GR. 40, LATEST REVISIONS.

4. STRUCTURAL STEEL TO MEET A.S.T.M. SPECIFICATION A-36, LATEST REVISION.

5. PROVIDE ¼" BED PLATES UNDER ALL STEEL COLS., & TO BE THE SAME SIZE IN PLAN AS THE BASE PLATES.

6. PROVIDE 4" DEEP CARDBOARD CARTON FORMS UNDER ALL GRADE BEAMS.

7. ALL COL. TO BEAM & BEAM TO BEAM CONN, TO BE ERECTED WITH A.S.T.M. A-325 HIGH STRENGTH BOLTS. ALL OTHER CONN. MAY BE ERECTED WITH STD. MACHINE BOLTS.

8. PROVIDE 6X6: 10/10 WIRE MESH FOR ALL CONC. SLABS ON GRADE UNLESS NOTED OTHERWISE.

9. STEEL JOISTS TO MEET STL. JST. INSTITUTE SPECIFICATIONS, INCLUDING BRIDGING & ACCESSORIES.

10. WELD STL. JST. TO THE STL. BEAMS, OR TO 4X6X¼ WELDING PLATES WITH 1-X8 WELDED ANCHOR ON THE BOT'M. OF EA. PLATE, IF STL. JST. BEAR UPON MASONRY WALLS. IF JOISTS BEAR INTO MASONRY WALLS PROVIDE STD. JOIST ANCHOR.

11. SEE MECH. DWGS. FOR EXACT DIM. OF OPNG'S & MECH. EQUIPT. DIM.

12. REINFORCING TIES IN PIERS & COLS. SHALL NOT BE HOOP TYPE, BUT PROVIDE SQUARE OR RECTANGULAR TIES.

13. PROVIDE 4 - #5 X3'-6"DOWELS FROM PIERS INTO GRADE BEAMS.

Fig. 16-28. Notes relating to the structural design of the building.

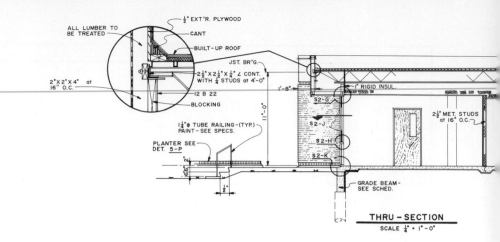

Fig. 16-30. A structural building section.

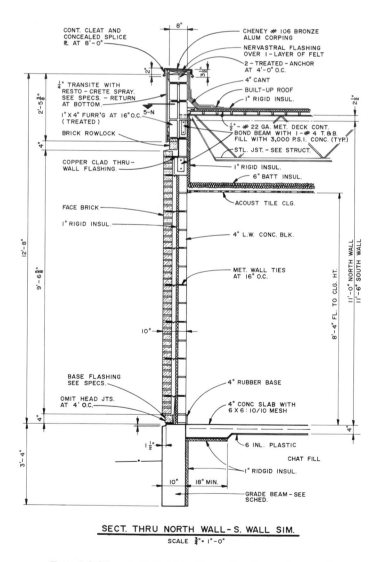

SECT. THRU NORTH WALL - S. WALL SIM.
SCALE ¾" = 1'-0"

Fig. 16-29. A typical wall section from the grade beam through the roof.

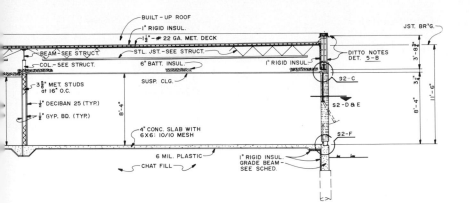

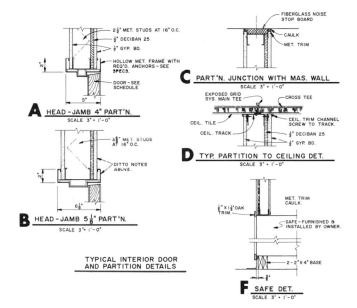

Fig. 16-31. Typical interior door and partition details.

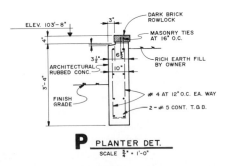

Fig. 16-32. A section through a concrete planter. Also see the floor plan in Fig. 16-8.

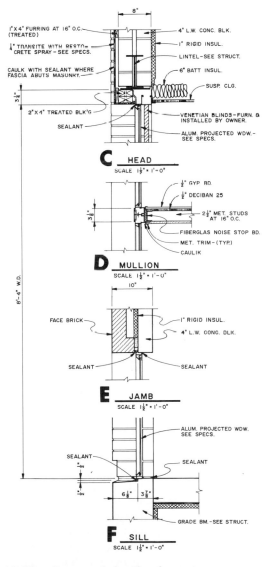

Fig. 16-33. Structural details of exterior entrance. Also see the elevation in Fig. 16-13.

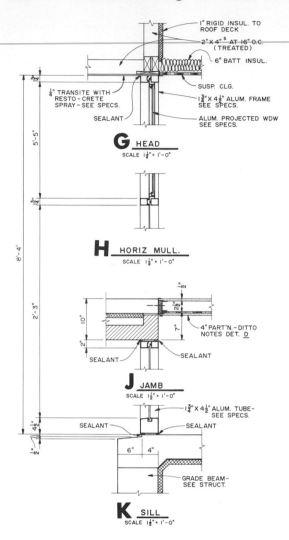

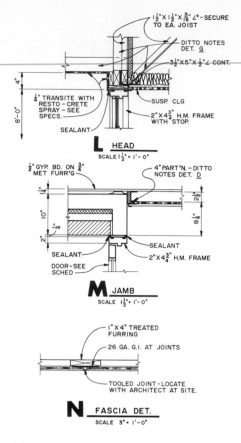

Fig. 16-34. Structural details of exterior window. Also see the elevation in Fig. 16-13.

Fig. 16-35. Structural details of the fascia.

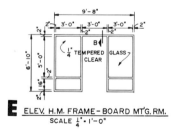

Fig. 16-36. Details of the door and glass wall of the board meeting room.

Environmental Impact Studies

When planning a large project, a study must be made to see how the project will influence the surrounding environment. How will it affect plant and animal life in the area? What impact will it have on property values? And how will it influence the lives of people living in the region?

Consider, for example, the situation when a dam is built across a river. A huge lake forms, flooding large areas of land. This greatly changes the environment for plants and animals as well as for the people living nearby.

Suppose a nuclear power plant is being considered. It could become an important source of power. But some nuclear plants release heated water into nearby streams or lakes. This could disturb the local ecology. Some people might feel somewhat uneasy living near the plant. Others may earn a living by working there. Property values could change. Both positive and negative possibilities must be examined when decisions are made about construction projects.

Special Design Considerations

Many of the design features of buildings cause problems for people who have physical disabilities. Such architectural barriers must be removed in existing buildings and access designed into new buildings so that all people can take an active part in social, recreational, educational, residential, and employment activities.

Legislation has been enacted to help meet special needs, and design standards have been formulated to assist architects. These are recorded in the publication, **Specifications for Making Buildings and Facilities Accessible to and Usable by Physically Handicapped People**, ANSI A117.1-1980. It is published by the American National Standards Institute, 1430 Broadway, New York, N.Y. 10018. State and local regulations must also be observed. These may vary from area to area.

Designing for Those with Mobility Restrictions

People with mobility restrictions include those who must rely upon wheelchairs, walkers, crutches, and other assisting devices for mobility. Since a wheelchair requires the largest space for moving and turning, the space required for using a wheelchair is usually adequate for people using other devices. The dimensions of a typical nonelectric wheelchair are given in Fig. 16-37. When a standard wheelchair is turned in a circle on its large wheels, the front castors make a circle with a 19-inch radius, and the foot platform forms a circle with a 32-inch radius. See Fig. 16-38. Electric wheelchairs

are usually slightly larger than standard wheelchairs and require more space for turning.

Entrances

1. Each building must have at least one principal entrance on an accessible route.
2. This entrance must be connected by an accessible route to public transportation, to parking and passenger loading zones, and to public streets and sidewalks.
3. Entrances shall connect by an accessible route to all accessible spaces within a building.

Doors

1. Revolving doors or turnstiles are not to be on the access path.
2. All doors must provide a minimum opening of 32 inches (813 mm) with the door at 90 degrees.
3. Double doors shall have at least one side that will provide a 32-inch (813 mm) opening.
4. Floor area must be provided for maneuvering at doors that are not automatic. This area should be level and clear. See Fig. 16-39.
5. When there are two doors in a series, the minimum space allowed between them is 48 inches (1220 mm) plus the width of any door swinging into the space. Both doors shall swing either in the same direction or away from the space between them. See Fig. 16-40.

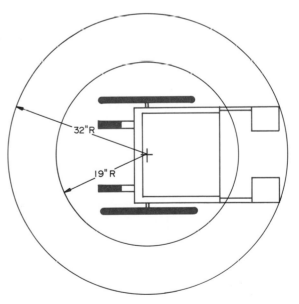

Fig. 16-38. Turning radius of the front castors and foot platform.

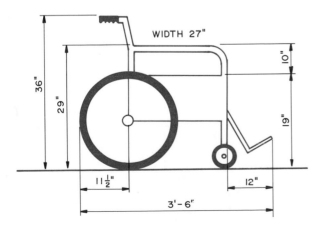

Fig. 16-37. Size of a standard wheelchair.

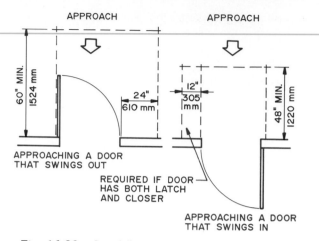

APPROACH   APPROACH

60" MIN. 1524 mm

24"
610 mm

12"
305
mm

48" MIN. 1220 mm

APPROACHING A DOOR
THAT SWINGS OUT

REQUIRED IF DOOR
HAS BOTH LATCH
AND CLOSER

APPROACHING A DOOR
THAT SWINGS IN

Fig. 16-39. Level floor area required in front of non-automatic swinging doors.

6. Thresholds must not exceed 3/4 inch (19 mm) in height for exterior sliding doors or 1/2 inch (13 mm) for other doors.
7. Door closers must open the door at least 70 degrees and take at least three seconds to close to within 3 inches (76 mm) of the latch.
8. The maximum opening force for exterior hinged doors is 8.5 lb/ft (37.8 N), and 5 lb/ft (22.2 N) for interior hinged, sliding, and folding doors.
9. Consider the use of power-operated doors. These are available as sliding or swinging units. They use electric or pneumatic sensing mats, photoelectric cells, motion detectors, keys, and switches. However, swinging power-operated doors can be dangerous for the visually handicapped. Without sight, it is difficult to anticipate the movement of the door.
10. Automatic doors must comply with the **American National Standard for Power Operated Doors**, ANSI A156.10-1979.

Halls

1. The minimum hall width to accommodate a single wheelchair is 3'-0" (915 mm). If two wheelchairs must pass, the hall must be at least 5'-0" (1524 mm) wide. When a wheelchair and a person must pass, the hall width should be 4'-0" (1220 mm).
2. A wheelchair requires a hall 3'-0" wide (915 mm) to turn a 90° corner. See Fig. 16-41. The minimum space needed to make a 180° turn around an obstruction is shown in Fig. 16-42. Making a U-turn requires an area 5'-0" (1524 mm) wide and 6'-6" (1982 mm) long.
3. The intersection of halls must allow the wheelchair to turn the corner.
4. The halls must provide at least one accessible route to all spaces within a building and one exterior entrance. Doors on this route must allow access.

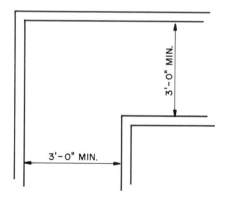

3'-0" MIN.

3'-0" MIN.

Fig. 16-41. A standard wheelchair requires a hall or walk 3 feet wide to turn a 90-degree corner.

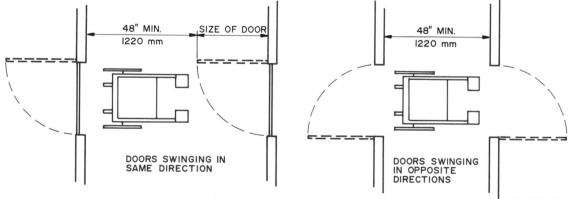

48" MIN.
1220 mm

SIZE OF DOOR

DOORS SWINGING IN
SAME DIRECTION

48" MIN.
1220 mm

DOORS SWINGING
IN OPPOSITE
DIRECTIONS

ANSI A117.1-1980

Fig. 16-40. Spacing of hinged doors in a series.

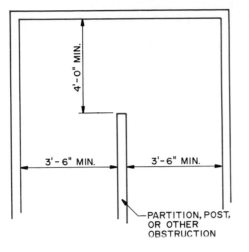

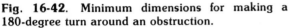

Fig. 16-42. Minimum dimensions for making a 180-degree turn around an obstruction.

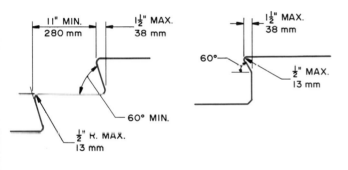

FLUSH RISER STAIR   ROUNDED NOSING STAIR

Fig. 16-43. Recommended stair design details.

5. Objects, such as cabinets, should not protrude more than 4 inches (101 mm) into halls.
6. Halls must have 6'-8" (2032 mm) clear headroom.
7. The floor surfaces should be stable, firm, and made of material that prevents slips.

Stairs

1. The maximum riser is 7 inches (177 mm) and the minimum tread is 11 inches (280 mm). The maximum nosing is 1½ inches (38 mm). See Fig. 16-43.
2. In a single flight of stairs, all riser heights and tread widths must be uniform in size.

Handrails

1. Place handrails on both sides of a stair, 30 inches (762 mm) high for adults and 24 inches (610 mm) for children.
2. Handrails should extend one foot beyond the riser at the top of the stair and one foot plus the width of one riser at the bottom. See Fig. 16-44.
3. Handrails must be continuous on stairs, ramps, and landings.
4. Handrails must clear walls by 1½ inches (38 mm).

Elevators

1. Small elevators are used in small commerical buildings.* See Fig. 16-45.
2. Elevators must be automatic and have a self-leveling feature that brings the car level with floor landings within a tolerance of ½ inch (13 mm).

---

*Small elevators are also useful in residences.

Fig. 16-44. Required handrail extensions at the top and bottom of a stair.

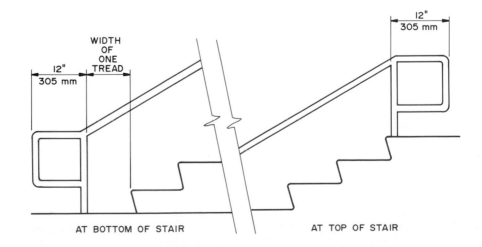

AT BOTTOM OF STAIR   AT TOP OF STAIR

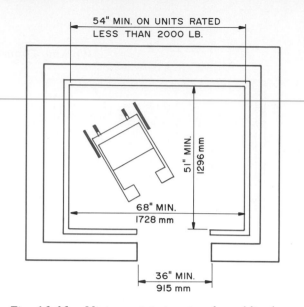

Fig. 16-46.   Minimum interior sizes for public elevators.

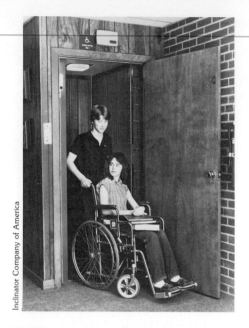

Fig. 16-45.   An elevator designed to accommodate a wheelchair.

3. Call buttons in the hall should be 42 inches (1067 mm) above the floor. The button should have a minimum diameter of ¾ inch (19 mm).

4. A visible and audible signal should be provided at each elevator entrance to indicate which car is answering the call. Audible signals either sound once for Up and twice for Down or have verbal annunciators that say "up" or "down." Visible signals should be 72 inches (1829 mm) above the floor and should be 2½ inches (64 mm) in their least dimension.

5. Elevator entrances shall have raised or indented floor designations on both jambs. They should be 60 inches (1524 mm) above the floor. The characters should be 2 inches (51 mm) high.

6. Elevator doors must open and close automatically. They must automatically reopen if the door is obstructed. Door reopening devices shall hold the door open for 20 seconds.

7. Elevator cars in public buildings shall provide room for wheelchair users to enter, maneuver within reach of the controls, and exit from the car. Acceptable door opening and inside sizes are given in Fig. 16-46.

8. Controls shall be located on the front wall of center opening doors or on the front wall or side wall of side opening doors. See Fig. 16-47.

9. Wheelchair lifts and stair lifts may be used if appropriate. These are more commonly used in residences. Refer back to Chapter 2.

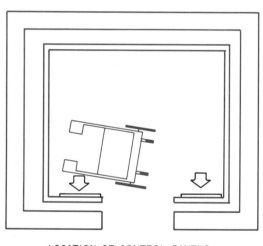

LOCATION OF CONTROL PANELS
FOR CENTER LOADING ELEVATOR

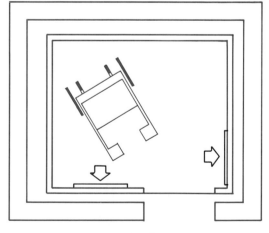

LOCATION OF CONTROL PANELS
FOR SIDE LOADING ELEVATOR

Fig. 16-47.   Location of elevator control panels.

Walks

1. Walks should have a non-skid surface.
2. A walk should be 4'-0" (1219 mm) wide. If two wheelchairs are to pass, the walk must be 5'-0" (1524 mm) wide.
3. Walks should be clear of any obstacles.
4. Walks should be level. If a change in elevation is necessary, provide a ramp.
5. All walks require a minimum of 6'-8" (2032 mm) clear headroom.

Ramps

1. The maximum slope of a ramp is 1:12.
2. The maximum rise for any one ramp is 30 inches (762 mm).
3. The minimum width of a ramp is 36 inches (915 mm).
4. Ramps having a rise of more than 6 inches (152 mm) or those that run longer than 72 inches (1829 mm) shall have handrails on both sides. Handrails for adults shall be 32 inches (813 mm) above the floor of the ramp, and those for children shall be 24 inches (610 mm) above the floor of the ramp.
5. The clear space between the handrail and a wall must be 1½ inches (38 mm).
6. If a ramp has a dropoff on the sides, 2-inch (51 mm) curbs must be built.
7. A 60-inch (1524 mm) level landing should be provided at the top and bottom of a ramp. A level area can be provided midway down a ramp.
8. A non-skid surface is required.

Curb Ramps

1. The minimum ramp width is 3 feet (915 mm), exclusive of flared sides. The maximum slope of the flare is 1:10 and the ramp 1:12. See Fig. 16-48.
2. Curb ramps should be placed at normal street crossings, where cars will not park in front of them.
3. Ramps should have a non-skid surface.
4. On the level area before a ramp descends, there should be a warning texture extending the full width of the ramp and flares. This gives a warning to those with visual impairments.

Parking

1. The minimum stall width is 8 feet (2439 mm), with an adjacent access aisle of 5 feet (1524 mm). See Fig. 16-49.
2. Parking stalls near the building should be clearly marked for the handicapped. The international symbol of access is shown in Fig. 16-50. It is always white and either blue or black.
3. Ramps should be used to provide access to the parking area if there is a difference in elevation.
4. Locate parking stalls so the person is in no danger from approaching autos and is clear of pedestrian traffic.
5. Locate the stalls so they can be reached without going between or behind other vehicles.

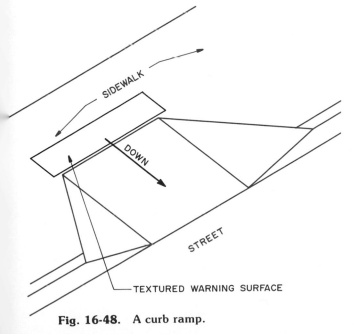

Fig. 16-48. A curb ramp.

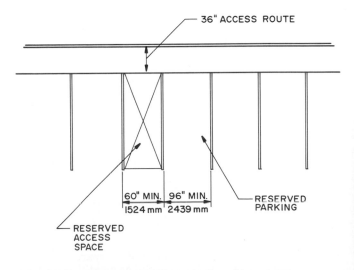

Fig. 16-49. Dimensions of required parking space.

Fig. 16-50. The international symbol used to indicate access to facilities.

6. The location of the parking area should be on the access route to the building. This should be the shortest route possible.
7. Passenger loading zones should provide a clear area 4 feet (1219 mm) wide and 20 feet (6096 mm) long beside the auto.

Drinking Fountains

1. Install drinking fountains so they are 36 inches (914 mm) above the floor. Have a hand-operated water control on the front or side of the unit. The spout should be on the front.
2. Wall-hung units should have a clear knee space of at least 27 inches (686 mm).

3. Wall-hung units need a clear floor area of 30" × 48" (762 mm × 1219 mm).

Rest Rooms

1. The height of water closets shall be 17 inches to 19 inches (432 mm to 483 mm), measured to the top of the seat.
2. The clear floor space for a water closet not in a stall is given in Fig. 16-51.
3. The clear floor space for a water closet in a stall is given in Fig. 16-52.
4. Grab bars should be installed as shown in Fig. 16-53.
5. Clearances for wall-hung lavatories are shown in Fig. 16-54.
6. The minimum clear floor space at a lavatory must be 30" × 48" (762 mm × 1220 mm). See Fig. 16-55.

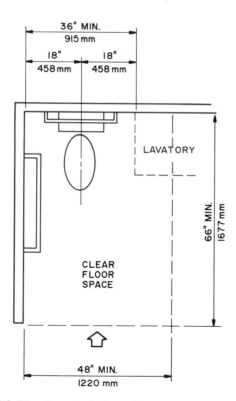

Fig. 16-51. Required clear floor space at a water closet not in a stall. Space can be located at the left or right. Note locations of grab bars.

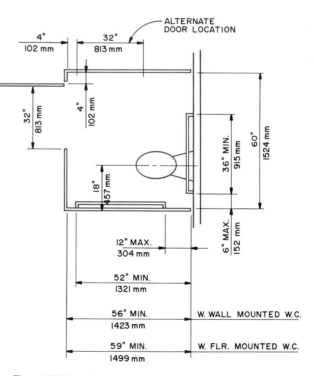

Fig. 16-52. Minimum dimensions for a stall.

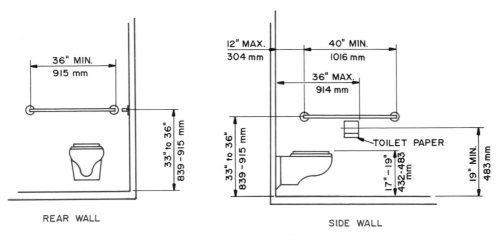

REAR WALL   SIDE WALL

Fig. 16-53. Location of grab bars.

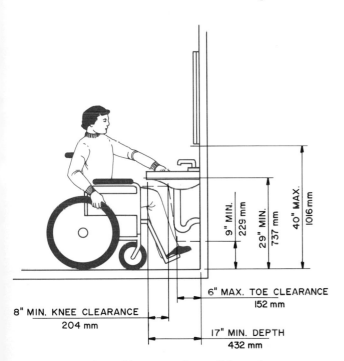

Fig. 16-54. Clearances for wall-hung lavatories.

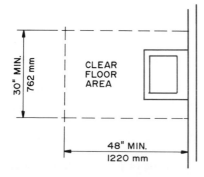

Fig. 16-55. Clear floor area required for wall-hung lavatories.

Designing for Those with Visual Impairments

Persons with restricted vision may be those who can see only large objects, such as furniture, or those who are completely unable to see. Most often a visually impaired person uses a cane by swinging it from left to right in an arc. The cane touches the ground just beyond the width of the person's shoulders. See Fig. 16-56. Another technique is to hold a cane diagonally across the body, with the cane tip touching the ground just outside one shoulder. In either case, obstacles and hazards are noticed only when they come within the range of the cane. If something is protruding into the path of a person who is visually impaired, the cane will detect it only if it is below 27 inches (686 mm) in height. Overhangs cannot be detected.

Following are recommended standards for planning for accessibility by the visually restricted:

1. Allow no protrusions greater than 4 inches (102 mm) into halls or walks, unless these protrusions are within 27 inches (686 mm) of the floor. See again Fig. 16-56.
2. Keep the minimum overhead clearance at 6'-8" (2032 mm).
3. Use a change in the surface of the floor to signify a change in areas. For example, a change from tile to an abrasive surface could indicate a danger area, such as a stair or ramp. Begin this new surface 2 to 4 feet (610 mm to 1219 mm) in front of the hazard.
4. Door push plates, handles, and knobs can be knurled as a warning that the door leads to a hazardous area.

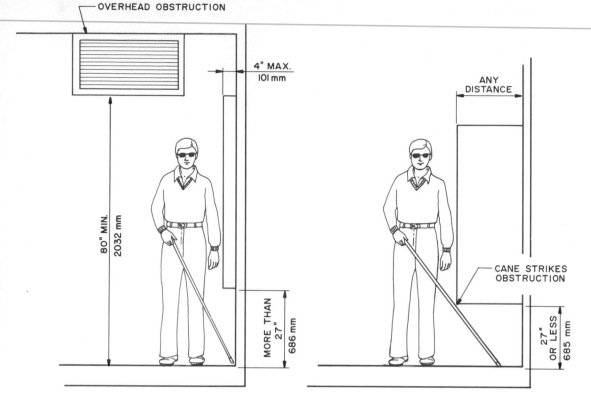

OVERHEAD OBSTRUCTION

4" MAX.
101 mm

80" MIN.
2032 mm

MORE THAN 27"
686 mm

ANY DISTANCE

CANE STRIKES OBSTRUCTION

27" OR LESS
685 mm

CANE TOUCHES WALL AND
PERSON CLEARS OBSTACLE

Fig. 16-56.   Restrictions on objects protruding into the walking space.

5. Consider the use of audible warning signals and directions.
6. Signs can be used if the letters are raised or in braille. Letters should be at least 5/8″ (16 mm) high. Braille characters should be placed to the left of printed characters. Place signs 4′-6″ to 5′-6″ (1372 mm to 1676 mm) above the floor.

Designing for Those with Hearing Impairments

1. Equip at least one telephone with an adjustable sound amplifier.
2. Use a teleprinter or other telephonic device to transmit printed messages via telephone lines.
3. Alarms, such as for fire, should be at a frequency that will attract people with partially restricted hearing.
4. Use lights as warnings for those who can hear nothing.
5. In an auditorium or theater, install a sound system that has earphone jacks with variable control.

Build Your Vocabulary

Following are terms that you should understand and use as a part of your working vocabulary. Write a brief explanation of each term.

accessibility
architectural barriers
audible signals
curb ramps
riser diagram

symbols legend
title block
title sheet
visible signals

Class Activity

If possible, examine a set of working drawings for a small commercial building in your area. Do the following as you refer to the drawings.

1. List the drawings shown on each sheet and the scale of each drawing.
2. Using specific examples, list ways in which this building design meets or does not meet the special needs of people with orthopedic impairments.

Chapter 17

Planning Merchandising Facilities

Each type of retail merchandising facility has special requirements that must be considered during the planning and designing phase. The planner should visit with those experienced in operating such a store and find out how the business is carried on, the best ways to display merchandise and the details of store operation. A study should be made to ascertain buying habits of customers, services they desire, what attracts attention and lures them into the store, and what things help close a sale. The planner is interested not only in the inside of the store, but in the outside as well. The entire facility should be planned as an integral unit.

A successful store requires careful consideration of three major elements.

1. A sales area that is efficient and attractive,
2. Service areas, such as a storage area for merchandise or an area in which to move items to the sales area, and
3. An exterior design that will attract customers into the store.

Following are general principles of planning which should be considered when planning retail merchandising outlets. Special planning considerations for selected types of stores follow the general principles.

An Efficient and Attractive Sales Area

Since the major measure of effectiveness of store planning is the sales record, the area in which goods are displayed and sales are concluded is the heart of the merchandising facility. All other areas exist only to facilitate the sale. The following general principles apply to planning the sales area:

1. The merchandise to be sold must be known before planning can begin. Merchandise is grouped into three types — impulse, convenience, and demand.

Impulse merchandise usually consists of luxury items or items dependent upon good display for sale. Such items are usually not sought after, but are bought by those who, upon seeing them, suddenly decide they want one. Scarves, jewelry, and cosmetics are impulse items.

Convenience merchandise consists of standard items that are popular and much used, such as sheets, food, or appliances.

Demand merchandise involves necessities that bring many people shopping. Customers look especially for these items. Suits, furniture, and prescribed medicines are demand items.

2. Some stores specialize in one type of item; for instance, a novelty shop mainly carries impulse items. In general, most stores predominately stock demand merchandise, with some impulse and convenience items. A person experienced in merchandising should classify items into their sales types for the person planning the store layout.
3. Demand merchandise is usually located far away from the entrance, since customers generally will persist until they find it.
4. Convenience merchandise is most often located midway between the entrance and the demand merchandise.
5. Impulse items are located near the store entrances. Here all customers pass them when they enter and leave the store. Much impulse merchandise is sold to demand and convenience buyers.
6. Customer traffic in the store must be carefully planned. Easy access from the store entrance to all sales sections is necessary.
7. Usually, the store is divided into shopping areas, with each area, in actuality, becoming a small specialty shop. For example, one area in

a department store may be devoted to men's clothing while another may display jewelry. These areas are reached by aisles. See Fig. 17-1.

8. Each specialty area should be planned as a separate shop, and the entire store plan should be developed by attractively grouping these areas.

9. The size and location of shopping aisles depends upon the size of the store and the type of merchandise. Most small retail outlets use one single, straight, center aisle extending the length of the store. If the store is fairly large, minor aisles should branch from the main aisle. In very large stores, several main shopping aisles may be used, with many minor aisles. In stores with heavy traffic in the main aisle, the minor aisles become the sales aisles, and the main aisle simply carries traffic.

10. The two patterns frequently used for placing aisles are the block system (or some variation) and the free-flow system.

The **block system** is illustrated in Fig. 17-2, top. It fits well with the columns in the structural system and is frequently used, but it has many disadvantages. Often it forces customers to go out of their way to the elevator or stairs. More importantly, it provides little flexibility. Some sales areas need more floor space than others. This is difficult to arrange with the block system. It is also difficult to rearrange fixtures and display seasonal merchandise.

The **free-flow system** is also illustrated in Fig. 17-2. Almost all trace of the block system of aisles is gone. Traffic can flow easily from area to area. Fixture design can be highly imaginative and flexible. Displays can be changed easily, and particular sales areas can receive more floor space during their sales season. Certainly, the space devoted to swimming suits in the summer should be adaptable to another use in the winter. Gloves and scarves can be displayed in this area merely by removing, adding, or rearranging fixtures. This system is recommended for most effective and economical use of space.

11. Aisle width should be determined by the capacity of the store entrances and the flow of people to them from elevators and stairs. The width of a main aisle is usually from 6'-0" to 11'-0" (1829 mm to 3353 mm), and minor aisles from 3'-0" to 4'-0" (914 mm to 1219 mm). Aisles for clerks need be only 2'-0" (610 mm) wide, unless additional space

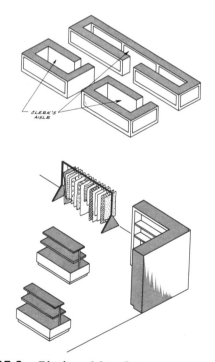

Fig. 17-1. Observe the division of merchandise into small departments or specialty areas. In the foreground is a seasonal display, with the steady demand items to the side and rear of the store.

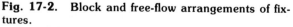

Fig. 17-2. Block and free-flow arrangements of fixtures.

Top: Block plan. Arrangement is frozen, since standard fixture is used to display all types of merchandise. Considerable space is consumed by the clerks' aisle inside the fixture.

Bottom: Free-flow arrangement. Rearrangement of displays is easily accomplished. Fixtures are designed to display merchandise. No space is devoted to a clerks' aisle.

is needed for opening deep drawers or other display or storage devices.

12. If a store has more than one level, stairs and elevators or escalators are necessary. If the store has only two levels, stairs may be satisfactory; however, some mechanical means of vertical transportation is necessary for three or more stories.

13. Stairs, escalators, and elevators must be carefully planned as a part of the traffic system. They must be easily accessible. Their width is usually determined by local building codes. In addition, they influence the placement of certain types of merchandise, in order that it may be viewed by the customer when ascending or descending.

14. Stairs in small stores are often located at the end of the building away from the entrance. This location causes customers to be exposed to considerable merchandise when passing the sales areas on the way to the stairs.

15. Escalators and stairs in large stores are usually located in the center of the sales area. Customers are drawn to the center of the store, regardless of the entrance used.

16. Escalators are installed in pairs. One unit handles the upward traffic and the other the down traffic. The number of units and their width depend upon the anticipated number of people expected to use them.

17. To insure even traffic flow and to get maximum use from the units, only one bank of elevators or escalators should be provided. Second sections located in another part of the store generally do not carry an equal share of the load.

18. The interior of a store should reflect the type of merchandise for sale. A food store should give the impression that it is a food store rather than the impression that it is a hardware store. Some food stores handle a considerable variety of merchandise, but it should be so displayed that it does not change the customer's initial reaction. If many things are sold in a store, such as a department store, each sales area should clearly reflect what it has for sale.

19. The atmosphere desired for the various sales areas should be considered. An area devoted to toys frequently has a carnival air, while expensive silver or furs require a quiet, restful atmosphere. This influences the location of the sales areas on the floor plan. Bargain areas or noisy sales areas should be carefully located so they do not disturb those areas requiring quieter surroundings.

20. Sales areas that feature related merchandise should be conveniently near one another. Men's ties should be close to the shirt area, and paint brushes near the paint. Again, merchandising experience is valuable in making decisions pertaining to relationship.

21. If sales areas are on more than one floor, special attention should be given to the needs and demands of customers. For example, if lamps are sold with furniture on the second floor, the shades should not be on the first floor.

22. A good view over the entire store from any point is necessary. This increases the store's attractiveness and helps customers locate the area they desire. High fixtures tend to block the overall view of the store.

23. Merchandising research has shown that most people turn to the right when they enter a store. This should be considered when planning sales areas. Attractive aisles to the left should be planned to pull some people that way.

24. The planning of each sales area on the total sales floor includes not only a consideration of the area in which to contact the customer and make the sale, but also a consideration of how much merchandise will be on display before the public and how much will be kept in reserve stock on the floor. The section of this discussion on service areas considers this in detail.

Merchandise Display Equipment

The sales areas also need effective means of displaying merchandise. This can be counters, racks, platforms, or any other similar device. Display fixtures can be used to store, protect, and display merchandise; however, few fixtures are able to serve all three functions at the same time. The merchandise to be handled determines which function will be emphasized. Jewelry cases usually combine all three, while less expensive merchandise is stored and displayed, but not protected.

25. A decision must be made concerning means of display. For example, what merchandise should be left open for customers to handle and serve themselves, and what should be kept out of sight in storage rooms? In exclusive clothing stores, the clerk selects the items to be shown the customer. In contrast, other stores may display the entire selection on open racks, permitting the customer to personally go through the entire stock on the floor.

26. Display fixtures should enable the maximum amount of merchandise to be available and seen, yet should not appear overcrowded.

27. Fixtures should be simple and unobtrusive. Their purpose is to display the merchandise. If the fixture attracts attention to itself, it detracts from the merchandise and reduces sales. The design of fixtures is a full-time occupation in itself.

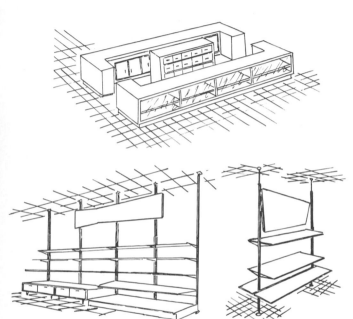

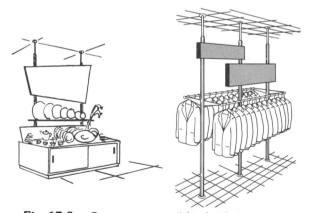

Fig. 17-3.   Common types of display fixtures.
   A — An island fixture with a stock storage cabinet inside the island. The customer sales counter can have display below or be used for storage.
   B — A wall fixture with the display to the floor.
   C — A freestanding fixture with storage below.
   D — A freestanding fixture with no provision for storage.
   E — An open rack for clothing display.

28. Fixtures should be flexible in use and easily moved to permit rearrangement of the sales floor. They should occupy a minimum of space, yet display a maximum of merchandise.

29. Some fixtures are used entirely for storage purposes. The top of the fixture serves as a counter, and the merchandise is removed from the fixture by the clerk. Racks are used to store and display hanging items such as coats; a top over the rack protects the items from dust. Other items, such as lawn mowers or large appliances, are displayed without fixtures.

30. Every display fixture should be designed to serve a particular type of merchandise. A store should not adopt a standard fixture of uniform size and height and then force merchandise to fit it.

31. Several types of fixtures are in common use — the island fixture, the wall fixture, and the freestanding fixture.

  The **island fixture** is a counter completely surrounding an area; the clerk serves from inside the island. A section of storage units also may be inside. See Fig. 17-3.

  The **wall fixture** is placed against a wall or partition. It may have shelves to the floor or have a base cabinet for storage, with display shelves above.

  The **freestanding fixture** provides customer access from all sides. It may have storage in the lower section or have open display shelves to the floor.

32. As fixtures are being planned, the need for chairs, mirrors, or tables should be considered. Areas such as shoe departments need these items.

33. Combined with fixtures and displays is the need for special utilities. A lamp department needs ample electrical outlets, while an appliance department needs water and sewer facilities.

34. Special facilities are needed to house the cash register and merchandise wrapping systems. Credit departments need a separate area. The type of fixtures used depends upon the system adopted.

35. After determining the size of the sales area needed for each specialty to be handled, the areas are related one to another, and aisles are planned to carry the customer traffic. Then the

details of each specialty area must be planned. The merchandise in each area must be carefully related to facilitate sales. In a jewelry area, the location of watch sales and watch repair must be considered. The planner must decide where to locate the sterling items, necklaces, and rings. Again, a knowledge of sales techniques for the particular merchandise is important.

36. Some specialty departments are large enough to warrant consideration of aisle usage within the one department. Also, each specialty department should consider the impulse, convenience, and demand items it has to display. Each sales area must be planned as carefully as is the entire sales floor and in more detail.

Lighting

37. The type of lighting used in the sales area varies with the type of merchandise displayed. The lighting in a clothing area is quite different from that in a jewelry area. Since the placement of merchandise in the sales area varies from season to season, the lighting system must be highly flexible.

38. The primary purpose of lighting is to improve the display of merchandise. Lighting fixtures should not attract attention, since this detracts from the items displayed and reduces sales.

39. The sales area requires sufficient overall illumination to provide normal visibility in the store. More intense, special lighting should be used on displays to call attention to the merchandise.

40. Colors influence lighting practices. The material used behind dark-colored merchandise should be light, but not so light that there is a severe contrast between it and the merchandise. Since light-colored merchandise has a high value of light reflection, it should be displayed against a rather dark background. For example, a white enamel washer would not be displayed to maximum advantage against a light-colored wall.

41. Special lighting is built into fixtures displaying items that require high-intensity illumination. Jewelry shows up best in well-lighted display cases. It might be drab and unattractive if shown under general illumination.

42. For merchandise needing natural light, such as fabrics, daylight fluorescent fixtures are frequently used.

43. While the recommended footcandles needed for selling varies with the type of merchandise

and the colors used, in general, the overall illumination should run from 25 to 100 footcandles (270 to 1090 cd/m^2*). High-intensity, display lighting may be as much as 200 footcandles (2150 cd/m^2*). Recommended footcandles can be obtained from architectural standards or lighting engineers' handbooks. Artificial light is usually preferred to natural light, since it is uniform and can be controlled easily.

44. The general illumination should be such that when the customers enter the store from the daylight, their eyes quickly and easily adjust to the light level inside the store.

45. The most effective and economical method of general illumination for a sales area is **direct illumination** from shielded or recessed light fixtures. This is a low-cost system. **Indirect lighting** is good for supplementary illumination, but it is not economical for general illumination.

46. If recessed, spot-style lights are used for general illumination, special lights to illuminate the ceiling are needed to give contrast between the sales floor and the ceiling. This combination effectively provides general illumination as well as special display lighting.

47. Recessed fluorescent fixtures provide good general illumination, as do recessed floodlights. These are economical systems.

Service Areas

Service areas provide facilities for storage of merchandise, receiving incoming stock, shipping outgoing items, moving stock about in the store, behind-the-scene's work, and store administration. In addition, they provide the space for the mechanical equipment that furnishes heating, cooling, light, and other needed utilities. These areas also provide eating, recreational, and lounge facilities for employees. The service areas generally should be hidden or should be a very inconspicuous part of the store operation.

The following factors should be considered when planning service areas for a merchandising facility:

1. A decision must be made early in the planning pertaining to the amount of reserve stock to be stored in the sales area to quickly replenish that sold from the display racks. Along with this should be a decision as to how much stock should be stored in larger storage areas on the same floor, but in storerooms hidden from public view.

2. The decision pertaining to storage of stock will vary with the type of merchandise and the philosophy of the merchandiser. Reserve stock areas located around the perimeter of the large sales floor generally are used. This allows easy access to stock and does not break the unified appearance of the sales floor.

3. The moving of merchandise to the sales floor should not interfere with the customer traffic patterns. The employee traffic patterns should be planned as carefully as those of the customers.

4. Outgoing merchandise should also be handled in a way that does not hamper sales.

5. Most merchandise is brought to and shipped from a store by truck. Facilities for docking, loading, and unloading are necessary. This area should be completely separate from the entrances used by customers and from the customer parking areas. The service area for merchandise should be adjacent to this docking area. Large stores frequently have ramps into basement areas for trucks. Trucks should not have to extend into public streets when being unloaded.

6. It is advantageous if docking facilities can be under a roof to provide protection during inclement weather.

7. A conveyor belt is a good way to transfer items from the receiving dock to the service area where they are marked and put into stock.

8. After merchandise is unloaded at the receiving dock, it should follow a predetermined traffic pattern as it is checked and marked, and then sent to storage, to the sales floor or to a customer. The receiving department needs tables, bins, racks, and carts to facilitate this process. A suggested receiving, marking, storage, and shipping area is illustrated in Fig. 17-4.

9. If the store is on more than one level, freight elevators are needed to move merchandise.

10. Consideration should be given to having a large, supplemental storage area where items may be kept until the storage areas on, or adjacent to, the sales floor can accommodate more stock. Stock could then be transferred to these forward storage areas when needed.

11. Additional storage areas are needed to handle off-season stock that was not sold during the season or that is acquired during the off-season. Stores handling large items, such as furniture or appliances, frequently have a separate warehouse for remote storage; items are moved to the sales building as needed.

12. Some items need special storage facilities. Frozen foods, fresh flowers, and furs are examples.

13. Usually, merchandise that has been sold and is to be delivered is wrapped on the sales floor and then travels by some system to a shipping room. Large stores have a system of conveyors or chutes to transport these items. In small stores, carts are frequently used.

14. Arrangements are necessary for providing space for delivery trucks to load. This generally should be kept separate from the receiving area, especially in large stores handling merchandise that sells rapidly.

15. Workrooms are necessary for service personnel to perform their duties. For example, if a store does alterations or if displays and posters are made by store employees, work space must be provided for these operations. The office and executive staff require space to handle the store accounts and purchasing. The maintenance staff needs space for equipment and supplies. Large stores frequently have

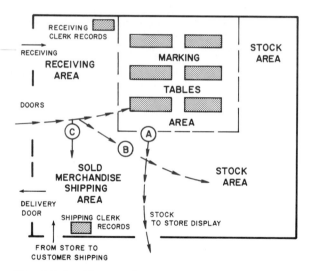

Fig. 17-4. This stock receiving, marking, storage, and shipping area is only a suggested plan to show the activities normally carried on in such an area. The exact layout will vary considerably from store to store, depending upon the specific type of merchandise handled and the size of the store.

A — The merchandise is received and uncrated if it is to be put on display or sent to storage areas on the sales floor; it is marked and sent to the sales floor.

B — Other items might be received and placed in stock for future uncrating and marking.

C — Items may be received and immediately placed in the sold-merchandise shipping area for delivery to waiting customers.

Various types of carts are used to move the merchandise. In large department stores, conveyor systems are used to transport items through this area. If the store is on several floors, freight elevators should be located in this area.

lounges, sick wards, and cafeterias for their many employees. Rest rooms are always a necessity.

16. Room is needed for the mechanical equipment used in heating, cooling, and operating the store. The amount of space needed varies with the systems used and the size of the store. This area is vital, however, and cannot be neglected in overall planning.

17. Mechanical equipment in small stores is usually located in a basement area or at the rear of the ground floor.

Exterior Design

The primary function of exterior design is to attract the customer into the store. This exterior should reflect the character of the store. Certainly, the exterior of a jewelry store should reflect a different character than a drugstore. A passerby should be able to know from the exterior what type of store it is. Some store fronts are so well designed that a sign is considered unnecessary. These are exceptions, however, and most stores use attractive signs as part of the front, to identify the store. Basically, a store front is an advertising medium.

Consider the following when designing the exterior of a store:

1. The **show window** is a major part of the exterior design and is used to display merchandise. The front may have floor-to-ceiling windows (open front) or smaller, framed windows, with backs blocking the customer's view into the store (closed fronts). Since "window shopping" is very popular, special attention should be given to the show window. It reveals the type of merchandise carried. In addition, it is a major "attention getter"; it promotes considerable impulse buying and accounts for a large percentage of the total store sales.

2. The size and construction of the show window depend upon the merchandise handled. A music store displaying pianos or television sets requires large open areas. A jewelry store needs smaller windows at such a height that viewing is easy.

3. Since many small stores carry merchandise of all sizes, consideration should be given to designing show windows that have great flexibility. For example, a window displaying major appliances should be easily converted to properly display small appliances, such as irons or toasters.

4. The show window is an integral part of the store front and should be designed so it blends architecturally with the exterior.

5. Provision should be made for control of sunlight on a show window. Because the window is so important to the sales program, it should never be blocked out by reflections. A properly designed **canopy** provides a satisfactory solution when glare is a problem. The canopy not only provides light and reflection control, but protects the prospective customer from inclement weather. It is frequently used to carry a sign identifying the store. A canopy, to be effective, should shade the sidewalk so the viewer stands in the darkened area; if the show window area has a higher light intensity, reflection will be reduced. If a canopy is too high, it will be ineffective. Frequently, a transom is built over a low canopy, thus permitting light to enter the show window. This provides natural lighting for the display area, yet shades the viewer.

6. Another way to reduce glare on a show window is to tilt the glass or to use curved glass. Glass surfaces reflect light at the same angle at which the rays strike the glass. Light reflected off the tilted glass can be directed away from the viewer's eyes and toward a light-absorbing surface. Common solutions to reflection control on show windows are shown in Fig. 17-5.

7. Easy access to the show window area is essential, so that displays can be set up.

8. Some stores use large front windows (open front) for display purposes, with no special, built-in show window. The interior of the store extends up to the exterior window. This enables the floor space occupied by window display to be converted easily to sales space whenever needed. It also allows flexibility in the use of display space.

9. Flexible lighting is necessary in show windows. Seldom is a display effective without dramatic lighting. It must be of greater intensity than interior lighting and must be more intense in the daytime than at night. The principles of lighting are the same as interior store lighting in that general illumination is necessary, along with special display lighting on certain items. Light sources should be shielded to avoid blinding the viewer and should be inconspicuous. For some displays, up to 500 footcandles (5400 cd/m^2) is necessary. Lighting should be positioned to avoid casting shadows inside the display area.

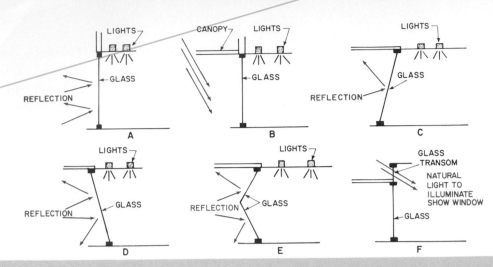

10. The color of interior materials lining a show window or other display areas is important. A white background reflects the light in the area, thus making it easier to view the merchandise. Dark colors absorb light and make viewing more difficult.

11. Show windows should be easily viewed by all passersby, including those in automobiles. The more directions and the greater the distance the display can be seen, the more valuable it is.

Since automobiles pass rapidly, the store front should attract attention instantly.

12. Many small stores can be built advantageously with an all-glass front, thus turning the entire store into a large show window. This is excellent planning, but the interior fixtures should be such that the view is not obstructed. Frequently, display cases and shelves are glass. These are used to reduce obstructions to view and to allow passage of light.

13. Consideration should be given to orienting a store to the weather. In northern latitudes, show windows which can receive most of the sun's rays are most effective. In areas of high winds, some provision must be made to protect the customer. A recessed entrance sometimes provides effective protection.

14. The heavy ornamentation of store fronts, which was common a few years ago, has been completely abandoned in favor of simplicity. Excess ornamentation is expensive and detracts from the exterior display of merchandise. An attractive exterior is shown in Fig. 17-6.

15. If a store is on a thoroughfare with heavy pedestrian traffic, room must be allowed along the front for those desiring to stop and view the merchandise. This could be provided by a recessed or slanted entrance. While this reduces inside floor space, the exterior display is of sufficient importance to warrant the loss.

Libbey-Owens-Ford Glass Company

Fig. 17-6. This commercial building illustrates effective use of the glass wall. It has character and appeal. The design is simple and uncluttered. Notice the ceiling spotlights providing general illumination.

This entrance also serves as an aisle for customers. This aisle is needed even if the store is built to the front property line.

16. The recessed store front increases the amount of window display space available. Since the volume of sales is influenced heavily by the show windows, this usually justifies the expense of using this type of front.

17. While the show window is the most important consideration when planning a store front, the exterior space above the window also holds advertising possibilities. This space is only effective for those viewing the store from a distance, since those carefully viewing the show windows seldom look above eye level. Because of this, the upper portion is usually a flat billboard with signs or slogans affixed. If this upper area also is devoted to show windows, they will compete with the street-level displays and be distracting. See Fig. 17-7.

18. A store on a corner has the advantage of having more room for display windows and more upper space for advertising. This space should be fully utilized in the exterior design.

19. The entrance doors are the final invitation to a customer to enter the store. They should appear to extend a welcome by allowing the person to see what is inside. Commonly used are full-length glass doors. All entrance doors should be easy to operate, and automatic doors are necessary in some stores.

20. Most stores should have double-entrance doors. Only extremely small stores can use a single door.

21. The door should fit into the interior customer traffic patterns. However, it generally is poor planning to place the door of a small store exactly on center. This creates equal-sized display windows on each side, thus limiting their flexibility. Ordinarily, one large window creates a better impression. This can be accomplished by placing the entrance off-center.

22. If at all possible, the level of the sales floor and the sidewalk should be the same. Ramps should be used if a difference exists. Steps should be avoided.

23. The design of signs identifying the store is a critical factor in overall exterior design. Generally, a simple sign is most effective and in best taste. The sign should always be lighted at night and should be easily read. The type of sign and the lettering used should be consistent with the character of the store.

Kawneer Company

Fig. 17-7. Notice the recessed entrance located off-center to provide a large show window. The second floor front has been converted into a simple billboard to provide a mounting place for the sign which has large and easily read lettering. A hood protects the show windows from excess glare and the customer from the weather.

24. Since most stores have year-round air conditioning and carefully controlled lighting systems, outside windows to admit light or air are unnecessary.

Shoe Stores

1. Most shoe stores must operate at a high sales volume. This requires both an exterior that attracts attention and displays to pull the customer into the store. For this reason, most shoe stores use the all-glass, open-front exterior. See Fig. 17-8.

Libbey-Owens-Ford Glass Company

Fig. 17-8. Display windows designed to attract attention and lure the customer inside. This is located off the mall in a large shopping center in Houston, Texas. The entire store and mall is air-conditioned.

2. The customer should be able to enter and leave without conflicting with sales personnel moving about to obtain merchandise.
3. A maximum number of chairs should be used. These should be attractive, durable, and as small as possible.
4. The floor should be carpeted, since it is necessary to protect the merchandise and customers may be walking in stocking feet.

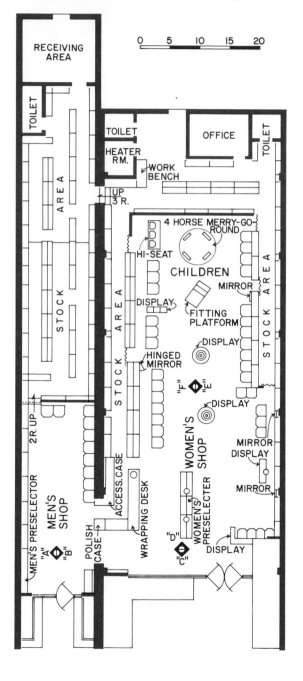

Fig. 17-9. The plan for a typical shoe store.

5. Shoes should be placed on open display. This gives the customer a chance to examine the merchandise.
6. The storage of shoes on the sales floor is difficult. Open shelving may be built along walls, and the boxes of shoes arranged on these. This shelving should be within the reach of the sales personnel. Attractive displays of shoes should be located at various points along the wall shelves. Reserve stock is best stored in a stock room off the sales floor.
7. Impulse items are important to the financial success of shoe stores. Items such as hose, shoe polish, belts, and ties are located near the entrance and sales desk. A typical plan is in Fig. 17-9.
8. Usually, a sales desk is located near the entrance. Here a clerk makes change, wraps the items, and asks about the purchase of impulse items nearby.
9. High-intensity lighting enhances the appearance of shoes.
10. Full-length mirrors are necessary for viewing the footware.

Jewelry Stores

1. Most jewelry stores are built with an all-glass, open front.
2. Considerable impulse buying is characteristic and should be given emphasis in developing the layout of the sales floor.
3. Luxury and dignity should be strived for, in both interior and exterior design. Everything should be in good taste.
4. Since most merchandise is small, the reserve stock area need not be large. Much can be kept in the display fixtures, yet out of sight.
5. Security from theft during both day and night is a major consideration. A vault and burglar alarm system are necessary.
6. Merchandise must be displayed with consideration to its security and protection. Self-service features are not useful, except for costume jewelry.
7. Usually, the jewelry store is departmentalized. Watch and jewelry repair, gem display, silver, china, fountain pens, rings and bracelets, and other items make logical groupings of merchandise. Notice the departments in Fig. 17-10. A gift department is shown in Fig. 17-11.
8. Small items of jewelry are displayed in glass counters, while silver, china, and other larger items are displayed on open shelving and tables.
9. The repair area is usually located in the rear of the store. This is a demand function, and this

location forces the customer to pass by a great many impulse items.

10. Service areas include: an office to handle sales details, a vault, private sales areas for the sale of exclusive merchandise, office space for normal business operation, and rest rooms for employees.

11. Incandescent lighting highlights gems very nicely. Displays of silver, pewter, or platinum merchandise need fluorescent lighting to show the items to their best advantage.

Fig. 17-11. A gift department in a jewelry store. The well designed fixtures plus adequate lighting enhance the display and invite customers to enter.

12. Quietness is characteristic of a luxury store. Use sound-absorbing materials on the ceiling and floor.

13. Mirrors should be placed to enhance a display. They also are necessary for customers to use in viewing their jewelry.

Furniture Stores

Some furniture stores offer very specialized merchandise, while others offer a complete line of home furnishings. Specialized stores might sell only contemporary furniture or Early American styles.

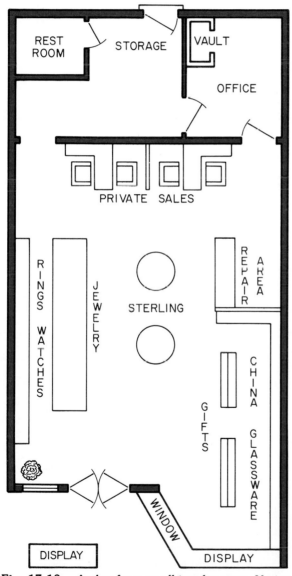

Fig. 17-10. A plan for a small jewelry store. Notice the location of the repair area which is a demand service. Impulse items as gifts, china, and glassware are exposed to all customers. Tables and chairs provide for customer comfort.

Fig. 17-12. A furniture store utilizing the open-front merchandising trend. The main store and adjoining show rooms all are open to public view.

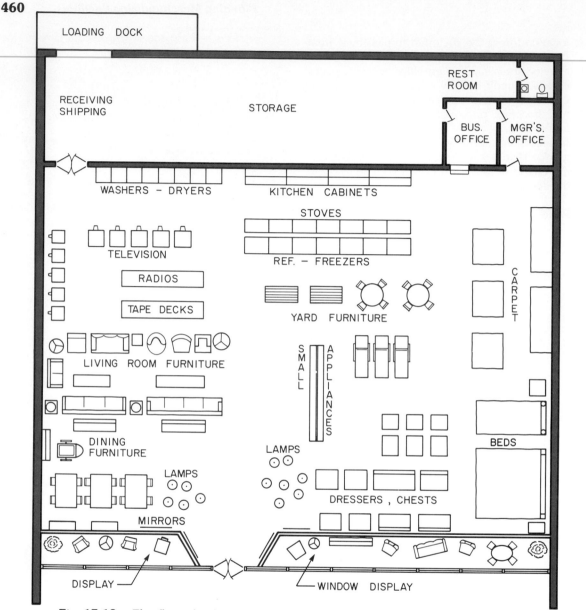

Fig. 17-13. This floor plan for a small furniture store suggests good locations for selected merchandise. The actual placement of individual items varies from day to day as they are sold and new stock is placed on the sales floor. Flexibility is needed to accommodate seasonal merchandise, such as porch and lawn furniture.

Furniture stores offering a broader line of merchandise stock all styles of furniture, as well as appliances, rugs, curtains, radios, television sets, bedding, lamps, china, and glassware. Sometimes an interior decorating service is offered.

When planning a furniture store, consider the following general principles:

1. Window display space is a vital part of the planning. Some of the store frontage should be devoted to show windows that are large enough to display entire rooms completely furnished. Window lighting is essential. See Fig. 17-12.

2. Consideration should be given to placement of impulse, convenience, and demand items. Large furniture pieces and rugs should be in the rear of the store. See Fig. 17-13. Radios, television sets, and appliances are convenience items and should be located between the demand and impulse items. Lamps and mirrors, as impulse items, are near the entrance.

3. Fixtures are few. A large, open floor area is needed. Furniture should be arranged in natural groupings, rather than simply lining up a long row of chairs or sofas.
4. Electrical outlets should be plentiful, so furniture can be easily rearranged.
5. General illumination is usually assisted by local lighting from lamps on display.
6. Considerable merchandise is displayed on the sales floor. In small stores, the sales floor may contain almost the entire stock on hand. Large stores may use upper floors for storage or may have a separate warehouse. A freight elevator is essential in stores with more than one level.
7. If a furniture store is on more than one level, the first floor generally contains living room furniture, lamps, china, radios, and television sets, while the second floor contains bedroom furniture, mattresses, and carpeting. If a third floor is used, various appliances are located here, as well as kitchen and outdoor furniture.
8. Space to store and prepare sold merchandise for delivery is necessary. An adequate loading and shipping platform is needed.
9. Space should be available to receive and uncrate merchandise for display on the sales floor.

Appliance Stores

The typical appliance store is usually departmentalized. Kitchen items are frequently grouped together, such as ranges, refrigerators, freezers, dishwashers, and kitchen cabinets. A second department is centered around laundry facilities such as washers, dryers, and ironers, while a third grouping includes radios, record players, and television sets. If small appliances are handled, these are grouped together.

Consider the following factors when planning an appliance store:

1. Fixtures are few in this type of store. Small appliances can be displayed on low, open shelves or tables. See Fig. 17-14. Major appliances and television sets are frequently placed upon low platforms.
2. Numerous electrical outlets are needed to provide flexibility of arrangement. Some appliances require 240-volt current. Television antennas are needed.
3. The small appliances are most often located near the entrance, with television sets and radios next. Major appliances generally are in the rear of the store.
4. If repair services are offered, a work space is needed with room for disassembly work. This

Fig. 17-14.   Small appliances can be effectively displayed on low, open fixtures.

type of work requires considerable space and may be located in an adjacent building or in the rear of the retail store. Space is also needed for parts storage and parking for service trucks.
5. The repair area needs all utilities to check the operation of repaired fixtures.
6. Adequate storage is needed for many large items.
7. Stores selling kitchen cabinets frequently build a model kitchen in the rear of the store.
8. Since many purchases are on credit, suitable office space and a place for customers to be interviewed and to make payments must be provided. This should be a part of the sales floor, yet should enable business transactions to be handled with a feeling of privacy.

Clothing Stores

Men's Clothing Stores

1. The atmosphere of a men's store is generally one of reserve and simplicity. See Fig. 17-15.
2. The interior should also reflect a pleasant, quiet, unhurried atmosphere.
3. The merchandise is rather stable from year to year and is easily divided into impulse (belts, ties, jewelry), convenience (shirts, hats, pajamas), and demand (suits, coats) items.
4. The merchandise is seasonal, and plans should be made for easy rearrangement of fixtures.
5. Good lighting on merchandise displayed on mannequins helps sales. General illumination should be adequate.
6. Fixtures should not be crowded together.

Fig. 17-15. A men's clothing store. Notice the simplicity of exterior design and the uncluttered display. The special spotlights in the ceiling along the glass display front highlight the window merchandise.

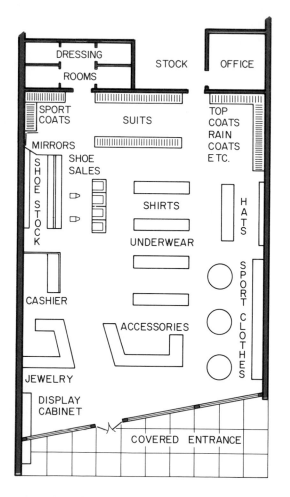

Fig. 17-16. Plan for a small men's store.

7. Dressing rooms and mirrors are necessary.
8. Workrooms for tailors are needed if the store does its own alterations.

A plan for a small store is given in Fig. 17-16.

Women's Clothing Stores

1. A women's clothing store contains a wide variety of items which can be logically grouped, such as hosiery, blouses, and lingerie. The demand merchandise includes coats, dresses, and suits. Typical convenience items are gloves, hosiery, and sweaters. Costume jewelry, handbags, and cosmetics are impulse items.
2. Most women's clothing stores rely heavily on personal service; however, some items could be somewhat self-service.
3. Flexibility in display fixtures is vital, since most items are seasonal and fashions change rapidly. See Fig. 17-17.
4. A striking exterior with a dramatic display area is necessary. Since mannequins are used, large display windows are necessary. See Fig. 17-18. The front should appear as a theatrical stage, with exciting sets and effective lighting. Exclusive shops can use a closed front with a few, framed, display windows, but this is not effective for stores relying on volume sales.
5. Dressing rooms are needed for customers to try on clothing. Full-length mirrors are also necessary.
7. Comfortable chairs are necessary in areas where expensive garments are shown.
8. If the store does alterations, a workroom is needed.
9. The usual offices, credit desk, sales desks, and rest rooms are required.

Fig. 17-17. These fixtures provide a high degree of flexibility. They can be raised and lowered or removed and replaced with other types of hanging rods, shelves, or cabinets. Even the mirrors can be easily moved to facilitate a rapid rearrangement of merchandise.

Natcor Corporation

H. Armstrong Roberts

Fig. 17-18. A women's clothing store with a center entrance. Notice the recessed display windows providing sidewalk space for customers viewing the display.

Fig. 17-19. A hardware and building supply store planned on two levels. The all-glass front transforms the store interior into a vast display. The interior lighting is especially effective.

Hardware and Building Supply Stores

The typical hardware store handles a considerable variety of items, many of which are very small and difficult to display to advantage. Other items, such as power tools, are large and require considerable floor space. Some stores also handle housewares, small appliances, sporting goods, garden tools, auto supplies, radios, and television sets. Such a store must be carefully departmentalized.

The typical building supply store carries a full line of hardware items, plus items needed for building

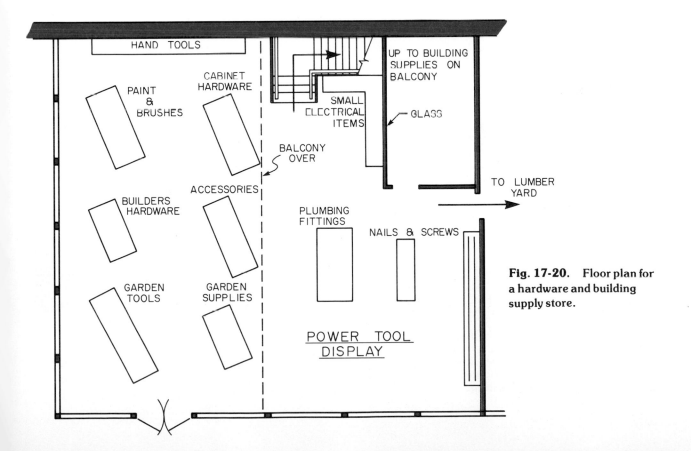

Fig. 17-20. Floor plan for a hardware and building supply store.

construction, such as plywood, lumber, lighting fixtures, doors, windows, and ceiling and wall materials. It seldom handles appliances and sporting goods. Figures 17-19 and 17-20 illustrate hardware and building supply stores.

When planning a store to merchandise hardware, consider the following general principles:

1. Impulse items include small electrical appliances, sporting goods, and garden tools and supplies. They can be effectively displayed along the main aisle carrying customer traffic.
2. Demand items, include paints, wallpaper, large electrical appliances, and large hardware items such as ladders.
3. Convenience items include such things as small hardware items (hinges, handles, nails), hand tools, china, cooking utensils, and cutlery.
4. Fixtures are usually the open type. Since much of the merchandise is not seasonal, special fixtures can be used. For example, fixtures to display electrical parts or small hardware items could be used. Tools are often hung on wall racks. See Fig. 17-21. Hinges, drawer pulls, and other hardware items may be displayed on wall panels, with the boxes of stock stored in cabinets behind these displays.

5. Open shelving along the exterior walls provides excellent display space for small appliances, cooking utensils, and paint. This shelving should be adjustable to accommodate items of various heights.
6. Open floor area is needed for seasonal items that are frequently handled, such as sleds and bicycles. These are ordinarily placed in front of the all-glass, open front.
7. Floors should be attractive, yet able to withstand hard wear.
8. General illumination should be excellent, since most merchandise on display usually is not specially lighted.
9. Colors should be kept light in order to "set off" the merchandise.
10. Most stores use the all-glass front and arrange items that are to be displayed in front of the window.
11. A receiving and stock room is necessary.

Service Stations

The services offered by stations vary considerably. Some simply sell gas and oil and handle a few accessories. Others include lubrication, car washing, tire and battery sales, and offer accessories such as seat covers, lights, and horns. Still others perform such minor repairs as tune-up and brake replacement. When a service station is being designed, the services to be offered must be clearly known. Following are some planning principles to consider.

The Site

When planning the site, remember that ingress and egress are vital to a good service station. These involve the approaches and driveways.

1. Maximum approach widths and their minimum distances from property lines are usually governed by city or state rulings. Average approach width is 30 feet (9144 mm) at the property line, with 10 to 15 feet (3048 mm to 4572 mm) a usual minimum distance from the edge of the approach to the property lines (at side and corner). Location of the approaches should give easy access to the islands, with no sharp turns necessary. See Fig. 17-22.
2. An inner drive width of 16 to 18 feet (4877 mm to 5486 mm) is most common. (This is the distance from the island's inside face to the building step.)

H. Armstrong Roberts

Fig. 17-21.   The hand tool and power tool section of the store in the floor plan in Fig. 17-20. Notice the wall panels for tool display, with stock stored on shelves below. The wall panels are lighted with fluorescent strips hidden behind a fascia strip.

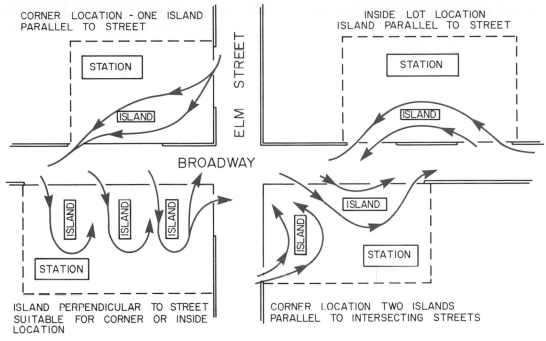

Fig. 17-22.　Typical traffic patterns relating customer flow to islands and street.

3. Grades are of the utmost importance. The service area around the islands should be nearly level to permit gasoline tanks to be filled completely, and yet should allow enough slope on the slab for drainage. A common slope is 1/8 inch per foot (3 mm per 305 mm) in the direction of the island and 1/4 inch per foot (6 mm per 305 mm) across the drives.

4. Surfacing adjacent to the islands where gasoline is dispensed should be concrete.

5. Minimum center-to-center spacing of gasoline dispensers is 5'-0" (1524 mm).

6. The island should provide water, compressed air, and window cleaning facilities. Some provide vacuum cleaning facilities.

7. It is essential that the approaches receive floodlighting so that they can be seen clearly by the approaching motorist.

The Building

The service station building should be attractive and should provide adequate facilities for operation. A typical plan is shown in Fig. 17-23. Normal elements to be considered in the building planning are: sales office, service stalls, work area, storage room, rest rooms, vending area, canopies, and building utilities (heating, plumbing, electrical).

8. The sales office should provide space for operator's desk and merchandise display.

Maximum visibility to and from the office is essential both for display of items in the office and for clear view of the driveway areas by the operator. Average office area required is 300 square feet (27.9 m²).

9. Service stalls are usually provided for car washing, lubrication, and (in the large stations) maintenance.

Phillips Petroleum Company

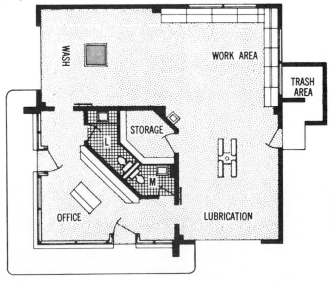

Fig. 17-23.　Plan for a contemporary service station.

10. The wash stall is provided with a mud sump to keep mud and oil from entering the sewers. Floor should slope from the perimeter of the stall a total of 2 to 3 inches (51 mm to 76 mm) to the sump for proper drainage.

11. The ceiling should be high enough to accommodate a car on the hydraulic lift at full height. Thirteen feet (3962 mm) from floor to ceiling is recommended. If trucks are serviced, the ceiling should be even higher.

12. Service stalls are usually 15'-0" wide by 27'-0" long (4572 mm wide by 8230 mm long). Their entrances have 10' × 10' overhead doors. Larger doors are necessary if trucks are to be serviced.

13. A locked storage area is necessary for merchandise items.

14. Rest rooms should be easily maintained. Tile floors and walls and good quality plumbing fixtures are best. Floor drains encourage frequent cleaning.

15. Formerly, rest rooms opened to the outside; however, some of the newer stations have inside rest rooms.

16. Space should be allowed for vending machines.

17. Canopies are installed on many stations. They provide shade and also protection from rain and snow. See Fig. 17-24.

18. Most codes require that any flame in the service stall area be 8 feet (2438 mm) above the floor to avoid igniting any gasoline vapors present. Gasoline vapors are heavier than air and settle to the floor area. Thus water heaters and furnaces (both oil- and gas-fired) are suspended from the ceiling or mounted on the wall above the 8-foot level.

19. There are no special plumbing requirements other than a mud sump or (in some cases) a grease trap.

Fig. 17-24. A contemporary service station.

20. Underground storage tanks are provided for regular, premium, and unleaded gasoline. Tanks should be of sufficient size to receive the full delivery of a transport. Tank capacity is usually 4000 to 6000 gallons.

Drugstores

The typical drugstore is much more complex than the name implies. Commonly, these stores offer food service, cigars and cigarettes, stationery, gifts, cosmetics, cameras, books and magazines, household items, school supplies, prescription service, and a wide variety of nonprescription medications. The exact items handled will vary from store to store.

Most drugstores rely heavily on self-service for most items; clerks merely help customers find items, give advice, and show items housed in protective cases, such as cameras.

Following are some general principles to be considered when planning a drugstore:

1. Strict adherence to placing impulse items near the entrance, demand items in the rear, and convenience items in between is necessary.

2. Most stores are arranged by departments. The prescription department, a demand department, should be located in the rear. See Fig. 17-25 for a typical plan.

3. The food service should be near the entrance and easily seen from the street.

4. Cosmetics tend to be impulse items and should be near the entrance. It is advantageous if they can be located opposite the food service.

5. Convenience items such as drugs, household items, and sundries should be located between the prescription department and the entrance. Typical fixtures for displaying the many items found in a drugstore are shown in Fig. 17-26.

6. If extensive food service is planned, a study of restaurant fixtures is necessary. See Chapter 19, Planning Food Stores, Restaurants, and Cafeterias.

7. Considerable area is needed for stock storage. This could be on a lower level or in the extreme rear.

8. The precription area should be open to public view and should be attractively arranged. Frequently, this area is protected by glass partitions.

9. Some drugstores use a closed or partially closed display window to advantage. The displays exhibited are simple and uncluttered. Customers should be able to see over the display into the store proper. Many stores use

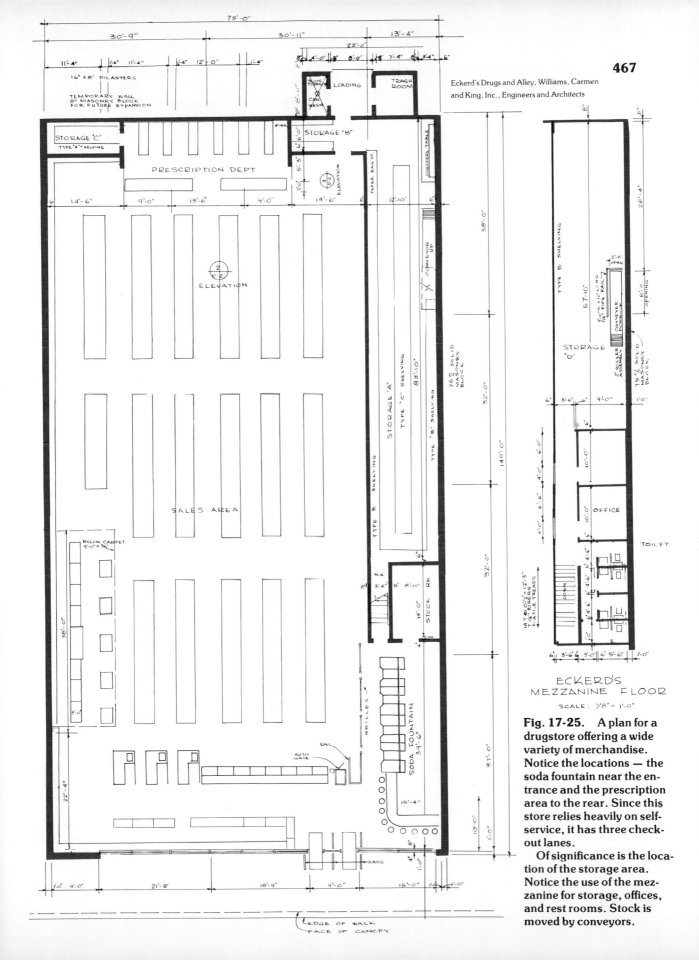

Eckerd's Drugs and Alley, Williams, Carmen
and King, Inc., Engineers and Architects

ECKERD'S
MEZZANINE FLOOR
SCALE: ⅛"= 1'-0"

Fig. 17-25. A plan for a drugstore offering a wide variety of merchandise. Notice the locations — the soda fountain near the entrance and the prescription area to the rear. Since this store relies heavily on self-service, it has three checkout lanes.

Of significance is the location of the storage area. Notice the use of the mezzanine for storage, offices, and rest rooms. Stock is moved by conveyors.

Fig. 17-26. Open island fixtures are commonly used to display merchandise in a drugstore. Notice the prescription center at the rear of the store.

Fig. 17-27. An open front drugstore designed to attract the customer by presenting the entire contents of the store as a display.

the all-glass, open front so that the entire interior is the store's "face." See Fig. 17-27.

10. Good general illumination is necessary. In display areas where merchandise is protected, each display case is lighted separately.

Build Your Vocabulary

Following are terms that you should understand and use as a part of your working vocabulary. Write a brief explanation of each term.

block aisle system	impulse merchandise
canopy	island fixtures
convenience merchandise	open fixtures
	reserve stock
demand merchandise	sales area
elevator	service area
escalator	show window
free-flow aisle system	wall fixtures

Class Activities

1. Make a checklist of the principles that should be observed in planning one store of a type discussed in this chapter. Select such a store in your community and secure permission to visit it. Then evaluate it by marking each item on the checklist as good, satisfactory, or poor. The checklist should include exterior design, as well as interior design and planning.

2. Visit one store of a type discussed in this chapter and make a freehand, scale sketch of the interior arrangement. In class, revise this to improve

the overall operation of the store, and draw the revised floor plan.

3. Build a cardboard model of a store interior (the building, with the roof and possibly one or two walls removed). Build scale models of typical store fixtures. Arrange these on the floor to develop the most satisfactory plan.

4. Visit a shopping area near the school. Find a store whose exterior details truly reflect its character. Report on these details. Explain why this exterior design is good. A photograph of the store would be useful in your report.

5. Build scale models of store fronts, illustrating the more frequently used types of display windows. Include provision for sun control and for customer protection from bad weather. Finish the project by adding a store sign that is in keeping with the exterior design and the type of merchandise handled.

6. Visit stores of a type covered in this chapter and examine their service areas until you find one that is well planned. Sketch a plan of this area and build a cardboard scale model, including storage shelves, bins, and conveyors. Prepare a report calling attention to the outstanding features of this service area.

7. Work with several other class members as a design team to research the data for a particular type of store. Make preliminary drawings, revise these, and draw a complete set of plans. As planning proceeds, you will need to study the material concerning structure in Chapters 16 and 23.

Chapter 18

Planning Offices, Banks, and Medical Clinics

The increasing costs of construction have caused designers of offices, banks, and medical buildings to focus attention upon the efficient use of space. Careful study of the activities to take place in an area has enabled designers to reduce the amount of space needed and, at the same time, to increase the efficiency of the workers.

This chapter discusses the principles to follow when designing office, banking, and medical facilities.

An Approach to Office Planning

The proper approach to office planning is to survey the needs, analyze these needs, and then apply the principles of planning to develop an effective layout. While exact procedures vary from company to company, the following steps suggest a logical approach:

1. Ascertain the personnel to be accommodated and what duties they perform.
2. List all needed equipment.
3. Develop an organizational chart of the workers, and chart the flow of work. See Fig. 18-1.
4. List all service facilities needed, such as rest rooms and storage rooms.
5. Determine the amount of floor area needed by each office activity.
6. Visit with those who will work in the office, and discuss their needs. If it is a new facility, visit similar facilities.
7. Provide for the future — anticipated needs or expansion.
8. Make a detailed layout of the office arrangement. Templates are useful.
9. Let those who will work in the facility review the plan.

Office Planning Principles

The following factors must be considered when planning office layouts:

1. A supervisor who is working with the employees generally should have a desk in the same room with them.
2. Employees within a unit doing the same type of work should be in the same room. Interrelated units of an office should be near each other. Transportation distance of work should be held to a minimum. If possible, work surfaces should be arranged so that each worker receives work from the person at the next desk. Refer again to Fig. 18-1.
3. Work in an office should follow a straight-line flow to minimize backtracking. **The locations of work stations should be dictated by work relationships.** This principle is most important, and all other considerations should be subordinate to it.
4. A large, open office area generally is better than the same space divided into smaller offices. The open office makes control and communication easier, provides better light and ventilation, reduces space requirements, facilitates a better flow of work, simplifies supervision, and eliminates partition costs.
5. Desks should face in the same direction. Placing desks back-to-back does not save space and often stimulates conversations and other time-consuming practices.
6. Large offices should be partitioned into related activities by file cabinets or low dividers.
7. Noisy operations, such as duplication or bookkeeping machines, should be separated from quieter clerical areas. Special acoustical treatment may be necessary.
8. All clerical desks should be the same size and color. This gives the office a more pleasant appearance and facilitates rearrangement.

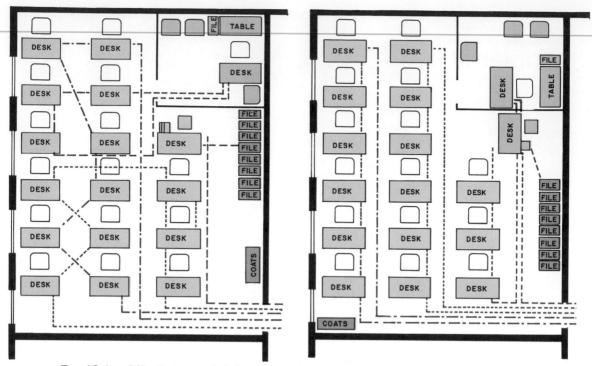

Fig. 18-1.   Office before and after a survey of work flow.
    Left: Office before work flow survey was made. Considerable time was being wasted because clerks had to leave work stations to forward work.
    Right: Office after survey established flow of work and clerks were seated to form a straight-line flow.

9. Desks are best arranged in straight lines to facilitate straight-line flow of work.

10. Clerical desks should be at least 30 inches (762 mm) apart, to allow sufficient space for the worker's chair. Maximum space is 48 inches (1219 mm). The ideal spacing is 36 inches (914 mm).

11. Primary aisles in a general clerical office should be at least 4'-0" (1219 mm). Crossover aisles should be 3'-0" (914 mm), and secondary aisles between desks should be 2'-0" (610 mm). These should be wider if heavy traffic is anticipated.

12. The most efficient arrangement of desks in a general clerical office is to place two desks side by side. This provides an aisle on each side. When more than two desks are arranged in this fashion, the spacing between the rows of desks should be increased to a 33-inch (838 mm) minimum. A more satisfactory spacing is 42 inches (1067 mm).

13. Desks should be arranged so that employees do not face the natural light. Light should come over the left shoulder.

14. Heavy equipment should be placed against walls or columns whenever possible, in order to avoid overloading the floor.

15. It is poor practice to locate private offices on the outside walls and clerical work stations in an inside area. This gives a boxed-in feeling. Clerical areas need some outside window area.

16. The size of office furniture varies considerably. That selected should be large enough to meet the needs of the worker.

17. Each office worker should have only the furniture needed to do the job.

18. Employees having considerable public contact should be near entrances to the office. Employees doing confidential work should be away from entrances.

19. The employee work station should be located near the equipment to be used.

20. Attractive, comfortable working conditions increase the efficiency of personnel.

21. Future needs for expansion or added services should be considered.

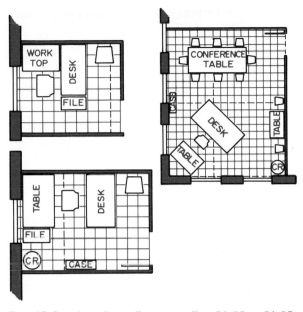

Fig. 18-2. A — A small private office 8'-0" × 9'-0".
B — A private office 9'-0" × 12'-0".

C — A large executive office with a small conference table requiring a room 16'-0" × 20'-0".

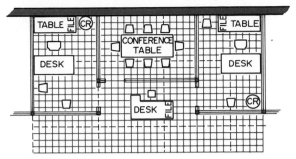
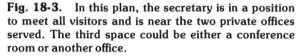

Fig. 18-3. In this plan, the secretary is in a position to meet all visitors and is near the two private offices served. The third space could be either a conference room or another office.

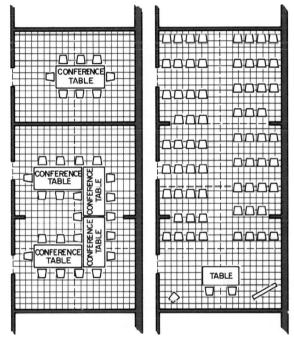

Fig. 18-4. A conference room 20'-0" × 45'-0". The use of folding doors enables this area to be used for three small conference rooms, one small and one large room, or the entire area may be opened for large meetings. The elimination of tables greatly increases the capacity of the room.

22. Doorless or open areas for coats and packages should be provided, with hanging rods for coats and shelves for hats and other personal belongings. These areas should be located in the space least desirable for work to be done.

23. Very few employees, including top executives, need private offices. The primary consideration is the need for privacy in the conduct of the job.

24. A person having few visitors can usually be accommodated in an 8'-0" × 9'-0" (2438 mm × 2743 mm) office, while one who has frequent visitors needs a 9'-0" × 12'-0" (2743 mm × 3658 mm) office. An executive who holds conferences in the office may require 300 to 400 square feet (27.9 m² to 37.2 m²). See Fig. 18-2.

25. The space allotted to a secretary depends upon the duties. Tasks other than taking dictation and typing, such as filing or greeting visitors, may be required.

26. An intercommunication system between executive and secretary is desirable.

27. The secretary's office should be designed to have an unobstructed view of the doorways, adequate file space, storage for office supplies, and seats for visitors. See Fig. 18-3.

28. Where frequent interviews with the public are required, such as in personnel offices, the use of interview cubicles should be considered.

Such cubicles need to be only large enough for the interviewer, the applicant, and a small desk or table.

29. The reception area should be pleasant and should favorably impress the visitor. Music, plants, paintings, drapes, and comfortable furniture contribute to this atmosphere.

30. The location of rest rooms near the reception area should be considered.

31. Small conference or interview rooms frequently eliminate the need for private offices.

32. In conference rooms accommodating up to thirty people, allow 20 to 25 square feet (1.9 m² to 2.3 m²) per person. Allow 10 to 15 square feet (0.9 m² to 1.4 m²) per person for rooms accommodating thirty to two hundred people. This does not allow for tables, but assumes chairs will be set in rows facing a rostrum. If tables are to be used, ascertain the table size and allow clearance for side chairs and an aisle behind the chairs. Selected plans are shown in Fig. 18-4.

33. Conference rooms can be located off the reception room if reduction of traffic in the office area is desired. See Fig. 18-3.

34. Inactive files should be removed from the general office to warehouse or storage space.

35. Active files should be located as near to those using them as possible. Some businesses prefer to use a central file room, Fig. 18-5.

36. The aisle space between files should be held to a minimum. Suggested aisle spacing follows:

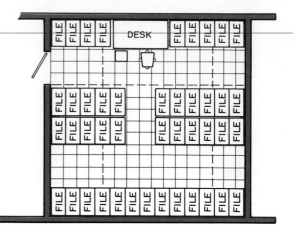

Fig. 18-5.   A centralized file room.

File Room	Drawers Per Person Working Therein	Aisle Space Suggested	
		Min. in. (mm)	Max. in. (mm)
Very Active	20 to 49	42 (1067)	45 (1143)
Active	50 to 74	36 (914)	41 (1041)
Semi-Active	75 to 150	31 (787)	35 (889)
Inactive	Over 150	24 (610)	30 (762)

37. Standard files require an allowance of 5 to 7 square feet (0.5 m² to 0.7 m²), and legal files require 6 to 8 square feet (0.6 m² to 0.7 m²).

38. Lighting requirements vary with the activity. Corridors need about 20 footcandles (215 cd/m²), conference rooms and reception rooms about 35 to 40 footcandles (380 to 430 cd/m²), general offices and secretarial areas from 50 to 75 footcandles (540 to 800 cd/m²), and drafting and business machine sections from 75 to 100 footcandles (800 to 1090 cd/m²).

39. Consideration should be given to the use of color. Color can reduce fatigue brought on by clerical duties. Bright colors or sharp contrasts should be used in halls, while a reception room should be painted in warm, inviting tones. In large, general clerical offices, the wall faced by the employees should be painted a light, cool color, while the other three walls should be a warm pastel, such as yellow or tan.

40. Year-round air conditioning is widely used and tends to increase employee efficiency.

41. Stairways and exits should be planned so the distance from any employee to the nearest exit is not over 150 feet (45 720 mm).

42. Rest rooms and drinking fountains should not be over 150 feet (45 720 mm) from offices.

43. Doors must be wide enough to permit easy movement of furniture.

44. In many offices, provision must be made for handling large quantities of mail.

45. If a large office is planned, column spacing must be considered. A minimum of 20 feet (6096 mm) between columns is recommended.

46. An office building plan is illustrated in Fig. 18-6. It shows needed work areas and offices for a group of lawyers. Examine this plan according to the office planning principles.

Data Processing Centers

The planning of a data processing center is an extremely complex assignment, involving the services of many people. Those concerned with the daily operation of the office or business divisions of a store or company must be consulted constantly. Experts from companies manufacturing data processing equipment must be deeply involved. Some of the factors to consider when planning such a facility follow:

1. The type of information to be processed must be ascertained. This could be data such as payrolls, materials handled, or invoices.

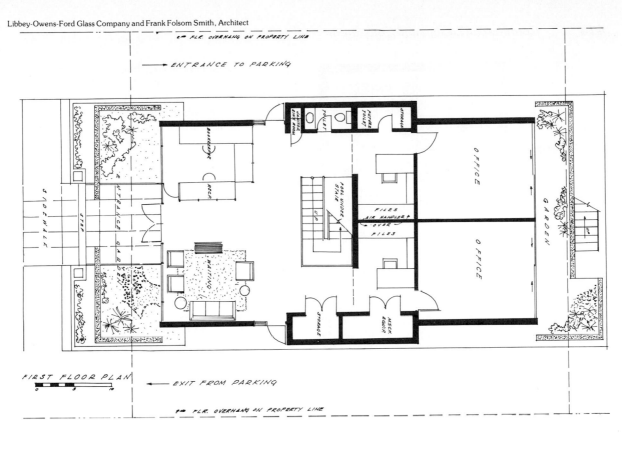

FIRST FLOOR PLAN

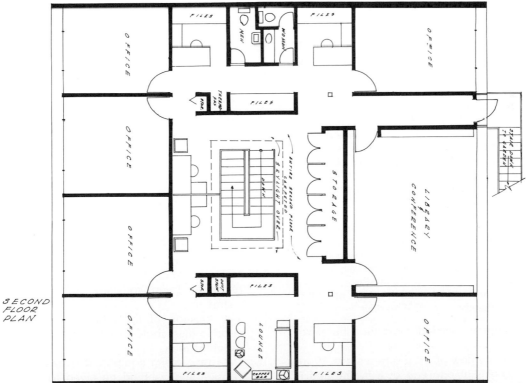

SECOND FLOOR PLAN

Fig. 18-6. Law office building in Sarasota, Florida. This building was designed to accommodate eight attorneys, with adjacent secretarial offices for each attorney. It is completely air-conditioned. The structure is steel-framed, with the second floor cantilevered to the property line. The first floor was set back to allow access to a parking area in the rear.

473

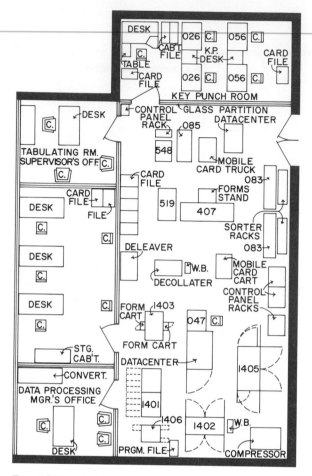

Fig. 18-7. A data processing center designed for the McLouth Steel Company. It handles payroll, tax information, check writing, labor distribution for cost analysis, mill order information, billing and invoicing, and sales analysis. The solid lines indicate machines and racks; the dotted lines indicate spaces needed for servicing the machine. This installation has equipment manufactured by the International Business Machines Corp. The machines are identified by number as follows:

1401 Data Processing System	**083** Sorter
1402 Card Read Punch	**407** Accounting Machine
1403 Printer	**519** Document Originating Machine
1405 Disk Storage Unit	**548** Interpreter
1406 Storage	**085** Collator
047 Tape to Card Printing Punch	**026** Printing Card Punch
	056 Verifier

2. The equipment needed must then be ascertained.
3. A plan for the flow of work is necessary. Work usually begins at one end of a facility and is carried in process to the other end. It is set up on a production-line basis. This type of installation is illustrated in Fig. 18-7.
4. The relationship between the data processing facility and other business operations must be ascertained. For example, if a company has a large accounting and statistical division, the data processing facility must be located so that employees involved with these other departments are a minimum distance from the facility.
5. Work aisles should be at least 3 feet (914 mm) wide, and aisles to carry heavy traffic 4 feet (1219 mm) wide.
6. Units providing work surfaces should be located within 4 feet (1219 mm) of the machine with which they are to be used.
7. Key punch machines should be at least 30 inches (762 mm) apart.
8. The key punch department should be isolated from the remainder of the operation. This arrangement facilitates temperature control and eliminates noise from the remainder of the installation.
9. Ceilings and walls should be acoustically treated. Walls are usually treated from the ceiling down to about 5 feet (1524 mm) from the floor.
10. Machine rooms require 40 footcandles (430 cd/m^2) of light at table level.
11. Electrical power requirements must be ascertained. A constant rate of power must be available in order for the machines to perform the high-speed computations accurately. A tremendous amount of power is required. A separate feeder line for the computers should be considered. Usually, transformers separate from those supplying the building are used for the facility.
12. Room temperature and humidity control is vital. The machines must be operated at the specific temperature recommended by the manufacturer.
13. Dust control is necessary. The room should be as "dust proof" as possible. A typical installation is shown in Fig. 18-8.
14. Machines should have sufficient space surrounding them to allow repairs and adjustments.
15. The location of the tabulating room supervisor's office should allow visual contact with the entire area.
16. Data processing equipment is heavy. Special consideration should be given to design the structure to carry the extra weight.

Fig. 18-8. A data processing system. Notice the soft general illumination and clean conditions.

Planning Banks

Bank planning varies considerably, depending upon the size of the bank and the services it offers. Areas must be planned for each activity, and attention must be given to the relationships that exist between these activities.

Planning Small Banks

The following planning principles should be considered when planning small banks:

1. The first consideration is to examine the site and locate the building, parking area, and drive-in facilities to best advantage. Then it is necessary to study the plan of operation of the bank and to ascertain the traffic patterns of the various employees and customers.
2. Adequate customer parking is vital. Easy entrance to the bank from the parking area is necessary. A covered entrance is of value for inclement weather.
3. In a small bank, each employee usually performs several functions. These must be considered when locating work areas. It is customary for tellers to work in the bookkeeping area during lull periods; therefore, the teller area and the bookkeeping area should be adjacent, allowing employees to move between the two quickly and with a minimum of steps. A small banking facility is illustrated in Fig. 18-9.
4. If a drive-in window is used, someone in the bookkeeping or teller area usually serves these customers; therefore, the drive-in window should be visible from these areas.
5. The tellers and employees in the bookkeeping and machine area should have visual contact

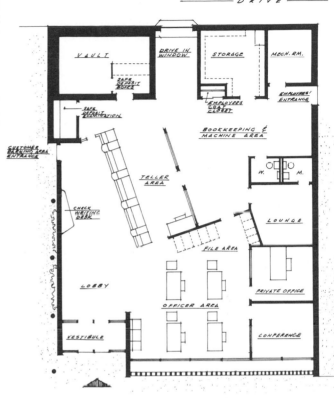

Fig. 18-9. This plan was developed to house an existing small bank.

Since a primary consideration in planning is the cost of the facility, the designer is limited to a building size the bank can afford. Within this floor area, the designer attempted to embody all of the desirable characteristics of bank planning that could be obtained within the cost limitations. Notice the relationships established between work areas, vault location and control, and the auto traffic.

with the vault area to control access to the vault.

6. In a small bank, the vault serves the dual purpose of money and record storage and safe-deposit box location. Some control of customer traffic to this area is necessary. Usually, a low partition with a gate is used to separate the lobby from the vault. Employees responsible for overseeing activities in the vault should have their work stations near this entrance.
7. The vault should have a ventilator to supply oxygen to the area in case someone is inadvertently locked inside.
8. An alarm system is important and should be planned by experts.

9. A hidden telephone should be installed in the vault.

10. A private cubicle near the vault, but outside it, is needed for customers to examine the contents of their safe-deposit boxes.

11. The number of safe-deposit boxes needed and their size must be determined. They are available in a variety of sizes.

12. The interior ceiling height of the vault is usually 8 feet (2438 mm).

13. The small bank generally has three types of customers — those seeing the general teller, those writing checks and making deposits, and those seeing bank officers concerning loans and notes. To reduce customer congestion, the check-writing and deposit-slip desk should be apart from the bank officers' area.

14. The bank officers' desks should be located away from the main lobby. This increases privacy for those discussing loans and notes.

15. Offices for bank officials are frequently clustered in a semi-private area. Desks may be separated by partitions if desired.

16. The bank president frequently desires a private office.

17. Several chairs near the bank officers' area are needed to seat waiting customers.

18. A small conference room is desirable. This can serve as a meeting place for bank directors and for private functions, such as the disposition of wills. A telephone should be in this area.

19. In some banks, officers also serve at the tellers' windows. In this case, the officers' area and tellers' area should be closely related.

20. The tellers need drawers, shelves, and cabinet space for storage of supplies used in their work. These should have locks for security.

21. A carefully planned, intercommunication system is necessary. It should be designed to keep conversations private. Especially important is a connection between the bank officials' area and the bookkeeping area.

22. Rest room facilities for employees are needed. These should be located so they are not in general view of the customer.

23. A small, employees' lounge is desirable. If space is at a premium, this area also could serve as a director's room or an extra office.

24. A storage room near the bookkeeping and machine area is needed. Tellers can store several days' supply of needed items at their stations.

25. A separate room for office machines and equipment is best if it can be centrally located.

26. Year-round air conditioning is vital.

27. An area for files should be planned. This should be near the department using the files.

28. The equipment to be accommodated in the bookkeeping and mechanical room must be determined so these areas can be planned. The typical, small bank has posting machines, microfilm equipment, adding machines, check-cancelling equipment, typewriters, and filing equipment.

29. Closets are needed for employees' coats and personal belongings.

30. Working areas, such as the teller area, bookkeeping and machine area, and bank officers' area, require 100 footcandles (1090 cd/m^2) of light. The lobby needs only a minimum of general illumination. Other areas, such as the check-writing table, can use accent lighting.

31. Acoustical treatment is vital, expecially in the machine area.

32. If the bank operates a small-loan office, provision should be made for easy access to this by customers after regular closing hours, since this area operates the full day.

Additional Considerations for Larger Banks

33. Larger banks face different problems in relating areas. They employ full-time tellers and bookkeepers, so easy access between these areas is not important. A good intercommunication system serves well here.

34. Drive-in banking windows offer special planning problems. A stock unit is shown in Fig. 18-10. Customers should be able to reach these windows easily from the street, without cutting across lanes of traffic or heavily traveled sidewalks. The window should be on the driver's side of the auto. See Fig. 18-11.

35. If several drive-in windows are used, space should be provided for automobiles to pass one another to reach alternate windows or to leave immediately upon completion of the business transaction. See Fig. 18-12.

36. Well placed signs are necessary to give drivers directions of traffic patterns to windows. Electric signs are available to tell a waiting customer when a window is empty.

37. Provision must be made to accommodate customers that are waiting for drive-in service. Waiting autos cannot be permitted to block traffic in the street. See Fig. 18-13.

Fig. 18-10. A stock drive-in window equipped with a microphone for communicating with the customer.

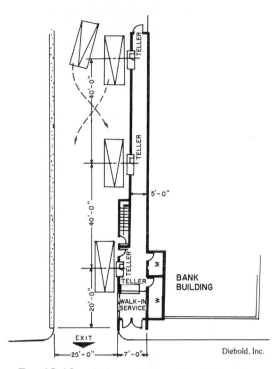

Diebold, Inc.

Fig. 18-12. A three-window drive-in facility. Notice the spacing between windows and the drive area, permitting customers to pass one another to reach unoccupied windows or to leave upon conclusion of their business.

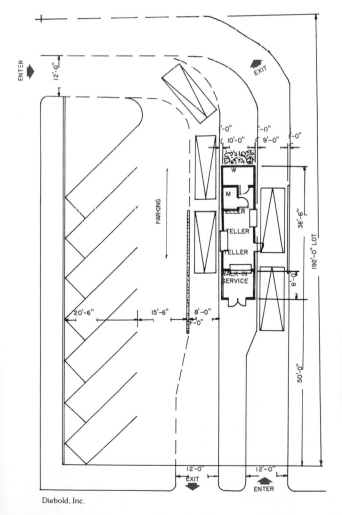

Diebold, Inc.

Fig. 18-13. A multiple-station drive-in facility with sufficient driveway space to accommodate a number of customers waiting for service.

Fig. 18-11. Plan of a two-station drive-in banking facility. Notice the walk-up window off a small lobby.

Fig. 18-14. A remote banking station utilizing closed-circuit television and pneumatic tube for transferring materials to a teller in the bank building.

38. A walk-up banking window serves customers who do not want to take time to enter the bank. Nearby parking is needed. Refer again to the plans shown in Figs. 18-11 and 18-12.

39. A drive-in banking unit located in a remote area can be installed where space does not permit the usual drive-in banking with teller service. This unit has closed-circuit television for contact with a teller. The materials are transferred from the unit to a teller via a pneumatic tube which the teller can operate from inside the bank building. See Fig. 18-14.

40. State banking laws regulate the distance that drive-in stations can be located from the bank building.

41. A night-deposit window should be considered. It should be easily accessible and well lighted.

42. The exterior styling of banks varies considerably. It usually reflects the geographic location of the bank and the attitude of the management. There is no one best style, and colonial as well as contemporary styling is successfully used.

Medical and Dental Offices and Clinics

A clinic is a building designed to provide facilities for two or more doctors or dentists. Patients are treated here, but usually are not kept overnight as is the practice in a hospital.

Two clinic plans are shown in Figs. 18-15 and 18-16.

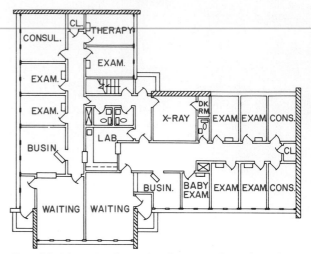

Fig. 18-15. The floor plan for an L-shaped medical clinic on one level.

Medical Offices and Clinics

As medical offices and clinics are being planned, the following principles should be considered:

1. The size of the waiting room depends upon the type of medical activity housed and the number of doctors using the room for their patients.

2. The duties of the receptionist must be known. In small clinics, the receptionist often serves as business office clerk as well. In this case, the reception and business areas should be combined in one room. See Fig. 18-17. Larger clinics have a separate business staff and require a special business office. This should be near the waiting room and reception area.

3. In large clinics, provision should be made for a lounge for employees.

4. Rest rooms for staff and patients are necessary. Easy access from the waiting room is desirable.

5. The equipment in a medical examination room varies according to the type of practice and the preferences of the doctor. The equipment normally includes an examination table, scale, lavatory, waste container, desk, two chairs, and a dressing booth. If a room is used for special examinations, space must be allowed for the needed equipment. This type of room is commonly used for consultation as well as examination. In a room used only for examination, the desk and chairs could be omitted. Figure 18-18 illustrates typical examination and consultation rooms, while Fig. 18-19 illustrates office suites for medical specialists.

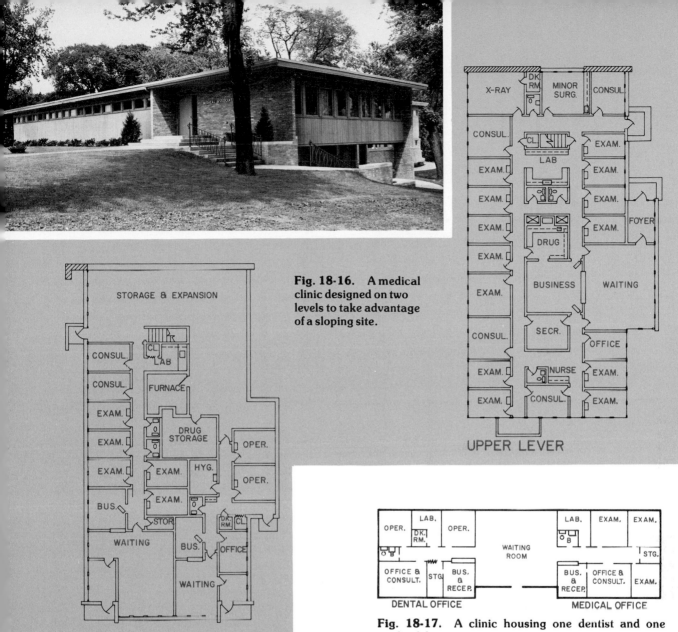

Fig. 18-16. A medical clinic designed on two levels to take advantage of a sloping site.

UPPER LEVER

LOWER LEVEL

Fig. 18-17. A clinic housing one dentist and one medical doctor.

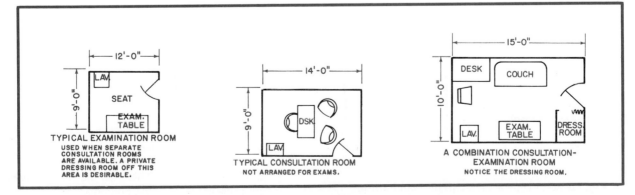

TYPICAL EXAMINATION ROOM
USED WHEN SEPARATE CONSULTATION ROOMS ARE AVAILABLE. A PRIVATE DRESSING ROOM OFF THIS AREA IS DESIRABLE.

TYPICAL CONSULTATION ROOM
NOT ARRANGED FOR EXAMS.

A COMBINATION CONSULTATION-EXAMINATION ROOM
NOTICE THE DRESSING ROOM.

Fig. 18-18. Examination and consultation rooms.

Marshall Erdman and Associates, Inc., Madison, Wisconsin, Designers and Builders of Medical Office Buildings

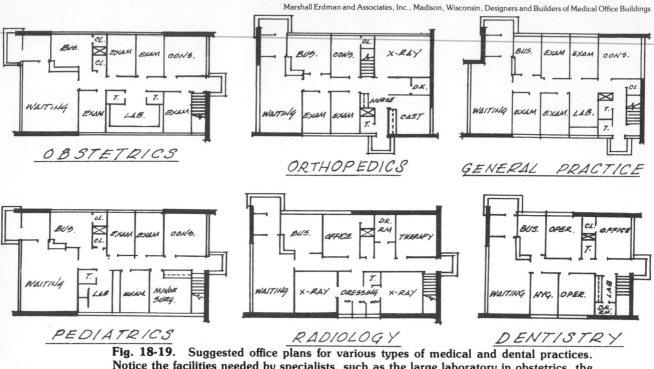

Fig. 18-19. Suggested office plans for various types of medical and dental practices. Notice the facilities needed by specialists, such as the large laboratory in obstetrics, the cast room in orthopedics, and X-ray and therapy facilities in radiology.

6. The walls of consultation and examination rooms should be soundproof.

7. In a very small clinic, one nurse frequently serves as a laboratory technician and receptionist, as well as the nurse. In such a situation, a combination business office and reception room is used, with the laboratory adjoining this area.

8. In clinics of moderate size, one person may serve as receptionist and handle the business functions. This frees the nurse for medical duties. In this case, the nurse's station and the laboratory could be combined so the nurse also can serve as a laboratory technician. A combined laboratory and nurse's station is illustrated in Fig. 18-15. A plan with these duties separated is given in Fig. 18-16.

9. The nurse's station should be located near the area used by the doctors served.

10. The nurse's station should be as near the waiting room as possible, in order to meet patients and take them to consultation or examination rooms.

11. The size of a laboratory and the equipment located there varies with the size of the clinic and the availability of large commercial laboratories in the area. A typical medical clinic laboratory is illustrated in Fig. 18-20.

12. A rest room adjoining the laboratory is helpful for taking specimens.

13. Figure 18-21 illustrates a typical X-ray laboratory, darkroom, and dressing room as found in medical clinics.

14. All sterilization should take place in one area. Sometimes this is a part of the laboratory. Provision should be made to vent the area to remove moisture.

15. Suggested space requirements for medical and dental clinics are shown in Fig. 18-22.

16. Provision must be made for storage of janitorial supplies. This can be in the least desirable portions of the building.

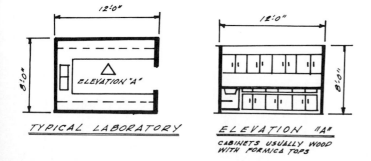

Fig. 18-20. Typical medical clinic laboratory.

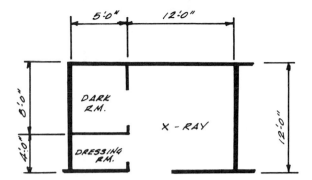

Fig. 18-21. Typical X-ray laboratory as found in a medical clinic.

Fig. 18-22.

Typical Medical and Dental Space Requirements[1]

Room	Square Feet	Square Meters[2]
Waiting room - medical (to seat 20 patients)	400	37
Waiting room - dental (to seat 6-8 patients)	100	9
Examining rooms - medical	100-140	9-13
Consultation rooms	100-140	9-13
Rest rooms	30	3
Nurse's station and workroom	80	7
Laboratory - medical (to handle EKG, BMR, and X-ray)	100-300	9-28
Laboratory - dental	75-100	7-9
Recovery room - medical	140	13
Recovery room - dental	40-50	4-5
X-ray room - medical	120-180	11-17
Darkroom - medical	50	5
Darkroom - dental	15-25	1-2
X-ray storage and viewing room	100-120	9-11
Patient dressing rooms	30-40	3-4
Minor surgery room - medical	130	12
Operatory - dental	80	7
Supply room (sterilized supplies, sterilizers, autoclave, utensil washing)	80-120	7-11
Employee lounge	150-175	14-16
Kitchen (for patients in recovery rooms and personnel lunches)	65	6
Linen storage	100-120	9-11
Janitor room	80-120	7-11
Furnace room	As Req.	As Req.
Library	100-120	9-11
Private doctor's office	100	9
Receptionist area	80-150	7-14
Business office	100-150	9-14

[1]Space requirements were developed by examining existing facilities. Requirements vary considerably, depending on type of practice and personal preference.
[2]Rounded off to nearest square meter.

17. The atmosphere of clinics should be one of friendliness and warmth.
18. Year-round air conditioning is very important. It is necessary that the air in each room be changed frequently and that the fresh-air supply be considerable. Odors must be removed. Patients must be comfortable at all times. Certain rooms, such as those in medical and dental surgery, require a 100 percent fresh-air supply.
19. Special plumbing fixtures are available for medical use. Manufacturers' catalogs should be studied to ascertain what is available and the advantages of each.

Dental Offices and Clinics

Many of the planning factors discussed for the medical clinic apply as well to the dental clinic. Following are considerations unique to dental clinic planning:

1. The waiting room can be considerably smaller than that required for medical clinics.
2. The receptionist should be able to talk with a patient without being overheard by those waiting. Usually, this is accomplished by separating the reception area from the waiting room.
3. The darkroom can be very small, since X-ray work is limited.
4. The dentist should have a private office.
5. Three operatories are most often preferred for a single dentist. The dentist, with an assistant, can use two, while a dental hygienist can use the third for X-ray, scaling, cleaning, and tissue examinations. A plan for an office for one dentist is given in Fig. 18-19.
6. It is desirable if the operatories can face the north to take advantage of this light. A typical operatory is shown in Fig. 18-23.
7. Operatories used for dental surgery require 100 percent fresh-air supply.
8. The operatories should be adjacent.
9. The operatory walls should be soundproof.
10. If space is available, a small room with a chair and couch is desirable for patients who need time to recover before leaving the building.
11. A small area is needed for a sterilizer. This is usually not much larger than a closet.
12. Suggested space requirements are shown in Fig. 18-22.
13. The first floor plan for a larger dental clinic is shown in Fig. 18-24. Notice the relationship the architect established between operatories and other facilities.

Courtesy of Dr. Gregory G. Rackauskas

Fig. 18-23.   A dental operatory.

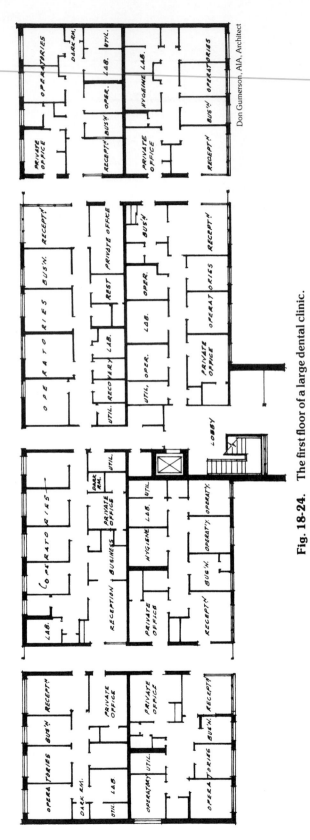

Don Gumerson, AIA, Architect

Fig. 18-24.   The first floor of a large dental clinic.

Doctors' Parks

With the rapid growth of suburban areas, the problem of providing convenient medical facilities has become a growing concern. One solution developed is the doctors' park. It is designed to furnish medical facilities for doctors specializing in various types of practice. The park is composed of a series of small buildings, each usually housing two to four doctors.

Parking areas and approaches to each building are individualized, and the patient can recognize the doctor's office and location.

Doctors' parks are usually developed in residential areas. Facilities can be expanded easily by the addition of other small buildings when the need warrants.

The buildings should be placed so that each preserves its individuality and has adequate, adjacent parking. Parking areas frequently are separated by plantings and grass plots to further individualize the facility.

Figures 18-25 and 18-26 illustrate doctors' parks.

482

Fig. 18-25. A doctors' park. Note the individual buildings and convenient parking.

Marshall Erdman and Associates, Inc.,
Madison, Wisconsin, Designers and Builders
of Medical Office Buildings

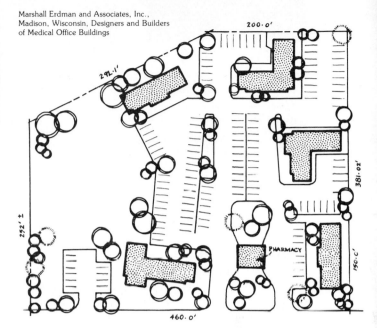

Fig. 18-26. A site plan for a doctors' park. Notice the provision for a pharmacy. The parking areas are nicely individualized, helping to give each building and the doctors practicing there an individual identity.

Build Your Vocabulary

Following are terms that you should understand and use as a part of your working vocabulary. Write a brief explanation of each term.

Medical
dental office
dental operatory
doctors' park
examination room
laboratory
medical clinic
medical office
reception area

Banks
drive-in window
vault

Offices
data processing
organization chart
work flow

Class Activities

1. Make a checklist of the planning principles that should be observed when planning an office, bank, or medical office or clinic. Then visit one of these and evaluate it by marking each item on the checklist as good, satisfactory, or poor. The checklist should include exterior design, as well as interior design and planning.
2. Visit an office, bank, or medical facility and make a freehand, scale sketch of the interior arrangement. In class, revise this to improve the overall operation and draw the revised floor plan.
3. Build a cardboard, scale model of the interior of one of the commercial buildings discussed in this chapter (the building with the roof removed and possibly one or two walls removed). Build to scale, typical fixtures and equipment. Arrange these on the floor plan to develop the most satisfactory plan.

4. Design a doctors' clinic. Work with several other class members as a team. One member of the team can design a dental office, and others medical offices for various types of medical specialists. An X-ray room and laboratory could be included. The individual plans will need to be coordinated by the team leader so they will logically fit together into one building. A waiting room will have to be planned. After the preliminary planning is finished, a set of working drawings should be completed. It will be necessary to consult other chapters concerning structure and exterior design.
5. If several teams have developed medical buildings, and each building accommodates several doctors, develop a plan to work these buildings into a doctors' park. The structures could be cut from blocks of pine (at a very small scale) and arranged on a three-dimensional model, with roads, parking, and shrubbery arranged around them.
6. Design a small branch bank office. Provide two drive-in windows and space for two walk-up teller's windows. Select a lot in a desirable location, and plan the building to fit the site. Give special consideration to parking and to traffic flow to the drive-in windows.

Chapter 19

Planning Food Stores, Restaurants, and Cafeterias

When planning either retail food stores or food service facilities, the designer faces a number of problems common to both. The exact solutions to these problems vary with the type of facility. In both facilities, food must be received, prepared, and stored. Refuse must be removed. Customers must be received, and money must be collected.

Some problems facing the designer are peculiar to the particular type of business. A food store must display its items. A restaurant must cook its food and keep it in readiness to be served. Restaurants also must handle dishes and soiled linens.

This chapter presents points to consider when solving the common problems and the special problems of each facility.

Food Stores

The typical food store must stock and display thousands of items. Some are perishable and others are fragile. Some foods are received in bulk and require preparation and packaging. Considerable effort is required in stocking the sales area, and the location of various products influences the sales of these items. The following principles must be considered when planning a food store:

1. The store front is the most important factor in the success of a food store. An attractive design is shown in Fig. 19-1.
2. The store front is usually the all-glass, open type. This helps attract customers, since they can see the entire inside display. See Fig. 19-2.
3. Customer traffic control is necessary from the time of entering until leaving. The entrance and exit should be separate, even in the small food store. They should open onto a lobby space.
4. The entrance and exit should be closely related to the parking area.
5. Adequate parking is a must.

Libbey-Owens-Ford
Glass Company

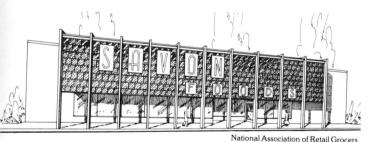

National Association of Retail Grocers

Fig. 19-1. This food store has a popular modern style.

The front of the building does not have ornamental brick or stone, so decorative grillwork is placed forward for "glamour." In this design, the grill screen is supported with I-beams that carry up to the roof line and extend back, forming a protective covering over the walk. The grill screen serves as a windbreak and creates interesting shadow panels on the building. Instead of grillwork, brick or color tiles may be used, but the thinner grill discourages birds nesting and is preferred over brick or tile.

Letters in store identification should be individual, boxed, and illuminated. Those selected should carry out the modern theme of the design.

Fig. 19-2. This food store is located in Fort Worth, Texas. The interior of the store is the show window. The glass panels are 8 feet high and 7 feet wide, stacked three tiers high 24 feet to the roof.

6. The parking lot should be designed so that 250 feet (76.2 m) is the maximum distance a customer must walk to reach the store.

7. Good general illumination is needed over the entire store. Using fluorescent fixtures reduces the heat damage to perishable merchandise.

8. A store must be attractive and must be arranged so that it allows ease of shopping and efficiency in stocking.

9. A study of the traffic flow of customers and employees is necessary.

10. All food stores should be completely air-conditioned.

11. The colors selected for walls, ceiling, and fixtures should be light or neutral.

12. Noise is a problem usually combatted with acoustical ceilings. Piped-in music helps to reduce attention to store noises and creates a pleasant atmosphere.

13. The interior walls should be made of a durable material that can be washed. Floors must be easy to clean and resilient enough for noise reduction and comfort in walking.

14. Economy of operation should be considered. This includes providing for a smooth work flow, allowing sufficient room for employees to work, and operating the store with a minimum number of employees.

15. Equipment selected must be dependable and must have sufficient capacity.

Steps in Food Store Planning

16. The first step in planning the store layout is to select the basic equipment for each department. The proper amount of equipment is based upon the estimated sales.

17. The building itself should not be planned until the floor layout is completed. The building is a shell to house the store and attract customers and should not influence the layout of the store. An exception would be limitations caused by the shape of the lot.

18. Each department should be planned to handle the anticipated sales. Based on averages, the following percentages serve as a guide for planning a balanced store:

Percent of Store Sales Dollar Volume		
Department	Range	Average
Grocery	57% to 77%	69%
Meat	16% to 29%	22%
Produce	5% to 10%	6%
Bakery	2% to 5%	3%
Frozen Food	3% to 5%	4.5%

National Association of Retail Grocers of the United States

19. Next, equipment must be selected for each department. This involves first a study of equipment offered by manufacturers and then the selection of equipment to be used. The capacity of the selected equipment and the sales capacity per week per lineal foot of equipment must be ascertained. This information is available from manufacturers. To illustrate this, some sales capacities might be:

Dollar Sales Per Lineal Foot of Display Cabinet Per Week	
Grocery	$ 60
Meat	250
Produce	60
Bakery	80
Frozen Food	50

20. Floor area allocation varies depending upon the designer. Following are data developed from a survey of existing stores:

Allocation of Floor Space for the Total Store			
Store Area	Range		
	High	Low	Median
Sales Area	86.7%	55.8%	68.3%
Back Room Area	37.7%	13.3%	28.6%
Front End	25.7%	4.5%	14.8%
Other	17.0%	.2%	6.4%

University of Missouri-Columbia and Agriculture Marketing Research Institute

The sales area is where the customer and the merchandise come together. The back room area is where items are stored and prepared for placement in the sales area. The front end is where the check-outs, cart storage, and manager's booth are located.

21. If the estimated weekly dollar sales in each department is divided by the sales capacity of the equipment selected, the number of lineal feet of equipment needed in each department can be found. To illustrate:

	Weekly Sales	Dollar Volume Per Lineal Foot	Lineal Feet of Display Needed
Grocery	$55,200	$ 60	920
Meat	17,600	250	70
Produce	4,800	60	80
Bakery	2,400	80	30
Frozen Foods	3,600	50	72

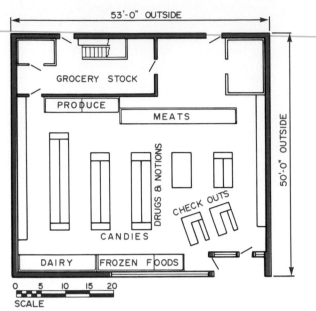

Fig. 19-3. A store of 2650 square feet. The traffic pattern is based on the premise that customers want to select meat first, produce second, and then groceries. Notice that the dairy is in the front corner, drawing customers to that area. The check-out stands are placed on an angle because this provides more room. Notice that drugs, notions, and candies are in view of the checker. A basement area is used for grocery storage, thus increasing space available for sales.

22. Lineal feet of display refers to single-face cabinets as found along a wall. Island-type fixtures provide display area on both sides, and each side counts separately in computations.
23. A commonly used guide for a sales objective is to plan for a minimum weekly sale of $18 per square foot of selling area. Therefore, a food store expecting to do $90,000 per week would have to have approximately 5000 square feet (464.5 m²) of selling area, plus storage and preparation areas. Well engineered food store layouts are illustrated in Figs. 19-3, 19-4, and 19-5.

Other General Planning Considerations

24. More merchandise can be displayed on multi-deck racks. Items such as frozen foods, cold cuts, smoked meats, chickens, produce, and dairy products can be displayed in two- and three-deck fixtures. These can be refrigerated if needed. See Fig. 19-6.
25. Some food retailers use high gondolas in the center of the store. This increases display area. These should never be so tall that they block the overall view of the store.

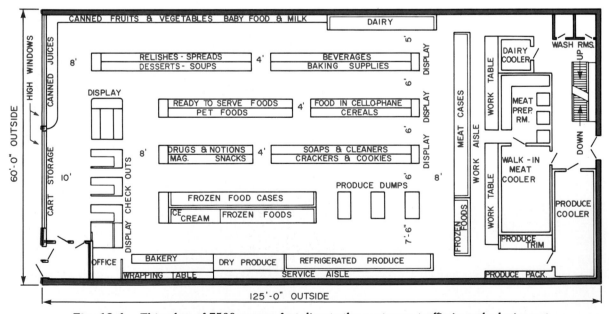

Fig. 19-4. This plan of 7500 square feet directs the customer traffic in a clockwise pattern. The produce and frozen foods are near the end of the shopper route. This plan has excellent meat and produce receiving, storing, and preparation areas. Groceries are stored in the basement. Notice the conveyor near the receiving doors. The manager's office is conveniently located.

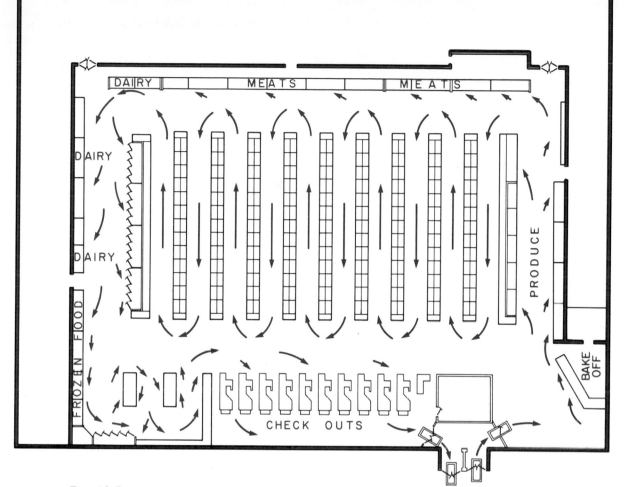

Fig. 19-5. This layout shows the best characteristics of a store layout. The building is divided into sales area 71% and back room 29%. The front end uses 9% of the sales area. The sales area is divided as follows: grocery sales — 83%, meat sales — 6% and produce sales — 11%.

Fig. 19-6. A multideck dairy bar. This fixture displays considerably more merchandise per lineal foot of cabinet than the single-deck type. This increases the sales capacity of the store without expanding the building.

Fig. 19-7. An attractive, well lighted produce department. Notice the use of mirrors to enhance the display. The department is attractively identified by a large sign and enhanced by a cluster of palm trees.

Two- and three-deck produce fixtures are available for both refrigerated and dry produce.

Tyler Refrigeration Corporation

Fig. 19-8. Frozen food and ice cream cases assembled in an island installation. These are available in many forms and provide for multilevel display if desired.

Mirawal, Birdsboro Corporation

Fig. 19-9. A service-type, fresh meat display. This unit requires personnel to wait upon each customer. Notice the durable, easily cleaned walls.

Examples of selected types of sales cases are shown in Figs. 19-7 through 19-9. Manufacturers' catalogs should be consulted for details as to size, electrical power, drains, refrigerants, and other needs which must be considered when planning the mechanical aspects of the building.

26. A $20,000 weekly volume is needed for each mechanical check-out station.
27. Aisles between gondolas should be about 4'-6" to 6'-0" (1372 mm to 1829 mm). The longer the aisle, the wider it should be.
28. Customers select grocery items comparatively rapidly. These aisles tend to clear rapidly. In the dairy and frozen food areas, customers take more time; therefore, wider aisles are needed. In meat and produce areas customers take considerable time to make selections. Since this causes congestion, the widest aisles should be in the meat and produce departments. In very small stores, a 4'-6" (1372 mm) aisle in meat and produce is adequate. Average-sized stores usually allow at least an 8'-0" (2438 mm) aisle.
29. Sufficient space should be allowed behind check-out stations to accommodate waiting customers and to permit normal traffic flow for shopping customers. This might be only 6'-0" (1829 mm) in a very small store. However, average stores should have 8'-0" to 10'-0" (2438 mm to 3048 mm).
30. Consideration must be given to technical problems such as refrigerant lines, electrical power, plumbing needs, and heating and air conditioning.
31. Rest rooms for employees are necessary.
32. The store manager needs space for an office. Usually, this is located to give an overall view of the store. The best location for the office is often at the front of the store.
33. Most food stores are departmentalized. Common groupings are: fresh fruits and vegetables, dairy products, baked items wrapped for self-service, meats, canned and packaged goods, sometimes a bakery and a delicatessen, liquor, and frozen foods. Some large stores include other merchandise areas, such as household items (dishes, pans, can openers), books, and potted plants.
34. Preparation areas should be planned, where meat and produce can be processed into the forms and quantities preferred by customers.

Restaurants

A restaurant is an establishment where food and refreshments are served to the public. There are numerous varieties of restaurants, each serving special functions. This discussion is limited to restaurants providing waiter service and to self-service cafeterias.

In initial planning of any food service, the following decisions must be made:

1. What type and number of customers are to be served? Are customers repeat or transient? How many must be seated at peak demand hours?
2. What type of service is to be offered? Will this be cafeteria service or waiter service? Will it be counter service and/or table service?

3. What hours of operation are planned? Will all three daily meals be available?
4. What type of menu is planned? How elaborate will be the offerings and quality of food? What price meals are planned?
5. What type of interior atmosphere is desired?
6. What type of service areas are planned? Will private and public dining areas be needed?
7. What facilities are needed for food preparation? What food receiving, storage, preparation, cooking, and service areas are needed?
8. What provisions are needed for dishwashing and waste disposal?
9. How large a staff is planned?
10. What is the total area needed to accommodate all these services?

There are many types of food-serving facilities, each having special requirements. The planning of a commercial cafeteria differs in many respects from that of a school cafeteria. A restaurant catering to transient trade differs from one serving luxury meals to repeat customers. The type of facility to be designed must be carefully studied, and decisions must be made to enable it to serve its function effectively.

General Considerations

11. The exterior and entrance area should reflect the character of the facility. The restaurant catering to a transient clientele frequently uses large glass areas to expose the busy, attractive interior, thus inducing people to enter. See Fig. 19-10. An exclusive, expensive restaurant is better characterized by less interior exposure, thus reflecting a private, more fashionable atmosphere.
12. The entrance lobby in restaurants handling a large volume of trade is usually a simple, open area designed to accommodate incoming and outgoing traffic with a minimum of congestion. In more exclusive restaurants, it can be attractively furnished to provide a place for customers to wait before they are seated in the dining area.
13. Provision should be made for customers' coats.
14. The customer must be able to move easily from the entrance to the dining area.
15. Rest rooms, coat rooms, and bars should be located near the entrance.
16. The cashier's counter is located near the exit.
17. The materials used on the floor and walls should be extremely durable and easy to clean. The ceiling should absorb sounds and reduce the noise level.

Northrop Architectural Systems

Fig. 19-10. A small restaurant using glass walls to expose the attractive interior.

18. Year-round air conditioning is vital.
19. Rest rooms provided for the employees should be separate from those for customers.
20. Subtle, indirect lighting enhances the atmosphere creating a relaxed mood for dining. The level of illumination should be low, with general illumination of the dining area being from 5 to 20 footcandles (54 to 215 cd/m^2).

Dining Area

21. The seating in the dining area should be highly flexible. The use of a variety of means of seating is recommended.
22. The seating capacity of the dining area must be related to the kitchen production capacity.
23. Round tables occupy less space than square tables. Since many people eat alone or in couples, several tables should be for two persons.
24. Customer and service aisles should be at least 3'-0" (914 mm) wide — more desirably, 5'-0" (1524 mm).
25. Rectangular tables should be spaced from 3'-0" to 4'-0" (914 mm to 1219 mm) apart. A table should be at least 2'-0" (610 mm) from the wall, if its chairs are backed against the wall. Circular tables should be at least 2'-6" (762 mm) apart.
26. While booths vary in size, the typical booth designed to accommodate four persons, seated two on each side, measures 4'-0" (1219 mm) long and 5'-4" (1626 mm) wide.

A booth to accommodate two persons facing each other commonly measures 2'-6" (762 mm) long and 5'-4" (1626 mm) wide. The aisle between a row of booths should be at least 3'-0" (914 mm). The backs of booths seldom should be higher than 4'-0" (1219 mm) or lower than 3'-6" (1067 mm).

27. Pedestal tables are being used increasingly. They are more convenient than the conventional type, and they make it easier for the floors to be cleaned.

28. In restaurants with waiter service, the customer traffic and service traffic use the same aisles. The service entrance to the dining area should be located well away from the customer's entrance.

29. Allow 10 to 12 square feet (0.9 m^2 to 1.1 m^2) per table seat, when the dining area is planned.

Kitchen Area

The proper planning of the kitchen area is vital to the successful operation of a restaurant. It is highly complex and requires the services of a food service consultant. Some planning principles follow:

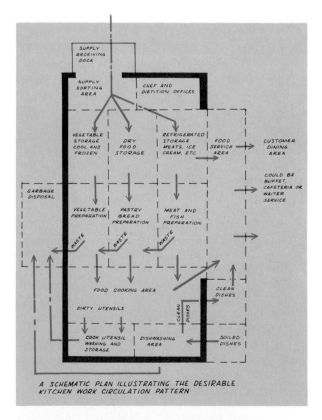

A SCHEMATIC PLAN ILLUSTRATING THE DESIRABLE KITCHEN WORK CIRCULATION PATTERN

Fig. 19-11.   Desirable flow of kitchen work.

30. The kitchen area involves the receiving of supplies, the storage of food, the preparation, cooking, and serving of this food, and the washing of dishes and utensils. A theoretical floor plan is presented in Fig. 19-11.

31. The size of the food preparation area and the equipment located there varies a great deal. A small establishment specilizing in short orders has a simple layout and receives many items, such as bread and pastry, from an outside supplier. Large restaurants and cafeterias have extensive facilities. The planner must know what types of foods are to be served and how the preparation is to be undertaken.

32. The storage of food supplies should be as near to the preparation area as possible.

33. Some foods require refrigeration or freezing, while others require dry storage. Food storage should be near the supply entrance.

34. The food preparation area is between the food storage and serving areas.

35. Each kitchen area should be distinctly separate, yet all must be closely related for efficient operation.

36. In a kitchen used with waiter service, the prepared food ready to be served should be in the following order: salads near the incoming waiters' door, then sandwiches and cold meats, followed by the hot foods. Desserts and pastry should be next, with beverages last, near the door opening into the dining area. See Fig. 19-12.

37. The serving area for the salads and cold meats needs mechanical cooling units.

38. The serving area for the hot foods needs heating units.

39. Clean dishes should be stored near the food serving area. It should require only a minimum of handling to get them here from the washing area.

40. In large kitchens, separate areas should be provided for the preparation of the various kinds of food. Usually meat, fish, vegetables, bread, and pastry are separated.

41. Doors to the kitchen should be located so that incoming waiters are clear of those going out. Two separate doors are generally provided; these are sometimes automatic.

42. The dietition and chef should have office space near the kitchen and should be a part of any intercommunication system.

43. A kitchen ventilating system should be provided to remove odors and fumes. Hoods are frequently used. This system is often separate from that used for the dining area.

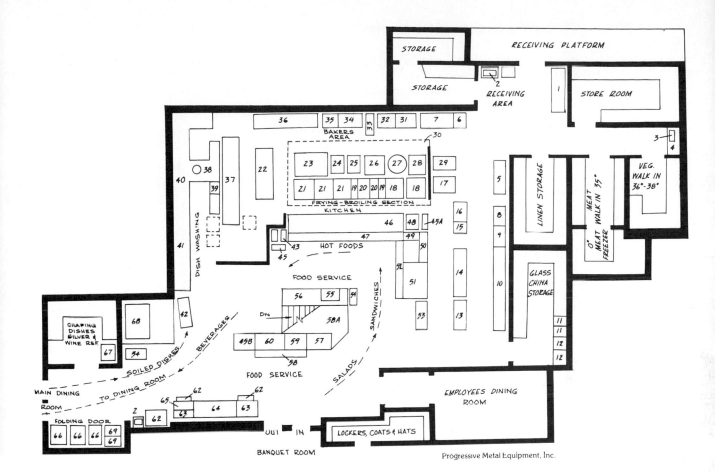

LIST OF EQUIPMENT

NO.	QUAN	NAME	NO.	QUAN.	NAME
1	1	RECEIVING TABLE	37	1	CONVEYOR DISHWASHER
2	1	LAVATORY	38	1	PRE-WASHER DISPOSAL
3	1	MEAT CHOPPER	39	1	GLASS WASHER
4	1	WORK TABLE	40	1	SET OF SOILED AND CLEAN DISH TABLES
5	1	SALAD REFRIGERATOR	41	1	CONVEYOR
6	1	VEG. PEELER	42	1	SOILED DISH TRUCK
7	1	VEG. SINK	43	1	SOUP WARMER
8	1	SALAD CUTTER & SHREDDER	44		
9	1	DISPOSAL UNIT WITH SPRAY	45	1	LOWERATOR BOWL
10	1	SALAD PREPARATION TABLE	45a	1	LOWERATOR PLATE
11	1	CLOSED DISH TRUCK	45b	1	LOWERATOR CUP & SAUCER
12	1	TRAY TRUCK	46	1	CHEF'S COUNTER
13	1	ICE CREAM CABINET WITH DIPPER WELL	47	1	HOT FOOD SERVICE
14	1	PASS THRU REFRIGERATOR	48	1	RADARANGE
15	1	SLICER	49	2	TOASTERS
16	1	TABLE	50	1	SANDWICH UNIT
17	1	COOK'S REFRIGERATOR	51	1	SALAD TABLE WITH OVERSHELVES
18	2	BROILERS	52	1	TRAY SLIDE AND SHELVING UNIT
19	2	SPREADERS	53	1	SALAD REFRIGERATOR
20	2	FRYERS	54	1	PORTABLE ICE TRUCK
21	3	RANGES	55	1	ROLL WARMER
22	1	POT AND PAN STORAGE CABINET	56	1	TRAY TABLE
23	1	ROAST & BAKE OVEN	57	1	BEVERAGE COOLER
24	1	PAN RACK	58	1	URN STAND, ICE BIN AND SHELF
25	1	PYRA STOVE	58a	1	STAIRWELL COVER
26	1	PORTABLE WORK TABLE	59	1	ICE TEA URN
27	1	STEAM KETTLE	60	1	COFFEE URN
28	1	STEAMER	61		
29	1	BAKER'S REFRIGERATOR	62	1	GLASS TRUCK
30	1	CANOPY	63	1	WATER FILLER
31	1	FOOD CUTTER	64	1	ICE CUBE MACHINE
32	1	WORK TABLE	65	1	WATER STATION WITH BINS
33	1	MIXER	66	1	HOT & COLD FOOD TRUCK
34	2	PORTABLE CANS	67	1	WINE REFRIGERATOR
35	1	BAKER'S TABLE	68	1	ICE FLAKER
36	1	POT AND PAN SINK	69	1	DISH TRUCK

Fig. 19-12. A food preparation area for a large restaurant.

Notice the dishwashing area is fairly close to the entrances from the dining room and banquet hall. Soiled dishes are put on large carts in the dining room and moved to dishwasher. The extensive storage area is located adjacent to the receiving platform.

The sandwich preparation area is near the sandwich serving area (52) and the salad preparation is near the salad service area (51). Note the traffic flow through the food service area from salads (51), to sandwiches (52), to hot foods (47), to beverages (58).

Of special value is the relationship established in the kitchen area between the ranges (21), fryers (20), and broilers (18). The baking area is small, but a close relationship exists between the oven (23), pan rack (24), baker's table (35), and mixer (33). Notice that the entire cooking facility is covered with a huge hood (30). The pot and pan storage (22) is adjacent to the washing area and the kitchen.

491

44. The kitchen should have about 20 footcandles (215 cd/m²) of light at the working level.

45. Floor drains are necessary, since the kitchen is completely washed down every day.

46. The floor should be on one level; differences in level are dangerous.

47. Kitchen walls, from ceiling to floor, should be washable. Tile is used most often. Floors are also tile.

48. Kitchen equipment should be easy to clean. Stainless steel equipment is commonly used. It should be located so the floor under it and the walls behind it can also be cleaned easily.

49. The dishwashing section should be near the door used by those bringing soiled dishes into the kitchen.

50. This area usually includes: shelves for holding trays of soiled dishes while they are unloaded onto a table, sinks for soaking, preflushing equipment to dispose of food particles, a dishwashing machine, clean-dish table, and racks for dish storage. If dishes are moved to storage nearer the entrance to the dining area, trucks are necessary by the clean-dish table. A small dishwashing area is illustrated in Fig. 19-13.

Cafeterias

A **cafeteria** is a restaurant designed to be almost entirely self-service. It usually caters to those desiring an inexpensive meal; therefore, the interior furnishings generally are not luxurious nor expensive. Such a food service needs a large volume of trade to succeed. Following are some factors to consider when planning a cafeteria:

51. The cafeteria should be as nearly self-service as possible.

52. The food service counter should be attractive and well lighted to display food to best advantage. Some areas need units to keep foods

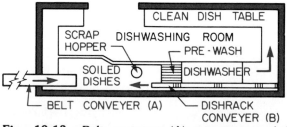

Fig. 19-13. Belt conveyor (A) transports soiled dishes from the cafeteria dining area to the soiled-dish table in the dishwashing room. Conveyor (B) carries the empty dish racks back to the soiled-dish table after they have passed through the dishwashing machine and the clean dishes have been removed.

Fig. 19-14. A food service counter designed to display the food to advantage. The clean lines of the installation plus effective use of lighting enhance the service considerably.

hot, and other foods must be kept cool. See Fig. 19-14.

53. Food in the service counter should be protected, yet should be readily available to the customer. In some cafeterias, hot foods are served, while salads, desserts, and drinks are self-service. Self-service is also used for many small items, such as silver, butter, bread, and salad dressing.

54. Two systems for paying checks are in use. In one system, the customer pays as he leaves the food service counter. This slows service. In the second plan, the customer receives a check at the end of the line and pays the cashier when leaving the cafeteria.

55. Six to eight persons per minute can pass through a single serving line and pay for their meals at a cashier's stand at the end of the line. If a checker is used and the cashier is located elsewhere, ten or eleven persons per minute can be handled in a single line. If greater capacity is needed, a second line must be planned.

56. Many plans are acceptable for the design of food service lines. Those in common use are shown in Fig. 19-15. Of great importance is a free flow of customers along the line so congestion is avoided.

57. A long food counter can be divided into sections. Usually, beverages and desserts are in a center section, with hot foods at one end and salads, sandwiches, and cold meats at the other end. This, in effect serves as two lines.

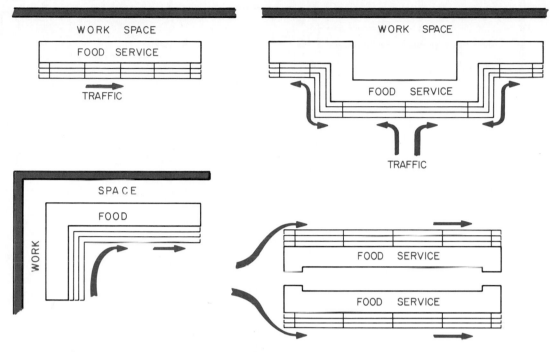

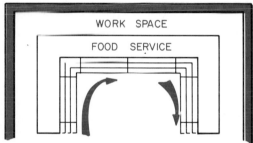

Fig. 19-15. Cafeteria food service patterns. Notice the work space needed for all plans to allow attendants to service the counters.

Plans for service to a single line are shown in the left portion of this illustration: top — straight line service, middle — L counter, bottom — inside U counter. The important factor is that the line flows smoothly and quickly.

Plans for service to two lines are shown in the right portion of this illustration: top — outside U counter, bottom — open counter. Duplicate food service is necessary for each line.

Those desiring a hot meal go to the end serving hot foods and progress to the beverages and desserts in the center and to a cashier. Those wanting a cold meal enter from the end displaying cold foods. See Fig. 19-16, page 494, for a split line serving hot meals and short orders.

58. Provision must be made for food preparation and delivery to the food counter. Also, a way must be provided to replenish the food service counters without interfering with the flow of customer traffic. This usually means that some work space is needed behind the food service counter.

59. The removal of dishes is a big problem. Frequently, a conveyor is used. In some cafeterias, the customers are expected to carry their dishes to a conveyor for removal to the washing area. In others, employees do this.

Class Activities

1. Make a checklist of the planning principles that should be observed when planning a food store, restaurant, or cafeteria. Visit one of these and evaluate it by marking each item on the checklist as good, satisfactory, or poor. The checklist should include exterior design as well as interior design and planning.
2. Visit a food store, restaurant, or cafeteria and make a freehand, scale sketch of the interior arrangement. In class, revise this to improve the overall operation, and draw the revised floor plan.
3. Build a cardboard scale model of the interior of one of the commercial buildings discussed in this chapter (the building with the roof removed and possibly one or two walls removed). Build,

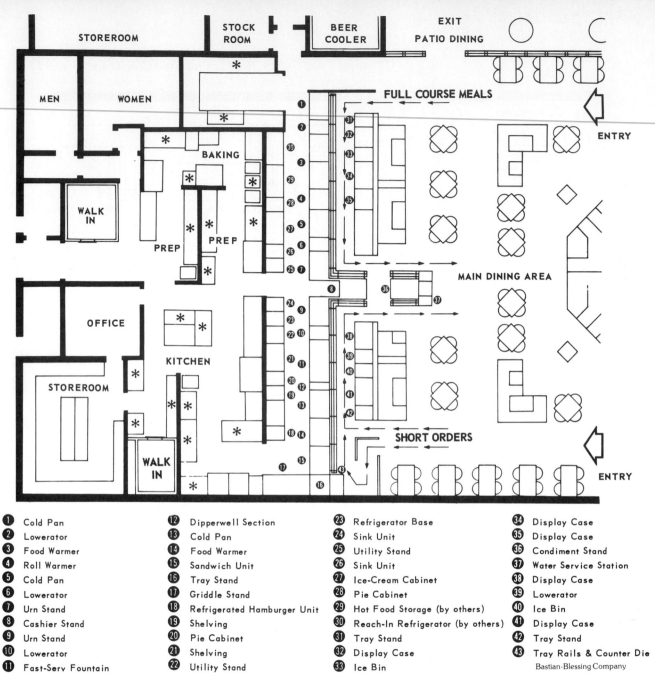

Fig. 19-16. Cafeteria serving layout for meals and short orders. From receiving, to preparation, to serving, this shows efficiency. There is one main serving line for full-course meals and one short-order line. Customers help themselves to cold drinks from shelves near serving counters. Total area is 7,509 sq. ft.; seating capacity is 250; 10 food serving employees suggested; 7 food preparation employees suggested; 7 table-cleaning and dishwashing employees suggested; average number of people served is 7 per line per minute.

1 Cold Pan	12 Dipperwell Section	23 Refrigerator Base	34 Display Case
2 Lowerator	13 Cold Pan	24 Sink Unit	35 Display Case
3 Food Warmer	14 Food Warmer	25 Utility Stand	36 Condiment Stand
4 Roll Warmer	15 Sandwich Unit	26 Sink Unit	37 Water Service Station
5 Cold Pan	16 Tray Stand	27 Ice-Cream Cabinet	38 Display Case
6 Lowerator	17 Griddle Stand	28 Pie Cabinet	39 Lowerator
7 Urn Stand	18 Refrigerated Hamburger Unit	29 Hot Food Storage (by others)	40 Ice Bin
8 Cashier Stand	19 Shelving	30 Reach-In Refrigerator (by others)	41 Display Case
9 Urn Stand	20 Pie Cabinet	31 Tray Stand	42 Tray Stand
10 Lowerator	21 Shelving	32 Display Case	43 Tray Rails & Counter Die
11 Fast-Serv Fountain	22 Utility Stand	33 Ice Bin	Bastian-Blessing Company

to scale, typical fixtures and equipment. Arrange these on the floor plan to develop the most satisfactory plan.

4. Design a complete food store, restaurant, or cafeteria. Work with several other class members as a team. For a food store, one team member could design the receiving and storage area, another the meat preparation and sales area, and others the general sales area.

For a restaurant or cafeteria, one could design the food receiving and storage area, another the food preparation area, and others could work on the serving and dining area. Both facilities need careful planning to provide adequate parking. The team leader will need to coordinate the efforts of the individual team members. It will be necessary to consult other chapters concerning structure and exterior design.

Chapter 20

Planning Motels and Parking Facilities

Providing accommodations for travelers has become a big business. Architects have devoted considerable study to the problem of providing motel units that offer the comfort guests demand and the attractiveness they expect. Providing parking facilities for shoppers has become a big business, too. The corner parking lot has given way to multistory parking structures costing thousands of dollars.

Both these facilities must be carefully planned to efficiently accommodate the largest number of customers at the lowest per-unit cost.

Motel Planning

The term **motel** has many different meanings. Generally speaking, it refers to rental units, usually for overnight guests. Most often, a motel is a long, low building. However, some motels are on two or more levels. See Fig. 20-1.

Motel/Motor Inn Journal

Fig. 20-1.   A courtyard view of multistory motel.

A trend is toward the use of **urban motor hotels.** These hotels are multistory buildings designed to receive casually dressed travelers. Automobiles are parked in the garage, located below the ground. Travelers can drive into the garage, register, and take an elevator directly to the floors where their rooms are located. This procedure is convenient, and it makes it unnecessary for travelers to make an unwanted appearance in the hotel lobby.

The motor hotel is frequently located downtown. It combines the informality of a motel with the downtown location of a hotel. See Fig. 20-2.

The planning of a motel involves the consideration of many factors. The location and size of the building site are of primary importance. Other factors to be considered include parking space, access roads, attractive appearance, and the designing of living units to meet the needs of the prospective guests. Demands upon motels vary, since some locations accommodate only overnight guests, and others often accommodate guests who spend a week or more.

Following is a discussion of factors important to consider in planning motels.

Site Utilization

Of primary importance is the selection of a building site in a location where the motel will prosper. This consideration is beyond the scope of this study. The designer, however, must make the most of the site selected. Some factors to consider in site utilization follow:

1. Guests must be shielded from noise.
2. Privacy for each living unit is important. The windows in a unit should not have to be constantly covered to maintain privacy.
3. Each living unit should have windows facing an attractive, private scene. A much used plan is to place the bath (with high windows) on the side facing the access road. From the other side of the unit, the occupant views a pleasant, landscaped area. See Fig. 20-3.

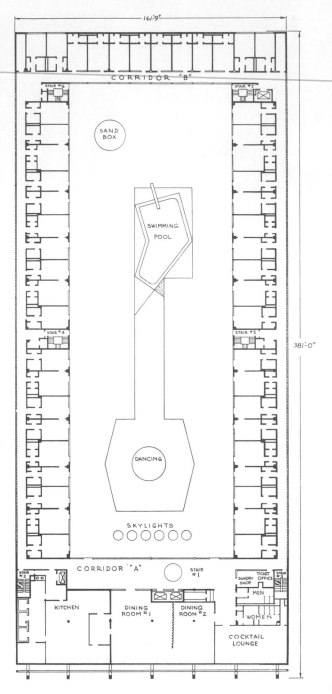

Fig. 20-2. Exterior view and floor plan of Sahara Motor Hotel, Cleveland, Ohio.

Note that all exterior walls are solid, and all units face on the garden park area. At the upper right, adjacent to stairway No. 7, is a private elevator for guests in swimsuits.

Rooms facing corridor **B** are for conferences, private meetings, and parties — all completely separated from guest rooms. Also, note that the dining rooms are separated by a folding partition. Areas may be combined into a large banquet room.

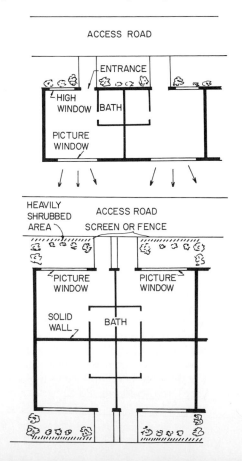

Fig. 20-3. Planning to provide a pleasant view.

Above: Good way to provide easy access to unit and yet offer a pleasant view. This plan is effective only for buildings one-unit wide.

Below: A way to provide a pleasant view for buildings two-units wide. Area between the unit and access road could be fenced and made into a private terrace. The parking area tends to be quite a distance from the unit.

Motel/Motor Inn Journal

Fig. 20-4. A motor inn designed to fit the climate and atmosphere of its geographic location. Notice the attractive sign. This facility is located in San Diego, California.

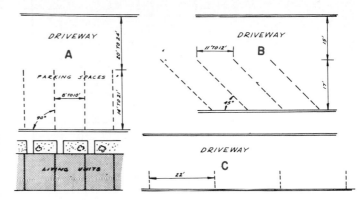

Fig. 20-5. Parking plans and space requirements.
   A. 90° parking allows space for one car per unit. Parking space width is about equal to width of one living unit.
   B. 45° parking requires more curb space, but less depth from curb to curb than perpendicular parking.
   C. Parallel parking space is about as long as two living units; therefore, it will not provide sufficient parking space.

4. A two-story building on a small site, reduces privacy and complicates parking and the moving of baggage; however, it does provide an economical structure to build and operate.
5. The entire site should be utilized as a part of the overall planning procedure. This would include such features as a swimming pool, children's play area, restaurant facilities, and a garden or park area.
6. The overall appearance should be one of unity and relaxation. It is necessary to put a pleasant "face" to the highway to encourage guests to stop for the night.
7. An attractive sign that invites the overnight guest is important. It should be located so it can be seen for some distance before the guest reaches the motel entrance. See Fig. 20-4.

Planning Factors Related to Parking

The parking of the guests' cars consumes considerable space and becomes quite a problem on a small site. Factors to consider when planning parking space follow:

8. Sufficient space should be provided for each car. This should be as near to the door of the unit as possible.
9. The parking space should be large enough to permit the doors on both sides of the car and the trunk to be opened.
10. Parking space should be provided for guests with second cars or boats or trailers.

11. Parking 45 degrees to the curb requires more curb space than a 90-degree parking angle, but the cars protrude less into the access road. Parking parallel to the curb takes considerable curb space and usually does not allow space for one car per unit. See Fig. 20-5.

Planning the Shape of the Motel

12. The shape of a motel is often dictated by the shape of the site. See Fig. 20-6. The U-shape and the L-shape have corners that usually are wasted space. These corners can be utilized for storage, furnaces, and water heaters.
13. The wasted corner in an L- or U-shaped building is frequently opened into a passageway, thus making all footage under roof rentable space. See Fig. 20-7.

Planning the Individual Unit

It is poor practice to design one good living unit and then repeat it until the required number of rental units are available. Motels should have a variety of rental units. These could vary from a small unit for one person to several units that could make a suite. The following factors should be considered when planning the individual units:

14. The unit should provide space for bathing, dressing, sitting, sleeping, and some storage. Cooking and dining facilities are provided in some units.

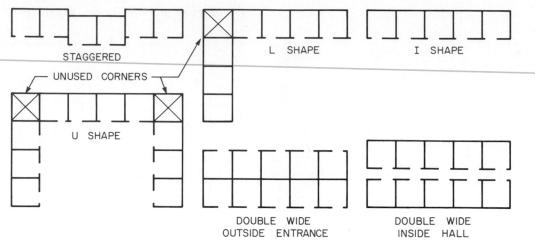

Fig. 20-6.   Typical motel building shapes.

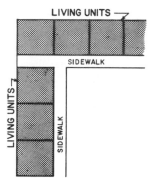

Fig. 20-7.   With the addition of a passageway, the corner becomes rentable, even though a car cannot be parked directly by the door to the unit.

Fig. 20-8.   A spacious room designed to accommodate four persons. It is large enough for guests who wish to stay for extended periods of time.

15. Some of the same area can be used for several activities. For example, a large portion of the sleeping area can be used as a part of the sitting area during the day. In some units, the beds make up into sofas and thus comprise the major portion of the sitting area.

16. Early in the planning, it is necessary to determine what furniture will be used in each unit. The area must be large enough to accommodate this furniture and still allow unrestricted movement about the room. (See Chapter 2, Planning the Individual Rooms.)

17. Two double beds are generally used instead of twin beds. See Fig. 20-8. Such an arrangement permits sleeping four persons. King size and queen size beds are popular but require more space.

18. Furniture should be arranged to facilitate cleaning the room and making the beds.

19. Inexpensive, wall-hung racks with hangers are adequate in most overnight rental units.

20. Each unit must have a private bath. However, proper design can allow several people to use the facilities at the same time. It is desirable to locate the water closet in a compartment by itself. See Fig. 20-9.

21. Units for rapid heating and cooling should be provided.

22. The heating or cooling for each unit should be controlled by the occupants of the unit. This calls for a flexible system. Chapter 12, Heating and Air Conditioning, discusses the various systems available.

Parking Facility Design

A basic decision that must be made when planning a parking facility is whether it will be **self-parking** or **attendant-operated.** Facilities utilizing

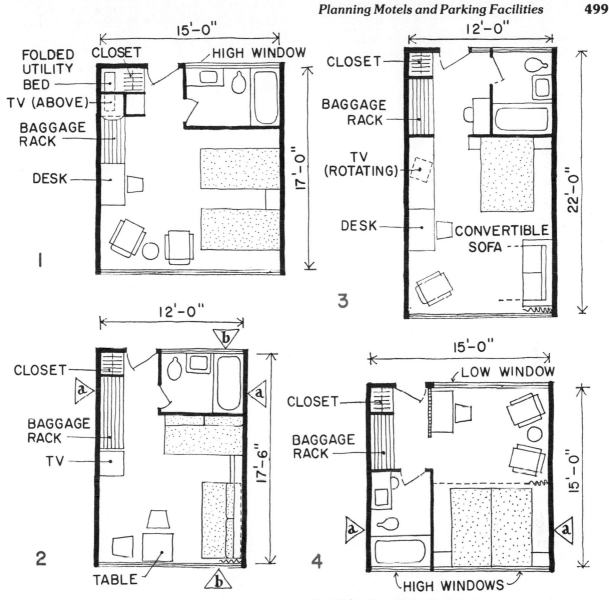

Reprinted from **Motels** by Baker and Funaro, Reinhold Publishing Company, by permission

1. The minimum depth in this plan is made up of the pair of twin beds with night table and clearances plus the bathroom which cannot be fitted into less than 5 feet. The minimum width is fixed by the bed length plus space for the chair grouping.

2. With sofa-beds set against the walls, the unit width can be reduced from 15 feet to 12 feet. The controlling factor is then a total of bathroom, plus passageway, plus closet and baggage rack depth.

3. A double bed and convertible sofa show clearly the difficulties of placing furniture of such a size in a rental unit of minimum dimensions. The convertible is more expensive than a sofa bed, which in turn is far more expensive than a folding utility bed. Yet most visitors will still consider the convertible to be no better than extra, emergency sleeping accommodations.

Fig. 20-9. Basic plans for various types of living units.

These almost-minimum room plans are arranged and furnished to fit the special requirements of one-night occupants (e.g. baggage racks instead of bureau drawers).

The bathroom shown is of conventional plan with a separate make-up table in the dressing area.

4. Unit width a is reduced by using a pair of twin beds conventionally placed at right angles to the wall, but recessed in an alcove. The beds are hidden during the day by a curtain drawn across to close off this alcove. At night, when this curtain is drawn back, the beds are in a spacious room with good through-ventilation. This plan, is well suited to a site plan where entrance and view are on the same side.

attendant parking require less space per car because the drivers are skilled. However, the increasing cost of labor seems to be offsetting this savings.

The two major objectives in designing a parking facility are efficient use of space and rapid handling of incoming and outgoing automobiles. All design considerations should implement these two objectives. Following are principles to consider when planning a parking facility.

Ramp Parking Facilities

1. The size of the auto stall must be decided. Since auto sizes vary a great deal, a size must be chosen that will accommodate the majority of cars. A study of popular lengths and widths currently on the road is necessary. Some experts recommend a stall 19'-0" (5791 mm) long by 9'-6" (2896 mm) wide.

2. The stall should be wide enough that auto doors can be opened with minimum of conflict with the next car. Also, the average driver must be able to park without difficulty.

3. The width of the aisles handling traffic flow on the parking deck varies with the width of the parking stall. The narrower the stall, the wider the aisle required. See Fig. 20-10.

4. Pedestrians generally use the auto traffic aisles to get to and from their parked autos. Special sidewalks take considerable space from the parking area and reduce the number of parking stalls available.

5. A decision must be made concerning the angle of the parking stall. A 60-degree or 90-degree stall angle is most often used. Convenience and speed of parking are the major considerations.

6. The angle of the parking stall influences the distance required from the curb to the rear edge of the parking stall as well as the number of stalls that can be located along a

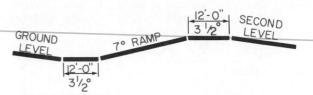

Fig. 20-11. Ramps require a lesser slope at the beginning and end to prevent autos from bumping the floor. This slope should be half the ramp angle.

wall. Design tables prepared by Baker and Funaro in their book **Parking** indicate that a 19'-0" (5791 mm) stall on a 30-degree angle requires 16'-6" (5029 mm) from curb to rear of stall, while a 90-degree stall requires 19'-0" (5791 mm). However 20 stalls 9'-6" (2896 mm) wide on a 60-degree angle require 382'-0" (116 434 mm) lineal wall length, while 90-degree parking accommodates the same number of stalls in 190'-0" (57 912 mm).

7. Consideration must be given to approaches from the street into the parking area and to fast and easy access to the street for cars leaving the facility.

8. Parking facilities on more than one level require ramps to transport autos to upper levels (or lower levels if it is an underground facility).

9. The minimum land requirement for a ramp-type facility is about 20,000 square feet (1860 m²).

10. The slope of the ramp is critical. If it is too steep, the front and rear of the auto will strike the pavement. A 7-degree slope is commonly accepted as maximum. The steeper the ramp, the less floor area consumed. See Fig. 20-11.

11. The floor-to-floor height of the multistory facility influences ramp length. The greater this height, the longer the ramp needed. A height of 8'-0" (2438 mm) is common. Figure 20-12 presents data for selected angles of ramps and floor-to-floor height.

Fig. 20-10.

Suggested Aisle Widths for Parking

Stall Width	60° PARKING		90° PARKING
	Stall Length to Curb (19' Stall)	Aisle Width[1]	Aisle Width[2]
8'-6"	20.7	18'-6"	25'-0"
9'-0"	21.0	18'-0"	24'-0"
9'-6"	21.2	18'-0"	24'-0"

[1]Aisle accommodates one-way traffic and provides room to back out of stall.
[2]Aisle accommodates two-way traffic and provides room to back into or out of stall.

Reproduced with permission from George Baker and Bruno Funaro, Parking, Reinhold Publishing Company, New York.

Fig. 20-12.

Ramp Length for Straight Ramps (In Feet)

Angle (Degrees)	Ramp Grade (Percent)	Floor-to-Floor Height		
		8'-0"	9'-0"	10'-0"
4	7	114	128	143
5	9	89	100	111
6	10	80	90	100
7	12	67	75	83
8	14	57	64	72

Reproduced with permission from George Baker and Bruno Funaro, Parking, Reinhold Publishing Company, New York

Fig. 20-14. A parking garage using helical ramps. The ramps are protected from the weather by plastic panels.

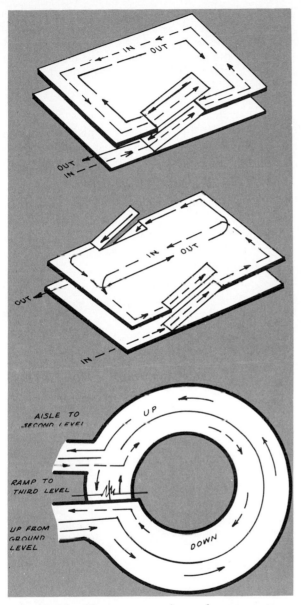

Fig. 20-13. Three commonly used ramp systems. There are many other ways of locating ramps.

13. In most cases, a single system of ramps in a self-parking garage is adequate to handle all traffic. See Fig. 20-13.

14. **Helical ramps** are used instead of straight ramps in some facilities. These can be one-way or two-way traffic aisles as shown in Fig. 20-13.

15. The diameter of the helical ramp is controlled by the turning radius of the auto. A two way helical ramp requires 42'-0" to 45'-0" (12 802 mm to 13 716 mm) **radius** to the outside edge of the outer lane. A one-way helical ramp could have a **diameter** of 30'-0" to 32'-0" (9144 mm to 9754 mm) to the outside edge of the lane. An exterior view of a helical ramp facility is shown in Fig. 20-14.

16. Attendant parking facilities with ramps require some means for allowing the employees to return rapidly to the entrance level. The autos are driven up the usual ramp system. Elevators or firepoles are commonly used to return attendants.

17. Provision for moving pedestrians from upper-level, self-service parking to the street is necessary. A two- or three-story facility usually includes stairways. Larger facilities have elevators or escalators. Even with an elevator, a stairway should be included for emergency use. Stairs should be located so the customer does not have to walk a great distance from the car to reach them.

18. Since parking customers do not arrive at an even pace, provision should be made for accommodating a number of waiting autos to prevent them from blocking traffic in the street.

12. Sufficient space is needed at the end of the ramp for the car to turn a corner. This is based upon the turning radius of the auto. A one-way aisle requires about 30'-0" to 32'-0" (9144 mm to 9754 mm), and a two-way aisle about 42'-0" to 45'-0" (12 802 mm to 13 716 mm).

Class Activities

1. Visit a motel. Try to envision problems facing a customer when checking in at the office, parking the car, carrying luggage into the room, and becoming comfortable in the room. List the good and poor planning features of the motel. Special consideration should be given to parking.

2. Visit a motel and determine its ability to withstand hard use. Prepare a report listing things that seem to have become unusually worn. Then explain what improvements you recommend by revising the plan or substituting a different material. For example, a motel owner may try to reduce costs by using plastic tile in the bath. Due to heavy use, it may be loose or may present a poor appearance. What would you do to remedy this problem?

3. Working with other students, plan a motel and adjoining food service. The work could be divided according to interest. Both units must attract the passing customer and entice him to stop. Refer to the previous chapter for food service planning.

4. Visit with attendants at a parking facility. Be familiar with the principles of planning such a facility so you can ask questions. Prepare a report listing the special problems they cite concerning the design and use of their installation.

5. Plan a two-story or higher parking facility to accommodate the automobiles driven daily to your school. This entails a careful count over a period of a week, plus selecting a suitable building site.

Chapter 21

Planning Shopping Centers

Today's shopper tends to purchase needed items from stores in close proximity to one another. A popular plan to meet these demands is the grouping of a variety of stores, so the customer can easily walk from one to another. Store owners have found their combined drawing power enables them to increase business, because a larger number of potential customers pass their doors.

Shopping Centers

A shopping center is a carefully integrated area of merchandising facilities. The types of merchandise to be offered must be ascertained before any other planning starts. This information must be obtained from a study of the shopping needs of the geographic region to be served. Then, a plan can be developed to meet all these needs. In this sense, therefore, a row of small stores located along a busy thoroughfare is not a true shopping center. Some of the principles to consider when planning a shopping center follow.

Preliminary Planning

1. Before a center is built, much preliminary planning must be done. Of first importance is an analysis of the potential marketing area. Usually, the center needs a well populated, residential area or a new, rapidly expanding area for maximum economic success.
2. The selection of the site is another preliminary planning factor. A careful study must be made of such features as the shape of the plot, orientation, access roads to the site, availability of utilities (sewers, water, gas, electricity), slope of the land and drainage possibilities, and soil conditions influencing construction costs.
3. Sites with unusual shapes cause planning difficulties and should be avoided.
4. A popular trend is the utilization of sloping sites, thus enabling a center to be planned with entrances on two levels.
5. The site selected should be large enough to accommodate a center that will meet the needs of the marketing area. If the area is growing, room must be allowed on the site for future expansion of the sales area and parking.
6. A study of subsurface soil conditions is strongly advised before a site is purchased. Unsuitable conditions may require excessively expensive foundations. This may make the site development cost prohibitive.
7. The building site is best if it is in one tract of land. Some sites with a minor street cutting through them can be developed, but special consideration of auto and pedestrian traffic is necessary. The street could be spanned by a pedestrian overpass. Sometimes stores are built on both sides of the street and are connected at several levels by passways built over the street. These enable customers to cross over the street without leaving the sales building.
8. Utilities should be at the site or close enough for economical connections.
9. As is true with any construction, zoning regulations should be studied to see that the site can be used as intended.
10. The site should have suitable access roads to handle the heavy flow of traffic. Heavily traveled, major thoroughfares are usually not the best choice; the additional traffic from the shopping center can cause severe difficulties.
11. After the site location is decided, the architect can begin preliminary building design and site layout drawings. As this is done, many other decisions need to be made, such as the types of stores to be in the center, their location in relation to other stores, the architectural styles to be used, and the locations of parking areas and malls.

503

Access Roads and Parking

12. The site layout must provide roads connecting the passing thoroughfares.

13. For large centers, traffic patterns should separate truck traffic from customer traffic. Very large centers may have subsurface shipping and receiving tunnels leading to basement receiving, shipping, and storage areas.

14. Large areas for customer parking are a must. Usually, these areas surround the shopping complex on all sides, thus reducing the distance a customer must walk. A distance of 300 to 400 feet (91.4 m to 121.9 m) from the parking area to the store is considered maximum.

15. The amount of parking space needed varies with the size and type of center. The most common plan for neighborhood centers is to allow 2 square feet of parking area for each square foot of building area (including upper floors, basements and mezzanine areas); for larger centers, allow 3 square feet of parking for each square foot of building.

    Each auto requires 300 square feet (27.9 m²) of area (including traffic aisles). If portions of the parking area are to be landscaped or if pedestrian walks are a part of the parking area, these need to be added to the allotment. Suggestions for arranging the parking area can be found in Chapter 20.

Fig. 21-1.   A shopping center with a mall. Notice the attractive plantings. Customers can move freely, with no fear of automobile traffic.

Owens-Corning Fiberglas Corporation

16. The parking area is best if it is all in one tract of land and not cut by cross-traffic streets.

17. If land is expensive, a multideck parking building can be used. Some stores provide parking area on the roof.

18. Traffic congestion can be relieved by the use of drive-in windows for services such as cleaning, laundry, and banking.

19. Raised pedestrian walks between the rows of parked autos generally are not used. Most people walk in the auto traffic aisles. The walks take considerable space, and the cost is prohibitive in areas where land is expensive.

20. Within the shopping center, pedestrian traffic is directed over walks, courts, and malls. See Fig. 21-1. These areas often have roofs for protection from sun and inclement weather.

21. The center should contain areas filled with plants, trees, and sculpture to create a pleasant atmosphere. Benches for rest and fountains for beauty are common.

Grouping the Stores

22. A basic principle in store grouping is that closely related stores, which can be of mutual benefit to one another, should be placed together. Success has been experienced by locating department stores and food stores next to each other, stores catering to women together, stores providing service and repairs together, and food stores and variety stores together. In general, doctors' offices are most satisfactorily located in a separate building. Second-floor locations seem to be inferior, except for businesses which have regular customers who park for short times, such as dance studios.

23. The center should be built around one or two large stores, such as department stores. The addition of small shops adds to the completeness of the center and gives it a balance of services. Frequently, restaurants and banks are included.

24. Large stores attract great numbers of customers. For this reason, they should be located so that customers, to reach them, must pass the many small shops in the shopping center. Their "pulling power" is used to provide a flow of customers to the smaller stores.

25. A small neighborhood center most often uses a food store, a junior department store, or a drugstore as a major customer attraction.

Other stores, such as cleaners, jewelers, and novelty shops, rely heavily upon the customers that the larger stores attract.

26. It is desirable to have more than one large store. These are placed in different sections of the center, so customers must pass the small stores.

27. If only one major store is planned, it is ordinarily located in the middle of the shopping complex, with the smaller stores grouped around it.

28. Store locations must be arranged to promote maximum customer traffic throughout the shopping center. The best way to accomplish this is to assign locations to each type of store.

29. It is desirable to have several stores of each kind in a center. This provides the opportunity for comparison shopping and keeps the store managers alert to competition.

30. Opinions vary on the best way to group stores. Some architects desire to group like stores together. For example, two food stores could be adjacent, with a bakery and/or other food services nearby. This has worked in actual practice, but it does cause some difficulties, such as parking congestion.

    Other designers prefer to separate the types of stores. Since most centers are a complex of buildings, these designers attempt to balance the types of stores in each building. For example, each building might include a jewelry store, clothing store, barber shop, and a shoe store.

31. Stores handling mainly impulse items are generally located on the portions of the mall with heavy traffic or on the main court. Stores handling demand items, such as appliance stores, can be located on the outskirts or outside walks.

32. In a residential area, it may be necessary to provide a buffer zone of trees and shrubs around the shopping center. This is done mainly to soften the impact of the buildings and parking areas.

Exterior Design

33. The architecture of the entire center should be integrated. The center should present a uniform facade, yet each store should exhibit an individual character.

34. Store signs, while they can reflect individuality, should be controlled by an overall plan to preserve the unity of the center.

35. Materials, colors, and other features should be used to individualize each store and to prevent it from being lost among the other stores.

36. Each store front should be designed to complement, rather than overshadow, its neighbor.

Illustrative Example

The following illustrates the application of some of the principles of planning a shopping center:

A plan for a neighborhood center is presented in Fig. 21-2, page 506. This center uses a variety store and a supermarket as its major customer attractions. A rear delivery area is provided, and a special road furnishes access to it. Of special interest is the area reserved for additional stores as the neighborhood grows. On the original plan, this area was not to be paved. Stores may also be expanded behind the center, if more floor area is needed.

The inclusion of a post office and bank, with drive-in and walk-up service, is another interesting feature.

The original architect's drawing and the plot plan are not shown, but they do include important information. The original architect's drawing includes the elevation data and the location and size of all utility lines. The location and point of connection of utilities to the service in the streets are also shown. The plot plan sheet includes utility details, such as sections through drainage pipes and plates for the parking area, sewer connections, and a catch basin. The buildings or groups of small shops are identified on the drawings with large letters. Each letter serves as a key to a complete set of working drawings. Since the center is too large to be drawn as one set, it is drawn as if it were eight buildings. The plot plan shows the relationship of these buildings with each other and with the site. The architects also developed an enlarged plan of the parking area.

A Look to the Future

The typical retailer today operates a small store as a part of a neighborhood shopping center or a large regional center. The store is one of a cluster of small shops built side-by-side. In the future, this typical shopping center may give way to the **one-stop shopping center.** This single store will handle a comprehensive variety of goods. Everything from drugs to furniture and food will be available in one store, thus offering the customer great convenience in shopping. Obviously, such a plan will greatly change the type of store facility that must be built.

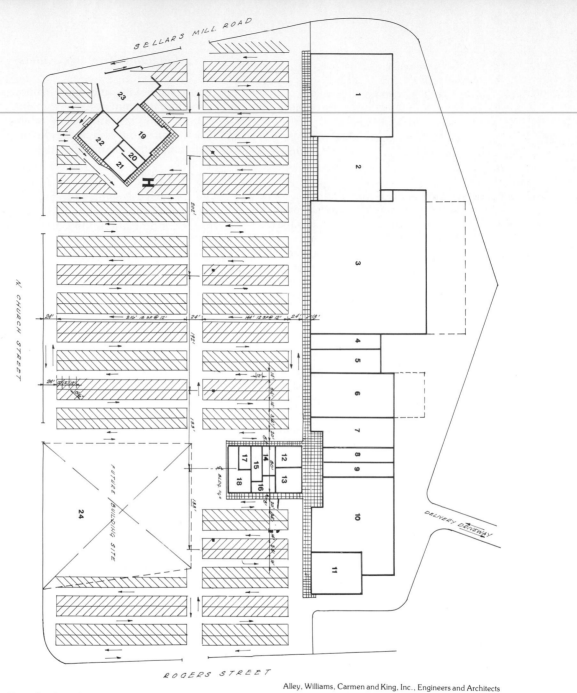

Alley, Williams, Carmen and King, Inc., Engineers and Architects

Fig. 21-2. An abbreviated plan for a small shopping center. This is the Cum Park Plaza Shopping Center, Burlington, North Carolina.

Key to Buildings:

1. Supermarket.
2. Hardware store.
3. Variety store.
4. Shoe store.
5. Ladies clothing.
6. Drugstore.
7. Clothing store.
8. Men's clothing.
9. Gift shop.
10. Dry goods.
11. Clothing store.
12. Dairy bar.
13. Family stamp store.
14. Barber shop.
15. Beauty shop.
16. Shoe store.
17. Bakery.
18. Specialty shop.
19. Post office.
20. Laundromat.
21. Dry cleaners.
22. Bank with drive-in window.
23. Post office truck maneuvering area.
24. Site for future building.

Fig. 21-3. A comprehensive, one-stop shopping center. In the future, this type of facility will replace the typical small shopping center.

Possible shell for the 300′ × 335′ one-stop shopping center with sales area on the second level. The ground floor of the building is open on three sides. The arches are decorative and serve to break up the immense front span. The skylight is not visible on the exterior of the building.

Cross section of the one-stop shopping center shows the relationship of the ground floor, main shopping floor, and balcony. There is sufficient space on the 80-foot-deep balcony for stores for men's and women's wear, children's apparel, shoes, jewelry, carpeting and drapery, TV-radio and hi-fi equipment, an optometrist's office, and a bank.

This floor plan, an overall 100,500 square feet, provides 68,750 square feet of sales area and 31,750 square feet of storage on the main floor. The balcony has 51,700 square feet for sales and 31,750 square feet for storage. Therefore, 65% of the total area is for merchandise display and sales, while 35% is for storage. Parking is on the ground level, beneath the store. The market totals 22,800 square feet.

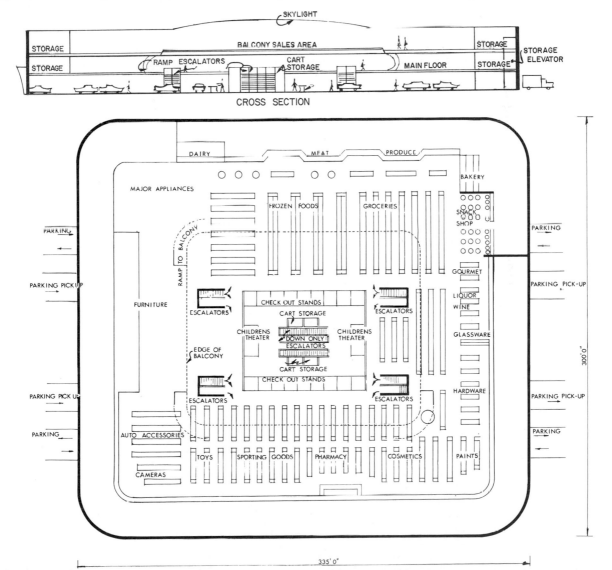

Beverly Willis, Industrial Designer

507

European stores are experimenting with one solution — the use of a shopping card system. All the items in the facility (except produce and meat) are for display only. The customer detaches a card from the rack containing the desired merchandise and presents this card at the check-out counter.

When entering the store, the shopper picks up a number card divided into four sections. One section is attached to produce selected and another to the selected meat. These purchases are sent by a conveyor to the stock room. The third section is presented to the check-out clerk, along with the collected merchandise cards. These are sent via a pneumatic tube to a basement stock room. Here the purchases are assembled, together with the meat and produce sent down. The customer gets into an auto, drives to a pick-up window, and presents the fourth section of the card as a claim check for all the purchases.

Such a system eliminates all problems of maintaining stock on the sales floor and removes the need for push carts, thereby reducing congestion. Figure 21-3 illustrates how this concept might work. Large multimillion dollar, regional centers may not follow this trend.

Class Activities

1. Visit a small shopping center.
    A. Ascertain if the parking area is adequate. What is the greatest distance any customer has to walk? Is this in keeping with planning suggestions? In your report, cite the data you collected to defend your decisions.
    B. Analyze the traffic pattern used to move autos from the street into the parking lot and provisions for autos to leave the area. Make suggestions for improvement. Illustrate these with a freehand drawing.
    C. Report on the exterior design of the center. Do the stores have an individual character, yet blend comfortably into the overall plan? Is the center attractive? Do signs detract?
    D. Examine provisions for receiving and shipping merchandise and removing trash. Prepare a report on problems existing here, and suggest means for improvement.
    E. Analyze the types of stores in the center. Do they meet the suggested principles for the selection of occupants? Consider the relationship of a store to its neighbor. Is the choice of adjoining stores sound? What changes would you suggest?
2. A possible class project would be to design a small shopping center. This is a large project and could involve every class member. A building site should be selected, and the grade ascertained. Then decisions must be made concerning the types of stores to include. Subgroups of students could begin to plan each store. Space allocations and final decisions concerning the overall size and shape of the center will need to be made. The completed plan could then be constructed as a three-dimensional model.

Chapter 22

Planning Churches

In the design of churches and synagogues, the needs and interests of many people are involved. Differences exist in the physical facilities needed by different religions. The building must provide areas for educational, recreational, and social activities, as well as the necessary worship centers. Since the building must be designed to house the functions of the church effectively, it should take on a form that is, in reality, a symbol to the world of the church and what it believes. This form must reflect the purpose of the church. Because the objectives vary from congregation to congregation within a denomination, they can be ascertained only by the minister and the congregation.

General Planning Considerations

1. When a church building is planned, the congregation must determine what activities are to take place and what type of facility is needed for them.

2. The architect must know the theology and history of the church and the meaning of the **liturgy** (rites). The design of the building reflects the spirit of the church. Solutions for different denominations are shown in Figs. 22-1 through 22-3.

3. A careful study of the site must be made before the building design is started.

California Redwood Association, Charles Warren Callister, Designer

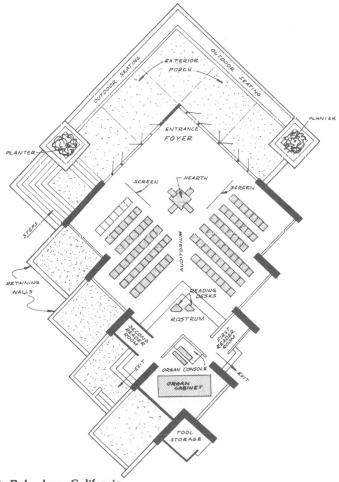

Fig. 22-1. First Church of Christ, Scientist, Belvedere, California.
Notice the special requirements for this denomination and the terms used to identify the parts of the church. The narthex is a foyer, the nave is an auditorium, and the chancel is a rostrum. Special provision must be made for the first and second readers.

Durham, Anderson and Freed, Architects

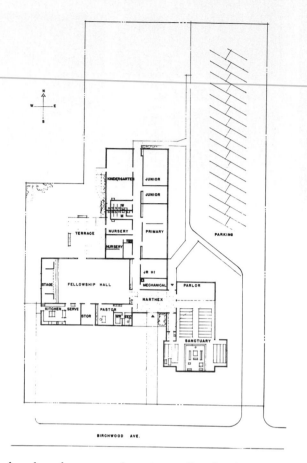

Fig. 22-2. A small Presbyterian Church in Bellingham, Washington.

The interior is lighted with cone-shaped spotlights and a skylight above the altar. The structural members are laminated wood arches, and the exterior is a blending of native materials.

The church office is compact, but contains the desirable features of an outer secretary's office, enclosed workroom, and a private office for the pastor. The kitchen and social hall are adjoining, and a storage room is readily available.

The narthex serves the fellowship hall and the sanctuary with equal effectiveness. The parlor serves to accommodate overflow congregations. The mechanical room is centrally located in order to serve all three wings of the building.

4. Exterior gardens and plantings should be in harmony with the building.
5. Churches and church school buildings should have year-round air conditioning.
6. Churches have a special heating problem. Some portions need to be kept at a comfortable temperature for daily activities, while other areas need not be fully heated, except for special times during the week. A zoned heating system is recommended.
7. Ramps instead of steps are of value wherever they can be used.
8. Lack of funds to complete the entire church plant occasionally requires that some rooms be used for several purposes. The sanctuary frequently serves also as a social hall and sometimes houses church school classes. Such measures should be only temporary.
9. The church office is a busy place. The principles of planning offices (Chapter 18) must be carefully observed.
10. The clergy should have private offices.
11. The church and clergy offices should be on the ground floor, near an exterior entrance.
12. Many churches have a social hall for congregational meetings and dinners. Frequently, this room can be divided into small rooms for use by church school classes.
13. The kitchen is next to the social hall. Most church kitchens are designed to prepare food and provide table service. (See Chapter 19.)
14. Storage is needed for folding tables and chairs.

Fig. 22-3. A Catholic Church in Alhambra, CA.
This building has a concrete roof surface over a plywood diaphragm. The reredos (partition wall behind the altar) is flagcrete.

J. Earl Trudeau, AIA, Architect

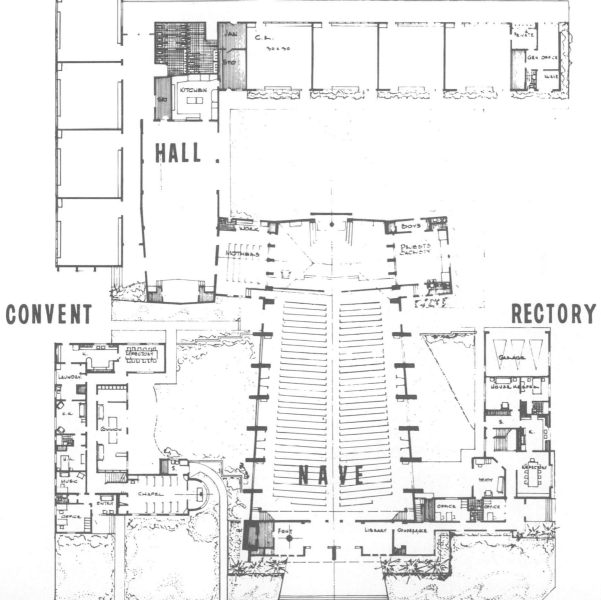

SCHOOL

HALL

CONVENT

RECTORY

NAVE

The Narthex

The **narthex** refers to the vestibule, foyer, or entrance hall of a church. The main entrance to the church opens into this area. It is essential to any church plan. In Fig. 22-4, a narthex is shown as it would be seen from the front entrance.

1. The narthex should contain 2 square feet (0.2 m²) of floor area for every seat in the nave.
2. Since this area serves as a gathering place before and after services, it should be separated from the nave with a sound-deadening wall and should have ceiling treatment to reduce noise.
3. The narthex should have special racks for coats. One foot (305 mm) of hanging space for every fifteen seats in the nave is recommended.
4. Rest rooms should be located conveniently off the narthex; however, their doors should not open directly off this area.
5. A sufficient number of outside doors should open off the narthex to enable rapid exit.
6. To reduce chilling drafts, a double set of outside doors can be used. This is not necessary in warm climates.
7. The doors into the church and nave should be wide enough to allow passage of pallbearers carrying a casket. Minimum width is 5½ feet (1676 mm).

The Nave

The **nave** is that portion of the church used for seating the congregation during the worship service. It adjoins the narthex and is entered from it. A nave in a contemporary church is shown in Fig. 22-5.

Pennsylvania Wire Glass Company, Carroll, Grisdale and Van Alen, Architects

Fig. 22-5.   A view of the nave and chancel of the Methodist Church, Linwood Heights, Pennsylvania.

1. An overall plan of traffic flow must be developed for the nave. A center aisle is almost universally used. It should be at least 5 feet (1524 mm) wide, since it is used for processionals and as a main exit upon conclusion of the service. Other aisles should be at least 4 feet (1219 mm) wide, to accommodate two persons side-by-side.
2. A major cross aisle should be located between the chancel and the front row of pews. A chancel is shown in Fig. 22-8.
3. The Lutheran Church suggests planning nave seating capacity for 50 percent of membership if this membership is over 400 persons. If under 400 but more than 175, seating capacity should be planned for 65 to 75 percent.
4. Building codes usually limit seating to fourteen persons per pew if both ends open onto aisles, or seven persons per pew if one end is against a wall and the other opens onto an aisle.
5. The average person occupies 20 inches (508 mm) of pew seat. If finances permit, 22 to 24 inches (559 mm to 610 mm) is a better allowance. Rows should be spaced 36 inches (914 mm) apart, from the back of one pew to the back of that in the next row. See Fig. 22-6 for suggested pew dimensions.

California Redwood Association, Mario Corbett, Architect

Fig. 22-4.   A view from the narthex into the nave of the Hope Lutheran Church, Colma, California.

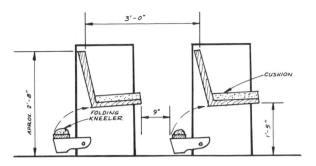

Fig. 22-6. Suggested pew dimensions.

6. The location of exits, columns, heat sources, and various obstructions must be considered when planning a seating layout.
7. The floor of the nave should be level.
8. It is desirable to design the structure so the nave is free of columns.
9. The nave should be shielded from all noises from other rooms or outside the building.
10. The communion rail, when used, is placed at the front of the nave and on the same floor level. The kneeling step in front is approximately 6 inches (152 mm) high, and the rail is about 24 inches (610 mm) high.

    It should accommodate at least 10 percent of the seating capacity of the nave. The center section should be removable to provide access to the chancel during the worship service. A minimum opening of 4 feet (1219 mm) is needed.
11. The chancel rail separates the chancel from the nave. It is on the floor level of the chancel. The rail may be solid or pierced and is usually 36 inches (914 mm) high.
12. At least 3 feet (914 mm) of open floor space should separate the communion rail and the chancel rail. Clergy must walk here during communion services.
13. While most churches in the Western World are rectangular in plan, some are beginning to follow the Eastern World plan of a round church with the altar in the center. This is commonly referred to as a **church-in-the-round.** The altar rail surrounds the altar on all sides. The cross is suspended from the ceiling above the altar. The pulpit is on one side, and the baptismal font is placed on the other. This, in effect, creates a religious amphitheater, with the congregation gathered around the altar. To enable everyone to see the raised altar, the nave floor must be sloped or stepped in tiers. A church-in-the-round is shown in Fig. 22-7.

A. The mass and form of the structure plus the textures of the materials selected combine to form this striking church exterior. It was designed in a round shape to accommodate the interior seating and altar.

B. The altar and surrounding communion rail support the theology of the Lord's Supper—the Lord's people gathered around the Lord's Table. Notice the specially designed, circular pews and the accent lighting of the pulpit and lectern. The cross is hung above the altar.

C. Of special interest is the pulpit shown on the left. Notice it is raised and of considerable prominence as compared to the lectern. The choir, organ, and organ pipes are strikingly accommodated on a balcony. The ceiling and walls of the choir space are plaster so they will reflect the music. General illumination is by spotlights in the ceiling. High-intensity illumination over the altar focuses attention upon it, and the altar is raised so that it is visible to the entire congregation.

Fig. 22-7. St. Luke's Episcopal Church, Dallas, Texas.

The Chancel

The **chancel** is reserved for the clergy to use to lead the worship service. See Fig. 22-8.

1. The floor of the chancel should be elevated above the floor of the nave.
2. The chancel should be designed to accommodate the ritual of the church.
3. There should be no physical separation between the chancel and nave. The chancel, which contains the altar, is the dominant feature of the church and the focal point of attention.
4. The choir and organ are sometimes placed in the chancel.
5. The **predella** is a part of the chancel. It is a platform, usually one step above the chancel floor, on which the altar is placed.

Fig. 22-8. A chancel. Notice that the chancel floor is raised two steps above the nave floor. The altar is on the predella which is three steps above the chancel floor, with an open altar rail of simple design. The predella allows ample room for the clergy to walk around the altar to perform the ritual. Behind the altar, rising to the ceiling, is a magnificent reredos.

6. The predella should be wide enough to permit the clergy to walk while administering holy rites. Usually 36 inches (914 mm) is necessary.

The Sacristy

Some denominations require two sacristies — a working sacristy and a clergy's sacristy. Non-liturgical churches usually do not have a clergy's sacristy. The **working sacristy** is a room used for the storage, preparation, and care of vestments, sacred utensils, paraments, and other items used in the chancel. The **clergy's sacristy** is a private room devoted to personal meditation. The clergy also don their robes and vestments here.

1. Storage cabinets and drawers are needed to store linens, vestments, and robes.
2. Special cabinets for storing communion trays, the chalice, missal stand, flower vases, and candlesticks must be designed.
3. A sink and counter space are needed for the preparation of floral arrangements.
4. Direct access to the chancel is necessary.
5. A substantial safe is necessary for the storage of silver utensils.
6. The clergy's sacristy should connect with the minister's study.

The Chapel

Consideration should be given to the inclusion of a **chapel**. Large churches frequently have a small worship area for special services or private occasions, such as weddings.

1. The planning of a chapel requires a thorough study of the types of services that are to occur there in order to accommodate them.
2. The maximum seating capacity for a chapel is usually 75 to 100 persons. Most are smaller.
3. The chapel should be as accessible as the main sanctuary for weddings, baptismals, and other services. It should be near the main entrance so the public may enter during the week for prayer and meditations.
4. Choir space is usually omitted.
5. A small electric organ is needed.
6. The chapel in Fig. 22-9 was designed as an independent structure, rather than as a part of a large church.

Choir and Organ Placement

The church choir serves three basic functions: to lead congregational singing, to add to the inspiration of the worship service, and to provide appropriate responses and transitions in the service.

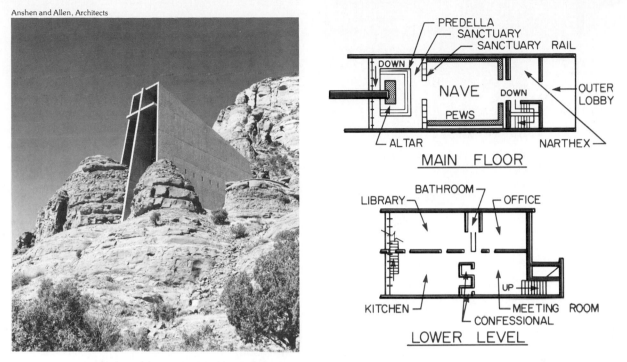

Anshen and Allen, Architects

PREDELLA
SANCTUARY
SANCTUARY RAIL

DOWN

NAVE   DOWN

PEWS

OUTER
LOBBY

ALTAR

NARTHEX

MAIN FLOOR

LIBRARY   BATHROOM   OFFICE

KITCHEN   MEETING ROOM
CONFESSIONAL

UP

LOWER LEVEL

Fig. 22-9.   Chapel of the Holy Cross, Sedona, Arizona.
   This Roman Catholic Church is situated in a magnificent setting, and the structure is well suited to the site. Notice the use of the architectural principles of mass, scale, surface, and texture of materials. The building shell is reinforced concrete, 12 inches thick, and forms the interior and exterior walls. Glare is reduced by using smoke-colored glass. The congregation is seated in pews along the side walls. Individual chairs are used to seat the remainder of the congregation.

1. The placement of the choir is limited to the chancel, a rear balcony, or an alcove off the side of the nave.
2. The choir should be closely grouped, and the organ should be located so that the organist has a close relationship with the choir.
3. Usually it is best to elevate the choir above the nave floor.
4. The choir members should be able to see and hear the clergy and to participate in the service.
5. The choir needs a room (called a **vestry**) in which members can put on their robes, secure their music and rehearse.
6. The size of the choir should be decided.

Furnishings

Any furnishings in the chancel that play no part in the practice or spirit of the liturgy should be discarded. Secular symbols should not be allowed to share honors with the symbols of God's presence in His church.

The final and lasting impression given by the interior of a church is mainly made by the furniture and accessories. The selection of pews, altar,

pulpit, communion table, sculpture, special windows, and other art pieces is greatly significant in the total picture. A structurally sound church building with a fine floor plan is incomplete if the furniture and accessories are poorly selected.

Lighting

Church lighting should enhance the beauty of the architecture and preserve the spiritual values of the ecclesiastical atmosphere. The lighting plan should be developed as the building plans are made. Lighting can be used to create different moods and to focus attention on important features. See Fig. 22-7.

1. The nave needs well distributed, low-intensity, general illumination. This should be variable from darkness to reading level by means of a dimmer system.
2. The chancel should receive 20 to 30 foot-candles (215 to 320 cd/m^2) of light.
3. The altar or communion table and the pulpit should receive 50 footcandles (540 cd/m^2) of light.

Fig. 22-10. Chapel of Manhattenville College, Purchase, New York.

Stained-glass windows are designed to complement the interior and exterior architecture. Notice the altar is lighted from the side by a window.

4. Two systems of indirect lighting are in common use. In the first, floodlights are concealed in coves along the upper part of the wall. The light is directed to the ceiling, where it is reflected to the floor. This requires powerful light sources, since much light is absorbed by the ceiling. In the second system, the light is aimed directly toward the surface it is to illuminate. This is done by putting the bulb inside a metal tube, which concentrates the light and sends it in one direction. As the light leaves the tube, it spreads into a cone shape.

5. Fluorescent lighting generally is unsatisfactory for use in the nave.

6. The lighting controls should be as simple as possible, since inexperienced persons most often operate them.

7. Colored light is used in church lighting. The changes should be subtle, rather than dramatic as in a theatre.

8. The altar, pulpit, lectern, choir, cross, and other specific areas need supplementary lighting.

9. The pulpit and lectern should be lighted from two sources, in order to reduce shadows and reflections.

10. In the past, church windows were formally and evenly spaced. However, windows in a contemporary church are placed wherever they will contribute to the effect desired. They may be concentrated in one wall or a portion of a wall. For example, windows may be located only in the chancel, to provide a dramatic lighting effect on the altar. See Fig. 22-10.

11. Careful use of natural light by means of precisely placed windows and skylights is a fine source of dramatic illumination. A skylight could be so located as to light an altar. See Fig. 22-5.

12. A congregation should never be seated so they face windows directly. The eyestrain caused by looking into such a bright source of light is considerable. An alternate plan is to place windows at the side of the chancel, thus removing them from the direct view of the congregation. See Fig. 22-10.

13. The halls and stairs of the church building require adequate illumination for night use. Classrooms, offices, and meeting rooms need a high-level, general illumination.

14. The architect can deliberately use lighting to cast shadows, creating a dramatic scene. All other shadows should be eliminated.

15. Exterior lighting that points up architectural features is necessary. See Fig. 22-11.

16. Exterior steps, sidewalks, and parking areas should be fully illuminated, and the church name should be visible at night.

Fig. 22-11. Dramatic night lighting enables St. Luke's Episcopal Church, Dallas, Texas, to communicate its character to passersby.

Art in Church Buildings

The integration of art in religious buildings is of great importance, and requires the services of experts. Generally, a portion of the overall budget should be designated for utilization of the visual arts. It is recommended that the architect be given overall control of the arts (sculpture, painting, murals, stained-glass windows) and that artists are employed to help plan these. See Figs. 22-10 and 22-12. This should be done as the structure is designed. Each art medium should serve a specific, preplanned purpose and should directly contribute to the **total** architectural picture of the church building.

The art should meet the requirements of the modern congregation. People no longer live as they did in past centuries. The Gothic and Romanesque Periods are past. Art should reflect the new age and should interpret the spirit of the present. It should contribute to the atmosphere of worship.

California Redwood Association, Shreve, Lamb and Harmon, Associates, Architects

Fig. 22-13. The Church of the Savior, Paramus, New Jersey.

Rambusch Decorating Company

Fig. 22-12. Sculpture design complements the exterior architecture of the Church of the Transfiguration, Collingswood, New Jersey.

Designing the Exterior

The styling of a church exterior will and should vary considerably from one congregation to another. The most frequent problem faced by the congregation is whether to copy an established style of the past or to design a building that reflects the purpose of the church today. The congregation and the architect should establish the needs of the church and plan the use of space to serve these needs. Once this is resolved in an interior plan, then the exterior should be considered. The exterior grows out of the plan and shelters it. It should be in tune with the times. The designer works with form, proportion, balance, scale, texture, color, light, and shade to achieve a structure that is an expression of the religious beliefs of the congregation. The building should have an effect upon the worshippers and passersby. It should stir the imagination and emotions and should reflect the purpose for which the structure was planned.

A church should reflect strength and courage. It should present a welcome atmosphere, yet be sincere, simple, and restrained. It should draw people inside by symbolizing harmony, peace, and dignity. See Fig. 22-13.

Build Your Vocabulary

Following are terms that you should understand and use as a part of your working vocabulary. Write a brief explanation of each term.

chancel	nave
chapel	predella
lectern	sacristy
liturgy	vestry
narthex	

Class Activities

1. Visit the church of your choice, and talk with the minister about the meaning of the liturgy. Report on the space and items needed in the various church ceremonies.
2. Plan a small church for the denomination with which you are most familiar. This project would be especially valuable if a congregation could be found which actually needed a new building. Ascertain the congregation size, and design the building to suit the needs and projected growth of the church.
3. Plan a small chapel. Many institutions — such as large churches, colleges, and hospitals — have a chapel available. If it is on a campus, it is usually a separate building. If it is a part of a large hospital, it is usually a room on the first floor. If possible, visit several chapels to obtain details.
4. Visit a church office. Record all the activities that must take place there. Then design a facility to handle these tasks. Consult Chapter 18 for principles of office planning.
5. The design of church furniture is an art. Try designing a church pew, chancel rail, altar, pulpit, and matching lectern.
6. Visit the ministers of several churches in which choirs are seated in different places. Find out what the ministers think are the advantages and disadvantages of each choir location.
7. Examine the exteriors of several churches. Select the church you feel has the most outstanding exterior styling. Write a paper giving the reasons you selected this church.
8. Make a survey of art inside and outside of church buildings. Report the type of art work found, its location in the building or on the site, and the subject matter represented.

Chapter 23

Construction Techniques for Commercial Buildings

The structural systems discussed in this chapter are based upon the use of stock members. No attempt is made to discuss the designing of structural members. This is an engineering function and is beyond the scope of this text. Small commercial buildings can be designed and built satisfactorily with stock members. Multistory buildings present engineering problems that are not considered. All discussion pertains to one-story structures. However, some illustrations show multistory applications of structural members.

The systems presented in this chapter are for light construction. If heavy or concentrated loads are involved or if the building will be subject to heavy impact loads or vibration of machinery, special design considerations are necessary. All of the prefabricated systems presented should be designed and built under the supervision of a competent architect or design engineer.

They are presented here for student use to provide a means of designing a structural system to house small commercial establishments. The student should realize that each business and building presents special design problems that can be solved only by professional architects and engineers.

All measurements are U.S. Customary standards or directly related to these standards. Therefore, no metric conversions will be presented. If conversions are desired, use the conversion tables presented in Chapter 5.

The method for ascertaining the load on a structural member is explained in Chapter 9. This process is not explained in this chapter.

Footings

The design of footings has been discussed in Chapter 8. The same principles apply to small commercial construction.

Pile Foundations

When a soil investigation reveals that it would be unwise to pour a foundation upon the soil, the building should be supported upon a pile foundation. The type of piling, its depth, size, and spacing vary with the building and soil characteristics. Pile types vary considerably, but fall generally into three groups — steel, concrete, and timber or some combination of these. While the design of a pile foundation is beyond the scope of this text, the following comments are informative.

Soil Investigations

Careful study of the character and physical properties of the rock or soil upon which a building is to be built is of great importance if the structure is to remain stable and structural damage avoided. This is true for light buildings, such as a service station, as well as large multistory buildings. In a soil investigation, specialists bore into the soil in the area where a building is to be constructed. The soil samples brought up provide the engineer with the information necessary to design the foundations. A structure can be properly designed above the ground, but be damaged due to the yielding of the soil supporting the structure. Damage is especially great when soil conditions permit one portion of the structure to settle more than another.

Many companies conduct soil investigations before they purchase property upon which to erect a building. This data gives the engineer an idea of the foundation costs.

The most frequently used procedure of foundation test borings is the **Gow** method. In this method, a small-diameter borehole is sunk into the ground. At each change of soil formation, or at more frequent intervals when desired, a sample tube is retrieved; it is opened so that the soil can be examined, classified and sealed in a sample jar.

Soiltest, Inc.

Fig. 23-1. Soil samples are taken at the building site and then tested to help determine the type of foundation that will be needed.

Raymond International, Inc.

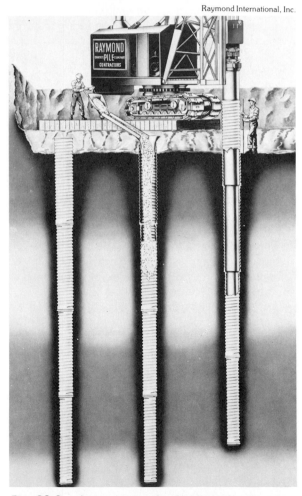

Fig. 23-2. A step-taper pile. On the right, the pile is being driven with a steel core, and in the center, it is being filled with concrete. At the left is a finished pile.

See Fig. 23-1. Where hard strata or ledge rock must be penetrated, a rotary or diamond drill is used.

Types of Piles

Step-taper piles consist of a hollow steel core that is in sections, each one inch larger in diameter than the next. The step-taper shells are helically corrugated to provide strength against ground pressure, and they are available in a variety of steel gauges and lengths. The joints between the sections are screw-connected. The pile is closed at the point by a steel plate, welded to the bottom driving ring. Usually, the pile is driven with a rigid steel core. After it is in place, it is inspected and then filled with concrete. See Fig. 23-2.

Also in use is a **concrete pile,** employing a concrete shell rather than one of steel. See Fig. 23-3. The concrete shells are threaded on a steel mandrel. The bottom section is closed by a solid concrete shoe, and joints between the sections are sealed by recessed steel bands.

After being driven the required depth, the mandrel is withdrawn and the shell is filled with concrete. During the driving, the mandrel bears directly on the concrete closure shoe, and the driving head engages the upper end of the concrete shell.

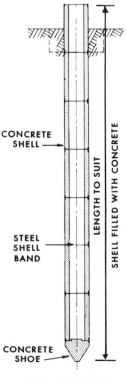

Fig. 23-3. Concrete-shell pile. The steel bands are used to seal the joints between the sections.

CONCRETE SHELL

STEEL SHELL BAND

CONCRETE SHOE

LENGTH TO SUIT

SHELL FILLED WITH CONCRETE

Raymond International, Inc.

Cast-in-place piles are also used extensively. In simple terms, a hole is drilled into the soil, and this is filled with concrete. One variation of this procedure is to insert a pipe into the hole and pump concrete grout through it, thus filling the hole from the bottom. See Fig. 23-4. This prevents dirt from falling into the hole, as frequently happens when concrete is poured from the top. These piles can be accurately cast close together. See Fig. 23-5.

Precast Concrete Structural Systems for Large Spans and Heavy Loads

Available from a number of companies are complete systems of precast concrete structural members. These are beams and girders, joists, roof and floor decking, and bents. See Fig. 23-6. Bents are frameworks which carry both lateral and vertical loads.

Concrete members are usually cast in a factory and transported to the building site where they are rapidly erected. They provide a fireproof system that requires little maintenance.

Intrusion Prepakt, Inc.

Fig. 23-4. Cast-in-place pile. The hole is being filled with concrete grout from a pressure pumping station. Notice the auger used to bore the piling hole.

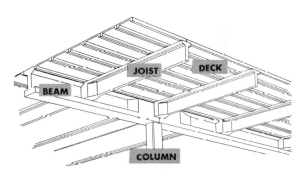
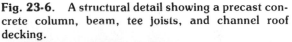

Fig. 23-6. A structural detail showing a precast concrete column, beam, tee joists, and channel roof decking.

Intrusion Prepakt, Inc.

Fig. 23-5. Cast-in-place concrete piles exposed by excavation for an addition. Notice the accurate placement. A close-up of the concrete piles is on the right.

The designer should carefully observe the recommended spans and the maximum allowable loads permitted on members of various sizes. The following descriptions illustrate those systems available from one company. Other systems with different capacities are available.

Precast Columns

Precast concrete columns are available in a wide variety of cross-sectional dimensions and lengths. They have a steel base plate welded to the main reinforcing members, which are cast integrally with the column. Holes are drilled in the base plate to receive anchor bolts that are cast into the foundation.

Construction crews erect these columns by raising them into position, lowering them to the foundation, placing the anchor bolts through the base plate holes, and tightening the anchor-bolt nuts. Below the base plate, 1 1/4 inches of nonshrinking grout is placed. Figure 23-7 gives stock column data.

Precast and Prestressed Beams and Girders

Precast and prestressed concrete beams and girders are available in unlimited lengths and sizes.

Fig. 23-7.

Design Data for Precast Concrete Columns*
Loads below heavy line are short column loads.

Column Size	Weight Per Foot In #	Round Columns Unsupported Column Height				
		8'-0"	9'-0"	10'-0"	11'-0"	12'-0"
6"	29	24.0	22.4	20.6		
7"	40	32.9	31.0	29.6	27.0	
8"	52	47.0	44.5	42.5	40.0	38.0
9"	66	63.5	61.0	58.5	56.7	53.0
10"	82	82.6	80.5	77.5	74.0	71.5
11"	99	92.0	92.0	89.0	86.5	83.0
12"	118	107.0	107.0	107.0	104.0	100.0

Column Size	Weight Per Foot In #	Square Columns Unsupported Column Height				
		8'-0"	9'-0"	10'-0"	11'-0"	12'-0"
6" × 6"	38	29.0	26.8	24.6		
7" × 7"	51	43.5	41.0	39.2	35.8	
8" × 8"	67	63.0	60.0	57.0	53.5	51.0
9" × 9"	84	83.0	79.0	76.0	73.0	69.0
10" × 10"	104	105.0	103.0	99.0	94.0	90.0
11" × 11"	126	116.5	116.5	112.0	109.0	104.0
12" × 12"	150	134.0	134.0	134.0	130.0	125.0

*Safe axial load is in kips.   Shlagro Steel Products Corporation

Fig. 23-8.

Safe Superimposed Loads for Prestressed Beams
(Lbs. per Lineal Foot)

Beam No.	Clear Span in Feet				
	15	20	25	30	35
[1]20-10	1940	1090	700		
20-15	2900	1630	1045	725	
20-20	3870	2175	1395	965	710
20-25	4840	2720	1740	1205	890
20-28	5410	3040	1950	1350	995
[2]24-10		1300	835	575	
24-15		1950	1250	870	640
24-20		2600	1665	1155	850
24-25		3250	2085	1445	1065
24-30		3900	2500	1735	1280
24-36		4680	3000	2080	1530

BEAM NO. 20-10 = 20" DEEP WITH 10 WIRES
$f_c = .4 f'_c$, INITIAL $f_s = .7 f'_s$, FINAL $f_s = .85$

$$\text{INITIAL } f_s, M = \frac{WL^2}{8}$$

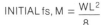

[1]20 Series Beams = 20" Deep × 12" Wide, 145 # Lineal Foot
[2]24 Series Beams = 24" Deep × 13" Wide, 190 # Lineal Foot

Southern Cast Stone Company, Inc.

They can be designed to accommodate all types of loadings. Usually, they are cast in an **I** or rectangular shape. See Figs. 23-8 and 23-9. A precast, prestressed beam requires a depth of only one-half to two-thirds that of ordinary concrete construction, thus allowing extra headroom and the spanning of longer distances. In a prestressed beam, the steel reinforcing members inside the beam are drawn to a predetermined tension before the concrete unit is cast around them.

Figure 23-10 illustrates why prestressing increases the load-carrying capacity of a member. Any beam (when loaded) tends to sag in the center, thus stretching the material at the bottom. This causes cracking and failure of the underside of the member. In a prestressed beam, the steel reinforcing is placed under tension as the concrete is poured. After the concrete sets, the tension is removed, thus compressing the concrete at the bottom of the beam. When a load is applied, the downward force tends to further compress the material. Since concrete resists compression, the beam can support the load without cracking.

The only way to overcome the tendency of an ordinary beam to crack is to increase its size. Therefore, a small prestressed beam can carry the same load as a large beam that is not prestressed.

Fig. 23-9.

Safe Loads for Rectangular Concrete Beams*

									Allowable Load Per Foot of Beam Kips
PRECAST CONCRETE f'c = 3000 #/□″									
Span	Width	Depth	Area	Weight Per Foot	Top Reinforcing	Bottom Reinforcing	Stirrups		
Feet	Inches	Inches	Square Inches	Pounds	Size	Size	Size	Spacing	
20	13	22½	292.5	320	#5	#8	#3	12″	1.71
	14	26	364.0	380	#5	#9	#3	12″	2.72
22	13	28	364.0	380	#5	#8	#3	12″	1.89
	17	28	476.0	496	#5	#9	#4	12″	3.11
25	17	28	476.0	496	#5	#9	#4	12″	1.76
	20	34	680.0	710	#5	#9	#4	12″	3.76

Shlagro Steel Products Corporation

*Safe load in kips.
Beams are designed for simple support.
Top reinforcing used for possible moment developed by handling.

LOAD

STEEL BAR

BEAM UNLOADED

CRACKS

BEAM LOADED CONCRETE IN TENSION

NOT PRESTRESSED

LOAD

BEAM UNLOADED

BEAM LOADED CONCRETE IN COMPRESSION

PRESTRESSED

Fig. 23-10. Visualization of load on a plain and a prestressed concrete beam.

Fig. 23-11.

Safe Superimposed Loads for Precast Concrete T-Joists
(Lbs. per Linear Foot)

	12-INCH HIGH UNITS						
Purlin No.	Clear Span in Feet						
	16	17	18	19	20	21	22
1219	411	365	309	275	236	207	
12T10	543	471	413	363	320	283	252
12T11	670	585	513	452	382*	327*	276*
12TC			530	469	415	370	330
12TC8						605	545

D = 12″ W = 12″ T = 3″ B = 3½″

$$f_c = 6000 \text{ psi}, \quad m = \frac{WL^2}{8}$$

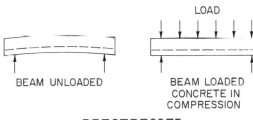

Southern Cast Stone Company, Inc.

Tectum Corporation

Precast Tee Joists

The precast joist, usually T-shaped, is placed between beams to carry the roof decking. See Fig. 23-11.

This structural system is suitable for supporting concrete slab roof decks, steel decking, wood-fiber decking, gypsum decking, and poured-in-place concrete and wood decking. A roof framed with these members and decked with wood-fiber planks is shown in Fig. 23-12. The planks form the finished ceiling, as well as the roof deck.

Precast Roof Decking Systems

Many concrete roof decking members are available, including channels, planks, and double tees.

Fig. 23-12. A roof framed with precast tee joists and decked with wood fiber planks.

Fig. 23-13.

Safe Superimposed Loads for Six-Inch Concrete Roof Channels
(Lbs. per Square Foot)

Unit No.	Clear Span in Feet				
	8	10	12	14	16
633	82	46			
644	162	97	62	40	27
655		159	105	73	51
666			155	108	79

$$f'c = 4000\,psi,\ fc = 1800\,psi,\ fs = 20{,}000\,psi,\ M = \frac{WL^2}{8}$$

Southern Cast Stone Company, Inc.

Fig. 23-14.

Safe Superimposed Loads for Double-Tee Concrete Roof Decking
(Lbs. per Square Foot)

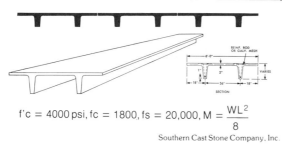

12-INCH HIGH UNITS									
Unit No.	DL lbs./ft^2	Clear Span in Feet							
		20	21	22	23	24	25	26	27
6T127	27	38	32	27					
6T128	27	59	51	44	38	32	28		
6T129	27	82	72	63	55	49	42	37	33
6T1210	28	103	91	80	71	63	56	49	44
6T1211	28	133	115	107	92	82	73	66	59

$$f'c = 4000\,psi,\ fc = 1800,\ fs = 20{,}000,\ M = \frac{WL^2}{8}$$

Southern Cast Stone Company, Inc.

Design data for some stock units manufactured by the Southern Cast Stone Co., Inc. are shown in Fig. 23-13 (channels) and Fig. 23-14 (double tees).

Precast Concrete Planks

Precast concrete tongue-and-groove planks are suitable for floors and for flat and sloped roof decking. Since they are reinforced with galvanized mesh on both sides, they are reversible, allowing either side to face up. This type of plank can be nailed to the supporting members. The planks are available with different structural capabilities. See Fig. 23-15 for design data and construction details.

Fig. 23-15.

Design Data and Construction Details for Precast Concrete Planks*
(Safe Loads in Lbs. per Square Foot)

Thick—Wgt.	2'-0"	3'-0"	4'-0"	5'-0"	6'-0"	7'-0"	8'-0"
2 " 14 lbs.	200	150	125	100	75		
2¾" 19 lbs.				150	105	75	60

*Planks are made to a maximum of 9'-0" in length with the allowable support spacing being 4'-0" or less.

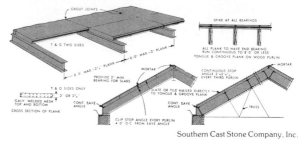

Southern Cast Stone Company, Inc.

A Precast Concrete Floor and Roof System

Another precast floor and roof system is manufactured by Flexicore Manufacturers Association. The two basic design considerations are span and load. The safe loads for one of the stock sizes are given in Fig. 23-16.

These units are set in place as shown in Fig. 23-17, and a concrete grout is placed between them. Shown in Fig. 23-18 is a typical section through a slab, with the keyslot grouted between it and the next slab. The load tables are based upon superimposed load, which includes the live load plus any dead load over and above the weight of the bare, grouted slab. A floor fill of at least 1 inch of concrete is placed over the slabs. This is a part of any superimposed load.

The roof decking usually consists of the application of rigid insulation with a built-up roof covering.

Flexicore units are precast as a monolithic unit. The steel reinforcement is prestressed, and the slabs are kiln-cured. Holes for pipes can be drilled with masonry drills or cut with a chisel. The slabs are adaptable to cantilevered construction, and overhangs up to 9 feet are practical.

These slabs can be used with masonry, reinforced concrete, or steel framing systems. The recommended minimum bearing of the slab on framing systems is: on steel, 2 inches; concrete, 3 inches; and masonry, 3 inches. A considerable saving in finishing costs can be achieved by painting the smooth interior surface of the slab.

The hollow cores can be used for running electrical conduit. An underfloor electrical system of metal raceways is available. These raceways are placed on top of the precast slab, and 1½ inches of concrete is poured as a floor topping. This enables electrical outlets to be located anywhere needed

Fig. 23-16.

Maximum Safe Superimposed[1] Working Loads for 6″ × 16″ Precast Concrete Decking
(Lbs. per Square Foot)

The table is based upon dead load plus grout of 61 lb. plf. or 45.8 psf.

Stirrups are needed for all loadings indicated in light figures above the heavy stepped dashed line.

The safe loads shown in light figures in shaded area produce deflection in excess of $1/360$ of the span.

Standard Designation	Simple Spans in Feet								
	8′-0″	10′-0″	12′-0″	14′-0″	16′-0″	18′-0″	20′-0″	22′-0″	24′-0″
S 99			253	174	122	87	62	43	29
S 94		364	238	163	114	80	56	39	25
S 90		344	225	153	106	74	51	34	
S 89		342	224	152	105	74	51	34	
S 84		320	208	141	97	67	46	30	
S 79		298	192	129	88	60	40	25	
S 74••	460	278	179	119	81	54	35		
S 70	434	262	168	111	74	49	31		
S 69	430	259	166	100	73	48	30		
S 65	399	239	152	99	65	42	25		
S 61••	377	225	142	92	60	38			
S 60	367	218	138	89	57	36	For Roof Slabs Only		
S 56	340	202	126	80	51	30			

[1]Includes the live load plus any dead load that is additional to the weight of the bare grouted slabs in place.

•• Indicates slabs with 2 rods for tensile steel. Data are for 6″ × 16″ Flexicore.

Flexicore Company, Inc.

Flexicore Company, Inc.

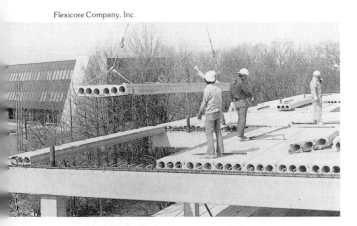

Fig. 23-17. Flexicore units being set into place to form a roof deck. Notice the hollow cores.

along the system, at any time during or after construction. This system is illustrated in Fig. 23-19. By running wiring through the core, circuits are made available in one direction; other circuits are run at right angles to these through the metal header ducts.

Construction Details

While exact details for construction with precast concrete structural systems vary, the following discussion illustrates a typical example for a small building.

Flexicore Company, Inc.

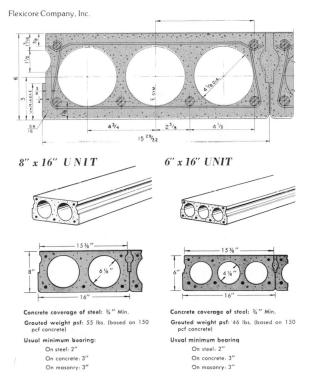

8″ x 16″ UNIT   **6″ x 16″ UNIT**

Concrete coverage of steel: ¾″ Min.

Grouted weight psf: 55 lbs. (based on 150 pcf concrete)

Usual minimum bearing:
   On steel: 2″
   On concrete: 3″
   On masonry: 3″

Concrete coverage of steel: ¾″ Min.

Grouted weight psf: 46 lbs. (based on 150 pcf concrete)

Usual minimum bearing
   On steel: 2″
   On concrete: 3″
   On masonry: 3″

Fig. 23-18. Details of Flexicore slab and installation. Notice that the keyslot space between slabs is grouted (upper right of this section).

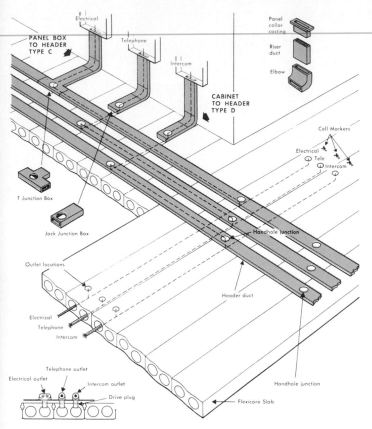

SECTION AT OUTLETS

Flexicore Company, Inc.

Fig. 23-19. Electrical system used with precast, hollow-core, concrete decking.

The footings are designed to support the calculated load, and then are poured. The foundation wall is constructed. Usually, this is poured concrete. Concrete blocks can be used, but the blocks upon which the joists will rest should be of solid masonry. This helps distribute the load over the wall.

The girder is set into place, and the floor joists are then placed according to the building plan. They are temporarily braced to hold them erect. The ends of the joists are permanently secured by filling between them with masonry bridging placed on top of the foundation. This could be brick, concrete block, or precast masonry units. See Fig. 23-20.

The forms for the cast-in-place floor slab are placed between the joists, and welded fabric wire is placed on top as reinforcement for the slab. See Fig. 23-21. If electrical conduit and plumbing are to be cast in the floor, they are installed now. The concrete is poured and finished. After it has attained sufficient strength, the forms are dropped and removed.

Where floor loads are expected to be heavier than the planned uniform load, the joists can be doubled. Double joists can be placed several inches apart to allow space for plumbing and heating ducts.

A 2-inch floor slab is adequate to carry the 100 pounds per square foot imposed load used in this system. The tops of the precast joists should be embedded $1/2$ inch into the slab.

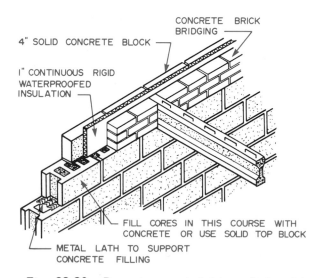

Fig. 23-20. Precast concrete joist as designed by Portland Cement Association. Notice recommended construction details.

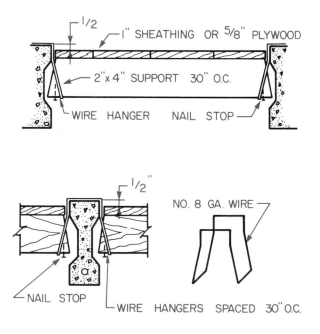

Fig. 23-21. One method of placing forms for pouring concrete slab over precast concrete joists.

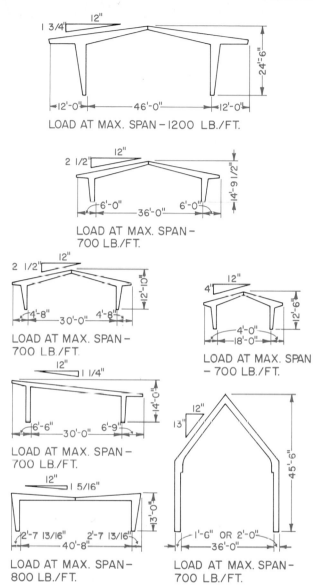

LOAD AT MAX. SPAN – 1200 LB./FT.

LOAD AT MAX. SPAN – 700 LB./FT.

LOAD AT MAX. SPAN – 700 LB./FT.

LOAD AT MAX. SPAN – 700 LB./FT.

LOAD AT MAX. SPAN – 700 LB./FT.

LOAD AT MAX. SPAN – 800 LB./FT.

LOAD AT MAX. SPAN – 700 LB./FT.

Fig. 23-22. Stock precast bents as manufactured by Southern Cast Stone, Inc. Notice the allowable load at maximum span in pounds per foot.

Precast Bents

A wide variety of precast concrete bents is available in stock sizes. Design data are given in Fig. 23-22. These bents are suited for channel slab decks, wood, steel, or any other decking material. They are fireproof and maintenance free. Precast concrete bents are spaced according to the load they must carry and the distance the proposed roof decking can span.

A layout for a small building and details of construction are illustrated in Figs. 23-23 and 23-24, pages 528 and 529. The details concerning the joining of the bents at the ridge are shown in Section 3. The bent first is bolted to the grade beam, as shown in the details, and then grouting is placed between it and the grade beam to secure it in a level position. The finished floor is poured over the connection and tie rod. The tie rod extends under the floor to the mating bent at the opposite side of the building. This serves to contain the tension put upon the bent coupling at the grade beam. See the cross section through the building for details.

Section 1 illustrates how the 6-inch, precast concrete channels are secured to the bents. The space between the channel ends is filled with concrete.

Bents are rapidly erected. Generally, a crane is used to raise each half section and hold it in place as it is leveled and secured.

Figure 23-25 shows the interior of a church with exposed, precast concrete bents and precast roof channels. These can be decorated or colored in several ways to contribute to the atmosphere of the building.

Monolithic Reinforced Concrete Construction

A monolithic reinforced concrete structure is one in which the concrete structural system is strengthened with steel reinforcing rods. Monolithic means that the concrete members are cast as a solid, continuous unit.

Southern Cast Stone Company, Inc.

Fig. 23-25.   Interior of a church with exposed precast bents and roof channels.

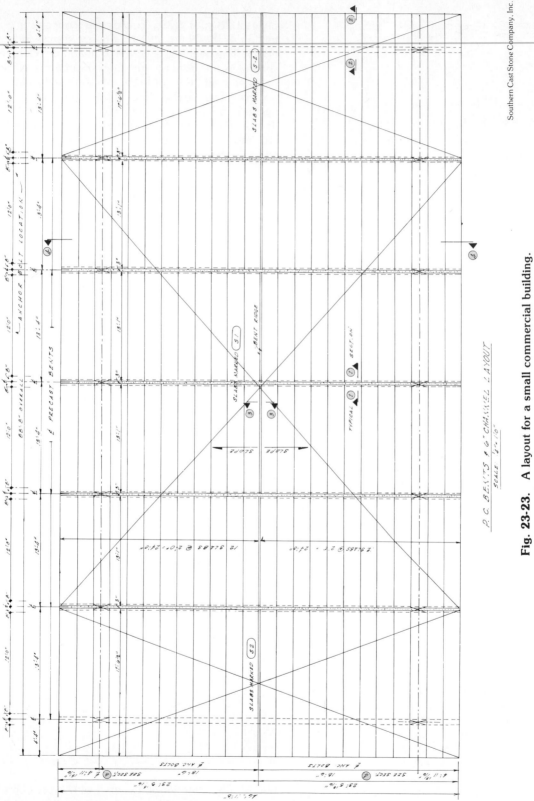

Southern Cast Stone Company, Inc.

Fig. 23-23. A layout for a small commercial building.

528

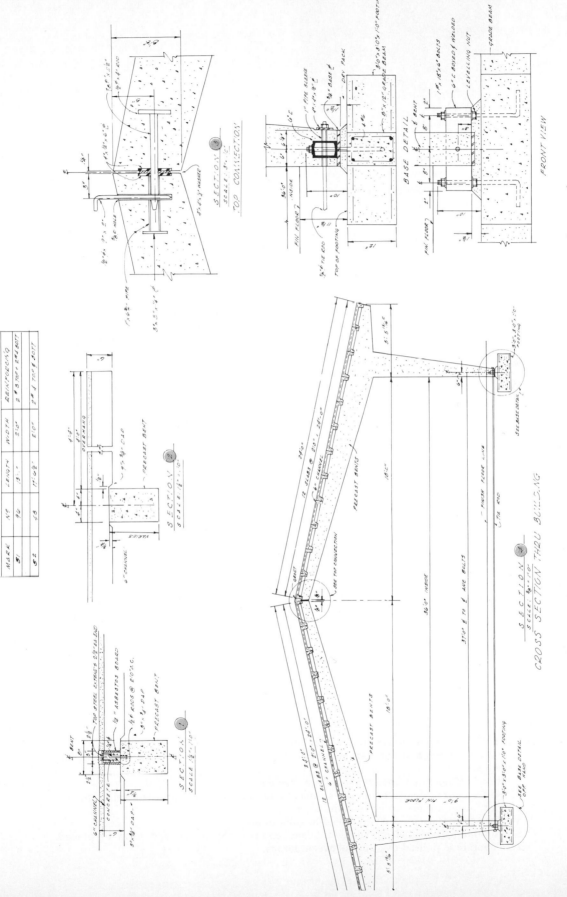

Fig. 23-24. Details for the same building (Fig. 23-23) using precast concrete bents and roof channels.

Fig. 23-26. This is a monolithic, cast-in-place, reinforced concrete structure.

As each floor is cast, the reinforcing steel and forms are moved up to form the next level. Notice the forms and shoring on the top level. The tower at the left is a lift for moving workers and materials.

Martin K. Eby Construction Company, Inc.

The concrete columns and beams are made by building forms on the job. The forms are in the places the columns and beams are wanted. Steel reinforcing rods are tied in place inside the forms. The concrete is poured into the forms to make the columns and beams into a single unit. See Fig. 23-26.

A much-used procedure is to pour monolithically the concrete joists and the floor or roof slab. The joists are generally arranged in one direction. They are formed on top of wood stringers that are supported by shoring. See Fig. 23-27 for erection procedures. The forms with steel reinforcing rods in place are shown in Fig. 23-28. This job is ready for the concrete to be poured.

After the concrete has gained proper strength, the shoring, stringers, and soffit boards are removed. The steel domes are removed by using

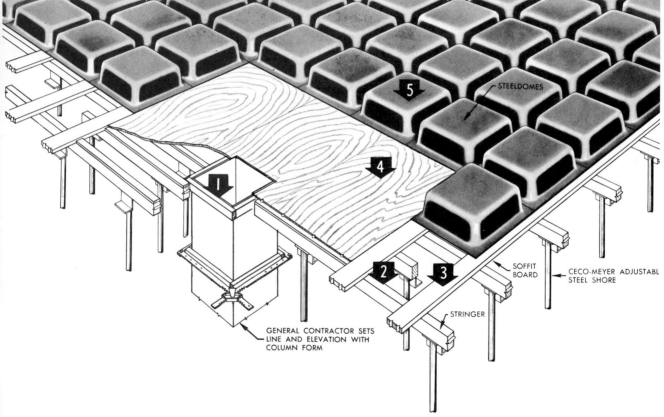

STEEL DOMES

SOFFIT BOARD

CECO-MEYER ADJUSTABL STEEL SHORE

STRINGER

GENERAL CONTRACTOR SETS LINE AND ELEVATION WITH COLUMN FORM

Ceco Corporation

Fig. 23-27. Erection procedure for monolithic, reinforced concrete construction.
1. Forms for columns, beams, and column capitals below the joist level are put into place.
2. Wood stringers and metal shores are erected.
3. Soffit boards are placed on the stringers and spaced to support the steel forms.
4. Flat forms are placed for all solid areas.
5. Steel forms are placed and nailed to soffit boards.
6. Plumbing sleeves, electrical outlet boxes, and reinforcing steel and mesh are placed.
7. Concrete is poured over and between domes.

compressed air. See Fig. 23-29. After they are cleaned and oiled, they can be used again.

The interior ceiling can be painted or sprayed with acoustical asbestos to present a pleasing appearance. One such ceiling is shown in Fig. 23-30. A suspended ceiling can also be installed.

Concrete is made from portland cement, water, sand, and gravel. The strength and appearance varies with the quantity and quality of these materials.

Concrete has good compressive strength. **Compressive strength** is the ability of a material to withstand forces tending to shorten it. See Fig. 23-31. The tensile strength of concrete is low. **Tensile strength** is the ability of a material to withstand forces tending to lengthen it. Steel has a high tensile strength. This is why steel reinforcing rods are added to structural concrete members. The concrete resists the compressive forces. The steel resists the tensile forces.

Ceco Corporation

Fig. 23-28. A roof ready to be poured. Notice the steel domes, the reinforcing rod in the spaces forming the ribs, and the wire mesh for the roof slab. In the center of the photo is an area without domes, which is to be a solid slab.

Ceco Corporation

Fig. 23-29. The steel domes are removed by air pressure without damaging the concrete. When the flanges of the steel domes are butted, a 6-inch rib is formed. If a wider rib is desired, a filler strip is placed between the flanges. This photo illustrates two wide ribs and several standard 6-inch ribs.

Ceco Corporation

Fig. 23-30. An exposed, concrete-joist ceiling sprayed with acoustical plaster and paint. Notice the suspended acoustical panels and spotlights.

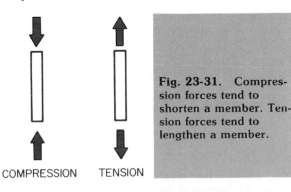

COMPRESSION TENSION

Fig. 23-31. Compression forces tend to shorten a member. Tension forces tend to lengthen a member.

Tilt-Up Concrete Construction

In tilt-up construction, reinforced concrete wall sections are cast in a horizontal position and then are lifted to a vertical position to form the walls of the building. The panels are framed with wooden forms, the reinforcement rods are tied into place and the concrete is poured. See Fig. 23-42. Electrical conduit may be cast in the panel. After the concrete is placed and screeded (leveled by passing a straightedge back and forth over the surface), it is permitted to stiffen. The forms are removed as soon as the concrete is firm enough to hold its shape, and the surface is trowelled and given its final finish. When the panel has set long enough to achieve sufficient strength to be lifted, it is tilted into place and temporarily braced. While there are many ways of lifting, a crane is most frequently used.

The reinforcing rods are extended outside the concrete panel, and the panels are erected with a gap between them. Vertical columns are cast in place at each opening between the panels. The reinforcing rods are treated so they will not adhere to the column, thus allowing some movement at the joint. See Fig. 23-46.

The panel reinforcement is the same as that used in cast-in-place walls. It could be bars or welded wire mesh. A quantity of small bars gives better crack control than the same weight of larger bars.

If openings are cast in the panel, extra reinforcement should be used on all sides of the opening.

Several types of joints are used between the panel and its supporting foundation. See Fig. 23-43. The most common procedure is to spread a layer of portland cement mortar on the foundation and tilt the wall on top of it. Some use of a premolded joint filler has been made. This can be sealed with portland cement mortar. Wedges are

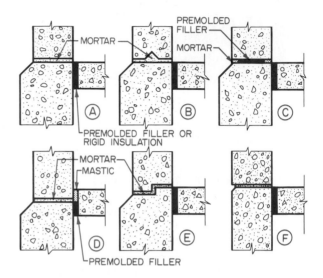
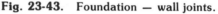

Fig. 23-43. Foundation — wall joints.

These are typical joints subject to many variations. **A** and **C** are the simplest and most commonly used. The offset from the floor level in **D** and the offset in the foundation or lower wall is sloped or offset slightly as shown in these sketches. Thus, there is no horizontal surface to catch the water and little possibility of leakage at this point than with any unit masonry wall.

Fig. 23-42. A tilt-up wall panel ready to be poured.

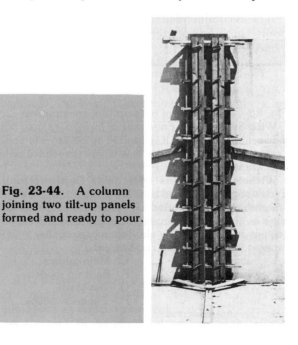

Fig. 23-44. A column joining two tilt-up panels formed and ready to pour.

sometimes placed on top of the foundation and the panel lowered on top of them. The wedges allow the panel to be leveled easily, and then they can be removed when the mortar in the joint sets.

Columns can be cast first and the panels fitted to them, or the panels can be erected first and the columns cast between them. The latter is most generally used. See Fig. 23-44. The column usually overlaps the panel. This hides any rough edges and makes variances in spacing unnoticed. It also hides differences in panel thickness. See Fig. 23-45.

Since movement between the column and the panels is necessary to allow expansion and contraction, provision for preventing bonding between the column and the panel must be made. This may be paper, felt, premolded joint filler, cork gasket, or some similar material. Even the reinforcing bars extending from the panels into the column must be coated to prevent bonding. An exception to this is in areas such as the West Coast, where earthquakes are likely to occur; the panels are joined together as firmly as possible.

The reinforcing bars usually extend from 2 to 6 inches from the panel into the column. They should extend into the column, just beyond the column reinforcement. See Fig. 23-46.

Tilt-up construction is adaptable to a wide variety of one-story buildings and has been used on multistory structures. Construction time is short, since the time-consuming framing necessary for poured-in-place walls is greatly reduced. Also, tilt-up construction tends to be less expensive.

The wall panels must be designed to meet two conditions. First, the panel should serve as the wall of the building and support the load of the roof. The design for this condition is the same as for a conventionally built reinforced concrete wall. Second, it must withstand the stresses placed upon it during the tilt-up operation.

The method to be used for the lifting process must be known before the bending stresses can be ascertained. With some pick-up systems, an analysis of stress becomes very involved. For small panels with a steel channel or angle bolted to the top edge, the panel should be designed as a simple slab that is supported on two ends. It is common practice to lift the panel by cables attached to the top edge. However, if the pick-up points are located one-quarter of the way down the panel from the top edge, the stresses are greatly reduced.

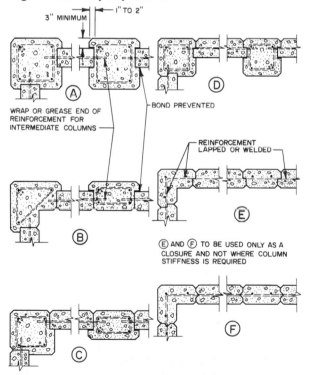

Fig. 23-45. Column — wall joints.

A and **D** are typical joints for use where movement at the joints is desired. They can also be used for rigid joints by lapping the reinforcement and omitting the bond-prevention material. Note that even where movement is desired at intermediate columns, the corner columns are bonded to the wall panels. V-joints should be used wherever the face of the wall is flush with the column.

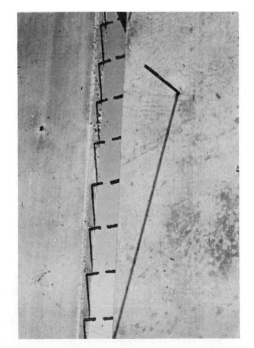

Fig. 23-46. Tilt-up panels in a vertical position ready to have a column formed and poured. Notice the reinforcing rods extending from the panel into the column.

Special consideration must be given to panels whose length is greater than their height. They tend to develop stresses longitudinally as well as horizontally.

If openings are cast into a panel, it is necessary to consider the effect of these when designing the panel to accommodate lifting stresses.

The most commonly used wall thickness is 6 inches. This usually meets structural requirements and is of sufficient strength that average-size panels can be lifted with minimum danger of cracking.

Panel design is also influenced by whether the walls are to be supported on a continuous footing or on the column footings only. If the panel extends from column footing to column footing, it serves as a deep beam, carrying its own weight plus the imposed weight from the roof. This also requires special consideration when designing the column footings, since they must carry considerably more load than with the continuous foundation. See Fig. 23-47.

Another design consideration is whether the tilt-up wall is to be load-bearing or non-load-bearing. If the roof is supported by the columns, walls can be of lighter construction.

Structural Steel Members

Following is a list of the most commonly used structural steel member classifications. Examples are shown in Fig. 23-48.

1. **American standard beams** are I-shaped members. They are identified by the symbol **S**.
2. **American standard channels** are identified by the symbol **C**.
3. **Miscellaneous channels** are special-purpose channel shapes other than the standard **C** shapes. They are identified by the symbol **MC**.
4. **Wide-flange shapes** are used for beams and columns. They are identified by the symbol **W**.
5. **Miscellaneous shapes** are identified by the symbol **M**. They are similar in cross section to the **W** shapes.
6. **Structural tees** are halves of **S, W,** and **M** shapes. They are identified by symbols indicating which member was used. The symbols are **ST, WT,** and **MT**.
7. **Angles** are identified by the symbol **L**. They have two legs which may or may not be of equal length. The legs are 90 degrees apart.
8. **Plates** and **flat bars** are rectangular members. The plate symbol is **PL**. The bar symbol is **BAR**. Square bars are shown by the symbol ▱. Round bars by the symbol ⊘.

The **S, C,** and **MC** shapes have tapered flanges. The **W** shapes generally have parallel flange surfaces. **M** shapes may have either parallel or tapered inner flange surfaces.

The drafting technician should know how to present data (billing) and show structural steel shapes on steel detail drawings. Recommended methods are given in Fig. 23-49. On drawings made to a

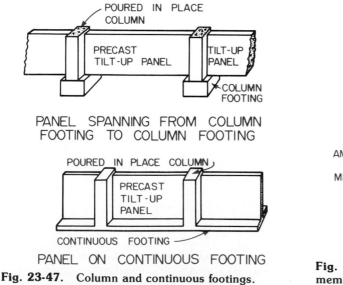

PANEL SPANNING FROM COLUMN FOOTING TO COLUMN FOOTING

PANEL ON CONTINUOUS FOOTING

Fig. 23-47.   Column and continuous footings.

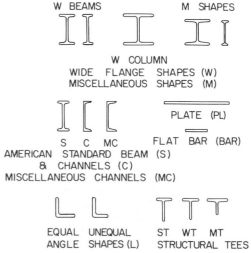

Fig. 23-48.   Shapes of commonly used steel members.

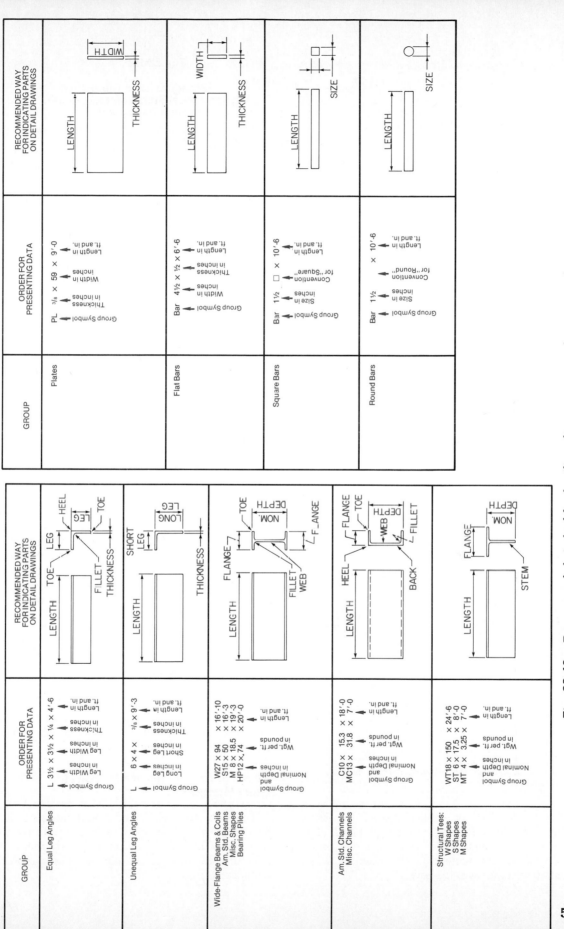

Fig. 23-49. Recommended method for detailing, dimensioning, and presenting descriptive data for structural steel members.

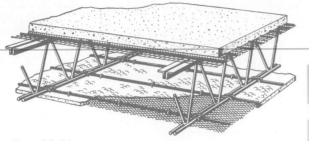

Fig. 23-50. Open-web steel joists with metal lath decking and a concrete slab. Notice the ribbed, metal lath and plaster ceiling.

Fig. 23-52.

Standard Load Table
Long span Steel Steel Joists, LJ-Series
18-inch Deep Joist

Joist Designation	Approx. Wt. in Lbs. per Linear Ft.	Safe Load* in Lbs. Between	Clear Opening or Net Span in Feet					
		21—24	26	28	30	32	34	36
18LJ02	13	8800	324	290	260	230	206	185
18LJ04	16	11500	423	367	321	283	251	224
18LJ06	19	15200	549	475	415	366	325	290
18LJ08	22	17000	639	595	538	474	421	376
18LJ10	27	18900	711	662	618	581	516	461
18LJ12	32	20700	776	722	675	634	597	565

*Loads are total safe, uniformly distributed, load-carrying capacities in pounds per lineal foot (live plus dead load).

Fig. 23-51.
Safe Loads for Open-Web Steel Joists

Total safe loads consist of live loads plus dead load plus weight of joist. Maximum deflection for tabulated spans and safe loads will not exceed 1/360th of the span. Tabulated safe loads are based on joists being properly braced laterally as required for standard joist construction. Design Working Stress: 20,000 psi.

Joist Type No.	Clear Span	Total Safe Load Pounds	Total Safe Load in Pounds Per Square Foot for Various Joist Spacing						
			*16"	24"	2'-6"	3'-0"	3'-6"	4'-0"	6'-0"
8S2	15'-0"	2333	117	78	62	52	45	39	—
10S2		2889	145	97	77	64	55	48	32
12S3		4400	220	147	117	98	84	73	49
14S4		5200	260	174	139	116	99	87	58
10S2	20'-0"	2167	81	54	43	36	31	—	—
12S3		3333	125	84	67	56	48	42	—
14S4		4900	184	123	98	82	70	61	41
16S5		6000	225	150	120	100	86	75	50
18S6		7200	270	180	144	120	103	90	60
20S6		7400	278	185	148	123	106	93	62
14S4	25'-0"	3920	118	79	63	52	45	39	—
16S5		5387	161	108	86	72	61	54	36
18S6	25'-0"	7200	216	144	115	96	82	72	48
20S6		7400	222	148	118	99	85	74	49
22S7		8000	240	160	128	107	91	80	53
24S8		9000	270	180	144	120	103	90	60
16S5	30'-0"	4489	113	75	60	50	43	38	—
18S6		6044	151	101	80	67	57	50	33
20S6		6556	164	110	88	73	63	55	36
22S7		8000	200	134	107	89	76	67	44
24S8		9000	225	150	120	100	86	75	50
18S6	35'-0"	5181	111	74	59	49	42	37	—
20S6		5619	121	81	64	54	46	40	—
22S7		7333	158	105	84	70	60	53	35
24S8		9000	193	129	103	86	73	64	43
20S6	40'-0"	4917	92	62	49	41	35	31	—
22S7		6417	120	80	64	53	46	40	—
24S8		8333	156	104	83	69	59	52	35

*Allowable uniform total load per foot of joist.　　　Bethlehem Steel Corporation

Structural steel members are made from a number of different steels. The type of steel used influences load-carrying characteristics. The most commonly used steel for general construction situations is designated as A-36. It is based on standards of the American Society for Testing Materials. It has a yield stress of 36 ksi. The abbreviation **ksi** means kips per square inch of cross section. A **kip** is 1000 pounds.

Open-Web Steel Joists

Open-web steel joists are welded assemblies of double-T structural members and bars. See Fig. 23-50. They are most commonly used for roof support and can span long distances. The joists are set in place, braced, and welded. Many kinds of roof decking can be used. The roof shown in Fig. 23-50 was formed by attaching paper-backed metal lath to the joists. Corrugated steel sheets could have been used. Concrete was poured on this. Steel reinforcement was built into the concrete. The bottom members of the joist hold the ceiling. Many types of hung ceilings could be used. In Fig. 23-50, metal lath was attached, and the ceiling was plastered.

The space between the ceiling and the roof is available to run electrical conduit and plumbing. Examples of the spans and loads typically available are given in Figs. 23-51 and 23-52.

Physical Properties of Structural Steel Members

The drafting technician should understand the basic physical properties of structural steel members and know the terms associated with them. The structural engineer constantly uses these and other technical terms.

scale of 1" = 1'-0" or smaller, it is not necessary to draw the toes rounded or show fillets. The thickness of the legs, webs, or flanges can be increased to suit the technician. If drawn to scale, almost no thickness would show.

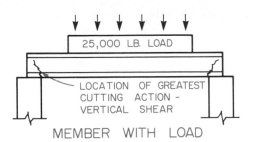

Fig. 23-53. Forces on structural members.

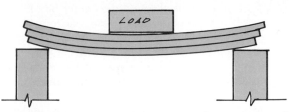

Fig. 23-54. Horizontal forces exist in a structural member under load. These tend to cause material to slide horizontally.

Stress

A member that is under **tensile stress** is under a force that tends to lengthen it. For example, if a steel rod with a cross-sectional area (**A**) of 2 square inches is hung from a beam and has a load of 15,000 pounds (**P**) hung from it, the computed tensile stress (f$_t$) would be 7500 pounds per square inch of the rod cross section.

$$f_t = \frac{P}{A} = \frac{15,000}{2.0}$$
$$f_t = 7500 \, lb./in.^2 \text{ or } 7.5 \, kips/in.^2$$

Compressive stress refers to forces acting upon a member which tend to shorten it. If the steel rod in the previous illustration were under compression, the compressive stress (f$_c$) would be 7500 pounds per square inch of the rod cross section.

Vertical shear refers to forces which tend to cause a member to fail by a cutting action at each support. In Fig. 23-53, a member is shown suspended between two supports. As a load is applied, the member tends to drop down between the supports, but is restrained by the resistance of the material from which the beam is made. The end of the member tends to remain on the supports. If sufficient load is applied, the member may break near the supports. This type of stress is often critical in short beams carrying very heavy loads. In beams, the vertical and horizontal shearing stresses generally cause a 45-degree fracture upon failure.

Horizontal shear is the tendency of the material in the structural member to slide horizontally. This can best be illustrated by stacking several boards between two supports. When a load is applied, the boards bend and tend to slide horizontally, thus becoming uneven on the ends. This same horizontal force is present in any beam under load. See Fig. 23-54.

On short spans, allowable beam loads may be limited by the shearing strength of the web, instead of the maximum bending stress allowed in the member. The shear limit of the web is indicated in the allowable uniform load tables of the American Institute of Steel Construction (AISC) **Manual of Steel Construction.** The loads in kips shown above the dark horizontal lines are the maximum allowable shear forces on the web. Selected tables are shown and discussed later in this chapter.

Shear also occurs in fasteners such as rivets or bolts. The engineer must ascertain the shear strength of these connectors, so they will not fail under load. The way shear force affects a rivet is illustrated in Fig. 23-55.

Stresses caused by **bending** are indirect stresses and, in this way, differ from tensile, compression, and shear stresses.

When a beam is supported at its ends and is loaded from above, the top portion is under compression and the lower portion is under tension. See Fig. 23-56. This occurrence also was illustrated in the discussion of prestressed concrete members.

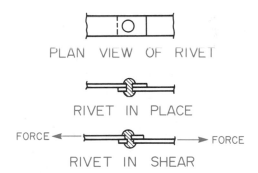
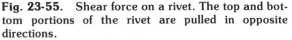

Fig. 23-55. Shear force on a rivet. The top and bottom portions of the rivet are pulled in opposite directions.

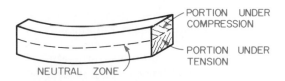

Fig. 23-56. Beam under bending stress. This beam is supported at the ends and loaded from above.

A horizontal plane through the center of a beam subjected to bending stress is a neutral zone, with compression stresses occurring above it and tension stresses occuring below. These stresses are greatest at the exterior or extreme surfaces of the beam, and they decrease as the center or neutral zone is approached.

The important point to consider when working with bending stress is the extreme fiber stress. In A-36 steel beams, for example, the allowable stress is 24,000 psi. The engineer must design a beam of sufficient cross section that the extreme fiber stress does not exceed this limit.

Elastic Limit, Yield Point, and Ultimate Strength

As a structural member is placed under a load, the member changes in length or shape. This change is referred to as **deformation.** If a member is under tension, the deformation occurring is a lengthening of the member; while if it is under compression, the deformation is a shortening. (When a member is subjected to bending, the deformation is called **deflection.** This is discussed in detail in a later section of this chapter.)

Up to a point, when a member is subjected to a load, the deformation is proportional to the stress. In other words, increasing the load causes a uniform and proportional deformation. This proportion continues until the member reaches its **elastic limit.** This is the point at which increased stress or load causes deformation to occur at a faster rate than that at which the load is increased. If a member is loaded to its elastic limit and then the load is removed, the member will return to its original length. If the elastic limit is exceeded, the member will have an increased length, rather than returning to its original length. This increase in length is called **permanent set.**

If the elastic limit is exceeded slightly, a small amount of deformation occurs, even though the load is not increased. The point at which this occurs is called the **yield point.** This is sometimes called **yield stress.** The symbol used is F_y. For A-36 steel, F_y is 36,000 psi. For all steels, the elastic limit and the yield point are close together.

As the load is increased beyond the elastic limit and the yield point, rapid deformation occurs. When the greatest stress is attained (just before beam failure), the **ultimate strength** of the member has been reached. Any stresses beyond the ultimate strength bring failure or rupture to the member. This is called the **breaking strength.**

Modulus of Elasticity

The **modulus of elasticity** of a material is its degree of stiffness. The greater the modulus of elasticity, the less deformation that occurs under load. The modulus of elasticity for various materials must be obtained from tables in architectural standards or manufacturers' publications. The modulus of elasticity for steel is 29,000,000 pounds per square inch.

Deflection

As a downward load is placed on a joist or beam, the member is subject to bending stresses. The deformation associated with these stresses is called deflection. **Deflection** is the vertical distance a beam moves from a straight line. It is often so slight that it is not visibly noticeable, but it is always present in a member under load.

The engineer must ascertain the deflection in a beam under the design loads. While the beam may be strong enough to sustain the stresses of bending, the deflection may be great enough to cause cracking of the plaster ceiling or the floor surface. Deflection is usually limited by building codes to $\frac{1}{360}$ of the span of the member.

The longer the span, the more likely that excessive deflection will occur. The deflection of a beam varies directly with the fiber stress and inversely with the depth of the beam. When a beam is selected, therefore, it is best to select one having the greatest practical depth, since depth reduces deflection. If two beams of similar cross-sectional area and capacity are under consideration, such as one 10″ × 12″ and one 8″ × 14″, the 8″ × 14″ will have less deflection because of its greater depth. Of course, the deeper beam tends to reduce ceiling height.

In tables giving allowable uniform load in kips for beams laterally supported, the deflection for the various spans is given for beams supporting the full, tabulated, allowable loads.

Using Safe Load Tables for Beams

The American Institute of Steel Construction **Manual of Steel Construction** gives design data for all common steel shapes, and many special designs. Included are tables giving allowable uniform loads, in kips, for beams laterally supported. Also given are tables showing dimensions for detailing and properties for designing members.

Fig. 23-58.

American Institute of Steel Construction

BEAMS — W shapes

W 8 $F_y = 36$ ksi

Allowable uniform loads in kips for beams laterally supported
For beams laterally unsupported, see page 2-84

Designation	W8	W8	W8	W8	W8	W8	W8	W8	W8	
Weight per Foot	35	*31	28	24	20	17	15	13	*10	
Flange Width	8	8	6½	6½	5¼	5¼	4	4	4	
L_c	8.5	8.4	6.9	6.9	5.6	5.5	4.2	4.2	4.2	
L_u	22.6	20.0	17.4	15.1	11.3	9.4	7.2	5.9	4.7	

Span in Feet	35	*31	28	24	20	17	15	13	*10	Deflection Inches
2								53.4		0.01
3							57.7	52.8	38.9	0.03
4					58.5	53.4	47.2	39.6	30.5	0.05
5			66.6	56.3	54.4	45.1	37.8	31.7	24.4	0.08
6	74.2	66.8	64.8	55.5	45.3	37.6	31.5	26.4	20.3	0.11
7	71.1	61.6	55.5	47.5	38.9	32.2	27.0	22.6	17.4	0.15
8	62.2	53.9	48.6	41.6	34.0	28.2	23.6	19.8	15.3	0.20
9	55.3	47.9	43.2	37.0	30.2	25.1	21.0	17.6	13.6	0.25
10	49.8	43.1	38.9	33.3	27.2	22.6	18.9	15.8	12.2	0.31
11	45.2	39.2	35.3	30.3	24.7	20.5	17.2	14.4	11.1	0.38
12	41.5	35.9	32.4	27.7	22.7	18.8	15.7	13.2	10.2	0.45
13	38.3	33.2	29.9	25.6	20.9	17.4	14.5	12.2	9.4	0.52
14	35.5	30.8	27.8	23.8	19.4	16.1	13.5	11.3	8.7	0.61
15	33.2	28.7	25.9	22.2	18.1	15.0	12.6	10.6	8.1	0.70
16	31.1	26.9	24.3	20.8	17.0	14.1	11.8	9.9	7.6	0.79
17	29.3	25.4	22.9	19.6	16.0	13.3	11.1	9.3	7.2	0.90
18	27.6	23.9								
19	26.2	22.7								
20	24.9	21.5								
21	23.7	20.5								
22	22.6	19.6								
		(23.6)							(23.5)	

For explanation of deflection see page 2-23

Properties and Reaction Values

	35	*31	28	24	20	17	15	13	*10
S in.³	31.1	27.4	24.3	20.8	17.0	14.1	11.8	9.9	7.8
V kips	37.1	33.4	33.3	28.2	29.3	26.7	28.8	26.7	19.5
R kips	**38.3**	**34.5**	**34.1**	**28.9**	29.3	**26.8**	**28.5**	**26.4**	19.2
R₁ kips	8.5	7.8	7.7	6.6	6.7	6.2	6.6	6.2	4.6
N, in.	3.4	3.4	3.4	3.4	3.5	3.5	3.5	3.5	3.6

Load above heavy line is limited by maximum allowable web shear.
Values of R in bold face exceed maximum web shear V.
* Tabulated loads for this shape are computed with the allowable stress (ksi) shown in parentheses at the bottom of the allowable load column.

American Institute of Steel Construction

Fig. 23-57.

BEAMS — S shapes

S 6-5-4-3 $F_y = 36$ ksi

Allowable uniform loads in kips for beams laterally supported

Designation	S6	S6	S5	S5	S4	S4	S3	S3
Weight per Foot	17.25	12.5	14.75	10	9.5	7.7	7.5	5.7
Flange Width	3⅜	3⅜	3¼	3	2¾	2⅝	2½	2⅜
L_c	3.8	3.5	3.5	3.2	3.0	2.8	2.6	2.5
L_u	9.9	9.2	9.9	9.1	9.5	9.0	10.1	9.4

Span in Feet	S6 17.25	S6 12.5	Defl.	S5 14.75	S5 10	Defl.	S4 9.5	S4 7.7	Defl.	S3 7.5	S3 5.7	Defl.
1	80.9		0.00	71.6		0.00	37.8		0.01	30.4	14.8	0.01
2	70.2	40.4	0.02	48.7	31.0	0.02	27.1	22.4	0.02	15.6	13.4	0.03
3	46.8	39.3	0.04	32.5	26.2	0.04	18.1	16.2	0.06	10.4	9.0	0.07
4	35.1	29.5	0.07	24.4	19.7	0.08	13.6	12.2	0.10	7.8	6.7	0.13
5	28.1	23.6	0.10	19.5	15.7	0.12	10.8	9.7	0.16	6.2	5.4	0.21
6	23.4	19.7	0.15	16.2	13.1	0.18	9.0	8.1	0.22	5.2	4.5	0.30
7	20.0	16.8	0.20	13.9	11.2	0.24	7.7	6.9	0.30	4.5	3.8	0.41
8	17.5	14.7	0.26	12.2	9.8	0.32	6.8	6.1	0.40			
9	15.6	13.1	0.34	10.8	8.7	0.40	6.0	5.4	0.50			
10	14.0	11.8	0.41	9.7	7.9	0.50						
11	12.8	10.7	0.50	8.9	7.2	0.60						
12	11.7	9.8	0.60									
13	10.8	9.1	0.70									

For explanation of deflection see page 2-23

Properties and Reaction Values

	S6 17.25	S6 12.5	S5 14.75	S5 10	S4 9.5	S4 7.7	S3 7.5	S3 5.7
S in.³	8.8	7.4	6.1	4.9	3.4	3.0	2.0	1.7
V kips	40.5	20.2	35.8	15.5	18.9	11.2	15.2	7.4
R kips	**54.1**	**27.0**	**56.7**	**24.6**	**36.9**	**21.8**	**38.9**	**18.9**
R₁ kips	12.6	6.3	13.3	5.8	8.8	5.2	9.4	4.6
N, in.	2.4	2.4	1.9	1.9	1.5	1.5	1.0	1.0

Load above heavy line is limited by maximum allowable web shear.
Values of R in bold face exceed maximum web shear V.

COLUMNS
W shapes

$F_y = 36$ ksi

W 8-6

TABLE I

Allowable axial loads in kips

$F_y = 36$ ksi with respect to least radius of gyration r_y

Effective length in ft. KL

Designation		W8		W8		W6			W6		
Nominal Depth and Width		8 × 6½		8 × 5¼		6 × 6			6 × 4		
Weight per Foot		28	24	20	17	25	20	†15.5	16	12	†8.5
	2	172	148	122	103	153	123	95	96	72	51
	3	168	144	118	100	150	120	93	92	68	48
	4	164	141	114	96	146	117	90	87	64	45
	5	160	137	109	92	142	113	87	81	60	42
	6	155	133	104	88	137	109	84	75	55	38
	7	150	128	98	83	132	105	81	69	50	34
	8	144	123	93	78	126	100	77	62	44	30
	9	138	118	86	72	120	96	73	54	38	25
	10	132	113	79	66	114	91	69	46	31	21
	11	125	107	72	60	108	85	65	38	26	17
	12	118	101	65	53	101	80	60	32	21	14
	13	111	94	56	46	94	74	55	27	18	12
	14	103	88	49	39	86	68	50	23	16	10
	15	95	81	42	34	78	61	45	20	14	
	16	86	73	37	30	70	54	39	18		
	17	78	66	33	27	62	48	35			
	18	69	59	29	24	55	43	31			
	19	62	53	26	21	49	39	28			
	20	56	47	24	19	45	35	25			
	22	46	39			37	29	21			
	24	39	33			31	24	17			
	26	33	28								

Properties										
Area A (in.²)	8.23	7.06	5.89	5.01	7.35	5.88	4.56	4.72	3.54	2.51
I_z (in.⁴)	97.8	82.5	69.4	56.6	53.3	41.5	30.1	31.7	21.7	14.8
I_y (in.⁴)	21.6	18.2	9.22	7.44	17.1	13.3	9.67	4.42	2.98	1.98
Ratio r_z/r_y	2.13	2.12	2.74	2.75	1.76	1.76	1.76	2.68	2.70	2.73
r_y (in.)	1.62	1.61	1.25	1.22	1.53	1.51	1.46	0.97	0.92	0.89
L_c (ft.)	7.0	6.9	5.6	5.6	6.5	6.4	6.4	4.3	4.3	4.2
L_u (ft.)	17.5	15.1	11.4	9.4	20.2	16.5	12.5	12.1	8.7	6.1
B_z } Bending	.339	.340	.347	.356	.441	.439	.456	.463	.489	.495
B_y } factors	1.246	1.259	1.683	1.771	1.308	1.328	1.412	2.156	2.376	2.486
a_z } Multiply	14.60	12.30	10.32	8.43	7.92	6.20	4.49	4.72	3.24	2.21
a_y } values by 10⁶	3.22	2.73	1.37	1.11	2.56	2.00	1.45	0.66	0.44	0.30

Heavy line indicates $Kl/r = 120$.   Values omitted for $Kl/r > 200$.
† Flange is non-compact.

For example, in Fig. 23-57, data are presented for S-beams from 3 inches through 6 inches. Data for 8-inch **W** shapes are given in Fig. 23-58, and Fig. 23-59 contains data for 6- and 8-inch **W** shapes used as columns. The axial load is given in Fig. 23-58. The **axial load** is one that is applied parallel with the length of the member.

Information across the tops of Fig. 23-57 and 23-58 gives the depth and width of the beam and its weight per lineal foot. Beneath this are the following symbols:

Lc — The maximum, unbraced length (in feet) of the compression flange, at which the allowable bending stress may be taken at $0.66 F_y$ (feet).

Lu — The same function as Lc, except the stress is $0.6 F_v$ (feet).

The column at the extreme right gives the deflection that will occur if the indicated load is placed on the beam at the listed span, in feet.

The bottom of the table gives data on properties and reaction values.

S in.^{3} — The elastic section modulus, (in.^{3}).

V kips — Statical shear on the beam, in kips.

R kips — Reaction (or concentrated transverse load) applied to the beam or girder, in kips.

R$_i$ kips — Increase in reaction (R), in kips, for each additional inch of bearing.

N$_e$ in. — Length at end bearing to develop maximum web shear, in inches.

Below these are data for high-strength steels, which are not considered here.

Factor of Safety

Most building codes have some regulations pertaining to a **factor of safety.** This number can be calculated by dividing the ultimate strength or yield of a material by the allowable or actual stress. For example, if the ultimate strength of a material is 66,000 pounds per square inch and the actual stress is 16,000 pounds per square inch, the factor of safety is 4.1. The larger the factor of safety, the smaller the allowable stress. Instead of specifying a safety factor, some building codes specify the allowable stress.

Steel Columns

A **column** is a structural member erected in a vertical position and subject to compression stresses parallel to its longitudinal axis. The most commonly used steel member is the wide-flange column. Where fireproofing is not required, the most effective shape for columns is the hollow cylinder. That is why the Lally columns used in light construction are round and hollow in cross section.

Design data for **W** shapes used as columns are in Fig. 23-59. Data across the top of the table give the nominal depth and width of the member and its weight per lineal foot. The body of the table is the allowable axial loads, in kips, for the various-sized members, at the effective length in feet (KL), with respect to the least radius of gyration. The column of figures at the extreme left is the effective length of the column.

Various properties are given at the bottom of the table as follows:

Area A (in^2) — Cross-sectional area of the column (in square inches).

I$_x$ (in.^{4}) — Moment of inertia of a section in inches^4 of the X-X axis.

I$_y$ (in.^{4}) — Moment of inertia of a section in inches^4 of the Y-Y axis.

Ratio r$_x$/r$_y$ — Ratio of: radius of gyration with respect to the X-X axis (r$_x$), and radius of gyration with respect to the Y-Y axis (r$_y$).

L$_c$ — Maximum, unbraced length (in feet) of the compression flange, at which the allowable bending stress may be taken at 0.66 F$_y$ (feet).

L$_u$ — Same function as L$_c$, except the stress is 0.6 F$_y$ (feet).

B$_x$ and B$_y$ — Bending factors, with respect to the X-X axis and Y-Y axis, for determining the equivalent axial load in columns subjected to combined loading conditions.

A$_x$ and A$_y$ — Components of amplification factor for solving Equation (7a — see **Manual for Steel Construction**), when bending is about the X-X axis and Y-Y axis respectively.

Many other types of columns are available. Most frequently found are round and square types. Design data for selected areas and lengths are given in Fig. 23-60.

Fig. 23-60

Allowable Loads for Concrete-Filled Steel Columns

Safe load is in kips (thousands of pounds).
Data apply to concentric loads. If eccentric loading is to occur, special tables must be consulted.

ROUND, EXTRA HEAVY COLUMNS							
Diam. of Column Inches	Weight Per Foot Lbs.	Unbraced Length of Column — In Feet					
		8	10	12	14	16	18
4	21	55	47	39			
4^1/$_2$	27	70	62	52	43		
5^1/$_2$	39	108	99	89	79	70	60
6^5/$_8$	56	159	148	137	126	115	104
8^5/$_8$	91	265	250	239	225	212	199

SQUARE COLUMNS							
Exterior Dimensions Inches	Weight Per Foot Lbs.	Unbraced Length of Column — In Feet					
		8	10	12	14	16	18
3 × 3 × ¼	16	37	30	24			
3½ × 3½ × ¼	20	50	43	37	30		
4 × 4 × ¼	26	64	57	50	43	36	
4 × 6 × ¼	36	85	76	66	57	48	39
4 × 8 × ¼	46	106	94	83	71	60	46
5 × 5 × ¼	37	94	86	80	72	64	57
6 × 6 × ¼	50	127	119	111	103	96	88

Lally Column Company

Fireproofing

One disadvantage of steel framing is that it loses its strength when subjected to heat. Steps are taken to overcome this. In buildings where fireproofing is required, steel columns are encased in a fireproof material such as metal lath covered with vermiculite plaster, or they have a concrete shell cast around them. See Fig. 23-61.

Steel beams are commonly fireproofed by installing concrete floors above and fireproof ceilings below. Such a ceiling is usually metal lath covered with one-inch, vermiculite plaster.

Consult an architectural standards book for other ways of fireproofing and the rating of the various methods. The common rating method is to report how many hours of fireproofing will protect the beam at both the indicated temperature and the type and thickness of the fireproofing.

Structural Steel Drawings

A structural steel framed building is shown in Fig. 23-62. The location, size, and type of each steel member is planned by an engineer. A drafting technician must understand the terms used before making structural drawings. Some of the common terms follow:

Martin K. Eby Construction Company, Inc.

Fig. 23-62.   A structural steel framed building.

Columns — Vertical steel members used to support the roof and floor. See Fig. 23-63.

Girders — Structural members running horizontally between columns. See Fig. 23-63.

Filler beams — Structural members running horizontally between girders. See Fig. 23-63.

Pitch — The distance between the center lines of fasteners or holes. See Fig. 23-64.

Gage line — A continuous center line passing through holes or fasteners. See Fig. 23-64.

Gage distance — The distance the gage line is from the back of the structural member. See Fig. 23-64.

Edge distance — The distance the first hole or fastener is from the end of the member. See Fig. 23-64.

Slope — An indication of the angle a member makes with the horizontal.

Gusset plates — Used on trusses to join the structural members. See Fig. 23-64.

Structural Steel Shapes

The most commonly used structural steel shapes are shown in Fig. 23-48. These members are made

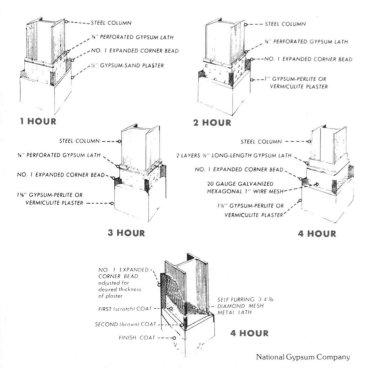

1 HOUR
STEEL COLUMN
¾" PERFORATED GYPSUM LATH
NO. 1 EXPANDED CORNER BEAD
½" GYPSUM-SAND PLASTER

2 HOUR
STEEL COLUMN
⅜" PERFORATED GYPSUM LATH
NO. 1 EXPANDED CORNER BEAD
1" GYPSUM-PERLITE OR VERMICULITE PLASTER

3 HOUR
STEEL COLUMN
¾" PERFORATED GYPSUM LATH
NO. 1 EXPANDED CORNER BEAD
1¾" GYPSUM-PERLITE OR VERMICULITE PLASTER

4 HOUR
STEEL COLUMN
2 LAYERS ½" LONG-LENGTH GYPSUM LATH
NO. 1 EXPANDED CORNER BEAD
20 GAUGE GALVANIZED HEXAGONAL 1" WIRE MESH
1½" GYPSUM-PERLITE OR VERMICULITE PLASTER

4 HOUR
NO. 1 EXPANDED CORNER BEAD adjusted for desired thickness of plaster
FIRST (scratch) COAT
SECOND (brown) COAT
FINISH COAT
SELF FURRING 3.4 lb DIAMOND MESH METAL LATH

National Gypsum Company

Fig. 23-61.   Typical ways of fireproofing a steel column.

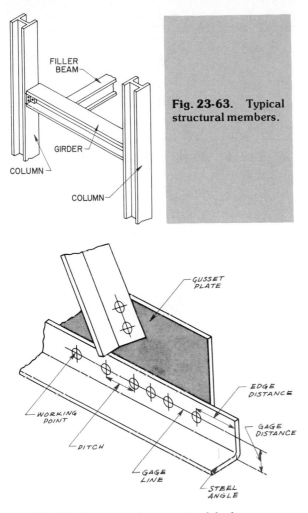

Fig. 23-63. Typical structural members.

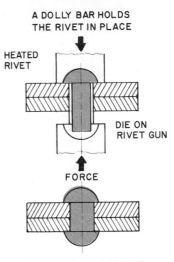

Fig. 23-65. Rivets are installed by heating the rivet, putting it in place, and forming a second head with a die on a rivet gun.

Fig. 23-64. Terms used in structural drafting.

in a wide variety of sizes. Tables give the maximum load each can carry.

Abbreviations and Symbols

A standard system of recording data in the form of notes on drawings is used. It was developed by the American Institute of Steel Construction. The system does not use inch or pound marks. They are understood as the symbol is read. Some of the most frequently used symbols are shown in Fig. 23-49.

Joining Steel Members

Structural steel members are joined by riveting, bolting, or welding. A **rivet** is installed by heating it to a cherry red color. It is inserted into a hole which is about 1/16 inch larger than the rivet diameter. A

head is formed on the straight end. See Fig. 23-65. The rivet is held in the hole with a **dolly.** It is a tool that fits over the head of the rivet. While one worker holds the dolly another forms the other head with a rivet gun.

When members are riveted together in a structural steel shop they are called **shop rivets.** When rivets are installed on the building site they are called **field rivets.** Shop rivets are shown on drawings as open circles. Field rivets are drawn as a solid black circle. See Fig. 23-66.

Structural bolts are widely used to erect steel structures. Special high tensile steel bolts are used. The selection of the proper size bolt is critical to structural soundness. Each bolt must be tightened to the proper tension. These data are given in the publication **Structural Joints Using ASTM A325 Bolts,** published by the American Institute of Steel Construction. Symbols for drawing bolts are in Fig. 23-66.

Welding is another way of joining steel members. The **fillet weld** is most commonly used. Specifications for welding are in the **Manual of Steel Construction** published by the American Institute of Steel Construction. Some welding details are shown in Fig. 23-85.

Spacing Bolts and Rivets

The gage line locates the center line of the fastener that is parallel with the length of the structural member. The pitch is the distance between the actual centers of each fastener. Refer again to Fig. 23-64.

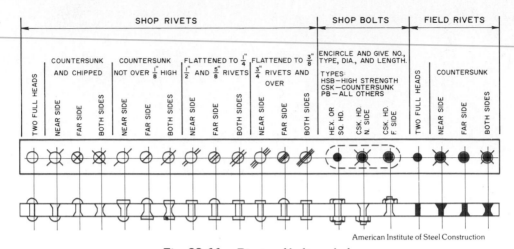

American Institute of Steel Construction

Fig. 23-66.   Rivet and bolt symbols.

American Institute of Steel Construction

LEG	8	7	6	5	4	3½	3	2½	2	1¾	1½	1⅜	1¼	1
G	4½	4	3½	3	2½	2	1¾	1⅜	1⅛	1	⅞	⅞	¾	⅝
G 1	3	2½	2¼	2										
G 2	3	3	2½	1¾										

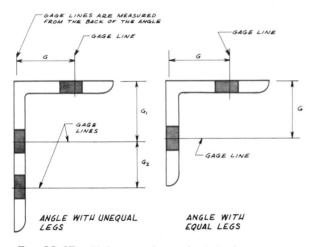

Fig. 23-67.   Hole gages for angles in inches.

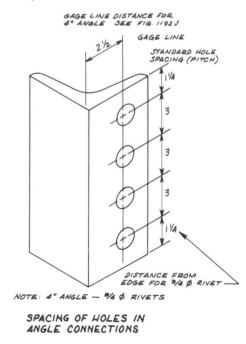

NOTE: 4" ANGLE — ¾ Ø RIVETS

SPACING OF HOLES IN ANGLE CONNECTIONS

Fig. 23-68.   Spacing of holes in angle connections.

The location of gage lines has been standardized for various connections. They vary with the length of the leg. In Fig. 23-67 are gage line distances for angle connections. The gage line is located by measuring from the back of the angle.

The pitch between holes of angle connectors is usually in 3-inch units. See Fig. 23-68. The angle connection is used to fasten a girder to a column.

The spacing of gage lines in **I** beams is shown in Fig. 23-69. The gage lines are measured from the center of the flanges. See Fig. 23-70. Spacing is usually 3″, 3½″, and 5½″. These holes are spaced along gage lines in 3-inch units.

For structural strength, holes must be kept specified distances from the edge of a member. The minimum distances are given in Fig. 23-71. Notice that they vary with the diameter of the rivet.

Standard Connections

Connections are used to fasten beams to other structural members. They are standardized by the

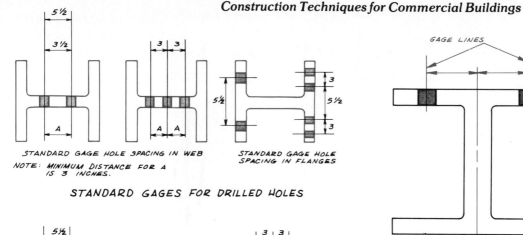

STANDARD GAGE HOLE SPACING IN WEB

NOTE: MINIMUM DISTANCE FOR A IS 3 INCHES.

STANDARD GAGE HOLE SPACING IN FLANGES

STANDARD GAGES FOR DRILLED HOLES

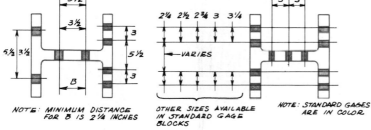

NOTE: MINIMUM DISTANCE FOR B IS 2¼ INCHES

OTHER SIZES AVAILABLE IN STANDARD GAGE BLOCKS

NOTE: STANDARD GAGES ARE IN COLOR

STANDARD GAGES FOR PUNCHED HOLES

American Institute of Steel Construction

Fig. 23-70. Gage lines are located from the center of I-beam flanges.

Fig. 23-69. Standard gages for punched holes.

Fig. 23-71.

Minimum Edge Distance (Inches) for Punched Holes

Rivet Diameter (inches)	In Sheared Edges	In Rolled Edge of Plate	In Rolled Edge of Structural Shapes*
½	1	⅞	¾
⅝	1⅛	1	⅞
¾	1¼	1⅛	1
⅞	1½	1¼	1⅛
1	1¾	1½	1¼
1⅛	2	1¾	1½
1¼	2¼	2	1¾

*May be decreased 1/8 inch when holes are near end of beam.

American Institute of Steel Construction

American Institute of Steel Construction. The two types commonly used are framed and seated. See Fig. 23-72. The **framed connections** are fastened to the side of the beam. The **seated connections** are fastened to the column and the beam rests on them.

Other types of connections are plates, tee and double-angle hangers, and bracket plates. See Fig. 23-73. Column connections are shown in Fig. 23-74.

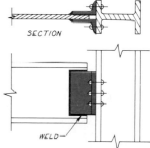

SECTION

TEE BEAM FRAMING

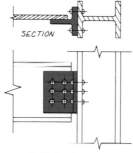

SECTION

WELD—

WELD BEAM CONNECTION

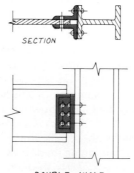

SECTION

DOUBLE ANGLE FRAMED CONNECTION

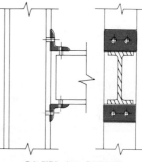

BOLTED AND RIVETED SEATED BEAM CONNECTIONS

Fig. 23-72. Beam connections.

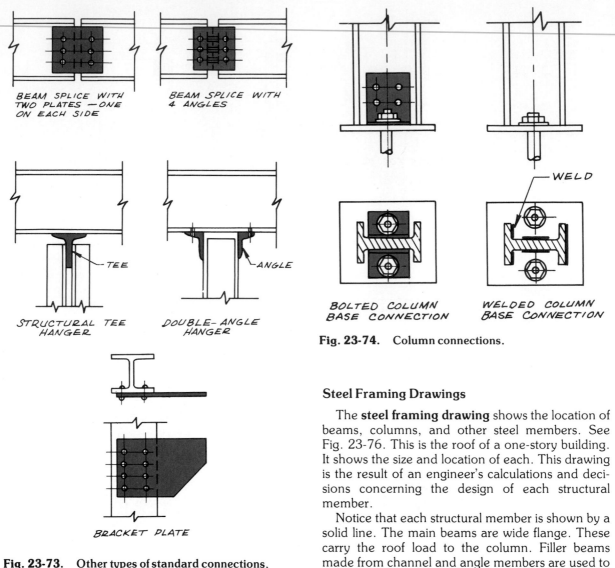

BEAM SPLICE WITH TWO PLATES — ONE ON EACH SIDE

BEAM SPLICE WITH 4 ANGLES

STRUCTURAL TEE HANGER

DOUBLE-ANGLE HANGER

TEE

ANGLE

BRACKET PLATE

Fig. 23-73. Other types of standard connections.

BOLTED COLUMN BASE CONNECTION

WELDED COLUMN BASE CONNECTION

WELD

Fig. 23-74. Column connections.

Steel Framing Drawings

The **steel framing drawing** shows the location of beams, columns, and other steel members. See Fig. 23-76. This is the roof of a one-story building. It shows the size and location of each. This drawing is the result of an engineer's calculations and decisions concerning the design of each structural member.

Notice that each structural member is shown by a solid line. The main beams are wide flange. These carry the roof load to the column. Filler beams made from channel and angle members are used to support the roof decking.

The size of each member is shown on the member. The actual length of each member is not shown. This is figured and shown on the shop drawings.

On the framing drawing, the location of each member is dimensioned to the center line of the member. The columns are also located.

Buildings several stories high have a separate framing plan for each floor plus one for the roof.

Erection Plans

The erection plan for the roof in Fig. 23-76 is shown in Fig. 23-77. **Erection plans** are really a type of assembly drawing. They show how the various members are put together.

Structural Steel Drawings

Three types of drawings are commonly used. These are a steel framing drawing, an erection drawing, and shop drawings. The sample drawings that follow are for the small building shown in Fig. 23-75.

Framing and erection drawings are usually made to the scale of $1/4'' = 1'-0''$. They can be made to a larger scale if desired. See Figs. 23-76 through 23-78. Shop drawings are usually drawn $3/4'' = 1'-0''$ to $1'' = 1'-0''$. See Fig. 23-79.

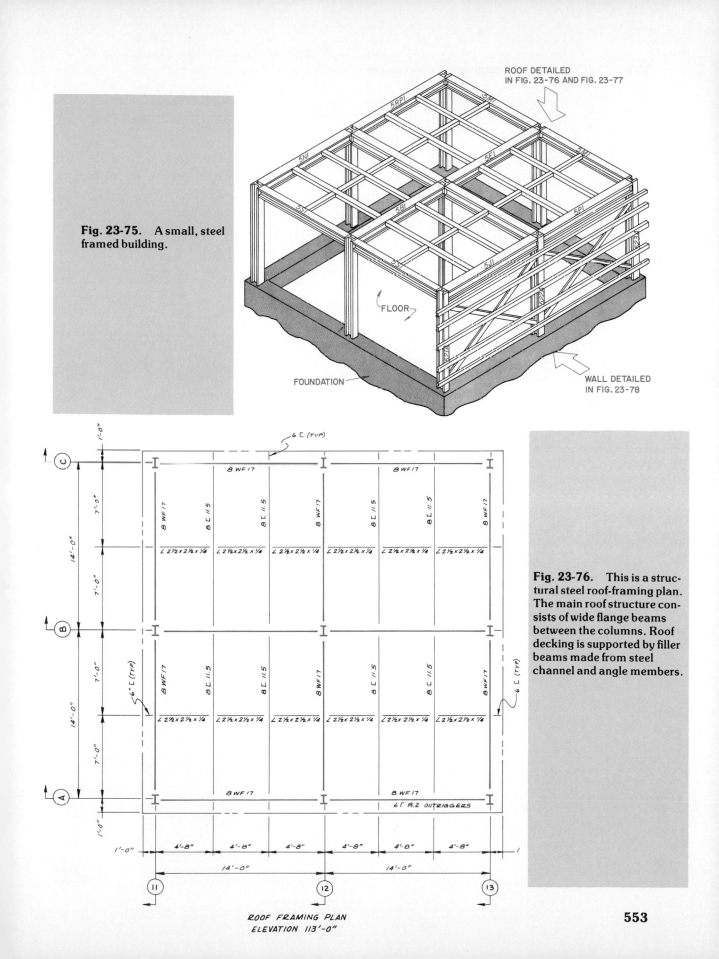

Fig. 23-75. A small, steel framed building.

ROOF DETAILED
IN FIG. 23-76 AND FIG. 23-77

WALL DETAILED
IN FIG. 23-78

FLOOR

FOUNDATION

Fig. 23-76. This is a structural steel roof-framing plan. The main roof structure consists of wide flange beams between the columns. Roof decking is supported by filler beams made from steel channel and angle members.

ROOF FRAMING PLAN
ELEVATION 113'-0"

553

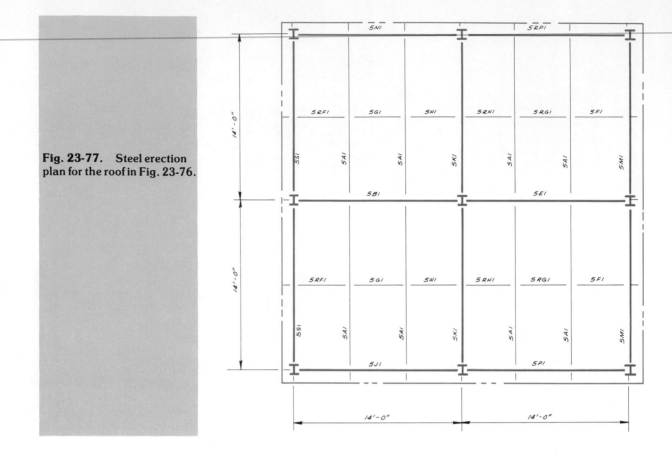

Fig. 23-77. Steel erection plan for the roof in Fig. 23-76.

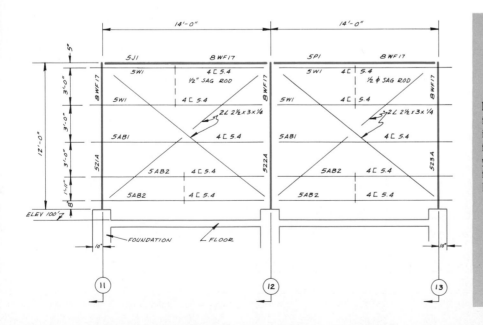

Fig. 23-78. An exterior wall-framing and erection plan. This is a combination plan with framing and erection data. Can you find this wall on the roof-framing and erection plans in Figs. 23-76 and 23-77?

Erection plans show those putting the framework together where each member belongs. Each member is made to the correct length in the structural shop. They are numbered. These numbers are shown on the erection plan. The construction crew finds a member with the number as shown on the plan and puts it in the place shown. They do not need to know the size data since they are only assembling the members.

The erection plan has only the dimensions locating the columns. Since each member is marked and all holes for assembly are drilled, no measuring is necessary. Many of the connections are shop-assembled.

Sometimes the framing plan and erection plan data are placed on one drawing. This can be done if the structure is not too complex. See Fig. 23-78.

If the building is several stories, an erection plan is made for each floor and the roof.

Wall Plans

If exterior walls are composed of structural steel members, framing and erection plans are made for each wall. The same principles apply here as just discussed for the roof plan. A combined framing and erection plan for one wall of the building in Fig. 23-75 is shown in Fig. 23-78.

Notice that the identification numbers for the columns appear on the wall plan. This does not appear on the floor or roof plan because they are not a part of that section of the structure.

Any structural member that appears in several drawings uses the same identification number on all drawings. In Fig. 23-78, beams 5J1 and 5P1 appear at the roof line. Find these on the erection drawing in Fig. 23-77.

This wall is to be covered with metal siding in large sheets. To support this steel, channel and angle members are used between the columns.

Shop Drawings

Shop drawings are detail drawings of each structural member. The drafting technician learns from the framing plan how structural members are joined and the design of connections.

Shop drawings show the size and length of the member. They detail the connections used. All hole locations are given. Dimensions must be complete because the member is made in the shop from this drawing. It must be possible for the member to be connected to the adjoining members in the field or shop with no additional machine work. Shop drawings include notes giving special instructions.

The scale used should produce a drawing large enough to be clear. A common scale is $3/4'' = 1'-0''$. Very long members do not have the length to scale. Break lines are not used. The member is simply drawn shorter but dimensioned true size. The height and width are always drawn to scale.

Each part of a structural system is given a number. This must appear on the shop drawing and erection drawing. It is used whenever the piece appears on a drawing. The number is painted on each member after it is made. If several members are identical, they are given the same number.

In Fig. 23-79, a shop drawing of a beam is shown. It is Beam 5J1 as shown in Figs. 23-75, 23-77 and 23-78. This beam has welded connections. The detail in Fig. 23-79 presents examples of how welded and bolted connections would be detailed.

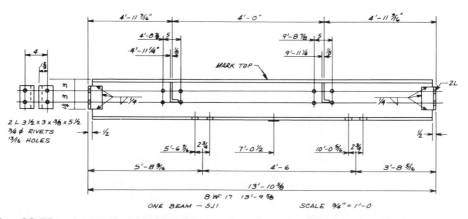

Fig. 23-79. A typical welded steel beam shop drawing. All parts of this type of drawing are to scale except the length of the beam. This is a beam from the building in Fig. 23-75.

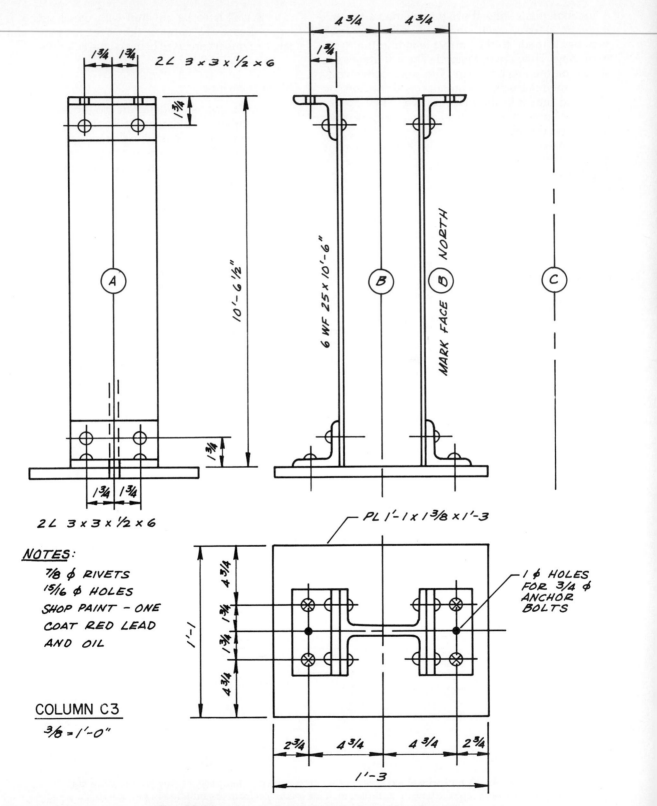

Fig. 23-80.   A typical riveted steel column shop drawing.

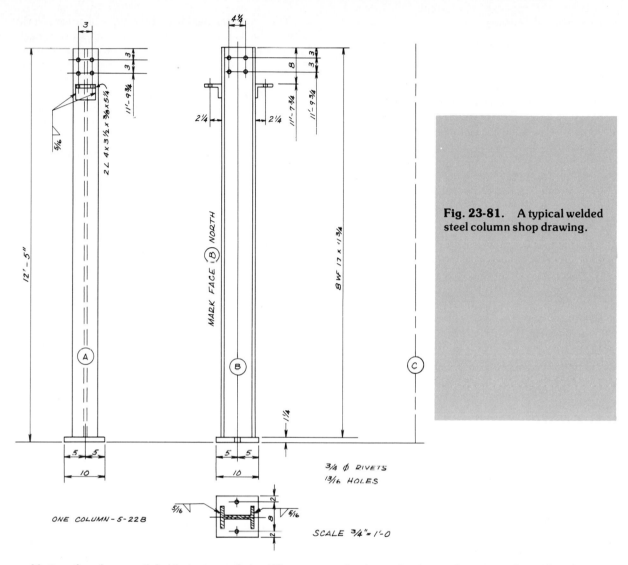

Fig. 23-81. A typical welded steel column shop drawing.

Notice the shop and field rivet symbols. Whenever possible, the structure is designed so holes line up on a common gage line. Each connection requires the location of gage lines, edge distance, and pitch.

Connections usually extend beyond the end of the beam. This set back distance is the difference between the overall length of the assembled member and the length of the beam alone with the connections. In Fig. 23-79, the beam alone is 13'-9 5/8" long. With the connections it measures 13'-10 5/8". This is the actual distance between the columns to which it is to be attached.

The left end view shows the angle connection holes and their dimensions. When the right end connections are the same, they need not be drawn. The beam is not shown in the end view.

A shop drawing of a riveted steel column is shown in Fig. 23-80. Usually, it can be described with views of two faces. The section view is actually a view from the top of the column.

The angle connections on this column are shop riveted to the column and plate. Notice that the top connections extend ½-inch beyond the end of the beam forming the column. The faces of columns are marked with a letter. The surface facing north is marked. This helps in the erection of the building. When two faces are the same, only one need be drawn. The other can be indicated with a center line and an identifying letter.

A welded steel column is illustrated in Fig. 23-81. All shop assembly was welded. The parts to be field-assembled have holes drilled for rivets or bolts. This column is the one in the center of the building shown in Figs. 23-75, 23-76, and 23-77.

Drawing Steel Trusses

Trusses are used to span distances too great for girders. Steel trusses are made from tees, wide flange beams, or angles.

Trusses must be carefully designed. The engineer must design them to withstand tension and compression stresses. The truss illustrated in Fig. 23-82 shows these stresses in a Fink truss. The common types of steel trusses are the Warren, Pratt, Fink, scissors, and bowstring. See Fig. 23-83.

The **Warren and Pratt trusses** are used to carry floor and roof loads. They are also good to use if the roof has to carry an extra load, such as an overhead crane.

The **Fink truss** is used for roofs where a steep slope is desired. It usually carries only the roof load.

The **bowstring truss** can span long distances. It gives the roof good slope.

The **scissors truss** is used for high-pitched roofs.

Drafting technicians must know the terms used in truss design. Refer again to Fig. 23-83. These are:

Chord — One of the principle structural members that is braced by web members.

Web members — Internal braces running between the chords.

Panels — The distances between two web members measured along a chord.

Span — The horizontal distance covered by the trusses with no support from columns or walls.

The engineer decides which type of truss will do the job best and then computes the stresses on each member. Next, a simple **design drawing** is made which shows tension and compression in pounds. See Fig. 23-84. Tension is shown by a plus (+) and compression by a minus (−). The engineer then selects structural members for each part of the

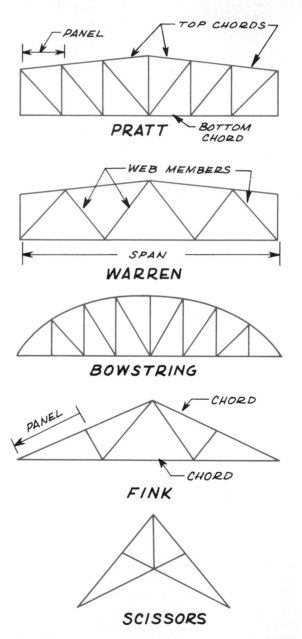

Fig. 23-83. Typical types of steel trusses. The type of truss selected for use depends on the special needs of the building.

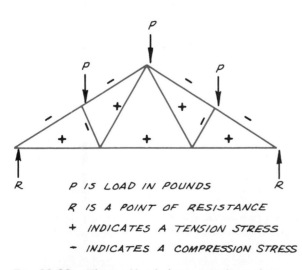

P IS LOAD IN POUNDS

R IS A POINT OF RESISTANCE

+ INDICATES A TENSION STRESS

− INDICATES A COMPRESSION STRESS

Fig. 23-82. The roof load places members of a truss under tension and compression stresses. The truss shown is a Fink truss.

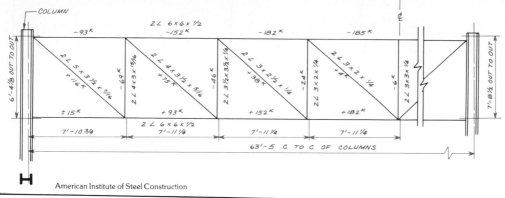

Fig. 23-84. A design drawing for a truss is prepared by the engineer. The shop drawing for this truss is in Fig. 23-86. The term "out to out" on the height dimensions refers to the outside surface at the top and bottom members.

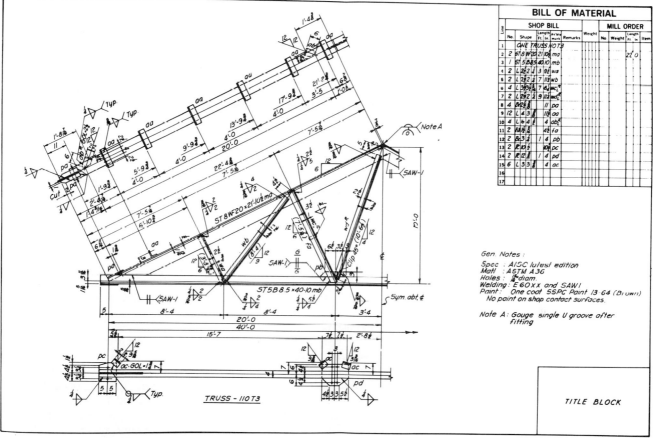

Fig. 23-85. A simple welded Fink truss. The web members are made of steel angles. The top chord is a wide flange T-beam. The bottom chord is a T-beam.

truss. These are recorded on the design drawing. The shop drawing is made from the data on this design drawing.

All information needed to make the truss is given in the **shop drawing.** Usually the truss is drawn in the assembled position. Individual pieces are seldom detailed separately since they are generally few in number and simple in design. A welded Fink truss is shown in Fig. 23-85. Notice the use of welding symbols. Ordinarily, only half a truss is drawn since the other half is identical. Study carefully the data in the bill of material.

In Fig. 23-86 is the shop drawing for the truss design shown in Fig. 23-84. Notice that the size of each piece is recorded. The gusset plates are fully dimensioned. Hole locations are indicated according to standard design practices. Some drawings can become very crowded. Figure 23-86 is such a

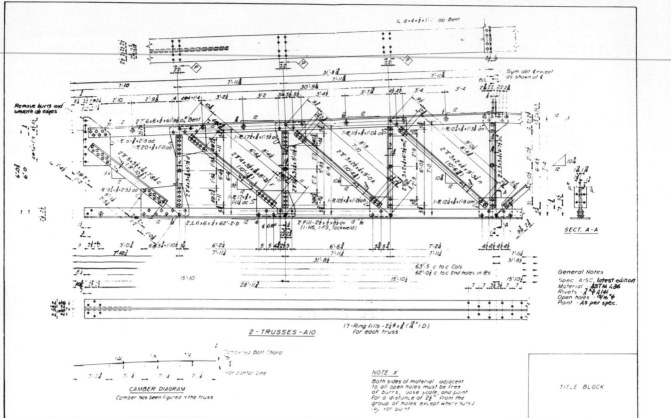

Fig. 23-86. A Pratt truss shop drawing. It is based on the design in Fig. 23-84. This is a riveted truss. All structural members are angles. Notice the large amount of dimensioning detail required.

drawing. The placement of dimensions must be very carefully planned. Such a drawing is very difficult to check; therefore the drafting technician must work very carefully. If there are details that can be clarified with a section, they are also drawn.

Generally, each piece of a truss is not identified with a number. The entire assembled truss is numbered. It appears on an erection drawing as a single member; therefore, it needs only a single identifying number. It is usually shop-assembled. An extremely large truss might be shipped to the site in two pieces which are field-joined.

Trusses are drawn to scale. All parts of the truss use the same scale. Details may be drawn to a larger scale.

Rigid Frame Drawing

The members of rigid frames are steel, girder-like members. They are assembled by welding, riveting, or bolting. They form the column and roof structure.

A drawing of a low rigid frame is shown in Fig. 23-87. The assembled form is drawn. Large scale

detail drawings are used to clarify assembly information.

Steel Roof Decking and Wall Panels

Numerous steel decking systems are manufactured. Some decking types are exposed to the weather, while others are used as forms for supporting a poured concrete roof or floor. Still others provide for electrical and heating systems to be built into the floor. Selected examples of these are illustrated in the following paragraphs.

Corrugated Panels

Design data for three types of corrugated roof and wall panels are given in Fig. 23-88. Steelbestos is a trade name for a steel sheet to which is applied an adhesive, a layer of asphalt-impregnated asbestos felt, and a plastic outer coating. The plastic coating is available in a variety of colors. The roof and wall panels of the power plant shown in Fig. 23-89 are made of corrugated steel.

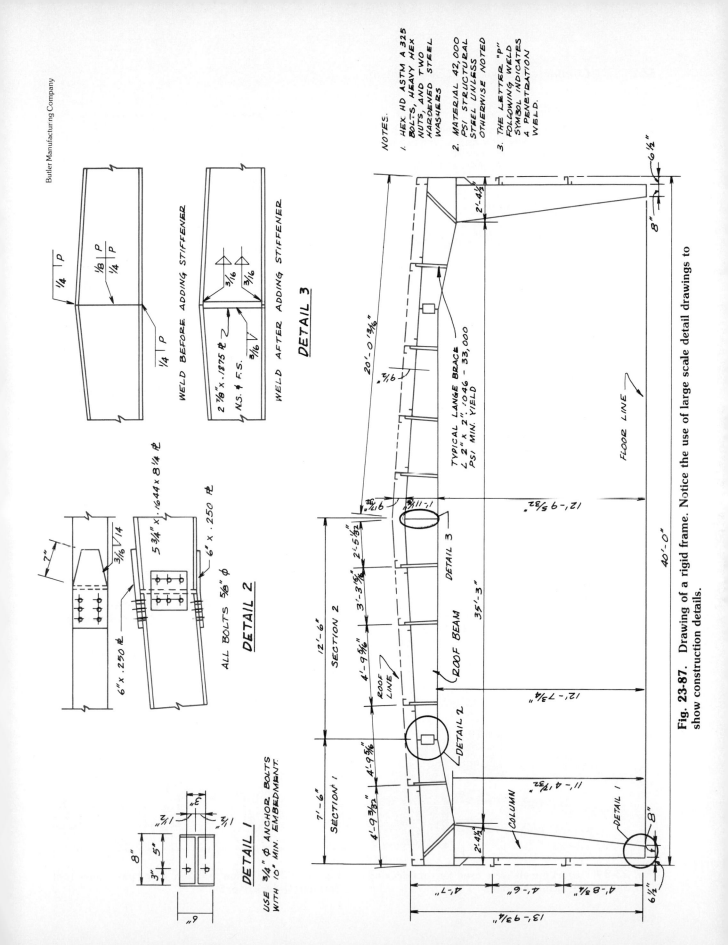

Fig. 23-87. Drawing of a rigid frame. Notice the use of large scale detail drawings to show construction details.

Fig. 23-88.

Design Data for Corrugated Panels
Maximum span-siding and panels — 20 lbs./sq. ft. total load; roofing — 40 lbs./sq. ft. total load.
Weights based on gross squares for roofing and siding and net squares for panels.

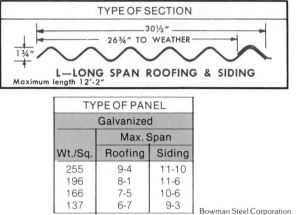

TYPE OF SECTION
← 30½" →
← 26¾" TO WEATHER →
1¾"
L—LONG SPAN ROOFING & SIDING
Maximum length 12'-2"

TYPE OF PANEL		
Galvanized		
	Max. Span	
Wt./Sq.	Roofing	Siding
255	9-4	11-10
196	8-1	11-6
166	7-5	10-6
137	6-7	9-3

Bowman Steel Corporation

Steel Roof Deck

Many types of steel roof decking are available. One type is illustrated in Fig. 23-90. The units have interlocking side joints. They are fastened to the supporting roof structure by arc welding. The safe loads for various spans are given in Fig. 23-91. After the steel is welded into place, it is covered with insulation board and a built-up roofing. See Fig. 23-92.

A decking designed for long, unsupported spans of roof is shown in Fig. 23-93. The safe load varies with the depth of the decking and the gauge of the steel. Refer again to Fig. 23-91. The decking sheets are welded to the structural steel frame. A built-up roof is applied over the decking, as shown in Fig. 23-94.

Bowman Steel Corporation

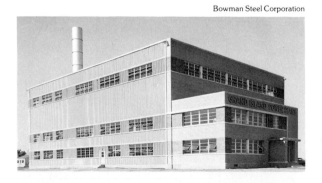

Fig. 23-89. Steel panels were used for roof decking and wall panels in this building.

H.H. Robertson Company

Fig. 23-90. Section through short-span, steel roof decking (Q-deck, No.3).

Fig. 23-91.

Allowable Loads for Steel Roof Decking[1]

LONG SPAN								
Type Span	5-75-18 A[2]	B[3]	5-75-16 A	B	5-75-14 A	B	5-75-12 A	B
16'-0"	84		144		183		261	
18'-0"	74		114		145		206	
20'-0"	67		92		117		166	
22'-0"	60		76		97		138	
24'-0"	50	50	64		81		116	
26'-0"			55	51	69	65	99	94
28'-0"			47	41	60	52	85	75
30'-0"			41	33	52	43	74	61
32'-0"			36	27	46	35	65	51

[1]Data are for **Q** deck No. 3 as manufactured by H.H. Robertson Co.

[2]Column A gives total, uniformly distributed loads (in lbs./sq. ft.) which unit will carry at a stress not to exceed 18,000 psi, premised on simple span ($\frac{1}{8}$ WL).

[3]Column B gives uniformly distributed loads per square foot on a simple span which will cause unit to deflect not more than ${}^{1}/240$th of span.

H.H. Robertson Company

H.H. Robertson Company

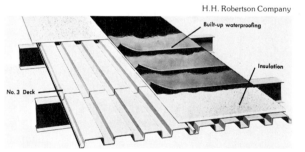

Fig. 23-92. Steel roof decking (Q-deck, No.3) with built-up roof.

H.H. Robertson Company

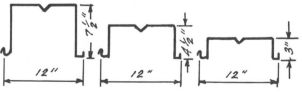

Fig. 23-93. Section through long-span, steel roof decking (Q-deck, No. 5).

H.H. Robertson Company

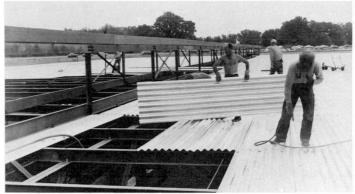

Fig. 23-94. Steel roof decking (Q-deck, No. 5) with built-up roof.

H.H. Robertson Company

Fig. 23-95. Exposed steel decking forms a ceiling.

Steel roof decking can be rapidly erected in any weather. The corrugated undersurface provides an attractive ceiling; however, suspended ceilings can be used. See Fig. 23-95.

Another type of steel floor and roof panel is designed to serve as a form for poured concrete roof or floor decks. It is coated with zinc to insure permanence. The zinc coating also unites with the cement in the concrete, thus forming a strong bond.

The steel sheets are welded to the structural steel frame. A special washer is placed on the sheet; the welder strikes an arc in the washer hole and burns through the sheet. This builds a plug weld from the structural steel member up into the washer, thus greatly strengthening the weld. See Fig. 23-96.

The steel sheets are then covered with an insulating concrete as shown in Fig. 23-97. The concrete serves as thermal insulation and as an effective vapor barrier, preventing moisture from penetrating the roof. This type of roof construction is lightweight. A built-up roof is applied over the insulating concrete.

The floor slab is structural concrete with a reinforcing mesh. See Fig. 23-98. The steel sheets may

Granco Steel Products Company

Fig. 23-96. Steel decking placed over structural steel frame and welded into place. The welded deck provides a working surface for the other trades.

Granco Steel Products Company

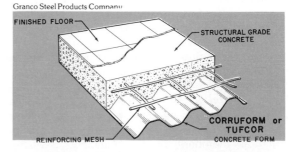

Fig. 23-97. Insulating concrete being poured and screeded over steel decking. This type of roof is erected rapidly.

Granco Steel Products Company

FINISHED FLOOR

STRUCTURAL GRADE CONCRETE

CORRUFORM or TUFCOR CONCRETE FORM

REINFORCING MESH

Fig. 23-98. A section through a floor with steel decking, a reinforced concrete slab, and a finished floor.

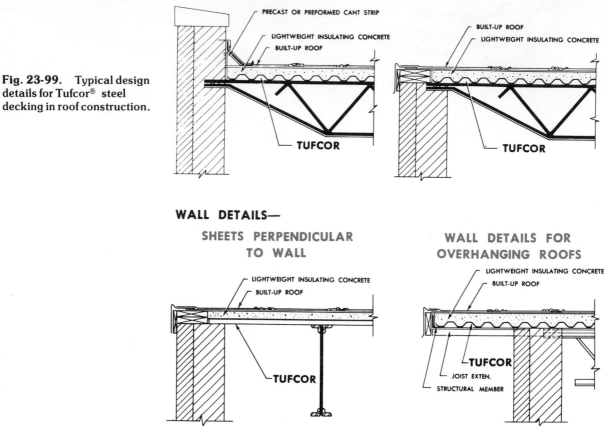

Fig. 23-99. Typical design details for Tufcor® steel decking in roof construction.

Granco Steel Products Company

Fig. 23-101.

Design Data for Steel Roof Decking and Insulating Concrete Slab

First consider the structural steel framing for the most economical purlin spacing; then select the type of decking; and finally choose the thickness of insulating concrete fill for desired **U** factor.

S signifies Standard Corruform and **T** Tufcor. Gauge of Tufcor is indicated by number preceding **T**

DECK SELECTION							
Uniform Design Live Load (lbs./sq. ft.)	Purlin Spacing O.C.						
	3'-0"	3'-6"	4'-0"	4'-6"	5'-0"	5'-6"	6'-0"
20	S	S	S	S	26T	26T	26T
30	S	S	S	26T	26T	26T	26T
40	S	S	S	26T	26T	26T	26T

Granco Steel Products Company

be used over precast concrete joists to serve as a permanent deck for a cast-in-place concrete floor.

While construction details vary, commonly used design details are illustrated in Figs. 23-99 and 23-100. Design data for roof and floor construction are presented in Figs. 23-101 and 23-102.

Cellular-Steel Floor Systems

A number of floor systems are manufactured using load-bearing, cellular-steel units for electrical distribution and heat and cooling distribution. In the system shown in Fig. 23-103, the cellular-steel units serve for load bearing and electrical distribution. The cellular-steel decking is placed upon the structural members of the building. See Fig. 23-104. On top of this, placed at right angles, is a crossover header. This carries wiring across the decking and permits it to extend into the cellular units, thus allowing electrical outlets and telephones to be located wherever needed. These outlets can be moved easily at any time.

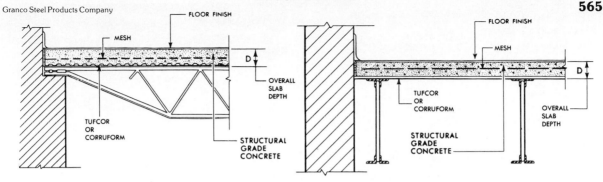

Granco Steel Products Company

FLOOR WITH STEEL CONSTRUCTION

PRECAST CONCRETE JOISTS

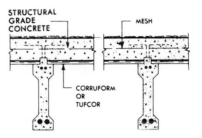

FLOOR WITH PRECAST CONCRETE JOISTS

Fig. 23-100. Typical design for steel decking in floor construction.

Fig. 23-102.

Design Data for Floor Construction with Steel Decking and Concrete Slab

Slab capacity is the safe superimposed load which slab will carry when reinforced as shown.
Form material symbols: **S** - Standard Corruform; **HD** -Heavy Duty Corruform.
Total slab weight and construction load equals 50 psf minimum, or greater.
Deflection under concrete load equals $^1/_{240}$ or less.

Depth of Slab	Slab Weight	Minimum Reinforcement Recommended		Span, O.C.					
				1'-6"	2'-0"	2'-6"	3'-0"	3'-6"	4'-0"
2½"	28	6 × 6 — 10/10	Form Material	S	S	S	S	S	
			Slab Capacity	365	193	113	70	44	
3"	35	6 × 6 — 10/10	Form Material	S	S	S	S	S	HD
			Slab Capacity	519	277	164	103	66	35

TABLE 1—FORM SELECTION TABLE AND ALLOWABLE SAFE SUPERIMPOSED LOADS ON FINISHED SLAB (psf)

For **2½" and 3" slabs.** For 3½" deep slab or over, see "long span construction" detail.

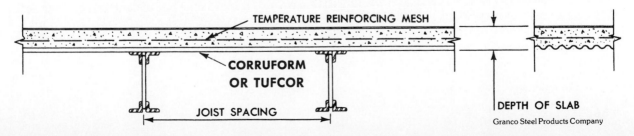

Granco Steel Products Company

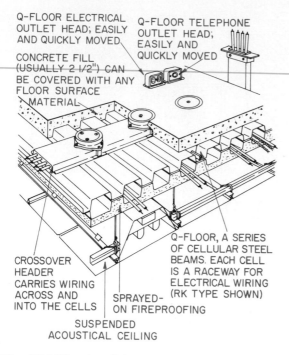

Q-FLOOR ELECTRICAL OUTLET HEAD; EASILY AND QUICKLY MOVED.

Q-FLOOR TELEPHONE OUTLET HEAD; EASILY AND QUICKLY MOVED

CONCRETE FILL (USUALLY 2 1/2") CAN BE COVERED WITH ANY FLOOR SURFACE MATERIAL

Q-FLOOR, A SERIES OF CELLULAR STEEL BEAMS. EACH CELL IS A RACEWAY FOR ELECTRICAL WIRING (RK TYPE SHOWN)

CROSSOVER HEADER CARRIES WIRING ACROSS AND INTO THE CELLS

SPRAYED-ON FIREPROOFING

SUSPENDED ACOUSTICAL CEILING

H.H. Robertson Company

Fig. 23-103. A cellular-steel floor used to carry a load and serve as an electrical raceway. Notice the sprayed-on fireproofing layer on the bottom of the steel.

Fig. 23-104. Cellular-steel, load-bearing units installed in a multistory building. This illustration shows both electrical-wiring cells and heat and cooling cells. A concrete floor is poured over the steel deck.

Fig. 23-105.

Design Data for Laminated Structural Members · Rigid Frame Units

SPAN		ROOF SLOPE 3:12				ROOF SLOPE 6:12				
	Loading Per Lineal Foot	Leg Height	Dimension			Loading Per Lineal Foot	Leg Height	Dimension		
			A	B	C			A	B	C
40′	600 lbs.	12′	$7'' \times 19^1/_2''$	10″	10″	600 lbs.	12′	$5^1/_4'' \times 19^1/_2''$	8″	10″
		16′	$7'' \times 21^1/_8''$	10″	11″		16′	$5^1/_4'' \times 21^1/_8''$	8″	10″
		20′	$7'' \times 21^1/_8''$	10″	11″		20′	$7'' \times 19^1/_2''$	10″	10″
	800 lbs.	12′	$7'' \times 22^3/_4''$	10″	11″	800 lbs.	12′	$7'' \times 19^1/_2''$	10″	10″
		16′	$7'' \times 24^3/_8''$	10″	12″		16′	$7'' \times 21^1/_8''$	10″	11″
		20′	$7'' \times 24^3/_8''$	10″	12″		20′	$7'' \times 22^3/_4''$	10″	11″
	1000 lbs.	12′	$7'' \times 26''$	10″	14″	1000 lbs.	12′	$7'' \times 21^1/_8''$	10″	11″
		16′	$7'' \times 27^5/_8''$	10″	14″		16′	$7'' \times 24^3/_8''$	10″	12″
		20′	$7'' \times 27^5/_8''$	10″	14″		20′	$7'' \times 26''$	10″	13″
60′	600 lbs.	12′	$7'' \times 27^5/_8''$	10″	16″	600 lbs.	12′	$7'' \times 22^3/_4''$	10″	12″
		16′	$7'' \times 29^1/_4''$	10″	15″		16′	$7'' \times 24^3/_8''$	10″	12″
		20′	$9'' \times 26''$	12″	13″		20′	$7'' \times 27^5/_8''$	10″	14″
	800 lbs.	12′	$9'' \times 27^5/_8''$	12″	16″	800 lbs.	12′	$7'' \times 27^5/_8''$	10″	16″
		16′	$9'' \times 29^1/_4''$	12″	15″		16′	$7'' \times 29^1/_4''$	10″	15″
		20′	$9'' \times 30^7/_8''$	12″	15″		20′	$9'' \times 27^5/_8''$	12″	14″
	1000 lbs.	12′	$9'' \times 30^7/_8''$	12″	20″	1000 lbs.	12′	$7'' \times 29^1/_4''$	10″	19″
		16′	$9'' \times 34^1/_8''$	12″	17″		16′	$9'' \times 27^5/_8''$	12″	14″
		20′	$9'' \times 34^1/_8''$	12″	17″		20′	$9'' \times 30^7/_8''$	12″	15″

Timber Structures, Inc.

566

Glued, Laminated Wood Structural Members

Data concerning butterfly, peaked, peaked and cambered, tapered, and simple straight beams are given in Chapter 9. In this chapter, structural members designed for use in larger buildings are discussed. The types presented include rigid frames; two-hinged or barrel arches; three-hinged arches; bowstring trusses; glued, laminated domes; and laminated wood post and beam construction.

Rigid Frames

The rigid-frame unit provides an economical method of achieving low-pitched roofs, while maintaining desired clearances. It accomplishes the same thing as the three-hinged arch, without the radius at the haunch. Each half bent is composed of separate leg and arm members, precision fabricated and joined by a patented haunch connection that maintains the correct roof slope under design loads. In Fig. 23-105, a rigid frame is illustrated and design data given for selected spans and roof slope. One of these units is shown in Fig. 23-106.

Two-Hinged Barrel Arches

The two-hinged arch, sometimes called the barrel arch, is adaptable for wide spans. It has been used for spans of 250 feet and can be designed for even wider spaces. There are three basic types — the foundation arch, the tied arch, and the buttressed arch. See Fig. 23-107. With **foundation arches,** the horizontal thrust is contained by foundation piers, with or without tie rods below the floor. **Tied arches** are supported by columns or bearing walls, with tie rods placed at the top of the column or wall. The use of tie rods is an important means of containing horizontal thrust. The horizontal thrust of **buttressed arches** is contained by concrete buttresses. Soil conditions and height of buttresses are determining factors as to fitness and cost.

Of course, special consideration must be given to the design of foundations, bearing walls, or buttresses to contain the horizontal thrust; however, this is not considered in this text. In Fig. 23-108, the spans, rise, and radius for selected arches are given, as well as data concerning horizontal thrust.

The arches are connected with purlins which support the roof decking. Since the loading on the purlins is considerably less than that on the arches, they are much smaller in cross section. Glued, laminated beams serve as purlins. Design data for safe

Fig. 23-106. A rigid frame with a slight concave curve to the roof arm, giving a graceful sloping roof.

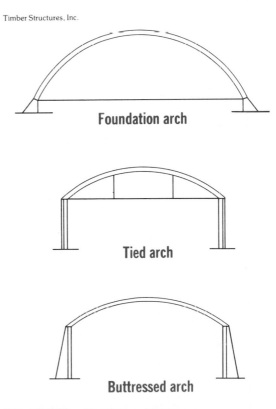

Foundation arch

Tied arch

Buttressed arch

Fig. 23-107.   Two-hinged arches.

Fig. 23-108.

Design Data for Two-Hinged Arches

The spacing of the arches depends upon imposed load. The load data given are per lineal foot of span.

Data are provided for horizontal thrust. This must be contained by the foundation buttresses or tie rods.

Data are based on conservative design criteria, for construction with suitable structural grade combinations, but are subject to local considerations and specific job requirements. Normal lateral support is assumed.

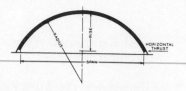

Span	Rise	Radius	SECTION SIZES REQUIRED IN INCHES					Maximum Horizontal Thrust per 100 lbs. of Design Load
			400	600	800	1000	1200	
DEPTH INCREMENTS BASED ON 3/4" LAMINATIONS								
50'	16'-8"	27'-1"	$3\frac{1}{4} \times 13\frac{1}{2}$	$5\frac{1}{4} \times 12\frac{3}{4}$	$5\frac{1}{4} \times 15$	$5\frac{1}{4} \times 16\frac{1}{2}$	$5\frac{1}{4} \times 18$	1750 lbs.
DEPTH INCREMENTS BASED ON 1 3/8" LAMINATIONS								
60'	20'-8"	32'-6"	$5\frac{1}{4} \times 13\frac{3}{4}$	$5\frac{1}{4} \times 16\frac{1}{2}$	$5\frac{1}{4} \times 19\frac{1}{4}$	$5\frac{1}{4} \times 20\frac{5}{8}$	$5\frac{1}{4} \times 22$	2100 lbs.
DEPTH INCREMENTS BASED ON 1 5/8" LAMINATIONS								
80'	26'-8"	43'-3"	$5\frac{1}{4} \times 17\frac{7}{8}$	$5\frac{1}{4} \times 22\frac{3}{4}$	$7 \times 22\frac{3}{4}$	$7 \times 24\frac{3}{8}$	$7 \times 24\frac{3}{8}$	2800 lbs.
100'	33'-4"	54'-2"	$5\frac{1}{4} \times 22\frac{3}{4}$	$7 \times 24\frac{3}{8}$	7×26	$7 \times 29\frac{1}{4}$	$7 \times 30\frac{7}{8}$	3500 lbs.
150'	50'-0"	81'-3"	$7 \times 29\frac{1}{4}$	$9 \times 30\frac{7}{8}$	$9 \times 35\frac{3}{4}$	$11 \times 35\frac{3}{4}$	11×39	5250 lbs.
180'	60'-0"	97'-6"	$9 \times 30\frac{7}{8}$	$9 \times 37\frac{3}{8}$	11×39	$11 \times 42\frac{1}{4}$	$11 \times 45\frac{1}{2}$	6300 lbs.

Span	Rise	SECTION SIZES REQUIRED IN INCHES					Maximum Horizontal Thrust per 100 lbs. of Design Load
		400	600	800	1000	1200	
60'	8'-0 7/16"	$3\frac{1}{4} \times 13$	$5\frac{1}{4} \times 13$	$5\frac{1}{4} \times 14\frac{5}{8}$	$5\frac{1}{4} \times 16\frac{1}{4}$	$5\frac{1}{4} \times 17\frac{7}{8}$	5640 lbs.
80'	10'-8 5/8"	$5\frac{1}{4} \times 14\frac{5}{8}$	$5\frac{1}{4} \times 17\frac{7}{8}$	$5\frac{1}{4} \times 19\frac{1}{2}$	$5\frac{1}{4} \times 21\frac{1}{8}$	$5\frac{1}{4} \times 22\frac{3}{4}$	7520 lbs.
100'	13'-4 3/4"	$5\frac{1}{4} \times 19\frac{1}{2}$	$5\frac{1}{4} \times 22\frac{3}{4}$	$7 \times 22\frac{3}{4}$	$7 \times 24\frac{3}{8}$	$7 \times 24\frac{3}{8}$	9400 lbs.
120'	16'-0 15/16"	$5\frac{1}{4} \times 22\frac{3}{4}$	$7 \times 24\frac{3}{8}$	7×26	$7 \times 27\frac{5}{8}$	$7 \times 29\frac{1}{4}$	11,280 lbs.

Timber Structures, Inc.

Timber Structures, Inc.

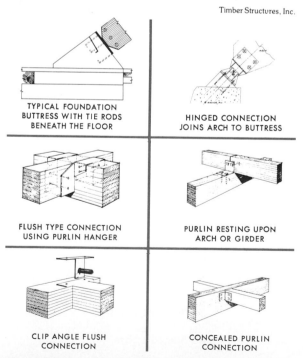

TYPICAL FOUNDATION BUTTRESS WITH TIE RODS BENEATH THE FLOOR

HINGED CONNECTION JOINS ARCH TO BUTTRESS

FLUSH TYPE CONNECTION USING PURLIN HANGER

PURLIN RESTING UPON ARCH OR GIRDER

CLIP ANGLE FLUSH CONNECTION

CONCEALED PURLIN CONNECTION

Fig. 23-109. Construction details for two-hinged arches.

loads are found in Chapter 9. Design details for this construction are shown in Fig. 23-109.

Three-Hinged Arches

Three-hinged arches add greatly to the interior beauty of a structure, and they are used where appearance requirements are most rigid, such as in churches and auditoriums. See Fig. 23-110 for one application of these units. There are three principal types — Tudor, Gothic, and continuous. Standard spans are from 30 feet to 100 feet, though wider spans can be designed. The principle types of three-hinged arches with details of construction are shown in Fig. 23-111. Construction is much the same as for the two-hinged arch, and purlins are used between the arches to carry the roof decking. Design data vary with the roof slope. In Fig. 23-112, page 570, design data for selected slopes and spans are given. Many other sizes are available, and manufacturers' catalogs should be consulted for a complete listing.

California Redwood Association

Fig. 23-110. A church using three-hinged, Tudor arches. Notice the purlins running perpendicular to the arches. The roof decking is a wood fiber product.

Bowstring Trusses

The bowstring truss is one of the most practical and most used of all truss types. It is designed to provide a high degree of fire safety and freedom from dimensional changes. While the design data in Fig. 23-113 includes spans up to 140 feet, units as wide as 250 feet are in use. When provided for in the design, these trusses can support an overhead monorail or other unusual loading. Another type of truss is illustrated in Fig. 23-114.

Timber Structures. Inc.

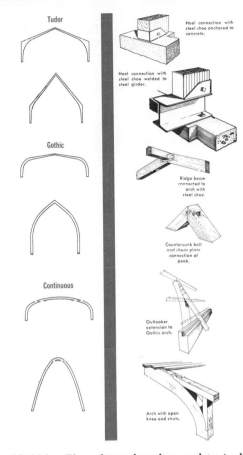

Fig. 23-111. Three-hinged arches and typical construction details.

Fig. 23-113.

Design Data for Typical Bowstring Trusses with Dimensions of Bearings[1]

Span	No. of Panels	Truss Height[2]	Roof Height	Arc Length[1]	Camber	Heel Width[3]	Length of Bearings[3]	Truss Weight
40'	6	5'- 9"	6'- 9$\frac{1}{2}$"	41'-10$\frac{11}{16}$"	1$\frac{1}{2}$"	6"	5"	950 Lbs.
60'	6	8'- 7$\frac{3}{4}$"	9'- 8$\frac{1}{4}$"	62'-10"	2$\frac{1}{2}$"	6"	8"	1530 Lbs.
80'	8	11'- 6$\frac{3}{8}$"	12'- 6$\frac{7}{8}$"	83'- 9$\frac{5}{16}$"	3"	6"	10"	2700 Lbs.
100'	8	14'- 4$\frac{1}{2}$"	15'- 5"	104'- 8$\frac{11}{16}$"	4"	6"	12$\frac{1}{2}$"	3710 Lbs.
120'	10	16'-11$\frac{3}{4}$"	18'- 0$\frac{1}{4}$"	125'- 8"	4$\frac{3}{4}$"	7$\frac{3}{4}$"	11$\frac{1}{2}$"	6230 Lbs.
140'	12	19'- 9$\frac{7}{8}$"	20'-10$\frac{1}{4}$"	146'- 7$\frac{3}{8}$"	5$\frac{1}{2}$"	7$\frac{3}{4}$"	13$\frac{1}{2}$"	8170 Lbs.

Heel connection uses heavy steel U-Strap and shear plates.

Web members are connected to chords with bolts and heavy steel straps. Shear plates are added when the stress requires.

Tim-Truss with heel connected to steel I-beam column.

[1]Tabular values are for average conditions. Individual designs may vary.
[2]Truss height is the vertical distance from bearing line to top of truss at mid-span. Roof height, as shown, is truss height plus depth of typical roof joists and thickness of 1″ nominal sheathing.
[3]Dimensions as determined for typical case of 40 lbs./sq. ft. total loading and 20-foot spacing, i.e., 800 pounds per linear foot of truss span. Heel width is also minimum pilaster width.

Timber Structures, Inc.

Fig. 23-112.

Design Data for Selected Three-Hinged Arches

Depth increments based on ¾ " laminations.
All sections shown are dictated by vertical loading.

Roof Slope	Wall Ht. (Ft.)	SPAN 40 FEET Vertical Loadings — Pounds per Lineal Foot of Span				
		400	600	800	1000	1200
3/12	10	5¼ × 9¾	5¼ × 13½	5¼ × 16½	5¼ × 20¼	7 × 17¼
	12	5¼ × 12	5¼ × 14¼	5¼ × 16½	5¼ × 20¼	7 × 18
	14	5¼ × 13½	5¼ × 16½	5¼ × 18¾	5¼ × 21	7 × 20¼
	16	5¼ × 14¼	5¼ × 17¼	5¼ × 20¼	7 × 18¾	7 × 21¾
	18	5¼ × 15	5¼ × 18¾	7 × 18	7 × 21	7 × 22½
6/12	10	5¼ × 10½	5¼ × 12¾	5¼ × 15	5¼ × 16½	5¼ × 18
	12	5¼ × 12	5¼ × 15	5¼ × 17¼	5¼ × 18¾	5¼ × 20¼
	14	5¼ × 13½	5¼ × 16½	5¼ × 18¾	5¼ × 21	7 × 19½
	16	5¼ × 14¼	5¼ × 17¼	5¼ × 20¼	7 × 19½	7 × 21
	18	5¼ × 15	5¼ × 18¾	5¼ × 21	7 × 21	7 × 21¾

Roof Slope	Wall Ht. (Ft.)	SPAN 60 FEET Vertical Loadings — Pounds per Lineal Foot of Span				
		400	600	800	1000	1200
3/12	12	5¼ × 16½	7 × 17¼	7 × 21	7 × 24¾	9 × 24
	14	5¼ × 18¾	7 × 19½	7 × 21¾	7 × 26¼	9 × 24¾
	16	5¼ × 20¼	7 × 21	7 × 24	7 × 27	9 × 26¼
	18	7 × 18¾	7 × 22½	7 × 25½	9 × 25½	9 × 27¾
	20	7 × 20¼	7 × 24	7 × 27¾	9 × 27¾	9 × 30
6/12	12	5¼ × 16½	5¼ × 20¼	7 × 19½	7 × 21¾	7 × 24
	14	5¼ × 18	7 × 18¾	7 × 21	7 × 24	7 × 26¼
	16	5¼ × 19½	7 × 20¼	7 × 23¼	7 × 26¼	9 × 24¾
	18	5¼ × 21	7 × 21¾	7 × 24¾	7 × 27¾	9 × 26¼
	20	5¼ × 21¾	7 × 22½	7 × 26¼	7 × 29¼	9 × 27¾

Roof Slope	Wall Ht. (Ft.)	SPAN 80 FEET Vertical Loadings — Pounds per Lineal Foot of Span				
		400	600	800	1000	1200
3/12	14	7 × 20¼	7 × 24	9 × 25½	9 × 30	9 × 34½
	16	7 × 22½	7 × 27	9 × 27¾	9 × 30¾	9 × 35¼
	18	7 × 24	9 × 25½	9 × 29¼	9 × 32¼	9 × 36
	20	7 × 24¾	9 × 26¼	9 × 30¾	9 × 33¾	11 × 33¾
	22	7 × 25½	9 × 27	9 × 31½	9 × 35¼	11 × 35¼
6/12	14	7 × 18¾	7 × 23¼	7 × 27	9 × 26¼	9 × 28½
	16	7 × 20¼	7 × 24¾	9 × 25½	9 × 28½	9 × 30¾
	18	7 × 21¾	7 × 26¼	9 × 27	9 × 30	9 × 32¼
	20	7 × 23¼	9 × 24¾	9 × 27¾	9 × 31½	9 × 34½
	22	7 × 24¾	9 × 25½	9 × 29¼	9 × 33	9 × 36

**tudor arches — 3-hinged
depth increments
based on ¾" laminations**

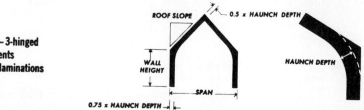

ROOF SLOPE   0.5 × HAUNCH DEPTH
WALL HEIGHT   HAUNCH DEPTH
SPAN
0.75 × HAUNCH DEPTH

Timber Structures, Inc.

Timber Structures, Inc.

Fig. 23-114. A lenticular-type truss over a supermarket. It spans **124** feet with chords extending **13′-6″** beyond the sidewalls to provide overhangs.

Timber Structures, Inc.

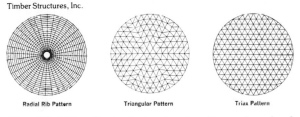

Radial Rib Pattern   Triangular Pattern   Triax Pattern

Fig. 23-115. Typical structural patterns for glued, laminated wood domes.

Domes

Glued, laminated structural domes have been built with clear spans up to 300 feet, and they are practical for even greater spans. With a tension ring to resist horizontal thrust, side walls need not be buttressed, and no tying members are needed. The ratio of rise to span is low. Since the members are in compression and need not be designed to resist

Timber Structures, Inc.

Fig. 23-116. A glued, laminated structural dome over a gymnasium, spanning **132′-6″**.

bending, section sizes are relatively small. Before the domes are raised into position, they are usually assembled in sections. Some typical construction patterns are shown in Fig. 23-115, and one application is shown in Fig. 23-116.

Drawing Laminated Wood Structural Systems

After the engineer produces the preliminary design of the structure, the drafting technician makes carefully dimensioned drawings. In Fig. 23-117 is a pictorial of parts of a wood structural

Fig. 23-117. Partially completed view of the frame of a laminated wood structure. Compare identified members with Figs. 23-118 and 23-121.

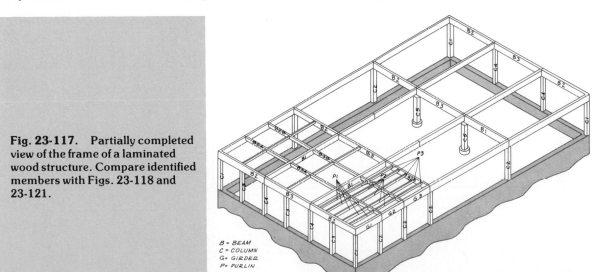

B = BEAM
C = COLUMN
G = GIRDER
P = PURLIN

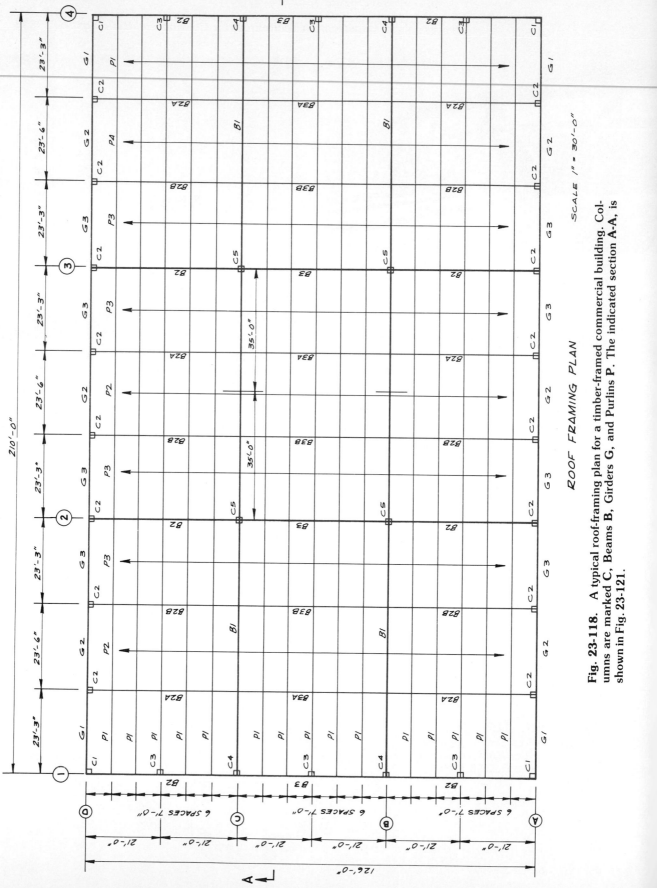

ROOF FRAMING PLAN

SCALE 1" = 30'-0"

Fig. 23-118. A typical roof-framing plan for a timber-framed commercial building. Columns are marked **C**, Beams **B**, Girders **G**, and Purlins **P**. The indicated section **A-A**, is shown in Fig. 23-121.

BEAM SCHEDULE

BEAM NO.	SIZE	LENGTH	NO. REQUIRED
B1	9 x 48 3/4	105'-0"	4

Fig. 23-119. Typical beam schedule.

COLUMN SCHEDULE

COLUMN	SIZE	LENGTH	NO. REQUIRED	TOP FAB	BOTTOM FAB
C1	9 x 9 3/4	25'-10 1/8	4	DET. T	CORNER DET.
C2	9 x 16 1/4	25'-10 1/8	16	DETAIL OF C2	TYP C1
C3	9 x 17 7/8	25'-10 1/8	6	DETAIL OF C1	TYP C1
C4	12 1/2 x 14 5/8	25'-10 1/8	4	DETAIL OF C1	COP AND ANCHOR
C5	12 1/2 x 16 1/4	24'-6 1/8	4	TYP INTERIOR	

Fig. 23-120. Typical column schedule.

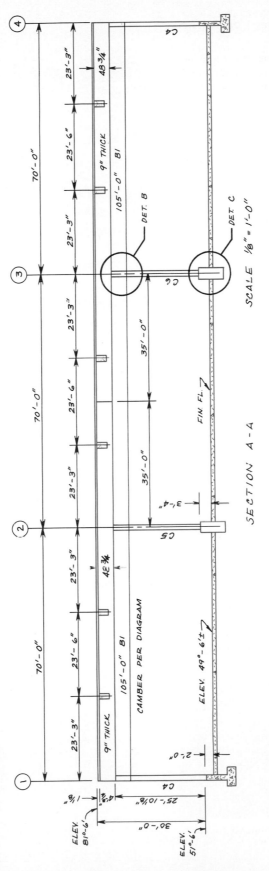

SECTION A-A

SCALE 1/8" = 1'-0"

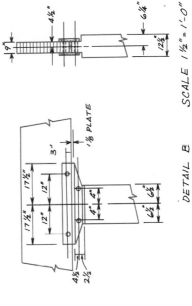

DETAIL B   SCALE 1 1/2" = 1'-0"

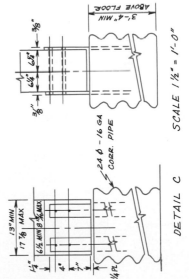

DETAIL C   SCALE 1 1/2" = 1'-0"

Fig. 23-121. This is the section (A-A) through the timber-framed commercial building indicated in Fig. 23-118. Typical large-scale construction details are shown.

573

system. Notice that the long spans are made with large beams. Shorter spans are made with smaller beams and girders. The areas between these have purlins to help support the roof deck. A complete roof-framing plan for this structure is shown in Fig. 23-118. It shows the columns, beams, girders, and purlins. Each of these is identified by a letter and a number. The exact size is given in schedules which are a part of the drawing. Examples are in Figs. 23-119 and 23-120. Notice that each member is dimensioned to its center line. The beams and girders are drawn much thicker than the purlins.

It is usually necessary to draw some sections to help clarify construction details. A typical example is shown in Fig. 23-121. It locates the structural members shown. The elevation of the foundation, roof, and finished floor are given.

The roof beams are designed with **camber**. This is a slightly convex (curved) condition. Engineers record the design data for the camber on the drawing.

Various special construction details are necessary. In Fig. 23-121 are some details showing how the columns are anchored to the foundation. Another shows how the beams are joined to the column.

Wood-Fiber Roof Planks

Numerous types of wood-fiber roof planks are available. In commercial construction, this type of decking is used with steel, wood, and concrete structural systems. In Figure 23-122, decking is being applied to open-web steel joists, and in Fig. 23-123, it is shown over a metal, box-section subpurlin. It is nailed to the steel joists in a nailing groove provided in the top member. Steel clips are used to secure it to box-section subpurlins. Design

Tectum Corporation

Fig. 23-123. **Wood-fiber decking applied to box-section sub-purlins.**

data for box-section subpurlins are given in Fig. 23-124. Another method of framing a roof is to use bulb-tees. Design data for these are given in Fig. 23-125. The space between the bulb-tee and the fiber deck is filled with a lightweight concrete grout, as shown in Fig. 23-126. See Fig. 23-127 for other methods used to secure wood-fiber roof decking to structural members.

This decking can be left exposed on the interior of a building, thus forming a finished ceiling as well as the roof decking and insulation. See Fig. 23-128.

Fig. 23-124.

Design Data for Box Section Subpurlins

Loads given are for three-span conditions. For one- and two-span conditions, multiply these by 9.

Deflections of spans in bold face type exceed 1/240 of span, but are less than 1/180 of span.

Total Loads PSF	Spacing O.C.	Spans 16 Ga.	Spans 18 Ga.	Spans 20 Ga.
30	32″	9′-9″	8′-7″	7′-1″
	36″	9′-2″	8′-2″	6′-8″
	42″	8′-6″	7′-7″	6′-2″
	48″	7′-11″	7′-0″	5′-9″
40	32″	8′-5″	**7′-6″**	6′-1″
	36″	7′-8″	6′-10″	5′-7″
	42″	7′-4″	6′-6″	5′-4″
	48″	6′-10″	6′-1″	4′-11″
50	32″	7′-7″	6′-8″	5′-6″
	36″	7′-1″	6′-4″	5′-1″
	42″	6′-7″	5′-10″	4′-10″
	48″	6′-2″	5′-6″	4′-6″

Tectum Corporation

Fig. 23-122. **Wood-fiber decking applied to open-web steel joists.**

Fig. 23-125.

Maximum Spans for Three-Span Bulb-Tees

All deflections should be checked, if limiting deflection is critical.
Tabulated spans below are calculated on the basis of Fs = 20,000 psi.
For high strength steels (Fs = 27,000 psi), multiply span below by 1.16.
For one- and two-span conditions, multiply spans below by 0.9.

Type of Bulb-Tee	Spacing O.C.	Loading in PSF						
		25	30	35	40	45	50	55
1 158	23³/₄″	7′-7″	6′-11″	6′-5″	6′-0″	5′-8″	5′-4″	5′-1″
	31³/₄″	6′-7″	6′-0″	5′-7″	5′-2″	4′-11″	4′-8″	4′-5″
1 168	23⁷/₈″	8′-11″	8′-2″	7′-7″	7′-1″	6′-8″	6′-4″	6′-0″
	31³/₄″	7′-9″	7′-1″	6′-6″	6′-1″	5′-9″	5′-6″	5′-3″
1 178	24″	10′-8″	9′-9″	9′-0″	8′-5″	7′-11″	7′-6″	7′-2″
	31⁷/₈″	9′-3″	8′-5″	7′-10″	7′-3″	6′-11″	6′-6″	6′-3″
1 200	24¹/₈″	12′-5″	11′-4″	10′-6″	9′-9″	9′-3″	8′-9″	8′-4″
	32″	10′-9″	9′-9″	9′-1″	8′-6″	8′-0″	7′-7″	7′-3″

Inland Steel Company

Fig. 23-126. Wood-fiber roof planks installed with metal bulb-tees. Notice the grout placed between the planks and around the bulb-tee.

Factory-Manufactured Commercial Buildings

Available from a number of manufacturers are small commercial buildings built of standard, preengineered components which are manufactured on a production basis. Some assembly is done in the factory. The components are fitted together on the job. See Fig. 23-129.

Such a structure can be rapidly erected with some saving in cost, since much on-the-site labor is eliminated due to factory cut and fitted components. These buildings can be easily expanded by adding on another unit; the components removed to attach the addition are utilized in the new structure. No demolition is necessary. See Fig. 23-130.

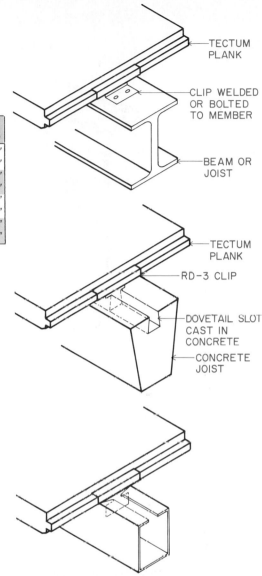

Fig. 23-127. Various types of metal clips are used to secure wood-fiber roof decking to structural members.

Tectum Corporation

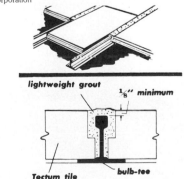

Fig. 23-128. Wood-fiber roof decking used as exposed ceiling, roof decking, and roof insulation.

Fig. 23-129. The factory-manufactured steel framework is rapidly assembled on the site.

Large, column-free, interior areas can be enclosed. See Fig. 23-131. The area is free of obstructions from wall to wall and from floor to roof peak.

Basically, the design procedure is simple. The footings and foundation are designed to support the necessary load. Since the steel frame carries the total load of the roof, heavy concrete footings are needed only at the uniformly spaced piers, not around the entire building. Factory-fabricated wall panels are designed to hang on the steel frame. If these are replaced with a conventional masonry wall, the footings for the wall need to support its weight.

Each company has a variety of standard framing systems available. See Fig. 23-132. Since some are better suited to certain types of buildings than others, a decision must be made concerning the system to be used. Design data for one system is given in Fig. 23-133. The needed span and roof pitch are two primary considerations.

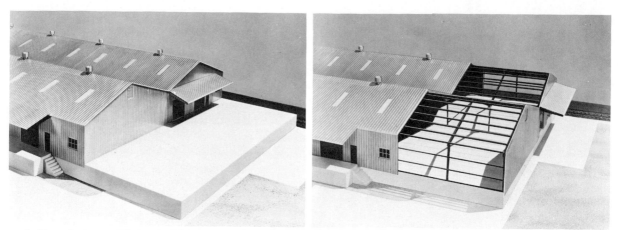

A. Foundation and floor of addition poured.

C. Steel frame and purlins frame the new addition.

B. Ends of units removed and placed in position for reuse on new addition.

D. Metal roof and siding secured to frame; building ready to be occupied.

Fig. 23-130. Building an addition to a factory-manufactured building.

Butler Manufacturing Company

Fig. 23-131. Interior of a warehouse built with low rigid frames and stamped metal roof and walls. Notice the wide, column-free floor area.

Butler Manufacturing Company

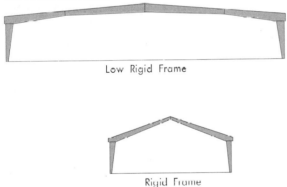

Low Rigid Frame

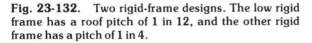

Rigid Frame

Fig. 23-132. Two rigid-frame designs. The low rigid frame has a roof pitch of 1 in 12, and the other rigid frame has a pitch of 1 in 4.

Another decision is the selection of exterior siding. Two metal types are available. One is a metal panel rolled in some form of ribbed or corrugated pattern. Generally, these panels are used on buildings which do not require insulation; however, they may be insulated after erection, if desired. A second type is a hollow metal panel, completely insulated when manufactured. The type illustrated in Fig. 23-134 is available in a variety of colors. The edge joint between the panels is a double tongue-and-groove with vinyl gaskets.

After the steel structure is erected, conventional materials, such as brick or stone can be used for the exterior wall. See Fig. 23-135. Since the total load of the structure is on the steel frame, the wall need only be self-supporting.

Fig. 23-133.

Data for Stock, Rigid Frame Steel Buildings

LOW RIGID FRAME (LRF) Clear Span Standard Widths										
	24	32	36	40	50	60	70	80	100	120
Eave Heights	10	10	10							
	12	12	12	12	12	12	12			
				16	16	16	16	16	16	16
				20	20	20	20	20	20	20
				24	24	24	24	24	24	24

	Building	Canopy	Width Extension
Bay lengths	18',20',21',24'	18',20',21',24'	18',20',21',24'
Total length	Unlimited	Unlimited	Unlimited
Roof slope	1 in 12	1 in 12	1 in 12
Design loadings[1]	2, 4, 6, 8	2, 4, 6, 8	2, 4, 6, 8
Widths		6', 10'	24', 48'

RIGID FRAME (RF) Clear Span Standard Widths										
	20	24	28	32	36	40	50	60	70	80
Eave Heights	10	10	10	10	10	10	10			
	12	12	12	12	12	12	12	12		
	14	14	14	14	14	14	14	14	14	
						16	16	16	16	16
						20	20	20	20	20
						24	24	24	24	24

	Building	Canopy	Width Extension
Bay lengths	18',20',21',24'	18',20',21',24'	18',20',21',24'
Total length	Unlimited	Unlimited	Unlimited
Roof slope	4 in 12	1 in 12	1 in 12
Design loadings[1]	1, 2, 4, 6	2, 4, 6	2, 4, 6
Widths		6', 10'	24', 48'

[1]Key to design loadings:
1 - Roof live load 10 psf or wind load 15 psf
2 - Uniform Building Code
4 - Southern Standard Building Code, Inland and Coastal
6 - Roof live load 30 psf and wind load 20 psf
8 - Roof live load 40 psf and wind load 20 psf

Butler Manufacturing Company

Curtain Walls

A **curtain wall** is a non-load-bearing wall erected on the exterior of a building to provide a barrier between the outer elements and the interior. More specifically, it is an entire enclosure element, which

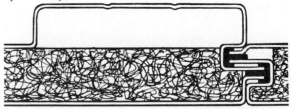

Fig. 23-134. Section through an insulated wall panel.

This is a hollow metal panel completely insulated. The double tongue-and-groove design creates a strong panel-to-panel joint. Double vinyl gaskets seal out moisture and wind.

Butler Manufacturing Company

Fig. 23-135.   A small manufacturing building assembled from stock low-rigid frames, with stone and metal panels as exterior siding.

is complete with exterior finish, interior finish, insulation, structural independence, and a means of attachment to a building. It is an engineered series of units designed for specified wind loads, structural requirements, insulating factors, and expansion and contraction, together with weathertightness. A curtain wall provides all the advantages of a masonry wall, plus the advantages of speedier erection, lighter weight, and better insulation. It has a long, maintenance-free life.

Curtain walls can be made of many materials. Combinations of extruded sections and insulation, patterned sheets, rigidized sheets, honeycombed panels, porcelain on aluminum or steel, or aluminum anodized sheets are used. The basic materials of aluminum, steel, stainless steel, bronze, cement-asbestos panels, plate glass, and plastic panels are used. Other systems use precast concrete, glass blocks, and ceramic-tile panels.

Windows are a prime component of these walls and must be engineered as a part of the entire unit. Generally, it is recommended that the designer use the window units suggested by the manufacturer of the curtain-wall system to be used.

Porcelain-Enamel Curtain-Wall Systems

Porcelain enamel is a hard, durable, glass-like coating applied to a metal base and produced by fusing to the base a carefully compounded mixture of mineral substances, such as cryolite, feldspar, quartz, borax, silica, tin, and zirconium oxides and clays.

The panels are available in a wide array of permanent colors and are resistant to acid and severe

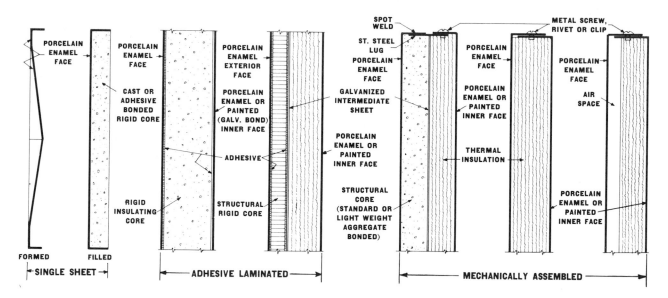

Single-Sheet Panels: These are metal units which form the exterior face and provide the required weather barrier. A back-up material is used to supply thermal insulation and fire resistance or to meet code requirements. The panel and the back-up material together comprise the composite type of curtain-wall construction.

Laminated Panels: In a laminated panel, assembly of the components is accomplished by use of

adhesives. Either flat or formed metal sheets or pans may be used.

Mechanically Assembled Panels: Exterior and interior components are fastened together by mechanical means. When the flanged pans are used back-to-back, they form a "box-type" panel with an interior space in which a variety of insulating materials may be held.

Fig. 23-136.   Types of porcelain-enamel panels used in curtain-wall construction.

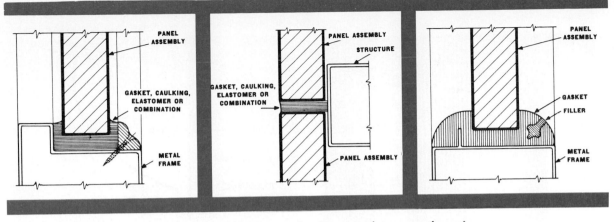

Fig. 23-137. Typical ways of sealing porcelain-enamel panels.

abrasive action. They can be cleaned by a simple washing. They are so durable that they can withstand severe atmospheric and climatic conditions. Another feature is their fireproof quality.

The two types of metal panels used in porcelain-enamel curtain-wall systems are steel and aluminum.

There are three types of porcelain-enamel panels in general use — the single-sheet panel, the laminated panel, and the mechanically assembled panel. See Fig. 23-136.

Porcelain-enamel curtain walls are classified into four types depending upon their appearance. These types are:

1. **Sheath** — No structural elements indicated;
2. **Grid** — Equal emphasis and expression given horizontal and vertical structural elements;
3. **Mullion** — Vertical structural elements emphasized;
4. **Spandrel** — Horizontal structural elements emphasized.

The panels may be sealed with extruded, neoprene rubber or polyvinyl-chloride weatherproofing gaskets. These seals have a longer life than caulking and can be installed faster. Caulking and elastomer compositions are also extensively used. The latter type develops a strong adhesive bond to the porcelain-enamel surface and has great flexibility and elasticity. It also has a longer life than usual caulking compounds. See Fig. 23-137.

Design Data

The design details shown in Fig. 23-138 pertain to an aluminum curtain-wall system as manufactured by Ceco Steel Products Corporation. Data are presented for a single-story building only; manufacturer's catalogs should be consulted for complete details. Other companies manufacture competitive systems.

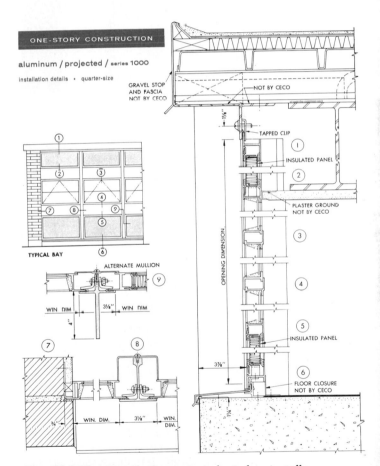

Fig. 23-138. Detail of a projected window installation.

In this single-story application, the curtain wall is anchored at head and sill. There is a drop ceiling, with an insulated panel covering structural members. This is standard window construction. Note the strong, simple mullion and cover (detail 8) and the alternate mullion (detail 9). Subsills may be simply applied to suit various masonry conditions. An insulated panel from floor to sill level conceals convectors, pipes, etc., from outside view.

579

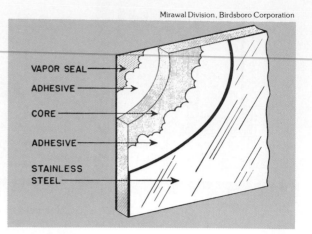

Fig. 23-140. A laminated, stainless-steel curtain-wall panel.

Fig. 23-139. Four Gateway Center, a building using stainless-steel panels and mullions.

Stainless-Steel Curtain Walls

Stainless-steel curtain-wall panels are in use on all types of buildings from single-story structures to skyscrapers. See Fig. 23-139. While single-thickness sheets of stainless steel have been applied as an outer covering on buildings, the most successful applications have occurred when laminated panels have been used. Several types are in use. These have a core made from mineral, vegetable, inorganic, or plastic materials. The core is then faced on both sides with cement-asbestos panels, and a veneer of stainless steel is bonded to them. These panels are usually custom-fabricated and may vary in thickness from 11/16" to 3", depending upon the design requirements.

Another type of veneer panel consists of a stainless-steel veneer laminated to a cement-asbestos or hardboard veneer. A variation of this is a panel with a rubber-impregnated, cork-fiber core.

This allows the panel to be bent around columns where a radius is needed. This radius can be as small as 6 inches. See Fig. 23-140.

Some types of stainless-steel panels can be field cut with a special power saw. Others require expensive cutting operations or must be returned to the factory to be cut.

Stainless steel requires no surface treatment, since it resists corrosion. It can be used in direct contact with brick, stone, or ceramics, since it is not affected by the alkalies and caustics in lime and mortar. Because it has a high tensile strength, it can be used in thinner gauges than other metals. Due to its mechanical properties, it can be stamped, deep-drawn, press-braked, or roll-formed into almost any design desired, thus allowing the architect great freedom of expression.

Mullions of stainless steel have the same long-wearing, lightweight qualities as the panels. They are frequently used with such panel types as porcelain enamel, as well as with stainless steel.

Copper and Copper-Alloy Curtain-Wall System

Copper and bronze curtain-wall panels, available in a variety of forms and shapes, exhibit all the desirable characteristics of other metal curtain-wall systems. They vary in color from bright red and yellow through dark brown and ebony. There are five architectural metals of this type in use — Copper, Red Brass, Architectural Bronze, Yellow Brass, and Nickel Silver.

They are available in the grid-type system, as discussed for the other metal panels. The mullions are extruded of the copper or copper alloy to be

Fig. 23-141. A building utilizing translucent plastic curtain walls. These walls admit a diffused, natural light.

used. Exact design details vary from the other systems, and manufacturers' catalogs should be consulted.

Translucent Plastic Curtain-Wall Panels

The translucent plastic, curtain-wall panel is a structural sandwich fabricated of two skins of glass fiber-reinforced, polyester plastic, laminated to an extruded aluminum gridwork core. Permanent lamination is accomplished by means of a rubber-based adhesive, which is activated by heat and pressure.

These panels offer full translucency with a wide range of color skins and color inserts. They admit an evenly diffused, natural light. See Fig. 23-141. These panels are available in 18 inch widths and in lengths of 8, 10, 12, and 20 feet. The standard thicknesses of the panels are 1½ and 3 inches. The standard skin is 2-ounce, glass fiber-reinforced, polyester plastic; however, heavier skins are available.

Cement-Asbestos Curtain-Wall Panels

Cement-asbestos curtain-wall panels are manufactured with a stable, unicellular, plastic insulation core. The cement-asbestos skins are adhered with epoxy resin to this core. See Fig. 23-142. The thickness of panels varies. This panel is faced with a .018-inch, polyester, glass fiber cloth face over the cement-asbestos board. These panels are available in a wide range of colors and in sizes up to 4'-0" by 12'-0".

Ceramic-Tile-Faced Curtain-Wall Panels

Curtain-wall panels faced with ceramic tile are available in two types — tile over a cast concrete panel and tile over a metal sandwich core.

The ceramic-tile facing is grouted with a specially formulated, weatherproof, flexible, latex grout. The tile is self-cleaning, requires no maintenance, and has a low-reflective surface.

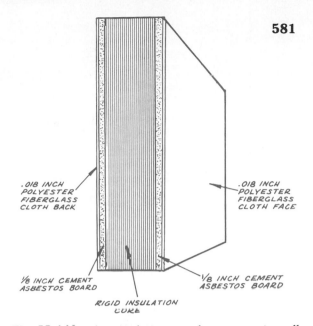

Fig. 23-142. A typical cement-asbestos curtain-wall panel with polyester fiber glass cloth face.

A section through a tile-on-concrete panel is illustrated in Fig. 23-143. The panel has a rigid-insulation core. A 1/8-inch clearance should be allowed between all edges of the panel, framing members, and framing stops.

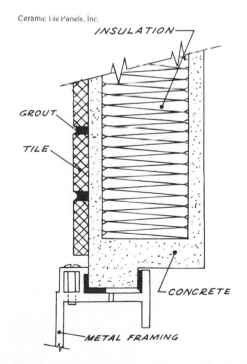

Fig. 23-143. A section through a reinforced concrete panel with a ceramic-tile facing. The interior of the panel is filled with a rigid insulation.

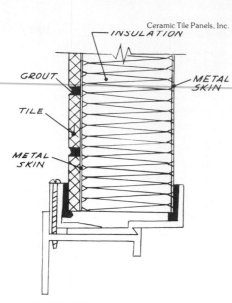

Ceramic Tile Panels, Inc.

Fig. 23-144. A section through a metal-skin panel with ceramic-tile facing.

Libbey-Owens-Ford Glass Company

Fig. 23-145. A two-story building with glass curtain walls. The horizontal and vertical framing members are extruded aluminum with clear glass window panels and heat-strengthened, colored glass spandrels.

Fig. 23-146. Details of one glass curtain-wall installation

Libbey-Owens-Ford Glass Company

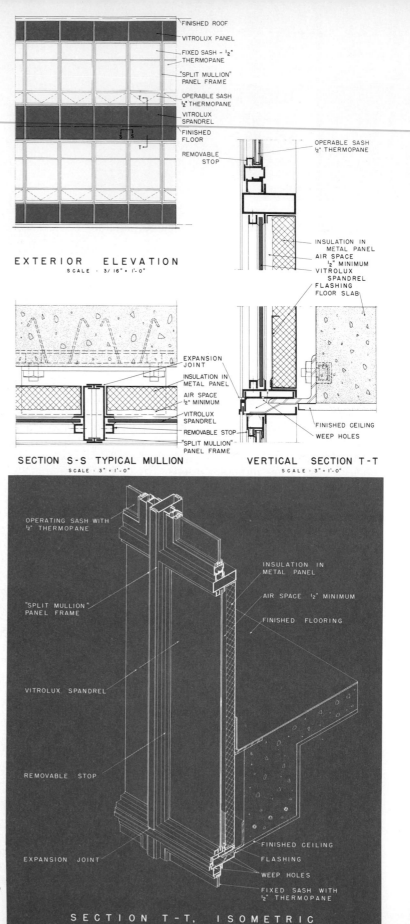

EXTERIOR ELEVATION
SCALE : 3/16" = 1'-0"

SECTION S-S TYPICAL MULLION
SCALE : 3" = 1'-0"

VERTICAL SECTION T-T
SCALE : 3" = 1'-0"

SECTION T-T, ISOMETRIC
SCALE : 1½" = 1'-0"

Tile-faced curtain-wall panels over a metal skin are available in a wide range of sizes and thicknesses. The metal skins are either aluminum or galvanized steel, bonded to a rigid insulation core. The tile is bonded to the metal skin. See Fig. 23-144.

Glass Curtain-Wall Panels

Glass curtain walls are made in many ways. One common system uses ¼-inch clear plate glass panels for wall areas where is is desired to see outside the building and makes the opaque spandrels from ¼-inch heat-strengthened, polished plate glass with ceramic color fire-fused on the back surface of the glass. This becomes an integral part of the glass panel. The color is sun-fast, and the panel resists weathering, crazing, and checking, exactly the same as clear plate glass does. See Fig. 23-145.

Glass curtain walls have the same advantages as mentioned for porcelain-enamel and stainless-steel curtain-wall systems. The opaque glass panels are available in a wide variety of colors allowing considerable variety of architectural expression. They are nonporous and nonabsorbent and resist most atmospheric acids and temperature changes.

The heat-strengthened panels are twice as strong and three times as resistant to thermal shock as ordinary plate glass. The standard maximum size for these panels is 48″ × 84″.

The panels are usually set in metal mullions and sealed with caulking or neoprene gaskets. The glass panel should have a ¼-inch clearance on all edges from the mullion to provide space for expansion.

A typical detail of a glass curtain-wall installation is illustrated in Fig. 23-146.

Glass-Block Curtain Walls

A wide selection of glass-block units for use in curtain-wall construction is available. These units are supplied in many sculptured and intaglio forms and in a variety of colors. They provide a wall that is low in maintenance cost, low in surface condensation, and high in insulation values. They can admit sunlight while excluding hot rays.

Since a wide array of shapes, colors, and sculptured surfaces is available, design opportunities are great.

Standard glass blocks are available in a number of stock sizes. These are commonly 6, 8, and 12 inches square and a 4 × 12-inch unit. Some are available with one dead air space inside, while others have two such spaces. The blocks are formed by fusing two sections of pressed glass together at elevated temperatures. The colored

Fig. 23-147. Curtain walls formed from 4″ × 8″ oval and 8″ × 8″ hourglass Intaglio glass blocks. This building is a library.

blocks have a fired-on, ceramic, enamel coating on one face.

Intaglio glass blocks are pressed, all-glass units in which the outer surfaces of each unit provide molded, translucent patterns, masked in a ceramic frit. They are available in 4″ × 8″ and 8″ × 8″ units.

Two of the stock units are shown in Fig. 23-147. The curtain walls are a combination of oval and hourglass units.

Also available are sculptured glass blocks. These are used, as are other glass blocks, for interior and exterior walls. They offer light transmission and insulation values.

Glass blocks can be cast in frames made from other materials. Blocks cast in a terra cotta frame are shown in Figs. 23-148 and 23-149.

Fig. 23-148. Interior view of glass-block curtain wall illustrated in Fig. 23-149.

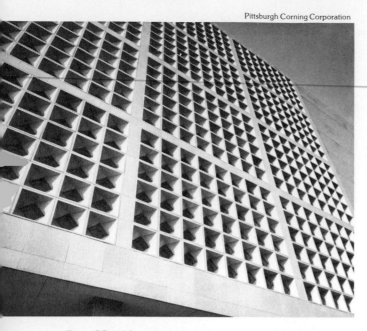

Fig. 23-149. Exterior view of curtain wall made from clear glass blocks, set in terra cotta framing to light the public foyer of a county courthouse. The glass blocks have fire-polished faces to allow vision through them. They are 4 inches thick, with a hollow interior being at a partial vacuum. They have an insulating value equivalent to an 8-inch masonry wall.

The glass-block panels require a reinforcement of galvanized-steel, double-wire mesh. This reinforcement should be embedded in the horizontal mortar joints on approximately 24-inch centers and in joints above and below all openings within the panel. The reinforcement should run continuously from end to end of the panels and should be lapped at least 6 inches when it is necessary to use more than one length in a joint. Expansion joints should not be bridged with reinforcement.

The panels are anchored to the walls or masonry columns with perforated, steel-panel anchor strips, 24 inches long and 1¾ inches wide. They should be crimped with expansion joints, so movement is possible. Usually, the panel anchors are placed 24 inches apart and in the same mortar joint as the wire-mesh, panel reinforcement.

The glass blocks are set in portland cement that has a waterproof ingredient added.

The blocks are cushioned at jambs, heads, and intermediate supports with oakum. Caulking should be a waterproof mastic. Installation details are shown in Fig. 23-150.

Precast Concrete Facing and Curtain-Wall Panels

Precast panels are available in a wide variety of sizes and surfaces. Manufacturers' catalogs should be consulted for information on these many kinds.

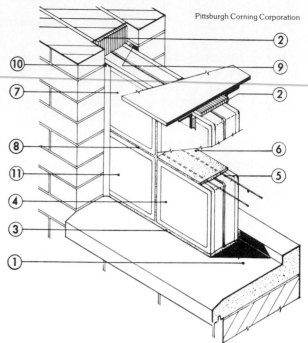

Fig. 23-150. Procedure for installing glass blocks.
1. Check that the sill area to be covered by mortar has a heavy coat of asphalt emulsion.
2. Adhere expansion strips to jambs and head with asphalt emulsion. Make sure expansion strip extends to sill.
3. When emulsion on the sill is dry, place the full mortar bed joint — do not furrow.
4. Set lower course of block. All mortar joints must be full and not furrowed. Steel tools must not be used to tap blocks into position.
5. Install panel reinforcements in horizontal joints where joints were required as follows:
   a. Place lower half of mortar bed joint. Do not furrow.
   b. Press panel reinforcement into place.
   c. Cover panel reinforcement with upper half of mortar bed, and trowel smooth. Do not furrow.
   d. Panel reinforcement must run from end to end of panels and, where used continuously, must lap 6 inches. Reinforcement must not bridge expansion joints.
6. Place full mortar bed for joints not requiring panel reinforcement. Do not furrow.
7. Follow above instructions for succeeding courses of blocks.
8. Strike joints smoothly while mortar is still plastic and before final set. At this time, rake out all spaces requiring caulking to a depth equal to the width of the space. Remove surplus mortar from the faces of glass blocks and wipe dry. Before mortar takes final set, tool joints smooth and concave.
9. After mortar sets, pack oakum tightly between glass-block panel and jamb and head construction. Leave space for caulking.
10. Caulk panels as indicated on details.
11. Final cleaning of glass-block faces should not be done until after final mortar set.

The George Rockle and Sons Company

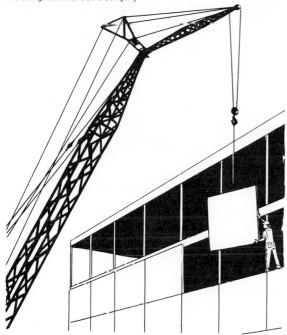

Fig. 23-151. One method of handling precast concrete facing and curtain-wall panels.

Precast facing units are most economically cast in sizes from 20 to 60 square feet and in 2-inch thicknesses. Larger units can be cast, if the thickness is increased according to steel-reinforcement requirements. While the weight varies according to the type of panel, 25 pounds per square foot is a common design weight for 2-inch thick panels. Exact weights must be ascertained from manufacturers' catalogs.

These panels are cast in mold boxes and are reinforced with welded, galvanized, wire mesh. The crushed aggregate, cement, and reinforcing mesh are compacted in the mold box with high-frequency vibrations to assure good bond and maximum strength and density.

The panels are designed for placement in a vertical position and must be stored, transported, and handled in this position. One method of handling panels is shown in Fig. 23-151.

The panels can be mechanically fastened to a steel or wood frame or to a masonry wall. Special fastening devices are manufactured to provide this connection. See Fig. 23-152.

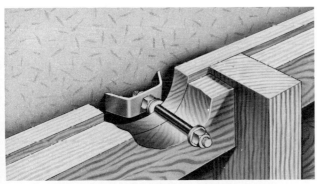

A. Precast panel secured to wood-framed wall.

C. Precast panel secured to metal framing.

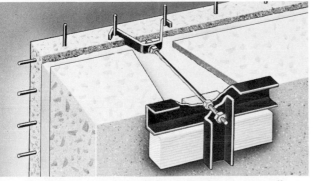

B. Precast panel secured to poured masonry wall.

D. Precast panel secured to brick wall.

Mo-Sai Associates, Inc.

Fig. 23-152. Methods of fastening precast concrete panels.

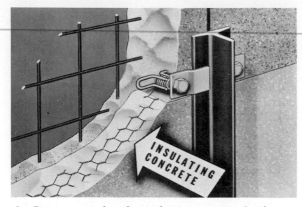

A. Precast panel with insulating concrete back-up and metal back strip.

B. Precast panel with sandwich-style, rigid insulation and metal back strip.

Mo-Sai Associates, Inc.

Fig. 23-153. Insulating precast concrete curtain walls.

Precast curtain-wall units are available with built-in insulation. Lightweight insulation concrete can be cast as a core of the unit, or various types of rigid insulation can be cast in the slab. See Fig. 23-153.

Build Your Vocabulary

Following are terms that you should understand and use as a part of your working vocabulary. Write a brief explanation of each term.

anodizing	grout
bending stress	kips
breaking strength	modulus of elasticity
caulking	mullion
compressive stress	permanent set
curtain wall	shear stress
deflection	tensile stress
deformation	ultimate strength
elastic limit	unicellular plastic core
expansion joints	vertical shear
glass fiber panels	yield point
grid system	

Problems for Study

1. A one-story structure is to have precast concrete columns having an unsupported length of 12'-0". The load on each column was computed to be 64 kips. What size round column should be used? What is the weight per foot of this column? What size square, precast column should be used? What is the weight per foot of this column? Which column would seem to be the more economical? Explain.

2. A building requires a prestressed concrete I-beam to span 35'-0" in the clear. It must carry a superimposed load of 800 pounds per lineal foot. What beam should be used? What should be the depth and width of the beam? How many prestressed wires must be cast in the beam?

3. A rectangular, precast beam is required to span 20 feet and carry a load of 1600 pounds per foot of beam. What beam would carry this load? What is the width and depth of this beam? What size reinforcing bars must be used?

4. Precast concrete tee joists are to be used to form the roof structure. They must span 20'-0" in the clear. The superimposed roof load is 285 pounds per lineal foot. What size joist must be used?

5. A roof is to be built of double-tee, precast roof decking. It must span 25'-0" in the clear. If the computed roof load is 40 pounds per lineal foot, what size decking unit should be used? What should be the width of the unit?

6. A roof is to be decked with precast concrete planks. The planks are to span 6'-0". The roof load is 70 pounds per square foot. What thickness plank must be used?

7. A building is to use precast concrete Flexicore units for floor and ceiling. The floor units span 12'-0" and the roof units, 22'-0". The floor loading is 150 pounds per square foot and the roof, 30 pounds per square foot. A 6" × 16" section is to be used. What are the stock units that should be used for each?

8. A round, extra-heavy steel column has an unbraced length of 12'-0". It must support a load of 85,000 pounds. What diameter column must be used?

9. A steel rod, 1½" in diameter, supports a tensile load of 20,000 pounds. Compute the tensile unit stress.

10. If the allowable tensile unit stress on a rod is 18,000 pounds per square inch and if the rod supports a load of 80,000 pounds, what must be its cross-sectional area? If the rod is rectangular with a thickness of ½ inch, what would be its width? If the rod is round, what diameter would be required?

11. If an 8 × 8 W8 35 beam spans 15'-0" with a uniform load of 30,000 pounds, what is the actual deflection?

    What is the maximum permissible deflection in inches? Will this beam carry the assigned load without excess deflection?

12. A concrete roof slab, 3½ inches thick, is to be poured over a corrugated steel decking. The decking spans 5'-0" between joists. What form material should be used?

13. Long-span, steel roof decking (as manufactured by H. H. Robertson Co.) is used with a built-up roof covering. It spans 30'-0" in the clear with a stress not to exceed 8000 psi. If it carries a load of 40 pounds per square foot, what stock decking should be used?

14. A building is to use corrugated, galvanized-steel panels for exterior siding. If the panels are to be 8'-0" high, what is the maximum span for steel panels of this gauge if used for roof decking?

15. A roof is to be framed of wood bowstring trusses that must span 100'-0". What are the height and weight of the truss?

16. A church is to be built using three-hinged wood arches. They must span 60 feet, with a roof slope of 6 to 12. The wall height is 14'-0", and the vertical loading is 800 pounds per lineal foot of span. What is the depth increment of the arch frame?

17. A two-hinged wood arch must span 80'-0" and carry a load of 400 pounds per lineal foot of span. What are the rise and radius of the arch? What thickness of lamination is used in constructing this arch? What is the total horizontal thrust activated by this member?

18. A roof is to be framed with steel box-section subpurlins and wood-fiber roof decking. If the roof load is 40 pounds per square foot and if the spacing is 32 inches on center, what is the maximum span for these subpurlins?

    If the same roof is to be framed with three-span, steel bulb-tees, what is the maximum span allowed for the bulb-tee members?

Class Activities

1. Explore your community and find commercial buildings using curtain-wall construction. Prepare a report giving the street address and name of the company occupying the building and identify the type of curtain-wall system used. Find as many different kinds as possible.

2. Examine manufacturers' catalogs and compile a list of companies manufacturing the various curtain-wall systems.

University of Minnesota, Minneapolis

Fig. 23-154. This innovative building design required the use of new and different construction techniques. The building extends seven stories underground, and encloses 150,000 square feet of office and classroom space. Sunlight is directed down to the lowest hallway through a 132-foot periscope using 30-inch twin tracking mirrors.

Apollo Computer Inc.

Sun Microsystems, Inc.

Fig. 24-6.   These Apollo and Sun high-performance workstations have the power of minicomputers but serve one person. They can be linked in a network.

Microcomputers are small-scale computers. Good examples are the IBM PC and PS/2 series and the Apple Macintosh series. The MicroVAX II by Digital Equipment Corporation (DEC) is a super-microcomputer. See Fig. 24-7.

Confusion exists in the industry as to the definitions of a personal computer, minicomputer, and mainframe. At one time there were clear distinctions among them. Today, though, there are microcomputers which are more powerful than mainframes of the past. Microcomputers were once considered single-user units, but today many offer multi-user capabilities. One thing is certain: the price of microcomputers continues to decline as the performance continues to rise. Because of this, in the future almost all computers will be based on microcomputer technology.

Microcomputer Components

CPU. The "brain" of a microcomputer is its microprocessor, a single chip which serves as the computer's CPU (central processing unit). All information passes through the CPU prior to its display on the monitor. Generally, the more powerful the CPU, the faster the computer.

A basic IBM PC/AT contains the Intel 80286 microprocessor (see Fig. 24-8), and an IBM PS/2 Model 80 contains the faster Intel 80386 chip. The Apple MAC II contains a Motorola 68020 chip. Both the 80386 chip and the 68020 chip are 32-bit microprocessors, whereas the 80286 chip is a 16-bit microprocessor.

Math coprocessors. Also referred to as numeric coprocessors, these speed the generation of graphics on the display screen. Since graphics information is stored as coordinates (which are numbers), the math chip helps to process this information faster. A graphic image will generate on the screen three to five times faster when a math coprocessor is resident in the computer. See Fig. 24-9.

Apple Computer, Inc.

Fig. 24-7.   An Apple Macintosh II computer.

Fig. 24-8. A close-up view of the 80286 microprocessor.

Intel Corporation

Fig. 24-9. This is the 80287 math coprocessor.

Intel Corporation

RAM. RAM (random access memory) is the temporary memory storage area in the computer. Data and programs are input, manipulated, and output here. Generally, the more RAM, the better. A microcomputer-based CADD system contains RAM ranging from 512,000 bytes (single characters) of information to 16 million bytes. Once the computer is turned off, all the information stored in RAM is lost. This is the reason for permanent storage devices.

Permanent storage devices. The most common devices for permanent storage are floppy diskettes and hard disks. Floppy diskettes are thin, flexible disks encased in plastic or paper. They are currently made in three diameters: 3½", 5¼", and 8". The 5¼" diskettes are the most popular, while the 8" are the least. High-capacity 3½" diskettes are quickly becoming popular and someday may replace the 5¼" diskettes. See Figs. 24-10 and 24-11.

Hard disks, also referred to as fixed disks, are many times faster than floppy diskette units and can store huge amounts of data. For instance, a 20 megabyte fixed disk stores over 55 times as much data as a floppy diskette used in a standard IBM PC. Unlike floppy diskettes, fixed disks are not removable from the disk drive unit.

Streaming tape systems are used to make backup files. Backup files are just that — they serve as a backup in case the files stored on the floppy diskette or fixed disk are lost or damaged. Some computers have built-in tape backup systems; others do not. See Fig. 24-12. Another popular practice is to put backup files on floppy diskettes, which are then stored in a safe place.

Fig. 24-11. These are 5¼" floppies. When not in use, they should be stored in a safe place away from heat, humidity, magnets, and electrical devices.

Fig. 24-10. This diskette drive can accommodate 3½" as well as 5¼" floppy diskettes.
Hewlett-Packard Company

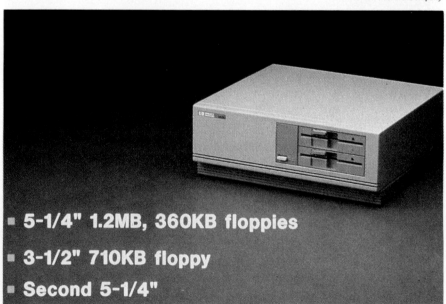

- 5-1/4" 1.2MB, 360KB floppies
- 3-1/2" 710KB floppy
- Second 5-1/4"

Standard ports. Input and output devices connect to the ports found at the rear of a microcomputer. Most input devices, such as a digitizer, and output devices, such as a pen plotter, connect to the standard RS-232 serial port. Most matrix and daisy wheel printers connect to the standard parallel port. See Fig. 24-13.

Fig. 24-12. This streaming tape unit is designed to mount in the host computer and can store up to 60 MB of data on a high-performance tape cartridge. A tape controller board is needed internally in the host computer to allow the tape backup unit to interface with the computer. *Colorado Memory Systems, Inc.*

Expansion slots. Most computers are built for expansion. Inside, there are several empty slots. Printed circuit boards, also referred to as cards, are plugged into these slots. New cards may be inserted to increase RAM, to add more serial ports, or to increase speed.

Monitors

Also referred to as displays, monitors vary greatly in purpose, size, quality, and price. Single-color (monochrome) monitors are popular for applications such as word processing. Color is more desirable for CADD, though monochrome monitors are often used for CADD too. Elaborate systems often contain two monitors: a large color monitor for the graphics and a small monochrome monitor for the text information. See page 588 for an illustration of a two-monitor system.

Resolution refers to the fineness of detail observable in the images on the monitor. Monitors of low resolution display coarse images. For instance, diagonal lines may appear as a set of stair-steps. As the resolution increases, the stair-step effect diminishes.

Resolution is measured in pixels. One pixel is one lighted dot on the screen. All monitors have an x number of pixels horizontally and a y number of pixels vertically on the screen. Low-resolution monitors have approximately 200-400 pixels horizontally and 100-200 vertically. Medium-resolution monitors are in the 640 x 400 range. High-resolution monitors have a resolution of 1000 x 800 pixels or higher. See Fig. 24-14.

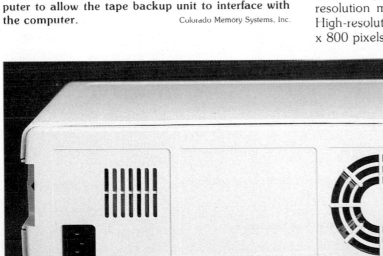

Fig. 24-13. The serial and parallel ports are located at the rear of the computer.

Radio Shack, a division of Tandy Corporation

Fig. 24-14. This 20″ monitor has a resolution of 1280 x 1024 pixels.
Mitsubishi Electronics America, Inc.

Hercules Computer Technology

Fig. 24-15. The Hercules Incolor Card permits the display of 3,072 software-definable characters.

The size of the monitor affects the resolution too. For instance, an image viewed on an 11″ monitor at 640 x 400 resolution will appear much better than the same image viewed on a 19″ monitor at the same resolution.

Graphics Adapters and Controllers

The printed circuit board that is connected to the graphics monitor is the graphics or display adapter, sometimes referred to as the graphics controller. Technically, a graphics adapter is a low-cost device, whereas a graphics controller has more capabilities and costs more. Like monitors, adapters and controllers are specified by resolution and color. An adapter or controller must have as good a resolution and color capability as the monitor to which it is connected in order for the monitor to display its own resolution and color capabilities. In other words, the monitor displays the lesser of the capabilities offered by the two. Ideally, the monitor and adapter or controller have the same capabilities. See Fig. 24-15.

Input and Pointing Devices

Keyboard. The keyboard offers a means for alphanumeric (letters and numbers) input. No device currently available is better for entering such information into a computer. However, to use it efficiently, you must know how to type. See Fig. 24-16.

Mouse. A mouse pointing device is used for picking points and objects that appear on the screen and for selecting screen menu items. A mouse is one of the least costly pointing devices. The mouse

Fig. 24-16. The computer keyboard is similar to a typewriter keyboard, but it has additional keys. The hand is pointing to keys which will move the cursor on the computer screen.
Texas Instruments

connects to the standard RS-232 serial port or into a special board mounted in one of the bus slots inside the computer.

Some mice contain a small roller ball on their underside. The ball turns as it moves across the table. The screen cursor tracks this motion. Other mice are optical. They contain a small light source that shines on a pad containing a metal grid. The light and grid combination controls the motion of the screen cursor. See Fig. 24-17.

Digitizer. A digitizer consists of a large, smooth surface and an attached cursor control. Digitizers, also referred to as tablets, are larger and more expensive than mice, but they offer more capabilities. For instance, a menu overlay can be attached to the digitizer. Commands on the menu are entered in the computer by "picking" them with the cursor control. This is much faster than entering commands from the keyboard. In addition, symbol libraries can be part of the tablet menu. See Fig. 24-18.

Digitizers are capable of tracing hardcopy drawings. This process is known as digitizing. A paper drawing is taped to the surface of the digitizer. In conjunction with the CADD commands and functions, coordinates are digitized from the drawing. Digitizing is often faster than recreating the drawing from scratch. Large digitizers are used to digitize large drawings such as maps of land developments. See Fig. 24-19.

A stylus or puck-type cursor control is used with a digitizer. A stylus is a pen-like object used for such functions as drawing lines and picking items from the tablet menu. See Fig. 24-20. The puck performs these functions too. In addition, it usually contains buttons. The buttons activate functions such as the keyboard's "return" key, or even commands such as "line," "erase," and "zoom." See Fig. 24-21.

Other, less popular pointing devices are available for CADD systems, such as **light pens, joysticks, function boxes, track balls,** and **dials.** See Figs. 24-22, 24-23, and 24-24.

Fig. 24-18. A digitizer with a menu overlay. The blank portion of the overlay is the screen pointing area. This area represents that portion of the monitor on which the drawings will appear. *Texas Instruments*

Texas Instruments

Fig. 24-17. The computer operator is using a mouse with a pad.

CalComp

Fig. 24-19. Digitizing a hardcopy drawing. Once the drawing is in the computer, it can be manipulated and altered just like any computer-generated drawing.

Fig. 24-20. Using a stylus.

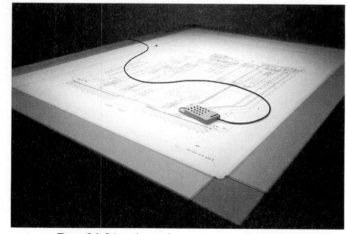

Fig. 24-21. A puck-type cursor control with pro-grammable buttons.

Fig. 24-22. A joystick.

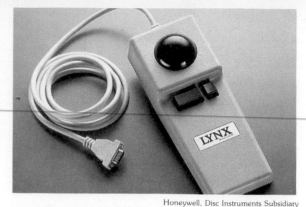

Fig. 24-23. This trackball was designed for use with an Apple Macintosh computer.

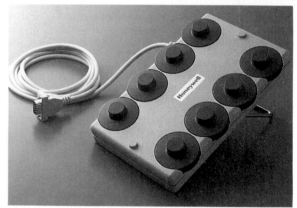

Fig. 24-24. The dial set provides manual control over image motion, orientation, and presentation on graphics work stations. It is particularly suited for three-dimensional modeling applications.

Graphics Output Devices

Pen plotters. Pen plotters are the most popular devices for producing quality CADD line drawings. There is a variety of plotter styles and sizes available. The plotter pens come in various widths and colors.

A pen plotter operates by movement of a pen and a medium, such as paper. Microgrip (the most common type) and drum style plotters move the paper on one axis while the pen moves across the other axis. Simultaneous movement results in angular or curved lines. See Fig. 24-25. Flatbed plotters operate on a different principle. The sheet remains stationary, and the pen moves across the entire sheet. Flatbed plotters are usually placed on tables, while microgrip and drum style plotters are mounted on floor stands. See Fig. 24-26.

much slower, usually between 2 and 8 inches per second. See Fig. 24-28. Ceramic tip pens are becoming popular. They provide nearly the same quality as liquid ink pens, and they run much faster and are not as messy. See Fig. 24-29.

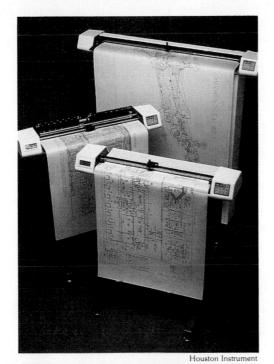

Houston Instrument

Fig. 24-25. Microgrip pen plotters.

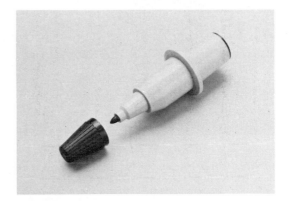

Fig. 24-27. A fiber-tip plotter pen. When the pen is not in use, the cap should be kept on to prevent drying of the tip.   Koh-I-Noor

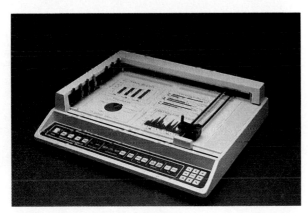

Fig. 24-26. A flatbed plotter.   Houston Instrument

Fig. 24-28. These liquid ink plotter pens are refillable.   Koh-I-Noor

Fig. 24-29. A ceramic tip pen.   Mile High Engineering Supply

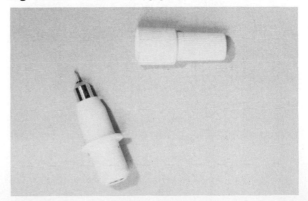

Pen plotters run at different speeds depending on the limitations of the plotter as well as the pens and media being used. Felt or plastic tip pens usually run at a high speed (15-25 inches per second) on paper or vellum. See Fig. 24-27. For much better quality, liquid ink pens are used on Mylar polyester film or vellum. These liquid ink pens should move

Fig. 24-30. An assortment of font cartridges is available for this laser printer. *Hewlett-Packard Company*

Laser printers. Laser technology offers a fast and quiet means for producing text and small graphics. Laser printers are especially popular in outputting images produced by the integration of CADD graphics with computer-aided publishing systems. See Fig. 24-30.

Thermal plotters, electrostatic plotters, and **ink jet printers** are other graphics output devices. Thermal plotters use heat to transfer lines and solid images to the media. See Fig. 24-31. Electrostatic plotters use an electrical charge. See Fig. 24-32. Ink jet printers spray a fine jet of ink onto the media. See Fig. 24-33.

Color impact printers. These offer another alternative for color graphics output. Generally, the quality is not as good as that produced by pen plotters, but impact printers cost less and are fast. They are especially good for preliminary draft work. See Fig. 24-34.

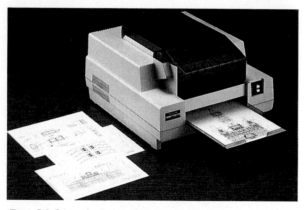

Fig. 24-31. A thermal printer/plotter. *CalComp*

Fig. 24-32. An electrostatic plotter. *CalComp*

Fig. 24-33. An ink jet printer. At the right is a color graphics rasterizer. Used together, the two provide a resolution of 120 dots per inch and a palette of 130,000 colors. *Tektronix*

Fig. 24-34. A color impact printer/plotter. *JDL*

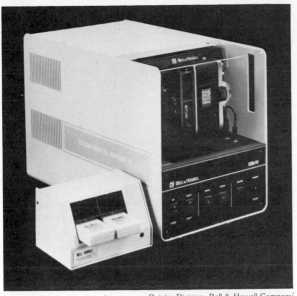

Quintar Division, Bell & Howell Company

Fig. 24-35. This film recorder produces color slides and prints from computer graphics.

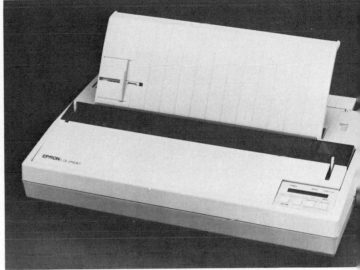

Epson America, Inc.

Fig. 24-37. This dot-matrix printer provides 324 characters per second in draft mode and 108 characters per second in near-letter-quality mode.

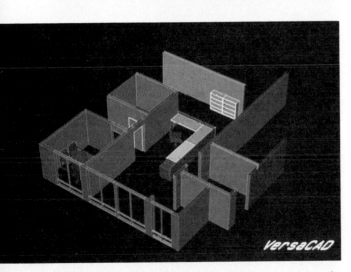

Fig. 24-36. This picture originated as a slide made with a film recorder.

Fig. 24-38. Output from a dot-matrix printer.

```
This text was output on a
matrix impact printer. Note
that the characters are made
up of small dots.
```

Film recorders. Film recorders produce 35-mm slides and color prints of the graphic image. They are an excellent means for producing graphics for presenting CADD to clients or to classrooms. See Figs. 24-35 and 24-36.

Text Output Devices

A surprising amount of text is generated from CADD systems. Examples include bills of materials, menu information, and printouts of settings such as layer information. Printers are used to produce hardcopy output of this text information.

Matrix impact printers. These operate on the same principle as the color impact printer mentioned earlier. Matrix impact printers are a fast and low-cost means of generating single-color text output. The characters (letters and numbers) they print are made up of small dots. See Figs. 24-37 and 24-38.

Daisy wheel printers. Also referred to as letter-quality printers, these provide print equal in quality to that of electric typewriters. However, they are much slower than matrix impact printers. Daisy wheel printers print 10-50 characters per second (cps), while matrix impact printers print at a speed of 50-200 cps or faster. See Figs. 24-39 and 24-40.

Fig. 24-39. A daisy wheel. The characters are at the end of the wheel's "spokes."

```
This text was output on a
letter quality printer.  Note
that the characters are made
of continuous lines.
```

Fig. 24-40. Output from a daisy wheel printer.

Typical CADD Commands and Functions

What follows is an overview of common CADD commands and functions. The discussion is not exhaustive, but it does provide a representative collection of features found in most popular CADD software.

Most of these commands and functions are applied while the CADD system is in the **graphics editor** mode. The graphics editor allows the user to create and modify drawings. Most CADD work is done in the graphics editor. The majority of the commands and functions discussed here are entered at the command prompt, as shown in Fig. 24-41.

CADD commands and functions are usually entered one of three ways: (1) typed using the computer keyboard, (2) picked from the screen menu using a pointing device such as a mouse or digitizer (tablet) cursor control, or (3) picked from the tablet menu using a mouse or digitizer cursor control.

The pointing device (mouse or digitizer cursor control) moves the cross-shaped cursor about the screen. It is used to pick points during geometric construction and to select drawing elements such as lines and text when editing. It is also used to select commands and functions from the screen menu.

Embedded in most of the commands and functions are options. For example, after entering the line command, the user will have the option of entering absolute coordinates (such as "4, 3") at the keyboard or picking a point anywhere on the screen with the pointing device. Another example: A circle can be drawn by specifying its center and radius or by specifying its center and diameter. These points and values can be entered with the keyboard or with the pointing device. These command options provide flexibility when creating geometric figures.

Drawing Commands and Functions

LINE — This is the most often used CADD command because drawings consist mainly of lines. Keyboard entry of X,Y coordinates creates a point on the screen. Entry of a second set of coordinates forms a line. For instance, entry of absolute points 2,2 and 5,5 would form a diagonal line. (See Fig. 24-42.) Pointing devices, such as a mouse or digitizer cursor control, draw lines quickly because of the speed by which they are moved by the hand. The user simply picks a point anywhere on the screen, moves the pointing device any desired distance and direction, and picks the second point. This creates a line on the screen. The process is called "rubber-banding" because the line stretches from the last point picked.

DOUBLE LINE — Floor plan walls are created with the double line function. Specific thicknesses of walls are entered by typing a value on the keyboard. On some systems, the double line function offers a facility for proper joining of wall intersections and corners. See Fig. 24-43.

ARC — Specifying three points forms an arc. Other options are also available, such as specifying the arc's center point and radius. As with most commands and functions, these points and values can be entered with the keyboard or with the pointing device. See Fig. 24-44.

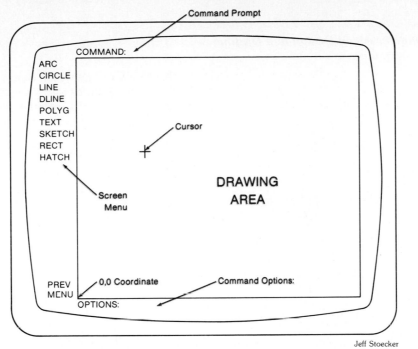

Jeff Stoecker

Fig. 24-41. When a CADD system is in the graphics editor mode, the computer screen will look similar to this. As the user creates drawings, they appear in the "drawing area."

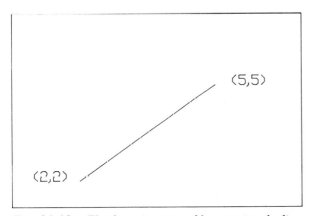

Fig. 24-42. This line was created by entering the line command and then entering the coordinates 2,2 and 5,5 in response to the computer prompts.

Fig. 24-43. The double line function can be used to create walls.

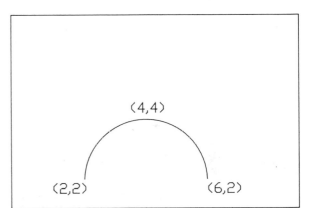

Fig. 24-44. This arc was formed by specifying the coordinates of the arc's start, center, and end points.

DRAG — The drag function makes it possible to dynamically drag an image on the screen. For example, after the start and center points of an arc have been specified, the arc forms on the screen. Moving the pointing device causes the arc to lengthen or shorten. When the desired length has been achieved, the user presses the pick button to "set" the image. Drag is usually embedded in other commands and functions, such as the arc command. Editing commands such as move and scale also utilize this impressive near real-time drag function. See Fig. 24-45.

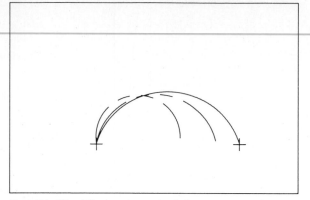

Fig. 24-45. The cross at the left represents the first pick point. The phantom lines represent the areas that form on the computer screen as the cursor control is moved. When the end point of the arc is picked (cross at right), the arc stops moving. Only the final arc remains on the screen.

CIRCLE — Specifying a center point and radius forms a circle. The radius is entered at the keyboard or is picked with the pointing device. A circle can also be formed by specifying a center point and diameter. See Fig. 24-46.

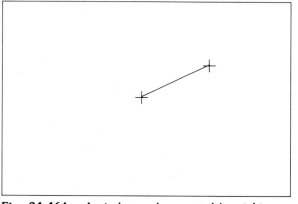

Fig. 24-46A. A circle can be created by picking a center point and a radius, as shown here.

Fig. 24-46B. This is the circle that formed when the radius was picked.

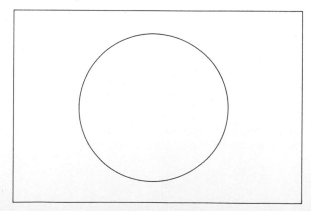

POLYGON — Entry of a center point, the number of sides, and a radius forms a regular polygon. The polygon is specified as being either inscribed within a hypothetical circle or circumscribed around it. See Fig. 24-47.

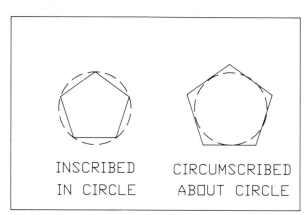

Fig. 24-47. Two methods of creating polygons.

SKETCH — While the sketch command is entered, movement of the pointing device creates a free-form line. See Fig. 24-48.

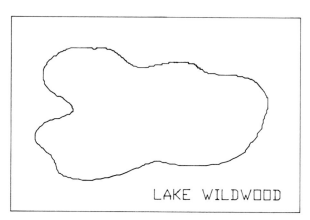

Fig. 24-48. The sketch command allows the creation of free-form lines. This feature is especially useful for landscaping and mapping applications.

TEXT — Text, such as notes and specifications, is entered at the keyboard. Text is left or right justified (aligned) or is centered from a specific point. Depending on the system, text is available in a variety of styles and fonts. For instance, text can be tall and thin or short and fat and can lean to the right or to the left. Popular fonts include Gothic and italic. See Fig. 24-49.

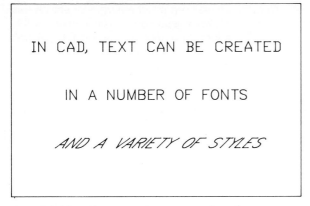

Fig. 24-49. Here are three examples of computer-generated text.

RECTANGLE — Entering a width and height forms a rectangle. Or, a corner is specified and the opposite corner is dragged into place.

HATCHING — Most systems provide a variety of hatch patterns. Examples include bricks, shingles, concrete, earth, insulation, and the standard cross-hatch pattern. Areas are hatched by selecting the boundaries with a pointing device. See Fig. 24-50.

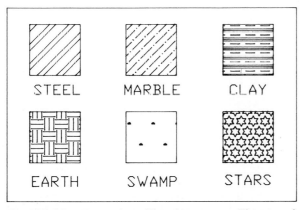

Fig. 24-50. Examples of hatch patterns. The size of the pattern can be scaled to suit the proportions of the drawing.

Editing Commands and Functions

Editing means changing existing text or graphics. The following commands and functions are typically used to edit drawings.

MOVE — Drawing elements (such as lines or circles) to be moved are selected with the pointing device. An arbitrary base point (handle) is chosen on or near the drawing elements and a destination point is picked. The drawing elements are moved to the new location. The drag option is issued to show dynamic movement of the objects. See Fig. 24-51.

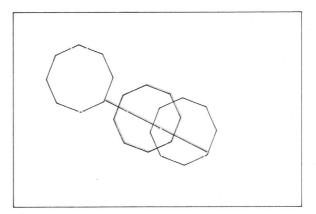

Fig. 24-51. This drawing illustrates how an object can be dynamically dragged to a new location on the screen.

WINDOW — With the window function, the user places a rectangle (window) around an object or group of objects. The window function is embedded into editing commands such as move, scale, and erase. Objects are often selected using the window function because several objects can be chosen in one step, saving time. See Fig. 24-52.

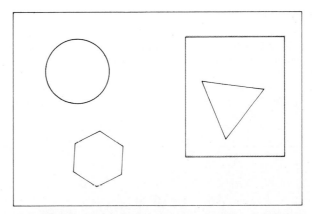

Fig. 24-52. A window has been placed around the triangle. The window function is especially useful when several objects need to be edited. All objects within the window are edited simultaneously.

COPY — This command creates duplicates of existing objects. See Fig. 24-53.

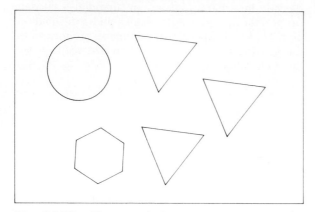

Fig. 24-53. The triangle from the previous illustration has been copied twice.

MIRROR — This command produces a mirror image of an object. For example, one side of a heart could be drawn and then mirrored to produce the other half. See Fig. 24-54.

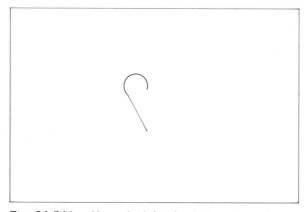

Fig. 24-54A. Here, the left side of a heart was drawn using the arc and line commands.

Fig. 24-54B. The right side of the heart was created using the mirror command. This assures that the two halves of the figure are symmetrical.

SCALE — This command enlarges or shrinks an object. A scale factor (such as 1.5) is entered, or the object is dragged to the proper size. See Fig. 24-55.

Fig. 24-55. The heart from Fig. 24-54B has been enlarged 1½ times using the scale command.

ROTATE — Objects rotate around a specified point with this command.

ERASE — Objects to be erased are selected with the pointing device. They then disappear from the screen. See Fig. 24-56.

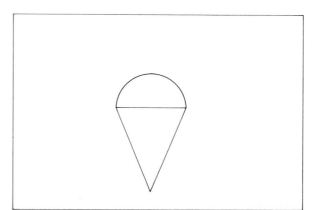

Fig. 24-56A. These drawings of an ice cream cone illustrate the erase command. After the user enters the erase command, the computer will prompt for object selection.

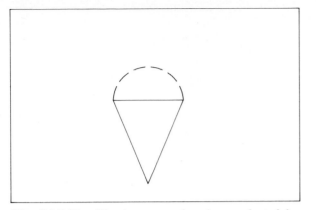

Fig. 24-56B. The ice cream has been selected for erasure. The computer has changed the solid line to a broken one to indicate exactly what will be erased.

Fig. 24-56C. Pressing the "RETURN" key caused the ice cream to disappear.

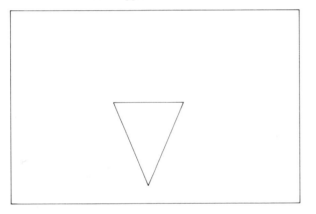

UNERASE — This command restores the graphic elements of a previously erased object.

STRETCH — Objects are stretched into place with this command. For example, an attached garage on a house can be made wider in one step. To do this, a window (rectangle) is placed around the portion of the garage to be stretched and a base point and a destination are chosen, similar to what is done with the move command. See Fig. 24-57.

Fig. 24-57A. The stretch command makes it possible to lengthen or widen one portion of a drawing.

Fig. 24-57B. Here, the right side of the figure from 24-57A has been lengthened. Note that the portion at the left remains unchanged.

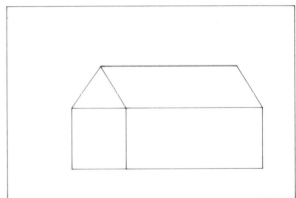

FILLET — This produces a rounded corner of a specified radius at a vertex. For example, the fillet command is used to round the corner of a sidewalk leading from the driveway to the front door of a house. See Fig. 24-58.

Fig. 24-58A. The corners of this rectangle can be rounded by using the fillet command.

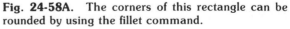

Fig. 24-58B. Here is the same rectangle, after the corners were filleted.

CHAMFER — This is similar to the fillet command, except the chamfer command inserts a chamfered corner at a vertex. A chamfer distance is specified at the keyboard. See Fig. 24-59.

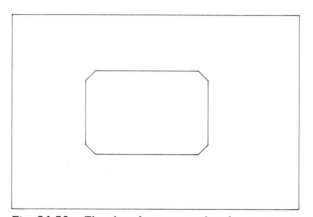

Fig. 24-59. The chamfer command makes it easy to produce chamfered corners.

EXTEND — Drawing elements such as lines are lengthened to a specified distance with the extend command.

ARRAY — This powerful command multiplies an object and arranges the objects in rows and columns or around a center point. For instance, a single chair in an auditorium can become 10,000 chairs in 200 rows and 50 columns with the array command. Distances between the objects are specified. See Fig. 24-60.

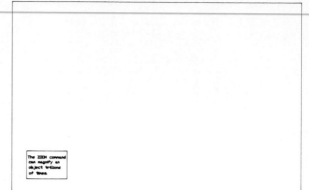

Fig. 24-60A. This sign could be duplicated using the copy command, but if many copies are needed, the array command is faster.

Fig. 24-60B. Here the array command was used to produce five rows and four columns of signs.

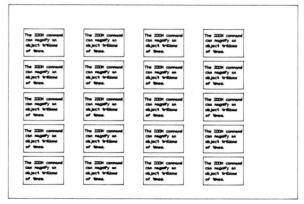

Display/Window Commands and Functions

ZOOM — This command magnifies objects on the screen. If a detail of an object is not currently seen on the screen, the zoom command can be used to make it appear. For instance, a drawing showing the planets of the solar system may be on the screen. A series of zooms will focus in on a crater on Earth's moon. Additional zooms inside the crater will fill the screen with a moon rock the size of a fist. Further zooms will display the rock's molecular structure. (Of course, someone must initially draw all these details in order to make these zooms possible.) See Fig. 24-61.

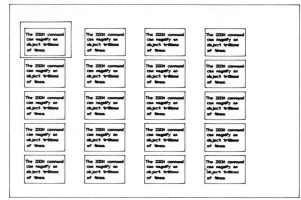

Fig. 24-61A. One of the signs in the array has been selected for magnification. A window has been placed around it.

Fig. 24-61B. The "zoomed" sign is much easier to read than the original.

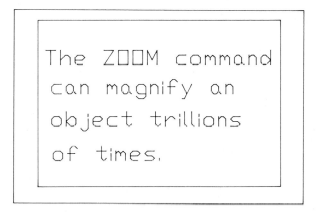

PAN — This function moves the display window from one location to another. For example, if the screen is zoomed in on the kitchen of a floor plan, it is possible to move to the living room while remaining at the same zoom magnification. This allows addition of detail without time-consuming zooms (zooming out of the kitchen, moving over to the living room, zooming in on the living room).

VIEW — With the view function, zoomed views are stored in memory for subsequent retrieval. For example, if an architect is about to leave the kitchen area of a floor plan but anticipates a need to return to the kitchen for editing purposes, the zoomed view of the kitchen is stored. View is an alternative to using the zoom and pan commands.

Dimensioning Commands and Functions

LINEAR — A linear dimension appears after picking the endpoints of a line within an object and picking the distance from the object to the new dimension line. This is known as semiautomatic dimensioning. Some CADD systems display the dimension after the user picks the object and the desired position of the dimension line. This feature speeds the dimensioning process. See Fig. 24-62.

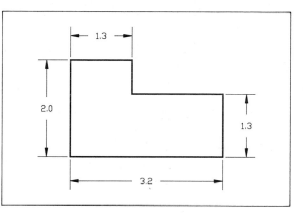

Fig. 24-62. For this figure, the CAD operator selected the lines to be dimensioned and the distance of the dimension lines from the object. The rest — measuring the lines, drawing the extension and dimension lines, and writing the numbers — was done automatically.

RADIAL — Radial dimensioning involves arcs and circles. The dimension appears after selecting the radial object and the desired placement of the dimension. See Fig. 24-63.

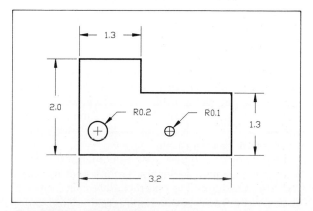

Fig. 24-63. Radial dimensioning.

ASSOCIATIVE DIMENSIONING — With associative dimensioning, the dimensions on a drawing automatically change when the drawing is scaled or stretched.

Layers, Colors, and Linetypes

Layers, also known as levels, are like transparent overlays. For example, a layer named WALLS may contain the walls of a floor plan. A layer named DIM may contain the dimensions, and a layer called NOTES may contain the text information (notes, specifications, and so on). See Fig. 24-64.

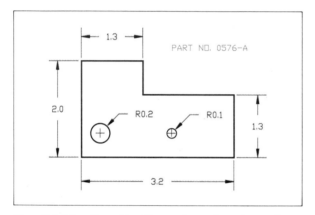

Fig. 24-64. For this illustration, the object lines were drawn on one layer, the dimensions on another, and the text on a third.

Layers are turned on or off. When on, the objects on the layer are displayed; when off, they are invisible. When plotting a drawing, any combination of layers can be on or off.

Colors are assigned to the layers. For example, the layer WALLS could be in green on the screen. The layer DIM could be in yellow and NOTES in red. Even if the CADD system does not include a color monitor, color assignments are still important. When plotting, pen numbers are assigned to colors. For instance, a .7-mm black pen may go with green, a .3-mm black pen with yellow, and a .3-mm red pen with red.

Like colors, linetypes, such as center and hidden lines, are assigned to layers. Colors and linetypes can also be assigned to objects individually, without regard to the layer on which they reside. This feature increases flexibility.

Symbols (Blocks)

Symbols, also referred to as blocks, make it unnecessary to repeatedly draw shapes, details, and parts. A symbol is a combination of drawing elements stored on disk and given a name. Symbols are retrieved by their name, scaled, rotated, and inserted into drawings as needed. Common examples in architecture are symbols for doors, windows, plumbing, appliances, furniture, electrical wiring and lighting, and landscaping symbols such as shrubs and trees. See Fig. 24-65.

A symbol is treated as a single element. For instance, only one selection is necessary when moving a symbol. The creation of a symbol is reversed with the explode function. After a symbol is exploded, the elements within the symbol can be edited one by one.

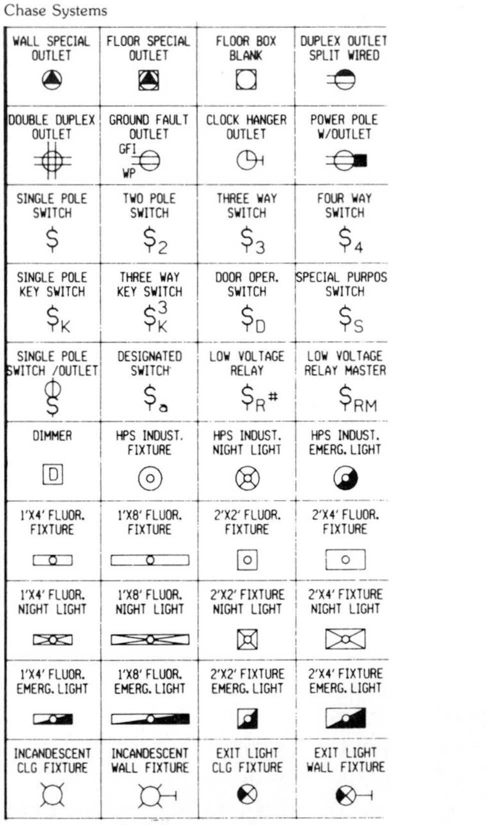

Fig. 24-65. These are electrical symbols from a commercially available symbol library. The symbols are on a digitizer tablet overlay and also are stored in the computer's memory. Touching a symbol on the overlay with the pointing device (stylus or puck) causes that symbol to appear on the computer screen.

Inquiry Commands and Functions

LENGTH — Length, also referred to as distance, measures the distance from one point to another. Two points are chosen with a pointing device. After selection of the second point, the distance between the two points appears on the screen in numerical units.

AREA — This command calculates the area of an enclosed space. An outline of the space is traced by picking points around the perimeter, forming a polygon with an unlimited number of sides. The area command also calculates the perimeter of an enclosed space.

PROPERTIES — This command, also called the list command, lists certain properties of objects. Examples are the circumferences of circles, the position (in absolute coordinates) of objects, and the layers on which they reside. This information appears when the properties function is issued and an object is selected.

TIME — The time function tracks and displays the time spent on a drawing.

FILES — The files function allows the user to access disk information such as the names of the drawings stored on the disk. It also provides a facility for renaming drawings or deleting unwanted drawings.

Drawing Aids

GRID — A grid is a visual drawing aid in the graphics editor. The grid function allows the user to set an alignment grid of dots on the screen at any desired spacing. The grid does not become part of the drawing. It simply provides a visual reference for distances and overall drawing size. See Fig. 24-66.

Fig. 24-66. A grid on the graphics editor makes it easier to judge distances when drawing.

SNAP — The snap function is similar to the grid function. Snap also creates a grid, but an invisible one. Although it cannot be seen, the effects of the snap grid become evident as the cursor is moved with the pointing device. When the snap function is in effect, it forces all points and selections to be placed on the points of the invisible grid. The snap spacing is determined by the user. The snap function is especially useful when laying out floor plan walls at modular increments such as 2'.

OBJECT SNAP — This function provides a means for snapping to specific points on an object. (The points do not have to be on a snap grid.) For instance, suppose it is necessary to draw an arc from the end of one straight line to the beginning of the next. With the object snap function, it is easy to locate the endpoints of the lines precisely. See Fig. 24-67.

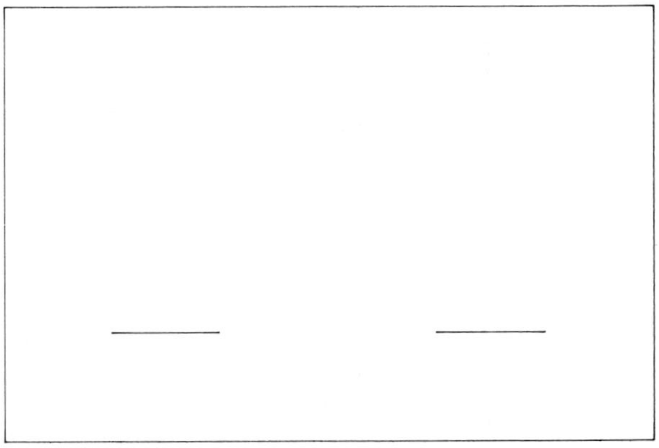

Fig. 24-67A. These two lines are to be joined by an arc.

Fig. 24-67B. The object snap function made it easy to begin and end the arc precisely at the endpoints of the lines.

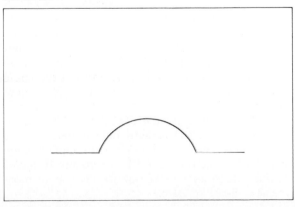

ISOMETRIC DRAWING — Most CADD systems offer isometric drawing capabilities. The isometric drawing is created in a manner similar to creating the isometric on paper. Some systems generate the isometric drawing automatically from the three orthographic views of the drawing. Other systems even generate orthographic views from an isometric view. See Fig. 24-68.

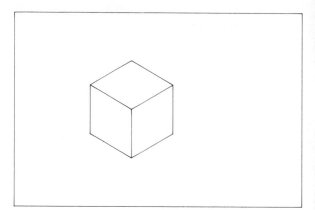

Fig. 24-68. An isometric drawing generated with CADD.

Plotting

Plotting is the production of hardcopy drawings. All designing and drawing on CADD systems is electronic in nature until the plotted drawing is produced.

The CADD software controls most of the plotting parameters. For example, the speed with which the pen should move on the paper, the relationship of color to pen number, the paper size, and the scale are all specified at the computer prior to plotting.

3-D Functions

Some CADD software is 2-D, some is 2½-D, while other software stores the graphic information in 3-D. The 2-D systems store graphic data with X and Y coordinates only. The 2½-D systems store coordinate data the same as 2-D systems, but they provide 3-D visualization of objects if the user enters a value for object thickness. The 3-D systems store graphic data using X, Y, and Z coordinates. These systems are more powerful, but they are often more difficult to learn.

WIREFRAME and HIDDEN LINE REMOVAL — The 2½-D and 3-D systems offer a feature called hidden line removal. This feature is applied when the object is viewed as a 3-D wireframe. Removal of the hidden lines makes the object appear more realistic. See Fig. 24-69.

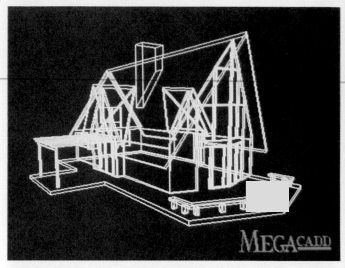

Fig. 24-69A. A wireframe drawing of a house.

Fig. 24-69B. The same drawing, but with hidden lines removed.   MEGA CADD, Inc.

SURFACE MODELING — The next step toward making an object look as real as possible is surface modeling or solids modeling. Surface modeling is the placement of a skin on the wireframe object. The skin may appear in a variety of colors depending on the capabilities of the hardware. Surface modeling provides realistic perspective drawings of buildings. See Fig. 24-70.

SOLIDS MODELING — Solids modeling is often confused with surface modeling because the two appear the same in many instances. However, solids modeling treats the object as though it were solid. If you were to cut a section from the solid model, it would not be hollow like a surface model. Solids modeling requires a lot of power and memory. Consequently, it is not yet common on a microcomputer, but it will be in the future.

Fig. 24-70. Surface modeling gives a more realistic appearance to drawings.   MEGA CADD, Inc.

SHADED SOLIDS MODELING — The most realistic 3-D view of an object is provided by shaded solids modeling. Shades of color and shadows are represented on and near the object according to a specified light source. For instance, if a light source is specified at the upper right corner and behind the object, shadows of the object appear at the lower left in front of the object.

Advanced Applications

Networking

Networking is linking two or more computers so that each can share resources from the other. For instance, a stair detail being designed on one workstation can be quickly transferred to another workstation for insertion into a building section drawing being designed on that workstation. Symbol libraries, menu files, and drawing files are all shared among workstations included in the network. Several workstations can be linked to a single plotter or printer, making it unnecessary to buy numerous output devices.

Symbol Libraries

As mentioned earlier, use of symbol libraries avoids repetition. The symbol libraries serve as an electronic template of shapes. Moreover, these shapes offer much more detail than shapes outlined from plastic templates. See Fig. 24-71.

Attributes and Bills of Materials

Information about an object's characteristics can be assigned to its symbol. For instance, a door symbol may contain information about the size of the door, style, material, swing, manufacturer, etc. The information does not have to appear on the drawing; it can be invisible. These items of information, called **attributes**, can be extracted from the drawing file to form a door schedule. The same is true for windows, appliances, and any other symbols. See Fig. 24-72. A bill of materials can also be generated from the attribute information. See Fig. 24-73.

Links with Other Software

Software for CADD connects to other software for special applications. For example, the bill of materials shown in Fig. 24-73 was produced with a program external to the CADD program. This section describes some of the software packages used to customize CADD systems and increase their capabilities.

Spreadsheets. These programs display information in cells made by rows and columns. Formulas are assigned to values contained in the spreadsheet, such as costs of building materials. As new values are input by the user, the computer makes the calculations that show how the changes affect overall costs. When a spreadsheet program is linked with a CADD program, cost estimates can be calculated, printed, and attached to the drawings.

The strength of **data base programs** is their ability to handle large amounts of data, sort the data, and provide specialized reports. For example, a large commercial building, such as a hospital, contains hundreds of electrical fixtures. Attributes from these fixtures, stored as individual records, can be kept in the data base and later extracted according to their field specification. The field specification may be fluorescent lighting, for instance, and a report on all fluorescent lighting fixtures would then be generated.

Word processors. Several word processors, such as Volkswriter and WordPerfect, link to popular CADD programs. While CADD is strong in editing graphics, word processors are strong in editing text. Since CADD attributes are text information, word processors are useful in formatting reports and correcting mistakes.

Autodesk, Inc.

Fig. 24-71.   The AutoCAD AEC tablet overlay contains numerous architectural symbols.

DOOR and HARDWARE SCHEDULE

DOOR MARK	DOOR SIZE	DOOR			FRAME			DOOR HARDWARE	REMARKS
		TYPE	MATERIAL	FINISH	TYPE	MATERIAL	FINISH	KEY CARD / HORN / N.R. B.B. HINGES / B.B. HINGES / HINGES / PANIC HDW. / OFFICE 2 3/4 B.S. / STORAGE / PASSAGE / BATHROOM / DUMMY / ROLLER CATCH / PUSH–PULL / D–BOLT W/T.B. / CLOSER / FLUSH BOLTS / SURFACE BOLTS / KICK PLATES / WALL STOP / FLOOR STOP / ELEC. STRIKE	
1	2 3'–0" x 6'–0" x 1–3/4"	B	H.M.	PAINT	II	H.METAL	PAINT		
2	3'–0" x 7'–0" x 1 3/4"	C	S.C. WOOD	VARNISH	II	H.METAL	PAINT		
3	3'–0" x 7'–0" x 1 3/4"	C	S.C. WOOD	VARNISH	II	H.METAL	PAINT		
4	2 3'–0" x 7'–0" x 1–3/4"	A	AL & GL	NONE	I	ALUM	NONE		

Fig. 24-72. A door and hardware schedule produced on a CADD system. *Joseph McRae, Dearcon Group Inc.*

```
Code          Size              Grade   Description
----------    ----------------  -----   ------------------------------------
door30        3'-0"             1       Interior Door
desk6         6'-0" X 29"       1       Office Desk
exchair       Normal            1       Executive Chair
window60      6'-0" X 8'-0"     3       Office Window
window40      4'-0" X 8'-0"     3       Office Window
execdesk      6'-0" X 3'-0"     1       Executive Desk
divid6        6'-0" X 3'-0"     1       Room Divider
desk4         4'-0" X 29"       1       Office Desk
sechair       Standard          1       Secretary Chair
armchair      Normal            1       Arm Chair
couch         Normal            1       Large Couch
lgchair       Normal            1       Chair
endtable      2'-6"             1       Small Endtable
lamp          100 Watt          1       Lamp
bookcase      5'-0"             1       4 Shelve Bookcase
file4         Standard          1       4 Drawer File Cabinet
```

Fig. 24-73. This bill of materials for an office was created from the attributes assigned to the components of the floor plan. *VersaCAD Corporation*

Parametrics. Parametrics is a fascinating concept for the efficient design, storage, and retrieval of drawings and symbols. With parametrics, a drawing is stored on disk, but the dimensions of the object remain variable. This means that when the object, such as a door symbol, is inserted on the screen, the user is prompted to enter design parameters such as the door type, size, material, swing, etc. It is possible to produce unlimited variations of the same basic door design. Storage of the object is very efficient too because only one drawing is stored on disk as opposed to hundreds.

Two popular software packages use parametric programming techniques. ACAD Partner, produced by Chase Systems (Westerville, OH), is a selection of symbol libraries for a variety of building applications. All symbols are contained on tablet menu overlays (templates) for easy retrieval. See Fig. 24-74. Synthesis, produced by Synthesis, Inc.

(Bellingham, WA), is a design program that operates with the AutoCAD computer-aided drafting package. An AutoCAD drawing, such as an elevation, is entered into Synthesis and a variable is assigned to each drawing dimension. These variables are later replaced by dimension values to produce a variation of the design. See Fig. 24-74. As with ACAD Partner, unlimited variations of the basic design can be produced.

Space diagrams. Architectural-specific CADD software enables the user to create space diagrams. The space diagrams are then converted to floor plan drawings.

The first step is to define individual spaces such as a kitchen, living room, bath, and bedrooms. These spaces are simple rectangular diagrams made up of single lines produced with CADD. Next, with the CADD move capability, the spaces are placed in any desired relationship to one

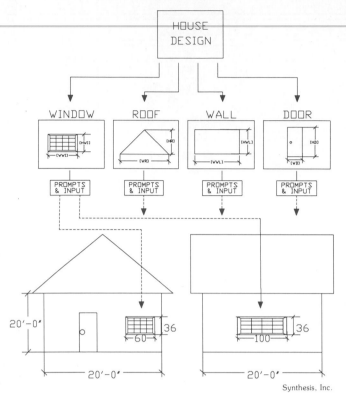

Fig. 24-74. As you can see on this diagram, Synthesis allows you to create variations of a basic design.

Synthesis, Inc.

Fig. 24-75. These two drawings were made using the AutoCAD computer drafting software and LAND-CADD landscaping software.   LANDCADD, Inc.

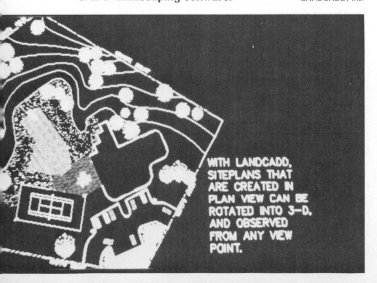

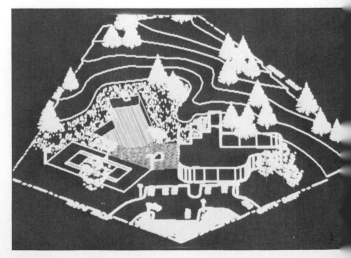

LANDCADD, Inc.

another. The desired interior and exterior wall thicknesses are entered. Lastly, the CADD system converts the schematics to walls and properly connects them. Some CADD systems can even show the floor plan in 3-D so that the user can evaluate the spaces in terms of volume.

Landscape design. CADD is a landscaper's dream come true. With CADD, landscape architects insert plant symbols such as shrubs and trees and other site amenities. Cut and fill earth work and grading calculations are also performed. In addition, landscape designs are linked to the CADD data base for performing quantity takeoffs and cost estimates at any phase of the design. See Fig. 24-75.

The LANDCADD software package, produced by LANDCADD (Franktown, CO), provides these capabilities and operates with AutoCAD. Its irrigation module calculates flow, velocity, and pressure loss and can automatically size pipe.

Structural engineering. *Finite element analysis* measures the physical and thermal stress within a mechanical member and determines the amount of pressure that can be exerted on a mechanical member before it deforms or breaks. Computer programs can perform the complex calculations required for such analysis.

Supersap software, produced by Algor Interactive Systems (Pittsburgh, PA), is a good example of such a program. It is available for both AutoCAD and VersaCAD. Supersap analyzes static and dynamic loads, pressure, thermal stress, constant acceleration and centrifugal loads, weight and center of gravity, and steady state heat transfer.

MSC/pal 2, produced by Macneal-Schwendler Corporation (Los Angeles, CA), is a micro-based finite element program for stress and vibration analysis of structures. It also interfaces with AutoCAD and VersaCAD.

Exchanging Information among CADD Systems

CADD communication standards play an important role in the industry-wide trend toward sending and receiving electronic CADD data. When drawing files from one CADD system are compatible with other CADD systems, they can be transferred easily and are often sent thousands of miles in just minutes through telephone lines. In addition, architects, engineers, and contractors are able to eliminate, or at least minimize, paper drawings and express mail costs.

The DXF file format is a de facto standard for translating files from one micro-based CADD system, such as AutoCAD, to another, such as VersaCAD. Both AutoCAD and VersaCAD contain facilities for translating and accepting files in the DXF format.

Autodesk (producers of AutoCAD) created the DXF format, known as the drawing interchange file format. DXF files are standard ASCII (text) files. They are easily translated to the formats of other CADD systems or submitted to other programs for specialized analysis.

IGES stands for the Initial Graphics Exchange Specification. IGES is an industry standard for interchange of graphic files between small- and large-scale CADD systems. AutoCAD and VersaCAD are also IGES compatible.

File translation via IGES or DXF from one CADD system to another is useful. However, each CADD system has some unique characteristics. Consequently, certain features, such as layers, blocks, text, and linetypes, are potential problem areas when translating files. For example, some CADD systems use only numbers for layer names and do not accept words such as WALLS or DIM for layer names. If a drawing file that uses words for layer names is translated to a system using only layer numbers, all of its layer names would change to numbers.

Programming Languages

The open architecture of CADD software, combined with the user programming languages contained in popular CADD software, contributes greatly to the growth and shape of the CADD industry. For instance, VersaCAD contains CPL (CAD Programming Language). Applications developers use CPL for development of add-on programs and custom menus.

One of the most significant programming developments is the AutoLISP programming language embedded in AutoCAD. AutoLISP is a version of the LISP programming language, which is often associated with knowledge-based systems and artificial intelligence applications. Literally hundreds of AutoLISP macros and programs are being developed. (A macro executes a series of inputs automatically.) These macros and programs enhance AutoCAD's drafting and design capabilities.

Looking Ahead

In the future, we can expect integrated CADD packages. This means that the software will provide not only graphics capabilities but also the following: parametrics, finite element analysis, bill of materials generation, word processing, spreadsheets, data base management, and even desktop publishing all in one software package.

Architects, engineers, and contractors will work from the same CADD data base. Cost estimates will be more accurate because all subcontractors will work from the same set of drawings and specifications. Building designs will improve, and construction will be more efficient.

Imagine this. You are an architect designing an apartment complex. At the initial design stage, the CADD system asks for the style of the building, the approximate square footage, and the number of living units. You enter the information. It then asks you for types and sizes of rooms, halls, wall thicknesses, windows, doors, appliances, etc. Considerations such as city building codes are already in the system, providing a knowledge base upon which to make design-related decisions. Good architectural design practices such as efficient plumbing systems (bathrooms back-to-back, for example) and people traffic flow considerations are in the system too.

The computer evaluates the information, and before your eyes, it draws an optimized floor plan according to your specifications. It also proposes an efficient use of floor space, suggests building materials, and calculates approximate cost per square foot.

The computer then asks whether or not you like the preliminary floor plan design and provides an opportunity for making alterations. Finally, you embellish the drawing with detail.

It sounds almost like science fiction, but you can expect this level of sophistication on a micro-based CADD system in the future. This incredible system of tomorrow will bring remarkable power and artificial intelligence capability well beyond present-day CADD technology. Just around the bend, a totally new dimension to the architectural design and drafting process awaits us.

Build Your Vocabulary

Following are terms that you should understand and use as part of your working vocabulary. Write a brief explanation of each term.

array command	networking
associative dimensioning	object snap function
attribute	pan command
data base program	parametrics
digitizer	pen plotter
drag function	RAM
editing	resolution
files function	snap function
floppy diskette	solids modeling
graphics editor	spreadsheet program
grid function	surface modeling
hard disk	symbol library
layer	view command
micro-based CADD	window function
mouse	zoom command

Class Activities

1. Visit a CADD facility of your choice and talk to the owner or manager about the effects that CADD has had on his/her business. Find out what kind of retraining was needed for employees. Report your findings to the class.

2. Interview an architect. Find out what changes CADD has made in the way that designers approach a construction project.

3. Sketch ten architectural shapes that you would include in a CADD symbol library.

4. Visit your city's planning commission or the office of a civil engineering firm. Report to the class about how CADD is used to develop site plans.

5. Examine college, junior college, and trade school catalogs in your school's media center or library. Determine what kind of CADD training is available. Calculate the costs involved in that training. Find out what high school courses are recommended for students who wish to major in CADD-related fields. Seek information about the occupational outlook for persons interested in CADD employment. Report your findings to the class.

Trademarks

The following products were discussed in this chapter:

ACAD Partner is a registered trademark of Chase Systems

Apollo is a registered trademark of Apollo Computer, Inc.

AutoCAD and AutoCAD AEC are registered trademarks of Autodesk, Inc.

AutoLISP is a trademark of Autodesk, Inc.

CADVANCE is a trademark of Calcomp

CAD Solutions is a registered trademark of Sigma Design

CPL is a program feature of VersaCAD

Cyber 205 is a trademark of Control Data Corporation

DataCAD is a trademark of Microtecture

IBM 4300 and IBM PC, and IBM PS/2 are registered trademarks of IBM Corporation

Intel 80286, Intel 80287, and Intel 80386 are trademarks of Intel Corporation

LANDCADD is a trademark of LANDCADD, Inc.

Lotus 1-2-3 is a registered trademark of Lotus Development Corp.

Macintosh is a trademark of Apple Computer, Inc.

MicroVAX II is a trademark of Digital Equipment Corp.

Motorola 68020 is a product of Motorola

MSC/pal 2 is a trademark of Macneal-Schwendler Corp.

Personal Architect is a trademark of Computervision

Personal Designer is a trademark of Computervision

RoboCAD is a registered trademark of Robo Systems Corporation

Sun is a trademark of Sun Microsystems, Inc.

Supersap is a trademark of Algor Interactive Systems

Synthesis is a trademark of Synthesis, Inc.

VAX 8600 is a trademark of Digital Equipment Corp.

VersaCAD/Architect is a product of VersaCAD Corporation

VersaCAD is a registered trademark of VersaCAD Corporation

Volkswriter is a registered trademark of Lifetree Software Inc.

WordPerfect is a registered trademark of Satellite Software International

architecture

Design
Engineering
Drawing

Careers

Chapter 25
Career Opportunities
in Architecture
and Related Fields

619

Chapter 25

Career Opportunities in Architecture and Related Fields

People working in architecture and fields related to it transform ideas into workable plans and plans into structures. This process of designing and building structures requires the talents and skills of many people. You could be one of these people, for many types of career opportunities exist in these areas.

Included in this chapter are discussions of some career possibilities. Use this information to begin an investigation of your own. Read the text and select the careers that interest you. Then write to the professional organizations suggested and request further information. The addresses given are current at the time of this writing. Also, consult professionals in the areas you select. Read professional publications. Discuss your plans with a career counselor. A career in architecture or a related field can be challenging as well as satisfying.

Architectural Design and Construction

The construction industry offers a wide variety of career choices. Occupations involving structure design and building are directly involved in the industry, and many other occupations are closely related. See Fig. 25-1.

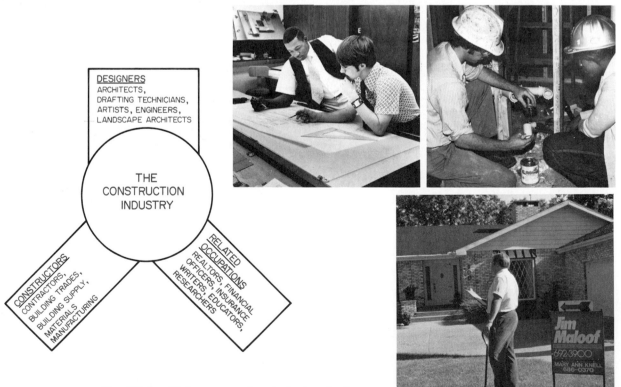

Fig. 25-1. Major occupational clusters in the construction industry.

Maloof Realty. Roger Bean

The group of occupations classified as designers includes architects, engineers, drafting technicians, and landscape architects. Occupations in the constructor group may be subdivided according to the type of construction involved. Basically, there are two types: building construction and heavy construction.

Building construction is the building of residences and commercial buildings. This is the type of construction that is discussed and illustrated in this book. **Heavy construction** is large scale construction. Building projects such as dams, canals, highways, underground construction, bridges, water control construction, piping systems, electrical power installations, and communications construction are all heavy construction projects. See Figs. 25-2 through 25-6.

Santa-Fe International Corporation

Fig. 25-2. Engineers, contractors, and operators of heavy equipment are some of the people involved in laying a pipeline.

Bureau of Reclamation, U.S. Department of Interior

Fig. 25-4. Workers in heavy construction are building a turbine which will be used to generate electricity in a power-generating plant connected with a large dam.

Mackinac Bridge Authority

Fig. 25-3. Building a bridge is a heavy construction project.

West Virginia Highway Department

Fig. 25-5. This tunnel will carry a highway through a mountain. After the engineers do the designing, a contractor for heavy construction sees that the work is done.

Fig. 25-6. The construction of a communications tower requires engineering design, construction skill, and courageous construction workers.

Design Careers

Career opportunities in architectural design are many and varied. A career may be selected that involves designing an entire building (architect), designing a mechanical system for a building (mechanical engineer), or, perhaps, interpreting designs into working drawings (drafting technician). Brief discussions of these and other career possibilities in architectural design follow.

Architect

Architects are involved in the basic planning, engineering, aesthetic design, and the drawing of plans and specifications. They are involved in contract negotiations and must understand business practices. They must be licensed to practice architecture. See Fig. 25-7.

Architects are prepared in universities. The program of study takes four to five years and involves extensive study in the fields of mathematics and science.

Information may be obtained by writing to:

The American Institute of Architects
1735 New York Avenue, N.W.
Washington, D.C. 20006

Fig. 25-7. The architect uses drafting as a basic design skill.

Architectural Drafting Technicians

Architectural drafting technicians work for architectural, engineering, or construction firms under the direction of an architect, engineer, or an experienced drafting technician. They become involved in the preparation of architectural drawings and may specialize in particular areas. Some technicians specialize in structural drafting. Others concentrate on the drafting of heating and air-conditioning systems, electrical systems, or plumbing systems. See Fig. 25-8.

Preparation is available in vocational schools and junior and senior colleges.

Information may be obtained by writing to:

American Institute for Design and Drafting
3119 Price Road
Bartlesville, Oklahoma 74003

Fig. 25-8. Architectural drafting technicians may work on a variety of architectural drawings.

Fig. 25-9. Landscape architects are responsible for all details of the landscaping of the site.

Fig. 25-10. Interior designers assist clients in making color and fabric decisions.

Landscape Architects

Landscape architects plan the arrangement of walks, terraces, shrubs, and trees on a building site. They do both residential and commercial work and supervise the grading, construction, and planting. Some states require that they be licensed. See Fig. 25-9.

Preparation is available in colleges and universities. Usually, four years' study is required to complete the bachelor's degree.

Information may be obtained by writing to:

American Society of Landscape Architects, Inc.
1750 Old Meadow Road
McLean, Virginia 22101

Interior Designers

Interior designers plan and supervise the design and arrangement of building interiors and furnishings. They assist in the selection of furniture, carpeting, draperies, wall covering, light fixtures, and art. They also supervise those who are doing the work, including the painting, paperhanging, and cabinetmaking. See Fig. 25-10.

Preparation is available in community and four-year colleges. Programs of study require from 2 to 5 years.

Information may be obtained by writing to:

American Institute of Interior Design
730 Fifth Ave.
New York, N.Y. 10019

Engineering

Various types of engineers participate in the architectural design process. Electrical, mechanical, and civil engineers are among those employed.

Electrical engineers design the electrical and communications systems. They become involved in overseeing the installation and testing of electrical and communications equipment.

Mechanical engineers are involved with the design and installation of mechanical systems. These systems include heating, air conditioning, and refrigeration.

Civil engineers design and supervise the construction of roads, harbors, bridges, sewage systems, and structural systems of buildings. Some are employed as supervisors of construction projects. See Fig. 25-11.

Many jobs require that engineers be licensed.

Information may be obtained by writing to:

Institute of Electrical and Electronic Engineers
American Society of Civil Engineers
American Society of Mechanical Engineers
345 East 47th Street
New York, N.Y. 10017

Fig. 25-11. A civil engineer supervises the construction of a building.

Fig. 25-12. A crew from a roofing firm that was subcontracted to do this part of the job.

Construction Careers

The following discussion is related to careers for those who are interested in the construction process which occurs after the building or other project has been designed.

General Contractor

The general contractor has the full responsibility for the completion of the project at a specified cost and time. This includes securing the materials and equipment, hiring the workers, and managing the actual building of the structure. A general contractor must understand all aspects of construction and use sound business practices.

Subcontractor

A subcontractor specializes in one kind of work. Examples are roofing, electrical wiring, heating and air conditioning, and excavation. Subcontractors are usually hired by general contractors and do much of the actual work on projects. See Fig. 25-12.

Project Manager

The project manager is an employee of the general contractor. This person is at the construction site and is responsible for the total construction project. Generally, the project manager has a college degree in engineering, architecture, construction technology, or business administration.

Project Engineer

The project engineer is assigned to the construction site and must report to the project manager. The project engineer is responsible for all engineering aspects of the construction.

Expediter

Expediters work in the main office of a general contractor. They see that items needed for construction are shipped to the site when needed. Expediters are responsible for keeping accurate records and reports. See Fig. 25-13.

Fig. 25-13. The expediter must keep accurate records of materials for many different jobs.

Fig. 25-14. A surveyor determines precise measurements and locations of land areas.

Surveyors locate precisely the corners and boundaries of pieces of land. In addition, they produce drawings of the site and provide legal descriptions. They are also involved in locating buildings, roads, and other such projects.

Surveyors must be licensed to practice. Usually, surveyors have completed a bachelor's degree in civil engineering.

Surveying aides may receive preparation in community college programs taking 1 or 2 years. Through experience, a person can progress into more responsible jobs on the survey team. See Fig. 25-14.

Information may be obtained by writing to:

American Congress on Surveying and Mapping
Woodward Building
733 15th Street, N.W.
Washington, D.C. 20005

THE FINISHING TRADES

Floor Coverer
Roofer
Terrazzo Worker
Glazier
Marble Setter
Plasterer
Lather
Paperhanger
Painter

THE MECHANICAL TRADES

Millwright
Elevator Constructor
Sheet-Metal Worker
Pipe Fitter
Plumber
Electrician

THE STRUCTURAL TRADES

Ornamental-Iron Worker
Stonemason
Rigger and Machine
    Mover
Reinforcing-Iron Worker
Cement Mason
Bricklayer
Boilermaker
Operating Engineer
Carpenter

Fig. 25-15. The skilled occupations in the construction industry.

Robert E. McKee, Inc.

Fig. 25-16. Employees in structural trades often work under hazardous conditions.

Skilled Occupations

The construction occupations make up the largest group of skilled workers in the work force. They can be divided into three large groups—the structural trades, the finishing trades, and the mechanical trades. See Fig. 25-15.

Preparation for employment is available through an apprenticeship program or by enrolling in a vocational program at the secondary or post-secondary level. See Figs. 25-16, 25-17, and 25-18.

An excellent source of career information is the **Occupational Outlook Handbook** published by the Superintendent of Documents, Washington, D.C.

The Architectural Firm

The staff of an architectural firm designs new buildings and related structures, improves existing buildings, and prepares drawings and contract documents for additions, renovations, and new facilities. While the actual organization varies from one firm to another, the following descriptions are typical.

LOF Libbey-Owens-Ford Company

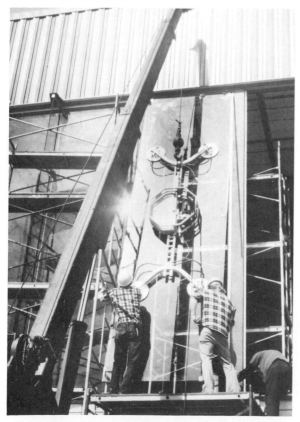

Fig. 25-17. Glaziers belong to one of the finishing trades.

Fig. 25-18. Plumbing is one of the mechanical trades.

Sole Proprietorship

A **sole proprietor** is the owner-manager of a business. Often an architect will establish an office and begin a business as an individual rather than joining an architectural firm. This person is the sole proprietor. He or she sets up the office, seeks clients, does the design work, does some or all the drafting, oversees construction, and handles all contracts and business operations. In short, the architect as the sole proprietor performs all the duties required to operate the business. He or she gets all the profits and takes all the losses.

As business increases, the architect may employ a drafter either part-time of full time. Also, he or she usually will employ engineers on a "by-the-job" basis to do some of the engineering work, such as the structural, mechanical, and electrical design. As the sole proprietor, the architect employs these specialists, reviews their work, and pays their fees. Then he or she pulls together the work of these various specialists into a complete set of working drawings and specifications. The architect may write the specifications or employ the services of an experienced specification writer.

If the sole proprietorship is successful and business increases, the architect will employ full-time architectural drafters and other architects. See Fig. 25-19. The business may become so large that it becomes difficult for one person to handle all the duties. The sole proprietor can then consider forming a partnership with one or more other architects and engineers.

Partnerships

A **partnership** is formed when a contract is entered into by two or more persons. In this contract, each partner agrees to furnish a part of the capital and, sometimes, the labor for a business enterprise. Each shares a predetermined proportion of the profits and losses of that business.

The partnership provides additional capital needed to enable the architectural firm to grow. These funds could permit the employment of additional staff and the purchase of equipment, such as computers, needed to expand the business. Additional capital also permits the hiring of people with special skills, such as mechanical and electrical engineers.

Each partner has a voice in running the business and the authority to act for the firm. Usually, a majority vote of partners is necessary to decide the main lines of action of the firm.

There are several types of partnerships. Those used most often in architectural firms are the active partner and the silent partner.

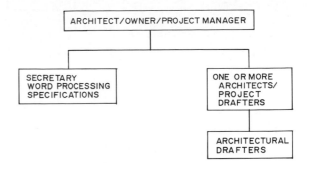

Fig. 25-19. A sole proprietorship architectural firm.

An active partner contributes financially to the firm as well as giving his or her full personal attention to the business. Active partners serve on a board of directors. As an architect or engineer, he or she could specialize in one area, such as seeking clients and handling contracts and business operations. This frees the other partners to concentrate on design and field work.

The silent partner has a financial investment in the business but takes no part in its operation. This partner has no voice in the management of the firm but is liable for any obligations the firm may contract. The existence of a silent partner is not secret. In architectural firms, a retired senior official of the company may be a silent partner.

Large Firms

A large architectural firm has a staff of specialists. While the organization varies from firm to firm, that shown in Fig. 25-20 is typical.

Role of Partners. Partners oversee the entire operation. As mentioned earlier, they often divide up the major responsibilities. But in order for a large partnership to function, many duties must be delegated to various staff members.

Teams. Often, a large architectural firm is divided into small teams which are much like the small sole proprietorship organization. The firm's organization would then be several small offices, each headed by a partner who is responsible for all projects assigned to that organization. Usually there is little interaction between offices.

Actual production of architectural projects is the responsibility of the **project manager.** The engineering staff also reports to him or her. The manager reports to a partner.

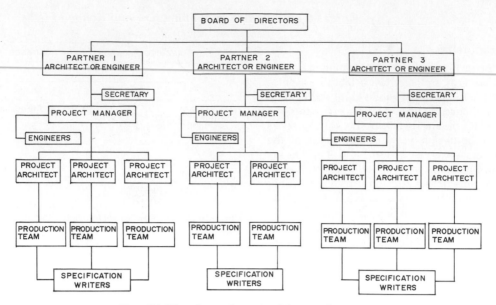

Fig. 25-20. An architectural firm with partners.

Drafters. Beginning drafters on the production team start at the bottom and progress in assigned responsibility as they acquire experience and show they can do the work. They usually start as assistants to the chief drafter and learn office procedures, detailing, and sources of information.

Newly employed experienced drafters undergo a brief orientation period and are then expected to handle a full load. Experienced drafters are expected to interpret design data and prepare architectural working drawings. In very large firms, drafters tend to specialize in an area such as electrical drafting.

The chief drafter is responsible for all drafting services and, possibly, specification writing and artistic work. A company might employ an **artist** to do color perspectives of projects and, sometimes, to also build architectural models. Besides overseeing these tasks, the chief drafter works with the architectural and engineering staffs. See Fig. 25-20.

Architects. When first employed, recently graduated architects start working as assistants to experienced, licensed architects. They learn office procedures and gain technical knowledge. After working in this manner for several years, they become qualified to take the state professional architect examination. The number of years of work experience required varies from state to state (approximately three years).

The chief architect is responsible for architectural services, and members of his or her staff may each be responsible for a limited area of design.

Engineers. A large firm often has electrical, civil, and mechanical engineers on the staff to do the structural, electrical, heating, air-conditioning, and refrigeration design. They give their design data to experienced drafters who complete the working drawings.

Support staff. Secretaries and a business office staff will be needed to handle purchasing, contracts with clients, payroll, and other business functions. Most firms will not have a full-time attorney but will retain these services on an "as-needed" basis.

The architectural firm may rent office space in a large building or own its own building. If it owns its own building, additional staff will be required to clean and maintain the facilities. This is an added business expense.

Build Your Vocabulary

Following are terms that you should understand and use as a part of your working vocabulary. Write a brief explanation of each term.

partnership
sole proprietorship

Class Activity

Arrange for the class to visit the offices of an architect operating as the sole proprietor and a large architectural firm. Prior to the visit, develop a list of questions to ask or observations to be made. Assign these questions or observations to various class members. Students will report their findings at the next class meeting.

Developments in Architecture and Construction

Chapter 26
Reproduction of Drawings

Chapter 27
Current Practices in Building Construction

Chapter 26

Reproduction of Drawings

Architectural working drawings are most commonly reproduced using diazo or blueprint processes or an electrostatic process called xerography. Drawings may also be reproduced from microfilm.

Using Diazo Processes

Diazo is a light-sensitive dye used to coat paper on which copies of original drawings can be made by using special processes. There are two diazo processes: dry and moist. These are used to produce two types of copies: whiteprints and intermediates.

Whiteprints

Whiteprints are paper copies of an original drawing made on vellum or plastic drafting film. The whiteprint paper has a **diazo** compound coating which is light sensitive.

To make a copy of a drawing, the diazo paper and the drawing are placed together and fed into a white-print machine. See Fig. 26-1. The whiteprint

paper is placed with the dye side up and the drawing is placed on top of it with the pencil or ink image up. They are fed together into the exposure unit of the machine where they are exposed to ultraviolet light. The light goes through the translucent drawing vellum and hits the diazo coating destroying it. The lines on the drawing block the light rays, preserving the coating. As the two sheets leave the exposure unit, they are separated. The diazo-coated sheet is fed into the developing unit.

In the **dry process,** the developer uses heated ammonia vapor to develop the remaining diazo lines, producing a colored image on a white background. Blue lines are most commonly used.

The **moist process** involves the same exposure procedure, but the exposed print is developed by running it over one or more rollers which are continuously coated with an alkaline coupler that develops the diazo coating. The moist print is partially developed as it leaves the machine. A few seconds after leaving the developing unit, it is dry and completely developed.

Intermediates

An **intermediate** is a copy of an original drawing that is developed by the diazo process on a special translucent paper. The intermediate copy is used instead of the original drawing to produce whiteprint copies. This saves wear and tear on the original drawing. Also, changes can be made on the intermediate, leaving the original unchanged.

Blueprinting

A **blueprint** is made on a chemically treated paper that turns blue when exposed to light and developed. The exposure process is the same as that used for making whiteprints. Exposed blueprint paper is developed by washing it in clear water and coating it with a solution of potassium dichromate.

Blu Ray, Incorporated

Fig. 26-1. This white printer produces colored lines on a white background.

Microfilm

A **microfilm** is a small photographic negative of a drawing or other information such as project specifications. The original drawing or document is photographed with a microfilm camera. The film is developed in a film processor. After the negative is developed, it is mounted in an aperture card. This card contains information identifying the drawing or document. When it is placed in a reader-printer, an enlargement of the image is displayed on a screen. The reader-printer will also produce enlarged whiteprints of the image on the negative. See Fig. 26-4.

Microfilm provides a way to store hundreds of large drawings in a small file cabinet. It also preserves originals because it can be used to produce copies. Microfilm can be sent to other companies and they can produce enlarged prints for their use.

Drawings to be microfilmed must be drawn carefully. Line widths must be uniform and dense. Ink drawings produce the best negatives. Lettering must be large enough to be reduced photographically and dense enough to be enlarged without losing clarity. Standard lettering sizes are shown in Fig. 26-5.

Fig. 26-5.

Minimum Lettering Sizes for Microfilm

METRIC (mm)		CUSTOMARY (in.)	
Size of Drawing	Letter Height	Size of Drawing	Letter Height
A4, A3	3.5	8.5 x 11 (A)	.125
A2	4.0	11 x 17 (B)	.125
A1, A0	4.5	17 x 22 (C)	.125
Titles	7.0	22 x 34 (D)	.156
Drawing number		34 x 44 (E)	.156
in title block	8.0	Titles	.250
		Drawing number	
		in title block	.250 (.312 over 17 x 22 in.)

3M Corporation

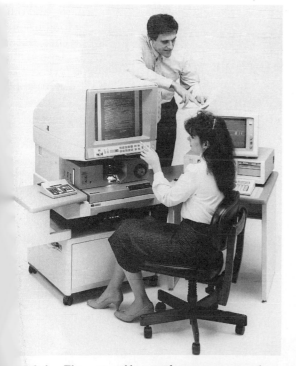

26-4. This microfilm reader-printer provides a [mea]n for reading images stored on microfilm and will [produ]ce an enlarged image on paper.

Build Your Vocabulary

Following are terms that you should understand and use as part of your working vocabulary. Write a brief explanation of each term.

blueprint	microfilm
diazo	whiteprint
intermediate	xerography

Class Activities

1. With the machine set on the same speed, make a number of whiteprints using several drawings with lines ranging from very light to very dark. If possible, prepare an ink drawing and make a whiteprint of it. Describe how line quality influences the quality of the whiteprint.

2. Using a good, dark drawing, run several whiteprints changing the speed each time. Mark the speed on each print. Describe your results and tell why speed influences the quality of the whiteprint.

The print must then be hung to dry. The areas exposed to light turn blue. The protected areas under the lines of the drawing wash away leaving white lines on a blue background.

Using Xerography

Architectural drawings can be reproduced by an electrostatic printer using a process called **xerography.** See Fig. 26-2. It is a dry process that produces black-line images on a white background.

The Xerographic Process

In the xerographic process, the printer has a selenium-coated plate that is given a positive electrical charge. When the drawing is fed into the machine, its image is projected through a lens onto this plate. The image is held there by the positive electrical charge. The rest of the plate loses the positive charge. A negatively charged powder called a toner is spread over the plate, sticking to the positive image. A sheet of paper with a positive charge is placed on the plate attracting the negatively charged powder. Heat fuses the powder to the paper, producing a copy of the original drawing.

Xerographic units will reproduce almost any image fed into them. The image does not have to be on translucent paper. It will make copies of whiteprints, typed material, magazine pages, and other opaque-based images. It will also make enlarged or reduced copies. For example, in architecture, a 24″ x 36″ drawing can be reproduced as an 8″ x 10″ drawing.

Fig. 26-2. This xerographic engineerir reproduce drawings and other documen 50″.

Transmission of Drawings

Architectural drawings, reports, ch such material can be transmitted ove xerography networks using telephor cable, or microwave transmission equipment sends the image which reproduced as a positive image by receiver. See Fig. 26-3.

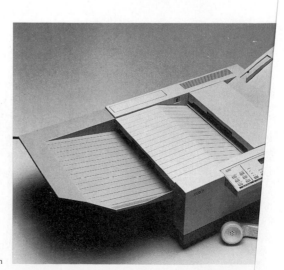

Fig. 26-3. A plain paper fax transmits drawings and uses thermal transfer technology to reproduce drawings transmitted via a telephone.

Fig
scre
pro

Chapter 27

Current Practices in Building Construction

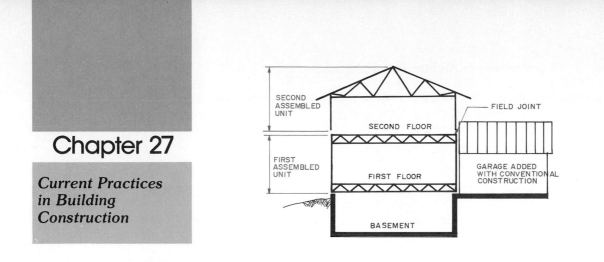

Many developments occurring in architecture and building construction today involve new applications or revised uses of current materials and products. As needs are recognized and new information becomes available, new materials and building techniques are developed. Following are some of the more important practices being applied today. But to stay well-informed on current developments, architectural designers and building contractors will also want to read construction magazines and examine available advertising literature.

Health and Safety

The health and safety of the occupants of buildings is a primary concern of the architect. There are many factors involved such as space utilization and material selection. Following are examples of the things that must be considered.

Fire Safety

Regulations pertaining to fire safety must be observed when designing residential and commercial buildings. The various organizations developing building codes also have **fire codes.** One publication, **The Uniform Fire Code,** sets out provisions necessary for fire prevention. It is sponsored jointly by the International Conference of Building Officials and the Western Fire Chiefs Association. The publication, **Uniform Fire Code Standards,** is a companion to **The Uniform Fire Code.** It contains the standards of the American Society for Testing and Materials and the National Fire Protection Association. **The National Fire Codes** is an eight-volume set that provides guidelines for code enforcement, inspection, design, and maintenance of buildings. Codes include information related to fire alarm systems, the fire rating of materials and assemblies of materials, fire detection systems, fire

escapes, fire-extinguishing systems, and fire resistance (passageways, roofs, wood, walls, floors, etc.). In addition, building techniques, such as fire stopping and fire separations, and information about flammable materials are covered in the codes.

Ventilation

Ventilation of living areas and storage spaces is necessary for health and safety. Some aspects of ventilation are covered in Chapters 1, 3, and 13. In addition, codes provide rigid regulations for various commercial buildings, such as public garages and auto repair shops. Also, detailed ventilation codes apply to heating system design.

Radon

Radon is a colorless, ordorless, tasteless, radioactive gas that occurs naturally in soil gas, underground water, and outdoor air. It exists at various intensity levels throughout the United States. Prolonged exposure to elevated concentrations of radon decay products has been associated with increases in the risk of lung cancer.

Soil gas enters homes through exposed soil in crawl spaces and through cracks and openings in slab-on-grade floors and below-grade basement walls and floors. The water in some private water wells contains radon. When this water is heated and agitated, as in a shower, it gives off small quantities of radon.

Construction Principles. Following are three construction principles to be observed to reduce the inflow of radon:

1. Design and construct the building to minimize pathways of soil gas to enter.
2. Design the building to maintain a neutral pressure differential between indoors and outdoors.
3. Provide facilities to remove radon from inside the building, such as a ventilation system.

Construction Techniques. Figure 27-1 shows ways to reduce the pathways that provide radon entry.

1. Place a 6-mil polyethylene vapor barrier under concrete slab floors. Overlap sheets 12 inches and seal the seams.
2. Avoid penetrating the walls and floor as much as possible. When it is necessary to run an item such as a pipe or wire or other item through a wall or floor, seal and tape the opening.
3. Place welded wire fabric in all concrete floors to minimize the size of cracks that may occur.
4. Seal joints between parts of a building, such as the foundation wall and footing. Use a flexible expansion joint material and cover this with a thick polyurethane caulking material. Heavily seal the outside of the foundation wall to provide waterproofing.
5. Provide a French drain (perforated pipe) around the edge of the slab or footing. Tie this into the ventilation system.
6. Pour the concrete slab and foundation monolithically.
7. Reinforced cast-in-place concrete foundations are less likely to crack than concrete block foundations.
8. Floor drains should run to daylight, a sewer, or a sump. Seal sumps at the top.
9. Seal hollow-block walls by filling the blocks with concrete or capping the walls with solid concrete blocks.
10. Coat interior surfaces with a good water-resistant coating.
11. Any heating ducts in a crawl space or below a concrete slab should have their joints tightly taped. When possible, avoid putting ducts in these places.
12. All doors and windows in basements should have airtight seals.

Figure 27-2 shows how to reduce the vacuum effect.

1. Install vents in the crawl space. Have at least 1 square foot of vent for each 150 square feet of under-floor area.
2. Seal all openings in the wood floor over the crawl space.
3. Seal all places where air could flow, such as around chimneys, plumbing chases, and stairs.
4. Provide an external air duct to supply air to the fireplace.

Figure 27-3 shows construction methods that help remove radon after the house is occupied.

1. Place 4 inches of gravel aggregate below concrete slab floors to facilitate the installation of a sub-slab ventilation.
2. Lay a continuous loop of perforated 4-inch pipe around the inside edge of the foundation footing. Vent this to the outside using a fan-driven vent.
3. Put 4-inch holes in the slab and run vent pipes up and out of the building.

Truss Systems

Truss systems are being designed that are changing methods of construction.

Truss-Framing System

The Truss-Framing System® is used for residential and light commercial building construction. The unit consists of a roof truss and floor truss joined by

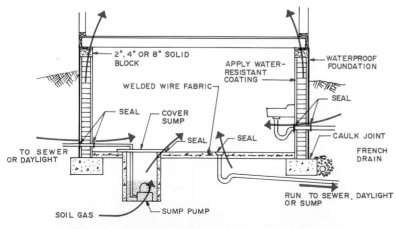

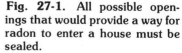

Fig. 27-1. All possible openings that would provide a way for radon to enter a house must be sealed.

studs. See Fig. 27-4. These are assembled in a factory and shipped to the site. They are erected on the foundation spaced two feet on center. The units are connected with sheathing, producing a strong wood-frame shell. See Fig. 27-5. Windows can be selected to fit between the studs. Larger windows require the studs be cut and a conventional header be installed as shown in Fig. 27-6.

Framing used in 2 x 3 and 2 x 4 stock which eliminates the need for large, more costly members such as 2 x 8 and 2 x 10 rafters and joists. In some cases where snow and wind loads are high, 2 x 6 wall studs are required.

A variety of designs can be used, including a two-floor Truss Frame. See Fig. 27-7.

Truss-Joist Floor System

A Truss Joist® is a structural wood member made using structural plywood webs and Microlam® laminated veneer lumber flanges. The web is glued into a groove in the flanges. A Truss Joist is used just as a solid wood joist is used. See Fig. 27-8. They have long-span capabilities and, on most residential designs, eliminate the need for a beam through the center of the house. Typical construction details are shown in Fig. 27-9. When a heavy load must bear upon a section of the floor the joists can be doubled. Recommended spans for selected joist sizes used in light construction are given in Fig. 27-10.

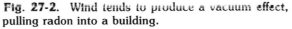

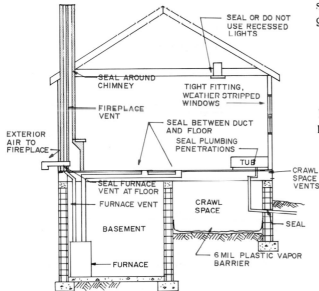

Fig. 27-3. These construction methods help reduce radon penetration into the building.

Fig. 27-2. Wind tends to produce a vacuum effect, pulling radon into a building.

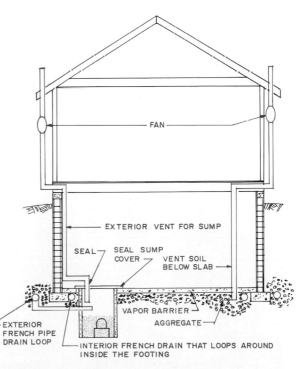

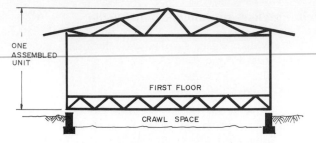

Fig. 27-4. A truss-framing system consists of a roof truss and floor truss joined by studs.

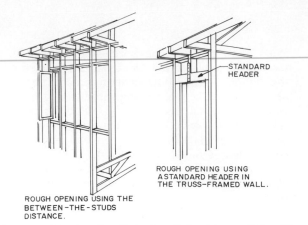

Fig. 27-6. Windows can fit between studs or openings can be made using headers.

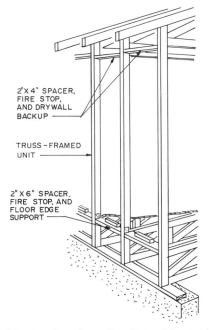

Fig. 27-5. Roof-floor truss units are spaced two feet on center and joined with sheathing.

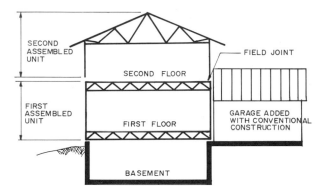

Fig. 27-7. A two-story structure can be built using two truss-framed sections that are joined during construction.

Fig. 27-8. TJI® [1] joists are available in long lengths to accommodate multiple spans.

[1]TJI® is a registered trademark of T.J. International, Boise, Idaho.

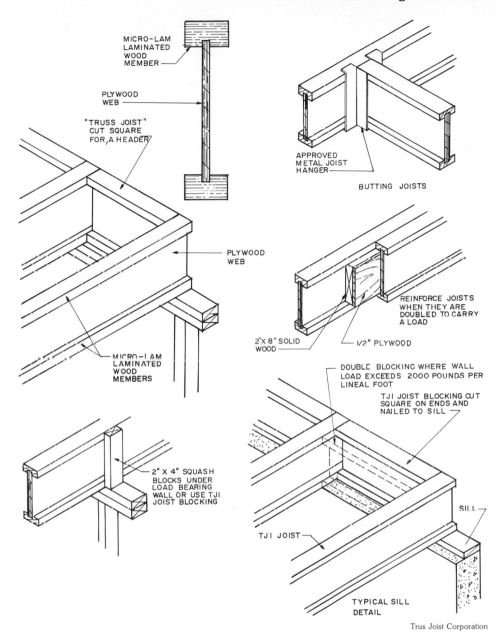

MICRO-LAM
LAMINATED
WOOD
MEMBER

PLYWOOD
WEB

"TRUSS JOIST"
CUT SQUARE
FOR A HEADER

PLYWOOD
WEB

MICRO-LAM
LAMINATED
WOOD
MEMBERS

2" X 4" SQUASH
BLOCKS UNDER
LOAD BEARING
WALL OR USE TJI
JOIST BLOCKING

APPROVED
METAL JOIST
HANGER

BUTTING JOISTS

REINFORCE JOISTS
WHEN THEY ARE
DOUBLED TO CARRY
A LOAD

2"X 8" SOLID
WOOD

1/2" PLYWOOD

DOUBLE BLOCKING WHERE WALL
LOAD EXCEEDS 2000 POUNDS PER
LINEAL FOOT

TJI JOIST BLOCKING CUT
SQUARE ON ENDS AND
NAILED TO SILL

SILL

TJI JOIST

TYPICAL SILL
DETAIL

Trus Joist Corporation

Fig. 27-10.

Spans for TJI® Floor Joists[1]

Fig. 27-9. Construction details for installing TJI® joists.

O.C. Spacing (inches)	Joist Depth			
	9 1/2	11 7/8	14	16
12	18'-2"	21'-9"	24'-8"	27'-3"
16	16'-6"	19'-9"	22'-5"	24'-10"
19.2	15'-6"	18'-6"	21'-1"	23'-4"
24	14'-4"	17'-1"	19'-8"	21'-7"
32	11'-4"	14'-4"		

[1]Based on a floor load of 40 lb/sq ft. and 10 lb. dead load.

Trus Joist Corporation

Wood Floor Trusses

Wood floor trusses are structural members that span long distances and carry reasonably high floor loads. Load data for selected joists from one manufacturer are given in Fig. 27-11. These joists will span the width of most residential designs, eliminating the beam commonly required in the center of the structure. Since they are trusses, they provide open spaces through which plumbing, electrical wires, and heating ducts can be run. This greatly facilitates the installation of such systems.

Fig. 27-11.

Span and Load Data for Truss-Type Floor Joists[1] Spaced 24″ O.C.

Clear span in feet	Joist Depth					
	12	14	16	18	20	22
14	120	118				
16	83	100				
18	66	79	92	105	117	120
20	53	63	74	85	95	103
22		52	61	70	79	88
24			51	59	67	74
26				50	57	63
28					49	55
30						47

[1]Pounds per square foot.

Alpine Engineered Products, Inc.

Typical design and construction details are in Fig. 27-12. The joists are usually spaced 24 inches on center.

Roof Windows

Roof windows, sometimes called skylights, provide natural light and ventilation through openings in the roof. They are often used instead of dormers and have the design advantage of maintaining the line of the roof. They can be set at eye level so it is possible to see out of the room as shown in Fig. 27-13. Or they can be placed high in the roof as shown in Fig. 27-14. The low windows are opened with a crank. This permits the window to be opened without removing the insect screen. Those set high in the roof require that a pole be used to turn the crank. High windows can also be equipped with an electronic control unit that opens and closes the window with a push-button keyboard.

Roof windows have a variety of ways to block the inflow of the sun when it is not desired. Some use a venetian blind while others use a solid blind that pulls over the window. See Figs. 27-14 and 27-15. An exterior sun-blocking awning is also available.

Roof windows are designed with the necessary flashing to provide a watertight installation. See Fig. 27-16. Design data are presented in Fig. 27-17.

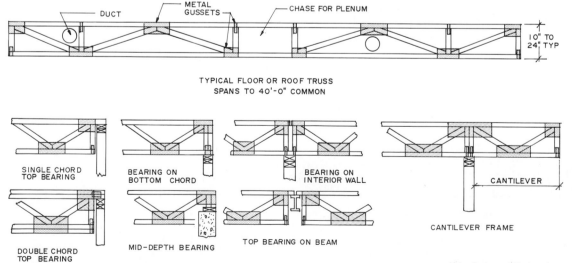

Fig. 27-12. Construction details for truss-type wood floor joists.

Fig. 27-13. These eye-level roof windows open from the sill and allow operation without removing the insect screen.

Fig. 27-14. These roof windows are placed high in the roof. Notice the blinds used to control the sun pulled partially down over the glass.

Fig. 27-15. These high roof windows provide light and ventilation. Notice how venetian blinds are used to control the influx of sunlight.

Fig. 27-16. Typical flashing details used on roof with thin roofing materials such as asphalt shingles. This design is suitable for roofs where the pitch is from 20° to 85°.

Fig. 27-17.
Design Data For Roof Windows

Velux-America, Inc.

Outside Frame W x H (inches)	30 5/8 x 38 1/2	30 5/8 x 55	44 3/4 x 46 1/2	27 1/2 x 46 1/2	21 1/2 x 38 1/2
Rough Opening W x H (inches)	31 1/8 x 39	31 1/8 x 55 1/2	45 1/4 x 47	28 x 47	22 x 39

639

Solar Greenhouses

Solar greenhouses provide a means for direct heat gain, the most common type of passive solar heating. The sun is admitted to the living area and heat is stored in massive masonry and concrete floors and walls. The heat can be moved to other parts of the building with ducts and low-velocity fans.

The greenhouse provides additional living space, opens the building to views of the outdoors, captures the sun during the day, and allows a view of the stars at night. It creates a beautiful atrium dining enclosure, or a room for family relaxation, or a place for a hot tub or sauna. It could also be a lovely garden room filled with flowers and other live plants.

There are many approaches to designing greenhouses. A manufactured unit, such as that shown in Fig. 27-18, can be purchased and added to a building. These are available in aluminum- and wood-framed units. The greenhouse could be integrated into the exterior wall of the building and provide natural light, ventilation, and heat to a large area such as a living room or great room. See Fig. 27-19. Sunscreens and high-performance tempered glass are available to shut out the sun in summer months. Thermostatically controlled ventilation systems automatically change the air as needed.

Build Your Vocabulary

Following are terms that you should understand and use as part of your working vocabulary. Write a brief explanation of each term.

fire codes
radon
roof window
Truss-Framing System®

Class Activities

1. Secure a copy of one of the fire codes and prepare a report detailing the requirements for one of the following:
   a. fire escapes
   b. fire alarm systems
   c. sprinkler systems
   d. storage of flammable materials
2. Contact a company in your area that tests for radon and arrange for a representative to come to your class and show how the tests are conducted. If this is not possible, perhaps your teacher can arrange for the school to buy a commercial radon testing kit. Then the class can conduct a test of the classroom.
3. In a book or magazine, find a section drawing of a building. Mark all the places that might permit radon to enter the building.

Fig. 27-19. This wood-framed solar greenhouse uses high-performance tempered glass to control temperature and withstand stresses.

Four Seasons Greenhouses

Four Seasons Greenhouses

Fig. 27-18. This greenhouse utilizes a high-strength aluminum alloy frame and high-performance glass.

Appendix A

Weight of Materials

Finish Materials	Pounds Per Square Foot
Cement finish, 1″	12
Fiberboard, 1/2″	0.75
Gypsum wallboard, 1/2″	2
Marble and setting bed	25
Plaster, 1/2″	4.5
Plaster on lath	10
Plywood, 1/2″	1.5
Tile, glazed wall, 3/8″	3
Quarry tile, 1/2″	5.8
Terrazzo 1″ or 2″ in stone concrete	25
Hardwood flooring, 25/32″	4
Wood block flooring, 3″ on mastic	15

Glass	Pounds Per Cubic Foot
Plate glass, 1/4″	3.3
Plate glass, 1/2″	6.6
Insulating glass, 5/8″ with air space	3.3
Wire glass, 1/4″	3.5
Glass block	18

Concrete	Pounds Per Cubic Foot
Plain cinder	108
Plain expanded slag	100
Plain expanded clay	90
Plain stone and cast stone	132
Reinforced cinder	111
Reinforced slag	138
Reinforced stone	150

Mortar and Plaster	Pounds Per Cubic Foot
Mortar, masonry	116
Plaster, gypsum, sand	105
Plaster, gypsum, perlite, vermiculite	55

Partitions	Pounds Per Square Foot
Wood stud, 2 x 4; gypsum board	8
Metal stud, 4″; gypsum board	6
Concrete block, lightweight, 4″; gypsum board	26
Concrete block, lightweight, 6″; gypsum board	35
Solid plaster, 2″	20
Solid plaster, 4″	32

Stone (Veneer)	Pounds Per Square Foot
Granite, 2″; 1/2″ parging	30
Granite, 4″; 1/2″ parging	59
Limestone, 6″; 1/2″ parging	55
Sandstone, 4″; 1/2″ parging	49
Marble, 1″	13
Slate, 1″	14

Structural Clay Tile	Pounds Per Square Foot
Hollow tile, 4″	23
Hollow tile, 6″	38
Hollow tile, 8″	45

Structural Facing Tile	Pounds Per Square Foot
Facing tile, 2″	14
Facing tile, 4″	24
Facing tile, 6″	34

Suspended Ceilings	Pounds Per Square Foot
Mineral fiber tile, 3/4″	1.5
Mineral fiberboard	1.4
Acoustic plaster on gypsum lath	10
Fiberglass panels	

Appendix B
Power Requirements

Heavy Duty Appliances[1]

Basic demand (lights, outlets)	4000 watts
Clothes washer	800
Dishwasher	1200
Range plus oven unit	8000
Oven (separate built-in)	4500
Range top (4 units)	5000
Clothes dryer	5000
Water heater	As rated (2500 average)
Refrigerator	500
Freezer	600
Disposal	400
Water pump	400
Attic fan	400
Electric bathroom heater	1300
Central heating (gas or oil)	700
Central heating (electric)	As rated
Room air conditioner	1200
Central air conditioner	As rated

Small Portable Power Tools[2]

Electric drill	400 to 750
Portable circular saw	1400 to 1600
Belt sander	600 to 800
Grinder	400 to 600
Pad sander	300
Router	600 to 1000
Electric motors	
1/3 horsepower	250
1/2 horsepower	375
1 horsepower	746

Small Electric Appliances[2]

Incandescent lights	40 to 150 watts
Fluorescent lights	15 to 60
Radio	40 to 75
Television	75 to 100
Sun lamp	250 to 400
Heat lamp	200 to 250
Heat pad	50 to 75
Electric blanket	150 to 200
Slide or movie projector	300 to 400
Portable electric heater	1200 to 1800
Portable fan	50 to 150
Sewing machine	50 to 100
Vacuum cleaner	250 to 700
Steam iron	600 to 1200
Hot plate (each burner)	500
Toaster	1000 to 1500
Coffee maker	600 to 1200
Waffle iron	1000 to 1200
Roaster	1200 to 1650
Rotisserie	1200 to 1650
Deep-fat fryer	1200 to 1650
Frying pan	1000 to 1400
Blender	250 to 400
Mixer	250 to 500
Grill	1000
Popcorn popper	525
Can opener	120
Portable oven/broiler	1500
Air purifier	100
Video disc recorder	30 to 50
Microwave oven	500 to 700

[1]These are minimum ratings required by the U.S. Department of Housing and Urban Development

[2]These are averages. Actual figures will vary with the appliance selected.

Appendix C

Glossary

Acoustical materials. Materials that will stop or decrease the passage of sound through walls and floors and will control sound within a space.

Altitude (of the sun). The angle formed by the rays of the sun and the earth's surface.

Ampere. The unit used to measure the rate of electrical flow.

Anchor bolt. A threaded rod embedded in the foundation. It is used to hold the sill to the foundation.

Angle iron. A piece of structural steel bent to form a 90° angle. Identified by the symbol **L**.

Array command. A CADD command which multiplies an object and arranges the objects in rows and columns or around a center point.

Associative dimensioning. A feature of some CADD software. It automatically changes the dimensions on a drawing when the drawing is scaled or stretched.

Atrium. An open area within a building, usually filled with plants.

Attic. The space between the ceiling joists and the roof rafters.

Attribute. Information about an object, such as its size, material, manufacturer, etc.

Awning window. A window that is hinged at the top and swings outward.

Azimuth. An angle measured from the north-south axis. For example, due east has an azimuth of 90°, while south has an azimuth of 180°.

Backfill. To replace the earth around a foundation.

Balloon framing. A type of residential framing in which one-piece studs extend from the foundation to the roof on a two-story building.

Balustrade. A series of balusters (posts) topped by a single rail, used to form the railing on a stair.

Baseboard. A wood finish strip placed along an interior wall where the wall meets the floor.

Base plate. A steel plate forming the bottom of a steel column.

Batt. A blanket-type insulation designed to be installed between studs or joists.

Batten. A narrow strip of wood used to cover joints between boards on exterior siding.

Beam. A horizontal structural member that supports a load.

Bearing wall. A wall that supports vertical loads as well as its own weight. Supports the floors and walls above it and the roof.

Bent. A rigid frame made of two vertical supporting members and a horizontal structural member.

Bevel siding. Wood siding boards that are thicker on one edge than on the other. Applied in such a way that the thick edge of one overlaps the thin edge of the next.

Bib (or bibb). An outside faucet that is threaded to allow a hose to be attached.

Blocking. Wood framing members placed between other members to add strength or to provide a nailing surface.

Blueprint. A copy of an original drawing that has white lines on a blue background.

Board measure. A system for specifying a quantity of lumber. One unit is one board foot, which is the amount of wood in a piece $1'' \times 12'' \times 12''$.

Box beam. A hollow structural member built up of plywood sheets and solid wood members.

Brick veneer. A layer of brick laid over the surface of a wood or concrete block wall.

Bridging. Rows of small diagonal braces nailed between floor joists to stiffen the floor.

BTU. British thermal unit. Used to measure heat gain and loss.

Building code. A series of legal requirements specifying design and construction details to ensure the health and safety of people.

Built-up beam. A beam made by joining several smaller members.

Built-up roof. A roof covering made of alternate layers of building paper or felt and asphalt, with a final layer of gravel.

CADD. Computer-aided design and drafting.

Cantilever. The extension of a structural member beyond its point of support.

Casement window. A window that is hinged on the side and swings outward.

Casing. The trim around interior door and window openings.

Caulking (or calking). A waterproof material used to seal joints or cracks.

Cavity wall. A masonry wall made of two or more layers of masonry units joined with ties but having air space between them.

Cement. A gray powdery material which when mixed with water will harden and adhere to masonry units.

Center to center. The distance from the center of one member to the center of the next. Abbreviated on drawings as O.C. or OC (on center).

Central heating. A system using a single source of heat, with ducts for heat distribution.

Chase. A vertical space in the face of a masonry wall, used for running pipes, ducts, and conduits.

Chimney. Vertical flue(s) used to draw off smoke and gases from furnaces and fireplaces.

Chord. A principal member of a roof truss.

Compressive strength. The ability of a material to withstand forces tending to shorten it.

Cornice. The part of the roof that extends out beyond the exterior wall.

Course. A single continuous row of bricks or stone.

Court. An open area surrounded partially by a building and/or walls. It has no roof.

Crawl space. An area between the floor joists and the earth under a building or part of a building where there is no basement.

Cripple. A stud that is less than the full height of the wall, such as one used over a door or window opening.

Curtain wall. An exterior wall that protects the interior from the weather but carries no structural loads.

Data base program. A computer program which enables the user to file and sort data according to field specification.

Datum level. Basic level used as a reference for reckoning heights when building.

Dead load. The weight of the materials used to build a structure. (Contrast Live load.)

Decibel. A unit used to measure the relative intensity or loudness of sound.

Deed. Written contract which conveys the title of ownership of a piece of real estate.

Deflection. The vertical distance a beam moves from a straight line as a result of being under the stress of a load.

Dehumidify. Reduce the amount of moisture in the air.

Diazo. A light-sensitive dye that is used to coat paper used in making copies.

Digitizer. An input device for a CADD system. It consists of a large, smooth surface and an attached pointing device.

Dormer. A small window structure projecting through a sloping roof.

Double glazing. A door or window pane made of two layers of glass that are sealed together. Dead air space is between the layers, providing insulation.

Double-hung window. A window having two sashes that move in a vertical direction.

Downspout. A vertical pipe used to carry rainwater from a roof gutter to the ground.

Drag function. A CADD function which makes it possible to dynamically drag an image on the computer screen.

Dressed lumber. Lumber that has been machined to standard sizes and smoothed.

Dry wall. Interior wall made of gypsum board panels instead of plaster.

Ducts. Pipes used to carry heated and cooled air to and from registers.

Easement. Limited right to use land belonging to someone else. For example, landowners may grant utility companies the right to install and service utility lines on their property.

Eave. The section of the roof that overhangs the exterior wall.

Editing. Changing existing text or graphics.

Elevation. Drawings showing the exterior of a building.

Ell. A wing of a building built at a right angle to the main part of the structure.

Excavation. A pit dug in the earth for the purpose of building a footing and foundation.

Expansion joint. A joint in a masonry or concrete unit, used to provide for expansion and contraction of materials due mostly to changes in moisture content and temperature.

Facade. The exterior face that is the front of a building. It is usually the most elaborate.

Face brick. High quality brick used in places that will be exposed to view.

Fascia. A vertical board nailed to the ends of rafters forming the face of the cornice.

Files function. A CADD function which allows the user to access disk information such as the names of the drawings stored on the disk.

Finish lumber. Good quality lumber used to form surfaces that will be painted.

Firebrick. A special brick that withstands high temperatures, used to line fireplaces.

Fire codes. Regulations pertaining to fire safety that must be followed when designing buildings.

Fire stop. Members installed inside walls and floors to keep fire from spreading.

Flashing. Material used in places such as at roof intersections and around windows and doors to keep water from leaking in.

Floor plan. The drawing showing the top view of one floor of a house, "cut through" to show locations of all important features such as windows and doors.

Floppy diskette. A thin, flexible disk encased in plastic or paper. It is used as a permanent storage device for computer data.

Flue. A pipe in a chimney through which furnace or fireplace gases and smoke pass.

Footing. A concrete pad upon which the foundation is set. It is broader than the foundation to distribute weight over a larger surface.

Frieze. A flat board that is on the top of the exterior wall where the siding meets the soffit.

Frost line. The depth to which soil freezes in a particular area.

Furring. Wood strips fastened to the walls and ceiling to form a straight surface for applying finished wall material.

Fuse. A soft metal link inserted in an electrical circuit. This metal will melt when an overload occurs.

Gable. The triangle area of an exterior wall at the end of a house from the top plate to the rafters.

Gable roof. A roof with two sides that slope in two opposite directions from a central ridge. Sometimes called an **A** roof.

Gambrel roof. A roof with the slope broken into two different planes on each side.

Girder. A horizontal structural member used to support walls and floor joists.

Grade. The elevation above sea level of the ground on a building site.

Graphics editor. In CADD, the computer mode which allows the user to create and modify drawings.

Grid function. A CADD function which allows the user to place an alignment grid of dots on the screen at any desired spacing. The grid provides a visual reference for distances and overall drawing size, but does not become part of the drawing itself.

Ground fault interrupter. An electrical device that detects current leaks in a circuit and shuts off the power to that circuit.

Grout. A thin cement mortar used for leveling masonry surfaces.

Gussett. A plywood or metal plate used to join the members of a truss and to add strength at each joint.

Gutters. Plastic or metal troughs at the lower ends of the roof for carrying away rainwater coming off a roof.

Gypsum board. Sheet material having a gypsum core laminated between layers of heavy paper. Also called plasterboard.

Half-timbered construction. Exterior wall construction using exposed heavy wood structural members with the spaces between them filled with masonry.

Hangers. Iron straps used to suspend pipes or joists.

Hardboard. Sheet material made by bonding wood fibers.

Hard disk. A permanent storage device for computer data. Also called fixed disk.

Head. The top frame of a door or window.

Header. A horizontal supporting member that spans an opening in a frame wall.

Headroom. The distance between stair treads and the ceiling above.

Hip rafter. A diagonal rafter that runs from the ridge to the corner intersection of the top plates of two walls.

Hip roof. A roof with four sides sloping away from the ridge.

House drain. The large horizontal sewer line that runs under the house and receives the waste from the soil pipes.

House sewer. The horizontal sewer line that runs outside the house from the house drain to the city sewer or septic tank.

Insulation. Materials used to stop or inhibit the passage of heat, cold, and sound. Most often used in walls, ceilings, and floors, but may be used in special ways such as wrapped around pipes.

Interior trim. The molding, casing, and baseboard used to finish walls and door and window openings.

Intermediate. A copy of the original drawing reproduced on translucent paper by the diazo process.

Jack rafter. A short rafter running between the roof ridge and a hip or valley rafter.

Jamb. The lining of a door or window opening.

Joist. A horizontal structural member used to support a floor or ceiling.

Kelvin (K). A metric measure of temperature.

Kip. 1000 pounds of deadweight load.

Knee wall. A short wall used in construction of the second floor of a one-and-one-half-story house.

Lally column. A round steel column with a square steel base and top plate. It is used to support beams and girders.

Laminated beam. A beam made by bonding several layers of wood.

Landing. A platform at the top or bottom of a stair or between flights of stairs.

Lath. Metal screening, wood strips, or solid gypsum panels used as the base upon which plaster walls are built.

Lavatory. A sink in a bathroom.

Layer. In CADD, an electronic equivalent to a transparent overlay. Also called a level.

Ledger. A wood strip nailed to the side of a wood beam and used to support joists meeting the beam.

Light. A single window pane.

Lintel. A horizontal structural member that spans the opening in a wall. It carries the weight of the material over the opening.

Live load. The weight of movable and variable loads placed on a building. (Contrast Dead load.)

Load-bearing walls. Walls that carry a load from above in addition to their own weight.

Lobby. An entrance area in a commercial building.

Loggia. A roofed, open gallery along the front or side of a building.

Lookout. A horizontal framing member that supports a roof overhang.

Louver. A unit with fixed or movable slats that are designed to regulate light and ventilation.

Mansard roof. A roof having two slopes on each of four sides. The top slopes are less steep than the lower slopes.

Mantel. A shelf attached to the wall above a fireplace, used for decorative purposes.

Masonry. Units made of concrete, stone, cement, brick, or tile.

Micro-based CADD. A CADD system based on a microcomputer.

Microfilm. A small photographic negative of a drawing.

Modular construction. The construction of buildings using units based on standard sizes.

Modular unit. A factory-built, finished section of a building, room-size or larger.

Module. A standardized unit of measure.

Modulus of elasticity. A measure of the degree of stiffness of a material.

Molding. Machined strips used for interior and exterior decoration.

Monolithic. Method of concrete construction in which all members are cast at the same time as a solid, continuous unit. For example, the footing and foundation may be cast monolithically.

Mortar. A mixture of portland cement, sand, and water, used to bond masonry units together.

Mortgage. A loan that is secured by pledging property as security for the debt.

Mouse. In a CADD system, a pointing device used for controlling the screen cursor and for picking items on the screen. Some mice operate on a tabletop; others require a pad.

Mullion. A large vertical member separating multiple windows or doors.

Muntin. Small bars, both horizontal and vertical, between the panes of glass in a window.

Networking. Linking two or more computers so that each can share resources from the other.

Newel post. A post, often decorative, that supports a handrail at the top or bottom of a stair or at a landing.

Nominal size. The size of lumber after it has been cut at the sawmill but before it has been planed smooth.

Nonbearing wall. A wall that carries no loads other than its own weight.

Nosing. The rounded edge of a stair tread that extends past the riser.

Object snap function. In CADD, a function which allows the user to place the cursor on specific points of an object.

Ohm. The unit of electrical resistance.

Orientation. Positioning of a building, to take advantage of the sun, wind, and view.

Outlet. An electrical unit that permits electrical current to be drawn from a circuit.

Overhang. The horizontal distance between the fascia and the exterior wall.

Pan command. In CADD, a command which moves the display window from one location on the drawing to another.

Panelboard. An electric box in which controls for electrical circuits are located.

Parametrics. A method of computer-aided design and drawing in which generic drawings stored in the computer can be made into many customized drawings by entering different dimension values.

Parapet. The part of an exterior wall that extends above the edge of the roof.

Parging. Applying a layer of mortar to the surface of masonry units to smooth or waterproof them.

Partition. An interior wall.

Partnership. A legal relationship in which two or more persons agree to furnish capital and sometimes labor for a business enterprise.

Patio. A paved, open area near a home. Sometimes called a terrace.

Penny (d). Term used to specify the length of a nail.

Pen plotter. A CADD output device which plots computer-generated drawings by movement of a pen and a medium, such as paper or vellum.

Perspective. A type of pictorial drawing in which the appearance of an object as drawn on a plane is similar to the appearance of the object in a natural three-dimensional state.

Pier. A vertical masonry unit, usually concrete, that supports a load from above.

Pilaster. A vertical masonry column constructed as part of a masonry wall but thicker than the wall itself, used to strengthen the wall or to support a beam.

Pile. Concrete, steel, or wood column driven into the ground and upon which a building is built.

Pitch. The angle of the roof from the ridge board to the plate, determined by dividing rise by span.

Plank. A long, flat wood member 2 to 4 inches (50 to 100 mm) thick and 6 or more inches (150 mm) wide.

Plaster. A mixture of portland cement, lime or gypsum, sand, and water used to cover walls and ceilings.

Plasterboard. See Gypsum board.

Plat. A drawing showing the layout of an area of land, such as that of a housing development.

Plates. Horizontal wood members located above and below studs in a wall.

Platform framing. A type of house framing where each story is built on top of the one below but framed independently. (Contrast Balloon framing.)

Plenum. A box on a furnace from which ducts run to outlets in various rooms being heated.

Plot. A piece of ground with specific dimensions upon which a building will be constructed.

Plumb. Absolutely vertical.

Plywood. Wood sheets made by gluing thin layers together with the grain in each layer at 90° to the previous layer.

Portland cement. A fine, gray material made from burning compounds of lime, silica, and alumina together. It is the bonding agent in concrete.

Post. A vertical structural member.

Post-and-beam construction. Wall and roof construction using posts and beams as the structural frame. Planks are applied transversely across the beams. Sometimes called post, plank, and beam construction.

Precast. Casting and curing concrete units before they become part of a building.

Prefabricated. Assembling units in a factory and shipping them to the site. For example, roof trusses are often prefabricated.

Prestressed. Casting concrete units in a mold using reinforcing steel that is under tension.

Purlin. A structural member that runs between and perpendicular to roof beams, rafters, or trusses. It is used to support sheathing.

Radiant heating. Heating by heat rays, without blowers or other air movement.

Radon. A colorless, odorless, tasteless, radioactive gas that occurs naturally in soil gas, underground water, and outdoor air. It can be a health threat.

Rafters. Structural members used for supporting the roof sheathing and the waterproof outer layers of the roof. Rafters run from the ridge to the exterior walls.

RAM (random access memory). The temporary storage area in a computer. When the computer is turned off, anything stored in random access memory is erased.

Rebars. See Reinforcing rods.

Register. A unit on the end of a heat duct, used to direct air into the room.

Reinforced concrete. Concrete with steel bars and/or mesh added to increase its strength.

Reinforcing rods. Round, steel reinforcing bars placed in concrete to increase tensile strength. Also called rebars.

Resolution. The fineness of detail observable in the images on a computer monitor. The higher the resolution, the finer the detail.

Retaining wall. A wall built to hold back soil.

Reveal. The depth between the wall surface and the door or window unit.

Ribbon. A wood member joined to studs to support the floor joists in balloon framing.

Ridge. The top peak of the roof where the sloping roof areas meet.

Rise. The vertical height of a roof.

Riser. The vertical part of a stair.

Roof window. Opening in a roof used to provide light and ventilation.

Rough opening. An unfinished opening in a frame wall, sized to accept a door or window unit.

Run. The horizontal distance covered between the roof ridge and the plate. Half the roof span. It is also the horizontal distance of a stair.

R-value. Measure of resistance to heat flow.

S shape. A standard structural steel beam shape.

Sandwich wall. A masonry wall with two adjacent panels separated by an insulating panel.

Sash. The movable frame of a window, containing the glass panes.

Scab. Wood pieces used to join other wood members. They are nailed across joints on the outside faces.

Scale. The use of proportional measurements, in which one distance represents another. Also, a drafting tool used to measure or lay out dimensions on a drawing.

Schedule. A list of parts, such as a window schedule.

Scratch coat. The first coat of plaster, applied directly to the lath.

Screed. A long straight board or rod used to level concrete in forms.

Section. A drawing showing interior features when an imaginary cut is made through a portion of a building.

Septic tank. A steel, concrete, or plastic tank placed in the ground to receive sewage.

Service entrance. The electric lines from the power company's pole to the building.

Set. The hardening of cement by hydration.

Setback. A zoning regulation limiting how close a building can be placed to the street.

Shakes. Handcut wood shingles.

Shear. A condition caused by forces that tend to produce an opposite but parallel sliding motion of the planes in a member.

Sheathing. The material placed directly on the exterior side of the studs.

Shed roof. A flat roof that slants in one direction.

Shim. A thin piece of material placed between two parts to level a surface or fill a void.

Shingles. Relatively small, thin pieces of material applied in an overlapping manner, used to weatherproof a roof or the exterior of a wall.

Shiplap. Siding boards with lapped joints along the edges applied over sheathing.

Shoring. Wood or metal braces used to provide temporary support for buildings during construction. Used to support forms for cast-in-place concrete frames.

Siding. The finish material forming the outside layer of an exterior wall.

Sill. The wood member bolted to the top of the foundation. Also the horizontal member of a window or door unit located below the door or window.

Skylight. See Roof window.

Slab. A strip of concrete, poured as a single piece. Often refers to a reinforced concrete floor.

Slope. Ratio between rise and run of a roof.

Snap function. In CADD, a mode in which the computer forces all points and selections to be placed on the intersections of (invisible) grid lines.

Soffit. The horizontal return from the fascia to the exterior wall.

Soil pipe. Part of the drainage system that receives waste from water closets.

Soil stack. The main vertical pipe into which waste flows from all fixtures.

Solar heat. Heat generated by the rays of the sun.

Sole. The horizontal member below a line of studs. Also called the bottom plate.

Sole proprietorship. A business owned and operated by one person.

Solids modeling. In CADD, a 3-D drawing method which treats objects as though they were solid.

Sound transmission class (STC). A means of specifying the ability of a material to resist the transmission of sound.

Span. The distance between two supporting members. Also the distance from one exterior wall to the other.

Spandrel. The wall area above a window.

Specifications. A written record of details relating to the design and construction of a building. It includes things that cannot be shown on the working drawings.

Splice. The joining of two members to form one piece.

Spreadsheet program. A computer program which displays information (such as financial data) in rows and columns. When the user changes a value in one location, the computer automatically adjusts the other values so that subtotals and totals will be accurate.

Stack. In plumbing, a general term referring to a vertical pipe.

Steel framing. A structural frame of steel members.

Stile. The vertical member of a door or window.

Stool. An interior horizontal member below a window.

Stop. A wood strip used to hold a window in place or against which a door closes.

Storm sewer. A sewer used to carry away surface water but not sewage.

Stress. A force, such as a load, acting upon a member.

Stressed-skin panels. Hollow, prefabricated, structural units built of plywood and solid lumber and used in floors, walls, or roofs.

Stringer. A side of a stair, cut to receive the treads.

Stucco. A cement mixture used as an exterior covering on buildings.

Stud. Vertical wood or metal units used in wall construction.

Subfloor. The floor material nailed directly to the floor joists.

Sump. A pit in a basement, in which water collects and is removed with a sump pump.

Surfaced lumber. Lumber that has been smoothed and sized by planing.

Surface modeling. In CADD, a 3-D drawing method which places a "skin" over a wireframe object.

Swale. A low area between two higher sloping sections of earth.

Symbol library. A collection of graphics stored on computer disk and inserted into drawings as needed.

Tensile strength. The ability of a member to withstand forces tending to lengthen it.

Termite shield. A metal shield on top of a foundation, used to block the passage of termites.

Thermal conductor. A material able to conduct heat.

Thermostat. A device that regulates the air temperature in a building by controlling the heating and air-conditioning unit.

Threshold. A beveled strip, usually wood or stone, that is fastened to the door sill and is directly below the door.

Tie. A structural member used to hold parts together.

Tilt-up construction. Concrete wall units cast in horizontal forms and lifted into position when cured.

Timber. Construction lumber larger than 4″ × 6″ (102 × 152 mm) in cross section.

Toenail. To drive nails at an angle at the end of one board into another board and fasten the two together.

Tongue-and-groove construction. Machined projection on a board (tongue) fits into groove in the next board.

Trap. A plumbing unit that keeps sewer gases from backing up into the house.

Tread. The horizontal part of the stair upon which a person steps.

Trim. See Interior trim.

Trimmer studs. Framing members forming rough openings in walls, floor, or roof.

Truss. A preassembled unit used for roof construction. It includes rafters, ceiling joists, and necessary bracing.

Truss-Framing System®. A wood framing system in which a single assembled unit contains the roof truss and floor truss connected by studs.

Underlayment. Floor covering used to provide a level surface for finish floor coverings such as carpet.

Valley. The intersection of two sloping roof sections.

Valley jacks. Rafters that extend from the ridge board to the valley rafter.

Valley rafter. A diagonal rafter running from the ridge board to the top plate, forming the intersection of two sloping roofs.

Valve. A device used to regulate the flow of water or gas.

Vapor barrier. A material that prevents the passage of moisture through walls and floors. Also called moisture barrier.

Vellum. Durable, transparent paper used for making architectural drawings.

Veneer. Placing a facing of one material over a structure of a different type. Example: placing a layer of brick over frame construction.

Vent. An opening to provide ventilation. In plumbing, a part of the pipe system used to keep the sewage system under atmospheric pressure.

Vent stack. The top section of the soil stack. It extends out through the roof.

Vestibule. An open area at an entrance to a building.

View command. In CADD, a function which allows zoomed views of a drawing to be stored in memory for subsequent retrieval.

Volt. Unit used to measure electrical pressure or force.

W shape. A wide flange structural steel member.

Wallboard. A general term used to refer to large rigid sheets used to cover interior walls. It can be made of wood fibers, gypsum, or other materials.

Waste pipe. A horizontal pipe used to carry waste from bath and kitchen fixtures to the soil pipe.

Waterproof. Prevents the passage of water.

Water table. The level of underground water.

Watt. A unit of electrical power.

Weather strips. Metal or fabric strips installed around the edges of doors and windows to reduce heat loss and gain.

Web members. Internal braces running between chords in a truss.

Weep hole. A small opening in the bottom of a wall that permits moisture to drain out.

Whiteprint. Copy of an original drawing reproduced on paper using the diazo process which produces a colored line drawing on a white background.

Winder. A triangular tread on a stair.

Window function. In CADD, a function which enables the user to select an object or group of objects on the display screen by drawing a rectangle around it. The objects within the window can then be moved, copied, erased, etc.

Working drawings. Drawings containing design details. The plans from which construction is done.

Xerography. An electrostatic process used to reproduce drawings by producing a black line on a white background using a positively charged selenum-coated plate and a negatively charged powder toner.

Zoning. Restrictions regulating the type and use of buildings that may be constructed in a specific area. In house planning, refers to dividing a house into areas according to activities performed.

Zoom command. A CADD command which magnifies objects on the display screen.

Appendix D
Abbreviations of Architectural Terms*

Following are standard abbreviations commonly used on architectural drawings. Notice that most are in capital letters. Also, if an abbreviation could be mistaken for a word, a period is placed after it.

Access panel . . . AP
Acoustic . . . ACST
Actual . . . ACT.
Addition . . . ADD.
Adhesive . . . ADH
Aggregate . . . AGGR
Air condition . . . AIR COND
Alternate . . . ALT
Alternating current . . . AC
Aluminum . . . AL
Amount . . . AMT
Ampere . . . AMP
Anchor bolt . . . AB
Angle . . .
Apartment . . . APT.
Approved . . . APPD
Approximate . . . APPROX
Architectural . . . ARCH.
Area . . . A
Asbestos . . . ASB
Asphalt . . . ASPH
At . . . @
Automatic . . . AUTO
Avenue . . . AVE
Average . . . AVG

Balcony . . . BALC
Basement . . . BSMT
Bathroom . . . B
Bathtub . . . BT
Beam . . . BM
Bearing . . . BRG
Bedroom . . . BR
Bench mark . . . BM
Between . . . BET.
Beveled . . . BEV
Blocking . . . BLKG
Blower . . . BLO
Blueprint . . . BP
Board . . . BD
Board measure . . . BM

Both sides . . . BS
Bottom . . . BOT
Bracket . . . BRKT
Brick . . . BRK
British thermal unit . . . BTU
Bronze . . . BRZ
Broom closet . . . BC
Building . . . BLDG
Building line . . . BL
Built in . . . BLT-IN
Button . . . BUT
Buzzer . . . BUZ
By . . . X

Cabinet . . . CAB.
Candela . . . cd
Caulking . . . CLKG
Cast concrete . . . C CONC
Cast iron . . . CI
Ceiling . . . CLG
Celsius . . . C
Cement . . . CEM
Center . . . CTR
Center line . . . CL
Center to center . . . OC
Centimeter . . . cm
Ceramic . . . CER
Circle . . . CIR
Circuit . . . CKT
Circuit breaker . . . CIR BKR
Circumference . . . CIRC
Cleanout . . . CO
Clear . . . CLR
Closet . . . CL
Coated . . . CTD
Cold water . . . CW
Column . . . COL
Combination . . . COMB.
Common . . . COM
Composition . . . COMP
Concrete . . . CONC

Concrete block . . . CONC B
Concrete masonry unit . . . CMU
Conduit . . . CND
Construction . . . CONST
Continue . . . CONT
Contractor . . . CONTR
Corrugate . . . CORR
Courses . . . C
Cross section . . . X-SECT
Cubic foot . . . CU FT
Cubic inch . . . CU IN.
Cubic yard . . . CU YD
Cylinder . . . CYL

Damper . . . DMPR
Dampproofing . . . DP
Dead load . . . DL
Decibel . . . DB
Degree . . . (°) DEG
Design . . . DSGN
Detail . . . DET
Diagonal . . . DIAG
Diagram . . . DIAG
Diameter . . . or DIA
Dimension . . . DIM.
Dining room . . . DR
Direct current . . . DC
Dishwasher . . . DW
Ditto . . . DO.
Division . . . DIV
Door . . . DR
Double . . . DBL
Double-hung . . . DH
Down . . . DN
Downspout . . . DS
Drain . . . DR
Drawing . . . DWG
Drinking fountain . . . DF
Dry well . . . DW
Dryer . . . D
Duplicate . . . DUP

*Reproduced from ANSI Y1.1-1972 with the permission of the publisher, The American Society of Mechanical Engineers.

Each . . . EA
East . . . E
Elbow . . . ELL
Electric . . . ELEC
Elevation . . . EL or ELEV
Elevator . . . ELEV
Emergency . . . EMER
Enamel . . . ENAM
Entrance . . . ENT
Equal . . . EQ
Equipment . . . EQUIP.
Estimate . . . EST
Excavate . . . EXC
Existing . . . EXIST.
Exterior . . . EXT

Fabricate . . . FAB
Family room . . . FAM R
Fahrenheit . . . F
Feet . . . (') FT
Feet board measure . . . FBM
Finish . . . FIN
Finished floor . . . FIN FL
Fire brick . . . FBRK
Fire extinguisher . . . F EXT
Fireproof . . . FPRF
Fireproof self closing . . . FPSC
Fixture . . . FIX.
Flashing . . . FL
Floor . . . FL
Floor drain . . . FD
Flooring . . . FLG
Fluorescent . . . FLUOR
Foot . . . (') FT
Footcandle . . . FC
Footing . . . FTG
Foundation . . . FDN
Full size . . . FS
Furnace . . . FURN
Furred ceiling . . . FC

Galvanize . . . GALV
Galvanized iron . . . GI
Garage . . . GAR
Gas . . . G
Gage . . . GA
Girder . . . G
Glass . . . GL
Glue-laminated . . . GLUELAM
Grade . . . GR
Grade line . . . GL
Gypsum . . . GYP.

Hall . . . H
Hardware . . . HDW
Hardwood . . . HDWD
Head . . . HD
Heater . . . HTR
Height . . . HT
Horizontal . . . HOR
Hose bib . . . HB
Hot water . . . HW
House . . . HSE
Hundred . . . C

Impregnate . . . IMPG
Inch . . . (") IN.
Incinerator . . . INCIN
Inside diameter . . . ID
Insulate . . . INS
Intercommunication . . . INTERCOM.
Interior . . . INT
Iron . . . I

Joint . . . JT
Joist . . . JST
Junior beam . . . M

Kelvin . . . K
Kilogram . . . kg
Kilowatt . . . kW
Kilowatt hour . . . kWh
Kip . . . K
Kitchen . . . KIT.
Kitchen cabinet . . . KC
Kitchen sink . . . KS

Laminate . . . LAM
Landing . . . LC
Laundry . . . LAU
Laundry chute . . . LC
Lavatory . . . LAV
Left . . . L
Length . . . LG
Length overall . . . LOA
Level . . . LEV
Light . . . LT
Linear . . . LIN
Linen closet . . . L CL
Live load . . . LL
Living room . . . LR
Long . . . LG
Louver . . . LV
Lumber . . . LBR

Main . . . MN
Manhole . . . MH
Manual . . . MAN.
Manufacturing . . . MFG
Mark . . . MK
Masonry opening . . . MO
Material . . . MATL
Maximum . . . MAX
Medicine cabinet . . . MC
Membrane . . . MEMB
Metal . . . MET.
Meter (water or gas) . . . M
Meter (metric) . . . m
Millimeter . . . mm
Minimum . . . MIN
Minute . . . (') MIN
Miscellaneous . . . MISC
Mixture . . . MIX.
Model . . . MOD
Modular . . . MOD
Motor . . . MOT
Molding . . . MLDG

Natural . . . NAT
Nominal . . . NOM
North . . . N
Not to scale . . . NTS
Number . . . NO.

On center . . . OC
Office . . . OFF
Opening . . . OPNG
Opposite . . . OPP
Overall . . . OA
Overhead . . . OVHD

Painted . . . PTD
Pair . . . PR
Panel . . . PNL
Parallel . . . PAR.
Part . . . PT
Partition . . . PTN
Passage . . . PASS.
Penny (nails) . . . d
Permanent . . . PERM
Perpendicular . . . PERP
Piece . . . PC
Plaster . . . PLAS
Plate . . . PL
Plumbing . . . PLMB
Point . . . PT

Pound . . . LB
Pounds per square inch . . . PSI
Precast . . . PRCST
Prefabricated . . . PREFAB
Preferred . . . PFD
Push button . . . PB

Quality . . . QUAL
Quantity . . . QTY

Radiator . . . RAD
Radius . . . R
Range . . . R
Receptacle . . . RECP
Reference . . . REF
Refrigerator . . . REF
Register . . . REG
Reinforce . . . REINF
Reproduce . . . REPRO
Required . . . REQD
Return . . . RET
Revision . . . REV
Revolutions per minute . . . RPM
Riser . . . R
Roof . . . RF
Roofing . . . RFG
Room . . . RM
Rough . . . RGH
Round . . . RD

Safety . . . SAF
Sanitary . . . SAN
S-beam . . . S
Scale . . . SC
Schedule . . . SCH
Self closing . . . SC
Second . . . (") SEC
Section . . . SECT
Select . . . SEL
Service . . . SERV
Sewer . . . SEW.
Sheet . . . SH
Sheathing . . . SHTHG

Shower . . . SH
Side . . . S
Siding . . . SDG
Sill cock . . . SC
Similar . . . SIM
Sink . . . S
Soil pipe . . . SP
South . . . S
Specification . . . SPEC
Square . . . SQ
Stairs . . . ST
Steam . . . ST
Standard . . . STD
Steel . . . STL
Stock . . . STK
Street . . . ST
Storage . . . STG
Structural . . . STR
Supply . . . SUP
Surface . . . SUR
Suspended ceiling . . . SUSP CLG
Switch . . . SW
Symmetrical . . . SYM
Symbol . . . SYM
System . . . SYS

Tar and gravel . . . T & G
Tangent . . . TAN.
Tarpaulin . . . TARP
Tee . . . T
Telephone . . . TEL
Television . . . TV
Temperature . . . TEMP
Terra-cotta . . . TC
Terrazzo . . . TER
Thermostat . . . THERMO
Thick . . . THK
Thousand . . . M
Through . . . THRU
Toilet . . . T
Tongue and groove . . . T & G
Total . . . TOT.
Tread . . . TR

Tubing . . . TUB.
Typical . . . TYP

Unfinished . . . UNFIN
Urinal . . . UR

Valve . . . V
Vent pipe . . . VP

Ventilate . . . VENT
Vent or ventilator . . . V
Vertical . . . VERT
Vestibule . . . VEST.
Vitreous . . . VIT
Volt . . . V
Volume . . . VOL

W-beam . . . W
Wall cabinet . . . WC
Washing machine . . . WM
Water . . . W
Water closet . . . WC
Water heater . . . WH
Waterproofing . . . WP
Watt . . . W
Weather-stripping . . . WS
Weatherproof . . . WP
Weep hole . . . WH
Weight . . . WT
West . . . W
Width . . . W
Window . . . WDW
Wire glass . . . WG
With . . . W/
Without . . . W/O
Wood . . . WD
Wrought iron . . . WI

Yard . . . YD

INDEX